# Integration and Bridging of Multiscale Bioengineering Designs and Tissue Biomechanics

Jun Liao • Joyce Y. Wong
Editors

# Integration and Bridging of Multiscale Bioengineering Designs and Tissue Biomechanics

Springer

*Editors*
Jun Liao
Department of Bioengineering
University of Texas at Arlington
Arlington, TX, USA

Joyce Y. Wong
Department of Biomedical Engineering
Boston University
Boston, MA, USA

ISBN 978-3-031-81742-7 ISBN 978-3-031-81743-4 (eBook)
https://doi.org/10.1007/978-3-031-81743-4



This Springer imprint is published by the registered company Springer Nature Switzerland AG
The registered company address is: Gewerbestrasse 11, 6330 Cham, Switzerland

# Preface

During the past decade, many innovative bioengineering approaches have been developed to fabricate functional tissues, biomaterials, and medical implants to restore organ/tissue functions. These bioengineering innovations include novel tissue engineering methods such as biomimicking scaffolds with in vivo temporal control and anatomic/physiologic construct design enabled by 3D bioreactor and 3D bioprinting techniques. Other examples include cell-scaffold optimization and integration to form complicated multiscale anatomies; organ decellularization/recellularization; tissue-derived extracellular matrix (ECM) scaffold/hydrogel applications; bioactive polymer-tissue hybrid composites; coordinated cell therapy; and modeling-guided biomaterial design/intervention. All these efforts share a common goal, i.e., to design and manufacture tissue constructs or implantable materials characterized by structural integrity, matched mechanical behavior, and sustained function at various hierarchical levels. It is critical to understand how nature uses hierarchical design spanning from molecules (amino acids) to structural ECM (collagen, elastin, and proteoglycans), to cells (various phenotypes), to tissue composites/combinations, and ultimately to functional organs/tissues. The knowledge of mechanistic integration of tissue behavior at various hierarchical levels to achieve optimal physiological functions often serves as an essential guide for bioengineering design and manufacture. The purpose of this book is to present the most recent, high-impact research accomplishments in bioengineered and natural hierarchical systems with clinical context. Our goal is to (i) highlight and emphasize the importance of integrating and bridging bioengineering designs at various tissue hierarchical levels, and (ii) provide a framework to encourage dialogue and collaboration among bioengineers, biomechanists, and clinicians.

In short, we hope this book provides not just insights but also inspiration by providing specific examples of applying multiscale approaches to clinically important problems. By understanding how nature uses tissue hierarchies to achieve optimal functions or by understanding how dysregulation of these processes leads to pathophysiology, we can apply these principles in bioengineering to bridge multiscale levels. The tissue hierarchy-based multiscale approach has the potential to achieve novel biomaterial designs, advance tissue engineering and regeneration, assist tissue function integration, improve high-fidelity computational modeling—aided by machine learning, and better design novel characterization tools/methodologies. Lastly, this book would not have been possible without the invaluable contributions, hard work, unwavering support, and patience of all the chapter contributors. It is with sincere gratitude that the editors thank them for their efforts and kind help during book preparation.

Arlington, TX, USA Jun Liao
Boston, MA, USA Joyce Y. Wong

# Contents

# Part I
# Structural Extracellular Matrices (ECMs) in Supporting Tissue Functions

# Structural Hierarchy in Collagenous Tissues and the Effects of Genetic Mutations

**Isabella J. Tan and Frederick H. Silver**

## 1 Introduction

### *1.1 Collagen as the Main Structural Component of Connective Tissues*

Collagen is the fundamental building block of connective tissues and extracellular matrices found throughout the human body. It is a fibrous protein renowned for its strength, durability, and structural significance [1]. In the context of this discussion, collagen plays an especially pivotal role in the skin [1]. The skin is a remarkable organ that serves multiple crucial functions in the human body and is influenced by both external and internal forces) (see Fig. 1). Beyond its role as a protective barrier against the external environment, the skin plays integral roles in temperature regulation, immune defense, and sensory perception [2]. Its distinctive molecular structure and assembly characteristics enable it to both store and disperse applied energy. This unique property not only confers mechanical stability to tissues but also allows it to form fibrils and fibers participating in integrin-mediated mechanotransduction [3]. Moreover, the skin significantly contributes to our physical appearance and sense of self-esteem. Therefore, the health and integrity of the skin are of paramount importance. Research indicates that approximately 45% of all deaths are linked to fibrotic diseases, underscoring the critical significance of understanding collagen structure and function [4].

I. J. Tan · F. H. Silver (✉)
Department of Pathology and Laboratory Medicine, Rutgers Robert Wood Johnson Medical School, Rutgers, the State University of New Jersey, Piscataway, NJ, USA
e-mail: ijt11@rwjms.rutgers.edu; silverfr@rutgers.edu

 
J. Liao, J. Y. Wong (eds.), *Integration and Bridging of Multiscale Bioengineering Designs and Tissue Biomechanics*, https://doi.org/10.1007/978-3-031-81743-4_1

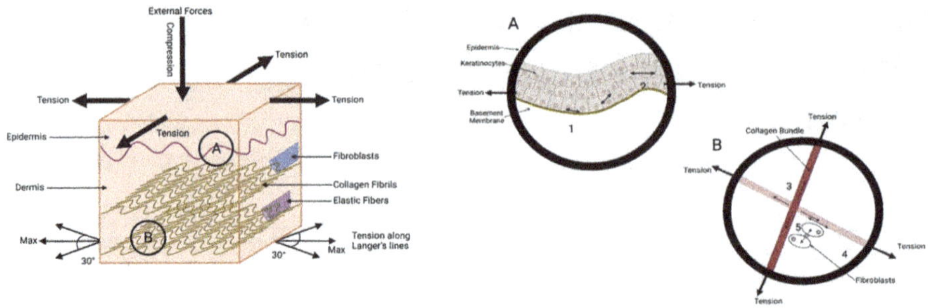


**Fig. 1** Diagram showing forces that exist in skin including gravity and skin tension. (Left) Illustration of the mechanical forces that exist in skin from gravity, and internal tension in both the epidermis and dermis. Note the tension in the epidermis (A) and collagen fibers (B) of the dermis maintain the skin microstructure and play a role in mechanotransduction. Mechanotransduction is the regulation of tissue synthesis and removal based on the mechanical forces exerted on a tissue. Collagen fibers in the dermis provide tension along Langer's lines while elastic fibers pull the collagen fibers taut in young skin. Cells provide forces that oppose the tension in the dermal collagen fibers. (Right) Diagram illustrating the tension in the epidermis (A) and in collagen fibers in the dermis (B). Tension in the collagen fibers in the dermis creates fibroblast tension almost perpendicular to the tension in the collagen fibers

**Fig. 2** Diagram of an overview of the collagen triple helix composed of three left-handed α-chain helices wrapped around each other to form a right-handed superhelix. Molecular modeling studies suggest that the collagen molecule is made of alternating rigid and flexible regions containing the amino acids proline and hydroxyproline (rigid) connected to flexible regions that lack the amino acids proline and hydroxyproline. The collagen molecule can therefore be considered a semi-flexible rod with nonhelical ends where crosslinks are located

## *1.2 Hierarchical Organization of Collagen*

A comprehensive understanding of the profound impact of collagen on skin health requires delving into its intricate hierarchical structure, which encompasses microfibrils, fibrils, fibers, and the formation of a biaxial network within the plane of the skin [5]. At its most foundational level, collagen consists of individual molecules (see Fig. 2) assembled into collagen fibrils, and these fibrils further amalgamate into more substantial structures referred to as collagen fibers [5]. However, the complexity of this hierarchical arrangement extends beyond mere fibers, culminating in the formation of elaborate networks that confer upon the skin its unparalleled mechanical properties and resilience [5]. This complex structural organization ensures that the skin functions as a biaxial network, granting it the remarkable ability to withstand a variety of mechanical stresses from multiple planes (Fig. 3).

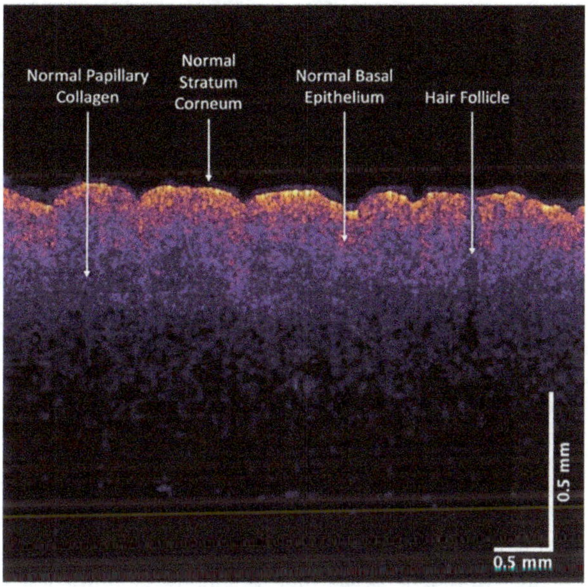


**Fig. 3** Color-coded optical coherence tomography (OCT) image of human skin. Skin is composed of a yellow stratum corneum, a germinating set of layers in pink, a basal epithelium in pink, and a papillary dermal component containing collagen and elastic fibers in blue. The papillary layer of skin contains predominantly collagen fibers. The collagen fibers are biaxially oriented allowing the skin to stretch within the plane and over joints without tearing

## *1.3 Importance of Proper Collagen Assembly and Organization for Tissue Function*

The maintenance of the skin's structural integrity and functionality is highly contingent on the proper assembly and organization of collagen, a process involving multiple steps. This assembly begins with the synthesis of individual peptide chains on fibroblast rough endoplasmic reticulum (RER), followed by subsequent post-translational modifications, including hydroxylation [6]. These modified chains are then transported through the Golgi apparatus and assembled within deep recesses in the cell membrane to form fibrils [6]. These fibrils continually grow as they are added to by other fibrils. To enhance collagen's structural integrity, crosslinking occurs through the formation of several types of linkages, including lysine-derived crosslinking, which modifies initially weak fibers into robust, viscoelastic materials [6]. This structural arrangement, characterized by collagen as a block copolymer with rigid regions connected to flexible ones, enables the storage of energy in the flexible regions, while the rigid regions prevent mechanical failure [7]. This conservation of the precise location of these flexible and rigid regions in the molecule, fibril, and fiber packing patterns ensures the maintenance of molecular properties, even when the molecules are quarter-staggered within the fibril structure [7]. Additionally, proteoglycans (PGs), such as decorin, attach to the positively stained collagen fibrils, facilitating slippage between fibrils during tensile stretching [8]. Collagen fibrils are heteropolymers, encompassing types I, III, and V collagen molecules and the sequence of assembly is precisely controlled within the cell (Table 1) [9].

**Table 1** Collagen assembly and organization in skin [9]

| Assembly and organization aspects | Description |
|---|---|
| Synthesis of peptide chains | Initiated on fibroblast RER |
| Post-translational modifications | Includes post-translational modifications, such as hydroxylation |
| Golgi apparatus transport | Chains processed through Golgi |
| Fibril formation | Assembly within cell membrane recesses |
| Fibril growth | Continuous growth by addition |
| Crosslinking for strength | Lysine-derived crosslinking for strength and elasticity |
| Block copolymer structure | Combines rigid and flexible regions |
| Conservation of structure | Precise alignment maintained |
| Proteoglycans and slippage | Proteoglycans (e.g., decorin) enable fibril movement during stretching |
| Heteropolymeric fibrils | Includes types I, III, and V collagen |

When collagen is correctly organized, it creates a dense and supportive framework that imparts strength and elasticity to the skin, complemented by a small quantity of elastic fibers that provide rebounding capability [10]. This organization equips the skin to withstand various mechanical stresses, including stretching and compression, while maintaining its resilience and suppleness. However, any disruption or irregularity in collagen's hierarchical organization can significantly impact skin health.

## 1.4 *Molecular and Fibrillar Structure of Collagen and Proteoglycans in Skin*

Understanding the molecular and fibrillar structure of collagen and proteoglycans within the skin is pivotal for comprehending both its normal function and the changes associated with skin disorders. Collagen molecules are characterized by their unique triple-helix structure, consisting of three polypeptide chains intertwined in a rope-like fashion [11]. This triple helix structure is primarily composed of the amino acids glycine, proline, and hydroxyproline, which contribute to its remarkable tensile strength [11]. Moreover, the hierarchical organization extends from these triple-helix molecules to the formation of collagen fibrils, where individual molecules are quarter-staggered to maximize the strength of the fibril and allow hole zones for mineral deposition in bone in the quarter staggered structure (Fig. 4) [11]. This intricate arrangement allows collagen to endure mechanical stress, such as stretching and compression, making it a crucial component of the skin's structural integrity.

Collagen and proteoglycans interact intricately to bolster the skin's ability to withstand mechanical forces. Proteoglycans, including decorin, ensure the correct spacing between collagen fibrils, enabling efficient transmission and dissipation of

**Fig. 4** Diagram illustrating the rigid and flexible sequences in the collagen molecule, the quarter-staggered packing pattern, and the end-to-end molecular crosslinks that stabilize collagen fibrils. Each collagen molecule is composed of a series of rigid (black vertical lines in **a**) and flexible helical components (white lines in **a**). These rigid and flexible bands are conserved in the quarter-staggered packing pattern that is found in collagen microfibrils and fibrils (**b**). Crosslinks shown in red at the end of the molecules form a continuous mechanical network that stores energy elastically and that prevents premature mechanical failure

forces while preventing excessive fibril stiffness [12]. These hydrophilic molecules also lock in moisture, maintaining skin hydration. Moreover, proteoglycans act as cushions against compressive forces, dispersing loads and shielding collagen fibrils during activities like walking or sitting [12]. This interplay between collagen and proteoglycans safeguards the skin's structural integrity and resilience through interactions with cell surface integrins found on fibroblasts and other cells in the skin (Fig. 5).

## 1.5 Changes Associated with Skin Disorders

Skin disorders, particularly those with genetic underpinnings, can lead to alterations in collagen structure and organization. Such changes often result in skin fragility, reduced elasticity, and impaired wound healing [13]. These alterations can manifest as heritable connective tissue disorders, which encompass a group of genetic conditions that affect the skin and other connective tissues.

## 1.6 Connection Between Collagen, Proteoglycans, and Skin Disorders

Collagen and proteoglycans within the skin are intricately connected, with proteoglycans serving as essential components of the extracellular matrix that surrounds collagen fibrils [13]. This matrix contributes to the skin's hydration, elasticity, and

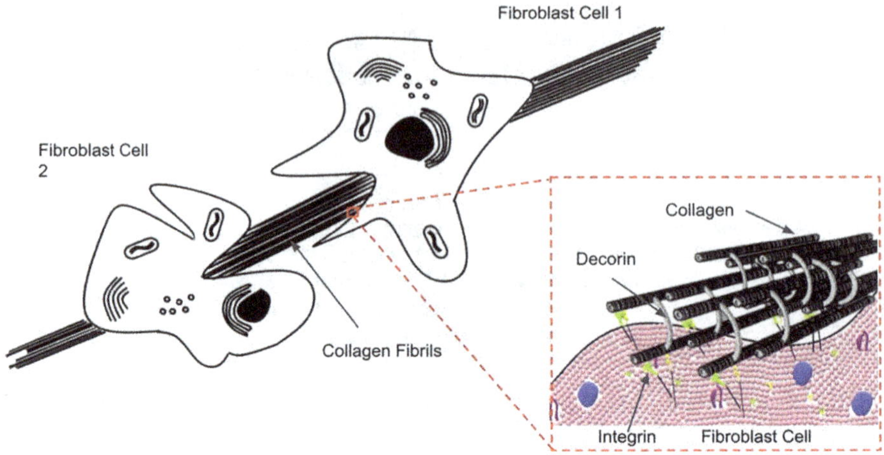


**Fig. 5** Diagram showing the relationship between collagen fibrils, decorin, and cell surface integrins. Fibroblasts sit on bundles of collagen fibrils that contain decorin on their surfaces. (Left) Fibroblasts sit on the surface of collagen fibrils and are attached via cell surface integrins. (Right) Integrins are attached to collagen fibrils and transfer external forces to the cell cytoskeleton and modify cell nuclear events through a process termed mechanotransduction. U-shaped decorin molecules are found on the surface of the collagen fibrils and allow the fibrils to slide by each other during stretching

overall mechanical properties. When genetic mutations or other factors disrupt this delicate balance, it can lead to the development of skin disorders characterized by structural abnormalities and functional deficits [13].

The significance of understanding the molecular and fibrillar structure of collagen and proteoglycans in the skin, along with the changes associated with skin disorders, cannot be overstated. Such knowledge not only deepens our appreciation of the skin's complexity but also holds the key to unraveling the mysteries of heritable connective tissue disorders. By delving into these intricate connections, researchers and clinicians aim to develop more effective diagnostic tools and targeted therapeutic interventions for individuals afflicted by these conditions. In the subsequent sections of this chapter, we will explore these concepts in greater detail, shedding light on the molecular mechanisms and clinical manifestations of skin disorders related to collagen and proteoglycan abnormalities.

## 2 Collagen: The Building Block of Skin

### *2.1 Collagen Types in Skin: Fibrous Versus Other Types*

In the context of the skin, collagen is a pivotal structural element, forming the backbone of the extracellular matrix (ECM). Notably, collagen in the skin is not

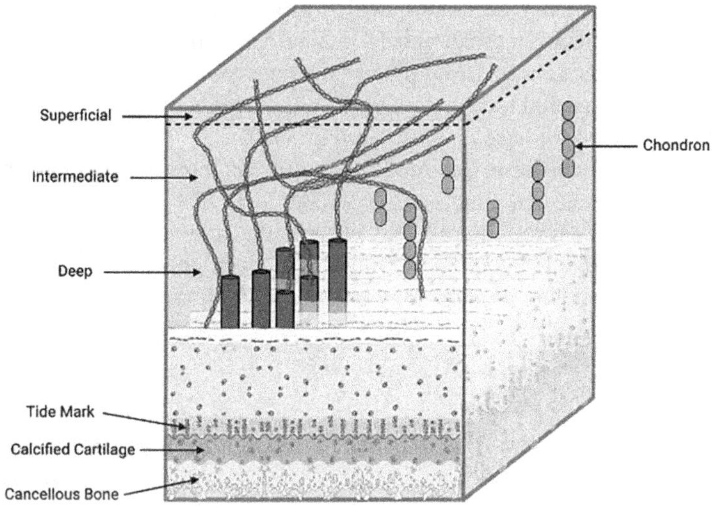


**Fig. 6** Diagrammatic representation of the structure of hyaline cartilage that lines joint surfaces. Hyaline cartilage which lines the surface of joints in long bones consists of a superficial zone, intermediate zone, and deep zone and contains collagen type II and proteoglycans. In the superficial zone, the collagen fibers are found approximately parallel to the surface very much like those seen in tendons. In the intermediate zone, the collagen fibers are found at angles of about ±45° to the cross-section of the tissue. Chondrons containing chondrocytes and PGs are found in vertical groups. In the deep zone, the collagen fibers are almost vertical and become calcified when they insert into the subchondral bone near the tide mark

a uniform entity but encompasses a diverse array of subtypes, each with unique characteristics and functions. These collagen types can be broadly categorized into two groups: fiber-forming and nonfiber-forming collagens, with each contributing distinctly to the skin's structure and function.

Fiber-forming collagens are responsible for providing tensile strength and structural integrity to the skin [14]. These collagens form organized, rope-like structures within the ECM. In the skin, the predominant fiber-forming collagens are type I and type III [14]. Type I collagen is the most abundant in the skin and is a hallmark of fibrous tissues [14]. It imparts remarkable tensile strength to the skin, ensuring its resilience and ability to withstand mechanical stress. Type III collagen, often found in association with type I collagen, plays a crucial role in maintaining the skin's pliability and flexibility [14]. It is particularly prevalent in areas requiring elasticity, such as the skin's dermal layers. The coexistence of type I and type III collagen in these regions ensures a balance between strength and suppleness.

In contrast, nonfiber-forming collagens are generally less abundant in the skin but still serve vital functions. Type II collagen, while less common in the skin, is prevalent in cartilage, where it forms the basis of the cartilage matrix [15]. This collagen type is vital for maintaining the health and function of cartilaginous tissues, as seen in joints and the nose (see Fig. 6) [15]. Type IV collagen is a major component of basement membranes, which are specialized ECM structures that separate the epidermis from the dermis [15]. These membranes play a pivotal role in the skin's barrier function, assisting in the exchange of nutrients and waste products between the epidermis and the dermis. Type IV collagen's unique properties support the structural integrity of basement membranes [15].

The collagen composition in the skin is diverse, tailored to meet the skin's mechanical, structural, and functional demands [16]. The balance between fiber-forming and nonfiber-forming collagens ensures the skin's strength, elasticity, and overall health [16]. Understanding the nuances of each collagen type is crucial for comprehending normal skin physiology and the alterations associated with various skin disorders.

## 2.2 *Structure and Organization of Collagen Fibers in Skin*

The organization of different collagen types in the skin contributes to its structural integrity and functionality. Table 2 reflects an overview of the organization of each collagen type and a summary of their roles in the skin.

The skin's collagen organization is highly structured and hierarchical [18]. Different collagen types have distinct roles in the skin. They provide tensile strength, support tissue repair, contribute to elasticity, and ensure the integrity of the skin's layers [18]. Collectively, these collagen types maintain the skin's structural integrity, making it a strong yet flexible and resilient protective barrier. Understanding the roles of collagen in the skin is essential for comprehending both normal skin function and the changes associated with collagen-related skin disorders.

**Table 2** Collagen types and their roles in the skin

| Collagen type | Organization | Roles in the skin |
|---|---|---|
| Type I | Thick fibers, densely packed [17] | Provides tensile strength and structural support [17] Predominantly found in the dermis, contributing to the skin's strength and integrity [17] |
| Type II | Forms fibrillar triple-helix molecules [6] | Crucial for cartilaginous tissues like joints and the nose, forming the cartilage matrix [6] |
| Type III | Thin fibers, loose network [6] | Supports tissue repair and elasticity. Found in the dermis, helping the skin return to its original shape after stretching [6] |
| Type IV | Mesh-like and sheet-like [6] | Mainly found in the basement membrane, separates the epidermis from the dermis [6]. Acts as a structural scaffold and supports the skin's barrier function [6] |

### 2.3 Importance of Collagen for Skin Strength, Elasticity, and Repair

Collagen's significance in skin cannot be overstated. It provides the skin with its structural framework, lending tensile strength that allows the skin to withstand mechanical stress. Collagen is essential for maintaining skin elasticity, enabling it to return to its original shape after stretching. Furthermore, collagen plays a pivotal role in wound healing, as it forms the scaffold for new tissue formation, ultimately ensuring the skin's repair and regeneration [18].

In the subsequent sections of this chapter, we will delve deeper into the molecular intricacies of collagen in the skin, exploring its role in skin disorders and the consequences of genetic mutations that affect collagen structure and function.

## 3 Role of Collagen in Mechanotransduction in Skin and Cancer

Collagen, as a fundamental component of the ECM, plays a pivotal role in mechanotransduction processes, particularly in integrin-modulated mechanotransduction [19]. Mechanotransduction is the conversion of mechanical cues or forces into biochemical signals, ultimately influencing cellular behavior [19]. Collagen's structural and mechanical properties are closely intertwined with mechanotransduction processes, impacting tissue homeostasis and disease progression [19]. In the context of skin and cancer, understanding how collagen structure is involved in mechanotransduction is crucial, both in health and in connective tissue diseases (Fig. 7).

Collagen's role in mechanotransduction is integral to understanding cellular behavior and its impact on both skin health and cancer progression. The modulation of collagen's structural properties influences the mechanotransduction pathways that govern cellular responses, providing opportunities for targeted therapies in skin-related conditions and cancer. The involvement of molecules like YAP, TAZ, beta-catenin, and WNT signaling underscores the complexity of these processes and their potential as therapeutic targets [21].

## 4 Proteoglycans: The Regulators of Skin Architecture

### 4.1 Definition and General Properties

Proteoglycans (PGs) are a family of complex molecules found in the extracellular matrix (ECM) of the skin and other connective tissues [23]. They are characterized by their unique structure, consisting of a core protein to which long chains of

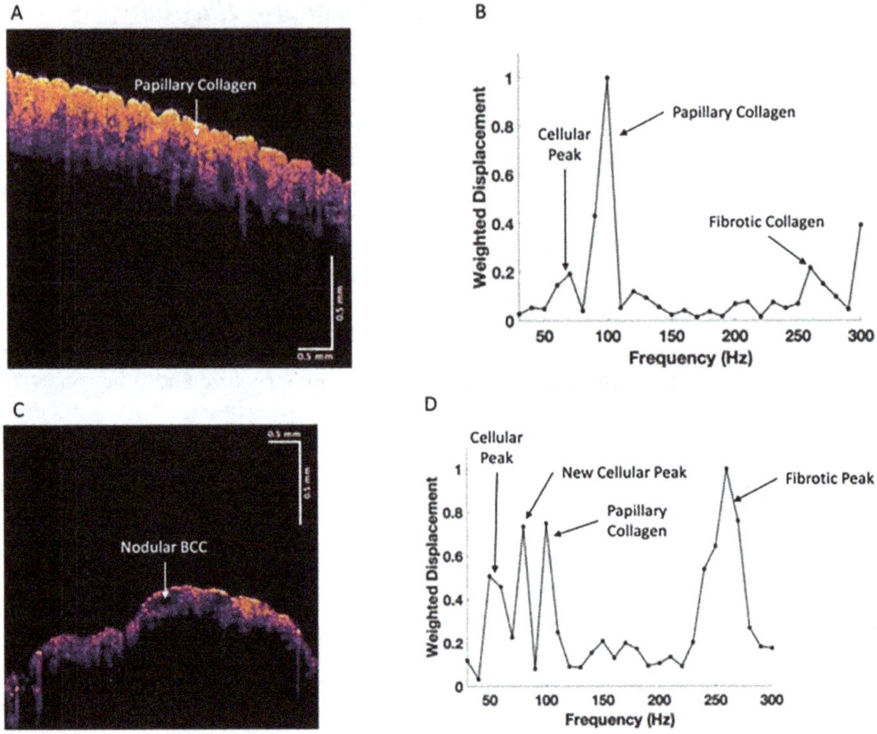


**Fig. 7** (**a**) Color-coded OCT image of normal skin showing the epidermis in yellow and pink and the papillary dermal collagen in blue. (**b**) Plot of weighted displacement versus frequency for normal skin showing the cellular (60–80 Hz), papillary collagen fibers (100), and a small fibrotic peak (250–260 Hz) obtained from vibrational optical coherence tomography studies on human skin. The stiffness of each tissue component is related to the square of the frequency. (**c**) Color-coded OCT image of a nodular basal cell carcinoma (BCC) showing the thin epidermis (yellow and pink) and the papillary dermis in blue. Note: the black oval spots are the location of the BCC lesions. (**d**) Plot of weighted displacement versus frequency for BCC showing the new cancer cellular peak (80 Hz), papillary collagen fibers (100–120 Hz), and new large fibrotic peak (250–260 Hz). Skin cancers appear stiffer than normal skin due to the presence of stiffer cells (80 Hz) and fibrotic tissue (250–260 Hz). The elastic modulus (stiffness) of each skin component is proportional to the square of the resonant frequency for each skin component (Table 3).

glycosaminoglycans (GAGs) are attached [23]. These GAGs are highly negatively charged due to the presence of sulfate groups, contributing to the hydrophilic nature of PGs [24].

### *4.2 Types of PGs in Skin and Their Locations*

In the skin, several types of proteoglycans are present, each with specific roles and distribution:

**Table 3** Collagen in skin and cancer mechanotransduction

| Aspect | Skin mechanotransduction | Cancer mechanotransduction |
|---|---|---|
| ECM stiffness and cellular response | Collagen's role in ECM contributes to mechanical properties. Changes in collagen density and organization impact ECM stiffness. Skin cells (fibroblasts, keratinocytes, immune cells) respond by modifying behaviors like migration, proliferation, and differentiation [19] | Tumors exhibit heightened ECM stiffness due to altered collagen. Cancer cells sense and respond by enhancing invasiveness and metastatic potential [20] |
| Collagen alignment and migration | Collagen fiber alignment plays a key role in cellular migration. Skin cells tend to align and migrate along collagen fibers. Critical during processes like wound healing, where efficient cell migration is required [19] | Collagen alignment induces cancer cell migration through contact guidance, providing a direct route for cancer cell invasion within the tumor microenvironment [20] |
| Skin aging | Changes in collagen structure and abundance contribute to skin aging. Reduced synthesis and increased degradation result in altered skin elasticity and strength. Mechanotransduction regulates genes related to collagen synthesis and degradation [19] | Altered collagen in cancer affects tissue stiffness. These changes impact cancer cell behavior and progression, including the regulation of genes associated with proliferation and invasiveness [20] |
| YAP and TAZ activation | YAP (yes-associated protein) and TAZ (transcriptional co-activator with PDZ-binding motif) are central to mechanotransduction. Mechanical cues, including ECM stiffness, regulate their activation. In stiff ECM, YAP and TAZ are activated and modulate gene expression [21] | YAP, TAZ, and beta-catenin are central players in the process of cancer tissue formation. In response to ECM stiffness, these molecules upregulate genes associated with cell proliferation, survival, and metastatic progression [21] |
| Beta-catenin and WNT signaling | Beta-catenin and WNT signaling pathways can become activated under conditions of ECM stiffness, modulating fibroblasts and keratinocytes to aid in the repair process [22] | Alterations in beta-catenin and WNT signaling, which are crucial in the regulation of cellular differentiation and survival, can contribute to developmental abnormalities and alterations in growth and homeostasis [22] |

1. Dermatan Sulfate Proteoglycans (DSPGs): These are found in the dermis and play a role in maintaining skin hydration and elasticity [25].
2. Heparan Sulfate Proteoglycans (HSPGs): HSPGs are present in both the epidermis and dermis. They are involved in various functions, including barrier formation and cell signaling [26].
3. Keratan Sulfate Proteoglycans (KSPGs): KSPGs are primarily located in the epidermis and contribute to the skin's barrier function [27].

### 4.3 Composition and Role of Proteoglycans in Skin

PGs are multifaceted molecules that are central to maintaining skin health and function. Comprising a core protein and one or more GAG chains, the unique variations in the lengths and compositions of these GAG chains result in a diverse array of proteoglycan structures [24]. In the skin, these core proteins are strategically embedded within the extracellular matrix (ECM), while the GAG chains extend outward, forming a gel-like environment [24]. This hydrated matrix envelops cells and collagen fibers, creating a dynamic support system.

Tendons are rich in a variety of PGs, each contributing to structural and mechanical properties. Among these PGs, decorin, a small leucine-rich proteoglycan, binds to the d band of positively stained type I collagen fibrils [24]. Additionally, hyaluronan, a high-molecular-weight polysaccharide, is present in tendons. Other small leucine-rich PGs found in tendons include biglycan, fibromodulin, lumican, epiphycan, and keratocan [28]. The amount of PGs associated with collagen fibrils in tendons decreases as fibril diameter and age increase [24]. PGs, particularly decorin and other glycoproteins in the ECM, have an essential role in normal collagen fibrillogenesis [29]. Decorin also assists in aligning collagen molecules and facilitating sliding during mechanical deformation.

Their role in preserving the skin's health and functionality is indispensable. Proteoglycans serve as moisture regulators by virtue of their highly hydrophilic GAG chains. These chains have an innate affinity for water molecules, and thus play a pivotal role in maintaining skin hydration, ensuring that the skin remains taut and well-hydrated.

Beyond this, proteoglycans contribute significantly to the skin's biomechanical properties. They enhance the skin's ability to stretch, which is crucial for withstanding mechanical stress and minimizing damage [25]. This resilience is particularly important given the skin's constant exposure to external forces and potential injuries, thereby underscoring the vital role of proteoglycans in sustaining the skin's structural integrity and overall health.

### 4.4 Interactions Between Collagen and Proteoglycans in the Skin Matrix

Collagen and proteoglycans are intimately connected within the skin matrix:

1. Collagen-Proteoglycan Complexes: Proteoglycans bind to collagen fibrils, forming complexes that contribute to the mechanical properties of the skin. These complexes help maintain the spacing and organization of collagen fibers, enhancing the skin's structural integrity [30].
2. Collagen-Related Disorders: Genetic mutations or abnormalities in proteoglycans can disrupt these interactions, leading to skin disorders. The loss of proper

collagen-proteoglycan associations can result in compromised skin structure and function [30].

In the upcoming sections of this chapter, we will delve deeper into the molecular details of proteoglycans in the skin and explore how changes in proteoglycan composition or function can lead to skin disorders and heritable connective tissue disorders.

## 5 Changes in Collagen and Proteoglycan Composition in Skin Disorders

### *5.1 Overview of Heritable Connective Tissue Disorders Affecting the Skin*

Heritable connective tissue disorders (CTDs) encompass a group of genetic conditions that affect various components of connective tissues, including collagen and proteoglycans. Many of these disorders manifest with skin abnormalities as a prominent feature. Table 4 illustrates an overview of some notable heritable CTDs affecting the skin.

### *5.2 Alterations in Collagen Structure and Synthesis Associated with Skin Disorders*

Skin disorders often involve changes in collagen structure and synthesis:

1. Increased Collagen Deposition: Some disorders, like scleroderma, lead to excessive collagen deposition in the skin, causing thickening and hardening [31].
2. Collagen Fragmentation: In conditions like dermatomyositis, collagen may undergo fragmentation, leading to skin inflammation and damage [37].
3. Abnormal Cross-Linking: Altered collagen cross-linking can affect skin elasticity, as seen in Ehlers-Danlos syndrome [33].

### *5.3 Abnormalities in Proteoglycan Metabolism and Distribution in Skin Disorders*

Proteoglycan metabolism and distribution can also be disrupted in skin disorders:

1. Reduced Hydration: Changes in proteoglycan composition can reduce skin hydration, contributing to dryness and vulnerability [23].

**Table 4** Skin disorders related to collagen and proteoglycan abnormalities

| Group | Disorder | Description of disorder |
|---|---|---|
| Collagen-related disorders | Scleroderma (systemic sclerosis) | Complex disorder characterized by abnormal collagen deposition in the skin and internal organs, resulting in skin thickening and tightness [31] |
| | Osteogenesis imperfecta (OI) | Genetic disorder causing brittle bones; skin fragility and easy bruising due to collagen synthesis abnormalities [32] |
| | Ehlers-Danlos syndrome | Group of disorders marked by skin hyperelasticity and joint hypermobility; associated with genetic mutations affecting collagen structure or processing [33] |
| | Epidermolysis bullosa | Rare genetic disorders causing blistering and skin fragility, often related to collagen and proteoglycan abnormalities [34] |
| Autoimmune and connective tissue disorders | Cutaneous lupus erythematosus | Lupus erythematosus affecting the skin, leading to various skin lesions, rashes, and photosensitivity; may involve autoimmune mechanisms affecting collagen and proteoglycans [35] |
| | Marfan's syndrome | Primarily affects the cardiovascular system and skeletal tissues but can also impact the skin, resulting in stretch marks and skin laxity related to fibrillin abnormalities [36] |
| | Dermatomyositis | Autoimmune disorder affecting the skin and muscles; characterized by a characteristic rash and perimysial inflammation in the skin [37] |

2. Altered Matrix Organization: Abnormal proteoglycan distribution can disrupt the skin's structural organization, affecting elasticity and barrier function [23]
3. Implications of Collagen and Proteoglycan Changes for Skin Integrity and Disease Progression

Collagen and proteoglycan alterations in skin disorders have significant implications for skin integrity and disease progression:

1. Skin Fragility: Changes in collagen and proteoglycan composition often lead to skin fragility, making the skin more susceptible to damage and injury [23].
2. Impaired Healing: Altered collagen and proteoglycans can hinder the skin's ability to heal properly, leading to chronic wounds and scarring [24].
3. Functional Deficits: Skin disorders associated with these changes can result in functional deficits, affecting mobility, sensation, and overall quality of life [24].

Understanding the molecular mechanisms underlying these alterations is crucial for developing targeted therapies aimed at mitigating the impact of collagen and proteoglycan changes in heritable connective tissue disorders affecting the skin.

# 6 Skin's Ability to Store, Transmit, and Dissipate Energy: Impacts of CTDs

One critical aspect of skin function is its ability to manage the energy applied to it. This energy can come in various forms, including mechanical forces, impact, and stress [38]. The skin's capacity to store, transmit, and dissipate this energy efficiently is vital for maintaining its structural integrity, resilience, and overall functionality [38]. However, this capacity can be significantly compromised in individuals with CTDs, in which structural proteins such as collagen and proteoglycans are impacted, which are essential for skin health.

## *6.1 Energy Storage in Healthy Skin*

Healthy skin possesses remarkable energy-absorbing capabilities, primarily attributed to its complex ECM composition [39]. Collagen, the predominant structural protein in the ECM, forms a dense network of fibrils that can store mechanical energy when subjected to deformation. Proteoglycans, with their hydrophilic properties, help maintain tissue hydration, enhancing the skin's ability to withstand stress. When energy is applied to the skin, such as during movement or minor mechanical stress, collagen and proteoglycans work in harmony to store this energy, preventing immediate tissue damage and maintaining skin integrity [39].

## *6.2 Energy Transmission and Dissipation*

Apart from energy storage, the skin also acts as a conduit for energy transmission and dissipation. During mechanical stress or impact, energy is propagated through the skin's layers, allowing it to be distributed across a broader surface area [39]. This minimizes localized damage and protects underlying tissues. Collagen's hierarchical organization, from molecules to fibrils to fibers, plays a pivotal role in facilitating this transmission while proteoglycans help maintain proper spacing and organization within the ECM [7].

## *6.3 Connective Tissue Disorders and Altered Energy Management*

CTDs encompass a spectrum of genetic conditions that disrupt the synthesis, assembly, processing, or stability of collagen and proteoglycans, profoundly impacting

**Table 5** Molecular alterations in connective tissue disorders

| Connective tissue disorder | Collagen mutations | Hyperelasticity | Impact on energy management |
|---|---|---|---|
| Ehlers-Danlos syndrome (EDS) | Mutations in collagen genes or genes encoding enzymes involved in collagen processing [33] | Collagen mutations lead to a disorganized and hyperelastic ECM [33] | Hyperelasticity in the skin alters its ability to store energy efficiently, rendering it vulnerable to tissue damage upon mechanical stress [33] |
| Osteogenesis imperfecta (OI) | Primarily results from mutations in collagen genes, especially type I collagen [32] | Skin is fragile and susceptible to energy impact [32] | Altered collagen structure makes the skin less effective at storing and transmitting energy, increasing the risk of bruising and tearing [32] |
| Epidermolysis bullosa (EB) | Mutation in the COL7A1 gene, which encodes type VII collagen [34] | Mutations disrupt structural stability, resulting in weakened cell-matrix interactions [34] | Inability to properly store and dissipate energy leads to blister formation and separation of skin layers, causing open sores and scarring [34] |

the skin's ability to store, transmit, and dissipate energy. Understanding the specific molecular changes underlying these alterations is crucial (Table 5).

The molecular changes in collagen and proteoglycans caused by CTDs have a profound impact on the skin's ability to manage mechanical energy. Alterations in collagen structure, cross-linking, and ECM organization lead to increased vulnerability to tissue damage, including bruising, tearing, and blistering [7]. Understanding these molecular mechanisms is fundamental for developing targeted therapies that can address the root causes of these disorders and improve clinical management in affected individuals.

## 6.4 *Implications and Research Directions*

Understanding the issues related to the skin's ability to manage energy and how CTDs affect these properties is essential for several reasons:

1. Diagnostic Significance: Altered energy management in the skin can serve as a diagnostic indicator for CTDs.
2. Treatment Development: Insights into these issues can guide the development of therapies aimed at improving the mechanical properties of the skin in individuals with CTDs.
3. Quality of Life: Addressing energy management challenges in CTDs can significantly enhance the quality of life for affected individuals by reducing symptoms and complications related to skin fragility.

# 7 Effects of Genetic Mutations on Collagenous Tissues

## *7.1 Genetic Mutations Affecting Collagen Synthesis, Assembly, Processing, Structure, and Function*

Genetic mutations can profoundly impact the synthesis, assembly, processing, structure, and function of collagen, which are essential for the integrity of various tissues, including the skin [40]. These mutations can occur in the genes encoding collagen or enzymes responsible for collagen post-translational modifications [40]. This section provides an overview of how genetic mutations can disrupt collagen's normal biology and lead to a spectrum of consequences, particularly in the context of skin disorders.

## *7.2 Consequences of Abnormal Collagen in Various Tissues and Skin Disorders*

Understanding the genetic basis of collagen-related disorders is crucial for accurate diagnosis, prognosis, and the development of targeted therapies. It also underscores the significant role of collagen in maintaining tissue integrity and function throughout the body (Table 6).

Understanding the genetic basis of collagen-related disorders is crucial for accurate diagnosis, prognosis, and the development of targeted therapies. Additionally, it highlights the interconnectedness of collagen with various tissues, emphasizing the importance of collagen in maintaining tissue integrity and function.

# 8 Effects of Genetic Mutations on Collagen-Related Skin Disorders

## *8.1 Introduction to Ehlers-Danlos Syndrome, Osteogenesis Imperfecta, and Other Connective Tissue Disorders*

Ehlers-Danlos syndrome (EDS) and osteogenesis imperfecta (OI) are prominent examples of heritable connective tissue disorders with significant skin involvement. EDS presents with a wide array of clinical features, including joint hypermobility, hyperelastic skin, atrophic scarring, easy bruising, and, in severe cases, vascular complications [33]. The underlying pathology in EDS revolves around abnormalities in collagen, whether it is related to its structure, processing, or expression [33]. Mutations in various collagen genes or enzymes responsible for collagen processing underpin different subtypes of EDS.

**Table 6** Collagen-related skin disorders and their consequences

| Category | Disorder and description |
|---|---|
| Skin disorders | Genetic mutations affecting collagen can result in heritable connective tissue disorders with significant skin involvement. Some examples include: |
| | *Ehlers-Danlos syndrome (EDS):* Group of disorders characterized by joint hypermobility and hyperelastic skin; mutations in collagen genes or processing enzymes weaken skin's integrity [33] |
| | *Osteogenesis imperfecta (OI):* Primarily associated with brittle bones, it can also affect the skin due to collagen gene mutations, resulting in fragile skin and increased bruising [32] |
| | *Epidermolysis bullosa (EB):* Rare disorders with skin blistering and fragility due to mutations in collagen genes or proteins impacting skin's structural stability [34] |
| Connective tissues | Collagen is a major component of connective tissues, and genetic mutations can have consequences beyond the skin, including: |
| | *Joint disorders:* Abnormal collagen can lead to joint hypermobility, pain, and dislocations, often seen in EDS [41] |
| | *Vascular disorders:* Collagen mutations can affect blood vessels, leading to conditions like aneurysms, dissections, and vessel fragility in vascular EDS [42] |
| | *Bone disorders:* OI, characterized by brittle bones, results from collagen mutations affecting bone strength and density [32] |
| Other tissues | Collagen mutations can impact various tissues, including the cornea (causing vision problems), the cardiovascular system (leading to heart-related issues), and the respiratory system [43] |
| Wound healing | Abnormal collagen can impair wound healing, causing chronic wounds, delayed tissue repair, and increased scarring |
| Systemic effects | Collagen mutations can have systemic effects, affecting multiple tissues and organs, resulting in complex clinical presentations [44] |

In contrast, OI primarily affects bone integrity, leading to frequent fractures, which are often associated with skin manifestations like fragile skin and easy bruising [32]. The genetic mutations in OI disrupt the synthesis of type I collagen, which plays a pivotal role in both bone and skin tissues [32].

These genetic mutations affecting collagen have profound molecular consequences. They can lead to the incorrect assembly of collagen, resulting in structurally abnormal collagen molecules. Moreover, post-translational modifications may be impaired, affecting collagen's functionality [32]. Abnormal collagen may also be more susceptible to degradation, leading to tissue fragility. Beyond this, mutations can disrupt collagen cross-linking and the process of fibrillogenesis, ultimately weakening connective tissues. Additionally, these mutations can interfere with the essential interactions between collagen and proteoglycans, which normally form a well-organized matrix in tissues. Such disruptions contribute to the disorganization of the extracellular matrix, compromising tissue integrity.

Understanding these molecular consequences is vital for comprehending the pathophysiology of collagen-related skin disorders and developing targeted therapies for individuals affected by these conditions.

## 9 Techniques for Studying Collagen and Proteoglycan Structure

### *9.1 Imaging Techniques to Visualize Collagen and Proteoglycans in Skin*

Several imaging techniques are employed to visualize collagen and proteoglycans in skin tissues. These techniques enable researchers and clinicians to gain insights into their spatial distribution, organization, and structural changes. Vibrational Optical Coherence Tomography (VOCT) is one such technique that combines optical coherence tomography with vibrational spectroscopy [45]. VOCT allows for noninvasive imaging of collagen and proteoglycans with high resolution and can provide valuable information about their structural arrangement within skin tissues [45].

### *9.2 Molecular Analysis Methods for Studying Collagen and Proteoglycan Composition*

Understanding the molecular composition of collagen and proteoglycans is crucial. Techniques such as mass spectrometry and proteomics can identify specific collagen types and post-translational modifications, providing insights into their composition [46]. Similarly, molecular assays like polymerase chain reaction (PCR) and gene expression analysis can quantify collagen-related gene expression and proteoglycan levels in skin samples [46].

### *9.3 In Vitro Systems for Investigating Skin Disorders Related to Collagen and Proteoglycan Abnormalities*

To study skin disorders associated with collagen and proteoglycan abnormalities, researchers utilize in vitro systems that mimic aspects of skin biology. These systems include:

1. Cell Culture Models Using Human or Animal Skin Cells: These models involve culturing skin cells, such as fibroblasts, keratinocytes, or melanocytes, in a

controlled environment. They allow researchers to manipulate and study specific cellular processes related to collagen and proteoglycan biology [47].

2. Organotypic Skin Models: Organotypic skin models are three-dimensional culture systems that aim to replicate the structure and function of native skin. They often incorporate multiple cell types and ECM components, including collagen and proteoglycans, providing a more physiologically relevant environment for studying skin disorders [48].
3. Bioengineered Scaffolds: These scaffolds are designed to mimic the extracellular matrix (ECM) of skin. They can be used to investigate cell-matrix interactions, tissue regeneration, and the impact of genetic mutations on collagen and proteoglycan organization [49].

Additionally, researchers can introduce genetic mutations into these in vitro systems to model specific collagen-related disorders. By using gene editing techniques like CRISPR-Cas9, they can create cellular models that carry the same mutations found in patients, allowing for the study of disease mechanisms and potential therapeutic interventions [50]. These in vitro systems provide valuable insights into the molecular and cellular aspects of skin disorders associated with collagen and proteoglycan abnormalities, aiding in the development of targeted treatments and interventions.

## 10 Therapeutic Approaches for Collagen Disorders

### *10.1 Current Treatment Options for Collagen-Related Skin Disorders*

Current treatment options for collagen-related skin disorders, such as Ehlers-Danlos syndrome (EDS), osteogenesis imperfecta (OI), and epidermolysis bullosa (EB), primarily focus on managing symptoms and improving patients' quality of life (Table 7).

### *10.2 Targeting Molecular Pathways Affected by Genetic Mutations*

Developing treatments that target the molecular pathways affected by genetic mutations in collagen-related skin disorders is an active area of research. This involves understanding the specific genetic mutations associated with each disorder and identifying potential therapeutic targets (Table 8).

**Table 7** Treatment options for collagen-related skin disorders

| Treatment | Description |
|---|---|
| Physical therapy | Recommended for managing joint hypermobility, improving muscle strength, and enhancing function in Ehlers-Danlos Syndrome (EDS) patients [51] |
| Pain management | Involves prescribing pain medications, nonsteroidal anti-inflammatory drugs (NSAIDs), or physical modalities to alleviate pain associated with collagen disorders |
| Wound care | Essential for patients with Epidermolysis Bullosa (EB), focusing on meticulous wound care. It includes using dressings and bandages to minimize friction and protect fragile skin [52] |
| Fracture management | Crucial for osteogenesis imperfecta (OI) patients. Bisphosphonates may be prescribed to increase bone density and reduce the risk of fractures [53] |
| Supportive care | Patients with collagen disorders often require orthopedic interventions, splints, braces, and assistive devices to manage mobility and joint issues |

**Table 8** Emerging therapeutic strategies for collagen disorders

| Strategy | Description |
|---|---|
| Enzyme replacement therapy | Considered in specific collagen disorders like certain types of osteogenesis imperfecta (OI) to correct underlying enzymatic defects [54] |
| Gene therapy | Gaining momentum as a potential treatment for collagen disorders, the approach aims to correct genetic defects by introducing functional copies of mutated genes into affected cells, targeting the genetic defect at its source [55] |
| Stem cell-based therapies | Stem cell therapies, including mesenchymal stem cell (MSC) treatments, hold potential for collagen disorders. MSCs can differentiate into various cell types, including fibroblasts, which produce collagen, offering the potential to replace or repair damaged tissues [56] |
| CRISPR-Cas9 and genome editing | Genome editing techniques like CRISPR-Cas9 offer precise DNA editing capabilities, potentially correcting genetic mutations associated with collagen disorders by directly modifying the affected genes [50] |
| Biologics and targeted therapies | Exploration of biologics and targeted therapies that can modulate specific molecular pathways involved in collagen disorders, addressing the underlying molecular mechanisms of these conditions [57] |

While current treatments for collagen-related skin disorders focus on symptom management, ongoing research is advancing in therapeutic approaches targeting the molecular pathways affected by genetic mutations. Emerging strategies, including gene therapy, stem cell-based therapies, and genome editing, offer hope for more effective treatments and potential cures in the future.

## 10.3 Viscoelasticity of Skin and Basal Cell Carcinomas

The skin's ability to store, transmit, and dissipate energy prevents tearing and premature mechanical failure of underlying organs and tissues. This function requires both elastic behavior (energy storage) and viscous behavior (energy dissipation). Elastic behavior is achieved at low strains by deformation of the elastic fiber network and at high strains by stretching and aligning the collagen fibers along the stretch direction [58–61]. The stiffness of the skin is the result of the contribution of cells, dermal collagen, blood vessels, and any fibrotic material present [60, 61]. Figure 8 illustrates that the elastic modulus (stiffness) of basal cell carcinomas increases over that of normal skin. This increased stiffness is associated with the presence of fibrotic tissue with a resonant frequency of 250 Hz.

Fibrosis is present in all skin cancers and other skin disorders as is seen in Fig. 9 for basal cell carcinomas. The ability of the skin to dissipate energy occurs by slippage of collagen fibers by each other during stretching [58]. This involves proteoglycans including decorin and is limited by tissue fibrosis [58, 59].

Energy dissipation occurs predominately at low strain rates in skin and other soft tissues as is shown in Fig. 10. Energy dissipation occurs via collagen fiber slippage and fluid rearrangement when the rate of loading is small. This prevents premature tearing of the skin. At high loading rates, the skin becomes stiff and tears easily since collagen fibril slippage and fluid rearrangement do not occur rapidly enough.

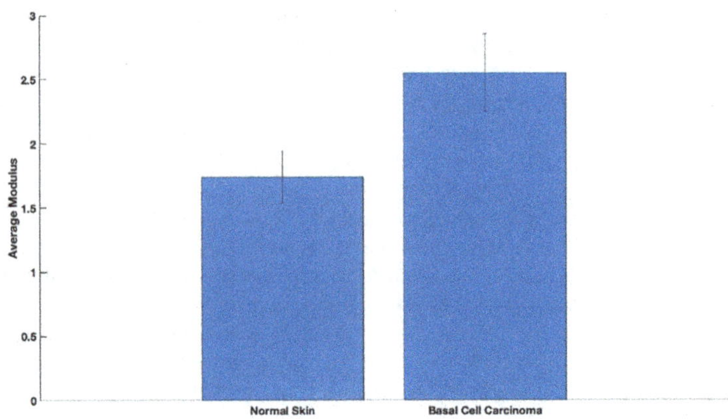


**Fig. 8** Plot of average elastic modulus for normal skin and basal cell carcinoma measured noninvasively using VOCT. The average modulus was calculated by measuring the relative contribution based on the displacement of each component in Fig. 9 (i.e., cells, dermal collagen, blood vessels, and fibrotic tissue) and weighing the modulus of each peak by the displacement contribution. This plot illustrates how tissue fibrosis stiffens tissues and leads to impaired function

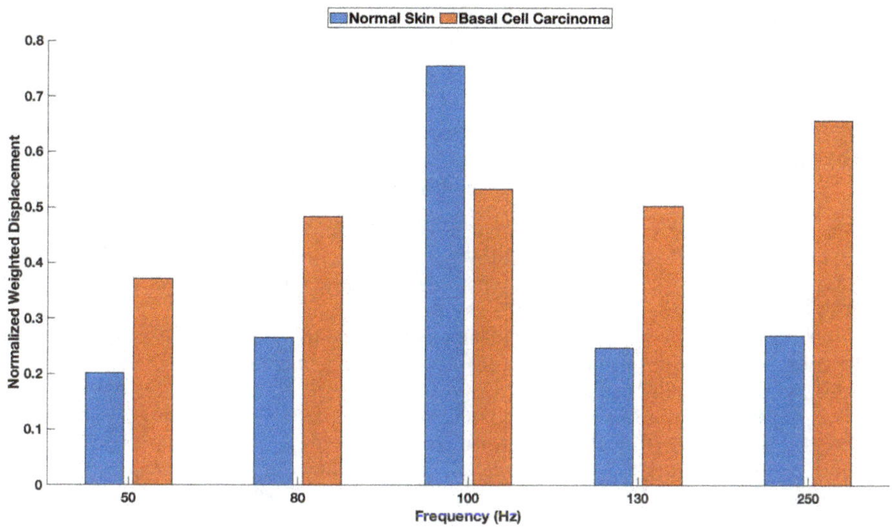


**Fig. 9** Bar plot of normalized weighted displacement versus frequency for normal skin and basal cell carcinoma measured noninvasively using VOCT. The displacement of each peak was normalized by dividing by the largest peak in each weighted displacement versus frequency plot [60, 61]. The peaks at 50 Hz (epithelial cells), 100 Hz (dermal collagen), and 250 Hz (fibrotic tissue) are all found in normal skin. The peaks at 80 Hz (cancer-associated cells), 130 Hz (thin tumor-associated blood vessels), and 250 Hz (fibrotic tissue) are found associated with all epithelial cancers [60, 61]. Fibrosis not only stiffens tissues it also leads to impaired skin function

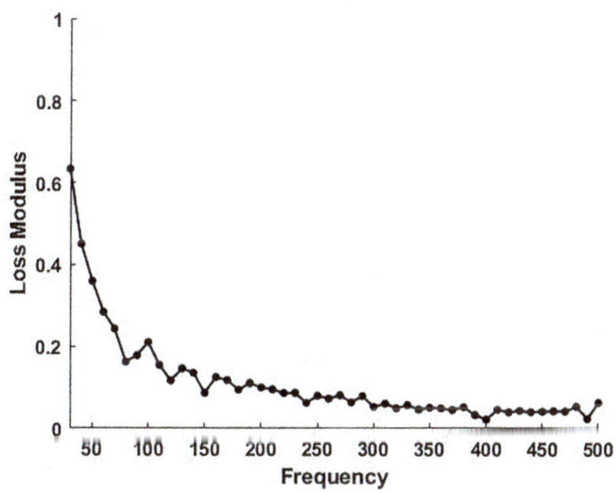


**Fig. 10** Loss modulus (energy dissipation) calculated as a fraction of the elastic modulus versus frequency in Hz (cycles per second) for human skin measured using VOCT. Note the loss modulus decreases as the frequency increases suggesting that skin dissipates more energy at low frequencies than at high frequencies. During impact loading at high frequencies, the skin offers little energy-dissipating ability leading to tissue trauma. At low frequencies, the skin dissipates large amounts of energy providing mechanical stability to tissues

**Acknowledgments** The authors would like to thank Tanmay Deshmukh for his assistance with the preparation of the figures. The figures were created under a license from BioRender.

## References

1. Zhao C, Xiao Y, Ling S, Pei Y, Ren J. Structure of Collagen. *Methods Mol Biol*. 2021;2347:17–25. https://doi.org/10.1007/978-1-0716-1574-4_2
2. Romanovsky AA. Skin temperature: its role in thermoregulation. *Acta Physiol (Oxf)*. 2014;210(3):498–507. https://doi.org/10.1111/apha.12231
3. Yin J, Zhang S, Yang C, et al. Mechanotransduction in skin wound healing and scar formation: Potential therapeutic targets for controlling hypertrophic scarring. *Front Immunol*. 2022;13:1028410. Published 2022 Oct 17. https://doi.org/10.3389/fimmu.2022.1028410
4. Wynn TA, Ramalingam TR. Mechanisms of fibrosis: therapeutic translation for fibrotic disease. *Nat Med*. 2012;18(7):1028–1040. Published 2012 Jul 6. https://doi.org/10.1038/nm.2807
5. Mienaltowski MJ, Birk DE. Structure, physiology, and biochemistry of collagens. *Adv Exp Med Biol*. 2014;802:5–29. https://doi.org/10.1007/978-94-007-7893-1_2
6. Ricard-Blum S. The collagen family. *Cold Spring Harb Perspect Biol*. 2011;3(1):a004978. Published 2011 Jan 1. https://doi.org/10.1101/cshperspect.a004978
7. Bretaud S, Nauroy P, Malbouyres M, Ruggiero F. Fishing for collagen function: About development, regeneration and disease. *Semin Cell Dev Biol*. 2019;89:100–108. https://doi.org/10.1016/j.semcdb.2018.10.002
8. Tonniges JR, Clark DL, Velleman SG. The Effect of the Wooden Breast Fibrotic Myopathy in Broilers on Fibrillar Collagen Organization and Decorin-Collagen Binding. *Avian Dis*. 2019;63(1):48–60. https://doi.org/10.1637/11985-102218-Reg.1
9. Halper J, Kjaer M. Basic components of connective tissues and extracellular matrix: elastin, fibrillin, fibulins, fibrinogen, fibronectin, laminin, tenascins and thrombospondins. *Adv Exp Med Biol*. 2014;802:31–47. https://doi.org/10.1007/978-94-007-7893-1_3
10. Theocharidis G, Connelly JT. Minor collagens of the skin with not so minor functions. *J Anat*. 2019;235(2):418–429. https://doi.org/10.1111/joa.12584
11. Shoulders MD, Raines RT. Collagen structure and stability. *Annu Rev Biochem*. 2009;78:929–958. https://doi.org/10.1146/annurev.biochem.77.032207.120833
12. Gillies AR, Lieber RL. Structure and function of the skeletal muscle extracellular matrix. *Muscle Nerve*. 2011;44(3):318–331. https://doi.org/10.1002/mus.22094
13. Huang J, Heng S, Zhang W, et al. Dermal extracellular matrix molecules in skin development, homeostasis, wound regeneration and diseases. *Semin Cell Dev Biol*. 2022;128:137–144. https://doi.org/10.1016/j.semcdb.2022.02.027
14. Kumar VA, Wang BK, Kanahara SM. Rational design of fiber forming supramolecular structures. *Exp Biol Med (Maywood)*. 2016;241(9):899–908. https://doi.org/10.1177/1535370216640941
15. Stukel JM, Goss M, Zhou H, Zhou W, Willits RK, Exner AA. Development of a High-Throughput Ultrasound Technique for the Analysis of Tissue Engineering Constructs. *Ann Biomed Eng*. 2016;44(3):793–802. https://doi.org/10.1007/s10439-015-1507-0
16. Franchi M, Trirè A, Quaranta M, Orsini E, Ottani V. Collagen structure of tendon relates to function. *ScientificWorldJournal*. 2007;7:404–420. Published 2007 Mar 30. https://doi.org/10.1100/tsw.2007.92
17. Varma S, Orgel JP, Schieber JD. Nanomechanics of Type I Collagen. *Biophys J*. 2016;111(1):50–56. https://doi.org/10.1016/j.bpj.2016.05.038
18. Sharma S, Rai VK, Narang RK, Markandeywar TS. Collagen-based formulations for wound healing: A literature review [published correction appears in Life Sci. 2022 May 15;297:120436]. *Life Sci*. 2022;290:120096. https://doi.org/10.1016/j.lfs.2021.120096

19. Pozzi A, Yurchenco PD, Iozzo RV. The nature and biology of basement membranes. *Matrix Biol*. 2017;57–58:1–11. https://doi.org/10.1016/j.matbio.2016.12.009
20. Bhattacharjee S, Hamberger F, Ravichandra A, et al. Tumor restriction by type I collagen opposes tumor-promoting effects of cancer-associated fibroblasts. *J Clin Invest*. 2021;131(11):e146987. https://doi.org/10.1172/JCI146987
21. Yui S, Azzolin L, Maimets M, et al. YAP/TAZ-Dependent Reprogramming of Colonic Epithelium Links ECM Remodeling to Tissue Regeneration. *Cell Stem Cell*. 2018;22(1):35–49.e7. https://doi.org/10.1016/j.stem.2017.11.001
22. Zhou Q, Lyu S, Bertrand AA, et al. Stiffness of Nanoparticulate Mineralized Collagen Scaffolds Triggers Osteogenesis via Mechanotransduction and Canonical Wnt Signaling. *Macromol Biosci*. 2021;21(3):e2000370. https://doi.org/10.1002/mabi.202000370
23. Carrino DA, Sorrell JM, Caplan AI. Age-related changes in the proteoglycans of human skin. *Arch Biochem Biophys*. 2000;373(1):91–101. https://doi.org/10.1006/abbi.1999.1545
24. Oh JH, Kim YK, Jung JY, Shin JE, Chung JH. Changes in glycosaminoglycans and related proteoglycans in intrinsically aged human skin in vivo. *Exp Dermatol*. 2011;20(5):454–456. https://doi.org/10.1111/j.1600-0625.2011.01258.x
25. Larjava H, Heino J, Krusius T, Vuorio E, Tammi M. The small dermatan sulphate proteoglycans synthesized by fibroblasts derived from skin, synovium and gingiva show tissue-related heterogeneity. *Biochem J*. 1988;256(1):35–40. https://doi.org/10.1042/bj2560035
26. Wang F, Wolfson SN, Gharib A, Sagasti A. LAR receptor tyrosine phosphatases and HSPGs guide peripheral sensory axons to the skin. *Curr Biol*. 2012;22(5):373–382. https://doi.org/10.1016/j.cub.2012.01.040
27. Mori M, Shrestha P, Sakamoto F, Yang LJ, Qin C, Tsujimura T. Histogenesis and possible mechanism of chondroid changes in mixed tumour of the skin: immunohistochemical evaluation of bone morphogenetic protein, glycosaminoglycans, keratin, vimentin and neuronal markers. *Arch Dermatol Res*. 1994;286(5):285–292. https://doi.org/10.1007/BF00387602
28. Pang X, Dong N, Zheng Z. Small Leucine-Rich Proteoglycans in Skin Wound Healing. *Front Pharmacol*. 2020;10:1649. Published 2020 Jan 28. https://doi.org/10.3389/fphar.2019.01649
29. Reed CC, Iozzo RV. The role of decorin in collagen fibrillogenesis and skin homeostasis. *Glycoconj J*. 2002;19(4–5):249–255. https://doi.org/10.1023/A:1025383913444
30. Junqueira LC, Montes GS. Biology of collagen-proteoglycan interaction. *Arch Histol Jpn*. 1983;46(5):589–629. https://doi.org/10.1679/aohc.46.589
31. Rosendahl AH, Schönborn K, Krieg T. Pathophysiology of systemic sclerosis (scleroderma). *Kaohsiung J Med Sci*. 2022;38(3):187–195. https://doi.org/10.1002/kjm2.12505
32. Deguchi M, Tsuji S, Katsura D, Kasahara K, Kimura F, Murakami T. Current Overview of Osteogenesis Imperfecta. *Medicina (Kaunas)*. 2021;57(5):464. Published 2021 May 10. https://doi.org/10.3390/medicina57050464
33. Riley B. The Many Facets of Hypermobile Ehlers-Danlos Syndrome. *J Am Osteopath Assoc*. 2020;120(1):30–32. https://doi.org/10.7556/jaoa.2020.012
34. Bardhan A, Bruckner-Tuderman L, Chapple ILC, et al. Epidermolysis bullosa. *Nat Rev Dis Primers*. 2020;6(1):78. Published 2020 Sep 24. https://doi.org/10.1038/s41572-020-0210-0
35. Petty AJ, Floyd L, Henderson C, Nicholas MW. Cutaneous Lupus Erythematosus: Progress and Challenges. *Curr Allergy Asthma Rep*. 2020;20(5):12. Published 2020 Apr 4. https://doi.org/10.1007/s11882-020-00906-8
36. Yuan SM, Jing H. Marfan's syndrome: an overview. *Sao Paulo Med J*. 2010;128(6):360–366. https://doi.org/10.1590/s1516-31802010000600009
37. O'Connell KA, LaChance AH. Dermatomyositis. *N Engl J Med*. 2021;384(25):2437. https://doi.org/10.1056/NEJMicm2033425
38. Laga AC, Larson A, Granter SR. Histopathologic Spectrum of Connective Tissue Diseases Commonly Affecting the Skin. *Surg Pathol Clin*. 2017;10(2):477–503. https://doi.org/10.1016/j.path.2017.01.012
39. Mackanic DG, Chang TH, Huang Z, Cui Y, Bao Z. Stretchable electrochemical energy storage devices. *Chem Soc Rev*. 2020;49(13):4466–4495. https://doi.org/10.1039/d0cs00035c

40. Kuivaniemi H, Tromp G. Type III collagen (COL3A1): Gene and protein structure, tissue distribution, and associated diseases. *Gene*. 2019;707:151–171. https://doi.org/10.1016/j.gene.2019.05.003
41. Grahame R. Joint hypermobility and genetic collagen disorders: are they related?. *Arch Dis Child*. 1999;80(2):188–191. https://doi.org/10.1136/adc.80.2.188
42. Sundel RP. Putting collagen back in the collagen vascular disorders. *Rheumatology (Oxford)*. 2011;50 Suppl 5:v60–v62. https://doi.org/10.1093/rheumatology/ker398
43. Sorkin N, Varssano D. Corneal collagen crosslinking: a systematic review. *Ophthalmologica*. 2014;232(1):10–27. https://doi.org/10.1159/000357979
44. Raisz LG. Pathogenesis of osteoporosis: concepts, conflicts, and prospects. *J Clin Invest*. 2005;115(12):3318–3325. https://doi.org/10.1172/JCI27071
45. Silver FH, Deshmukh T, Kelkar N, Ritter K, Ryan N, Nadiminti H. The "Virtual Biopsy" of Cancerous Lesions in 3D: Non-Invasive Differentiation between Melanoma and Other Lesions Using Vibrational Optical Coherence Tomography. *Dermatopathology (Basel)*. 2021;8(4):539–551. Published 2021 Dec 13. https://doi.org/10.3390/dermatopathology8040058
46. Zhang Y, Zhang N, Wei Q, et al. MiRNA-320a-5p contributes to the homeostasis of osteogenesis and adipogenesis in bone marrow mesenchymal stem cell. *Regen Ther*. 2022;20:32–40. Published 2022 Mar 28. https://doi.org/10.1016/j.reth.2022.03.001
47. Bennink LL, Li Y, Kim B, et al. Visualizing collagen proteolysis by peptide hybridization: From 3D cell culture to in vivo imaging. *Biomaterials*. 2018;183:67–76. https://doi.org/10.1016/j.biomaterials.2018.08.039
48. Shalom-Feuerstein R, Serror L, De La Forest Divonne S, et al. Pluripotent stem cell model reveals essential roles for miR-450b-5p and miR-184 in embryonic corneal lineage specification. *Stem Cells*. 2012;30(5):898–909. https://doi.org/10.1002/stem.1068
49. Pang Y, Wang X, Ucuzian AA, et al. Local delivery of a collagen-binding FGF-1 chimera to smooth muscle cells in collagen scaffolds for vascular tissue engineering. *Biomaterials*. 2010;31(5):878–885. https://doi.org/10.1016/j.biomaterials.2009.10.007
50. Uehara H, Zhang X, Pereira F, et al. Start codon disruption with CRISPR/Cas9 prevents murine Fuchs' endothelial corneal dystrophy. *Elife*. 2021;10:e55637. Published 2021 Jun 8. https://doi.org/10.7554/eLife.55637
51. Reychler G, De Backer MM, Piraux E, Poncin W, Caty G. Physical therapy treatment of hypermobile Ehlers-Danlos syndrome: A systematic review. *Am J Med Genet A*. 2021;185(10):2986–2994. https://doi.org/10.1002/ajmg.a.62393
52. Kirkorian AY, Weitz NA, Tlougan B, Morel KD. Evaluation of wound care options in patients with recessive dystrophic epidermolysis bullosa: a costly necessity. *Pediatr Dermatol*. 2014;31(1):33–37. https://doi.org/10.1111/pde.12243
53. Rossi V, Lee B, Marom R. Osteogenesis imperfecta: advancements in genetics and treatment. *Curr Opin Pediatr*. 2019;31(6):708–715. https://doi.org/10.1097/MOP.0000000000000813
54. Bacon S, Crowley R. Developments in rare bone diseases and mineral disorders. *Ther Adv Chronic Dis*. 2018;9(1):51–60. https://doi.org/10.1177/2040622317739538
55. Schindeler A, Lee LR, O'Donohue AK, Ginn SL, Munns CF. Curative Cell and Gene Therapy for Osteogenesis Imperfecta. *J Bone Miner Res*. 2022;37(5):826–836. https://doi.org/10.1002/jbmr.4549
56. Botor M, Fus-Kujawa A, Uroczynska M, et al. Osteogenesis Imperfecta: Current and Prospective Therapies. *Biomolecules*. 2021;11(10):1493. Published 2021 Oct 10. https://doi.org/10.3390/biom11101493
57. Hellmich B, Henes JC. Biologika bei Kollagenosen und Vaskulitiden [Biologics for connective tissue diseases and vasculitides]. *Internist (Berl)*. 2022;63(2):143–154. https://doi.org/10.1007/s00108-021-01249-w
58. Silver FH, Freeman JW, DeVore D. Viscoelastic properties of human skin and processed dermis. *Skin Res Technol*. 2001;7(1):18–23. https://doi.org/10.1034/j.1600-0846.2001.007001018.x.
59. Silver FH, Kelkar N, Deshmukh T. Molecular Basis for Mechanical Properties of ECMs: Proposed Role of Fibrillar Collagen and Proteoglycans in Tissue Biomechanics. *Biomolecules*. 2021;11:1018.

60. Silver FH, Deshmukh T, Gonzalez-Mercedes M. Use of Vibrational Optical Coherence Tomography to Noninvasively Differentiate Between Benign Fibrotic Tissue and Fibrosis Associated with Melanoma, *21st Century Pathology*. 2023;3:139.
61. Silver, FH, Deshmukh, T, Ryan N, Romm A, Nadiminti, H. "Fingerprinting" Benign and Cancerous Skin Lesions Using Vibrational Optical Coherence Tomography: Differentiation among Cancerous Lesion Types Based on the Presence of New Cells, Blood Vessels, and Fibrosis. *Biomolecules* 2022;12:1332. https://doi.org/10.3390/biom12101332

# Contribution of Proteoglycans to the Multiscale Structure and Function of Soft Tissues

**Béla Suki and Matthew A. Nugent**

## 1 Introduction

Soft tissues are a key part of the body, and they contribute to the functionality of virtually all organs. The composition of soft tissues includes various cells, and a supporting framework, called the extracellular matrix (ECM). The organization of the ECM is often hierarchical giving rise to a multiscale functional behavior [1]. A salient example of such multiscale behavior is provided by the tendon which transmits mechanical force from muscle to bone at the macroscale [2]. To be efficient, the tendon needs to have a relatively high stiffness, loosely defined as the elastic resistance against deformation. This biomechanical function at the macroscale is a consequence of the stiffness and hierarchical organization of collagen starting from the triple helix of tropocollagen through microfibrils, fibrils, and fibers all the way up to the tissue level [1]. On the other hand, the stiffness of fibrils must also provide proper mechanical milieu for the resident cells that maintain the tendon [3]. The organization of the ECM in other tissues is similarly multiscale although not necessarily as hierarchical as that of the tendon. For example, in contrast to the cylindrical organization of the tendon, collagen fibrils in the skin form a random network which allows a high tear resistance [4].

While different tissues in the body have varying ECM composition and structure supporting tissue-specific function, the multiscale organization of virtually all soft tissues is regulated by a class of molecules called the proteoglycans (PGs) [5]. PGs are comprised of a core protein with covalently attached glycosaminoglycan (GAG)

B. Suki (✉)
Department of Biomedical Engineering, Boston University, Boston, MA, USA
e-mail: bsuki@bu.edu

M. A. Nugent
Department of Biological Sciences, Towson University, Towson, MD, USA

 
J. Liao, J. Y. Wong (eds.), *Integration and Bridging of Multiscale Bioengineering Designs and Tissue Biomechanics*, https://doi.org/10.1007/978-3-031-81743-4_2

chains. PGs are known to play roles in many important cellular processes including cell signaling, migration, mechanotransduction (conversion of mechanical cues to biochemical signals), maintenance, and repair [6–9]. In particular, cells secrete and deposit PGs into the ECM allowing them to bind to many ECM proteins which regulate fibrillogenesis and general ECM assembly during development and repair. For example, decorin, a small leucine-rich proteoglycan (SLRP), decorates collagen fibrils at regular intervals to control fibril diameter as well as help maintain a proper interfibrillar distance [10, 11]. Such a strategic role in organizing the microscale structure in turn affects tissue function at the macroscale. In this sense, SLRPs operating at the level of fibrils can be said to fine-tune the multiscale biomechanical function of soft tissues including, among others, tendon [12], blood vessels [13], skin [14], heart [15], cornea [16], and lung [17].

The question then arises whether PGs directly contribute to the multiscale biomechanical function of the ECM. The answer to this question is not clear and the issue has been debated for many years without a clear resolution. Various studies reported contradictory results regarding PGs' and/or their GAG side chains' potential in transmitting mechanical forces between neighboring fibrils. For example, the compressibility of cartilage is governed by the negative charge distribution of proteoglycans [18], more specifically aggrecans that belong to the class of aggregating proteoglycans. Lung tissue stiffness at low lung volumes is certainly influenced by the compressive and shear resistance of PGs [17]. In engineered elastin sheets, the PG network controls failure strain but not stress [19]. In addition, the residual stress in blood vessels has also been associated with PGs [20]. On the other hand, depletion of GAGs seems to have little effect on the viscoelastic properties of and hence force transmission in tendon [21]; instead, GAGs may facilitate the sliding of collagen fibrils on each other which may protect fibrils from rupture under supraphysiological loads [22]. Hence, it appears that PGs' participation in biomechanical function is tissue specific. Whether their role is also multiscale in those instances where they influence the biomechanical properties of specific tissues has rarely been addressed.

Beyond the biomechanical function, PGs have many other physiological and biological roles both in health and disease. A key function is to provide proper hydration to the ECM and its resident cells. Interestingly, PG hydration can also influence biomechanical function [17, 23]. PGs are also known to bind to many proteins such as growth factors, cytokines, cell surface receptors, fibronectin, elastin, and collagens [24]. For example, the wide array of decorin's binding partners forms an interesting interaction network [25] that controls many cell and ECM functions. Since PGs can bind to receptors and their cognate ligands, their binding is often considered the primary triggering event of cell signaling cascades. Hence, they play critical roles in tissue and organ development [13, 26–28] and also contribute to aging [29–31] and many disease pathologies [3, 5, 7, 30]. Although PGs' biological activity is limited to the molecular scale, the end result of this activity occasionally percolates to the macroscale which can therefore be considered multiscale.

The purpose of this chapter is to review PGs' general contribution to the structure-function relationship of soft tissues in the context of multiscale behavior. Since PG structure has been reviewed in many excellent reviews [3, 6, 7], here we will focus on the role of PGs in tissue function from both the biomechanical and biological points of view. We will attempt to achieve this via specific examples given that PG function, especially biomechanical, remains somewhat controversial and not easily generalizable. We will start with a general description of the multiscale biomechanical properties of soft tissues, especially how microscale mechanisms govern macroscale behavior. Next, we will describe the biophysical and biological properties and activities of PGs in various tissues such as blood vessels, cornea, and the lung. Following a short discussion of how PGs contribute to diseases from a multiscale perspective, we will conclude the chapter by attempting to answer questions such as why multiscale properties matter and what are the possible roles of PGs in the future of tissue engineering and regenerative medicine.

## 2 General Multiscale Properties of Soft Tissues

The popular notion of multiscale structures is best demonstrated via the hierarchical organization of tendon (Fig. 1), a soft tissue that links bones and muscles allowing the stabilization of joints and the transmission of muscular forces to bones [32]. The structure spans an enormous length range of about seven orders of magnitude from the amino acids at the subnanometer scale, through the alpha chains of the collagen triple helix, the fibrils, the fibers, and the fascicles to the tendon at the mm scale. At each level, the organization follows an approximate helical structure stabilized by hydrogen bonds and crosslinks within and between collagen molecules. The hierarchical nature of the tendon can be traced to thin, elongated self-similar objects at a given length scale winding around each other to produce a thin elongated but larger object at the next length scale. For example, the collagen molecule is a result of three left-handed alpha chains forming a right-handed superhelix. At the next level, microfibrils are composed of four or five collagen molecules at a given cross-section twisting around each other in a right-handed manner such that these microfibrils interdigitate with their neighbors [33]. The microfibrils in turn go on to form full fibrils which are arranged into fibers and so on. As illustrated in Fig. 1, PGs are located at the scale of fibrils attaching to them at regular intervals mostly in the gap region of the fibrils' D periods. This hierarchical organization in other soft tissues often stops at the level of fibrils or fibers which in turn become the structural units of various extended two-dimensional (2D) or three-dimensional (3D) networks. Basement membranes can be considered as 2D type IV collagen networks whereas the skin is a typical example of a quasi 2D network of type I collagen fibers. There are soft tissues with more complex organization such as the lung parenchyma which consists of two intertwined and mechanically interacting networks, the type I-type III collagen fibrils and elastic fibers.

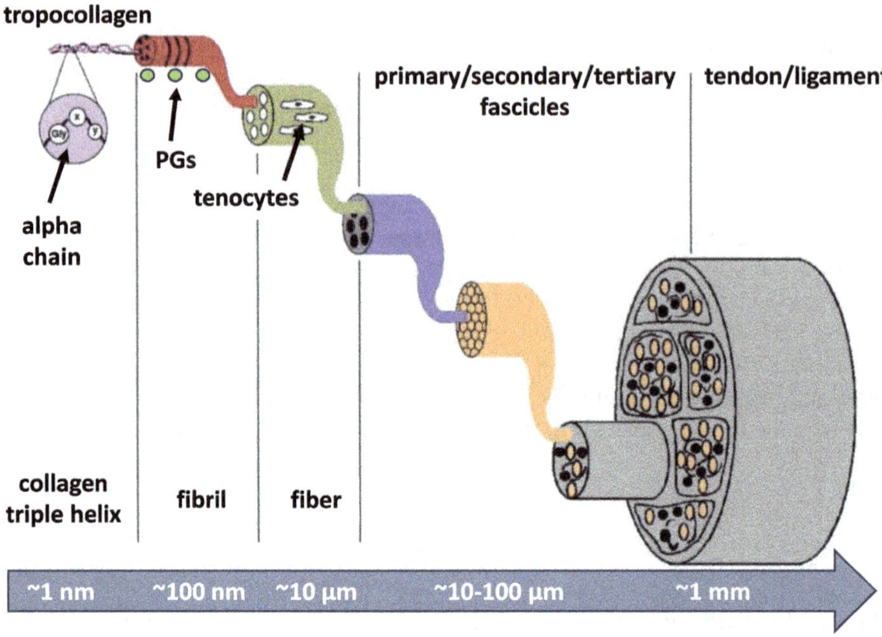


**Fig. 1** Multiscale hierarchical structure of tendons and ligaments starting at the scale of alpha chains through collagen triple helix, fibrils, fibers, and fascicles to the organ. Proteoglycans (PGs) are located in between fibrils and operate at the length scales between tens and hundreds of nm. (Modified with permission from [32])

What functional roles can such complex multiscale organizations play in general soft tissue behavior? The answer is tissue specific. Tendon, for instance, transmits force generated by the muscle. For this purpose, the best shape for the tendon is a cable-like structure. The internal organization of the cable depends on the functional demands on the tendon such as maximum load, maximum speed, and minimum dissipation during force transmission. For this, a nearly rigid cable with a very high stiffness, defined as the change in stress (force over area) divided by change in strain (relative change in length, area, or volume) in the axial direction, is suitable. However, the tendon also has another important function, namely, storing elastic energy during load transfer to augment muscle performance [34]. The ability to store elastic energy requires the tendon to be stretchable, the extent of which is determined by the stiffness of the tendon. Stiffness in turn is a function of Young's modulus (stiffness in uniaxial deformation) of the basic unit, the collagen molecule, and the way the units are assembled. As we have just seen, the organization of tendon and most other soft tissues is multiscale and, hence, stiffness at the organ level is an inherently multiscale concept in biology. The stiffness of a collagen molecule ranges from 3 to 9 GPa and that of the fibril is between 0.4 and 0.9 GPa [35]. The stiffness of different tendons in various species exhibits an enormously wide range of values from about 50 MPa to 1.6 GPa [36]. Although it seems difficult to find order in these data, it is interesting to note that another type of mechanism may help better understand such variability. The fibrils and fibers within the tendon are not completely straight; instead, there are microscopic crimps and waves along the fibrils and fibers (Fig. 2) with a distribution of waviness [37]. When the tendon is stretched, these waves become straighter suggesting they act as shock absorbers. The general assumption is that a wavy fibril does not contribute to the macroscopic force until it becomes straight which can be utilized to develop multiscale computational models of tendon mechanics [38]. Thus, a recruitment phenomenon occurs during tendon stretching that gives rise to a nonlinear stress-strain curve at low strains because as more fibrils become straight, their added contribution to force development gradually increases the macroscopic stiffness (Fig. 3). Such a variable stiffness may be in part a consequence of strain amplitude, strain rate, loading history, and tendon type. More importantly, since the tendon operates in the low-strain regime, the systematic variation of stiffness along the stress-strain curve can be exploited so that the tendon can efficiently respond to different loading conditions. This is clearly a functional multiscale behavior.

While the tendon needs to transmit force in a single direction, other soft tissues such as skin or lung parenchyma are stretched biaxially or triaxially, respectively. Furthermore, the high stiffness of tendon would not work for the skin as it would likely fail to serve the purpose of easy locomotion since skin deformation is affected by joint movement [39]. Similarly, too high stiffness for the alveolar walls in the lung would impair normal breathing. Nature has evolved a different structure for such cases, the random network of fibrils (Fig. 2, #5). Since fibrils in a random network can fold and align with the direction of loading before becoming stretched, the associated stiffness can be much smaller than that of the tendon. For example, the stiffness of the skin ranges from 4 to 15 MPa depending on the location and direction of stretching relative to the Langer lines [40], whereas the stiffness of the lung parenchyma is between 1 and 5 kPa and that of single walls is between 30 and 70 kPa [41]. At low strains, elastin appears to dominate the stiffness of both skin [42] and lung tissue [43]. Elastin's mechanical behavior is characterized by a nearly perfectly linear stress-strain curve without much dissipation [1]. On the other hand, the random network of collagen fibrils exhibits a nonlinear stress-strain behavior which is due to fibril alignment, another recruitment process. However, the fibrils themselves are also wavy; hence, following alignment, fibril straightening takes over to continue extending the nonlinear behavior. The two sequential recruitment processes, alignment and straightening, are in a sense analogous to the fibril straightening followed by the fiber crimp elimination in tendon. Hence, in the skin and the lung as well as other soft tissues, we see a multiscale functional mechanical behavior.

An interesting consequence of fibril alignment is its role in the superior tear resistance of the skin. A small cut in the skin does not propagate as a crack in hard tissues such as bone where large-scale alignment is not possible; instead, alignment orients fibrils perpendicular to the cut effectively stopping the mechanical conditions for crack propagation [4]. Thus, recruitment as a dynamic process operating on a multiscale structure can serve different functional purposes in different tissues.

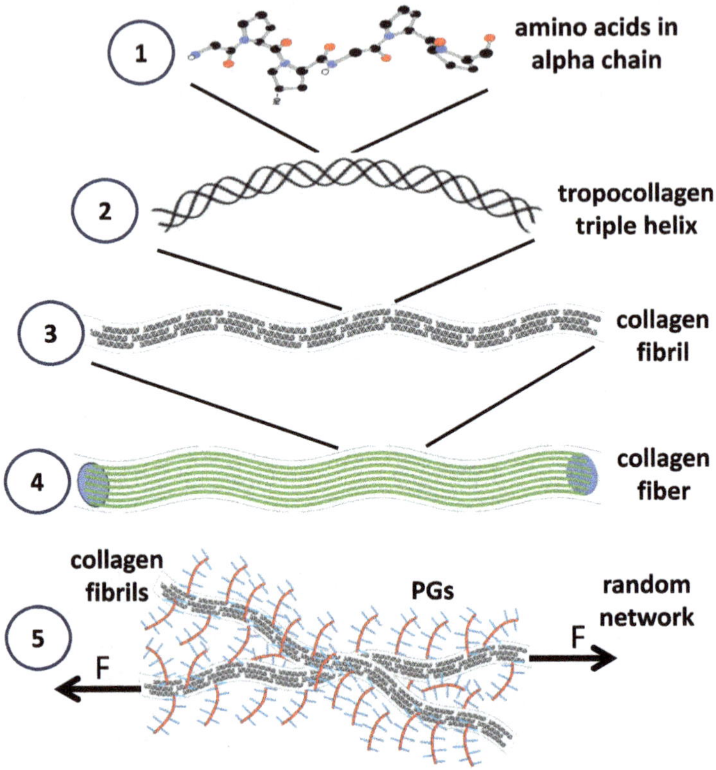


**Fig. 2** Multiscale dynamic structure and corresponding function. Wavy structures are shown at the alpha chain (1), tropocollagen (2), collagen fibril (3), and collagen fiber (4) levels. In various tissues, fibrils and fibers further organize to form fascicles (not shown) or random networks (5). In the presence of force (F), fibrils and fibers align with the direction of applied force

It has been argued that recruitment is in fact a general principle in biology that is inherently multiscale ranging from the tissue scale all the way down to the semiflexible molecules of the ECM (Fig. 2). Specifically, "during evolution, the ECM and its components may have been fine-tuned for optimal performance within the recruitment region because in this region the stiffness of the system can adaptively change so as to match its response to best satisfy function under widely different external loading conditions" [1].

What possible roles may PGs play in this multiscale scenario? From the structural point of view, as we pointed out in the introduction, PGs have the ability to fine-tune the hierarchical organization of collagen fibrils by modulating fibril diameter and spacing [10, 11]. Knockout studies provide clear evidence that the absence of PGs such as decorin, fibromodulin, or biglycan leads to grossly altered fibril shape and ECM organization [10, 44, 45]. Hence, PGs can indirectly influence the multiscale structure of tissues which in turn affect function. In general, larger-diameter fibrils

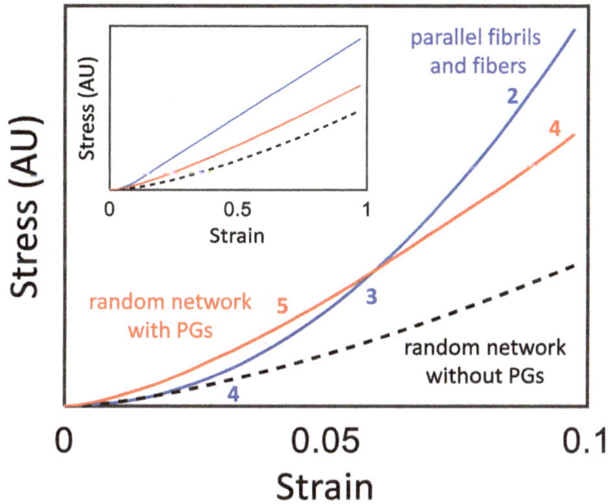


**Fig. 3** Low-strain regime of stress-strain curves of a parallel fibril system (blue) vs. a random network with (red) or without (black dashed) PGs. The numbers represent the alignment or straightening of wavy structures corresponding to those in Fig. 2, providing a recruitment-based nonlinear behavior. Inset: Simulations showing the linear region of the stress-strain curves, where the collagen molecules and the alpha chains become stretched

have lower stiffness and hence a lower bending modulus [46]. Hence, it may be expected that the heterogeneous fibril diameters caused by the lack of certain PGs would alter crimp morphology and recruitment. An important question is: Do PGs also contribute to the multiscale function of fully developed soft tissues? In tendon, enzymatically depleting GAGs did not seem to affect crimp morphology [47] implying no direct GAG involvement in mature tendon function. However, an electron microscopic study demonstrated that loading mitral valve chordae altered the angle between fibrils and PGs implying that PGs play a role in load transfer [48]. A more recent study using GAG depletion also arrived at a similar conclusion [49]. In an attempt to reconcile these observations, we can consider the response of PGs and GAGs to changing demands. For example, moderate running in young adult rats reduced crimp waviness allowing earlier recruitment and stronger force transmission that was accompanied by increased productions of type I collagen and PGs by tenocytes [12]. This intriguing finding demonstrates that stretch-induced mechanotransduction fine-tunes the crimp structure and the recruitment-associated function. More generally, we point out that the amount and types of PGs produced by tissue-specific fibroblasts are sensitive to mechanical stretch [12, 50, 51] as well as the stiffness of the underlying ECM [52, 53]. Thus, tissue distribution of PGs responds to the mechanical environment indicating that PGs play a role in the adaptive tuning of function. In addition, in the random network of collagen fibrils in lung parenchyma, alignment of fibrils was found to depend on solution tonicity and hence the swelling state of PGs [17]. Such mechanical interactions between PGs and collagen influence fibril recruitment at low strains. A similar example is the cornea where PGs are critical to controlling collagen fibril thickness and spacing (see below). It is interesting to note that digesting GAG chains results in significantly softer tensile properties [16]. Thus, the question of whether PGs have any role in multiscale function is not fully resolved and is clearly tissue specific.

## 3 Biological Properties of PGs with Multiscale Implications

PGs are complex macromolecules that are comprised of a protein core with one or more covalently attached linear glycosaminoglycan chains (Fig. 4). Proteoglycans are often categorized based on the specific GAG chains attached to the core protein: chondroitin sulfate (CS), dermatan sulfate (DS), keratan sulfate (KS), or heparan sulfate (HS). However, there are also instances where multiple GAG types can be attached to the same core protein, making mixed GAG PGs. Each GAG class is defined by the chemical properties of the repeating disaccharide unit that

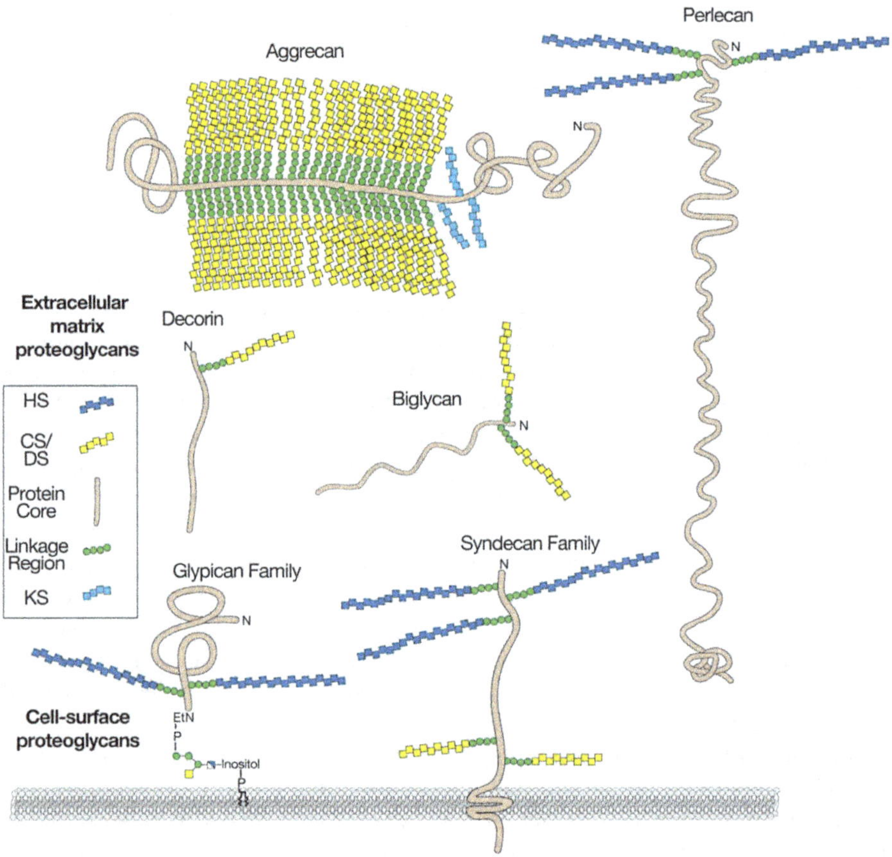


**Fig. 4** Structure of the major PG types. PGs are comprised of a core protein (light brown) with one or more covalently attached GAG chains (dark blue, heparan sulfate; yellow, chondroitin sulfate/dermatan sulfate; light blue, keratan sulfate). Membrane PGs, syndecan, and glypican families either span the plasma membrane (type I membrane proteins) or are linked by a glycosylphosphatidylinositol anchor. Extracellular matrix proteoglycans, aggrecan, biglycan, decorin, and perlecan are usually secreted. (This figure and information in the legend are from Reference [54] with permission (https://creativecommons.org/licenses/by-nc-nd/4.0/))

contains an amino sugar, either glucosamine (N-acetylated or N-sulfated) or *N*-acetylgalactosamine, and a galactose or uronic acid (glucuronic acid or iduronic acid). All PG-associated GAGs carry a high negative charge density due to the presence of sulfate and carboxyl groups. The negative charge contributes to the ability of GAGs to attract and bind water to control tissue hydration, as well as to bind proteins as a mechanism to control extracellular matrix structure and cell signaling in response to growth factors and hormones.

GAG fine structure is extremely heterogeneous as a result of structural modifications within the specific disaccharide units through the action of biosynthetic enzymes in the Golgi leading to variations in chain length and regions of high and low sulfation among other changes [54, 55]. This structural diversity is particularly noteworthy in HS where more than 18 biosynthetic enzymes produce as many as 32 distinct disaccharide units to make this class of compounds one of the most information-dense in nature [56–58]. The fine structure of HS, for example, can be further modified in the pericellular environment by endosulfatases and heparanases [59]. These modifications are tissue and condition specific, allowing HS structure to rapidly adjust to the demands placed on the tissue. These changes in HS fine structure alter protein binding specificity, which can lead to multiscale changes in tissue structure and response to mechanical force. For example, degradation of HS with the enzyme heparinase in endothelial cells reduced the adhesion strength of cells to the ECM and inhibited shear stress-induced cellular alignment and recruitment of phosphor-focal adhesion kinase in the direction of flow [9]. Similarly, disrupting HS biosynthesis by silencing modifying enzymes leads to suppression of flow-induced signaling and motility in vascular smooth muscle cells [60, 61]. At the tissue scale, transgenic mice overexpressing heparanase show increased arterial thickness, cellular density, and mechanical compliance, and the extent of heparanase expression correlated with neointimal thickening in response to vascular injury [62]. Hence, even though HS is not a mechanical component of tissue per se, altering its level and structure can modulate the macroscale mechanical properties of cells and tissues.

While the chemical structure of the specific PG core protein and associated GAG chains plays a significant role in defining function, sub-cellular location has been proposed to be a more effective means of classifying PGs [63]. Only one PG, serglycin, is found within cells packed within the granules of mast cells where it functions to store proteases. A large class of cell surface-associated proteoglycans, predominantly those containing HS chains, have been shown to regulate cellular response to external stimuli such as growth factors by orchestrating receptor activation and signaling. These HSPGs are linked to the cell surface through transmembrane core proteins or through glycosyl-phosphatidyl-inositol (GPI) anchors. In both cases, the PGs can be endocytosed or shed from the cell surface through the action of pericellular enzymes [64]. Internalization of HSPGs has been shown to modify the trafficking and processing of growth factors and their receptors and to inhibit histone acetyltransferase activity [65–68]. Shedding and enzymatic modification of cell surface-associated HS have been associated with changes in activities including inhibition or promotion of cancer development and progression [69–71].

The PG core proteins direct where PGs will be localized within the extracellular and pericellular environment and they also interact with other proteins to contribute to PG function. The number of core proteins is relatively small with ~40 having been identified [72]. Consequently, PG core proteins can be grouped by genetic homology and the presence of specific protein modules [73]. The syndecans are noted for having a transmembrane domain that embeds them in the plasma membrane, whereas the glypicans contain a GPI linkage region. Another family is the small leucine-rich PGs (SLRPs) that are characterized by the presence of leucine-rich repeats flanked by cysteine clusters [63]. The core proteins of SLRPs bind other proteins in the ECM, such as collagens, to control fiber diameter, spacing, and higher-order tissue structure [74]. Another family of core proteins that include aggrecan, versican, neurocan, and brevican is noted for its large size and ability to bind to hyaluronan to create large aggregates in the ECM [72]. These PGs have CS as their major GAG and were the first studied in detail because of their abundance and ease of isolation. The aggregating PGs modulate hydration and the overall physical structure of the ECM. Some core proteins, such as versican, are also subject to alternate mRNA splicing to give rise to several isoforms with distinct activities [75]. Thus, the diversity of core protein structure dictates the function of PGs through direct interactions with other active biomolecules as well as through their role in defining the localization of the PG.

While PG core proteins play critical roles in PG function, significant attention has been focused on the role of the GAG chains in mediating the wide range of activities of these complex molecules. In particular, the ability of the highly diverse HS chains to bind and modulate proteins, and in turn control cell response to external stimuli, has been an area of intense research. HS is essential for embryonic development [76] and is required for the function of all adult physiological systems. HS structure and expression can change rapidly during development [76–78], indicating that alterations in HS might be a key signal of functional changes in cells and tissues. A bioinformatics analysis of the HS interactome identified 435 human proteins that interact with HS or the structurally related heparin [79]. We will discuss two examples to illustrate the range of HS activity: the fibroblast growth factor (FGF) family and fibronectin.

The FGFs represent a large family of related proteins that bind to HS and have a wide range of biological activities including regulation of cell proliferation, migration, differentiation, and tissue morphogenesis [80, 81]. The involvement of HS in regulating FGFs represents a classic mechanism by which HS can modulate protein-protein interaction [82]. HS has been shown to function as a co-receptor that stabilizes the binding of FGF to its cognate receptors on cell surfaces through interactions with both proteins. The HS-FGF-receptor complexes signal cells to respond to FGFs in a robust manner, whereas the absence of HS leads to limited cell response [83]. The HS-FGF-receptor complexes also appear to have a distinct fate once internalized within cells leading to changes in signal persistence and ultimate cell response [65, 67]. While the structural requirements for HS to bind FGFs have been described and receptor specificity has been defined for the FGF system's 23 ligands and 7 receptor variants [84, 85], the HS structural requirements for interaction with FGF receptors or how distinct domain organization within HS can modulate FGF-FGFR complex formation remain to be defined. Thus, on a molecular scale, HS-containing PGs can modulate protein-protein interactions, which can then translate to a dramatic change in tissue response. For example, reduced or altered HS within the blood vessel wall has been shown to alter the vascular response to injury resulting in the development of thicker neointimal plaque, and increased thrombosis [62, 86–88]. The presence and structure of HS have also been shown to control the movement of HS-binding proteins like FGFs and vascular endothelial growth factors (VEGFs) to ultimately control the activity range of these bioactive proteins [89–92]. This molecular scale interaction can then lead to dramatic changes in tissue morphology during development and after injury [76, 93].

HS-containing PGs can also manipulate the function of the ECM. The ECM is principally comprised of proteins and polysaccharides. Besides the structural components such as fibrous collagens and elastin, a prevalent dynamic component is the glycoprotein fibronectin (Fn) that is found in increased quantities at sites of active tissue remodeling, such as during wound healing and development [94, 95]. Fn is also present at increased levels and has been associated with diseases such as cancer and atherosclerosis [96–99]. A function of Fn within the ECM is linked to its ability to bind to multiple proteins and to coordinate cell-ECM interactions via cell surface receptors called integrins. Interestingly, the conformation of individual Fn molecules can be modulated by interaction with HS, and these structural changes propagate to higher length scales in supermolecular Fn fibers [98, 100–102]. Altered Fn fiber structure controls cell migration and the response of cells within tissues to mechanical stimulation. One important consequence of the HS-mediated change in the Fn structure is to modify the ability of Fn to bind bioactive proteins. For example, it has been shown that HS can dramatically increase the binding of VEGFs and platelet-derived growth factor (PDGF) to Fn, allowing Fn to direct the activity of these growth factors [89, 103, 104]. Indeed, the ability of HS to control PDGF deposition to Fn is critical to proper gastrulation leading to tissue and organ development [104, 105]. Hence, changes in HS levels and fine structure can modify protein-protein interactions at their own length scale that in turn can percolate across all scales from the molecular through tissue level to the organization of the entire body plan.

There are also other examples where non-HS-containing PGs can have a dramatic multiscale impact on cell, tissue, and organ structure. The most well described is the ability of PGs to coordinate with structural proteins in the ECM through direct molecular interactions to create a functional structure. An interesting example is the role of KS, CS/DS-containing SLRPs in the cornea [106]. The cornea is a uniquely transparent tissue and this light transmissive process is the hallmark of its function. The ability of light to pass through the cornea is a consequence of a very tightly controlled structural arrangement of collagen fibrils in a lattice-like structure. Heterotypic fibrils of principally type I and V collagen are arranged in fibrils with a narrow distribution of diameters to form regular lattices which are further organized into stacked lamellae that are arranged at 90° intervals between lamellae. In this way, light scattering caused by one fibril must appropriately interfere with light scattered by adjacent fibrils in a precise manner to reduce scattering in all but the forward direction. The fibril spacing that allows such scattering is the result of interactions between PGs and the collagen fibrils [107, 108]. The five major SLRPs present in the cornea are lumican, keratocan, decorin, osteoglycin, and biglycan. Their function includes controlling collagen fibrillogenesis through binding of the core protein and the fibril whereas the GAG chains control spacing between fibers and lamellae. Not surprisingly, several studies have revealed that defective SLRPs lead to blindness in humans. For example, mutations in decorin are linked to stromal dystrophy and corneal opacity at birth [109]. Furthermore, mutations in the keratocan core protein have been shown to cause cornea plana, a congenital defect in cornea curvature, and decreased visual acuity [110]. As evidence of the role of the GAG chains, mutations in the carbohydrate sulfotransferase 6 gene that catalyzes the transfer of sulfate to KS also lead to corneal dystrophy [111]. Altered corneal curvature results from changes in the biomechanical properties such as the stiffness of the cornea. Indeed, the corneal structure must also satisfy biomechanical constraints because it must be able to withstand ocular pressure. Thus, although these multiscale mechanisms of SLRPs within the cornea have been extensively studied using electron microscopy and gene knockout studies in mice [106, 108], revealing the macroscopic functional implications of these mutations awaits further investigation.

## 4 PGs and Diseases

The ability of PGs to bind to water and other biological macromolecules within the extracellular matrix and on cell surfaces at the molecular scale can impact multiscale ECM and tissue function. Thus, it is no surprise that when the multiscale roles of PGs are disrupted disease ensues [7]. For example, the glypicans have been reported to play a role in several types of cancer including hepatocellular carcinomas, gliomas, and bladder carcinomas, and have been identified as key biomarkers to guide therapy [112]. The proteolytic shedding of syndecans has been implicated in mediating angiogenesis and vascular permeability in response to VEGF and hepatocyte growth factor, which promotes tumor growth and metastasis. Syndecans are also directly involved in inflammatory disease [113]. Upregulation of syndecans during inflammation enhances leukocyte adhesion to endothelial cells and transmigration into tissues. Thus, diseases associated with uncontrolled inflammation such as chronic obstructive pulmonary disease (COPD), rheumatoid arthritis, and vascular disease can be linked to altered syndecan expression [114]. For example, the ECM within vascular lesions is characterized by a shift from low PG and high elastic fiber content to a state of high PG levels leading to the formation of advanced lesions where elastic fibers have been replaced by collagen [115]. This shift leads to increased vascular stiffness, a precursor to vascular disease, the most significant being atherosclerosis, which is the number one cause of death worldwide.

PGs have long been known to play important roles in the development of the central nervous system, where they direct guidance factors to allow for proper axonal development and are also associated with neurodegenerative disease [116]. Plaques containing β-amyloid peptides in Alzheimer's disease contain high levels of PGs, mostly HS-containing PGs. HSPGs have been demonstrated to bind to β amyloid peptides to induce core deposits that propagate to plaque [117]. A similar mechanism has been proposed for the role of HSPGs in promoting Tauopathies through direct binding of HS to Tau leading to modulation of Tau secretion, internalization, and aggregation [118]. More recently, a novel mechanism for HSPGs has emerged in mediating infection by the SARS-CoV-2 virus where interaction between the spike protein on the virus surface with HS facilitates binding of the virus to its cognate receptor to gain entry into host cells [119]. Several genetic disorders are caused by mutations in the genes for GAG biosynthetic enzymes leading to a family of orphan diseases characterized by connective tissue and skeletal abnormalities, and impairment of multiple organs [120].

From a more biomechanical perspective, PGs' multiscale influence may be exploited for the purpose of treating diseases. For example, in the lung, PGs mechanically support the collagen network which allows them to contribute to basic mechanical properties such as Poisson's ratio, characterizing the lateral contraction of the alveolar network during uniaxial stretch [17]. This has major implications for COPD in which enzymatic damage to elastic [121] and collagen [122] fibers weakens the alveolar septal walls. Uneven mechanical stresses including those due to gravity can rupture these fibers and alveolar walls [123, 124]. Tissue rupture in turn leads to airspace enlargement, the characteristic structural feature that incessantly progresses in COPD lungs. Interestingly, PGs and GAGs have been shown to slow this rupture-induced disease progression in mice by taking up some of the mechanical stresses and hence protecting the fibers from rupture [125]. Indeed, treating mouse lungs with hyaluronan, a distinct GAG that does not have a protein core, was found to reduce the amount of cigarette smoke-induced airspace enlargement [126, 127]. Furthermore, when COPD patients were treated with aerosolized hyaluronan in a pilot clinical trial, sputum and plasma levels of desmosine and isodesmosine, elastin crosslinks representing degradation products, were statistically significantly reduced without side effects [128]. Although the results were attributed to hyaluronan's ability to coat and directly protect elastic fibers from digestion at the molecular level, it is quite likely that hyaluronan also played a multiscale biomechanical protective role by lowering mechanical stresses on the fibers. Thus, GAGs have the ability through their molecular scale interactions as well as macroscale biomechanical function to slow the progression of COPD, the third most deadly disease in the world [129].

## 5 Summary and Outlook

Many physiological phenomena as well as tissue and organ properties are multiscale in nature. A prime example is tissue stiffness which often arises from collagen or elastin molecules forming microfibrils, fibrils, fibers, and a network of fibers. Stiffness at the macroscale emerges from both the physical properties of its molecular constituents and their network organization. Fine-tuning the robustness of such multiscale structures for proper biological function has occurred over evolutionary time scales. Arguably, however, most diseases are also multiscale in the sense that pathology starts at the molecular scale perhaps due to a single site mutation whose effects gradually grow and percolate to the organ level which manifests in clinically observable symptoms of the disease. Such processes are examples of emergent behavior that can lead to the breakdown of the healthy multiscale structure or function over much shorter time scales than the human lifespan.

PGs and GAGs have traditionally been considered constituents of the ECM at the molecular scale. In this brief chapter, we have given examples and evidence that PGs are complex macromolecules that have the ability to control molecular scale events with far-reaching consequences in terms of length scale providing them with a multiscale attribute that ultimately controls tissue and organ structure and function. This ability of PGs to influence macroscopic function, such as tissue organization and stiffness, through binding water and interacting with other molecules at the molecular scale within the ECM is an intriguing example of how emergent properties can arise unexpectedly from lower-scale interactions. While the classic structural proteins such as the collagens form multiscale structures themselves, PGs and GAGs are unable to form such structures. Their multiscale function can be traced back to how they control cells that produce structural proteins as well as how they directly influence the processes that generate multiscale structures such as fibrillogenesis. Mechanical stresses operating at the macroscale in turn feed back to cells via mechanotransduction that may regulate PG production. This general scheme is illustrated in Fig. 5. Taken together it is clear that PGs, through their ability to engage in myriad interactions through their core proteins and most prominently through the highly charged GAG chains, have the ability to influence a broad range of macroscopic properties of tissues, organs, and organisms. The lack or excess of these molecular-scale interactions can cause a bifurcation in tissue growth, organization, and function culminating in disease. Alternatively, exploiting the unique ability of PGs to control and regulate multiscale tissue structure and behavior has the potential to advance tissue engineering and regenerative medicine approaches to solving major human diseases.

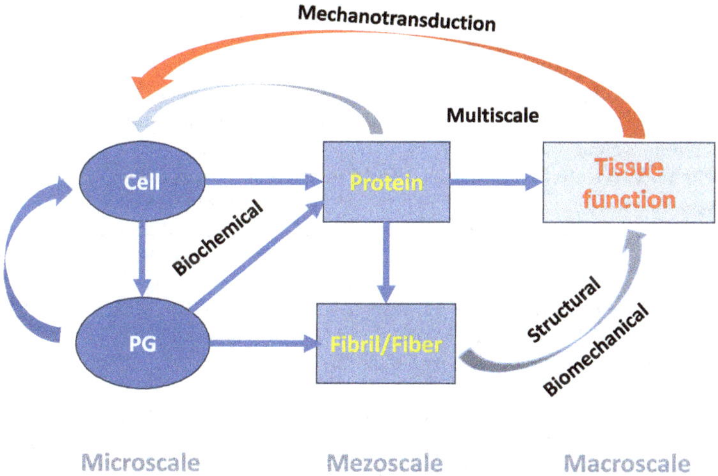


**Fig. 5** Multiscale role for PGs in soft tissue structure and function. PGs bind to growth factors and cytokines to modulate their action on cells and to structural proteins in the ECM to control fibrillogenesis. PGs further influence the higher-order arrangement of fibrils into fibers to define the structure and biomechanical properties of tissues, which feed back to modify cellular activity through mechanotransduction. Disruption of the role of PGs at any of these stages, through altered biosynthesis, degradation, or molecular structure, can lead to a broad range of disease states

# References


1. **Suki B**. *Structure and Function of the Extracellular Matrix: A Multiscale Quantitative Approach*. Academic Press, 2021.
2. **Maceri F, Marino M, and Vairo G**. An insight on multiscale tendon modeling in muscle-tendon integrated behavior. *Biomech Model Mechanobiol* 11: 505–517, 2012.
3. **Siadat SM, Zamboulis DE, Thorpe CT, Ruberti JW, and Connizzo BK**. Tendon Extracellular Matrix Assembly, Maintenance and Dysregulation Throughout Life. *Adv Exp Med Biol* 1348: 45–103, 2021.
4. **Yang W, Sherman VR, Gludovatz B, Schaible E, Stewart P, Ritchie RO, and Meyers MA**. On the tear resistance of skin. *Nat Commun* 6: 6649, 2015.
5. **Iozzo RV, Goldoni S, Berendsen AD, and Young MF**. Small Leucine-Rich Proteoglycans. In: *The Extracellular Matrix: an Overview*, edited by Mecham RP. Berlin, Heidelberg: Springer Berlin Heidelberg, 2011, p. 197–231.
6. **Schaefer L, and Iozzo RV**. Biological functions of the small leucine-rich proteoglycans: from genetics to signal transduction. *J Biol Chem* 283: 21305–21309, 2008.
7. **Xie C, Schaefer L, and Iozzo RV**. Global impact of proteoglycan science on human diseases. *iScience* 26: 108095, 2023.
8. **Frey H, Schroeder N, Manon-Jensen T, Iozzo RV, and Schaefer L**. Biological interplay between proteoglycans and their innate immune receptors in inflammation. *FEBS J* 280: 2165–2179, 2013.
9. **Moon JJ, Matsumoto M, Patel S, Lee L, Guan JL, and Li S**. Role of cell surface heparan sulfate proteoglycans in endothelial cell migration and mechanotransduction. *J Cell Physiol* 203: 166–176, 2005.

10. **Danielson KG, Baribault H, Holmes DF, Graham H, Kadler KE, and Iozzo RV**. Targeted disruption of decorin leads to abnormal collagen fibril morphology and skin fragility. *J Cell Biol* 136: 729–743, 1997.
11. **Iwasaki S, Hosaka Y, Iwasaki T, Yamamoto K, Nagayasu A, Ueda H, Kokai Y, and Takehana K**. The modulation of collagen fibril assembly and its structure by decorin: an electron microscopic study. *Arch Histol Cytol* 71: 37–44, 2008.
12. **Franchi M, Torricelli P, Giavaresi G, and Fini M**. Role of moderate exercising on Achilles tendon collagen crimping patterns and proteoglycans. *Connect Tissue Res* 54: 267–274, 2013.
13. **Thieszen SL, and Rosenquist TH**. Expression of collagens and decorin during aortic arch artery development: implications for matrix pattern formation. *Matrix Biol* 14: 573–582, 1995.
14. **Lochner K, Gaemlich A, Sudel KM, Venzke K, Moll I, Knott A, Stab F, Wenck H, Doring O, Bottger M, and Gallinat S**. Expression of decorin and collagens I and III in different layers of human skin in vivo: a laser capture microdissection study. *Biogerontology* 8: 269–282, 2007.
15. **Ross CJ, Laurence DW, Richardson J, Babu AR, Evans LE, Beyer EG, Childers RC, Wu Y, Towner RA, Fung KM, Mir A, Burkhart HM, Holzapfel GA, and Lee CH**. An investigation of the glycosaminoglycan contribution to biaxial mechanical behaviours of porcine atrioventricular heart valve leaflets. *J R Soc Interface* 16: 20190069, 2019.
16. **Hatami-Marbini H**. On the Mechanical Roles of Glycosaminoglycans in the Tensile Properties of Porcine Corneal Stroma. *Investigative Ophthalmology & Visual Science* 64: 3–3, 2023.
17. **Cavalcante FS, Ito S, Brewer KK, Sakai H, Alencar AM, Almeida MP, Andrade Jr JS, Majumdar A, Ingenito EP, and Suki B**. Mechanical Interactions between Collagen and Proteoglycans: Implications for the Stability of Lung Tissue. *J Appl Physiol* 98: 672–679, 2005.
18. **Buschmann MD, and Grodzinsky AJ**. A molecular model of proteoglycan-associated electrostatic forces in cartilage mechanics. *J Biomech Eng* 117: 179–192, 1995.
19. **Ritter MC, Jesudason R, Majumdar A, Stamenovic D, Buczek-Thomas JA, Stone PJ, Nugent MA, and Suki B**. A zipper network model of the failure mechanics of extracellular matrices. *Proc Natl Acad Sci U S A* 106: 1081–1086, 2009.
20. **Azeloglu EU, Albro MB, Thimmappa VA, Ateshian GA, and Costa KD**. Heterogeneous transmural proteoglycan distribution provides a mechanism for regulating residual stresses in the aorta. *Am J Physiol Heart Circ Physiol* 294: H1197–1205, 2008.
21. **Fessel G, and Snedeker JG**. Evidence against proteoglycan mediated collagen fibril load transmission and dynamic viscoelasticity in tendon. *Matrix Biol* 28: 503–510, 2009.
22. **Rigozzi S, Muller R, Stemmer A, and Snedeker JG**. Tendon glycosaminoglycan proteoglycan sidechains promote collagen fibril sliding-AFM observations at the nanoscale. *J Biomech* 46: 813–818, 2013.
23. **Eisenberg SR, and Grodzinsky AJ**. Swelling of articular cartilage and other connective tissues: electromechanochemical forces. *J Orthop Res* 3: 148–159, 1985.
24. **Gandhi NS, and Mancera RL**. The structure of glycosaminoglycans and their interactions with proteins. *Chem Biol Drug Des* 72: 455–482, 2008.
25. **Gubbiotti MA, Vallet SD, Ricard-Blum S, and Iozzo RV**. Decorin interacting network: A comprehensive analysis of decorin-binding partners and their versatile functions. *Matrix Biol* 55: 7–21, 2016.
26. **Schwartz NB, and Domowicz MS**. Roles of Chondroitin Sulfate Proteoglycans as Regulators of Skeletal Development. *Front Cell Dev Biol* 10: 745372, 2022.
27. **Puri S, Coulson-Thomas YM, Gesteira TF, and Coulson-Thomas VJ**. Distribution and Function of Glycosaminoglycans and Proteoglycans in the Development, Homeostasis and Pathology of the Ocular Surface. *Front Cell Dev Biol* 8: 731, 2020.
28. **Small DH, Mok SS, Williamson TG, and Nurcombe V**. Role of proteoglycans in neural development, regeneration, and the aging brain. *J Neurochem* 67: 889–899, 1996.

29. **Dunkman AA, Buckley MR, Mienaltowski MJ, Adams SM, Thomas SJ, Satchell L, Kumar A, Pathmanathan L, Beason DP, Iozzo RV, Birk DE, and Soslowsky LJ**. Decorin expression is important for age-related changes in tendon structure and mechanical properties. *Matrix Biol* 32: 3–13, 2013.
30. **Hardingham T, and Bayliss M**. Proteoglycans of articular cartilage: changes in aging and in joint disease. *Semin Arthritis Rheum* 20: 12–33, 1990.
31. **Lee DH, Oh JH, and Chung JH**. Glycosaminoglycan and proteoglycan in skin aging. *J Dermatol Sci* 83: 174–181, 2016.
32. **Laurent C**. 13 - Micromechanics of Ligaments and Tendons. In: *Multiscale Biomechanics*, edited by Ganghoffer J-F Elsevier, 2018, p. 489–509.
33. **Orgel JP, Irving TC, Miller A, and Wess TJ**. Microfibrillar structure of type I collagen in situ. *Proc Natl Acad Sci U S A* 103: 9001–9005, 2006.
34. **Biewener AA**. *Tendons and Ligaments: Structure, Mechanical Behavior and Biological Function*. New York: Springer, 2008, p. xviii, 506 p.
35. **Gautieri A, Vesentini S, Redaelli A, and Buehler MJ**. Hierarchical structure and nanomechanics of collagen microfibrils from the atomistic scale up. *Nano Lett* 11: 757–766, 2011.
36. **LaCroix AS, Duenwald-Kuehl SE, Lakes RS, and Vanderby R, Jr.** Relationship between tendon stiffness and failure: a metaanalysis. *J Appl Physiol (1985)* 115: 43–51, 2013.
37. **Franchi M, Fini M, Quaranta M, De Pasquale V, Raspanti M, Giavaresi G, Ottani V, and Ruggeri A**. Crimp morphology in relaxed and stretched rat Achilles tendon. *J Anat* 210: 1–7, 2007.
38. **Fang F, and Lake SP**. Modelling approaches for evaluating multiscale tendon mechanics. *Interface Focus* 6: 20150044, 2016.
39. **Choi SH, Lee BJ, and Lee SY**. A study on skin mobility according to joint movement: Variations in mobility according to joint motion range and correlation and influence with hydrica composition. *Skin Res Technol* 29: e13288, 2023.
40. **Kalra A, Lowe A, and Al-Jumaily AM**. Mechanical Behaviour of Skin: A Review. *Journal of Material Sciences & Engineering* 5: 1–7, 2016.
41. **Kim JH, Schaible N, Hall JK, Bartolak-Suki E, Deng Y, Herrmann J, Sonnenberg A, Behrsing HP, Lutchen KR, Krishnan R, and Suki B**. Multiscale stiffness of human emphysematous precision cut lung slices. *Sci Adv* 9: eadf2535, 2023.
42. **Oxlund H, Manschot J, and Viidik A**. The role of elastin in the mechanical properties of skin. *J Biomech* 21: 213–218, 1988.
43. **Setnikar I**. [Origin and significance of the mechanical property of the lung]. *Arch Fisiol* 55: 349–374, 1955.
44. **Tillgren V, Morgelin M, Onnerfjord P, Kalamajski S, and Aspberg A**. The Tyrosine Sulfate Domain of Fibromodulin Binds Collagen and Enhances Fibril Formation. *J Biol Chem* 291: 23744–23755, 2016.
45. **Corsi A, Xu T, Chen XD, Boyde A, Liang J, Mankani M, Sommer B, Iozzo RV, Eichstetter I, Robey PG, Bianco P, and Young MF**. Phenotypic effects of biglycan deficiency are linked to collagen fibril abnormalities, are synergized by decorin deficiency, and mimic Ehlers-Danlos-like changes in bone and other connective tissues. *J Bone Miner Res* 17: 1180–1189, 2002.
46. **Yang L, van der Werf KO, Fitie CF, Bennink ML, Dijkstra PJ, and Feijen J**. Mechanical properties of native and cross-linked type I collagen fibrils. *Biophys J* 94: 2204–2211, 2008.
47. **Franchi M, De Pasquale V, Martini D, Quaranta M, Macciocca M, Dionisi A, and Ottani V**. Contribution of glycosaminoglycans to the microstructural integrity of fibrillar and fiber crimps in tendons and ligaments. *ScientificWorldJournal* 10: 1932–1940, 2010.
48. **Liao J, and Vesely I**. Skewness angle of interfibrillar proteoglycans increases with applied load on mitral valve chordae tendineae. *J Biomech* 40: 390–398, 2007.
49. **Al Makhzoomi AK, Kirk TB, Dye DE, and Allison GT**. Contribution of glycosaminoglycans to the structural and mechanical properties of tendons - A multiscale study. *J Biomech* 128: 110796, 2021.

50. **Benallaoua M, Richette P, Francois M, Tsagris L, Revel M, Corvol M, Poiraudeau S, Savouret JF, and Rannou F**. Modulation of proteoglycan production by cyclic tensile stretch in intervertebral disc cells through a post-translational mechanism. *Biorheology* 43: 303–310, 2006.
51. **Gupta V, Tseng H, Lawrence BD, and Grande-Allen KJ**. Effect of cyclic mechanical strain on glycosaminoglycan and proteoglycan synthesis by heart valve cells. *Acta Biomater* 5: 531–540, 2009.
52. **Zhang YH, Zhao CQ, Jiang LS, and Dai LY**. Substrate stiffness regulates apoptosis and the mRNA expression of extracellular matrix regulatory genes in the rat annular cells. *Matrix Biol* 30: 135–144, 2011.
53. **Olivares-Navarrete R, Lee EM, Smith K, Hyzy SL, Doroudi M, Williams JK, Gall K, Boyan BD, and Schwartz Z**. Substrate Stiffness Controls Osteoblastic and Chondrocytic Differentiation of Mesenchymal Stem Cells without Exogenous Stimuli. *PLoS One* 12: e0170312, 2017.
54. **Merry CLR, Lindahl U, Couchman J, and Esko JD**. Proteoglycans and Sulfated Glycosaminoglycans. In: *Essentials of Glycobiology*, edited by Varki A, Cummings RD, Esko JD, Stanley P, Hart GW, Aebi M, Mohnen D, Kinoshita T, Packer NH, Prestegard JH, Schnaar RL, and Seeberger PH. Cold Spring Harbor (NY): 2022, p. 217–232.
55. **Gottschalk J, and Elling L**. Current state on the enzymatic synthesis of glycosaminoglycans. *Curr Opin Chem Biol* 61: 71–80, 2021.
56. **Whitelock JM, and Iozzo RV**. Heparan sulfate: a complex polymer charged with biological activity. *Chem Rev* 105: 2745–2764, 2005.
57. **Xu D, and Esko JD**. Demystifying heparan sulfate-protein interactions. *Annu Rev Biochem* 83: 129–157, 2014.
58. **Sarrazin S, Lamanna WC, and Esko JD**. Heparan sulfate proteoglycans. *Cold Spring Harb Perspect Biol* 3: 2011.
59. **Masola V, Bellin G, Gambaro G, and Onisto M**. Heparanase: A Multitasking Protein Involved in Extracellular Matrix (ECM) Remodeling and Intracellular Events. *Cells* 7: 2018.
60. **Nikmanesh M, Cancel LM, Shi ZD, and Tarbell JM**. Heparan sulfate proteoglycan, integrin, and syndecan-4 are mechanosensors mediating cyclic strain-modulated endothelial gene expression in mouse embryonic stem cell-derived endothelial cells. *Biotechnol Bioeng* 116: 2730–2741, 2019.
61. **Shi ZD, Wang H, and Tarbell JM**. Heparan sulfate proteoglycans mediate interstitial flow mechanotransduction regulating MMP-13 expression and cell motility via FAK-ERK in 3D collagen. *PLoS One* 6: e15956, 2011.
62. **Baker AB, Groothuis A, Jonas M, Ettenson DS, Shazly T, Zcharia E, Vlodavsky I, Seifert P, and Edelman ER**. Heparanase alters arterial structure, mechanics, and repair following endovascular stenting in mice. *Circ Res* 104: 380–387, 2009.
63. **Iozzo RV, and Schaefer L**. Proteoglycan form and function: A comprehensive nomenclature of proteoglycans. *Matrix Biol* 42: 11–55, 2015.
64. **Purushothaman A, Uyama T, Kobayashi F, Yamada S, Sugahara K, Rapraeger AC, and Sanderson RD**. Heparanase-enhanced shedding of syndecan-1 by myeloma cells promotes endothelial invasion and angiogenesis. *Blood* 115: 2449–2457, 2010.
65. **Sperinde GV, and Nugent MA**. Mechanisms of FGF-2 intracellular processing: A kinetic analysis of the role of heparan sulfate proteoglycans. *Biochemistry* 39: 3788–3796, 2000.
66. **Buczek-Thomas JA, Hsia E, Rich CB, Foster JA, and Nugent MA**. Inhibition of histone acetyltransferase by glycosaminoglycans. *J Cell Biochem* 105: 108–120, 2008.
67. **Hsia E, Richardson TP, and Nugent MA**. Nuclear localization of basic fibroblast growth factor is mediated by heparan sulfate proteoglycans through protein kinase C signaling. *J Cell Biochem* 88: 1214–1225, 2003.
68. **Stewart MD, Ramani VC, and Sanderson RD**. Shed syndecan-1 translocates to the nucleus of cells delivering growth factors and inhibiting histone acetylation: a novel mechanism of tumor-host cross-talk. *J Biol Chem* 290: 941–949, 2015.

69. **Beauvais DM, and Rapraeger AC**. Syndecans in tumor cell adhesion and signaling. *Reprod Biol Endocrinol* 2: 3, 2004.
70. **Vlodavsky I, Beckhove P, Lerner I, Pisano C, Meirovitz A, Ilan N, and Elkin M**. Significance of heparanase in cancer and inflammation. *Cancer Microenviron* 5: 115–132, 2012.
71. **Jung O, Trapp-Stamborski V, Purushothaman A, Jin H, Wang H, Sanderson RD, and Rapraeger AC**. Heparanase-induced shedding of syndecan-1/CD138 in myeloma and endothelial cells activates VEGFR2 and an invasive phenotype: prevention by novel synstatins. *Oncogenesis* 5: e202, 2016.
72. **Couchman JR, and Pataki CA**. An introduction to proteoglycans and their localization. *J Histochem Cytochem* 60: 885–897, 2012.
73. **Iozzo RV, and Murdoch AD**. Proteoglycans of the extracellular environment: clues from the gene and protein side offer novel perspectives in molecular diversity and function. *FASEB J* 10: 598–614, 1996.
74. **Kalamajski S, and Oldberg A**. The role of small leucine-rich proteoglycans in collagen fibrillogenesis. *Matrix Biol* 29: 248–253, 2010.
75. **Watanabe H**. Aggrecan and versican: two brothers close or apart. *Am J Physiol Cell Physiol* 322: C967–C976, 2022.
76. **Perrimon N, and Bernfield M**. Specificities of heparan sulphate proteoglycans in developmental processes. *Nature* 404: 725–728., 2000.
77. **Thompson SM, Connell MG, Fernig DG, Ten Dam GB, van Kuppevelt TH, Turnbull JE, Jesudason EC, and Losty PD**. Novel 'phage display antibodies identify distinct heparan sulfate domains in developing mammalian lung. *Pediatr Surg Int* 23: 411–417, 2007.
78. **Izvolsky KI, Shoykhet D, Yang Y, Yu Q, Nugent MA, and Cardoso WV**. Heparan sulfate-FGF10 interactions during lung morphogenesis. *Dev Biol* 258: 185–200, 2003.
79. **Ori A, Wilkinson MC, and Fernig DG**. A systems biology approach for the investigation of the heparin/heparan sulfate interactome. *The Journal of biological chemistry* 286: 19892–19904, 2011.
80. **Beenken A, and Mohammadi M**. The FGF family: biology, pathophysiology and therapy. *Nat Rev Drug Discov* 8: 235–253, 2009.
81. **Itoh N**. The Fgf families in humans, mice, and zebrafish: their evolutional processes and roles in development, metabolism, and disease. *Biol Pharm Bull* 30: 1819–1825, 2007.
82. **Eswarakumar VP, Lax I, and Schlessinger J**. Cellular signaling by fibroblast growth factor receptors. *Cytokine Growth Factor Rev* 16: 139–149, 2005.
83. **Nugent MA, and Iozzo RV**. Fibroblast growth factor-2. *Int J Biochem Cell Biol* 32: 115–120, 2000.
84. **Ornitz DM, Xu J, Colvin JS, McEwen DG, MacArthur CA, Coulier F, Gao G, and Goldfarb M**. Receptor specificity of the fibroblast growth factor family. *The Journal of biological chemistry* 271: 15292–15297, 1996.
85. **Zhang X, Ibrahimi OA, Olsen SK, Umemori H, Mohammadi M, and Ornitz DM**. Receptor specificity of the fibroblast growth factor family. The complete mammalian FGF family. *The Journal of biological chemistry* 281: 15694–15700, 2006.
86. **Edelman ER, and Karnovsky MJ**. Contrasting effects of the intermittent and continuous administration of heparin in experimental restenosis. *Circulation* 89: 770–776, 1994.
87. **Han RO, Ettenson DS, Koo EW, and Edelman ER**. Heparin/heparan sulfate chelation inhibits control of vascular repair by tissue-engineered endothelial cells. *Am J Physiol* 273: H2586–2595, 1997.
88. **Nugent MA, Nugent HM, Iozzo RV, Sanchack K, and Edelman ER**. Perlecan is required to inhibit thrombosis after deep vascular injury and contributes to endothelial cell-mediated inhibition of intimal hyperplasia. *Proc Natl Acad Sci U S A* 97: 6722–6727, 2000.
89. **Martino MM, and Hubbell JA**. The 12th-14th type III repeats of fibronectin function as a highly promiscuous growth factor-binding domain. *Faseb J* 24: 4711–4721, 2010.
90. **Zisch AH, Lutolf MP, and Hubbell JA**. Biopolymeric delivery matrices for angiogenic growth factors. *Cardiovasc Pathol* 12: 295–310, 2003.

91. **Dowd CJ, Cooney CL, and Nugent MA**. Heparan sulfate mediates bFGF transport through basement membrane by diffusion with rapid reversible binding. *J Biol Chem* 274: 5236–5244, 1999.
92. **Fannon M, Forsten-Williams K, Zhao B, Bach E, Parekh PP, Chu CL, Goerges-Wildt AL, Buczek-Thomas JA, and Nugent MA**. Facilitated diffusion of VEGF165 through descemet's membrane with sucrose octasulfate. *Journal of cellular physiology* 227: 3693–3700, 2012.
93. **Lin X, and Perrimon N**. Developmental roles of heparan sulfate proteoglycans in Drosophila. *Glycoconj J* 19: 363–368, 2002.
94. **Patten J, and Wang K**. Fibronectin in development and wound healing. *Adv Drug Deliv Rev* 170: 353–368, 2021.
95. **Zollinger AJ, and Smith ML**. Fibronectin, the extracellular glue. *Matrix Biol* 60-61: 27–37, 2017.
96. **Dinesh NEH, Campeau PM, and Reinhardt DP**. Fibronectin isoforms in skeletal development and associated disorders. *Am J Physiol Cell Physiol* 323: C536–C549, 2022.
97. **Lin TC, Yang CH, Cheng LH, Chang WT, Lin YR, and Cheng HC**. Fibronectin in Cancer: Friend or Foe. *Cells* 9: 2019.
98. **Mezzenga R, and Mitsi M**. The Molecular Dance of Fibronectin: Conformational Flexibility Leads to Functional Versatility. *Biomacromolecules* 20: 55–72, 2019.
99. **Speziale P, Arciola CR, and Pietrocola G**. Fibronectin and Its Role in Human Infective Diseases. *Cells* 8: 2019.
100. **Hubbard B, Buczek-Thomas JA, Nugent MA, and Smith ML**. Heparin-dependent regulation of fibronectin matrix conformation. *Matrix biology : journal of the International Society for Matrix Biology* 34: 124–131, 2014.
101. **Hubbard B, Buczek-Thomas JA, Nugent MA, and Smith ML**. Fibronectin Fiber Extension Decreases Cell Spreading and Migration. *J Cell Physiol* 231: 1728–1736, 2016.
102. **Mitsi M, Hong Z, Costello CE, and Nugent MA**. Heparin-mediated conformational changes in fibronectin expose vascular endothelial growth factor binding sites. *Biochemistry* 45: 10319–10328, 2006.
103. **Mitsi M, Forsten-Williams K, Gopalakrishnan M, and Nugent MA**. A catalytic role of heparin within the extracellular matrix. *J Biol Chem* 283: 34796–34807, 2008.
104. **Symes K, Smith EM, Mitsi M, and Nugent MA**. Sweet cues: How heparan sulfate modification of fibronectin enables growth factor guided migration of embryonic cells. *Cell Adh Migr* 4: 507–510, 2010.
105. **Smith EM, Mitsi M, Nugent MA, and Symes K**. PDGF-A interactions with fibronectin reveal a critical role for heparan sulfate in directed cell migration during Xenopus gastrulation. *Proc Natl Acad Sci U S A* 106: 21683–21688, 2009.
106. **Massoudi D, Malecaze F, and Galiacy SD**. Collagens and proteoglycans of the cornea: importance in transparency and visual disorders. *Cell Tissue Res* 363: 337–349, 2016.
107. **Knupp C, Pinali C, Lewis PN, Parfitt GJ, Young RD, Meek KM, and Quantock AJ**. The architecture of the cornea and structural basis of its transparency. *Adv Protein Chem Struct Biol* 78: 25–49, 2009.
108. **Lewis PN, Pinali C, Young RD, Meek KM, Quantock AJ, and Knupp C**. Structural interactions between collagen and proteoglycans are elucidated by three-dimensional electron tomography of bovine cornea. *Structure* 18: 239–245, 2010.
109. **Bredrup C, Knappskog PM, Majewski J, Rodahl E, and Boman H**. Congenital stromal dystrophy of the cornea caused by a mutation in the decorin gene. *Invest Ophthalmol Vis Sci* 46: 420–426, 2005.
110. **Pellegata NS, Dieguez-Lucena JL, Joensuu T, Lau S, Montgomery KT, Krahe R, Kivela T, Kucherlapati R, Forsius H, and de la Chapelle A**. Mutations in KERA, encoding keratocan, cause cornea plana. *Nat Genet* 25: 91–95, 2000.
111. **Musselmann K, and Hassell JR**. Focus on molecules: CHST6 (carbohydrate sulfotransferase 6; corneal N-acetylglucosamine-6-sulfotransferase). *Exp Eye Res* 83: 707–708, 2006.

112. **Wang S, Qiu Y, and Bai B**. The Expression, Regulation, and Biomarker Potential of Glypican-1 in Cancer. *Front Oncol* 9: 614, 2019.
113. **Czarnowski D**. Syndecans in cancer: A review of function, expression, prognostic value, and therapeutic significance. *Cancer Treat Res Commun* 27: 100312, 2021.
114. **Gopal S**. Syndecans in Inflammation at a Glance. *Front Immunol* 11: 227, 2020.
115. **Wight TN**. A role for proteoglycans in vascular disease. *Matrix Biol* 71–72: 396–420, 2018.
116. **Masu M**. Proteoglycans and axon guidance: a new relationship between old partners. *J Neurochem* 139 Suppl 2: 58–75, 2016.
117. **Snow AD, Cummings JA, and Lake T**. The Unifying Hypothesis of Alzheimer's Disease: Heparan Sulfate Proteoglycans/Glycosaminoglycans Are Key as First Hypothesized Over 30 Years Ago. *Front Aging Neurosci* 13: 710683, 2021.
118. **Zhu Y, Gandy L, Zhang F, Liu J, Wang C, Blair LJ, Linhardt RJ, and Wang L**. Heparan Sulfate Proteoglycans in Tauopathy. *Biomolecules* 12: 2022.
119. **Nugent MA**. The Future of the COVID-19 Pandemic: How Good (or Bad) Can the SARS-CoV2 Spike Protein Get? *Cells* 11: 2022.
120. **Mizumoto S, Ikegawa S, and Sugahara K**. Human genetic disorders caused by mutations in genes encoding biosynthetic enzymes for sulfated glycosaminoglycans. *J Biol Chem* 288: 10953–10961, 2013.
121. **Fukuda Y, Masuda Y, Ishizaki M, Masugi Y, and Ferrans VJ**. Morphogenesis of abnormal elastic fibers in lungs of patients with panacinar and centriacinar emphysema. *Hum Pathol* 20: 652–659, 1989.
122. **Imai K, Dalal SS, Chen ES, Downey R, Schulman LL, Ginsburg M, and D'Armiento J**. Human collagenase (matrix metalloproteinase-1) expression in the lungs of patients with emphysema. *Am J Respir Crit Care Med* 163: 786–791, 2001.
123. **Kononov S, Brewer K, Sakai H, Cavalcante FS, Sabayanagam CR, Ingenito EP, and Suki B**. Roles of mechanical forces and collagen failure in the development of elastase-induced emphysema. *Am J Respir Crit Care Med* 164: 1920–1926, 2001.
124. **Mondonedo JR, and Suki B**. Predicting Structure-Function Relations and Survival following Surgical and Bronchoscopic Lung Volume Reduction Treatment of Emphysema. *PLoS Comput Biol* 13: e1005282, 2017.
125. **Takahashi A, Majumdar A, Parameswaran H, Bartolak-Suki E, and Suki B**. Proteoglycans maintain lung stability in an elastase-treated mouse model of emphysema. *American journal of respiratory cell and molecular biology* 51: 26–33, 2014.
126. **Cantor JO, Cerreta JM, Ochoa M, Ma S, Chow T, Grunig G, and Turino GM**. Aerosolized hyaluronan limits airspace enlargement in a mouse model of cigarette smoke-induced pulmonary emphysema. *Exp Lung Res* 31: 417–430, 2005.
127. **Cantor JO, Cerreta JM, Ochoa M, Ma S, Liu M, and Turino GM**. Therapeutic effects of hyaluronan on smoke-induced elastic fiber injury: does delayed treatment affect efficacy? *Lung* 189: 51–56, 2011.
128. **Cantor J, Ma S, and Turino G**. A pilot clinical trial to determine the safety and efficacy of aerosolized hyaluronan as a treatment for COPD. *Int J Chron Obstruct Pulmon Dis* 12: 2747–2752, 2017.
129. **Stolz D, Mkorombindo T, Schumann DM, Agusti A, Ash SY, Bafadhel M, Bai C, Chalmers JD, Criner GJ, Dharmage SC, Franssen FME, Frey U, Han M, Hansel NN, Hawkins NM, Kalhan R, Konigshoff M, Ko FW, Parekh TM, Powell P, Rutten-van Molken M, Simpson J, Sin DD, Song Y, Suki B, Troosters T, Washko GR, Welte T, and Dransfield MT**. Towards the elimination of chronic obstructive pulmonary disease: a Lancet Commission. *Lancet* 400: 921–972, 2022.

# On Cardiac Elastin and Its Biomechanical Functions

**Jun Liao and Xiaodan Shi**

## 1 Structural Extracellular Matrices: Collagen, Elastin, and Proteoglycans

Cells are the smallest fundamental unit of living organisms. However, tissues/organs cannot achieve their structural organization/integrity, biomechanical behavior, and specific biological functions only via cells. The cells, either one type or several types, must be organized into a three-dimensional (3D) structure specific to each tissue/organ, and this 3D organization and structural support are realized by interweaving individual cells with structural extracellular matrices (ECMs) [1–4]. The three major structural ECMs are collagen, elastin, and proteoglycans [5–7]. Together with other ECM proteins and macromolecules, such as fibronectin, laminin, tenascin, and soluble signaling molecules, the structural ECMs build tissues/organs into structurally and mechanically stable entities and enable a dynamic biological environment that is able to regulate cellular behavior and thus achieve tissue-specific physiological functionalities [3, 4, 8, 9].

*Collagen* Among the essential ECM components, collagen is the most abundant structural protein identified in the human body and has 28 genotypes with functions specific to cell origins and tissue types [10]. Among all collagen types, Type I collagen is the most prominent (90% of the body's collagen) and serves as the fibrous scaffolding material to form cardiac muscles, skeletal muscles, skin, ten-

J. Liao (✉)
Department of Bioengineering, University of Texas at Arlington, Arlington, TX, USA
e-mail: jun.liao@uta.edu

X. Shi
Department of Bioengineering, University of Texas at Arlington, Arlington, TX, USA

Department of Biomedical Engineering, University of North Texas, Denton, TX, USA
e-mail: xiaodan.shi@unt.edu

 
J. Liao, J. Y. Wong (eds.), *Integration and Bridging of Multiscale Bioengineering Designs and Tissue Biomechanics*, https://doi.org/10.1007/978-3-031-81743-4_3

dons, ligaments, blood vessels, lungs, bladders, etc., providing the most important tensile mechanical resistance and strength essential to the biomechanical functions of those tissues [11, 12]. Type III collagen, with a finer structure and network formation tendency, is the second most abundant fibrous collagen often co-existing with Type I collagen as an assistant feature [13, 14]. As the major load-bearing structural protein in tissues, Type I and Type III are fibrous collagen and share a similar nature-designed hierarchical structure spanning from amino acid chains, collagen molecules, collagen fibrils, collagen fibers, and collagen fiber network [15–18]. Using tendon as an example, the collagen molecule is composed of three helical amino acid chains (e.g., Type I as heterotrimer; Type III as homotrimer) with a very high aspect ratio (~280 nm in length, ~1.4 nm in diameter) [18, 19]. In the next hierarchical level, the thin, rod-like collagen molecules self-assemble via covalent bonds to form long fibrils, i.e., collagen fibrils with a periodic pattern called D-period (or D-spacing), and this self-packing occurs accurately like "crystallinity." [15, 17, 18, 20] The discontinuous, very long collagen fibrils, mediated by small proteoglycans, then form collagen fibers with a diameter of ~1 μm [21–27]. Finally, collagen fibers form collagen networks, which take on a 3D arrangement specific to each tissue. As observed in tendon tissues (Fig. 1a), the collagen fibers take wavy pathways and have an overall alignment along the longitudinal direction to form a tendon fascicle at the next level, and eventually, many fascicles build a tendon. In planer tissues, such as skins and heart valve leaflets, the collagen fibers are aligned multi-directionally, often with various orientations/alignments specific to their layered anatomy [28–33].

Not only are they being constructed in a hierarchical manner, but the mechanical properties of the collagen structural entities at various scale levels are also drastically different. For example, the tensile modulus was measured to be 2.9 GPa for collagen molecules in bovine Achilles tendon [34], 96 MPa and 430 MPa for collagen fibrils in porcine mitral valve leaflet and bovine Achilles tendon, respectively [34, 35], and 54 MPa for separated collagen fibers in rabbit patellar tendon [36]. Although collagen fibers have very high stiffness (tensile modulus) and strength (failure strength), nature's design smartly packs them into a wavy fiber formation. This wavy formation of strong collagen fibers is the underlying mechanism that the soft tissues, such as the tendons/ligaments, fascia, skin, blood vessels, skeletal muscles, and heart muscles, can bear large mechanical loads (load bearing capacity), and at the same time can be deformed to large strains at lower stress levels (tissue extensibility) [16, 35, 37–44]. We can use the following analogy to summarize the structural-functional relation of collagen for an intuitive understanding. *Fibrous collagen in soft tissues behaves like wavy "**stainless steel wires,**" which enables flexible deformation in the physiological loading range via an uncrimping mechanism while providing mechanical stiffness and strength to soft tissues via a gradual fiber straightening/recruitment mechanism.*

*Proteoglycans* Proteoglycans are highly glycosylated proteins made of core proteins with covalently attached glycosaminoglycan (GAG) side chains and exist in nearly all tissues [45, 46]. Proteoglycans can be categorized into two major

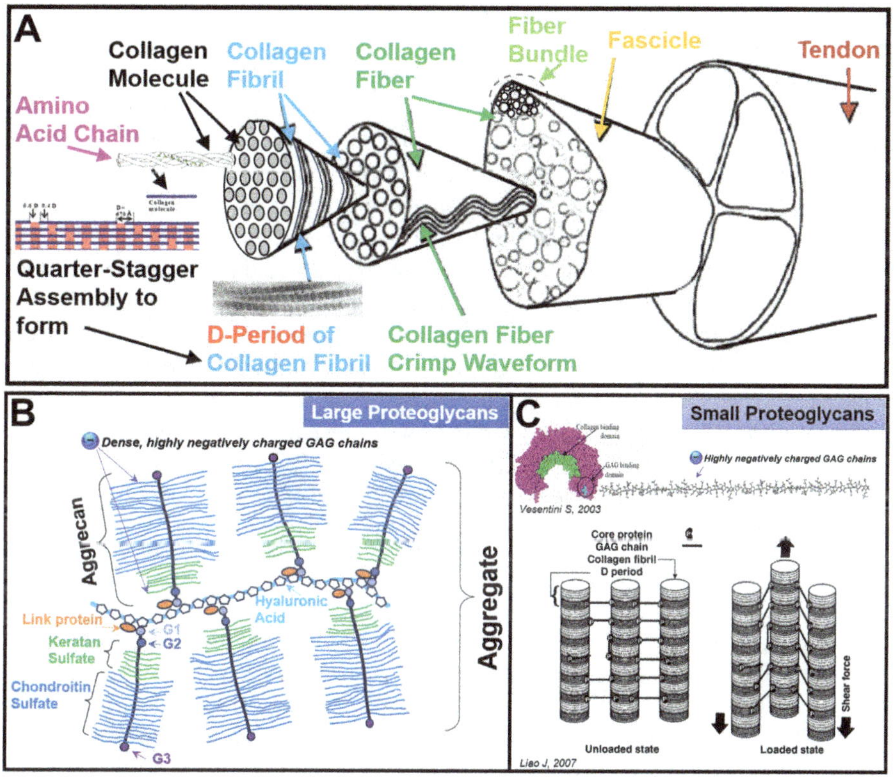


**Fig. 1** (**a**) Nature-designed hierarchical structure of fibrous collagen, spanning from amino acid chains, collagen molecules, collagen fibrils, collagen fibers, to collagen fiber network. (**b**) Hierarchical structure of aggrecans; aggrecans are large proteoglycans existing in cartilaginous tissues, in which abundant, highly negatively charged GAG side chains exist. (**c**) Small proteoglycan has a GAG chain covalently connecting with two horseshoe-shaped core proteins that can bind to specific sites of collagen fibril D-period, serving as interfibrillar linkages

types: (i) large proteoglycans, such as aggrecans (Fig. 1b) that provide shock absorbing, lubrication, and space-filling functions in cartilaginous tissues (e.g., articular cartilage and intervertebral disc) [46–51]; (ii) small proteoglycans (Fig. 1c) that interconnect/bind collagen fibrils into collagen fibers and contribute to viscoelastic properties of soft connective tissues such as tendons, ligaments, and heart valves [25, 26, 52–55]. GAG chains are lengthy linear polymers made of repeating disaccharides. According to their unique core disaccharide units, GAGs can be classified into four primary groups, i.e., heparin/heparan sulfate, chondroitin/dermatan sulfate, keratan sulfate, and hyaluronic acid [56, 57]. In the disaccharide unit of GAGs (except hyaluronic acid), one or both of the sugars contain one or two sulfate residues, giving GAG chains a high negative charge density and hence strong hydrophilic (water-attracting) properties [45, 56, 57]. Note that hyaluronic acid also achieves strong hydrophilic properties due to its polar molecules [56, 57].

In aggrecan, abundant, highly negatively-charged GAG sides chains (chondroitin sulfate and keratan sulfate) attach to a core protein to form a bottle-brush shape, and many of the "bottle-brush" units assemble along a long hyaluronic acid chain and thus build the aggregates that are able to attract and trap a lot of water due to very high hydrophilicity and GAG chain density [58, 59]. In small proteoglycans, the GAG chain covalently connects to two horseshoe-shaped core proteins that can bind to specific sites of the D-period of collagen fibril, and these structural features enable the small proteoglycans to bridge across collagen fibrils as interfibrillar linkages [25, 26, 53, 60]. *In short, for large proteoglycans, dense, highly negatively charged GAG chains attract and trap a lot of water in a way similar to "**gelatin,**" consequently making the tissues hydrated and able to provide osmotic swelling pressure/lubrication, resist repetitive compressive mechanical loading, and absorb shock energy, such as aggrecans in cartilage* [47, 50]. *For small proteoglycans, one can understand them as collagen interfibrillar "**bridging links,**" which are able to mediate collagen fibril dimensions and participate in force transformation between discontinuous collagen fibrils via shear lag mechanism (model)* [25, 26, 35, 54, 60–62].

*Elastin* Elastin is the major elastic protein in the human body and plays important roles in most of the dynamic soft tissues that experience large mechanical stretching and recoiling, such as the ligaments, lungs, bladder, blood vessels, and heart [63–66]. During elastogenesis, which mainly occurs at early developmental stages, soluble tropoelastin crosslinks and binds to fibrillin-rich microfibrils to form elastin fibers [67–69]. The amorphous tropoelastin core does not have specific shapes/regular patterns and is surrounded by microfibrils (Fig. 2a, b); the microfibrils serve to align, orient, branch, grow, and mesh elastin fibers into highly entangled networks [70]. Elastin has mechanical properties that are sharply different from collagen, i.e., a much lower stiffness when compared with collagen, very large extensibility (extremely stretchable), high elasticity capable of recoiling back to the initial reference status, and minimal energy dissipation during stretching-recoiling cycles [71–74]. The unique combination of elastin's primary, secondary, and tertiary structures encodes two fundamental mechanisms, which favor large deformation, high elasticity, and resilience. One is the *conformational entropy-driven recoiling* mechanism, i.e., stretch introduces order to the entangled, disordered peptide chains and decreases the entropy, hence generating the recoiling driven by entropic force (Fig. 2c); the other is the *hydrophobic effect-driven recoiling* mechanism, i.e., stretch increases the exposure of elastin's predominantly hydrophobic residues to water, hence driving the recoil by hydrophobic repealing and net thermodynamic force (Fig. 2d) [75–81].

*As an intuitive analogy, in dynamic soft tissues, elastin can be viewed as a network of "**entangled rubber bands,**" which can be stretched very far at a relatively low stress level, allowing great extensibility/large deformation, being able to bounce back/recoil to its original configuration, and storing and releasing the*

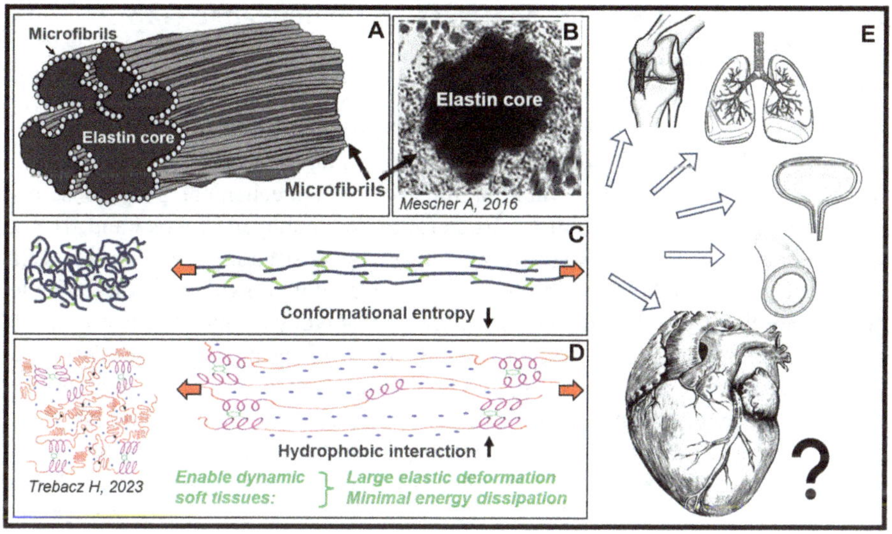


**Fig. 2** (**a**) During elastogenesis, soluble tropoelastin crosslinks and binds to fibrillin-rich microfibrils to form elastin fibers. (**b**) TEM image of tropoelastin core and microfibrils. Two fundamental mechanisms enable large deformation, high elasticity, and resilience of elastin fibers: one is (**c**) the conformational entropy-driven recoiling mechanism, and the other is (**d**) the hydrophobic effect-driven recoiling mechanism. (**e**) There is an apparent knowledge gap in cardiac elastin

*elastic energy efficiently during cyclic loading and unloading.* For most of the dynamic soft tissues, such as ligaments, lungs, bladder, blood vessels, and heart valves, the anatomical distributions, microstructural characteristics, and mechanical roles of elastin have been well studied [56–63]. There has been an accumulation of knowledge on how elastin contributes to optimal tissue behavior under physiological conditions and how elastin abnormalities cause pathophysiological behavior under diseased conditions. For instance, the rich elastin fibers in the lung alveolar sac allow for flexible expansion of the alveolar sac during inhalation and effective recoiling during exhalation. Triggered by smoking, exposure to pollutants, or chronic inflammation, the degeneration/loss of elastin fibers in the alveolar wall results in permanent enlargement of lung alveoli and emphysema, a debilitating chronic obstructive pulmonary disease [82–84]. In blood vessels, elastin fibers play critical roles in vascular mechanics by assisting wall expansion, optimal elasticity, resilience, and energy-efficient expansion and recoiling during cardiac cycles, hence accommodating the flow and distribution of blood to the human body [85–87]. Genetic mutation such as Marfan's syndrome can cause elastin fiber defects, degradation, and insufficiency, which drastically weaken the mechanical properties of blood vessels; as an example, loss of the protective mechanism of healthy elastin networks in the aorta results in accumulated microstructural damage, irreversible plastic deformation, wall thinning, and eventually aortic aneurysm that threatens the life of patients [88–91].

## 2 Cardiac Extracellular Matrices and Knowledge Gap on Cardiac Elastin

Like other soft tissues, collagen, elastin, and proteoglycans are the three major structural ECMs that provide the heart with unique anatomic architecture, cell assembly and organization, structural integrity and mechanical properties, and cardiac functions. Among the three structural ECMs, collagen (Types I and III) has been studied intensively due to its dominant role in constructing the muscular walls of the four heart chambers. Wrapping around the lengthy individual cardiomyocyte, the collagen fibers are able to (i) mesh many cardiomyocytes into long heart muscle fibers with or without branches, and (ii) assemble many heart muscle fibers into arrays of muscle fiber bundles. The cardiac collagen networks are delicate/robust and have key functions in housing individual cardiomyocytes inside unique "honeycomb" lacunae, tethering and transferring force among cardiomyocytes during systolic contraction, and preventing excessive myocardial fiber stretch during diastolic expansion [92–96]. During the pathological progression of myocardial infarction (MI), collagen continues playing a critical role and experiences pathological alterations, including (i) a breakdown of collagen content due to increased matrix metalloproteinase (MMP) activity in the necrotic phase [97–99], (ii) a large amount of collagen fiber deposition (Type I) forming stiff MI scar tissues in the fibrotic phase [97, 100–102], followed by (iii) a slowing down of collagen deposition and the crosslinking/stabilization of collagen fibers in the remodeling phase [100, 101, 103]. As for proteoglycans in the heart, the research has been more focusing on their involvement in cardiac matrix remodeling and fibrosis in diseases such as MI and ventricular pressure overloading [104–106]. The small proteoglycans are confirmed to function as the bridging and organizers of collagen fibrils, and the membrane-bound proteoglycans are the mediators of cardiac fibroblast signaling, a key player in matrix remodeling of the heart [104, 107–109].

In reviewing the literature, there is an apparent knowledge gap in cardiac elastin. Although several studies indicate that elastin might be important in cardiac mechanics [93, 110, 111], its detailed spatial distributions, microstructures, and mechanical contributions in the heart remain poorly understood when compared to what is known about elastin in other soft tissues. The previous knowledge on cardiac elastin was derived from genetic and pathophysiological research on conditions such as Williams syndrome and Marfan syndrome [112–114]. As in Williams syndrome, mutations in the elastin production genes cause deficiency or abnormal deposition of elastin, which could lead to structural defects in the heart, such as ventricular septal defects and hypertrophy of the ventricles [115–117]. New research has shed light on the potential biomechanical roles of cardiac elastin in heart function [93, 118–120]. Studies showed that overexpression of elastin could delay ventricular dysfunction by preventing scar expansion and stabilizing tissue structure in an infarcted heart [118, 119]. Moreover, it was believed that changes in collagen-to-elastin ratio, rather than collagen alone, were associated with stiffening of cardiac tissues due to aging and certain cardiac pathologies [93, 120]. There was also evidence that the loss of contractile function in native or diseased myocardium might partly be caused by the reduction in elastin content [98, 121, 122]. Microstructural knowledge on cardiac elastin was also limited in the literature. It was reported in a previous rat histological study that elastin was abundant in both the epicardial and endocardial layers of both ventricles and atria, and there was no apparent difference in their staining patterns [123]. A recent study using two-photon microscopic analysis showed that elastin fibers aligned with collagen fibers in the epicardium of porcine hearts, while elastin fibers in the endocardium took on a diffuse pattern [110]. In the nonvascular regions of the myocardium, the elastin fibers were reported as sparsely observed [93, 123].

It is thus important to understand, in greater depth, the spatial distributions, microstructural characteristics, and biomechanical roles of cardiac elastin in the heart. Note that we refer to cardiac elastin as the elastin associated with the heart wall tissues, not the vascular elastin in the coronary blood vessels. Our studies revealed that cardiac elastin can be categorized as epicardial elastin in the heart epicardium, endocardial elastin in the heart endocardium, myocardial interstitial elastin in the myocardium, and elastin fibers associated with cardiac Purkinje cells. In this chapter, we covered our recent major discoveries on cardiac elastin and its biomechanical roles in heart function. Based on a pig heart model, (i) the spatial distribution and network morphology of cardiac elastin, including epicardial elastin, endocardial elastin, myocardial interstitial elastin, and Purkinje cell-associated elastin, were described, and (ii) biomechanical roles of various cardiac elastin were discussed, with a focus on the epicardial prestraining protective mechanism. Lastly, (iii) we presented a preliminary investigation of cardiac elastin in the human heart, which verified the existence of prestraining phenomenon, indicating the importance of future studies on this topic and the potential translational impact of cardiac elastin research.

## 3 Spatial Distribution and Network Morphology of Cardiac Elastin

### *3.1 Anatomy of Ventricular Wall: Epicardium, Myocardium, and Endocardium*

A brief review of the anatomy of the ventricular wall is necessary to set the stage. The heart's ventricular wall has three anatomically distinguishable layers: the epicardium, myocardium, and endocardium. The epicardium (also known as visceral pericardium) is the outermost thin layer, and the major constituents are elastin fibers and collagen fibers [93, 110, 123–125]. The myocardium is the thick middle layer that fuses immediately with the epicardium and forms the main body of the heart wall. As we mentioned above, the myocardium contains mainly heart muscle fibers, which are delicately bound together by a network of Type I and III collagen, and the amount of elastin fibers in the myocardial ECM is believed to be relatively sparse [93, 124]. The innermost layer of the heart wall is the endocardium, which is also made of elastin fibers and collagen fibers [93, 123, 124].

## 3.2 *Epicardial Elastin: Elastin Network in the Epicardial Layer*

Porcine hearts (Yorkshire, ~6-month-old) were used for histological study. Epicardial, myocardial, and endocardial samples were dissected from the left ventricle anterior wall of the native porcine hearts. For endocardial samples, since the size and morphology of the trabeculae ridges varied greatly, we chose the regions between the trabeculae ridges that were relatively smooth in order to get a more consistent representation of the endocardial surface. Movat's pentachrome staining was used to reveal tissue microstructures, in which elastin was stained black, collagen stained yellow, heart muscle cells stained red, and Purkinje cells stained pale pink.

In the epicardial layer (Fig. 3a, b), elastin fibers (black color) were found to associate with collagen (yellow color), and the thickness of the epicardial layer

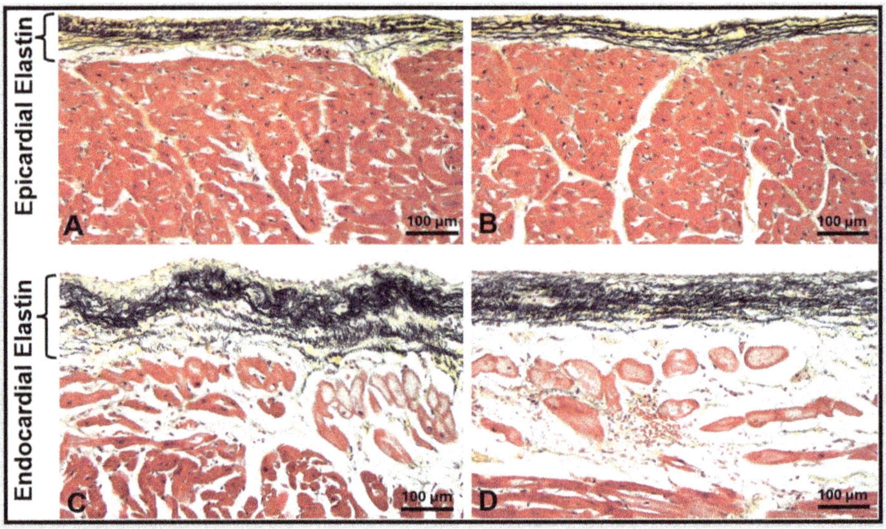


**Fig. 3** Histological images of the LV cross-sections of the native porcine hearts showed the microstructure of the epicardial layer (**a**, **b**). Epicardial elastin fibers covered a thickness range of 50–200 μm. The epicardial elastin fibers exhibited thick and distinguishable fibrous structure, with a wavy and meshed configuration. Microstructure of the endocardial layer (**c**, **d**): endocardial elastin fibers covered a thickness range of 100–250 μm. The endocardial elastin fibers were finer and formed much denser networks, with a higher degree of waviness when compared to the epicardial elastin network. Movat's pentachrome staining: collagen in yellow, elastin fibers in black, cardiac muscles in red, and the Purkinje fibers in pale pink. Scale bar = 100 μm

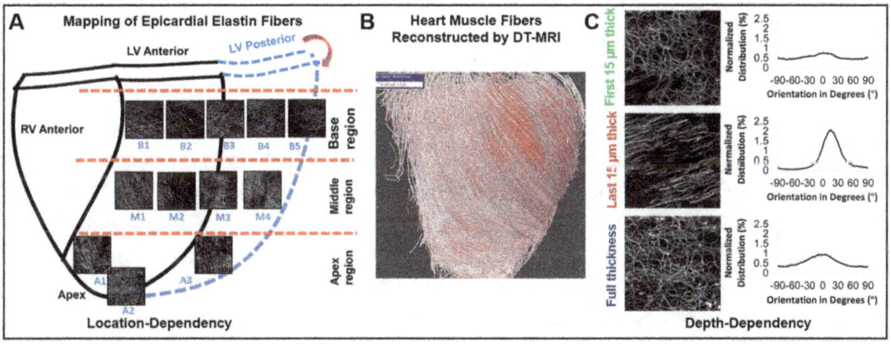


**Fig. 4** (**a**) The overall distribution of the epicardial elastin network in three regions (base, middle, and apex) of the left ventricle was obtained using the 3D laser scanning confocal microscopy (LSCM). (**b**) The multilayered helical structure of the heart muscle fibers revealed by our previous diffusion-tensor magnetic resonance imaging (DT-MRI) study [155]. (**c**) Layered epicardial elastin network in the apex region of the left ventricle. Top panel: projections of the first 15 μm thick layer; middle panel: projections the last 15 μm thick layer; bottom panel: projections of the full thickness of the sample. Scale bar = 100 μm

varied from ~50 μm to ~200 μm (Fig. 3a, b). The epicardial elastin fibers exhibit a thick and distinguishable fibrous structure, with a wavy and meshed configuration (Fig. 3a, b). In the epicardial histological images, the long elastin threads showed the fibers with in-plane orientation, and the short elastin strands showed the fibers with out-of-plane orientation, indicating the multidirectional orientation and depth-dependency of the epicardial elastin network [126].

The epicardial layer samples were dissected from various anatomical locations of the fresh porcine hearts, and the epicardial elastin fiber network was imaged by Laser Scanning Confocal Microscopy (LSCM) via elastin autofluorescence [126]. The epicardial elastin fiber network throughout the left ventricle demonstrated a location-dependent, with the overall elastin fiber orientation associating with the helical myocardial fiber orientation that spanned from the base to the apex (Fig. 4a, b). In the base region, elastin fibers showed a microstructure of multidirectional reinforcement and an overall trend wrapping circumferentially. In the middle region, elastin fibers exhibited an obvious oblique orientation pointing downwards. When approaching the apex, both the anterior regions and posterior regions showed a downward orientation and convergent to the tip of the apex. The epicardial elastin fiber network also showed depth dependency at the base, middle, and apex regions. Using the apex tip as an example (Fig. 4c), the deeper layer (the last 15 μm) had an alignment that as a continuity bridged the anterior wall fibers to the posterior wall fibers. However, the elastin fibers on the surface of the apex tip (the first 15 μm) exhibited a network of randomly meshed wavy elastin fibers, with a pore-like structure.

### 3.3 Endocardial Elastin: Elastin Network in the Endocardial Layer

The elastin network in the endocardial layer showed significantly different microstructural features compared to the elastin network in the epicardial layer (Fig. 3a–d). We observed that (1) the endocardial layer was thicker (~100–250 μm), and the endocardial elastin fibers were finer and more wavy, formed much denser networks, and had a higher elastin-collagen ratio; (2) the in-plane oriented long elastin fiber threads and the out-of-plane orientated short elastin fiber threads revealed that the endocardial elastin also had depth dependency in fiber alignment and orientation; and (3) the endocardial layer had areas where the layer was more undulating and areas where the layer was relatively smooth, indicating location dependency (Fig. 3c, d).

### 3.4 Myocardial Interstitial Elastin: Elastin Fibers in the Myocardium

With Movat's pentachrome staining, myocardial interstitial elastin fibers were detectable, appearing as finer fibers with a sparse existence. We observed that the fine interstitial elastin fibers in the myocardium were mostly associated with the perimysial collagen, the collagen network wrapping around and interconnecting individual heart muscle bundles (Fig. 5a, b). In-between muscle bundles, parallel-aligned elastin fibers (black color) were observed being embedded in yellow collagen (perimysial collagen), with a locally aligned, "hinge"-like morphology (Fig. 5c, b). An abundant number of interstitial elastin fibers were also found in the collagen network that connects the blood vessels to the surrounding myocardium, and this part of interstitial elastin fibers was more associated with the anatomical anchoring of the coronary blood vessels (Fig. 5e, f).

Note that the histological imaging was not able to detect the ultrafine elastin in the endomysial collagen (types I and III) that ensheathes each individual muscle fiber [127, 128]. In order to assess the ultrafine elastin fibers associated with the endomysial collagen, higher-resolution imaging techniques like transmission electron microscopy (TEM) or super-resolution light/laser microscopy via immunohistochemical staining must be used. Up to now, only Robinson et al. reported in 1983, via TEM observation, that the ultrafine elastic fibers, along with collagen fibers, interconnect individual heart muscle fibers, and the ultrafine elastin fiber helically winds around myocytes [127, 128]. Indeed, the presence of elastin in the myocardium might not be trivial as it can be found in the following sources: (i) fine interstitial elastin fibers that interconnect the heart muscle bundles (Fig. 5a–d), (ii) ultrafine interstitial elastin fibers that surround the individual muscle fiber [127, 128], (iii) interstitial elastin fibers that help to anchor the coronary blood vessels to the surrounding tissues (Fig. 5e, f), and (iv) vascular elastin in the coronary arteries/arterioles/veins that are plenty in the myocardium (coronary vasculature).

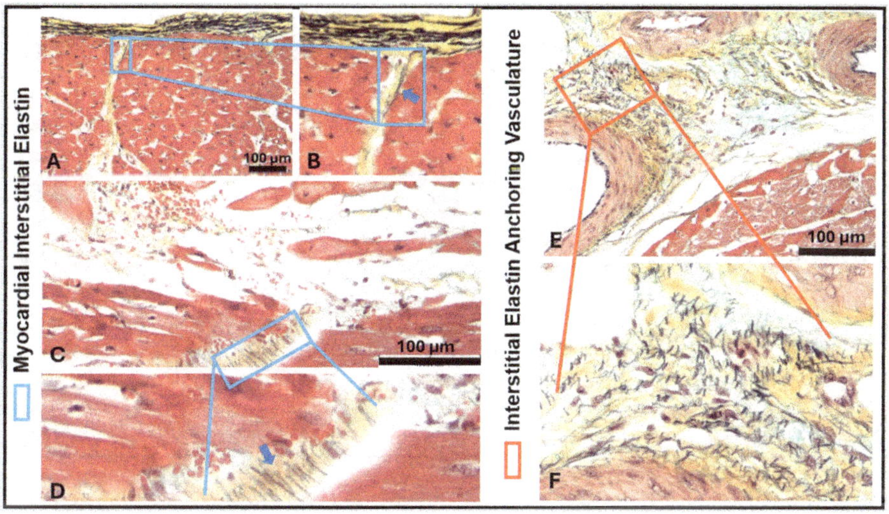


**Fig. 5** Histological images of the LV myocardium cross-sections of the native porcine hearts, with myocardial interstitial elastin fiber morphology revealed. The interstitial elastin fibers in the myocardium were found to co-exist with the perimysial collagen to bind the cardiomyocyte bundles (**a**, **b**), and some of the interstitial elastin fibers even showed locally aligned, hinge-like structures to connect the adjacent cardiomyocyte bundles (**c**, **d**). Arrows in (**b**) and (**d**) labeled the detailed interstitial elastin fiber structure. The histological image featured blood vessels and elastin associated with blood vessels. In the collagen network that meshed the blood vessels to the surrounding myocardium, abundant interstitial elastin fibers existed, and this part of interstitial elastin fibers was more associated with the anatomical anchoring of the coronary vasculature. Movat's pentachrome staining: collagen in yellow, elastin fibers in black, and cardiac muscles in red. Scale bar = 100 μm

## 3.5 *Elastin Fibers Associated with Purkinje Fibers and Purkinje Cells*

Located at the subendocardial region of the heart's ventricles (beneath the endocardium) and branching into the myocardium, Purkinje fibers are responsible for electrical conduction by sending electrical impulses to cardiomyocytes in the ventricles of the heart [129]. Studies of Purkinje fibers have mostly focused on their spatial distribution in the heart, electrophysiological properties, and cellular and biological characteristics of Purkinje cells. For example, it was found that the Purkinje cells are larger and have more mitochondria, but have fewer myofibrils than cardiomyocytes, and they are connected by desmosomes and gap junctions without intercalated discs, all in favor of their function in electrical conductivity [130, 131]. To describe their cellular morphologies at various anatomical locations, Pennisi *et al.* imaged the slow tonic myosin heavy chain in Purkinje fibers and reported the cellular morphologies of (i) the subendocardial Purkinje fibers, (ii) the branch point

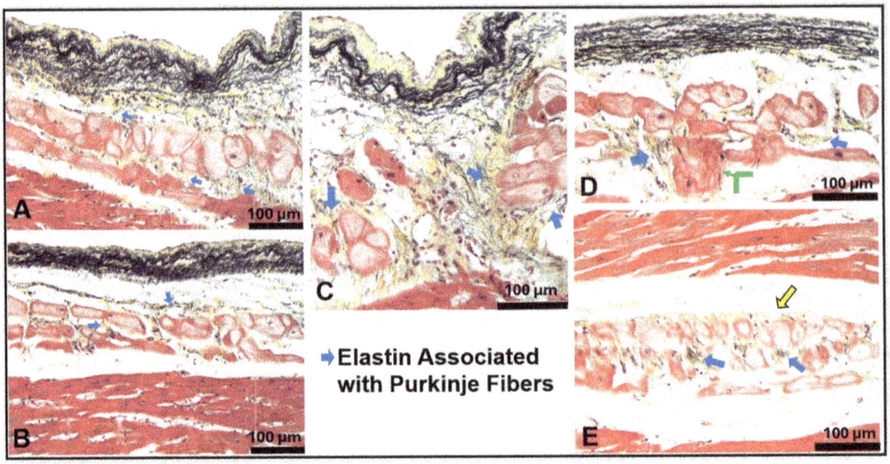


**Fig. 6** Histological images of the LV endocardial cross-sections (**a**, **b**, **c**, **d**) and LV myocardium cross-section (**e**) of the native porcine hearts showed that fine elastin fibers (black), along with collagen network (yellow), closely associated with the Purkinje cells, implying that this ECM association could be essential in organizing cardiac Purkinje cells into "fibrous" and "branching" morphologies. We noticed the fine elastin fibers (black), meshed with the collagen network (yellow), morphologically wrapping around groups of Purkinje cells and scattering in between the Purkinje cells (labeled by blue arrows). In (**a**, **b**, **c**), we captured the elastin fibers and collagen that were associated with subendocardial Purkinje fibers. In (**d**), we captured the elastin fibers and collagen associated with the branch point from subendocardial Purkinje fibers (green arrow in **d** shows where T-shape branching happens). In (**e**), we observed the elastin fibers (blue arrows) and collagen associated with the intramural Purkinje fibers (yellow arrow). Movat's pentachrome staining: collagen in yellow, elastin fibers in black, cardiac muscles in red, and the Purkinje fibers in pale pink. Scale bar = 100 μm

extended from the subendocardial Purkinje fibers, and (iii) the intramural Purkinje fibers (situated in and surrounded by the myocardium) [132].

Our study captured how structural ECMs were associated with Purkinje fibers and Purkinje cells, revealing important microstructural details. We found that the fine elastin fibers, meshing with collagen, morphologically wrapped around the groups of Purkinje cells along the longitudinal direction (Fig. 6a, b) and the groups of Purkinje cells in the cross-sectional view (Fig. 6c). Furthermore, the fine elastin fiber arrays, with local alignment, frequently existed between the Purkinje cells, and these elastin fiber arrays always associated with yellow-stained collagen (Fig. 6a–e). Judging from their morphology, these interstitial elastin and collagen were likely responsible for binding the Purkinje cells and allowing the formation of fibrous structures. Our study provided strong evidence to this point by capturing the elastin fibers and collagen (i) that were associated with the subendocardial Purkinje fibers (Fig. 6a–d), (ii) that were associated with the branch point extended from subendocardial Purkinje fibers (Fig. 6d, green arrow, T-shape branching), and (iii) that were associated with the intramural Purkinje fibers (Fig. 6e, yellow arrow).

# 4 Biomechanical Contributions of Cardiac Elastin in Epicardium, Endocardium, Myocardium, and Purkinje Fibers

## *4.1 Epicardial Residual Stress Revealed by Curling Phenomenon of Heart Surface Strip*

When dissecting heart surface tissues that consist of the epicardial layer and cardiac muscle, we discovered the bi-layered left ventricle (LV) surface strip always curled toward the epicardial side (Fig. 7a). When the LV surface strip was dissected off the heart, the contraction of the epicardial layer bends the surface strip, creating a curling morphology. This surface strip curling phenomenon reveals the existence of residual stress in the epicardial layer. Residual stress and strain in an organ are the stress and strain that remain in the tissue after all external physiological loads are removed, while certain local anatomical integrity is still preserved. Residual stress and strain in blood vessels are well-known examples of this; a single radial cut to the blood vessel ring segment can release the residual stress in a blood vessel wall. After the radial cut, the closed ring configuration cannot be maintained, and the ring opens up, forming an arc conformation [133]. The degree of the arc opening reflects the amount of residual stress and strain in the blood vessel wall, and a quantification called "open angle" can be defined and measured as the angle formed between two lines, which extend from the tips of the opened ring to the midpoint of the inner arc of the opened ring [133].

It has also been shown that the left ventricular wall of rats has residual stress. Omens and Fung demonstrated, that when a radial cut was applied to an equatorial slice of a rat heart, the left ventricle exhibited an open angle-like morphology (~30°), which indicates residual stress throughout the ventricular wall [134]. Based on a finite growth framework with fixed-point iteration, Genet et al. proposed that the residual stress of the ventricular wall is likely caused by the organ's growth [135]. Jobsis et al. continued the residual stress research on the left ventricle by examining the gross opening angle of the porcine left ventricle wall, and showed that the disruption of the epicardium reduced the gross opening angle of the left ventricular wall, indicating the likely contribution of the epicardium to the ventricular wall residual stress [110].

## *4.2 Curling Angle Measurements and Residual Stress Distribution on the Intact Heart*

Both the circumferential and longitudinal heart surface strips were dissected and measured (Fig. 7a). Native epicardial strips of ~15 mm × 5 mm × 1.5 mm, an optimal dimension to demonstrate curling, were dissected out from seven locations

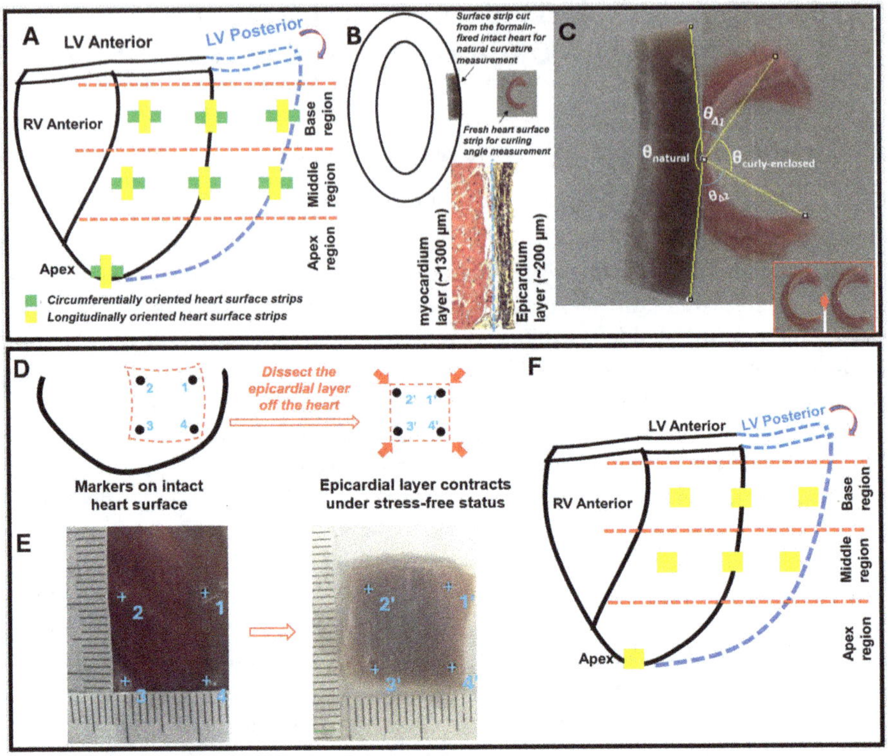


**Fig. 7** *Curling Angle Measurement*: Protocol to characterize the curling phenomenon of the native heart surface strip. (**a**) Sample preparation plan. Circumferentially oriented heart surface strips and longitudinally oriented heart surface strips were dissected from basal anterior, basal lateral, basal posterior, middle anterior, middle lateral, middle posterior, and apex locations. (**b**) Schematic illustration showing the curling phenomenon of the heart surface strips dissected from the native intact heart. The bi-layered LV surface strip, dissected from the fresh native heart, always curls toward the epicardial side (inserted picture on the right) revealing residual stress in the epicardial layer. The inserted picture on the left shows an LV strip dissected from the formalin-fixed heart which preserves the curvature of the heart surface as a comparison. Histology at the bottom shows the heart surface strip has a bi-layered structure consisting of the elastin-dominant epicardial layer (~200 μm) and a myocardial layer (~1300 μm). (**c**) Four types of angles were defined: $\theta_{\text{curly-enclosed}}$ as the surface curly enclosed angle from native LV strips after being dissected and reaching a stable curling status; $\theta_{\text{natural}}$ is the angle defined by the natural curvature contour measured from the formalin-fixed heart tissue strip; $\theta_{\text{reflex}}$ represents the reflex angle of $\theta_{\text{curly-enclosed}}$; $\theta_{\Delta}$ is the total curling angular change: $\theta_{\Delta} = \theta_{\Delta 1} + \theta_{\Delta 2} = \theta_{\text{reflex}} - \theta_{\text{natural}}$. Inserted picture in Fig. 1c demonstrates two important steps: (i) outlining the upper edge of the epicardial surface with two endpoints; (ii) finding the midpoint. *Epicardial Prestrain Measurement*: Schematic illustration shows the protocol to estimate the prestrains of the epicardial layer on the unpressurized intact heart. (**d**) By putting four squarely arranged markers on the epicardial surface of the unpressurized intact heart, we could obtain the in situ marker dimensions while the prestrains existed. After dissecting the same region of the heart, the marker dimensions of the stress-free epicardial layer could be measured for the calculation of the prestrains. (**e**) Markers tagged from ImageJ for in situ unpressurized intact heart (left panel) and stress-free epicardial sample (right panel). The ruler with the actual size was printed on paper. The paper ruler could be placed onto the left

from three regions of the LV of the fresh porcine hearts (base anterior, base lateral, base posterior, middle anterior, middle lateral, middle posterior, and apex). Histology showed the heart surface strip had a bi-layered structure consisting of the elastin-dominant epicardial layer (~200 μm) and a myocardial layer (~1300 μm) (Fig. 7b). Furthermore, formalin-fixed intact porcine hearts were used to estimate the natural surface curvature of the heart (Fig. 7b) [136]. To measure the degree of curling, after dissection, each fresh heart surface strip was immersed in 1X phosphate buffered saline (PBS) in a petri dish to create a free-floating and stress-free status. Digital pictures of each strip (side view) were taken right after the full occurrence of curling, which took place in a time frame of one minute. Side views of the strips from the formalin-fixed intact hearts were also imaged to get the natural surface curvatures. ImageJ (NIH, Bethesda, MD) is used to analyze the images. To estimate the total curling angular change ($\theta_{\Delta}$) of the fresh heart surface strip by comparing with the natural curvature of the heart, four types of angles were defined (Fig. 7c): (i) $\theta_{\text{curly-enclosed}}$ as the curly enclosed angle of the curling surface strip dissected from the fresh native heart; $\theta_{\text{curly-enclosed}}$ was defined by two straight lines connecting two endpoints of the strip upper edge to the midpoint of the strip; (ii) $\theta_{\text{natural}}$ as the angle defined by the natural curvature contour of the heart surface via the same quantification method; (iii) $\theta_{\text{reflex}}$ represented the reflex angle of $\theta_{\text{curly-enclosed}}$ defined by Eq. 1; and (iv) $\theta_{\Delta}$ was thus the total curling angular change as defined by Eq. 2.

$$\theta_{\text{reflex}} = 360^\circ - \theta_{\text{curly-enclosed}} \tag{1}$$

$$\theta_{\Delta} = \theta_{\Delta 1} + \theta_{\Delta 2} = \theta_{\text{reflex}} - \theta_{\text{natural}} \tag{2}$$

The curling of LV surface strips provided a characterization that was sensitive to the status of epicardial residual stress. Fresh native heart surface strips along the circumferential direction (CD) (Fig. 8b) and longitudinal direction (LD) (Fig. 8e) showed various degrees of curling in different anatomical locations. Heart surface strips from the formalin-fixed heart were also quantified to measure the natural surface curvature of the intact hearts (Fig. 8a, d). For the CD, the total curling angular change revealed that (i) larger curling happened in the basal lateral location (B2) ($\theta_{\Delta} = 103.2 \pm 23.7^\circ$) when compared with the basal anterior location (B1)

**Fig. 7** (continued) ventricle surface, perfectly fitting the left ventricle surface curvature (left panel). In this way, the readings of the real marker dimensions on the unpressurized intact heart can be accurate, without being skewed by using a straight ruler for a curved surface. By immersing the isolated epicardial layer in the 1X PBS in a petri dish, we obtained the new marker dimension of the epicardial layer under stress-free status (right panel). Rulers with actual size were printed on the transparent film and then placed beside the stress-free epicardial layer sample (right panel). (**f**) Sample dissection plan for estimating prestrains of the epicardial layer on the unpressurized intact heart

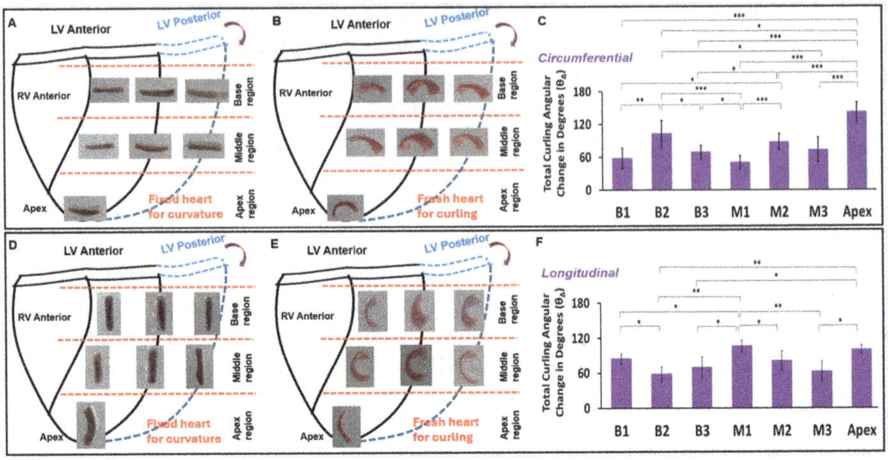

**Fig. 8** Native heart surface strips show the curling phenomenon. All seven anatomical locations of the left ventricle, i.e., basal anterior, basal lateral, basal posterior, middle anterior, middle lateral, middle posterior, and apex, are shown. (**a**) Circumferentially oriented heart surface strips dissected from the formalin-fixed intact heart; 10% buffered formalin solution fixation was able to preserve the natural curvature of the intact heart. Images are side views of the dissected formalin-fixed strips, showing the circumferential natural curvatures of the heart at seven anatomical locations. (**b**) Circumferentially oriented heart surface strips dissected from the unpressurized native intact heart. Images are side views of the dissected native strips, showing various degrees of curling toward the epicardial side at seven anatomical locations. (**c**) Total curling angular change ($\theta_\Delta$) in degrees for circumferentially dissected heart surface strips. (**d**) Longitudinally oriented heart surface strips were dissected from the formalin-fixed intact heart; 10% buffered formalin solution fixation was able to preserve the natural curvature of the intact heart. Images are side views of the dissected formalin-fixed strips, showing the longitudinal natural curvatures of the heart at seven anatomical locations. (**e**) Longitudinally oriented heart surface strips dissected from the unpressurized native intact heart. Images are side views of the dissected native strips, showing various degrees of curling toward the epicardial side at seven anatomical locations. (**f**) Total curling angular change ($\theta_\Delta$) in degrees for longitudinally dissected heart surface strips. Note that *, **, and *** donate the statistical significance between different anatomical locations of four types of angles (one-way ANOVA) when $p < 0.05$, $p < 0.01$, and $p < 0.001$, respectively

($\theta_\Delta = 60.0 \pm 18.2°$) and basal posterior location (B3) ($\theta_\Delta = 69.1 \pm 12.1°$); (ii) the larger curling happened in the middle lateral location (M2) ($\theta_\Delta = 87.6 \pm 14.4°$) when compared with the middle anterior (M1) ($\theta_\Delta = 50.0 \pm 12.3°$) and middle posterior (M3) ($\theta_\Delta = 72.9 \pm 22.8°$); and (iii) the apex location had the largest circumferential curling ($\theta_\Delta = 142.0 \pm 17.5°$) (Fig. 8c). For the LD, (i) the curling was larger in the basal anterior location (B1) ($\theta_\Delta = 85.0 \pm 8.8°$) when compared with the basal lateral location (B2) ($\theta_\Delta = 58.0 \pm 12.3°$) and basal posterior location (B3) ($\theta_\Delta = 69.6 \pm 17.8°$); (ii) the curling in the middle anterior location (M1) ($\theta_\Delta = 106.2 \pm 9.8°$) was larger than the middle lateral location (M2) ($\theta_\Delta = 81.2 \pm 14.4°$) and middle posterior location (M3) ($\theta_\Delta = 62.3 \pm 18.1°$); and (iii) again, the curling in the apex location was very large ($\theta_\Delta = 100.1 \pm 7.4°$) (Fig. 8f). The heart surface strip curling clearly revealed the existence of various amounts of residual stress in the epicardial layer. In other words, the epicardial layer is under tension (prestrained) when it covers the unpressurized intact heart in its in situ 3D configuration.

## 4.3 *Pre-tensioned Elastin Network as the Main Structural Contributor to Epicardial Residual Stress and Prestraining*

We designed a histological assessment to compare the epicardial microstructure from the unpressurized intact heart with that from dissected, curled surface strips to identify structural contributors of the surface strip curling. For the intact heart surface strip, the intact heart was fixed in a 10% buffered formalin solution for 72 h. The dissected heart surface strip, which preserved the natural curvature of the heart surface, was then embedded in paraffin and sectioned into 5 μm slices for Movat's pentachrome staining (Fig. 9b). To obtain the curled surface strip for histology, we let the freshly dissected native heart surface strip be immersed in 1X PBS in a petri dish and waited for the full curling occurrence to achieve a free-floating and stress-free status. The PBS solution was then replaced with a 10% buffered formalin solution for histological tissue fixation. We found that the formalin fixation was able to preserve the curled surface strip configuration, and the curled sample was then embedded in paraffin and sectioned into 5 μm slices for Movat's pentachrome staining (Fig. 9e).

Histological image of the epicardium from the unpressurized intact heart showed that (i) the elastin fibers (black color) exhibited a low degree of waviness, and the network had a pre-tensioned morphology and (ii) the collagen fibers (yellow color) that were associated with the elastin fibers demonstrate a moderate degree of waviness (Fig. 9c). On the contrary, the microstructural observation on the curled epicardial layer demonstrated that (i) the elastin pre-tensioned morphology disappeared, and the elastin fibers (black color) showed a recoiled, highly wavy morphology, indicating the release of tension; (ii) the collagen fibers (yellow color) associated with the elastin fibers also showed a morphology that was more wavy than that on the intact heart (Fig. 9f). The mechanism of the surface strip curling can thus be attributed to the pre-tensioned elastin network. Once a surface strip is dissected from the heart, the boundary condition that restricts the elastin network from contraction vanishes, and consequently the pre-tensioned elastin network contracts and bends the bi-layered surface strip, releasing the epicardial residual stress and prestraining. We noted that, on the unpressurized intact heart, the epicardial elastin network still showed a lower degree of waviness, and collagen fibers associated with the elastin had a moderate degree of waviness (Fig. 9c). Those configurations remind us that the intact heart is still in an unpressurized status, and both elastin and collagen fiber networks will be straightened and gradually recruited for load bearing once the LV wall is pressurized and deformed during the diastolic phase.

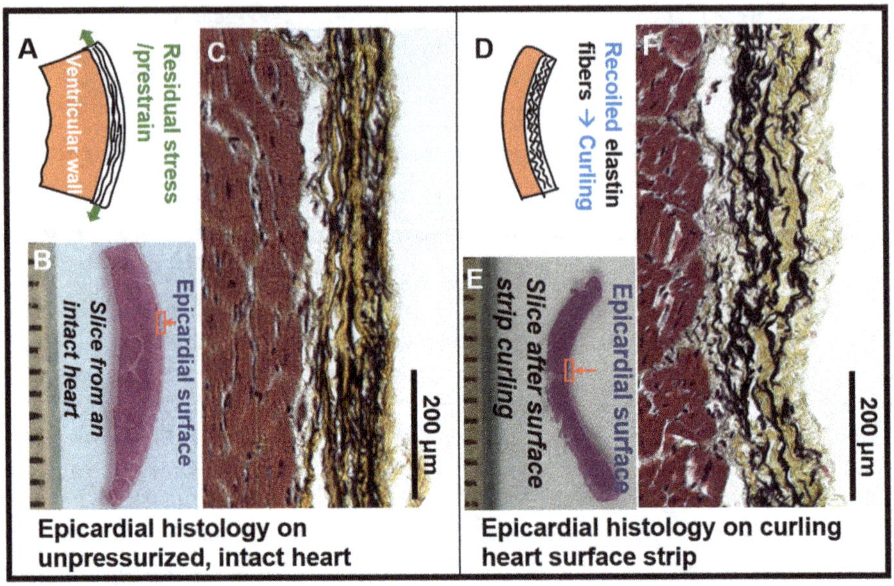


**Fig. 9** A histological study to compare the epicardial microstructure on the unpressurized intact heart to the epicardial microstructure on the curled heart surface strip. Movat's pentachrome staining was performed, in which elastin was black color, collagen yellow color, and muscle red color. (**a**) Schematic illustration shows the existence of prestraining/residual stress. (**b**) Histological slice from an intact heart. (**c**) On the intact heart, the elastin network in the epicardium appears less wavy, with a pre-tensioned morphology. (**d**) Schematic illustration shows that the recoiled elastin fibers bend the bi-layered surface strip. (**e**) Histological slice obtained after full curling of the native heart surface strip. (**f**) On the curled surface strip, the elastin pre-tensioned morphology disappears and the elastin network appears recoiled and has more waviness, indicating the release of residual stress

## *4.4 Regional Prestrains and Residual Stresses of the LV Epicardial Layer*

*Quantifying Epicardial Prestrains on the Unpressurized Intact Heart* By placing four squarely arranged markers on the epicardial surface of the unpressurized intact heart using a minimal amount of cyanoacrylate glue, we obtained the in situ marker dimensions while the epicardial prestrain existed (Fig. 7d). The actual size paper rulers were placed onto the LV surface, perfectly fitting the LV surface curvature. In this way, the readings of the real marker dimensions on the intact heart were accurate, without being skewed by using a straight ruler for a curved surface (Fig. 7e). After the in situ measurement, the epicardial layer of each location was dissected, gently removed of the attached heart muscles, and immersed in 1X PBS in a petri dish to ensure a free-floating and stress-free status (Fig. 7e). Rulers with actual size were printed on a transparent film and then placed beside the stress-free epicardial layer sample (Fig. 7e). Digital pictures were then taken

**Table 1** Estimated prestrains and residual stresses of the epicardial layer on the unpressurized intact heart

| $N = 5$ | $\varepsilon_X$ | $\sigma_{X(\text{residual})}$ (kPa) | $\varepsilon_Y$ | $\sigma_{Y(\text{residual})}$ (kPa) |
|---|---|---|---|---|
| Basal anterior (B1) | $-0.30 \pm 4.25\%$ | 0 | $8.59 \pm 3.62\%$ | 2.39 |
| Basal lateral (B2) | $4.66 \pm 0.44\%$ | 8.95 | $6.75 \pm 1.85\%$ | 3.05 |
| Basal posterior (B3) | $3.01 \pm 4.58\%$ | 4.68 | $7.72 \pm 4.38\%$ | 4.48 |
| Middle anterior (M1) | $-3.51 \pm 3.07\%$ | 0 | $8.72 \pm 6.70\%$ | 4.29 |
| Middle lateral (M2) | $7.09 \pm 3.43\%$ | 12.09 | $10.70 \pm 1.36\%$ | 7.88 |
| Middle posterior (M3) | $5.50 \pm 3.10\%$ | 2.89 | $8.31 \pm 4.55\%$ | 2.06 |
| Apex | $4.89 \pm 1.72\%$ | 12.24 | $8.84 \pm 3.37\%$ | 3.89 |

The residual stresses ($\sigma_{X(\text{residual})}$, $\sigma_{Y(\text{residual})}$) for each anatomical location were predicted by the prestrains ($\varepsilon_X$ and $\varepsilon_Y$) using the adjusted biaxial stress-stretch curves

to record the marker dimensions. For each heart, seven epicardial layer sample locations (~10 mm × 10 mm × 1 mm) were used to map the left ventricle (Fig. 7f). To calculate the prestrains (in the form of Green strain), we interpolated the displacements of all four markers from stress-free (the reference configuration) to in situ states of the sample using the first-order shape function for quadrilateral element [137].

The average prestrains ($\varepsilon_X$ and $\varepsilon_Y$) for each anatomical location were calculated (Table 1). Several observations were summarized as follows: (i) for nearly all the anatomical locations, the prestrains of both the circumferential and longitudinal directions ($\varepsilon_X$ and $\varepsilon_Y$) were positive, except for $\varepsilon_X$ in the basal anterior location and middle anterior location; the positive prestrains on the unpressurized intact heart mean that the epicardial layer was under tension and positive residual stresses existed. (ii) We found that $\varepsilon_Y$ was much larger than $\varepsilon_X$ in the basal anterior location and middle anterior location. (iii) $\varepsilon_X$ was slightly smaller than $\varepsilon_Y$ in the basal lateral location and middle lateral location, while both $\varepsilon_X$ and $\varepsilon_Y$ were large. (iv) $\varepsilon_X$ was smaller than $\varepsilon_Y$ in the basal posterior location and middle posterior location. (v) For the apex location, both $\varepsilon_X$ and $\varepsilon_Y$ were relatively large.

Note that the fresh porcine hearts subjected to prestrain measurements were at an unpressurized status after the captive bolt stunning and exsanguination (draining of blood) slaughtering procedure. After captive bolt stunning, the significantly thicker left ventricular myocardium in healthy animals squeezes the blood out of the ventricles as the heart muscle goes into rigor, which accounts for the slight contracture of the heart after death when rigor mortis has occurred [138–140]. The existence of epicardial prestrains on those slightly contracted hearts demonstrates that the observed prestraining is not caused by artifacts. On the other hand, if the slight contracture of the arrested hearts does not happen, the prestrains will be measured with slightly larger numbers.

*Biaxial Mechanical Testing and Residual Stress Estimation* After prestrain measurement, the samples dissected from the seven locations of the heart (Fig. 7f) were used for mapping the biaxial mechanical properties of the epicardial layer. One axis of the square sample (~10 mm × 10 mm × 1 mm) was aligned with the circumferential direction (CD) of the heart and the other with the longitudinal direction (LD). A biaxial mechanical testing system was used to capture the tissue behavior under physiologically relevant loading conditions [96, 137, 141, 142]. Each side of the square sample was mounted onto four stainless steel hooks that were attached to two loops of 000 polyester sutures. Four fiducial graphite markers, affixed to the center of the square sample, were monitored via a CCD camera to capture the real-time stretch ratios. Immersed in 1X PBS at 37 °C, the sample was preconditioned biaxially by imposing 10 cycles of 60 N/m Lagrangian membrane tension. After preconditioning, an equibiaxial tension protocol of $T_{CD}$:$T_{LD}$ = 60:60 N/m was performed. All the testing was performed with a 0.5 g tare load applied as a prerequisite for the software load control protocol. In order to accurately estimate the residual stresses, we adjusted the biaxial curves to real 0 g load reference status using a protocol described in our paper [137]. The compensatory strains, $\varepsilon_{X(0.5)}$ and $\varepsilon_{Y(0.5)}$, were estimated, and the adjusted curves starting from the 0 g load reference status were obtained.

With the adjusted biaxial stress-strain data (Examples in Fig. 10a) and the prestrain data (Table 1), we could estimate the residual stresses for each anatomical location. In Fig. 10a, red dashed lines label the circumferential residual stresses ($\sigma_{X(residual)}$) predicted by the circumferential prestrains ($\varepsilon_X$), and blue dashed lines label the longitudinal residual stresses ($\sigma_{Y(residual)}$) predicted by the longitudinal prestrains ($\varepsilon_Y$). The detailed estimations of $\varepsilon_X \rightarrow \sigma_{X(residual)}$ and $\varepsilon_Y \rightarrow \sigma_{Y(residual)}$ for each anatomical location are listed in Table 1. The mapping of residual stresses on the left ventricle showed that residual stresses did exist and varied with different anatomical locations. We found that (i) both the basal anterior location and middle anterior location showed longitudinal residual stress of several kPa but predicted negligible circumferential residual stress; (ii) the basal posterior location had both the circumferential residual stress and longitudinal residual stresses of several kPa; the middle posterior location also had comparable circumferential and longitudinal residual stresses around serval kPa. However, both the basal lateral location and middle lateral location showed a pattern of circumferential residual stress being larger (around 10 kPa) than the longitudinal residual stress. The apex region showed a pattern similar to the laterals, i.e., a large circumferential residual stress around 10 kPa, but a longitudinal residual stress of several kPa. Considering the geometry of the intact heart, the circumferential direction of the lateral locations as well as the apex location experience large curvature. The relatively large circumferential residual stresses ($\sigma_{X(residual)}$) of the basal lateral location, middle lateral location, and apex might be correlated with the large curvature of the heart surface.

## 4.5 *Effect of Epicardial Prestraining on Ventricular Mechanics*

We took a computational modeling and simulation approach to determine the contribution of the epicardial layer to ventricular mechanics. It is not feasible to

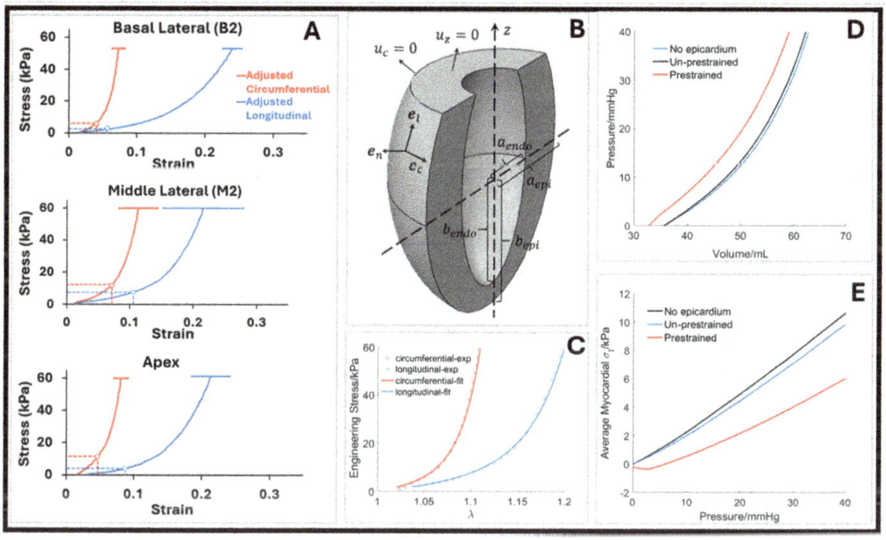


**Fig. 10** (**a**) Circumferential residual stresses $\sigma_{X(\text{residual})}$ (red dashed lines) and longitudinal residual stresses $\sigma_{Y(\text{residual})}$ (blue dashed lines) predicted by prestrains ($\varepsilon_X$ and $\varepsilon_Y$). Basal lateral, middle lateral, and apex locations were presented as examples. Adjusted biaxial stress-strain curves were used for the quantitative estimation of the regional residual stresses. Red solid curve: stress-strain curve along the circumferential direction with error bar (red solid bar); blue solid curve: stress-strain curve along the longitudinal direction with error bar (blue solid bar). FE modeling of the effect of epicardial prestrain. (**b**) An illustration of the model geometry. The left ventricle is modeled as a truncated ellipsoid, characterized by five parameters: the semi-axes lengths of the endocardial ($a_{endo}$, $b_{endo}$) and the epicardial ($a_{epi}$, $b_{epi}$) surface, and the distance between the apex and the basal plane ($h$). To denote the material orientations, we employ the local coordinate system whose base vectors are $\boldsymbol{e}_c$, $\boldsymbol{e}_l$, and $\boldsymbol{e}_n$, the circumferential, the longitudinal, and the transmural normal vector. On the basal plane,we prescribe $u_z = 0$, and on its outer edge $u_c = 0$ (in the circumferential direction). (**c**) The fitted biaxial stress-stretch curve of the middle lateral sample. (**d**) The cavity volume and (**e**) the averaged myocardial first principle stress, plotted against the cavity pressure. The blue line shows the case without epicardial prestrain, and the red curve shows the case with epicardial prestrain, and the black line shows the case without epicardium

surgically disrupt or peel off the epicardial layer from an in vivo beating human heart or animal heart in order to reveal its mechanical role due to ethical and physiological issues, as well as technical difficulties. However, with the computational models, we could alter the status of epicardial prestrained confinement so that the biomechanical role of the epicardial layer could be revealed.

For simplicity, we used a truncated ellipsoid to represent the shape of the porcine left ventricle (Fig. 10b). The lengths of the semi-axes of the endocardial surface are $a_{\text{endo}} = 15$ mm and $b_{\text{endo}} = 39$ mm, while those for the epicardial surface are $a_{\text{epi}} = 30$ mm and $b_{\text{epi}} = 47$ mm. The distance between the apex and the base plane is $h = 78$ mm. These geometrical parameters of the left ventricle were obtained by averaging the measurements from four porcine hearts (~6-month old, Yorkshire). The myocardium part consisted of 16,326 quadratic tetrahedral solid elements, and the epicardium was modeled as a shell covering the epicardial surface with 4430 quadratic shell elements with a thickness of 1 mm, consistent with the overall thickness of the epicardial samples we dissected for biaxial mechanical testing. To model the passive mechanical properties of the myocardium, we adopted the Holzapfel-Ogden model [143, 144], and the parameters we used are adopted from Goktepe et al. [145] For epicardium, we used the Holzapfel-Ogden model by fitting the biaxial tension data of the middle lateral sample. The fitted stress-strain curve was compared against the experimental data in Fig. 10c, which showed good agreement. The prestrains we applied came from the middle lateral sample (Table 1) and were introduced through a pre-factor in the deformation gradient. On the endocardial surface, we applied a pressure that increased from 0 to 40 mmHg. The calculations were performed using the open-source package FEBio [146].

The LV cavity volume-pressure curve predicted from the FE simulation showed that, with the presence of the epicardial layer alone, the cavity volume barely differed from the pristine ventricle model, and the average myocardial first principle stress only decreased slightly. Yet when prestrain was applied, a nontrivial shift (up to ~13%) toward smaller volume and lower average myocardial first principle stress emerged (up to ~30%) (Fig. 10d, e). This observation indicated that the epicardial layer provided additional confining and protection to the LV wall by contracting against myocardial tension. Moreover, the significant differences between the prestrained and the un-prestrained cases implied that the majority of the confinement effect actually comes from the prestraining instead of the epicardial layer alone. Usually, the parietal pericardium sac is associated with an important function of limiting ventricular filling and preventing high diastolic pressure [147], and our results indicate that such capability can also come from epicardial prestraining. If prestraining is absent, the confining capability of the epicardial layer to reduce the cavity volume and average myocardial first principle stress would be significantly impaired even if the epicardial layer itself is intact.

## 5 Potential Biomechanical Functions of Endocardial Elastin and Myocardial Interstitial Elastin

*Biomechanics of Endocardial Elastin* Knowing that (i) the epicardial layer acts like a prestrained "balloon" that wraps around the heart to provide additional resistance against ventricular diastolic expansion and ventricular wall protection by reducing myocardial stress; (ii) the pre-tensioned elastin network in the epicardial layer was the main structural contributor to the epicardial prestraining and residual stress, we speculate that the endocardial layer, rich in elastin, might also play an important role of interface mechanics in cardiac function. When compared to the epicardial layer, the endocardial layer was thicker (~100 μm – ~250 μm) and had a higher elastin-collagen ratio and denser elastin fiber network; moreover, endocardial elastin fibers were finer and more wavy than the epicardial elastin fibers (Fig. 3). The abundance of endocardial elastin hints that, similar to the epicardial layer, the endocardial layer possibly possesses prestraining and residual stress. Moreover, the drastic differences in elastin network microstructures between the endocardial layer and the epicardial layer might lead to different interface mechanics, which are likely to be associated with various deformations experienced by the endocardial and epicardial surfaces during ventricular expansion and contraction. Using an open-chest porcine model, Howard-Quijano et al. quantified the global longitudinal, radial, and circumferential strains of hearts [148]. They reported that regionally endocardial strain was greater than epicardial strain in both longitudinal direction and circumferential direction [148]. Apparently, more in-depth studies are needed to reveal the underlying structural-functional reasons that cause the elastin networks to adapt to layer-specific microstructure/morphology, thus serving different interface mechanics.

*Biomechanics of Myocardial Interstitial Elastin* Myocardial interstitial elastin were fine fibers associated with the perimysial collagen that wraps around the heart muscle bundles (Fig. 5). The co-existence of the interstitial elastin fibers with the perimysial collagen indicated the functional roles of perimysial elastin fibers in participating in the bound and connection of heart muscle bundles. The locally aligned, hinge-like elastin fiber array between heart muscle bundles suggests possible mechanical contribution of interstitial elastin fibers in (i) participating force transfer among heart muscle bundles during diastolic wall expansion and (ii) assisting recoil during the systolic phase by releasing the stored elastic energy. The ideal inter-muscle bundle connection should have both collagen network and interwoven elastin fibers, where the collagen fibers provide mechanical strength, and the elastin fibers facilitate large deformation and recoiling that are important for cardiac loading and unloading cycles. In the absence of elastin fibers, the collagen-only network would have an inferior performance in mediating the dynamic inter-muscle bundle connection due to the loss of elasticity components and capabilities in handling large deformation and recoiling. In fact, the mechanical role of interstitial elastin in mediating ventricular wall function is consistent with prior studies that investigated how elastin content/expression affects heart wall function as discussed in Section 2. It is worthy to quote those observations here: (i) The overexpression of elastin in an infarcted heart efficiently prevented scar expansion, stabilized tissue structure, and delayed ventricular dysfunction [118, 119, 149]; (ii) the loss of contractile function in the native or diseased myocardium might partially result from the reduction in the elastin content of the cardiac tissues [93, 98, 118, 121, 122, 150, 151]. Furthermore, considering the importance of myocardial interstitial elastin, future studies will need to focus on investigating the amount and microstructure of the ultrafine elastin fibers that are associated with the endomysial collagen to ensheathe each individual muscle fiber. In short, the anatomy, microstructure, and structure-function relation of the perimysial elastin and endomysial elastin, as well as their degenerations in response to aging/diseases, are currently a knowledge gap, and filling this gap is an imperative task in cardiac mechanics.

## 6 Potential Functional Roles of Purkinje Cell-Associated Elastin

Our histological study revealed that the cardiac elastin fibers, along with the collagen network, are closely intertwined with the Purkinje cells. Deducing from the morphology, these interstitial elastin and collagen could participate in organizing cardiac Purkinje cells into "fibrous" and "branching" features, enabling mechanical integrity, and serving as a protective mechanism when the electrically conductive Purkinje fibers experience large deformations during cardiac cycles. The evidence that the elastin fibers and collagen were associated with the subendocardial Purkinje fibers (Fig. 6a–d), the branch point from the subendocardial Purkinje fibers (Fig. 6d), and the intramural Purkinje fibers (Fig. 6e) warrants future study on the microstructures and compositions of the Purkinje fiber ECM. Knowledge of Purkinje fiber ECMs is essential not only for understanding the anatomy of Purkinje fibers but also for determining the ECM compositions, amounts, and structural characteristics to be incorporated into Purkinje fiber tissue engineering and regeneration.

Clinicians and bioengineers have always dreamed of restoring functional heart conduction pathways. A variety of conductive biomaterials have been investigated for cardiac tissue engineering, including graphene, carbon nanotube, gold-based nanomaterials, and polyaniline and polypyrrole, showing both potentials and limitations [152, 153]. Recently, Tracy et al. reported their research to reprogram human adipogenic mesenchymal stem cells (hADMSCs) to form Purkinje cells and bioprint those cells into a Purkinje network, in which pluronic acid is used as the sacrificial material and type I collagen as the structural material [154]. Via this approach, the bioprinted Purkinje networks demonstrated continuous syncytium, expressed gap junction protein Connexin 40, and showed membrane potential changes when electrical stimulation was applied [154]. We envision that future tissue-engineered or bioprinted Purkinje fibers/network could benefit by including both the collagen and elastin components as structural supporting materials. This nature-optimized combination can simultaneously provide extensibility, mechanical strength, and over-stretch protection while minimizing energy dissipation during dynamic loading. All those qualities are much needed for the engineered Purkinje fibers/network when they are eventually integrated into the ventricular wall and subjected to cyclic large deformation in vivo.

## 7 Epicardial Prestraining Phenomenon Verified on the Healthy Human Hearts

We obtained healthy human hearts (Fig. 11a, rejected for transplant) from the Southwest Transplant Alliance through an organ procurement organization research protocol (#NC228). We dissected the heart surface strip (epicardial layer + myocardium) and successfully observed the curling phenomenon. The curling phenomenon of the

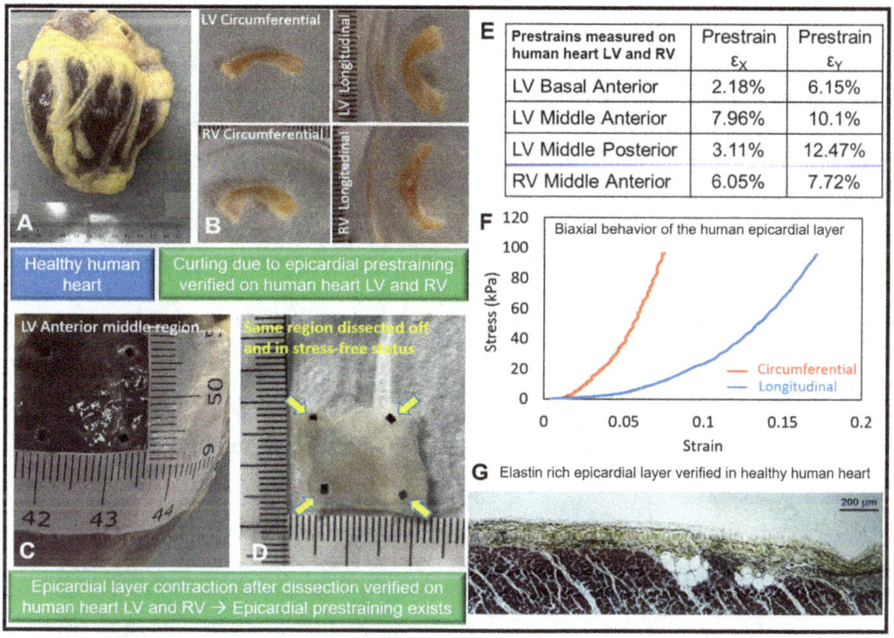


**Fig. 11** Epicardial prestraining was *verified* on *a healthy human heart* (**a**). Curling phenomenon (**b**) and epicardial prestraining (**c**, **d**) were observed on both LV and RV of the healthy human heart. (**e**) Prestrains ($\varepsilon_X$ and $\varepsilon_Y$) measured on human heart LV and RV. x: Circumferential, y: longitudinal. (**f**) Biaxial behavior of the human epicardial layer. (**g**) Elastin-rich epicardial layer was verified in healthy human heart. Black: elastin, yellow: collagen, red: muscle

surface strip, due to the contraction of the prestrained epicardial layer, was verified on both the left ventricle (LV) and right ventricle (RV) of the healthy human heart (Fig. 11b). We also carried out epicardial layer dissection to quantify the amount of prestrains of the epicardial layer on the human heart. It was found that, if only the epicardial layer was dissected off the human heart, the epicardial layer contracted in the free-floating and stress-free status (Fig. 11c, d), similar to that was observed on the porcine hearts. Epicardial prestrains measured on the human heart LV and RV showed values similar to those reported from the porcine heart study (Fig. 11e). The biaxial mechanical behavior of the human epicardial layer (Fig. 11f) was found consistent with the trend of the porcine epicardial layer (Fig. 10a) [137]. Our study verified the epicardial prestraining phenomenon on both the LV and RV of the healthy human hearts. Lastly, Movat's pentachrome histological staining confirmed that the human epicardial layer is rich in elastin (Fig. 11g), consistent with the natural ECM optimization of dynamic tissues, in which collagen provides mechanical strength and elastin fibers facilitate long-range deformability, elastic passive recoil, and efficiency in storing and releasing energy during dynamic loading and unloading cycles.

# 8 Final Remarks

In this chapter, we have discussed the detailed spatial distributions, network morphology, microstructural characteristics, and mechanical contributions of cardiac elastin across the heart ventricles. We showed that (i) epicardial elastin had both location- and depth-dependency and a correlation with the helical heart muscle fiber architecture; (ii) when compared to the epicardial layer, the endocardial layer was thicker, with endocardial elastin fibers being finer, denser, and more wavy, indicating different interface mechanics; (iii) in the myocardium, the interstitial elastin fibers co-existed with the perimysial collagen to bind and connect the heart muscle bundles; and (iv) fine elastin fibers, along with the collagen network, were closely associated with the Purkinje cells, and this ECM association could be essential in organizing cardiac Purkinje cells into "fibrous" and "branching" morphologies (summary illustration in Fig. 12). The combination of collagen and elastin is an

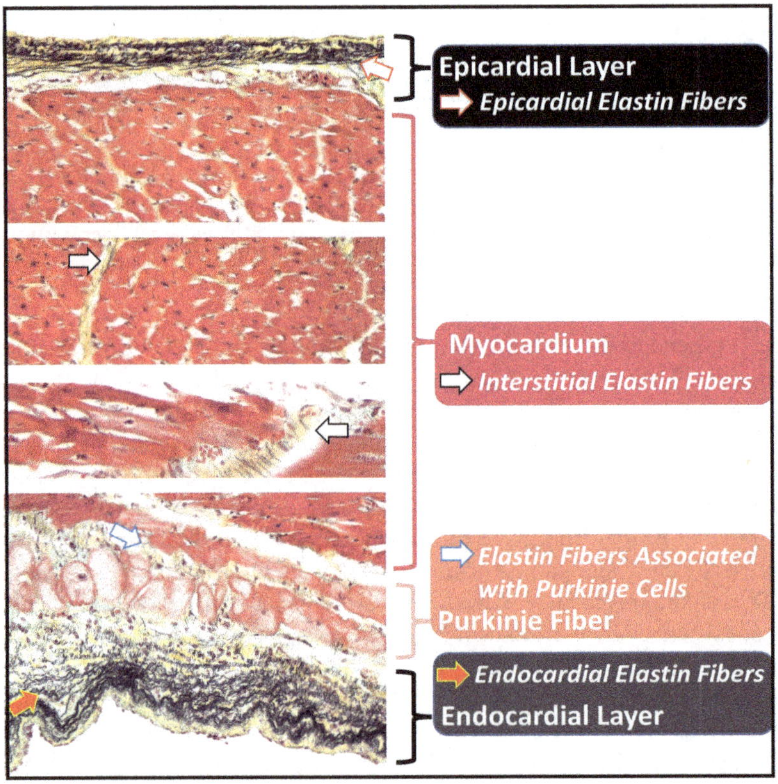


**Fig. 12** Illustration summarizes the spatial distributions and network morphology of the epicardial elastin fibers, endocardial elastin fibers, myocardial interstitial elastin fibers, and Purkinje cell-associated elastin fibers across the porcine left ventricle wall

optimal natural design for ECMs, in which the collagen fibers provide mechanical strength to the ventricular wall, and the elastin fibers enable large deformation during diastolic expansion and facilitate recoiling during systolic contraction. In more detail, the mechanical roles of cardiac elastin are to achieve long-range deformation (extensibility), robust recoiling (resilience), and high elasticity (minimal energy dissipation) during cardiac loading and unloading cycles. Using the porcine model, we demonstrated that the epicardial layer, rich in elastin, acts like a prestrained "balloon" that wraps around the heart. We further demonstrated that it is the prestraining of the epicardial layer that provides additional resistance against ventricular diastolic expansion and ventricular wall protection by reducing myocardial stress, serving as an interface protective mechanism for the heart. Lastly, we verified that the epicardial layer of the healthy human heart has abundant elastin, and the epicardial prestraining phenomenon also exists in the human heart. These discoveries have many implications for better understanding cardiac physiology and pathophysiology, as well as potential applications to facilitate biomaterial designs, tissue engineering/3D printing scaffolds, and bioprosthetic interventions for heart disease treatments.

**Acknowledgments** This work was supported in part by R15HL159599, R01HL157050, R15HL140503, R01EB022018, and T32 HL134613 from NIH, and UT STARS.

## References

1. Ahn, S., Lee, K. Y., Parker, K. K. & Shin, K. Formation of Multi-Component Extracellular Matrix Protein Fibers. *Sci Rep* **8**, 1913, https://doi.org/10.1038/s41598-018-20371-8 (2018).
2. Dzobo, K. & Dandara, C. The Extracellular Matrix: Its Composition, Function, Remodeling, and Role in Tumorigenesis. *Biomimetics (Basel)* **8**, https://doi.org/10.3390/biomimetics8020146 (2023).
3. Karamanos, N. K. *et al.* A guide to the composition and functions of the extracellular matrix. *FEBS J* **288**, 6850–6912, https://doi.org/10.1111/febs.15776 (2021).
4. Franchi, M., Piperigkou, Z., Mastronikolis, N. S. & Karamanos, N. Extracellular matrix biomechanical roles and adaptation in health and disease. *FEBS J* **291**, 430–440, https://doi.org/10.1111/febs.16938 (2024).
5. Alpesh Patel, B. F., Martin Sandig, Kibret Mequanint. Elastin Biosynthesis: The Missing Link in Tissue-Engineered Blood Vessels. *Cardiovasc Res* **71**, 40–49, https://doi.org/10.1016/j.cardiores.2006.02.021 (2006).
6. Paula Buijtenhuijs, L. B., Andre A Poot, Willeke F Daamen, Toin H van Kuppevelt, Pieter J Dijkstra, Rob A I de Vos, Lotus M Th Sterk, Bob R H Geelkerken, Jan Feijen, Istvan Vermes. Tissue Engineering of Blood Vessels: Characterization of Smooth-Muscle Cells for Culturing on Collagen-And-Elastin-Based Scaffolds. *Biotechnol Appl Biochem* **39**, 141–149, https://doi.org/10.1042/BA20030105 (2004).
7. Jin Gao, P. C., Robert Nerem, Yadong Wang. Co-expression of Elastin and Collagen Leads to Highly Compliant Engineered Blood Vessels. *J Biomed Mater Res A* **85**, 1120–1128, https://doi.org/10.1002/jbm.a.32028 (2008).
8. Kular, J. K., Basu, S. & Sharma, R. I. The extracellular matrix: Structure, composition, age-related differences, tools for analysis and applications for tissue engineering. *J Tissue Eng* **5**, 2041731414557112, https://doi.org/10.1177/2041731414557112 (2014).

9. Pompili, S., Latella, G., Gaudio, E., Sferra, R. & Vetuschi, A. The Charming World of the Extracellular Matrix: A Dynamic and Protective Network of the Intestinal Wall. *Front Med (Lausanne)***8**, 610189, https://doi.org/10.3389/fmed.2021.610189 (2021).
10. Linsenmayer, T. in *Cell Biology of Extracellular Matrix: Second Edition* 7–44 (Springer, 1991).
11. Amirrah, I. N. *et al.* A Comprehensive Review on Collagen Type I Development of Biomaterials for Tissue Engineering: From Biosynthesis to Bioscaffold. *Biomedicines***10**, https://doi.org/10.3390/biomedicines10092307 (2022).
12. Gelse, K., Poschl, E. & Aigner, T. Collagens—structure, function, and biosynthesis. *Adv Drug Deliv Rev***55**, 1531–1546, https://doi.org/10.1016/j.addr.2003.08.002 (2003).
13. Singh, D., Rai, V. & Agrawal, D. K. Regulation of Collagen I and Collagen III in Tissue Injury and Regeneration. *Cardiol Cardiovasc Med***7**, 5–16, https://doi.org/10.26502/fccm.92920302 (2023).
14. Namba, T. *et al.* Regulation of fibrillar collagen gene expression and protein accumulation in volume-overloaded cardiac hypertrophy. *Circulation***95**, 2448–2454, https://doi.org/10.1161/01.cir.95.10.2448 (1997).
15. Fung, Y. C. *Biomechanics: Mechanical Properties of Living Tissues*. 2nd edn, (Springer Verlag, 1993).
16. Silver, F. H., Freeman, J. W. & Seehra, G. P. Collagen self-assembly and the development of tendon mechanical properties. *Journal of biomechanics***36**, 1529–1553 (2003).
17. Chapman, J. A. & Hulmes, D. J. S. in *Ultrastructure of the Connective Tissue Matrix* (eds A. Ruggeri & P. M. Motto) 1–33 (Martinus Nijhoff, 1984).
18. Hodge, A. J. & Petruska, J. A. *Recent studies with the electron microscope on ordered aggregates of the tropocollagen molecule*. Vol. 289–300 (Academic Press, 1963).
19. Nimni, M. E. The molecular organization of collgen and its role in determining the biophysical properties of the connective tissues. *Biorheology***17**, 51–82 (1980).
20. Silver, F. H., Freeman, J. W. & Seehra, G. P. Collagen self-assembly and the development of tendon mechanical properties. *J Biomech***36**, 1529–1553 (2003).
21. Scott, J. E. Supramolecular organization of extracellular matrix glycosaminoglycans, in vitro and in the tissues. *Faseb J***6**, 2639–2645 (1992).
22. Grande-Allen, K. J., Griffin, B. P., Ratliff, N. B., Cosgrove, D. M. & Vesely, I. Glycosaminoglycan profiles of myxomatous mitral leaflets and chordae parallel the severity of mechanical alterations. *J Am Coll Cardiol***42**, 271–277 (2003).
23. Barber, J. E. *et al.* Mechanical properties of myxomatous mitral valves. *J Thorac Cardiovasc Surg***122**, 955–962, https://doi.org/10.1067/mtc.2001.117621 (2001).
24. Dahners, L. E., Lester, G. E. & Caprise, P. The pentapeptide NKISK affects collagen fibril interactions in a vertebrate tissue. *J Orthop Res***18**, 532–536 (2000).
25. Liao, J. & Vesely, I. A structural basis for the size-related mechanical properties of mitral valve chordae tendineae. *J Biomech***36**, 1125–1133, https://doi.org/10.1016/s0021-9290(03)00109-x (2003).
26. Liao, J. & Vesely, I. Skewness angle of interfibrillar proteoglycans increases with applied load on mitral valve chordae tendineae. *J Biomech***40**, 390–398, https://doi.org/10.1016/j.jbiomech.2005.12.011 (2007).
27. Liao, J. & Vesely, I. Relationship between collagen fibrils, glycosaminoglycans, and stress relaxation in mitral valve chordae tendineae. *Ann Biomed Eng***32**, 977–983 (2004).
28. Silver, F. H. *Biological Materials: Structure, Mechanical Properties, and Modeling of Soft Tissues*. (New York University Press, 1987).
29. Billiar, K. L. & Sacks, M. S. A method to quantify the fiber kinematics of planar tissues under biaxial stretch. *J Biomech***30**, 753–756 (1997).
30. Gilbert, T. W. *et al.* Fiber kinematics of small intestinal submucosa under uniaxial and biaxial stretch. *Journal Biomechanical Engineering* (in-press).
31. Kunzelman, K. S. *et al.* Differential collagen distribution in the mitral valve and its influence on biomechanical behaviour. *J Heart Valve Dis***2**, 236–244 (1993).
32. Grashow, J. S., Yoganathan, A. P. & Sacks, M. S. Biaxial stress-stretch behavior of the mitral valve anterior leaflet at physiologic strain rates. *Ann Biomed Eng***34**, 315–325 (2006).

33. May-Newman, K. & Yin, F. C. Biaxial mechanical behavior of excised porcine mitral valve leaflets. *Am J Physiol***269**, H1319–1327 (1995).
34. Sasaki, N. & Odajima, S. Elongation mechanism of collagen fibrils and force-strain relations of tendon at each level of structural hierarchy. *Journal of Biomechanics***29**, 1131–1136, https://doi.org/10.1016/0021-9290(96)00024-3 (1996).
35. Liao, J., Yang, L., Grashow, J. & Sacks, M. S. The relation between collagen fibril kinematics and mechanical properties in the mitral valve anterior leaflet. *J Biomech Eng***129**, 78–87, https://doi.org/10.1115/1.2401186 (2007).
36. Miyazaki, H. & Hayashi, K. Tensile Tests of Collagen Fibers Obtained from the Rabbit Patellar Tendon. *Biomedical Microdevices***2**, 151–157, https://doi.org/10.1023/A:1009953805658 (1999).
37. Gentleman, E. *et al.* Mechanical characterization of collagen fibers and scaffolds for tissue engineering. *Biomaterials***24**, 3805–3813 (2003).
38. Viguet-Carrin, S., Garnero, P. & Delmas, P. The role of collagen in bone strength. *Osteoporosis international***17**, 319–336 (2006).
39. Uitto, J. Connective tissue biochemistry of the aging dermis: age-related alterations in collagen and elastin. *Dermatologic clinics***4**, 433–446 (1986).
40. Chavoshnejad, P., Foroughi, A. H., Dhandapani, N., German, G. K. & Razavi, M. J. Effect of collagen degradation on the mechanical behavior and wrinkling of skin. *Physical Review E***104**, 034406 (2021).
41. Fung, Y. & Liu, S. Determination of the mechanical properties of the different layers of blood vessels in vivo. *Proceedings of the National Academy of Sciences***92**, 2169–2173 (1995).
42. Liao, J., Joyce, E. M. & Sacks, M. S. Effects of decellularization on the mechanical and structural properties of the porcine aortic valve leaflet. *Biomaterials***29**, 1065–1074, https://doi.org/10.1016/j.biomaterials.2007.11.007 (2008).
43. Liao, J., Yang, L., Grashow, J. & Sacks, M. S. Molecular orientation of collagen in intact planar connective tissues under biaxial stretch. *Acta Biomater***1**, 45–54, https://doi.org/10.1016/j.actbio.2004.09.007 (2005).
44. Liao, J. & Sacks, M. S. in *Advances in Heart Valve Biomechanics: Valvular Physiology, Mechanobiology, and Bioengineering* (eds M. S. Sacks & J. Liao) 81–104 (Springer 2018).
45. Iozzo, R. V. & Schaefer, L. Proteoglycan form and function: A comprehensive nomenclature of proteoglycans. *Matrix Biol***42**, 11–55, https://doi.org/10.1016/j.matbio.2015.02.003 (2015).
46. Yanagishita, M. Function of proteoglycans in the extracellular matrix. *Acta Patholigica Japonica***43**, 283–293 (1993).
47. Knudson, C. B. & Knudson, W. in *Seminars in cell & developmental biology.* 69–78 (Elsevier).
48. Pedlar, J. Biochemistry of glycosaminoglycans in the skin and oral mucosa of the rat. *Arch Oral Biol***29**, 591–597, https://doi.org/10.1016/0003-9969(84)90127-4 (1984).
49. Raman, R., Sasisekharan, V. & Sasisekharan, R. Structural insights into biological roles of protein-glycosaminoglycan interactions. *Chem Biol***12**, 267–277, https://doi.org/10.1016/j.chembiol.2004.11.020 (2005).
50. Roughley, P. J. & Lee, E. R. Cartilage proteoglycans: structure and potential functions. *Microscopy research and technique***28**, 385–397 (1994).
51. von Figura, K. [Biochemistry and function of glycosaminoglycans of the skin]. *Hautarzt***27**, 206–213 (1976).
52. Readioff, R., Geraghty, B., Kharaz, Y. A., Elsheikh, A. & Comerford, E. Proteoglycans play a role in the viscoelastic behaviour of the canine cranial cruciate ligament. *Front Bioeng Biotechnol***10**, 984224, https://doi.org/10.3389/fbioe.2022.984224 (2022).
53. Scott, J. E. & Thomlinson, A. M. The structure of interfibrillar proteoglycan bridges (shape modules') in extracellular matrix of fibrous connective tissues and their stability in various chemical environments. *J Anat***192 (Pt 3)**, 391–405, https://doi.org/10.1046/j.1469-7580.1998.19230391.x (1998).
54. Wu, J. *et al.* Viscoelastic shear lag model to predict the micromechanical behavior of tendon under dynamic tensile loading. *J Theor Biol***437**, 202–213, https://doi.org/10.1016/j.jtbi.2017.10.018 (2018).

55. Ahmadzadeh, H., Freedman, B. R., Connizzo, B. K., Soslowsky, L. J. & Shenoy, V. B. Micromechanical poroelastic finite element and shear-lag models of tendon predict large strain dependent Poisson's ratios and fluid expulsion under tensile loading. *Acta Biomater***22**, 83–91, https://doi.org/10.1016/j.actbio.2015.04.035 (2015).
56. Casale, J. & Crane, J. S. in *StatPearls* (2024).
57. Prydz, K. Determinants of Glycosaminoglycan (GAG) Structure. *Biomolecules***5**, 2003–2022, https://doi.org/10.3390/biom5032003 (2015).
58. Pomin, V. H. & Mulloy, B. Glycosaminoglycans and Proteoglycans. *Pharmaceuticals (Basel)***11**, https://doi.org/10.3390/ph11010027 (2018).
59. Han, E. H., Chen, S. S., Klisch, S. M. & Sah, R. L. Contribution of proteoglycan osmotic swelling pressure to the compressive properties of articular cartilage. *Biophys J***101**, 916–924, https://doi.org/10.1016/j.bpj.2011.07.006 (2011).
60. Redaelli, A. *et al.* Possible role of decorin glycosaminoglycans in fibril to fibril force transfer in relative mature tendons—a computational study from molecular to microstructural level. *J Biomech***36**, 1555–1569, https://doi.org/10.1016/s0021-9290(03)00133-7 (2003).
61. Eisner, L. E., Rosario, R., Andarawis-Puri, N. & Arruda, E. M. The Role of the Non-Collagenous Extracellular Matrix in Tendon and Ligament Mechanical Behavior: A Review. *J Biomech Eng***144**, https://doi.org/10.1115/1.4053086 (2022).
62. Massoudi, D., Malecaze, F. & Galiacy, S. D. Collagens and proteoglycans of the cornea: importance in transparency and visual disorders. *Cell Tissue Res***363**, 337–349, https://doi.org/10.1007/s00441-015-2233-5 (2016).
63. Steven G Wise, S. M. M., Anthony S Weiss. Engineered tropoelastin and elastin-based biomaterials. *Adv Protein Chem Struct Biol***78**, 1–24, https://doi.org/10.1016/S1876-1623(08)78001-5 (2009).
64. B Vrhovski, A. S. W. Biochemistry of tropoelastin. *Eur J Biochem***258**, 1–28, https://doi.org/10.1046/j.1432-1327.1998.2580001.x (1998).
65. Senior, R. M., G. L. G., Mecham, R. P. Chemotactic responses of fibroblasts to tropoelastin and elastin-derived peptides. *J Clin Invest***70**, 614–618, https://doi.org/10.1172/jci110654 (1982).
66. Silva, S. S. *et al.* in *Comprehensive Biomaterials II* (ed Paul Ducheyne) 228–252 (Elsevier, 2017).
67. Baumann, L. *et al.* Clinical Relevance of Elastin in the Structure and Function of Skin. *Aesthet Surg J Open Forum***3**, ojab019, https://doi.org/10.1093/asjof/ojab019 (2021).
68. Wagenseil, J. E. & Mecham, R. P. New insights into elastic fiber assembly. *Birth Defects Res C Embryo Today***81**, 229–240, https://doi.org/10.1002/bdrc.20111 (2007).
69. Daamen, W. F., Veerkamp, J., Van Hest, J. & Van Kuppevelt, T. Elastin as a biomaterial for tissue engineering. *Biomaterials***28**, 4378–4398 (2007).
70. Godwin, A. R. F. *et al.* The role of fibrillin and microfibril binding proteins in elastin and elastic fibre assembly. *Matrix Biol***84**, 17–30, https://doi.org/10.1016/j.matbio.2019.06.006 (2019).
71. Ritz-Timme, S., I. L., Collins, M. J.. Aspartic acid racemization: evidence for marked longevity of elastin in human skin. *Br J Dermatol***149**, 951–959, https://doi.org/10.1111/j.1365-2133.2003.05618.x (2003).
72. Oxlund, H., J. M., Viidik, A.. The role of elastin in the mechanical properties of skin. *J Biomech***21**, 213–218, https://doi.org/10.1016/0021-9290(88)90172-8 (1998).
73. Mercer, R. R., J. D. C. Spatial distribution of collagen and elastin fibers in the lungs. *J Appl Physiol (1985)***69**, 756–765, https://doi.org/10.1152/jappl.1990.69.2.756 (1990).
74. Anna Waterhouse, S. G. W., Martin K C Ng, Anthony S Weiss. Elastin as a nonthrombogenic biomaterial. *Tissue Eng Part B Rev***17**, 93–99, https://doi.org/10.1089/ten.TEB.2010.0432 (2011).
75. Gray, W. R., L. B. S., Foster, J. A.. Molecular model for elastin structure and function. *Nature***246**, 461–466, https://doi.org/10.1038/246461a0 (1973).
76. Hoeve, C. A. & Flory, P. J. The elastic properties of elastin. *Biopolymers***13**, 677–686, https://doi.org/10.1002/bip.1974.360130404 (1974).

77. Carvajal, M. *et al.* Dynamics in natural and designed elastins and their relation to elastic fiber structure and recoil. *Biophys J***120**, 4623–4634, https://doi.org/10.1016/j.bpj.2021.06.043 (2021).
78. Muiznieks, L. D., Weiss, A. S. & Keeley, F. W. Structural disorder and dynamics of elastin. *Biochem Cell Biol***88**, 239–250, https://doi.org/10.1139/o09-161 (2010).
79. Rauscher, S. & Pomes, R. Structural disorder and protein elasticity. *Adv Exp Med Biol***725**, 159–183, https://doi.org/10.1007/978-1-4614-0659-4_10 (2012).
80. Jamhawi, N. M., Koder, R. L. & Wittebort, R. J. Elastin recoil is driven by the hydrophobic effect. *Proceedings of the National Academy of Sciences***121**, e2304009121, https://doi.org/10.1073/pnas.2304009121 (2024).
81. Trębacz, H. & Barzycka, A. Mechanical Properties and Functions of Elastin: An Overview. *Biomolecules***13**, https://doi.org/10.3390/biom13030574 (2023).
82. Maclay, J. D. *et al.* Systemic elastin degradation in chronic obstructive pulmonary disease. *Thorax***67**, 606–612 (2012).
83. Mecham, R. P. Elastin in lung development and disease pathogenesis. *Matrix biology***73**, 6–20 (2018).
84. Seadler, B. D., Toro, F. & Sharma, S. in *StatPearls* (2024).
85. Clifford, P. S. *et al.* Spatial Distribution and Mechanical Function of Elastin in Resistance Arteries. *Arteriosclerosis, Thrombosis, and Vascular Biology***31**, 2889–2896, https://doi.org/10.1161/ATVBAHA.111.236570 (2011).
86. Wagenseil, J. E. & Mecham, R. P. Vascular Extracellular Matrix and Arterial Mechanics. *Physiological Reviews***89**, 957–989, https://doi.org/10.1152/physrev.00041.2008 (2009).
87. Wang, R., Yu, X. & Zhang, Y. Mechanical and structural contributions of elastin and collagen fibers to interlamellar bonding in the arterial wall. *Biomechanics and Modeling in Mechanobiology***20**, 93–106, https://doi.org/10.1007/s10237-020-01370-z (2021).
88. Seeburun, S. *et al.* Insights into elastic fiber fragmentation: Mechanisms and treatment of aortic aneurysm in Marfan syndrome. *Vascular Pharmacology***153**, 107215, https://doi.org/10.1016/j.vph.2023.107215 (2023).
89. Okamura, H. *et al.* Assessment of Elastin Deficit in a Marfan Mouse Aneurysm Model Using an Elastin-Specific Magnetic Resonance Imaging Contrast Agent. *Circulation: Cardiovascular Imaging***7**, 690–696, https://doi.org/10.1161/CIRCIMAGING.114.001658 (2014).
90. López-Guimet, J., Andilla, J., Loza-Alvarez, P. & Egea, G. High-Resolution Morphological Approach to Analyse Elastic Laminae Injuries of the Ascending Aorta in a Murine Model of Marfan Syndrome. *Scientific Reports***7**, 1505, https://doi.org/10.1038/s41598-017-01620-8 (2017).
91. Weismann, C. G., Hlebowicz, J., Akesson, A., Liuba, P. & Hanseus, K. Comprehensive Characterization of Arterial and Cardiac Function in Marfan Syndrome-Can Biomarkers Help Improve Outcome? *Front Physiol***13**, 873373, https://doi.org/10.3389/fphys.2022.873373 (2022).
92. Williams, C. & Black III, L. D. in *Biomaterials for Cardiac Regeneration* 1–35 (Springer, 2015).
93. Fomovsky, G. M., Thomopoulos, S. & Holmes, J. W. Contribution of extracellular matrix to the mechanical properties of the heart. *J Mol Cell Cardiol***48**, 490–496 (2010).
94. Borg, T. & Caulfield, J. in *Federation proceedings.* 2037–2041.
95. Sussman, M. A., McCulloch, A. & Borg, T. K. Dance band on the titanic biomechanical signaling in cardiac hypertrophy. *Circulation research***91**, 888–898 (2002).
96. Wang, B. *et al.* Structural and biomechanical characterizations of porcine myocardial extracellular matrix. *J Mater Sci Mater Med***23**, 1835–1847, https://doi.org/10.1007/s10856-012-4660-0 (2012).
97. Forrester, J. S., Diamond, G., Parmley, W. W. & Swan, H. J. Early increase in left ventricular compliance after myocardial infarction. *J Clin Invest***51**, 598–603 (1972).
98. Sato, S., Ashraf, M., Millard, R. W., Fujiwara, H. & Schwartz, A. Connective tissue changes in early ischemia of porcine myocardium: an ultrastructural study. *J Mol Cell Cardiol***15**, 261–275 (1983).

99. Etoh, T. *et al.* Myocardial and interstitial matrix metalloproteinase activity after acute myocardial infarction in pigs. *Am J Physiol Heart Circ Physiol* **281**, H987–994 (2001).
100. Fishbein, M. C., Maclean, D. & Maroko, P. R. The histopathologic evolution of myocardial infarction. *Chest* **73**, 843–849, https://doi.org/10.1378/chest.73.6.843 (1978).
101. Richardson, W. J., Clarke, S. A., Quinn, T. A. & Holmes, J. W. Physiological Implications of Myocardial Scar Structure. *Compr Physiol* **5**, 1877–1909, https://doi.org/10.1002/cphy.c140067 (2015).
102. Holmes, J. W., Yamashita, H., Waldman, L. K. & Covell, J. W. Scar remodeling and transmural deformation after infarction in the pig. *Circulation* **90**, 411–420 (1994).
103. Holmes, J. W., Borg, T. K. & Covell, J. W. Structure and mechanics of healing myocardial infarcts. *Annu. Rev. Biomed. Eng.* **7**, 223–253 (2005).
104. Christensen, G., Herum, K. M. & Lunde, I. G. Sweet, yet underappreciated: Proteoglycans and extracellular matrix remodeling in heart disease. *Matrix Biology* **75–76**, 286–299, https://doi.org/10.1016/j.matbio.2018.01.001 (2019).
105. Beetz, N. *et al.* Ablation of biglycan attenuates cardiac hypertrophy and fibrosis after left ventricular pressure overload. *Journal of Molecular and Cellular Cardiology* **101**, 145–155, https://doi.org/10.1016/j.yjmcc.2016.10.011 (2016).
106. Wang, X., Lu, Y., Xie, Y., Shen, J. & Xiang, M. Emerging roles of proteoglycans in cardiac remodeling. *International Journal of Cardiology* **278**, 192–198, https://doi.org/10.1016/j.ijcard.2018.11.125 (2019).
107. Schaefer, L. & Iozzo, R. V. Biological Functions of the Small Leucine-rich Proteoglycans: From Genetics to Signal Transduction*. *Journal of Biological Chemistry* **283**, 21305–21309, https://doi.org/10.1074/jbc.R800020200 (2008).
108. Mavroidis, M. & Capetanaki, Y. Extensive Induction of Important Mediators of Fibrosis and Dystrophic Calcification in Desmin-Deficient Cardiomyopathy. *The American Journal of Pathology* **160**, 943–952, https://doi.org/10.1016/S0002-9440(10)64916-4 (2002).
109. Lunde, I. G., Rypdal, K. B., Van Linthout, S., Diez, J. & González, A. Myocardial fibrosis from the perspective of the extracellular matrix: Mechanisms to clinical impact. *Matrix Biology* **134**, 1–22, https://doi.org/10.1016/j.matbio.2024.08.008 (2024).
110. Jöbsis, P. D. *et al.* The visceral pericardium: macromolecular structure and contribution to passive mechanical properties of the left ventricle. *American Journal of Physiology-Heart and Circulatory Physiology* **293**, H3379–H3387 (2007).
111. Benjamin, E. J. *et al.* Heart Disease and Stroke Statistics-2019 Update: A Report From the American Heart Association. *Circulation* **139**, e56–e528, https://doi.org/10.1161/CIR.0000000000000659 (2019).
112. Chowdhury, T. & Reardon, W. Elastin mutation and cardiac disease. *Pediatric cardiology* **20**, 103–107 (1999).
113. Perou, M. Congenital supravalvular aortic stenosis. A morphological study with attempt at classification. *Archives of pathology* **71**, 453 (1961).
114. O'connor, W. *et al.* Supravalvular aortic stenosis. Clinical and pathologic observations in six patients. *Archives of pathology & laboratory medicine* **109**, 179–185 (1985).
115. Yuan, S. M. Congenital heart defects in Williams syndrome. *Turk J Pediatr* **59**, 225–232, https://doi.org/10.24953/turkjped.2017.03.001 (2017).
116. De Rubens Figueroa, J., Rodríguez, L. M., Hach, J. L., Del Castillo Ruíz, V. & Martínez, H. O. Cardiovascular spectrum in Williams-Beuren syndrome: the Mexican experience in 40 patients. *Tex Heart Inst J* **35**, 279–285 (2008).
117. Phomakay, V., Gossett, J. M., Kaplan, P. B., Swearingen, C. J. & Collins, R. T., 2nd. Ventricular Hypertrophy on Electrocardiogram Correlates with Obstructive Lesion Severity in Williams Syndrome. *Congenit Heart Dis* **10**, 302–309, https://doi.org/10.1111/chd.12194 (2015).
118. Mizuno, T., Yau, T. M., Weisel, R. D., Kiani, C. G. & Li, R.-K. Elastin stabilizes an infarct and preserves ventricular function. *Circulation* **112**, I–81 (2005).
119. Li, S. H. *et al.* Elastin overexpression by cell-based gene therapy preserves matrix and prevents cardiac dilation. *Journal of cellular and molecular medicine* **16**, 2429–2439 (2012).

120. Mujumdar, V. S. & Tyagi, S. C. Temporal regulation of extracellular matrix components in transition from compensatory hypertrophy to decompensatory heart failure. *J Hypertens***17**, 261–270 (1999).
121. Henderson, B. C. *et al.* Reversal of systemic hypertension-associated cardiac remodeling in chronic pressure overload myocardium by ciglitazone. *International journal of biological sciences***3**, 385–392 (2007).
122. Cheng, X. W. *et al.* Elastolytic cathepsin induction/activation system exists in myocardium and is upregulated in hypertensive heart failure. *Hypertension***48**, 979–987, https://doi.org/10.1161/01.HYP.0000242331.99369.2f (2006).
123. Farquharson, C. & Robins, S. P. The distribution of elastin in developing and adult rat organs using immunocytochemical techniques. *Journal of anatomy***165**, 225 (1989).
124. Williams, C. & Black III, L. D. in *Chapter 1: The Role of Extracellular Matrix in Cardiac Development. Book: Biomaterials for Cardiac Regeneration* (eds Erik J. Suuronen & Marc Ruel) 1–35 (Springer, 2015).
125. Shi, X. *et al.* Spatial distribution and network morphology of epicardial, endocardial, interstitial, and Purkinje cell-associated elastin fibers in porcine left ventricle. *Bioactive Materials***19**, 348–359, https://doi.org/10.1016/j.bioactmat.2022.04.019 (2022).
126. Shi, X. *et al.* Spatial distribution and network morphology of epicardial, endocardial, interstitial, and Purkinje cell-associated elastin fibers in porcine left ventricle. *Bioactive materials***19**, 348–359 (2023).
127. Robinson, T. F., Cohen-Gould, L. & Factor, S. M. Skeletal framework of mammalian heart muscle. Arrangement of inter- and pericellular connective tissue structures. *Lab Invest***49**, 482–498 (1983).
128. Robinson, T. F., Cohen-Gould, L., Factor, S. M., Eghbali, M. & Blumenfeld, O. O. Structure and function of connective tissue in cardiac muscle: collagen types I and III in endomysial struts and pericellular fibers. *Scanning Microsc***2**, 1005–1015 (1988).
129. Garcia-Bustos, V. *et al.* A quantitative structural and morphometric analysis of the Purkinje network and the Purkinje-myocardial junctions in pig hearts. *J Anat***230**, 664–678, https://doi.org/10.1111/joa.12594 (2017).
130. Stephenson, R. S. *et al.* High resolution 3-Dimensional imaging of the human cardiac conduction system from microanatomy to mathematical modeling. *Sci Rep***7**, 7188, https://doi.org/10.1038/s41598-017-07694-8 (2017).
131. Atkinson, A. J. *et al.* Functional, anatomical, and molecular investigation of the cardiac conduction system and arrhythmogenic atrioventricular ring tissue in the rat heart. *J Am Heart Assoc***2**, e000246, https://doi.org/10.1161/JAHA.113.000246 (2013).
132. Pennisi, D. J., Rentschler, S., Gourdie, R. G., Fishman, G. I. & Mikawa, T. Induction and patterning of the cardiac conduction system. *Int J Dev Biol***46**, 765–775 (2002).
133. Fung, Y. What are the residual stresses doing in our blood vessels? *Annals of biomedical engineering***19**, 237–249 (1991).
134. Omens, J. H. & Fung, Y.-C. Residual strain in rat left ventricle. *Circulation research***66**, 37–45 (1990).
135. Genet, M. *et al.* Heterogeneous growth-induced prestrain in the heart. *Journal of biomechanics***48**, 2080–2089 (2015).
136. Weed, B. C. *et al.* Stress state and strain rate dependence of the human placenta. *Ann Biomed Eng***40**, 2255–2265, https://doi.org/10.1007/s10439-012-0588-2 (2012).
137. Shi, X. *et al.* Epicardial prestrained confinement and residual stresses: a newly observed heart ventricle confinement interface. *J R Soc Interface***16**, 20190028, https://doi.org/10.1098/rsif.2019.0028 (2019).
138. Grandin, T. Euthanasia and slaughter of livestock. *J Am Vet Med Assoc***204**, 1354–1360 (1994).
139. Jerlström, J. *When and what determines the death of an animal? A study investigating the heart activity during slaughter of farm animals.*, Swedish University of Agricultural Sciences.
140. Maxie, G. *Jubb, Kennedy & Palmer's Pathology of Domestic Animals*. 6th edn, 186 (Elsevier 2015).

141. Wang, B. *et al.* Myocardial scaffold-based cardiac tissue engineering: application of coordinated mechanical and electrical stimulations. *Langmuir* **29**, 11109–11117, https://doi.org/10.1021/la401702w (2013).
142. Wang, B. *et al.* Fabrication of cardiac patch with decellularized porcine myocardial scaffold and bone marrow mononuclear cells. *J Biomed Mater Res A* **94**, 1100–1110, https://doi.org/10.1002/jbm.a.32781 (2010).
143. Eriksson, T. S. E., Prassl, A. J., Plank, G. & Holzapfel, G. A. Influence of myocardial fiber/sheet orientations on left ventricular mechanical contraction. *Math Mech Solids* **18**, 592–606 (2013).
144. Holzapfel, G. A. & Ogden, R. W. Constitutive modelling of passive myocardium: a structurally based framework for material characterization. *Philos T R Soc A* **367**, 3445–3475 (2009).
145. Goktepe, S., Acharya, S. N. S., Wong, J. & Kuhl, E. Computational modeling of passive myocardium. *Int J Numer Meth Bio* **27**, 1–12 (2011).
146. Maas, S. A., Ellis, B. J., Ateshian, G. A. & Weiss, J. A. FEBio: Finite Elements for Biomechanics. *J Biomech Eng-T Asme* **134** (2012).
147. Katz, A. M. *Physiology of the heart*. 4th edn, (Lippincott Williams & Wilkins, 2006).
148. Howard-Quijano, K. *et al.* Left ventricular endocardial and epicardial strain changes with apical myocardial ischemia in an open-chest porcine model. *Physiol Rep* **4**, https://doi.org/10.14814/phy2.13042 (2016).
149. Mizuno, T., Mickle, D. A., Kiani, C. G. & Li, R.-K. Overexpression of elastin fragments in infarcted myocardium attenuates scar expansion and heart dysfunction. *American Journal of Physiology-Heart and Circulatory Physiology* **288**, H2819–H2827 (2005).
150. Raya, T. E. *et al.* Serial changes in left ventricular relaxation and chamber stiffness after large myocardial infarction in rats. *Circulation* **77**, 1424–1431 (1988).
151. Gupta, K. B., Ratcliffe, M. B., Fallert, M. A., Edmunds, L. H., Jr. & Bogen, D. K. Changes in passive mechanical stiffness of myocardial tissue with aneurysm formation. *Circulation* **89**, 2315–2326, https://doi.org/10.1161/01.cir.89.5.2315 (1994).
152. Mousavi, A. *et al.* Multifunctional Conductive Biomaterials as Promising Platforms for Cardiac Tissue Engineering. *ACS biomaterials science & engineering* **7**, 55–82, https://doi.org/10.1021/acsbiomaterials.0c01422 (2021).
153. Dong, R., Ma, P. X. & Guo, B. Conductive biomaterials for muscle tissue engineering. *Biomaterials* **229**, 119584, https://doi.org/10.1016/j.biomaterials.2019.119584 (2020).
154. Tracy, E. P. *et al.* 3D Bioprinting the Cardiac Purkinje System Using Human Adipogenic Mesenchymal Stem Cell Derived Purkinje Cells. *Cardiovascular engineering and technology* **11**, 587–604, https://doi.org/10.1007/s13239-020-00478-8 (2020).
155. Zhang, S. *et al.* The correlation of 3D DT-MRI fiber disruption with structural and mechanical degeneration in porcine myocardium. *Annals of biomedical engineering* **38**, 3084–3095 (2010).

# Complex Hierarchical Biomechanics of Dura Mater and Engineering Considerations in Graft Design

Mohammad Tabatabaei, Abigail S. Aplin, Justin D. Hilliard, and Lakiesha N. Williams

## 1 Introduction

### *1.1 Overview of the Dura Mater and Its Hierarchical Structure*

The dura mater, as the outermost layer of the meninges, plays a pivotal role in maintaining the protective environment of the brain. This integrity is largely attributed to the distinctive components of the dura mater, including its collagenous structure, vascular framework, and nerve structures [1, 2]. Among these elements, the collagenous structure is significant for its load-bearing capabilities, making it the dominant mechanical support within the meninges. However, trauma, disease, or neurosurgical procedures can disrupt the complex hierarchical structure and integrity of the dura mater, potentially compromising the brain's environment. This raises the pressing need for suitable dural grafts, a challenge that persistently captivates and drives the research of scientists, engineers, and clinicians.

M. Tabatabaei · A. S. Aplin · L. N. Williams (✉)
J. Crayton Pruitt Family Department of Biomedical Engineering, University of Florida, Gainesville, FL, USA
e-mail: lwilliams@bme.ufl.edu

J. D. Hilliard
Department of Neurosurgery, University of Florida, Gainesville, FL, USA

 
J. Liao, J. Y. Wong (eds.), *Integration and Bridging of Multiscale Bioengineering Designs and Tissue Biomechanics*, https://doi.org/10.1007/978-3-031-81743-4_4

## *1.2 Challenge in Creating Suitable Substitutes for Repairing Disrupted Dura Mater*

The dura mater encompasses intricate structural and functional elements critical to its role in the central nervous system. A notable function of the dura mater is its integral role in the brain's immune response. It incorporates various types of immune cells, acting as a reservoir for these cells when they are needed. The location-dependent architecture and heterogenous fibrous distribution, along with its ability to remodel, highlight the dura mater's versatility and resilience [3]. These unique features present challenges in creating effective dura mater grafts. A graft must not only replicate the dura mater's structural integrity but also emulate its diverse functional capabilities, making the task of developing a suitable graft far more complex than creating a simple lyophilized sheath.

Dural grafts, commonly used in neurosurgery, utilize either synthetic or naturally derived materials. Synthetic options, such as polyglycolic acid (PGA) and polylactic acid (PLA), offer biocompatibility and adjustable mechanical properties, approximating the robustness of the native dura mater. In contrast, naturally derived materials like collagen, fibrin, and chitosan provide excellent biological integration due to their bio-mimicry. While these materials may facilitate cell adhesion and proliferation, they may not match the mechanical strength of their synthetic counterparts [4, 5]. Ultimately, the choice between synthetic and natural materials in dural grafts hinges on specific clinical requirements. It underscores the critical need for multifaceted and versatile approaches in neurosurgical applications and tissue engineering. By deepening our understanding of the dura mater's structure and function, we can further refine these approaches, bringing us closer to creating more effective and reliable dura mater grafts.

## *1.3 Importance of the Dura Mater in Maintaining Mechanical Integrity for the Brain*

The structure of the dura mater extends from the outermost end, connecting to the periosteal layer of the inner cortex of the skull, to the innermost end, integrating with the arachnoid and then pia mater. This complex structure invites a multiscale approach for studying the dura mater—encompassing macroscale, mesoscale, and microscale perspectives. Beyond its structural importance, another fundamental characteristic of the dura mater lies in its watertight seal around the brain, maintaining cerebral spinal fluid (CSF) in its appropriate location (intradural). It plays a crucial role in preventing the leakage of CSF in both cranial and spinal regions, thereby maintaining a critical physiological balance within the central nervous system (CNS).

# 2 Structure and Function of the Dura Mater

## 2.1 Dura Structure

### 2.1.1 Collagenous (Fibrous) Structure

The meninges are composed of intricate layered membranes, which serve a dual purpose, functioning both as pathological and mechanical barriers (Fig. 1). They stand as the brain's primary defense, guarding it against external threats and potential damage [6]. This complex structure envelops the brain, adhering securely to the cranial vault and the base of the skull. Within the meninges are four distinct infoldings: the Falx cerebri, Tentorium cerebelli, Falx cerebelli, and Diaphragma sellae [7]. These infoldings are recognized for their roles in separating and cushioning different parts of the brain. Cranial meninges allow for the maintenance of homeostasis of the CNS as the layer provides mechanical, immunological, and vascular support to brain parenchyma [8].

Among the three layers of the meninges—the dura mater, arachnoid, and pia mater—the dura mater (Latin for "tough mother") is distinguished by its fibrous and robust nature. This toughness, essential for its role as the outermost layer, offers reinforced protection to the brain. Comprised of fibers, proteoglycans (PGs), and glycoproteins, its unique makeup, with elements of collagen, elastin, and minor

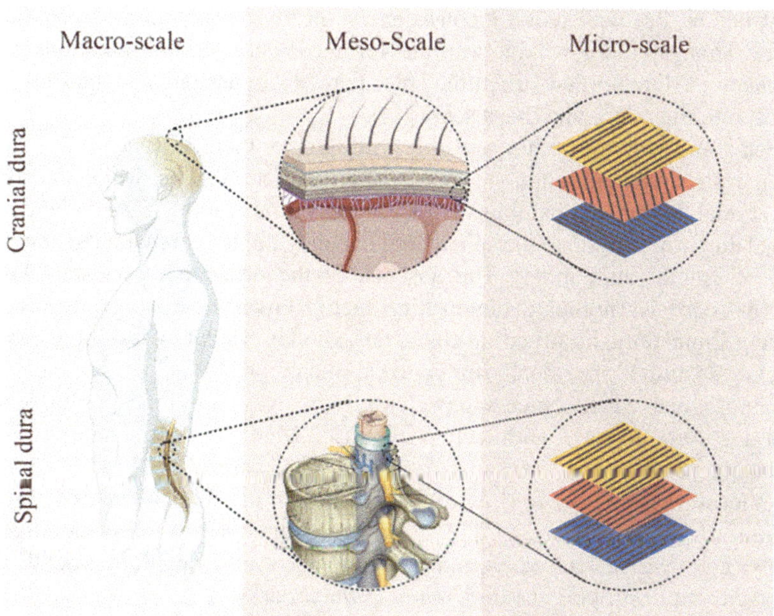


**Fig. 1** An intricate depiction of the hierarchical structure of the dura mater, encompassing macroscopic, mesoscopic, and microscopic scales

fibers like fibronectin, ensures structural integrity and vital cellular interactions [9]. There are two layers of dura mater, the outer periosteal layer (faces skull) and the inner meningeal (faces brain) [10]. The layers are tightly fused together and collectively referenced as dura mater. The layers are loosely attached in areas where they are separated by venous sinuses. Additionally, a meningeal lymphatic vessel runs parallel to the dural sinus separating the layers [11].

Delving into collagen's significance reveals it as a foundational element in many vertebrate structures. In biomechanics, the importance of collagen is heavily emphasized. Its ubiquitous presence in various tissues offers them the resilience to bear significant loads. There are five types of collagens, in tissues such as bones, tendons, and intervertebral discs much of their robustness is owed to a collagenous matrix. Within the dura mater, type I collagen is predominantly found. Type I collagen comprises 90% of the body's collagen and has specific mechanical properties thanks to its densely packed nature [12].

The connection between structure and function is a prevailing theme in biological systems, with the dura mater being a testament to this principle. The architecture of the dura mater is an area of keen interest to researchers, given the implications of its functionality. Data such as anisotropy and protein content have been employed to provide insights into the organization of fibers, with techniques like immunoassays and mass spectroscopy facilitating these investigations.

A notable contribution to this field is the study by Hamman et al. [13]. Using human cadavers, the collagenous architecture of the dura mater was quantified. Their findings revealed an anisotropic nature of the cranial dura mater's collagenous structure, but with noted inconsistencies in fiber orientation across different samples. This gap in knowledge warrants further insight and investigations into the importance of dura's microstructure. This sparked discussions and spurred further research into the micro and meso scale [14].

When comparing the cranial and spinal dura mater there exists an important distinction in structure and function (Fig. 1). The spinal dura mater, while an extension of the cranial dura mater, exhibits distinct architectural features. The collagen fibers in spinal dura run parallel to the spinal cord's length, contrasting with the anisotropic nature of cranial dura mater. The absence of the protective periosteal layer in the spinal cord is another distinguishing factor. Given neurosurgical procedures involving spinal dura, its structural characterization is critical to designing enhanced synthetic substitutes specifically for spinal applications [15].

Technological advancements have paved the way for a deeper understanding of the dura mater's architecture in vitro. Methods like scanning electron microscopy (SEM) [16], transmission electron microscopy (TEM) [17], atomic force microscopy (AFM) [18], confocal/multi-photon confocal microscopy, [19] and light-scattering microscopy [13] have been employed in these investigations. Notably, noninvasive tools are gaining prominence, allowing for in-depth explorations without destroying the dura mater in the process.

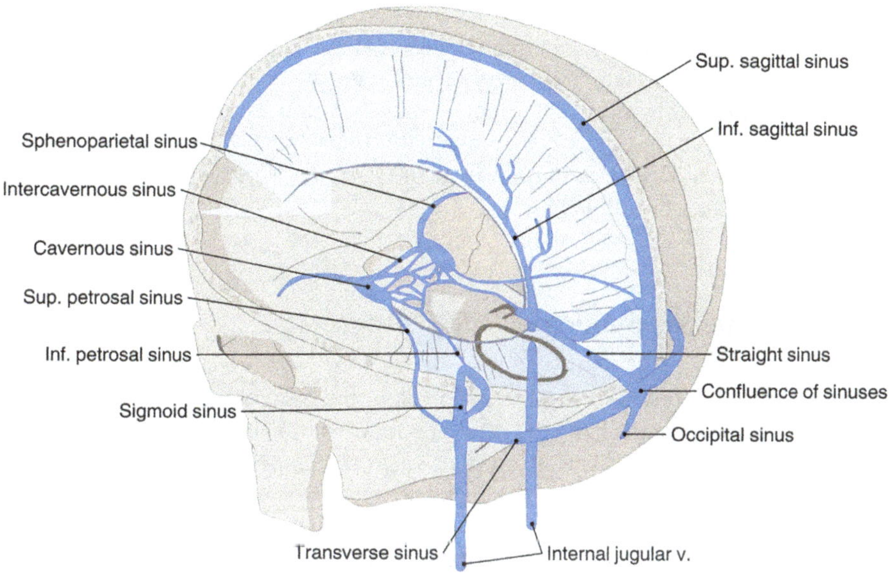


**Fig. 2** Schematic of dural venous sinuses in human

### 2.1.2 Dural Venous Sinuses: Structure and Function

Numerous venous sinuses are situated within the meninges, including the superior sagittal sinus, inferior sagittal sinus, straight sinus, transverse sinuses, sigmoid sinuses, occipital sinus, cavernous sinuses, superior petrosal sinus, inferior petrosal sinus, the confluence of sinuses (torcular herophili), intercavernous sinuses, sphenoparietal sinus, and the basilar plexus (Fig. 2). The dural venous sinuses serve two primary functions: (1) they collect deoxygenated blood from the brain and return it to the cardiovascular system and (2) they gather CSF from the subarachnoid space and drain it into the venous system via the sagittal sinuses. Structurally, the sinuses are valveless, allowing fluids to move bidirectionally. This system of sinuses further enhances the anisotropy of the dura mater and the complexity of designing the multilayered structure by providing separation in the layers making up the dura mater.

### 2.1.3 The Meningeal Lymphatic System. The Immune Response Centers

The meningeal lymphatics network supports CSF drainage and regulates fluid homeostasis in the brain. The network drains macromolecules and immune cells from the CNS, serving as a primary controller of tissue fluid balance. It purifies the tissues, hindering the buildup of potentially harmful substances like cellular metabolites or debris resulting from damaged or necrotic cells. The lymphatics are located in the lateral and basal parts of the skull and are the preferred route for CSF uptake and drainage to cervical lymph nodes. The flow of interstitial fluid (ISF) within the CNS takes place within the brain parenchyma. The presence of critical lymphatic structures and venous sinuses within the dura mater suggests that its condition and integrity can significantly impact overall human health. Any circumstances that would compromise or threaten the dura's homeostasis could diminish the brain's overall health.

## 3 Multiscale Mechanics of Dura Mater

This section categorizes the dura mater into the microscale, mesoscale, and macroscale to better understand the structure–function relationship of the meninges. While research largely focuses on the macroscale, understanding the dura's hierarchical structure requires an integrated approach across all levels. Insight into the hierarchical nature of dura aids further in the development of effective dural grafts, meeting the varied needs necessitated by its roles and interactions within the cranial environment.

### *3.1 Length Scales Of Dura Mater: Micro, Meso, and Macro*

Substitutes for damaged meningeal layers (such as dura mater) are common in clinical practice. This raises a question: What physical characteristics, including mechanical and geometric attributes, are pertinent to consider when defining the dura mater's hierarchical structure? Addressing this requires examination of the dura and meninges across macro, meso, and micro scales. A multiscale evaluation is essential to deepen our understanding of brain–skull interactions, particularly the transmission of forces between the brain and the skull. Three distinct interactions are involved: the brain and meninges interface, the three meningeal layers' interfaces, and the meningeal–skull interface. While the latter is significant due to its relative rigidity and role in morphogenesis, the interactions between the brain and meninges, as well as within the meningeal layers, are garnering increased attention from scientists. Understanding these interactions is key to revealing the structure and function at these interfaces and crucial exchanges within the brain.

#### 3.1.1 Microscale

The intricate architecture of meningeal tissues reveals significant variability across its thickness. Notably, while the dura mater is characterized by a dense collagenous composition, other layers like the subarachnoid space are instrumental in influencing the viscoelastic properties of the dura, which are crucial for dissipating

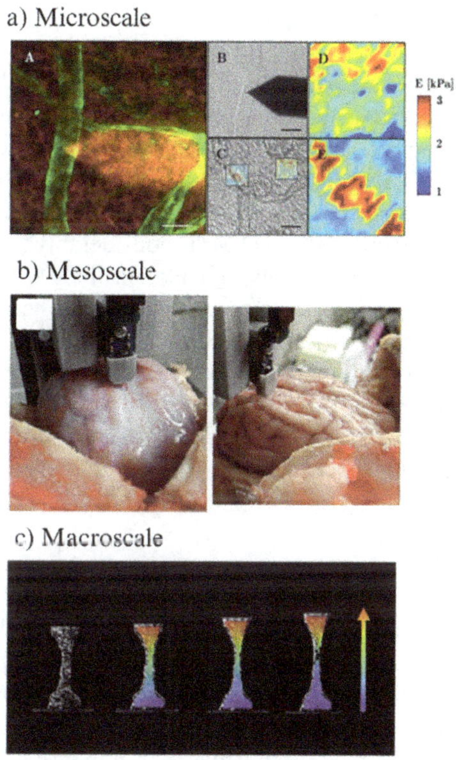


**Fig. 3** Illustration of mechanical properties at different scales: (**a**) microscale [20], (**b**) mesoscale [21], and (**c**) macroscale [1]

dynamic loads. Techniques such as atomic force microscopy (AFM) have been employed to discern details at the microscopic level. AFM experiments previously showed distinct mechanical differences associated with capillaries in veins and arteries. However, due to the invasive nature of tissue exposure and the potential for structural disruption during sectioning, research at this scale has been limited. The micrometric scale addresses the smallest hierarchical tissue elements, collagen molecules, and small fibrils (Fig. 3a).

AFM has been instrumental in capturing quantitative imagery of the dural structure. Sikora et al. [18] leveraged AFM to examine the structural properties of embryonic tissue, demonstrating that AFM can yield more quantitative results alongside imaging on delicate tissues. Although scanning electron microscopy (SEM) is widely used, AFM is preferred by some researchers due to the dura's fragility. Investigations of dried, excised dura mater revealed highly aligned collagen structures, with the periosteal layer displaying a distinct architecture compared to the meningeal layer. Microanalysis facilitates a deeper understanding of the complexity and heterogeneity inherent in collagenous tissues like the dura mater. Fabris et al. [20] explored the microstructure of the pia-arachnoid complex and assessed the heterogeneity of this complex through mechanical evaluation via nanoindentation. While such methodologies require direct tissue contact and possible damage, they offer valuable insights into tissue properties.

### 3.1.2 Mesoscale

Mesoscale studies are particularly beneficial in exploring the brain–meninges interface (Fig. 3b). A crucial benefit of mesoscale analysis is preserving the native structure. There is a notable scarcity of scientific literature in this area. Research conducted by MacManus et al. [21] provides valuable insights on the mesoscale dura structure. Their study of postmortem porcine specimens revealed that the dura-arachnoid mater (DAM) is considerably more rigid than the underlying brain tissue. This rigidity is instrumental in the DAM's ability to mitigate stress and strain, highlighting its protective function. The mesoscale analysis was conducted using a precise indentation technique of $70 \pm 3.6$ μm. Similarly, L. Qian [22] investigated the pia-arachnoid complex (PAC) through indentation methods, validating the PAC's protective role in porcine models. These studies are not only pivotal in understanding the mesoscale properties of meningeal layers but also provide insights that could be analogous to human neuroanatomy.

### 3.1.3 Macroscale

The meninges, characterized by their multilayered structure, exhibit a hierarchical organization within each layer. The meninges, as the brain's protective layer, display elastic, hyperplastic, and viscoelastic behaviors. This complexity necessitates the use of in vitro techniques for evaluating mechanical properties under various loading conditions (Fig. 3c). It is critical that the development of materials for dural substitutes physically and mechanically aligns with the properties of the surrounding dura mater at the macroscale. Research in this area, particularly focusing on the cranial and spinal properties of the meninges, is limited compared to other collagenous tissues. This scarcity of research underscores the need for a deeper understanding of the dura mater's mechanical properties to engineer effective substitutes. The mechanical properties of native dural tissue and dural grafts are a key area of focus. A systematic review [23] of current research methodologies and focuses revealed that while methodological differences pose challenges for direct comparisons, these studies are invaluable for the synthesis and optimization of dural graft materials. Mechanical properties have been evaluated across various species, including rodent [24], ovine [25], porcine [17], bovine [26], and human [27]. These comparative analyses provide critical insights for the development of dural substitutes, highlighting the need for materials to closely mimic the native dura mater's mechanical behavior.

## *3.2 Meninges Elastic Behavior*

The dura mater's elasticity, its ability to resist and recover from deformation, is a fundamental aspect of its mechanical properties. Research indicates a diverse range of elastic moduli within the dura mater, illustrating its varying resistance to deformation under different loads. For example, longitudinal Dura Arachnoid Complex (DAC) elastic moduli have been reported at 22.4 MPa in cervical, 38.1 MPa in thoracic, and 36.6 MPa in lumbar spinal levels [28]. Such variations reveal that the dura mater is not a uniform structure throughout the spinal cord. To analyze this mechanical behavior, researchers have applied models like the neo-Hookean hyper-elastic model. Additionally, studies on the cranial dura mater demonstrate nonlinear elastic characteristics. The dura mater is recognized as a heterogeneous material, though there is an ongoing debate about the nature of this heterogeneity and contradictions regarding its anisotropy levels [29]. Recent findings also indicate that the heterogeneity of cranial dura is not solely structural but also linked to varying protein contents in different regions [30]. To establish a correlation between structural and mechanical properties, researchers like J. Niestrawska et al. have employed microstructurally motivated material models for assessing dura mater morpho-mechanics [29]. These models are essential in understanding the local differences in the dura mater, thereby contributing to a more comprehensive understanding of its mechanical behavior.

## 4 Biomaterials in Dura Mater Grafts

Cranial trauma, tumor invasion, cerebrovascular disease, and neurosurgery can all result in defects of the dura mater [31]. Such defects warrant and necessitate repair. The ideal dural graft should prevent CSF leakage with a watertight seal, have similar strength and flexibility as human dura, minimize scar formation and inflammation, and reduce post-craniotomy adhesion risk [32]. Additionally, as time in the operating room correlates with the risk of post-operative infection, it is important for grafts to have ease of handling for efficient application. If dural repair is insufficient, CSF leakage can cause intracranial hypotension that can result in chronic headaches, wound breakdown, and life-threatening complications such as subdural hematomas and brain herniation [33–35]. Because of the intricate nature of the dura mater, there is no universally accepted graft material. Consequently, there is a fundamental requirement to investigate the dura mater at various length scales. The complexity of the meninges, comprising three distinct layers with unique properties, necessitates a comprehensive approach in dural graft research. Protasoni et al. [16, 36, 37] identified five layers in the human cranial dura mater using SEM, each with varying thickness, orientation, and structure. This discovery complicates the selection of suitable graft materials. The dura mater's strength often leads to its conflation with the meninges.

Dural defects, arising from trauma [36], tumor invasion [37], cranial or spinal surgeries [38], or medical interventions, pose significant clinical challenges. These defects can lead to complications such as seizures [39], meningitis [40], cerebrospinal fluid leakage [41] cerebral herniation [42], pseudomeningocele [43], and infections [44]. To address these issues, clinicians use dural grafts in surgical interventions, aiming for materials that offer ease of handling, match the biomechanical properties of native dura, promote new tissue growth, minimize scar formation, and avoid adhesion to adjacent tissues [45].

Historically, materials for clinical intervention have been categorized into biological sources (autologous, allogeneic, and xenogeneic), natural polymeric, and synthetic polymeric substitutes for dura mater reconstruction. The evolution of dural grafts over the past century has been marked by an increased understanding and refinement of these materials. The quest for the ideal dural substitute has been complicated by challenges associated with autogenic and allogenic grafts, including the need for an additional surgical site for harvesting the graft, lack of customizability, and the risk of Creutzfeldt-Jakob disease (CJD) from allografts, leading to the exploration of a broader spectrum of biological and synthetic materials [31].

The absence of a universally accepted standard for dural replacement underscores the necessity for a versatile approach to graft selection, tailored to the specific surgical procedure. A systematic review by Azzam et al. [38], covering the period from 2007 to 2017, highlighted trends in graft usage: allografts were preferred for decompression in Chiari malformations, while xenografts were more commonly used in decompressive hemicraniectomy for trauma scenarios. Synthetic grafts, meanwhile, were predominantly chosen for tumor surgeries over biological alternatives.

From a materials science perspective, the hierarchical structure and biological complexity of the native dura demand that ideal synthetic grafts replicate all biological and morpho-mechanical characteristics of the dura mater. The multiscale structure and mechanics of grafts play a pivotal role in facilitating cell and tissue regeneration. Minimal mechanical characterization falls short in capturing the biophysical intricacies of the dura, which is composed of several layers, each with distinct properties and roles. The dura mater functions as a critical interface, where parameters such as multiscale mechanical properties (elasticity and viscoelasticity), porosity, anisotropy, and structural thickness are crucial. In certain contexts, the biological significance of a graft is important, while in others, its biomechanical integrity is paramount. This comprehensive approach to developing and selecting dural substitutes is essential in surgical practice, where the goal is to mimic the native dura's function as closely as possible, ensuring successful clinical outcomes and patient recovery. An added layer of complexity is attributed to the many closure techniques, from suturing to glues to overlays. This area has garnered attention and resulted in multiple studies to assess not only grafts but their use in conjunction of different closure methods.

## *4.1 Biophysical Properties of Different Graft Types*

In the realm of neurosurgery, the repair and reconstruction of the dura mater are very important for patient recovery and the prevention of complications such as

CSF leakage. Various types of dural substitutes are employed to meet this need, each derived from different sources and possessing unique properties tailored to specific clinical requirements. This section explores the distinct categories of dural substitutes—autologous, allogeneic, xenogeneic, and synthetic—highlighting their applications, advantages, and the scientific advancements that enhance their efficacy and safety in clinical settings.

### 4.1.1 Autologous Grafts

Autologous dural substitutes, derived from the individual undergoing treatment, include tissues such as the pericranium, fascia lata, fat, and muscle. The primary advantage of using autologous tissue is the significantly reduced risk of immune rejection and the absence of risk for disease transmission [46]. Regarded as the gold standard due to their inherent compatibility, autologous grafts face limitations related to the availability of suitable donor tissue and the need for an additional surgical site for tissue harvesting [47].

Among the options for tissues, the fascia lata has mechanical properties (1–8 MPa) that contribute to its utility in dural repair [46, 47]. Pericranium is noted for its isotropic behavior, mirroring the extension and strain patterns observed in the longitudinal dura mater under yield and failure conditions. Despite its lower ultimate stress and tangent modulus compared to the longitudinal dura, the pericranium remains a viable tissue to cover dural defects [48, 49]. However, there have been reports of ossification of pericranial grafts, indicating a potential complication that necessitates careful consideration [50]. Fat, particularly from the abdominal region, has been explored as a substitute, showing promising outcomes [51]. Its primary advantage includes its soft consistency, which ensures minimal adhesion. Yet, a significant drawback is the potential for watertight seal failure and subsequent CSF leakage [51, 52]. Comparative studies reveal the final elastic modulus of human abdominal fat to be approximately 12.1 kPa [53], markedly lower than the dura mater's 61 MPa [1], underscoring the material's different mechanical properties.

### 4.1.2 Allogeneic Grafts

Allogeneic dural substitutes are derived from donors within the same species who are genetically distinct from the recipient [54]. These grafts, obtained from either cadaveric or living donors, undergo rigorous processing to reduce immunogenicity, aiming to minimize the risk of immune rejection and disease transmission [55]. Such processing is crucial not only for ensuring the grafts' safety but also for maintaining their initial mechanical integrity, despite concerns that such treatments might alter their structural properties. However, there is no conclusive evidence to support these concerns.

Clinically, various allogeneic sources have been utilized, including the amniotic membrane ($E = 6.8 \pm 0.22$) [1, 54] and acellular human dermis. The use of the amniotic membrane has proven effective and safe, demonstrating excellent watertightness and fostering an environment conducive to fibrocyte infiltration and subsequent healing of dural mater defects [38]. Notably acellular scalp, exhibiting biomechanical properties ($E = 20.0 \pm 9.0$ MPa), mirrors the native tissue closely under uniaxial tensile stress. This similarity suggests that the mechanical and structural integrity of the human dermis is largely retained following the acellularization process [55, 56].

A comprehensive study by A. Yahanda et al., encompassing 781 patients over a span from 2001 to 2019, indicated that complication rates between dural autografts and nonautologous grafts were comparable following posterior fossa decompression for Chiari Malformation Type I/Syringomyelia (CM-I/SM) [38]. This finding underscores the potential efficacy of allogeneic grafts as a viable option for dural repair, challenging the perceived superiority of autologous grafts and highlighting the advancement in allogeneic graft processing and application in clinical settings.

### 4.1.3 Xenogeneic Grafts

Xenogeneic dural substitutes, sourced from nonhuman species, are treated extensively to ensure compatibility with the human immune system. This process involves removing cellular elements and diminishing immunogenicity through methods such as glutaraldehyde stabilization and freeze-drying [54, 56]. Despite these modifications, there is an inherent risk of immune reactions, and the long-term integration and functionality of xenogeneic grafts may not fully align with those derived from autologous or allogeneic sources. The literature on the efficacy and outcomes of various dural substitutes is limited, making it challenging to definitively recommend one material over another.

Nevertheless, xenografts have been notably utilized in decompressive hemicraniectomy for the evacuation of acute subdural hematomas more frequently than other graft types [38]. Among the xenogeneic options, grafts derived from equine, porcine, and bovine sources are commonly available. Specific examples include the equine pericardium membrane [57] and porcine and bovine pericardium [58, 59], with studies documenting the thickness and mechanical properties of these materials post-processing, illustrating their robustness and potential for clinical application. Porcine small intestinal submucosa [60] represents another xenogeneic material utilized, demonstrating the broad spectrum of available xenografts. Although there may be slight mismatches in physical properties compared to autologous and allogeneic grafts, the theoretical advantage of an unlimited material supply positions xenografts as a valuable resource, provided they can be safely processed for human transplantation (Fig. 4).

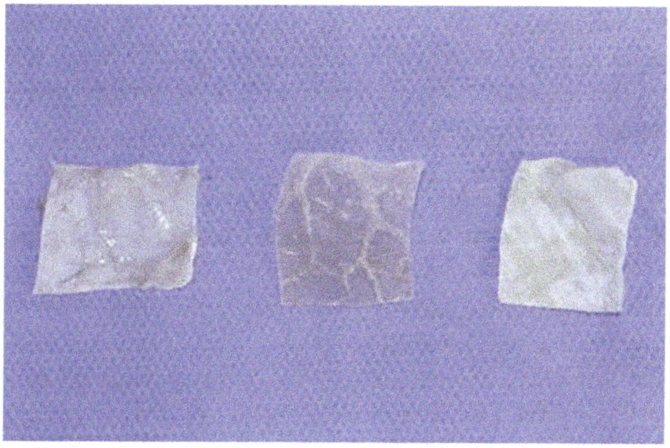

**Fig. 4** Examples of blocks of tissue samples used for the study. From left to right a pericranium, human dura mater, and temporalis fasci [61]

### 4.1.4 Natural Polymeric Substitutes for the Dura Mater

Natural polymeric substitutes are derived from organic, naturally occurring polymers. These materials are sourced from the environment and may include proteins (such as collagen [62] or gelatin [63]), polysaccharides (such as chitosan [64] or alginate [65]), or other biopolymers that are biocompatible and biodegradable. Natural polymeric substitutes are favored for their biological compatibility and the ability to promote cell adhesion, proliferation, and integration with the host tissue [66]. They are designed to mimic the extracellular matrix, providing a scaffold that supports tissue regeneration and healing. However, their mechanical properties and degradation rates must be carefully considered to match the requirements of dural repair, ensuring that the graft maintains structural integrity until the native tissue has adequately healed. In the case of collagen—the preferred biopolymer—the quick degradation and limited mechanical strength of pure collagen scaffolds have led to significant advancements, including chemical cross-linking and the incorporation of various bioactive polymeric compounds. Such innovations aim to enhance durability and functional integration with native tissues.

Commercial dural implants, including TissuDura® (Medtronic, MN, USA), Durepair® (Medtronic, MN, USA), and Duregen® (Integra, NJ, USA), derive collagen from diverse sources such as bovine and equine collagen. These different origins and manufacturing methods result in varied mechanical properties across products. Among the commercial dural implants, Durepair® is noted for closely replicating the mechanical behavior of the native dura [67]. Despite the scarcity of literature detailing their mechanical behaviors, retrospective studies have confirmed their high efficacy in clinical settings [44, 68, 69]. Beyond serving as mere substitutes, these materials also function as sealants, enhancing the mechanical and adhesive properties of the graft [70]. Furthermore, novel applications of these materials, either alone or in conjunction with other grafts like collagen or fibrin-based adhesives, have emerged. An intriguing example is the combination of autologous fat grafts with fibrin adhesive materials, presenting a promising avenue for dural replacement [49]. This multifaceted approach underscores the adaptability and potential of natural polymeric substitutes in neurosurgical applications, marrying traditional grafting techniques with cutting-edge biomaterial science for enhanced dural repair outcomes.

### 4.1.5 Synthetic Polymeric Substitutes for Dural Mater Reconstruction

While naturally derived materials boast high biocompatibility, they often face limitations due to restricted sources, rapid degradation, and insufficient mechanical strength [71]. Consequently, the development of synthetic materials that can be easily molded into various forms is crucial. Synthetic polymeric substitutes are laboratory-synthesized materials not found in nature, offering a modern solution to the drawbacks of traditional materials.

These engineered polymers include polytetrafluoroethylene (ePTFE) [72], polyurethane [73], polyglycolic acid (PGA) [74], polycaprolactone (PCL) [75, 76], and poly(L-lactic acid) (PLLA) [77]. They provide precise control over mechanical properties, degradation rates, and porosity, catering to specific needs in dural reconstruction. For instance, Lia et al. [78] developed a triple-layer dural structure incorporating PLLA, aiming to mimic the mechanical properties, structural features, biocompatibility, and watertight sealing necessary for effective dural repair.

Notably, synthetic options like Neuro-Patch represent nonabsorbable choices utilized in clinical settings. However, studies indicate that nonabsorbable synthetic substitutes may elevate the risk of infections [79]. Conversely, Cho et al. [77] examined the structural, mechanical, and in vivo responses of electrospun synthetic fibers of PLLA, demonstrating Safe-dura's acceptable mechanical strength with an average tensile strength of $3.18 \pm 0.13$ MPa and an average elongation of $41.6 \pm 5.27\%$.

Synthetic materials offer the dual advantage of replicating the native dural structure and enhancing mechanical properties. They can be absorbable or nonabsorbable and are frequently combined with natural materials to optimize dural substitutes. Moreover, the consistency, purity, and scalable production of these synthetics are beneficial, though their biocompatibility and immune response potential must be carefully managed through modifications with bioactive molecules or coatings, ensuring better integration with host tissue (Fig. 5).

## 5 Physical Characterization of the Dura Mater: CSF Sealing Function

The dura mater is responsible for retaining CSF in the brain in a watertight manner. When the dura mater is interrupted, the watertight seal needs to be reestablished. Over the past 40 years, there has been an ongoing evaluation of the effectiveness of

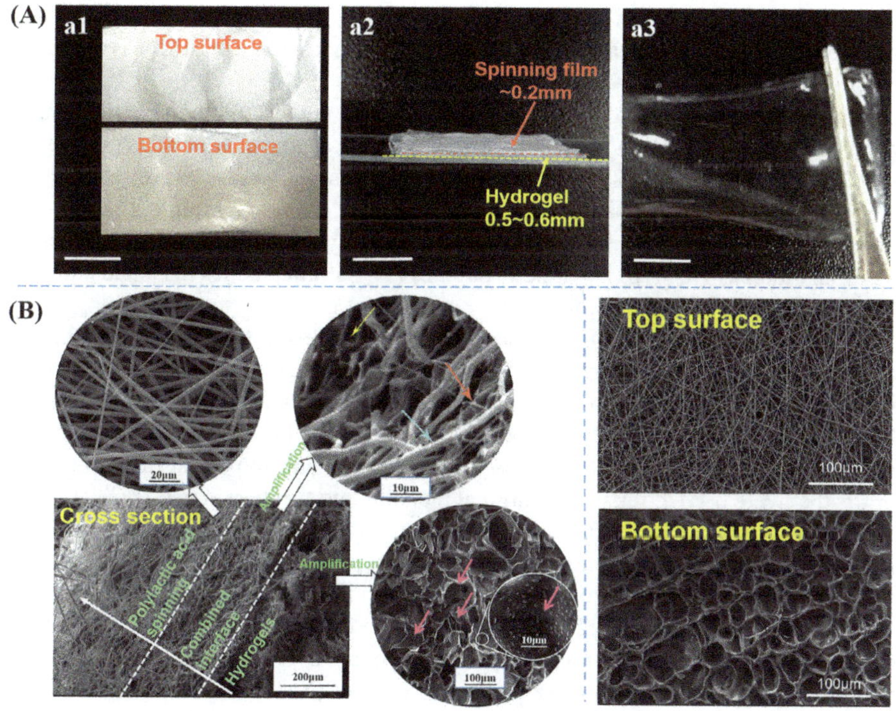


**Fig. 5** The macro and micro appearances of the triple-layered composites. (**a**) Macro appearances of the triple-layered composites, including (a1) the top and bottom surfaces; (a2) the cross-section; and (a3) hydrogels. scale bar: 1 cm. (**b**) SEM images of the top and bottom surfaces and the cross-section of composites. PLLA and CS/PLLA electrospinning fibers and hydrogels in the combined interface are indicated by blue, red, and yellow arrowheads, respectively. The acellular SIS powders evenly distributed within the hydrogel are indicated with red arrowheads in the SEM images of the hydrogel [78]

sealing techniques for dura mater following surgical intervention [31, 80]. Iatrogenic durotomies occur when neurosurgeons are forced to cut through the dura mater to access their region of interest inside the brain. Traumatic dural tears may occur from skull fractures that lacerate the dura mater and interrupt the watertight seal.

Depending on the region, neurosurgeons utilize sutures to repair the dura mater, however primary repair may not be watertight, especially in the setting of trauma where intracranial pressures are often excessive [81]. To address these issues, sealants and hydrogels are often applied overtop of the dural repair. For larger defects where suturing is not viable, overlays (both synthetic and natural) have been utilized.

Excision of the dura mater layer can cause CSF leakage which leads to a decrease in intracranial pressure (ICP) which can cause other complications [82]. Cerebrospinal fluid is found in the subarachnoid space just below the dura mater [83] and flows around the brain, ultimately returning to the venous sinuses, deep to the dura mater. Symptoms of CSF leakage include chronic headaches from intracranial hypotension, CSF otorrhea or rhinorrhea, nausea, neck pain or stiffness, tinnitus, and dizziness [84].

Animal models, in vitro models, and retrospective studies of human patients have been conducted to better understand which method or material is better suited for recreating the watertight seal. There has been a vast array of closure techniques, study findings, and professional opinions throughout the years. In vivo, [31, 85] and in vitro [86] models have largely been used to study the effectiveness of different materials such as glues [87], synthetic patches [5], natural patches [31], and suturing techniques [86] for direct comparison.

Retrospective studies [88–90] have allowed surgeons to inquisitively reflect on their techniques and the effectiveness of patches/glues, though many prove inconclusive due to the unique nature of each case. A systematic review conducted by Achinger et al. looked at CSF leakage occurrences in patients who underwent dural repairs in the infratentorial region [88]. The group looked at different dural grafts that were used to repair the surgical incision into the dura mater. By evaluating what is known about the grafts and their incidence of CSF leakage following surgical interventions, the need is evident for "larger prospective controlled studies" where the methodologies would allow for direct comparison like in in vitro studies.

In vitro models have largely been used to directly compare materials. Ebel, et al. developed a "standardized" in vitro model to compare over a dozen closure methods: suturing techniques, hydrogels, sealants, and patches [86]. The group compared the materials side-by-side to determine which one would provide a better "watertight seal" under pressure and controlled conditions. In vitro studies fail to include the reintegration of the material into the native dura mater, or the cellular interactions.

In vivo animal models have been useful in evaluating the cellular response to implants and sealants in situ along with the performance of the materials under pressure of CSF. Canine and Rabbit animal models have been used to mimic the surgical techniques used in humans. Nunamaker et al. looked at using alginate to seal the dura mater following electrode implantation in a rabbit model [85]. Islam et al. looked at using acellular dermal grafts along with suture techniques to repair the surgical incision [91].

Overall conclusions cannot be drawn from any one study or type of investigation. While the consensus is that of optimizing the prevention of CSF leakage among clinical and pre-clinical studies, there is no set benchmark for "watertight" in situ. There are many questions yet to be answered thus warranting longitudinal pre-clinical studies of CSF leakage based on graft type. The complete understanding of the multiscale, microscale architecture of the dura mater "watertighness" which prevents CSF from passing through it is unknown.

## 6 Conclusions: Final Considerations in Graft Design

This chapter has highlighted the multifaceted complexity of the dura mater, emphasizing the importance of mechanics, structure, immunogenicity, and clinical translatability in developing dura mater grafts. There are many considerations and options for surgeons in replacing dura mater during surgery. Figure 6 shows the decision factor tree some surgeons use to determine using native dura or a graft.

Prior to using any new implant material, it is important to understand the inflammatory and biocompatible response of the host [92, 93]. One of the most important factors is the need for the implant tissue to have a suitable environment for the growth and infiltration of the host cells. Growth of host cells within the implant tissue indicates acceptance of the implant by the host body and these cells also aid in the healing process. Despite extensive research, problems with dural replacement grafts include cerebrospinal fluid leakage, graft integration, and handleability [94, 95] still exist. In addition to the suitability of the material being considered for the scaffold design being governed by its biocompatibility, the mechanical integrity, biodegradability, and scaffold architecture must also be of primary consideration [94]. Understanding the structure–function relationships of dural grafts is crucial for their effective clinical translation. The involvement of engineers, scientists, and clinicians in researching and translating optimal dura mater substitutes is paramount.

Future research should include longitudinal studies both in vitro and in vivo. Various aspects need consideration in dura testing. While several in vitro studies have focused on fibroblast cells, a key question regarding the biological health of the implant remains: How do native brain cell populations integrate with the dura, and

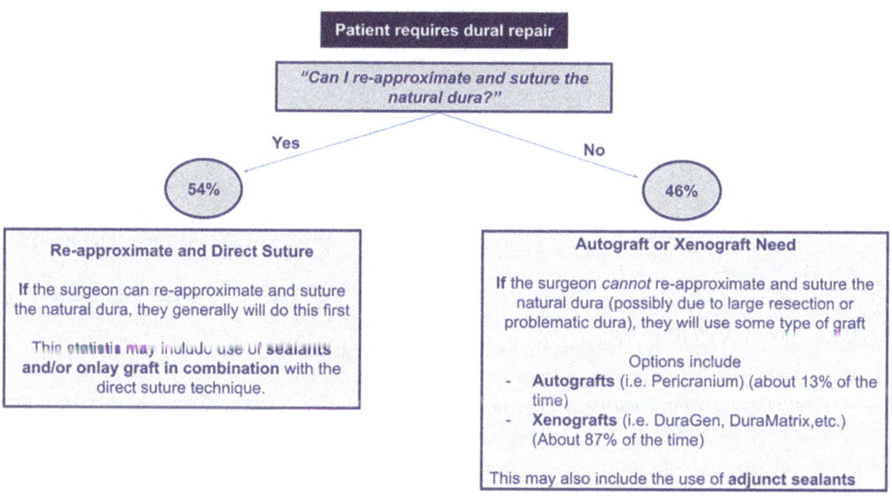


**Fig. 6** Surgeon decision tree for dural repair options. (Adapted from Integra Limit Uncertainty)

what are the adverse environmental changes cells experience upon graft exposure? Longitudinal studies are needed to assess the optimal stress state for mechanical testing of dura, sealant optimization based on graft type, and the feasibility of sutures and sealants in reducing leakage. In vivo comparative analysis should be conducted in both acute and chronic settings to determine changes in the graft over periods exceeding 6 months. Rodent models are useful for extrapolating these critical components of graft suitability. However, a primary potential challenge with using rodent models is that rodents have a small epidural space, which limits the use of a variety of clinically approved implants for testing. Larger animal models (e.g., rabbits and canines) have greater-sized epidural spaces that will enable testing the use of a broader range of commercially available cranial dural grafts. Studies using larger animal models can provide a more anatomically relevant space in the cranial cavity and offer a 1–1 match for dura placement in clinical settings. Acute and chronic in vitro and in vivo studies are essential for advancing dura graft design.

In addition to the concerns outlined earlier in this chapter, future research and development in dura graft design should prioritize native tissue regeneration, graft biodegradability, patient-specific design, and minimally invasive replacement. The regenerative properties of tissue-engineered grafts may facilitate seamless integration with native dura, mitigating the extreme immune response. Biodegradable grafts could gradually be replaced by native dura tissue, reducing the need for foreign materials in the body. Given the variation in dura's structural integrity across the brain, 3D bioprinting could enable the rapid printing of dura's soft, collagenous matrix based on the location of the injury. Ultimately, to minimize operating room time and open exposure of the brain, it may be possible to drill a small millimeter-sized burr hole in the skull and insert injectable multi-material hydrogel into the damaged area to cover the missing dura and serve as a graft.

## References

1. J. Zwirner, M. Scholze, J. N. Waddell, B. Ondruschka, and N. Hammer, "Mechanical properties of human dura mater in tension–an analysis at an age range of 2 to 94 years," *Sci Rep*, vol. 9, no. 1, p. 16655, 2019.
2. V. Kekere and K. Alsayouri, *Anatomy, Head and Neck, Dura Mater*. StatPearls Publishing, Treasure Island (FL), 2023. [Online]. Available: http://europepmc.org/abstract/MED/31424885
3. F. M. Cozzi, L. Ampie, M. T. Laws, and D. A. Brown, "The role of the dura mater in cerebral metastases," *Neurosurg Focus*, vol. 55, no. 2, p. E17, 2023.
4. W. Liu *et al.*, "In vivo Evaluation of Fibrous Collagen Dura Substitutes," *Front Bioeng Biotechnol*, vol. 9, p. 628129, 2021.
5. Y. Ramot *et al.*, "Local tolerance and biodegradability of a novel artificial dura mater graft following implantation onto a dural defect in rabbits," *Toxicol Pathol*, vol. 48, no. 6, pp. 738–746, 2020.
6. A. G. Clarke, "The Anatomy of the Meninges," *Postgrad Med J*, vol. 20, no. 220, pp. 74–78, 1944, https://doi.org/10.1136/pgmj.20.220.74.
7. V. Kekere and K. Alsayouri, "Anatomy, head and neck, dura mater," 2019.

8. K. Alves de Lima, J. Rustenhoven, and J. Kipnis, “Meningeal immunity and its function in maintenance of the central nervous system in health and disease,” *Annu Rev Immunol*, vol. 38, pp. 597–620, 2020.
9. E. M. Culav, C. H. Clark, and M. J. Merrilees, “Connective tissues: matrix composition and its relevance to physical therapy,” *Phys Ther*, vol. 79, no. 3, pp. 308–319, 1999.
10. J. Khan, “Anatomic basis of neurologic diagnosis.” AAN Enterprises, 2010.
11. A. Louveau *et al.*, “Structural and functional features of central nervous system lymphatic vessels,” *Nature*, vol. 523, no. 7560, pp. 337–341, 2015.
12. M. D. Shoulders and R. T. Raines, “Collagen structure and stability,” *Annu Rev Biochem*, vol. 78, pp. 929–958, 2009.
13. M. C. J. Hamann, M. S. Sacks, and T. I. Malinin, “Quantification of the collagen fibre architecture of human cranial dura mater,” *The Journal of Anatomy*, vol. 192, no. 1, pp. 99–106, 1998.
14. J. Zwirner, M. Scholze, B. Ondruschka, and N. Hammer, “What is considered a variation of biomechanical parameters in tensile tests of collagen-rich human soft tissues?—Critical considerations using the human cranial dura mater as a representative morpho-mechanic model,” *Medicina (B Aires)*, vol. 56, no. 10, p. 520, 2020.
15. D. Chauvet, A. Carpentier, J.-M. Allain, M. Polivka, J. Crépin, and B. George, “Histological and biomechanical study of dura mater applied to the technique of dura splitting decompression in Chiari type I malformation,” *Neurosurg Rev*, vol. 33, pp. 287–295, 2010.
16. M. Protasoni *et al.*, “The collagenic architecture of human dura mater,” *J Neurosurg*, vol. 114, no. 6, pp. 1723–1730, 2011.
17. A. Sharma, J. Liao, and L. N. Williams, “Structure and mechanics of native and decellularized porcine cranial dura mater,” *Engineered Regeneration*, vol. 4, no. 2, pp. 205–213, 2023.
18. A. Sikora and A. Kędzia, “A quantitative comparison of dura mater tissue structures measured with atomic force microscopy,” *Advances in Clinical and Experimental Medicine*, vol. 21, no. 4, pp. 487–493, 2012.
19. D. R. Walsh *et al.*, “Mechanical characterisation of the human dura mater, falx cerebri and superior sagittal sinus,” *Acta Biomater*, vol. 134, pp. 388–400, 2021.
20. G. Fabris, Z. M. Suar, and M. Kurt, “Micromechanical heterogeneity of the rat pia-arachnoid complex,” *Acta Biomater*, vol. 100, pp. 29–37, 2019.
21. D. B. MacManus, B. Pierrat, J. G. Murphy, and M. D. Gilchrist, “Protection of cortex by overlying meninges tissue during dynamic indentation of the adolescent brain,” *Acta Biomater*, vol. 57, pp. 384–394, 2017.
22. L. Qian, S. Wang, S. Zhou, Y. Sun, and H. Zhao, “Influence of pia-arachnoid complex on the indentation response of porcine brain at different length scales,” *J Mech Behav Biomed Mater*, vol. 127, p. 104925, 2022.
23. Q. Pearcy, J. Tomlinson, J. A. Niestrawska, D. Möbius, M. Zhang, and J. Zwirner, “Systematic review and meta-analysis of the biomechanical properties of the human dura mater applicable in computational human head models,” *Biomech Model Mechanobiol*, vol. 21, no. 3, pp. 755–770, 2022.
24. J. T. Maikos, R. A. I. Elias, and D. I. Shreiber, “Mechanical properties of dura mater from the rat brain and spinal cord,” *J Neurotrauma*, vol. 25, no. 1, pp. 38–51, 2008.
25. S. S. Shetye, M. M. Deault, and C. M. Puttlitz, “Biaxial response of ovine spinal cord dura mater,” *J Mech Behav Biomed Mater*, vol. 34, pp. 146–153, 2014.
26. Y. Jin, K. H. Yang, and A. I. King, “Mechanical properties of bovine pia–arachnoid complex in shear,” *J Biomech*, vol. 44, no. 3, pp. 467–474, 2011.
27. D. De Kegel, J. Vastmans, H. Fehervary, B. Depreitere, J. Vander Sloten, and N. Famaey, “Biomechanical characterization of human dura mater,” *J Mech Behav Biomed Mater*, vol. 79, pp. 122–134, 2018.
28. P. Sudres *et al.*, “Tensile mechanical properties of the cervical, thoracic and lumbar porcine spinal meninges,” *J Mech Behav Biomed Mater*, vol. 115, p. 104280, 2021.
29. J. A. Niestrawska *et al.*, “Morpho-mechanical mapping of human dura mater microstructure,” *Acta Biomater*, vol. 170, pp. 86–96, 2023.

30. D. R. Walsh *et al.*, "Regional mechanical and biochemical properties of the porcine cortical meninges," *Acta Biomater*, vol. 80, pp. 237–246, 2018.
31. X. Bi *et al.*, "Applications of materials for dural reconstruction in pre-clinical and clinical studies: advantages and drawbacks, efficacy, and selections," *Materials Science and Engineering: C*, vol. 117, p. 111326, 2020.
32. S. F. Danish, A. Samdani, A. Hanna, P. Storm, and L. Sutton, "Experience with acellular human dura and bovine collagen matrix for duraplasty after posterior fossa decompression for Chiari malformations," *J Neurosurg Pediatr*, vol. 104, no. 1, pp. 16–20, 2006.
33. E. Mazgajczyk, K. Scigala, M. Czyż, W. Jarmundowicz, and R. Bedzinski, "Mechanical properties of cervical dura mater.," *Acta Bioeng Biomech*, vol. 14, no. 1, 2012.
34. L. Chiapparini *et al.*, "Headache and intracranial hypotension: neuroradiological findings," *Neurological Sciences*, vol. 25, pp. s138–s141, 2004.
35. N. Adeeb, M. M. Mortazavi, R. S. Tubbs, and A. A. Cohen-Gadol, "The cranial dura mater: a review of its history, embryology, and anatomy," *Child's Nervous System*, vol. 28, pp. 827–837, 2012.
36. G. Singhal, S. Atri, S. Suggala, D. Jaluka, S. Singhal, and A. Shrivastava, "Growing skull fractures; pathogenesis and surgical outcome," *Asian J Neurosurg*, vol. 16, no. 03, pp. 539–548, 2021.
37. S. M. Barber *et al.*, "Cerebrospinal fluid leaks after spine tumor resection: avoidance, recognition and management," *Ann Transl Med*, vol. 7, no. 10, 2019.
38. D. Azzam *et al.*, "Dural Repair in Cranial Surgery Is Associated with Moderate Rates of Complications with Both Autologous and Nonautologous Dural Substitutes," *World Neurosurg*, vol. 113, pp. 244–248, 2018, https://doi.org/10.1016/j.wneu.2018.01.115.
39. H. Wang *et al.*, "Seizure after cranioplasty: incidence and risk factors," *Journal of Craniofacial Surgery*, vol. 28, no. 6, pp. e560–e564, 2017.
40. I. S. Kourbeti *et al.*, "Infections in patients undergoing craniotomy: risk factors associated with post-craniotomy meningitis," *J Neurosurg*, vol. 122, no. 5, pp. 1113–1119, 2015.
41. S. I. Moskowitz, J. Liu, and A. A. Krishnaney, "Postoperative complications associated with dural substitutes in suboccipital craniotomies," *Operative Neurosurgery*, vol. 64, no. suppl_1, pp. ONS28–ONS34, 2009.
42. Y. Motoyama *et al.*, "Risk of brain herniation after craniotomy with lumbar spinal drainage: a propensity score analysis," *J Neurosurg*, vol. 130, no. 5, pp. 1710–1720, 2018.
43. S. P. Norrdahl, T. L. Jones, P. Dave, D. S. Hersh, B. Vaughn, and P. Klimo, "A hospital-based analysis of pseudomeningoceles after elective craniotomy in children: what predicts need for intervention?," *J Neurosurg Pediatr*, vol. 25, no. 5, pp. 462–469, 2020.
44. T. D. McCall, D. W. Fults, and R. H. Schmidt, "Use of resorbable collagen dural substitutes in the presence of cranial and spinal infections—report of 3 cases," *Surg Neurol*, vol. 70, no. 1, pp. 92–96, 2008.
45. F. Yu, F. Wu, R. Zhou, L. Guo, J. Zhang, and D. Tao, "Current developments in dural repair: a focused review on new methods and materials," *Front Biosci (Landmark Ed)*, vol. 18, no. 4, pp. 1335–1343, 2013.
46. B. Weber, M. Y. Emmert, R. Schoenauer, C. Brokopp, L. Baumgartner, and S. P. Hoerstrup, "Tissue engineering on matrix: future of autologous tissue replacement," in *Seminars in immunopathology*, Springer, 2011, pp. 307–315.
47. G. Sabatino *et al.*, "Autologous dural substitutes: a prospective study," *Clin Neurol Neurosurg*, vol. 116, pp. 20–23, 2014.
48. T. Zeng, M. Wang, Z. Xu, M. Ni, and L. Gao, "Autologous Free Fascia Lata Can Be Used as Dura Graft in the Salvage Treatment of Recalcitrant Postcraniotomy Intracranial Infection Caused by Multidrug-Resistant Gram-Negative Bacteria," *Infect Drug Resist*, pp. 5667–5677, 2022.
49. M. Kirilova-Doneva and D. Pashkouleva, "The effects of age and sex on the elastic mechanical properties of human abdominal fascia," *Clinical Biomechanics*, vol. 92, p. 105591, 2022.
50. D. A. Hoover and A. Mahmood, "Ossification of autologous pericranium used in duraplasty: Case report," *J Neurosurg*, vol. 95, no. 2, pp. 350–352, 2001.

51. H. Di Vitantonio *et al.*, "Dural repair using autologous fat: Our experience and review of the literature," *Surg Neurol Int*, vol. 7, no. Suppl 16, p. S463, 2016.
52. C. A. Bohoun, T. Goto, H. Morisako, A. Nagahama, Y. Tanoue, and K. Ohata, "Skull base dural repair using autologous fat as a dural substitute: an efficient technique," *World Neurosurg*, vol. 127, pp. e896–e900, 2019.
53. N. Alkhouli *et al.*, "The mechanical properties of human adipose tissues and their relationships to the structure and composition of the extracellular matrix," *American Journal of Physiology-Endocrinology and Metabolism*, vol. 305, no. 12, pp. E1427–E1435, 2013.
54. E. A. Gruskin and R. Harten, "Sourcing animal and human tissue for implant use," in *Extracellular Matrix-derived Implants in Clinical Medicine*, Elsevier, 2016, pp. 119–138.
55. D. Dufrane, O. Cornu, C. Delloye, and Y.-J. Schneider, "Physical and chemical processing for a human dura mater substitute," *Biomaterials*, vol. 23, no. 14, pp. 2979–2988, 2002.
56. J. Pařízek, P. Měřička, J. Špaček, S. Němeček, P. Eliáš, and M. Šercl, "Xenogeneic pericardium as a dural substitute in reconstruction of suboccipital dura mater in children," *J Neurosurg*, vol. 70, no. 6, pp. 905–909, 1989.
57. R. Centonze, E. Agostini, S. Massaccesi, S. Toninelli, and L. Morabito, "A novel equine-derived pericardium membrane for dural repair: A preliminary, short-term investigation," *Asian J Neurosurg*, vol. 11, no. 03, pp. 201–205, 2016.
58. Y. Seo, J. W. Kim, Y. S. Dho, T. Chowdhury, S. Kim, and C.-K. Park, "Evaluation of the safety and effectiveness of an alternative dural substitute using porcine pericardium for duraplasty in a large animal model," *Journal of Clinical Neuroscience*, vol. 58, pp. 187–191, 2018.
59. D. Arbeiter, N. Grabow, Y. Wessarges, K. Sternberg, and K.-P. Schmitz, "Suitability of porcine pericardial tissue for heart valve engineering: biomechanical properties," *Biomedical Engineering/Biomedizinische Technik*, vol. 57, no. SI-1-Track-S, pp. 882–883, 2012.
60. D. J. Totten, N. F. Manzoor, K. L. Yancey, R. J. Yawn, D. S. Haynes, and A. Rivas, "Comparison of small intestinal submucosal graft and autologous tissue in prevention of CSF leak after posterior fossa craniotomy," *J Neurol Surg B Skull Base*, vol. 82, no. 06, pp. 695–699, 2021.
61. R. Morales-Avalos *et al.*, "Characterization and morphological comparison of human dura mater, temporalis fascia, and pericranium for the correct selection of an autograft in duraplasty procedures," *Surgical and radiologic anatomy*, vol. 39, pp. 29–38, 2017.
62. F. Esposito *et al.*, "Collagen-only biomatrix as a novel dural substitute: examination of the efficacy, safety and outcome: clinical experience on a series of 208 patients," *Clin Neurol Neurosurg*, vol. 110, no. 4, pp. 343–351, 2008.
63. H. Kawai *et al.*, "Effectiveness of a new gelatin sealant system for dural closure," *Neurol Res*, vol. 36, no. 10, pp. 866–872, 2014.
64. K. C. Wu *et al.*, "A tough bioadhesive hydrogel supports sutureless sealing of the dural membrane in porcine and ex vivo human tissue," *Sci Transl Med*, vol. 16, no. 739, p. eadj0616, 2024.
65. E. A. Nunamaker, K. J. Otto, and D. R. Kipke, "Investigation of the material properties of alginate for the development of hydrogel repair of dura mater," *J Mech Behav Biomed Mater*, vol. 4, no. 1, pp. 16–33, 2011.
66. R. Song, M. Murphy, C. Li, K. Ting, C. Soo, and Z. Zheng, "Current development of biodegradable polymeric materials for biomedical applications," *Drug Des Devel Ther*, pp. 3117–3145, 2018.
67. C. Kizmazoglu *et al.*, "Comparison of biomechanical properties of dura mater substitutes and cranial human dura mater: an in vitro study," *J Korean Neurosurg Soc*, vol. 62, no. 6, p. 635, 2019.
68. Z. N. Litvack, G. A. West, J. B. Delashaw, K. J. Burchiel, and V. C. Anderson, "Dural augmentation: part I—evaluation of collagen matrix allografts for dural defect after craniotomy," *Neurosurgery*, vol. 65, no. 5, pp. 890–897, 2009.
69. J. Tews *et al.*, "TachoSil dural reconstruction in extracranial-intracranial bypass surgeries," *J Neurol Surg A Cent Eur Neurosurg*, vol. 80, no. 01, pp. 39–43, 2019.
70. A. Kinaci, A. Algra, S. Heuts, D. O'Donnell, A. van der Zwan, and T. van Doormaal, "Effectiveness of dural sealants in prevention of cerebrospinal fluid leakage after craniotomy: a systematic review," *World Neurosurg*, vol. 118, pp. 368–376, 2018.

71. K. Yamada *et al.*, "Development of a dural substitute from synthetic bioabsorbable polymers," *J Neurosurg*, vol. 86, no. 6, pp. 1012–1017, 1997.
72. F. J. Attenello, M. J. McGirt, G. L. Garcés-Ambrossi, K. L. Chaichana, B. Carson, and G. I. Jallo, "Suboccipital decompression for Chiari I malformation: outcome comparison of duraplasty with expanded polytetrafluoroethylene dural substitute versus pericranial autograft," *Child's Nervous System*, vol. 25, pp. 183–190, 2009.
73. D. E. Sakas, K. Charnvises, L. F. Borges, and N. T. Zervas, "Biologically inert synthetic dural substitutes: appraisal of a medical-grade aliphatic polyurethane and a polysiloxane-carbonate block copolymer," *J Neurosurg*, vol. 73, no. 6, pp. 936–941, 1990.
74. K. Hida *et al.*, "Nonsuture dural repair using polyglycolic acid mesh and fibrin glue: clinical application to spinal surgery," *Surg Neurol*, vol. 65, no. 2, pp. 136–142, 2006.
75. A. Sanpakitwattana, W. Suvannapruk, S. Chumnanvej, R. Hemstapat, and J. Suwanprateeb, "Cefazolin loaded oxidized regenerated cellulose/polycaprolactone bilayered composite for use as potential antibacterial dural substitute," *Polymers (Basel)*, vol. 14, no. 20, p. 4449, 2022.
76. T.-Y. Shih, J.-D. Yang, and J.-H. Chen, "Synthesis, characterization and evaluation of segmented polycaprolactone for development of dura substitute," *Procedia Eng*, vol. 36, pp. 144–149, 2012.
77. M. Cho, K. M. Shim, S. S. Park, S. S. Kang, K. Jang, and S. E. Kim, "Evaluation of Biocompatibility and Healing Properties of Dural Substitutes Produced by Electrospinning Technology," *In Vivo (Brooklyn)*, vol. 38, no. 3, pp. 1119–1126, 2024.
78. J. Liao, X. Li, W. He, Q. Guo, and Y. Fan, "A biomimetic triple-layered biocomposite with effective multifunction for dura repair," *Acta Biomater*, vol. 130, pp. 248–267, 2021.
79. M. Malliti, P. Page, C. Gury, E. Chomette, F. Nataf, and F.-X. Roux, "Comparison of deep wound infection rates using a synthetic dural substitute (neuro-patch) or pericranium graft for dural closure: a clinical review of 1 year," *Neurosurgery*, vol. 54, no. 3, pp. 599–604, 2004.
80. R.-P. Dong, Q. Zhang, L.-L. Yang, X.-L. Cheng, and J.-W. Zhao, "Clinical management of dural defects: A review," *World J Clin Cases*, vol. 11, no. 13, p. 2903, 2023.
81. A. Diab *et al.*, "Efficacy and safety of polyethylene glycol dural sealant system in cranial and spinal neurosurgical procedures: Meta-analysis," *Surg Neurol Int*, vol. 12, 2021.
82. H. Ishikura *et al.*, "Risk factors for incidental durotomy during posterior open spine surgery for degenerative diseases in adults: a multicenter observational study," *PLoS One*, vol. 12, no. 11, p. e0188038, 2017.
83. M. Severson, C. G. Schaurich, and M. K. Strecker-McGraw, "Cerebrospinal fluid leak," in *StatPearls [Internet]*, StatPearls Publishing, 2023.
84. L. D'Antona *et al.*, "Clinical presentation, investigation findings, and treatment outcomes of spontaneous intracranial hypotension syndrome: a systematic review and meta-analysis," *JAMA Neurol*, vol. 78, no. 3, pp. 329–337, 2021.
85. E. A. Nunamaker and D. R. Kipke, "An alginate hydrogel dura mater replacement for use with intracortical electrodes," *J Biomed Mater Res B Appl Biomater*, vol. 95, no. 2, pp. 421–429, 2010.
86. F. Ebel *et al.*, "A standardized model for in vitro testing of sutures and patches for watertight dural closure," *J Neurosurg*, vol. 136, no. 5, pp. 1485–1494, 2021.
87. T. Mukai, N. Shirahama, B. Tominaga, K. Ohno, Y. Koyama, and K. Takakuda, "Development of watertight and bioabsorbable synthetic dural substitutes," *Artif Organs*, vol. 32, no. 6, pp. 473–483, 2008.
88. K. G. Achinger and L. N. Williams, "Trends in CSF leakage associated with duraplasty in infratentorial procedures over the last 20 years: a systematic review," *Crit Rev Biomed Eng*, vol. 51, no. 2, 2023.
89. U. Hani, F. Aziz, and M. S. Shamim, "Value of achieving a watertight dural closure, and the use of dural sealants after supratentorial cranial surgery," *J Pak Med Assoc*, vol. 70, no. 8, p. 1470, 2020.
90. S.-H. Jeon *et al.*, "Watertight sealing without lumbar drainage for incidental ventral dural defect in transthoracic spine surgery: a retrospective review of 53 cases," *Clin Spine Surg*, vol. 30, no. 6, pp. E702–E706, 2017.

91. S. Islam, K. Ogane, H. Ohkuma, and S. Suzuki, "Usefulness of acellular dermal graft as a dural substitute in experimental model," *Surg Neurol*, vol. 61, no. 3, pp. 297–302, 2004.
92. R. Khorramirouz *et al.*, "A novel surgical technique for a rat subcutaneous implantation of a tissue engineered scaffold," *Acta Histochem*, vol. 120, no. 3, pp. 282–291, 2018.
93. S. Mirsadraee *et al.*, "Biocompatibility of acellular human pericardium," *Journal of Surgical Research*, vol. 143, no. 2, pp. 407–414, 2007.
94. M. R. MacEwan, T. Kovacs, J. Osbun, and W. Z. Ray, "Comparative analysis of a fully-synthetic nanofabricated dura substitute and bovine collagen dura substitute in a large animal model of dural repair," *Interdisciplinary Neurosurgery*, vol. 13, pp. 145–150, 2018.
95. T. A. Barbolt, M. Odin, M. Léger, L. Kangas, J. Holste, and S. H. Liu, "Biocompatibility evaluation of dura mater substitutes in an animal model," *Neurol Res*, vol. 23, no. 8, pp. 813–820, 2001.

# Part II
# Applications in Orthopedic Bioengineering

# Advances in Hard Tissue Engineering: The Impact of Multiscale Bone Scaffold Designs via Customized Additive Manufacturing

**Juncen Zhou, Negar Akrami, Lukasz Witek, Yunzhi Peter Yang, Jie Shen, and Donghui Zhu**

## 1 Introduction

Categorizing human organs into hard and soft tissues sheds light on their structure and function. Hard tissues, known for their rigidity and durability, are vital for structural support, protection, and stress loading. In contrast, soft tissues, characterized by their flexibility, are essential for various bodily functions. In the realm of medical treatment, particularly focusing on orthopedic implants, hard tissues often require fixation or replacement when natural regeneration is impossible. Such devices include pins, screws, nails, and/or plates, which are used to stabilize fractures, enabling natural healing or serving as permanent fixtures in severe cases. Depending on the clinical scenario, these devices may be removed after healing or remain in place, demonstrating their adaptability in treating hard tissue conditions. When

J. Zhou · N. Akrami · D. Zhu (✉)
Department of Biomedical Engineering, Stony Brook University, Stony Brook, NY, USA

L. Witek
Biomaterials Division, NYU Dentistry, New York, NY, USA

Hansjörg Wyss Department of Plastic Surgery, NYU Grossman School of Medicine, New York, NY, USA

Department of Biomedical Engineering, NYU Tandon School of Engineering, Brooklyn, NY, USA
e-mail: lukasz.witek@nyu.edu

Y. P. Yang
Department of Orthopaedic Surgery, Stanford University School of Medicine, Stanford, CA, USA
e-mail: ypyang@stanford.edu

J. Shen
Department of Orthopaedic Surgery, Washington University School of Medicine, St. Louis, MO, USA
e-mail: shen.j@wustl.edu

 
J. Liao, J. Y. Wong (eds.), *Integration and Bridging of Multiscale Bioengineering Designs and Tissue Biomechanics*, https://doi.org/10.1007/978-3-031-81743-4_5

natural healing is not an option due to severe damage or medical conditions, these devices take on the role of a permanent replacement for the compromised or missing bone. Nonbiodegradable replacements, commonly used in hip replacements or jaw reconstructions, are designed to remain in the patient indefinitely. Yet, in some rare instances, critically sized bone defects have the potential to heal and regenerate with the help of bioactive bone grafts/devices [1]. The feasibility of replacing hard tissue with synthetic grafting contrasts with the complexities of doing so for soft tissues, underscoring the different treatment approaches required for each.

Autologous bone grafts are the traditional benchmark for treating bone defects, but unfortunately, they are associated with limitations and risks due to the surgical nature (e.g., harvesting) of the procedure. A viable alternative is alloplastic grafts, which present a favorable solution for clinicians when designed with suitable mechanical properties and compatibility with bone tissue [2]. However, challenges continue to arise in clinical scenarios, particularly due to the intricate architecture of bone [3]. Such complexity necessitates bone scaffold designs to reflect the complex architecture, leading to significant challenges in the design and production process. To effectively manage these challenges, a multi-scale approach is essential. Such an approach would involve designing scaffolds that are not only accurate in their overall form but also detailed at the micrometer level, ensuring that they mimic the complex microstructures of native bone. This level of detail is fundamental for the scaffold's integration with the body's tissues. Moreover, the wide range of bone defects, resultant from trauma, tumors, or infections, differ significantly in anatomical location, shape, and size. Such diversity necessitates a precision medicine approach, where each scaffold is individually tailored to meet the patient's particular needs [1].

Additive manufacturing (AM), commonly referred to as 3D printing (3DP), has emerged as a potential solution/modality in the field of bone repair. This innovative technique builds objects layer by layer, allowing for the creation of complex structures guided by computer-aided design (CAD) [4]. AM/3DP harnesses medical imaging data to construct bone scaffolds with intricate designs and high precision. It offers a highly customizable fabrication process, tailoring the scaffold's contour and porous structure to the material choice and post-treatment according to individual patient requirements. Therefore, AM's attributes position it as a powerful instrument in delivering precise treatments for bone defects [1].

This chapter will focus on the critical role of customization in fabricating bone scaffolds and treating bone defects. The chapter will begin by highlighting the necessity for customized AM bone scaffolds to address issues and then provide a detailed account of the AM bone scaffold production process. The protocol begins with diagnosing and imaging bone defects, which informs clinicians and engineers when devising scaffolds with suitable materials and shapes. After the design phase, this concept is converted into a tangible scaffold through a rapid, accurate, and versatile method. While customization presents challenges, it is also important to unlock the potential of bone scaffold applications, and AM is the most viable strategy to address these complexities. Ultimately, the chapter aims to impart an in-depth knowledge of customized AM bone scaffolds to clinicians and engineers, delving into the importance of customization at each step of the fabrication process, and transitioning the field from theoretical concepts, pre-clinical experimentation to practical, clinical applications.

## 2 Current Advances in Additive Manufacturing Scaffolds for Addressing Bone Defects

### *2.1 Regeneration and Reconstruction: The Dual Goals of Customized Additive Manufacturing Scaffolds in Bone Therapy*

When addressing bone defects through intervention, implant strategies are typically centered on two aspects: *(i) regeneration*, which aims to restore the biological and mechanical functions of the bone, and *(ii) reconstruction*, which focuses on maintaining the bone's appearance or functionality. It is important to note that the quality and size of bones are greatly affected by factors including age, gender, and overall health. This is crucial to understand, as bone defects commonly result from trauma or tumors. Consequently, bone defects present a wide variety of geometries, differing significantly from one individual to another. In contrast to standard implants with fixed sizes and shapes, custom AM bone scaffolds offer a significant advantage. They are specifically designed to fit individual defect profiles, based on defect assessments [1]. The correlation between these custom AM scaffolds and the unique bone defects they are designed to address is depicted in Fig. 1.

#### 2.1.1 Role of Customized AM Scaffolds in Influencing Different Stages of the Bone Regeneration Process

Injuries resulting from trauma or diseases, such as osteoporosis, osteopenia, and bone cancer result in bone defects, often requiring surgical intervention. For major defects, natural bone healing processes are typically insufficient, underscoring the necessity for scaffolds. These implants need to integrate flawlessly with the existing bone structures, facilitating regeneration—a process that is both complex and extended in duration. This regeneration process encompasses (i) an initial inflammatory stage marked by hematoma formation, (ii) the generation of bone, characterized by the development of fibrocartilaginous callus, bony callus, and revascularization, and (iii) the remodeling of bone, during which osteoblasts and osteoclasts remodel the bony callus [5, 6]. In each of these stages, customized AM bone scaffolds provide distinct benefits.

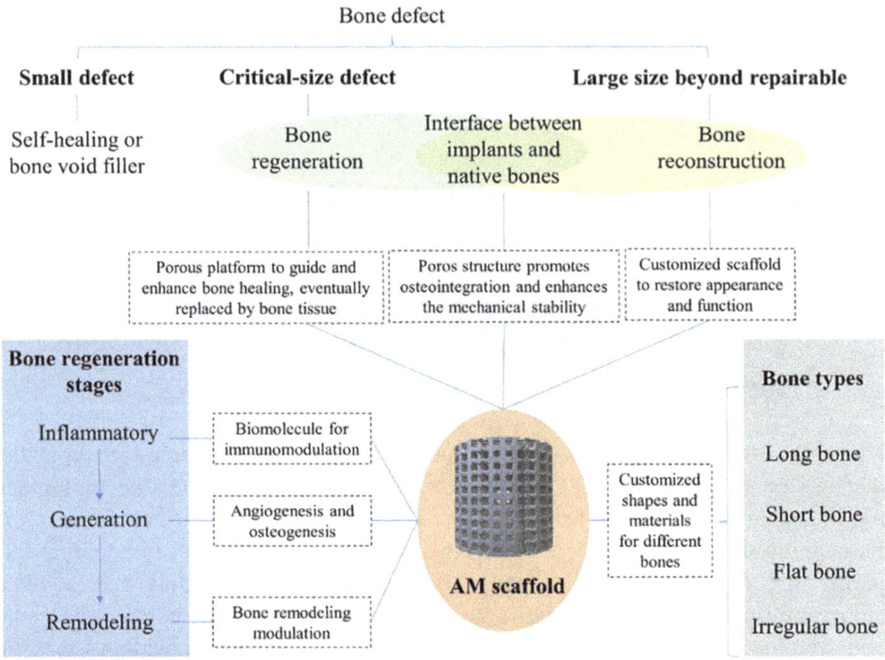


**Fig. 1** The association between the customized AM scaffold and its role in addressing bone defects [1]

I. Initial Stage of Inflammation

When a device/graft is inserted into a defect, it initiates a series of inflammatory responses. These responses can be influenced by using customized AM bone scaffolds. A common way to control these reactions is to encourage macrophage polarization towards the M2 type rather than M1. 3D-printed scaffolds, fabricated from various materials, have been demonstrated to be effective in directing the polarization of macrophages. Examples include scaffolds composed of PLGA modified by human umbilical cord mesenchymal stem cells (hUCMSC)-derived extracellular matrix [7], scaffolds composed of nanoscale bioactive glass [8], as well as scaffolds containing various cells with bone morphogenic protein-4 (BMP-4) [9]. These structures have been shown to facilitate M2 macrophage polarization and facilitate the repair of bone defects. The role of tumor necrosis factor (TNF)-α, known for its potential to hinder bone regeneration at elevated levels, has also been examined. Studies have shown that 3D-printed scaffolds incorporating Atsttrin within an alginate/hydroxyapatite (HAP) matrix can diminish TNF-α levels, thereby improving bone defect healing [10]. Furthermore, scaffolds made from 3D-printed poly (propylene fumarate) (PPF) have yielded positive results in terms of inflammation management, and β-tricalcium phosphate (β-TCP) scaffolds have been effective in reducing the expression of inflammation-related genes [11].

II. Stage of Bone Generation

In the stage of bone generation during healing, 3DP scaffolds have shown significant potential in improving both blood vessels and bone formation. For example, scaffolds made of 3DP titanium have effectively enhanced collagen mineralization and promoted in situ osteogenesis and angiogenesis [12]. This effect is also seen in 3DP ceramic scaffolds [13]. Scaffolds composed of β-tricalcium phosphate (β-TCP) integrated with mesoporous bioactive glass have outperformed β-TCP scaffolds in terms of bone and blood vessel formation [14]. In a different study, β-TCP scaffolds were enriched with MgO and $SiO_2$, releasing $Mg^{2+}$ and $Si^{4+}$ ions that stimulated osteogenesis and angiogenesis [15]. Additionally, the influence of pore structure on bone formation in calcium-deficient hydroxyapatite (CDHA) scaffolds has been investigated [16].

III. Final Stage of Bone Remodeling

The last stage of bone healing, the reshaping of new bone to match its original structure and adapt to mechanical load, is governed by the activity of osteoblasts and osteoclasts. Research on how this remodeling process interacts with 3DP scaffolds is limited, for example, one study used medical-grade polycaprolactone (PCL) and β-TCP scaffolds to examine biomineralization at the interface between soft tissue and bone [17]. This particular study demonstrated that bone near the native cortical bone is remodeled by basic multicellular units (BMUs). Another study examined a composite scaffold fabricated from PLGA/TCP/icariin using a rabbit model, observing the simultaneous progression of bone remodeling and scaffold degradation [18]. Another investigation focused on a 3DP PCL scaffold, which was combined with aspirin liposomes and bone-forming peptide-1 (BFP-1), demonstrating the scaffold's ability to facilitate bone remodeling, particularly through the PI3K/Akt signaling pathway [19].

### 2.1.2 Impact of Customized AM Scaffolds on Bone Reconstruction

In certain cases, the focus is centered more on bone reconstruction than regeneration. This is especially accurate for bones that impact physical appearance, such as maxillofacial and cranial bones, and bones near joints, like the humerus and acetabular bones. The use of custom 3D-printed scaffolds has proven to be effective in replicating the appearance and function of native bone. These scaffolds, with their porous design, aid in integration seamlessly with the native bone. In clinical settings, 3D printing has shown its efficacy in repairing extensive bone defects. This technique is employed in procedures like cranioplasty, either for creating prosthetic molds or for skull implants [20–23]. Additionally, in regions near joints, 3D-printed scaffolds have been applied successfully to address substantial segmental bone defects in the tibia [24], in revision surgeries of the acetabulum [25], and in treating severe defects of the humerus where traditional prosthetics are not suitable [26].

## 2.2 *Clinical and Pre-clinical Studies on AM Scaffolds Customized for Various Bone Categories*

Bones are ordinarily classified into four distinct categories: (i) long, (ii) short, (iii) flat, and (iv) irregular bones. Each type is susceptible to different kinds of defects, which exhibit unique features and challenges. As a result, AM of bone scaffolds designed for specific defects and customized to have distinct structures and properties [1]. Clinical and pre-clinical trials involving custom-made AM scaffolds for different bone categories are exemplified in Fig. 2.

### 2.2.1 Long Bone Studies

Long bones consist of trabecular (e.g., spongy) and cortical (e.g., compact) tissues, surrounding the central bone marrow. The primary function of a bone is primarily

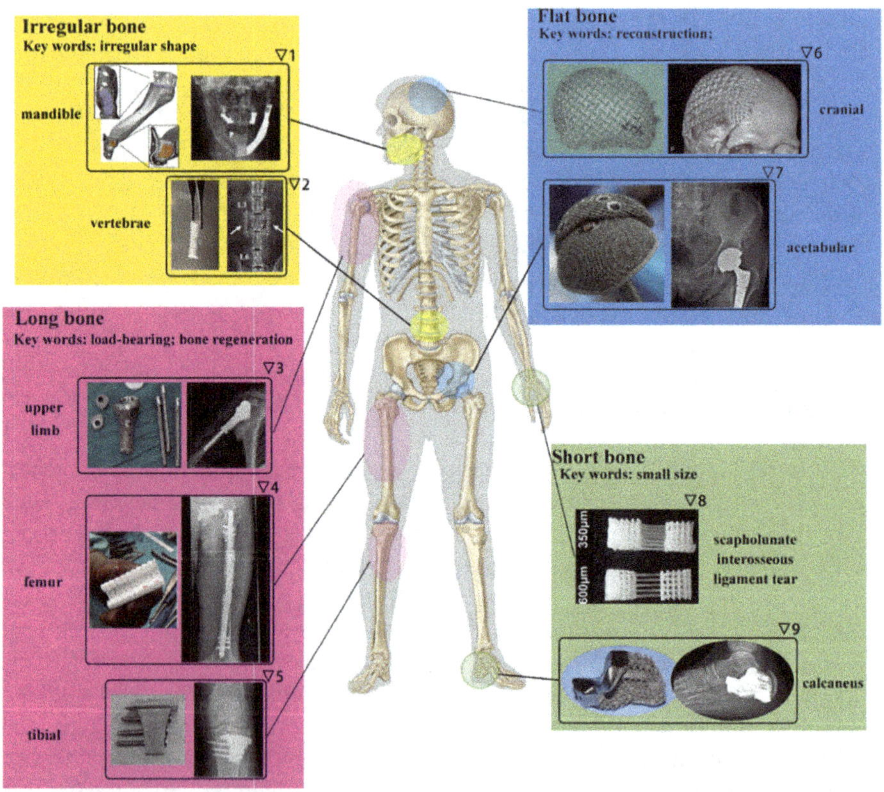


**Fig. 2** Illustration of the clinical applications/studies of customized AM scaffolds applied to various bone types. (Images modified from references: [1], ∇1 [27], ∇2 [28], ∇3 [26], ∇4 [29], ∇5 [30], ∇6 [20], ∇7 [25], ∇8 [31], ∇9 [32])

determined by the ratio of cortical to trabecular tissue. Cortical bone, composed of osteons, plays a crucial role in bending resistance, whereas trabecular bone is key in resisting compression. These structures incorporate vital elements such as osteocytes and Haversian canals, which contain blood vessels and nerves [33, 34]. In cases of significant segmental long bone defects that necessitate scaffold support for repair, these scaffolds need to demonstrate superior mechanical, biodegradable, biological, and architectural properties. The architecture of the scaffolds is critical, requiring porosity and interconnected designs that facilitate and stimulate the regeneration of the defect. Due to the necessity for optimizing architecture and characteristics like porosity to fit individual patient needs, AM stands out as an ideal method for this specialized type of fabrication.

Defects of the upper limb (e.g., humerus), arising from high-impact trauma or bone tumors, are addressed in procedures like total shoulder arthroplasty. In one particular case, a patient with chondrosarcoma who had an unstable proximal humeral replacement prosthesis underwent a replacement with a custom 3D-printed titanium alloy shoulder prosthesis, which was fabricated using electron beam melting technology, featuring 60% porosity [26]. Following 1 year post-op, the prosthesis demonstrated to be maintaining its stability and functionality. In another study, sheep with critical humerus models were tested with 3D-printed titanium-mesh scaffolds of varying stiffness. These implants exhibited lower stress shielding, facilitating early bridging and enhanced endochondral bone formation [35]. Additionally, a different study focused on the biological fixation between tendon and prosthesis implanted in rabbit humerus heads, using Ti-6Al-4V scaffolds fabricated using selective laser melting, featuring varied pore sizes [36]. It has been previously established that a pore size of 527.15 μm was most conducive for tendon growth. Two other research efforts delved into the potential of in situ bone repair in humerus defect models using 3D printing technologies [37, 38].

Long bones in the extremities, particularly the tibia and femur, are often the focus for evaluating 3D-printed scaffolds. In an instance of a femoral shaft fracture, a customized 3D-printed hybrid scaffold fabricated from medical-grade polycaprolactone and tricalcium phosphate was effectively employed, resulting in substantial bone fusion and interconnection following 1 year post-op [29]. In a separate case, five patients with femoral defects utilized customized 3D-printed micro-porous prostheses. These patients experienced swift weight-bearing capabilities and positive hip and knee joint assessments [39]. Extensive research on 3D-printed scaffolds has been conducted, examining aspects such as but not limited to the prevention of bacterial infections, enhancing vascular ingrowth, and localized gene therapy applications [40–42].

3D-printed scaffolds have demonstrated favorable outcomes in addressing tibia defects. For instance, a patient who underwent a procedure to remove giant cell tumors benefited from a 3D-printed porous implant augmented with an autograft. Such an approach has led to satisfactory limb functionality after 29 months [30]. In another case, a patient who presented with a tibia fracture received a custom 3D-printed titanium scaffold and experienced no functional limitations post implantation [24]. Research has also delved into various materials for 3D-printed scaffolds, such as beta tri-calcium phosphate (β-TCP). Β-TCP has been found effective in guiding the directional regeneration and remodeling of bone defects [43]. Additional studies identified scaffolds with a precisely controlled pore size of 400 μm provided the most effective capacity for bone formation [44]. Furthermore, the potential of 3D-printed polymeric scaffolds, augmented with compounds and cells such as alendronate, rhBMP-2, and MSCs, have been investigated for enhancing bone regeneration [45, 46].

### 2.2.2 Short Bone Studies

Short bones are composed of primarily spongy bone surrounded by a layer of compact bone. These bones, such as the carpals and tarsals, are small and typically cubic in shape. In common injuries, such as tears of the scapholunate interosseous ligament, scaffolds that replicate the natural tissue structure are created using AM. These multiphasic scaffolds, which include bone-ligament-bone components, promote bone growth when integrated with mesenchymal stem cells [31].

In a specific research study, a personalized 3D-printed implant made from a solid titanium-based alloy, enhanced with a mesh for ligament attachment, has been used to repair the calcaneus, a tarsal bone that bears much of the foot's weight, affected by tumor-related damage. This approach not only preserved the site where the Achilles tendon attaches but also permitted patients to walk without assistance [32]. In addressing defects in the calcaneus, the Masquelet technique, blending 3D printing with the induced membrane method, has been an effective solution. In a particular study, this technique involved using custom 3D-printed molds to shape bone cement, achieving a repair that closely matched the original bone morphology [47].

### 2.2.3 Flat Bone Studies

Flat bones, similar to short bones, also comprise a spongy bone nestled between two layers of compact bone. However, their distinct characteristic, such as broad plates, allow them to protect underlying tissues and provide attachment points for muscles. Among flat bones, the cranial and acetabular bones are particularly susceptible to defects arising from cancer, trauma, and a range of other health issues [1].

In the context of hip revision surgeries, trabecular titanium cups with high porosity, fabricated through electron beam melting (EBM), have been effectively employed for acetabular revisions [25, 48]. This method has yielded promising mid-term results, such as enabling early patient movement and weight-bearing capabilities, with no recorded failures in modularity [25]. Particularly for intricate acetabular bone defects, which are common in both initial and revision total hip arthroplasty, the use of 3D-printed porous augments has been advantageous. In experiments conducted using a swine model, these augments, made from Ti6Al4V powders, exhibited favorable biomechanical properties and compatibility with tissues [49].

Cranioplasty, the process of reconstructing the skull bone, aiming to restore its natural contours, is challenging due to the skull's curved surface, intricate geometries of the maxillofacial region, and the varying sizes of bone defects that depend on the individual patient's condition [1].

Numerous scaffold designs have been developed for the treatment of cranial defects. Clinical applications have included the use of polymethylmethacrylate (PMMA) and polyetheretherketone (PEEK) implants, which are fabricated using 3D printing technology to produce molds customized for individual patients [50–54].

A different method involves directly utilizing implant materials in 3D printing, guided by 3D-computed tomography (3D-CT) data. Titanium (Ti)-based meshes, chosen for their chemical and biological stability, have been employed for this purpose. These meshes have shown effectiveness in cranioplasty, especially for patients with extensive calvarial defects and recurring infections [22]. In a comparative analysis, no substantial differences were observed in post-surgical complications between 3D-printed Ti mesh and autologous bone flap, indicating that the choice of material should be specific to each patient [23]. Notably, combining 3D-printed Ti porous implants with calcium phosphate fillers has yielded encouraging outcomes, potentially reducing failure rates, shortening surgery time, and enhancing infection resistance [55, 56]. Additionally, using 3D-printed scaffolds in cranioplasty treatment may offer a cost-effective alternative to conventional implant techniques [57, 58].

### 2.2.4 Irregular Bone Studies

Bones with irregular geometrical configurations, such as vertebrae and mandible (lower jaw), consist of a layer of compact bone, which encases spongy bone. Commonly, implants are needed to treat defects associated with these types of bones [1].

Frequently, Ti6-Al4-V mandibular prostheses, which are tailor-made for a patient, are fabricated using EBM and implanted to repair a defect in the mandible effectively. Such prosthesis may feature porous designs at both the upper and lower ends and has been demonstrated in multiple patients to remain complication-free throughout a 9-month monitoring period [27]. Alternatively, magnesium-substituted β TCP and wollastonite scaffolds have been tested in a rabbit model for mandibular defects, where they exhibited encouraging osteogenic properties [59, 60]. Scaffolds composed of PLA/HAP/β-TCP and PCL/β-TCP have been effectively employed for repairing mandibular defects in canine models, while PLGA/HAP scaffolds were utilized for mandibular bone defect repair in rats, both demonstrating favorable outcomes for mandibular regeneration [61–63]. Naturally derived/sourced polymeric materials, such as chitosan, gelatin, and alginate, have been explored as potential sources for use as scaffolds. Scaffolds composed of alginate/TCP/HAP have been demonstrated to offer structural stability and cytocompatibility in vitro as well as exhibit antibacterial properties [64]. Comparably, scaffolds fabricated from nHAP/chitosan/gelatin have also been shown, in a swine model, to effectively enhance mandibular bone regeneration [65].

Surgical fusion of the spine, involving the merger of adjacent vertebrae using bone grafts, is a common modality for treating a variety of spinal ailments. Researchers have explored utilizing 3D-printed scaffolds composed of HAP, PLGA, and human demineralized bone matrix (DBM) in a rat model for posterolateral spinal fusion [66]. These scaffolds demonstrated remarkable fusion ratings as well as an increase in the expression of genes associated with osteogenesis, highlighting the significance of the scaffold's architecture and geometry. The most effective design for osteointegration and fusion was associated with the larger pore size and struts aligned at 45° [66]. Another study indicated that PLGA/β-TCP composite scaffolds with salvianolic acid B successfully enhanced bone fusion by promoting angiogenesis and osteogenesis [67]. The field has also delved into metallic biomaterials, as evidenced by a study where 3D-printed interconnected titanium alloy scaffolds enriched with HAP were used in a sheep model. These scaffolds showed enhanced osteogenic capabilities compared to those without HAP [68].

# 3 Process of Creating Customized Additive-Manufactured Bone Scaffolds

While additive-manufactured (AM) bone scaffolds offer substantial benefits in treating bone defects, their production involves a complicated, multi-faceted process that requires collaborative efforts. The necessity for each scaffold to be specifically adapted to meet the individual needs of patients underscores the importance of customization at every stage of the process. In this section of the chapter, the four principal components of the design and manufacturing process, *(i) imaging of bone defects, (ii) selection of materials, (iii) design of the scaffold, and (iv) methodology of fabrication*, will be explored and the respective significance of customization at each stage of the process [1].

## *3.1 Imaging of Bone Defects for Customization*

The first step in creating a patient-defect-specific scaffold is to acquire imaging of the patient's bone defect. Accurate imaging is crucial for achieving excellent integration at the interface between the tissue and scaffold, adapting mechanical needs, and tailoring functions of the scaffold. Such accuracy is vital when treating defects caused by tumors or trauma, which often have unique shapes. By analyzing in-depth medical imaging, clinicians and engineers can assess the anatomical features of the defect and create a scaffold that meets the unique specifications [1].

### 3.1.1 Overview of the Imaging Procedure

For obtaining of 3D imaging data, magnetic resonance imaging (MRI) and computed tomography (CT) are two of the most frequently utilized processes due to their capability of obtaining isotropic or near-isotropic data sets [69]. MRI has the advantage of no risk of radiation exposure and detailed visualization of soft tissue structures. However, artifacts resulting from patient movement can hinder its effectiveness in acquiring thin slices along the z-axis [70, 71].

Within the field of orthopedics, CT scans, noted for their high-contrast quality, are primarily used for image post-processing, often in conjunction with MRI data. An advanced variant, multirow detector computer tomography (MDCT), can capture exceptionally thin slices under 1 mm, significantly improving the process of 3D printing image post-processing [72–74]. A frequently utilized technique, cone beam computed tomography (CBCT), is remarkably effective in the oral and maxillofacial regions [75], offering lower radiation exposure and detailed 3D volumetric data in axial, sagittal, and coronal planes [76]. Irrespective of the imaging modality, the resulting data are stored as Digital Imaging and Communications in Medicine (DICOM) files, which are used for further processing.

The processing of DICOM files involves utilizing post-processing software to extract 3D projections from selected regions of interest. This involves using a thresholding voxel intensity value to identify bone. The resulting contours of these projections are converted using computer-aided design (CAD) software. The CAD data are saved in the Stereo Lithography (STL) format, a widely recognized standard for 3D files [77]. Engineers can modify the STL file to construct a personalized 3D-printed scaffold using 3D printers (Fig. 3).

Recent advancements in bone regeneration involve the transformation of 2D CT scans or X-rays into 3D-printed devices (e.g., scaffolds), capitalizing on the role of imaging for scaffold customization. These results are achieved using algorithms to convert 2D imaging into precise, modular 3D structures. This methodology allows for the direct creation of patient-specific scaffolds from conventional medical imaging, providing a streamlined and accurate solution for addressing bone defects, particularly those with complex geometries resulting from trauma or tumors. This innovation underscores the evolving integration between digital imaging and additive manufacturing, providing a more accessible pathway to customized bone repair solutions [78].

### 3.1.2 Imaging Techniques for Orthopedic Hardware

The presence of orthopedic hardware, in particular metallic implants, can negatively impact the clarity of CT and MRI scans. However, imaging with these devices is

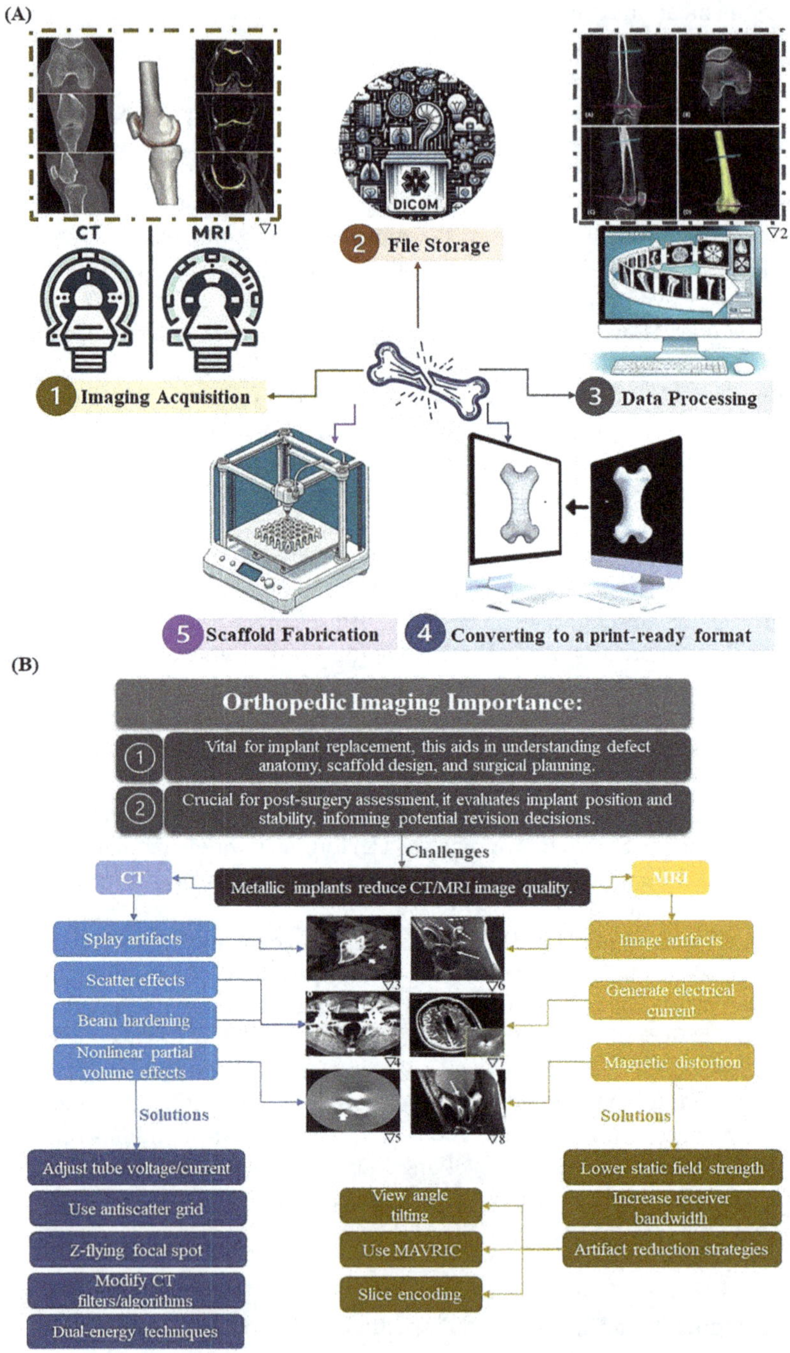


**Fig. 3** (**a**) Overview of the imaging procedure for creating customized AM bone scaffolds and (**b**) imaging techniques for orthopedic hardware. (Images modified from references: ▽1 [89], ▽2 [90], ▽3 [79], ▽4 [91], ▽5 [92], ▽6 [93], ▽7 [94], ▽8 [95])

often necessary, particularly in replacing dysfunctional implants. In these situations, imaging offers important information about the defect's anatomy, the scaffold's design, and the surgical approach. Additionally, imaging following surgery is important to assess the placement and integration of the implant, aiding medical professionals in deciding if additional adjustments are necessary [1].

The quality of CT scans can be compromised by the presence of metallic hardware, leading to issues such as beam hardening, splay artifacts, scatter effects, and nonlinear partial volume effects. The specific alloy composition, geometry, and dimensions of the implants further influence the extent of potential artifacts [79]. Several strategies can and have been employed to address these effects, each tailored to mitigate the specific challenges encountered. Adjusting the voltage and current can help minimize beam hardening, but this must be carefully balanced with the need to limit patient exposure to radiation and potential reduction in soft tissue contrast [80]. Splay artifacts may be reduced by applying a z-flying focal spot, and scatter artifacts can be decreased using an anti-scatter grid [81]. Other methodologies such as modifying CT reconstruction filters, algorithms [82, 83], and using dual-energy techniques [84] have been explored to elude artifacts.

When using MRI in the presence of metallic implants, an electrical current can be created, leading to magnetic distortion and artifacts in the images. The severity of these artifacts is contingent on the device's magnetic susceptibility [79]. Strategies to diminish metal-related artifacts in MRI have included opting for a lower static field strength [85], enhancing the receiver bandwidth [86], and implementing metal artifact reduction sequences. Techniques like view angle tilting, multi-acquisition variable-resonance image combination (MAVRIC), and slice encoding for metal artifact correction effectively mitigate these issues [87, 88].

## 3.2 *Selecting Appropriate Materials for Customized AM of Bone Scaffolds*

### 3.2.1 Traditional Materials for Printing Processes

Materials used to fabricate bone scaffolds should possess specific biological characteristics, such as biocompatibility, osteoconductivity, osteoinductivity, mechanical strength, and suitability for processing. These materials are generally divided into four categories: metals, polymers, ceramics, and composites. Additionally, integrating cells and drugs into the printing materials can confer distinct functions (Fig. 4). Selecting the most ideal materials for customized scaffolds depends on various aspects, such as the specific location and physiological characteristics of the bone defect, the intended role of the scaffold (esthetic or functional), its projected longevity and biodegradation (resorption) rate, as well as any specialized functionalities (i.e., antibacterial properties or osteogenesis promotion). The following sections will concisely overview the conventional materials employed in custom scaffold fabrication [1].

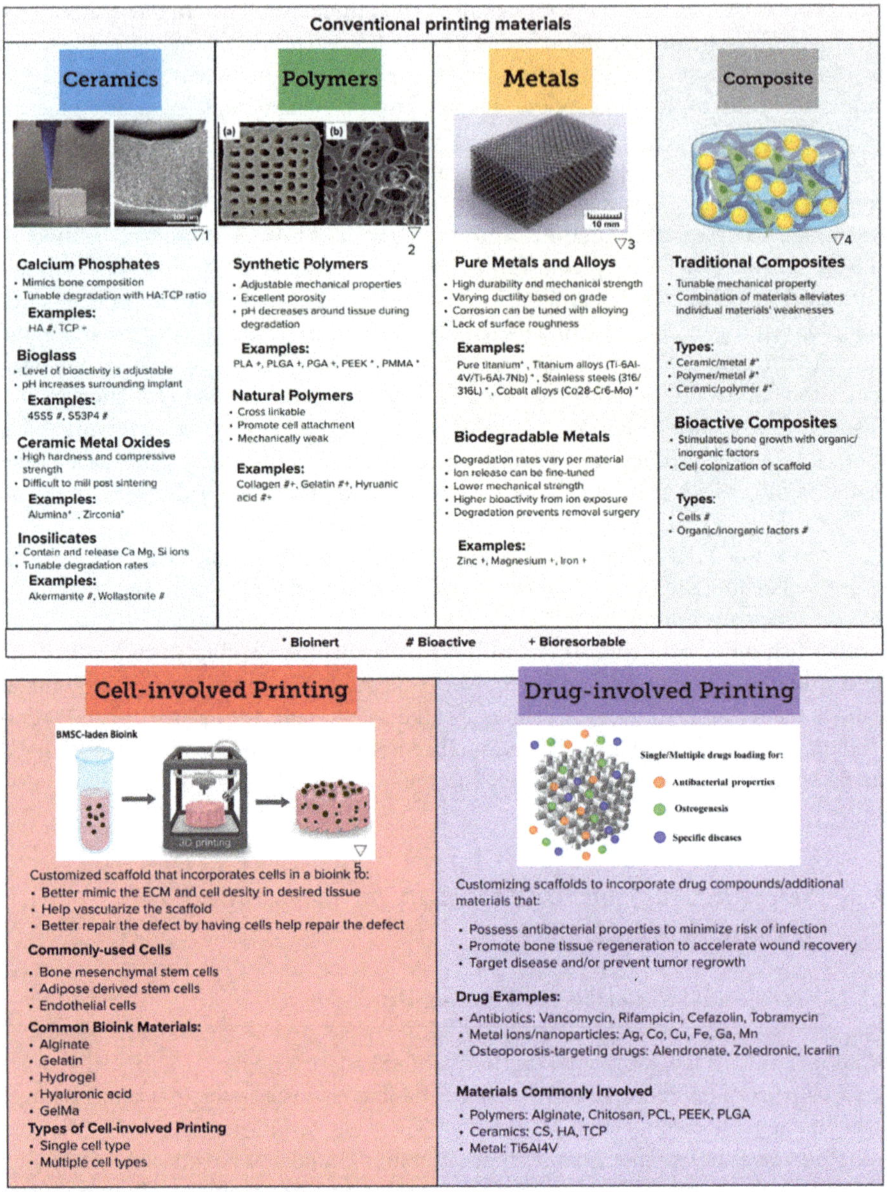


**Fig. 4** Materials selection framework for customized AM scaffolds: (**a**) various application environments; (**b**) characteristics of various material types and their corresponding application scenarios. (Images modified from references: [1], ∇1 [96], ∇2 [97], ∇3 [98], ∇4 [99], ∇5 [100])

### Ceramic Materials

For additive manufacturing of ceramic materials, there are two primary classes of ceramics: bioactive, which interacts with biological systems; and bioinert, which remains neutral in its environments. Ceramics are commonly known for their superior biocompatibility, and their elemental compositions having the capacity to be tailored to attain desired properties. In addition, the ceramics provide resistance to compression and corrosion, but their natural brittleness results in inferior fracture strength relative to other materials [101].

I. Bioactive Ceramic Materials

Bioactive ceramics, known for their biodegradability and ability to promote osteoconductivity and osteoinductivity, include materials such as tricalcium phosphate (TCP) [102, 103], calcium sulfate (CS) [104, 105], hydroxyapatite (HAP) [106, 107], akermanite [108], diopside [109, 110], and bioglass [111, 112].

TCP and CS are bioresorbable materials commonly used to treat bone defects, especially in load-bearing areas, because their resorption rates match the pace of bone healing. Both materials are available in two distinct phases, α and β, with varying degradation rates, notably quicker than Hydroxyapatite (HAP). β-TCP is often mixed with HAP to create biphasic TCP, facilitating a more controlled degradation rate. On the other hand, α-TCP, with its lower density and stability, is better suited for cement-like applications [113, 114]. Combining CS with either β-TCP or HAP in bone structures results in improved stability and resistance to pressure. This is attributed to the development of calcium-deficient hydroxyapatite (CDHAP) and dihydrate CS crystal formations, a combination extensively studied for its effectiveness in healing osteogenic defects [115]. Studies on varying mixtures of CS/HAP and CS/β-TCP powders for 3D printed scaffolds indicate that a ratio of 25:75 wt.% coarse HAP to CS powder mixture is ideal for achieving superior compressive strength, enhanced wettability, and ideal pore sizes for cellular adherence. Introducing dopants, such as specific metal ions or compounds, into these ceramic materials has significantly improved their physical and biological properties, such as osteoinduction [116]. Research on β-TCP scaffolds doped with Fe-Si has demonstrated a notable increase in mechanical strength, improved mineralization, and augmented angiogenesis relative to 100% β-TCP scaffolds. β-TCP scaffolds doped with $SiO_2$ and ZnO have also shown favorable results in terms of densification, degradation rates, sustained stability, and the formation of interconnected bone tissue [117, 118]. Ongoing research into the specific surface geometries and porosities of TCP and CS-based scaffolds has yielded promising findings regarding their biocompatibility and osteogenic effectiveness [103, 119, 120].

Hydroxyapatite (HAP), a naturally occurring bioactive ceramic in bone tissue, is employed in AM to fabricate bone void-filling scaffolds. Recent investigations have revealed that scaffolds composed entirely of HAP are exceptionally biocompatible and adept at replicating the complex, interconnected pore structures and topography found in natural bone matrixes [107, 121]. For instance, one study utilized digital light processing (DLP) technology with 20 nm HAP powder and a photopolymer to produce HAP scaffolds. The resulting post-sintered scaffolds effectively supported the growth and adhesion of MC3T3-1 cells. Despite these advantages, HAP's inherent brittleness and limited resorption and degradation capabilities could pose challenges in clinical use [122]. Such limitations may increase the likelihood of fractures near the site of the implant and might require the use of permanent implants rather than enabling the full replacement of the bone tissue [114]. However, these issues are often addressed by combining HAP with other composite materials, thus improving its overall suitability for medical applications.

Materials like akermanite, diopside, and wollastonite, which belong to the inosilicate category, are noted for their significant bioactivity. This is mainly due to their ability to release ions such as Ca, Mg, or Si, enhancing mineralization, osteogenesis, and angiogenesis. For instance, akermanite frameworks developed using direct-ink writing and augmented with 15% or 30% bioglass have increased compressive strength. This enhancement is vital for lowering melting temperatures and reducing the need for pressurization and sintering while preserving the porous structure. These modified frameworks have been observed to encourage angiogenic stimulation, foster bone regeneration, as well as maintain long-term stability and osteogenic adhesion, especially in areas of spinal fusion [108, 123]. Likewise, diopside, when amalgamated with bioglass, has been found to possess superior compressive strength and angiogenic properties, making it suitable for orbital implant applications. Concurrently, using DLP, studies have experimented with different ratios of diopside and wollastonite to fabricate durable, high-resolution scaffolds with extensive porosity and a diamond lattice configuration. These developments signal a promising trajectory for future innovations in this domain [110, 124]. Similarly, the integration of 15% akermanite into β-tricalcium phosphate scaffolds, fabricated via DLP, not only enhances mechanical strength but also significantly improves bioactivity and cell response. This advancement underscores the scaffold's ability to support enhanced cell adhesion, proliferation, and osteogenic differentiation, marking a promising direction for applications in bone repair and regeneration [125].

Bioactive glasses, composed of elements naturally found in the body, such as $SiO_2$, CaO, $Na_2O$, and $P_2O_5$, are ideal for use in bio-ceramic scaffolds. These materials feature a dense surface with a negative charge, which is effective in attracting serum proteins. Additionally, the elevated pH levels near the implant areas confer antimicrobial advantages [126, 127]. In specific applications, bioglass enriched with 2% $Fe_2O_3$ has been utilized to create scaffolds for replacing alveolar bone, with results indicating strong biocompatibility and increased activity of osteogenic markers [128].

Bioactive ceramic-based scaffolds are highly effective for bone regeneration due to favorable osteogenesis and their capacity to resorb. Such scaffolds facilitate the body's natural bone repair processes, ultimately aiding in the recovery of movement and the ability to bear weight. Adjustments in the material's properties enable the tuning of both degradation rates and mechanical strength, making these ceramics versatile for diverse uses, from load-bearing to nonload-bearing applications [1].

II. Bioinert Ceramic Materials

Bioinert ceramics, specifically alumina and zirconia, are widely used materials in manufacturing nonresorbable implants via AM. Alumina, historically one of the first bioceramics to be clinically adopted, is notable for its minimal wear, long-term stability, and high compressive strength. Studies have demonstrated that alumina surfaces structured with micropillars can positively influence osteogenic responses of human mesenchymal stem cells (hMSCs) [129]. Although alumina's strong mechanical characteristics are ideal for applications requiring load-bearing, its inherent brittleness is a limiting factor. Moreover, alumina often requires the addition of bioactive surface coatings or biomolecules to prevent its bioinert nature and enhance its compatibility with the biological environment [130].

Zirconia, a bioinert metal oxide ceramic type, is extensively employed in fabricating hip and dental implants. Its structural stability is notably enhanced when blended with certain oxides, such as MgO and CaO. In particular, the addition of 3% mol yttria (3Y-TZP) is prevalent, resulting in the production of tetragonal zirconia polycrystal (TZP) [131]. Advancements in material jetting systems for the 3D printing of implants have significantly accelerated the production process, enabling the rapid creation of ceramics that possess properties characteristic of the TZP range [132]. Furthermore, comparative studies between additive and subtractive manufacturing methods for materials like 3Y-TZP and alumina-toughened zirconia (ATZ) have shown minimal differences in their physical properties [131].

Bioinert ceramics are known for their high mechanical strength and minimal wear, making them well-suited for long-lasting scaffold applications, particularly in cases where natural bone healing is not possible. However, to avoid the issue of stress shielding, it is crucial for these scaffolds to closely emulate the mechanical properties of the bone tissue they are replacing [1].

### Polymer Materials

Biodegradable polymers are a widely utilized class of materials in applications such as bone and dental fixation, and in certain scenarios in tissue engineering. These materials need to be biocompatible to prevent any adverse autoimmune responses or biological harm. Polymers can be further divided into synthetic and natural types [133].

In the 3D printing domain, a variety of synthetic polymers are prominently used, such as poly (lactic acid) (PLA), poly (glycolic acid) (PGA), polylactic-co-glycolic acid (PLGA), polyurethane (PU), and polycaprolactone (PCL) [133, 134]. On the other hand, natural polymers, which are categorized into proteins and polysaccharides, are known to enhance cell adhesion and function [135]. Natural polymers that are regularly employed include collagen, gelatin, silk fibroin, chitosan, alginate, and hyaluronic acid.

Although clinical trials focusing on orthopedic uses are still in their early stages, both in vitro and in vivo studies have shown promising results. For example, a study utilized a PCL scaffold to preserve the height of the alveolar ridge following tooth extraction, which facilitated bone healing and improved maintenance of ridge height [136]. Whereas, an independent investigation compared the effectiveness of polymethyl methacrylate (PMMA) in overdentures, concluding that 3D-printed versions yielded more favorable outcomes for patients [137].

In the field of orthopedics, the application of 3D-printed soft polymers is slightly restricted, primarily due to mechanical properties. Currently, 3D-printed polymeric scaffolds are primarily limited to nonload-bearing areas and minor defects, as their mechanical strength is lower than that of bone tissues. Synthetic polymers are particularly suited for nonload-bearing (e.g., cranial region). Although, natural polymers, which form a part of the extracellular matrix (ECM) of bone tissue, can potentially mend minor load-bearing defects [1].

### Metal Materials

#### I. Bioinert Metal Materials

With respect to bone scaffold development, metallic implants, primarily produced using additive manufacturing (AM), provide durable and strong solutions. Metals (i.e., stainless steel, cobalt-chromium, and titanium alloys) are used in about 70% of these implants due to their superior mechanical strength and ability to resist fractures [138].

Titanium-based alloys, such as Ti-6Al-4V and Ti-6Al-7Nb, are known for their exceptional durability and mechanical strength. Nonetheless, an issue with certain alloys is the potential release of harmful substances due to corrosion, which can trigger immune reactions and result in chronic inflammation [139]. To address this concern, there is an increasing focus on alloys incorporating elements such as Mo, Nb, Ta, and Zr. These are gaining popularity due to their diminished corrosion and reduced stress shielding effects, thereby improving biocompatibility [140].

The inherent bioinert characteristics of solid metal implants often require the addition of bioactive coatings to improve their integration with adjacent tissue. A popular choice for coatings has been Hydroxyapatite (HAP), known to aid in bone tissue regeneration and osteointegration [141]. In addition, advances in AM now allow for precise control over the porosity of metals, which is believed to promote cell growth and adhesion [142–144]. For example, in research conducted by Taniguchi et al., selective laser melting was used to determine the most effective pore size in Ti and $TiO_2$ scaffolds. This preclinical, rabbit model, study concluded that a 600-um pore size is ideal for maintaining compressive strength, fostering osteoconductivity, and achieving osseointegration [145]. Moreover, recent advancements have brought forward the utilization of alloys like W4-Mg and Fe-Mn in fabricating metallic scaffolds.

#### II. Biodegradable Metals

Biodegradable materials offer advantages over permanent metal alloys for bone tissue regeneration aimed at reinstating function. These materials are recognized for promoting osteogenic and angiogenic activities through ion release. In scenarios where the materials are used in load-bearing applications, the differences in their degradation rates and mechanical properties need to be taken into consideration. Modifying these materials with substances that resist degradation and applying specific post-treatment techniques could be essential to manage their degradation kinetics appropriately [1].

Biodegradable metals serve as durable support for bone tissue throughout the healing process and gradually degrade, negating the necessity for further surgical procedures [146]. Their stiffness is significantly higher than scaffolds made from polymers, making them better suited for load-bearing applications [145].

Zn, Fe, and Mg are common, biodegradable, metals incorporated into scaffolds, each offering varied degradation rates. Specifically, Mg alloys have been developed to produce scaffolds that are not only degradable and biocompatible but also feature an open-pore structure that supports bone growth [147]. Due to their porosity, magnesium alloys exhibit an increased surface area, which can lead to quicker degradation. Studies on Mg alloys containing 4 wt.% Y have shown they degrade more slowly than pure magnesium scaffolds while maintaining biocompatibility [148]. Comparative in vivo research on 3D-printed scaffolds made from Ti, Mg, and Zn has further emphasized the viability of these materials [149].

### Composite Materials

Composite materials, encompassing polymer/metal, ceramic/polymer, and metal/ceramic blends, exhibit various characteristics, including bioactivity, degradation features, and mechanical strength. For example, the mixture of calcium phosphate with polyester has shown favorable results in promoting osteogenic differentiation and cellular proliferation [147]. Moreover, the integration of polycaprolactone (PCL) with calcium sulfate and pearl powder has been demonstrated to facilitate bone repair in a rabbit study [150]. Additionally, recent advancements have introduced 3D-printed scaffolds combining PLA, TCP, and CeNPs, offering superior osteogenic properties for effective bone regeneration. This breakthrough underscores the transformative role of nanocomposites in advancing tissue engineering [151].

Composites made from polymer and metal are noted for their enhanced structural strength, a benefit derived from the metallic components. For example, titanium/polymer composites have been shown to improve cell growth and adhesion [152, 153]. Alternatively, ceramic and metal blends, exemplified by magnesium-coated β-TCP, demonstrate significant levels of bioactivity and foster strong cell adhesion [154].

Sophisticated composites that incorporate molecular factors such as components of the extracellular matrix (ECM), inorganic minerals, and growth factors are being developed for custom bone regeneration implants [155]. Noteworthy instances include PLA composites augmented with carbohydrate particles and PEEK composites combined with calcium hydroxyapatite [156, 157]. Additionally, hydrogels augmented with minerals have yielded encouraging osteogenic outcomes [158, 159].

Materials such as demineralized bone matrices (DBM) and hydroxyapatite (HAP) are favored due to their rich content of growth factors and osteogenic agents [160]. Composite materials that include these substances have been associated with lower levels of inflammation and superior bone fusion outcomes [161, 162]. Additionally, printing implants with PLA combined with nano-sized HAP exhibit adjustable mechanical properties, indicating a high capacity for customization [163]. This adaptability in the properties of traditional ceramics, derived from such combinations, explains their broad application across numerous clinical conditions.

### 3.2.2 Cell-Involved Bioprinting

3D bioprinting marks a significant innovation in the production of custom bone implants, allowing the incorporation of living cells within the scaffold during the printing process. This modality provides a significant advantage by addressing the limitations of standard 3D-printing approaches, which often encounter difficulties in uniformly integrating cells after the object has been printed [164]. In the design process of a scaffold utilizing bioink, it is essential to evaluate the architecture of the tissue that is being substituted, select the proper cell type, and choose a bioink that is biocompatible and possesses ideal printing characteristics (e.g., suitable viscosity and sufficient mechanical durability) [165].

In 3D bioprinting, stem cells are commonly employed due to their versatility in differentiating into multiple cell types. The technique of extruding bioink, comprised of elements such as rat bone marrow stem cells (rBMSCs), nano silicate (nSi), gelatin, and alginate, is utilized to fabricate scaffolds that replicate bone extracellular matrix (ECM), promoting osteogenesis and addressing critical bone defects [166]. Further research has demonstrated the significance of silicon (Si)/silicate in improving stem cell survival within the bioink. This is achieved by incorporating silica and calcium hydroxide nanoparticles, which release Si ions into the hydrogel, thereby helping to maintain the stemness of mesenchymal stem cells (MSC) [167].

Advancements have led to the formulated bioinks devoid of nanoparticles by mesenchymal stem cells (MSCs) with compounds like fibrinogen, type A gelatin, hyaluronic acid, and glycerol. Such frameworks have the potential to evolve into bone, exhibiting preliminary hypertrophic properties that enhance vascularization and bone generation [168]. The selection of materials and the addition of particular drugs play a crucial role in guiding the differentiation and migration of MSCs. For example, scaffolds printed with nifedipine-loaded ethosome and laponite have been observed to support bone healing by stimulating osteogenic differentiation and encouraging the migration of bone marrow stem cells (BMSCs) [169]. In urgent care settings, the advancement of in situ printing bioinks is particularly significant, allowing immediate on-site tissue repair. These bioinks, when integrated with bone cement, can be shaped manually and 3D printed onto the patient's wound for swift tissue repair [170]. However, the effectiveness of mesenchymal stem cells (MSCs) in this context is restricted to repairing more small defects, primarily due to challenges related to achieving sufficient vascularization post implantation [171]. To address this issue, the scientific community has explored 3D printing techniques that involve a mixture of cell types, such as MSCs alongside endothelial cells, aiming to promote vascularization. Innovative approaches, including dual ink systems and interlaced printing methods, have been developed to produce scaffolds with pre-vascularized channels or to ensure a balanced cell distribution throughout the scaffold [172–174].

Incorporating cell spheroids in the bioprinting process can markedly augment intercellular communication [175]. This is exemplified when stem cells and human endothelial cells are amalgamated with mineral oil, leading to the formation of cell spheroids through oil-water interaction dynamics. Embedding these spheroids within supports composed of decellularized ECM and β-tricalcium phosphate yields a vascularized bone scaffold characterized by pronounced osteogenesis and angiogenesis.

Further illustrating the cutting-edge advancements in this field, a recent study introduces a 3D-printed hydroxyapatite scaffold functionalized with stem cell-derived small extracellular vesicles (sEVs). By employing fusion peptides for targeted sEV delivery, the scaffold demonstrated improved osteogenic activity, biocompatibility, and effective bone repair in rat models, marking a significant step forward in regenerative bone solutions [176].

Scaffolds that include living cells exhibit considerable promise for medical applications due to their bioactivity and customizable mechanical characteristics. Highlighting their capacity, a pivotal clinical case involved using a patient's bone marrow stromal cells within a 3D-printed scaffold for reconstructing a cleft alveolus, which showed encouraging results [177]. Yet, the presence of bioactive elements such as cells in scaffolds may restrict their application to smaller-scale defects. Despite such limitations, continued exploration in merging cells with bioinks is advancing toward customized therapeutic options for patients, demonstrating significant potential for future developments.

### 3.2.3 Drug-Involved Printing

Apart from their core uses in bone repair and regeneration, 3D-printed bone scaffolds offer the flexibility to be customized for individual patient needs. For instance, when an implant poses a substantial risk of infection, the scaffold can be engineered with anti-infective features to minimize bacterial infections specific to the patient's condition [1].

Incorporating anti-infective agents directly into the printing materials is one method to enhance scaffolds with antimicrobial properties. For instance, AM scaffolds made of tricalcium phosphate (TCP) and infused with vancomycin have been developed to allow precise control of drug release rates. This is achieved by altering the method of drug incorporation and integrating it with polymers [178].

Likewise, scaffolds composed of polycaprolactone (PCL) and enriched with the antibiotic rifampicin were produced via deposition printing at 60 °C, a technique that safeguards the antibiotic's efficacy against bacteria. Such innovations have proven effective in blocking the proliferation of pathogens such as *Escherichia coli* (*E. coli*) and *Staphylococcus aureus* (*S. aureus*) [179].

Aside from utilizing antibiotics, applying metal ions as anti-infective elements has garnered attention. For instance, PCL scaffolds that were combined with bioactive glass and gallium (Ga) exhibited antibacterial effects against methicillin-resistant Staphylococcus aureus (MRSA) and E. coli [180]. In a similar approach, silver nanoparticles (AgNps) were generated within a PCL solution via an in situ reduction process and subsequently used to create PCL/AgNps filaments for printing scaffolds [181]. Additionally, attaching anti-infective compounds to scaffold surfaces after printing offers another avenue for enhancing antimicrobial activity. For example, PLGA/HAP scaffolds were chemically modified to attach hydroxypropyl trimethyl ammonium chloride chitosan (HACC), showcasing the dual benefits of infection control and bone healing in an infected bone defect [40]. Furthermore, treating polymer scaffolds by immersing them in antibiotic solutions post manufacturing has also been employed to give them anti-infective characteristics [182, 183].

Individuals with osteoporosis are predisposed to a higher risk of bone fractures and defects, necessitating treatments that not only employ bone scaffolds but also include specific interventions for osteoporosis management. Osteoporosis-targeting medications such as alendronate, zoledronic, and icariin have been integrated into AM bone scaffolds created from materials like Ti6Al4V, PCL, and calcium phosphate [184–187]. In a notable research effort, freeze-dried platelet-rich plasma was utilized to enhance 3D-printed Ti6Al4V porous scaffolds to improve osseointegration in an osteoporotic animal model [188].

Following surgery to remove bone tumors, there is a dual need for bone regeneration and elimination of residual tumor cells. To address this, anti-cancer compounds such as doxorubicin hydrochloride and soy isoflavones have been incorporated into scaffolds fabricated from materials like TCP and a combination of PLGA/TCP [189, 190]. These specialized scaffolds are designed to facilitate bone healing and act against the activity or recurrence of tumors. Incorporating the drug at the time of printing may require a cryogenic setting to preserve the drug's efficacy [190]. Additionally, the use of photothermal effects from metal ions to attack tumors has been explored [191]. Metals, including Cu, Fe, Mn, and Co, were integrated into bioactive glass-ceramic (BGC) scaffolds. In vitro studies indicated that these scaffolds could effectively destroy tumor cells by triggering hyperthermia, leading to a significant decrease in tumor growth in vivo.

Outside of these specific instances, the needs of patients dealing with different conditions or unique health situations can be analyzed by medical professionals. Customizing scaffolds to align with the distinct requirements of each patient may substantially elevate the effectiveness of clinical interventions [1].

## 3.3 *Topographical Design for Customized AM Bone Scaffolds*

Once the material for the scaffold has been selected, establishing the topographical design of the AM bone scaffold becomes important. Imaging of bone defects provides essential information on the shape and dimensions of the scaffold's external contour. However, the configuration of its macro/microstructures is pivotal in enabling tissue integration, as well as in promoting the growth of blood vessels and nerve ingrowth [192].

### 3.3.1 Overall Impact of Parameters on Topographical Design

Various crucial aspects in the topography design, including porosity, the size of pores, and the overall pores configuration, have been the subject of in-depth research and critique [193–196], and the impacts of these elements are summarized in Table 1. However, when the customization of topography for a specific patient is undertaken, the foremost consideration is the site of implantation. This discussion delves into the considerations for designing topographies for trabecular and cortical bones, the two fundamental types of bone.

### 3.3.2 Designing Topography for Trabecular Bone Regeneration

Cortical bone significantly contributes to the overall mechanical strength of bone tissues, while trabecular bone facilitates a network for red bone marrow. Consequently, efforts in regenerative medicine for trabecular bone are primarily focused on replicating its porous nature. This bone layer exhibits a porosity between 50% and 90% and compressive strengths ranging from 4 to 12 MPa [40, 215]. Research predominantly utilizes a standardized topography for trabecular scaffolds, featuring repetitive cellular structures and uniform porosity [40, 41, 216, 217]. These scaffolds' porosity spans from 14% to 86%, with compressive strengths ranging from 10 to 200 MPa, underscoring that increased porosity generally leads to reduced compressive strength. Various scaffold architectures have been explored for their effects on the scaffolds' mechanical integrity and promoting bone tissue integration [202, 204, 205]. However, an optimal architectural standard remains undetermined, as it relies on various factors within clinical applications.

Mimicking the microstructure of natural trabecular bone represents an alternative approach to designing trabecular scaffolds, offering a solution by incorporating the patient's own bone structure into the design process. This approach was exemplified by utilizing the human femoral head, where micro-CT imaging and digital post-processing created a detailed template [197]. By overlaying the template multiple times, designers achieved porosities ranging from 15% to 70% and compressive moduli from 2579 to 3693 MPa. In vitro studies indicated good osteoblast viability and increased differentiation on these scaffolds. A similar method was employed

**Table 1** Factors in the design of scaffold topography [197–212]. (Images modified from references: [1], ∇1 [211], ∇2 [213], ∇3 [214])

| Porosity<br>the percentage of void volume in the whole scaffold | Value in literature | | |
|---|---|---|---|
| | materials | porosity | results |
| 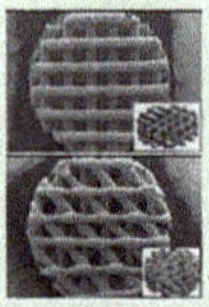 ∇1<br>**The porosity of one scaffold can determine its mechanical properties, which will influence the quality of new growth bones; High porosity can be beneficial for bone and blood vessel ingrowth.** | Ti-6Al-4V | 15-70% | The compressive moduli range 2579-3693 MPa, and the highest porosity leads to the most significant osteoblast differentiation and local factor production [197]. |
| | sodium alginate / gelatin / cellulose | 44-95% | Water-based cellulose nanocrystal scaffolds with gradient porosity have been developed [198]. |
| | Gelatin / carboxymethyl chitin / hydroxyapatite | Macroporosity and Microporosity, up to 90% | The introduction of bulk microporosity increases swelling and enhances the attachment, proliferation, and mineralization of human umbilical cord mesenchymal stem cells [199]. |
| | polycaprolactone (PCL) | Macroporosity of 60%, microporosity of 40% | Microscale porosity leads to a 3-fold increase in surface area and enhanced protein adsorption. The bone formation in the macro/micro pores is at a similar level [200]. |
| | α-TCP | 40% for the gradient scaffold, 51% for the constant architecture | The new bone volume in the scaffold with constant architecture compared with the gradient scaffold after seven months [201]. |
| | polylactic acid (PLA) | 30%, 50%, 70% | Scaffolds with lower porosity possess smaller pore sizes and better mechanical performance [202]. |

| Pore size<br>the diameter of the sphere pore; the side length of the square-shaped pore; in the case of an irregular shape pore, the average size will be used | Values in literature | | |
|---|---|---|---|
| | Materials | Pore size | Results |
| | Ti6Al4V | 400, 600, 800 μm | The pore size of 600 μm benefits the ingrowth of bone into pores [203]. |
| | PCL, PCL-HAP | 200, 500, 900 μm | The compressive moduli, yield stresses, and strains of scaffolds with different pore sizes have no significant difference [204]. |
|  ∇2<br>**The pore size plays a vital role in implant-induced osteogenesis since it can affect mass transfer, cell adhesion, and migration.** | Sr-HT-Gahnite | 390 – 590 μm | Authors suggest that the pore size in the range of 390 – 590 μm can enhance bone formation, and alternating large and small pores will increase the volume and functionality of new bone [205]. |
| | PLA | 500, 750, 1000 μm | Scaffolds with a pore size of 750 μm are suitable for osteoblasts' proliferation, metabolic activity, and the osteogenic differentiation of human mesenchymal stem cells [206]. |
| | PCL | 148-652 μm | The pore size dominates the mechanical properties of PCL scaffolds, while the strain rates are independent of pore size [207]. |
| | PCL | 200, 300, 400, 500, 600 μm | The time for cells to brighten a pore increases linearly with the overall pore size. The authors propose a reaction-diffusion model of tissue growth that qualitatively presents the tissue's evolution [208]. |

| Pore architecture<br>properties of the pore, including the pore shape, pore curvature, and pore gradient | Values in literature | | |
|---|---|---|---|
| | Materials | Architecture | Results |
| | pure/magnesium-doped calcium silicate | single-/double-layer printing | Single-layer printing scaffolds showed higher osteoconduction in the early stage, while the double-layer printing scaffolds had stronger osteogenic capacity in the late stage [209]. |
|  ∇3<br>**Different pore shapes present various responses to cell proliferation and migration; The pore curvature affects cell contraction and adhesion; The pore gradient makes the scaffold microstructure fit the native bone tissue.** | PCL/PCL-HAP | vertical uniform/gradient | All scaffolds have compressive moduli within the range of trabecular bone, and the compressive properties of gradient scaffolds are mainly affected by the weakest section of the gradient [204]. |
| | PCL | pores with a curvature control | The growth rate of tissue into pores depends on the local curvature of the substrate [208]. |
| | PLA | orthogonal/isometric design | The orthogonal design presents better mechanical properties [210]. |
| | Ti6Al4V | four scaffold designs with similar porosity and pore size | The diamond lattice unit is the best structure for the ingrowth of bone, contributing to the slight internal fluid velocity difference and the long fluid flow trajectory in the scaffold [211]. |
| | Mg-substituted wollastonite | Tailorable pore height | The pore height of 200 μm retards the bone tissue ingrowth, while 450 and 600 μm can facilitate the ingrowth of bone tissue and enhance bone regeneration and remodeling [212]. |

using templates from trabecular bone in the distal radius [218]. This research probed how various 3D printing settings, such as resolution, boundary conditions, support materials, STL mesh refinement, and duplication parameters, affect the scaffolds' mechanical strengths. This meticulous control over the printing process ensures the preservation of the bone's microarchitecture, optimizing both its biological function and structural support. The trabecular bone's intricate structure was further refined using Voronoi tessellation with design software, creating porous scaffold models [219]. Processing micro-CT scans of the L3 human vertebra into 2D Voronoi cells and 3D isotropic porous models facilitated the production of models for 3D printing, mirroring trabecular bone's histomorphometric traits. Critically, these scaffold models' mechanical and fluid dynamics properties could be predefined early in the Voronoi design stage, offering significant promise for clinical uses. The biocompatible polymer UHMWPE was also created into trabecular scaffolds employing the Voronoi technique [220], showcasing the method's versatility.

### 3.3.3 Designing Topography for Cortical Bone Regeneration

When designing scaffolds for regenerating cortical bone without mechanical reinforcement, the topographical approach mirrors that used for trabecular bones, focusing on porosity to facilitate bone tissue development and reconstruction within cortical areas. However, topographical adjustment is essential when these scaffolds are to be placed in areas subjected to significant mechanical stress. The primary objective is to engineer scaffolds that not only support porous frameworks to enhance bone regrowth but also match the mechanical strength inherent to cortical bone [1].

Improving the stiffness of scaffolds can be accomplished by minimizing the pore size to beam thickness ratio (PO:BT) [221]. The correlation between the PO:BT ratio and porosity changes across different topographical designs, allowing for selecting a lower PO:BT ratio while still achieving the required porosity. Additionally, incorporating more vertical beams within the scaffold design is another strategy to enhance stiffness [221]. Advanced architectural scaffold designs have also been introduced to elevate mechanical strengths without diminishing porosity levels. An example of this is glass-ceramic scaffolds that incorporate strontium-doped hardystonite grains, submicron gahnite clusters, and a glass phase within a hexagonal layout, achieving a compressive strength of 110 MPa, comparable to that of cortical bone [222]. This hexagonal structure provides superior compressive and flexural strength compared to other designs like curved, rectangular, and zigzag patterns.

Similar to techniques applied to trabecular bone, the concept of printing structures that mimic bone presents a promising path for developing customized bone scaffolds. An innovative approach detailed in one study introduced a method for producing scaffolds with a radial-gradient, bone-like structure via extrusion-based three-dimensional (EB-3D) printing. This technique, inspired by the intricate Koch snowflake fractal design, yields scaffolds with enhanced radial porosity and superior mechanical characteristics, surpassing the typical limitations of standard EB-3D printers in producing scaffolds with graded functionalities [223]. Further research demonstrates the production of scaffolds that simulate the Haversian system, utilizing digital laser processing (DLP) 3D printing technology. This process enables precise control over the scaffold's structural attributes, which influence its mechanical strength and porosity, thereby closely mimicking the natural structure of bone. These studies highlight the effectiveness of biomimicry in the development of structured, functional biomaterials, marking a significant advancement in the field of tissue regeneration [224].

Metallic materials are prime choices for load-bearing implants due to their mechanical properties and resistance to fatigue, allowing for more versatile topographical designs in scaffolds. Bone scaffolds created from nonbiodegradable metals, such as tantalum, titanium, and nitinol, are engineered with various structures to improve osteogenesis and osteointegration [211, 225, 226]. These metallic scaffold designs can be customized to provoke specific biological responses, with mechanical strengths aligned with that of natural cortical bone. In contrast, scaffolds made from bioresorbable metals, particularly magnesium alloys, have their mechanical and degradation behaviors significantly influenced by their topographical architecture [227, 228]. Opting for smaller pore sizes in these designs is advantageous for reducing hydrogen gas release and maintaining the scaffold's structural strength [229].

## *3.4 Material-Dependent Fabrication Method Selection for Custom AM Bone Scaffolds*

AM encompasses a wide array of techniques, many of which are utilized in creating orthopedic scaffolds. Each technique is based on distinct mechanisms, making them appropriate for different types of applications (Fig. 5). Selecting an appropriate AM technique for bone scaffold production requires the evaluation of multiple aspects, including the scaffold's material, accuracy, production time, surface quality, and requirements for post-processing [230]. When it comes to scaffolds for individual patients, the choice of manufacturing technique is significantly influenced by the type of material used in the scaffold. Consequently, this discussion delves into AM techniques according to their efficacy with different scaffold materials, emphasizing the essential aspect of customization.

### 3.4.1 Methods Used for the Fabrication of Scaffolds from Ceramic Materials

The scaffolds highlighted in this overview are primarily made from ceramic components, which are subjected to sintering to reinforce their mechanical strength.

I. SLS Method

Selective laser sintering (SLS) offers significant benefits, as it employs a technique where raw ceramic powder particles are heated to just below their melting point, facilitating the creation of scaffolds from ceramics such as hydroxyapatite (HAP) and β-tricalcium phosphate (β-TCP) through solid-state bonding [231]. For example, a composite scaffold of biphasic calcium phosphate/molybdenum disulfide

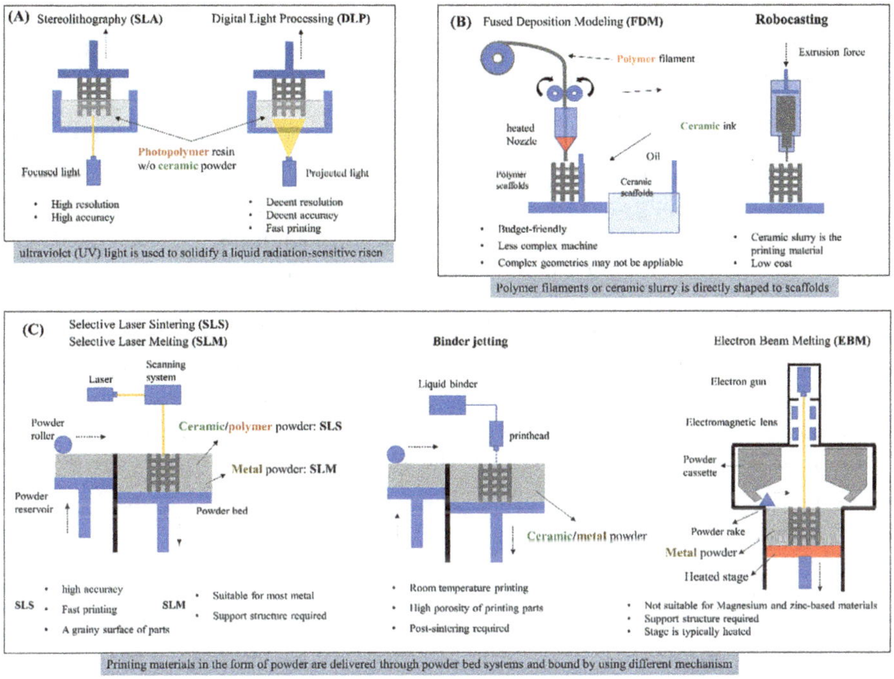


**Fig. 5** The various manufacturing techniques for personalized AM scaffolds. (**a**) Techniques utilizing UV light solidification (on the left: selective laser stereolithography (SLA), on the right: digital light processing (DLP)). (**b**) Techniques that employ direct deposition methods (on the left: fused deposition modeling (FDM), on the right: robocasting). (**c**) Techniques that operate based on powder bed fusion with distinct mechanisms for binding powder (on the left: selective laser sintering (SLS)/selective laser melting (SLM), in the middle: binder jetting, on the right: electron beam melting (EBM)) [1]

reinforced with polyamide12, created through SLS, shows enhanced mechanical strength, biocompatibility, and osteoconductivity [232]. The process's flexibility is highlighted by the ability to alter its characteristics, including laser power and scanning speed, thereby adjusting the scaffold's chemical composition and microstructural properties [233, 234]. This level of customization allows for the precise engineering of scaffolds to meet specific functional requirements. Moreover, SLS streamlines the scaffold fabrication process by removing the necessity for a subsequent sintering step, making it an efficient option for rapidly producing customized ceramic scaffolds.

II. SLA/DLP Methods

For fabricating ceramic scaffolds, stereolithography (SLA) and digital light processing (DLP) present effective methods, utilizing a process that hardens a mixture of photopolymers and ceramic materials under light exposure. Subsequent to this light-induced solidification, a sintering stage is applied to integrate the ceramic components fully, allowing for a high degree of customization. The primary distinction between SLA and DLP resides in their performance trade-offs: SLA is characterized by its superior resolution, albeit at a slower fabrication rate; in contrast, DLP achieves quicker production times with a precise compromise.

The adaptability of SLA and DLP in scaffold fabrication is showcased through the wide range of printing suspensions (e.g., resins) that can be utilized. These suspensions, comprising various ceramic materials along with additional substances like dispersants and defoaming agents, significantly improve the biological and mechanical characteristics of the scaffolds [235, 236]. The incorporation of various additives into the ceramic mixture throughout the SLA/DLP processes further allows for an enhanced level of property customization within the scaffolds [236].

The fabrication of ceramic scaffolds through SLA and DLP is characterized by a two-phase sintering method. Initially, a lower-temperature sintering step is applied to eliminate photopolymers, followed by a higher-temperature stage designed to enhance the mechanical strength of the scaffolds. The meticulous control of sintering temperatures is important for preserving the material's structural phase [237, 238]. Innovations in formulating aqueous suspensions have also emerged, addressing the issues of excessive viscosity and the environmental impact associated with conventional suspensions [239].

In summary, SLA and DLP stand out for their capability to customize ceramic scaffold characteristics precisely. This is achieved through the flexible incorporation of materials, management of resolution and fabrication speed, and maintenance of the material phase via customized sintering strategies.

III. Binder Jetting Method

Distinct from SLA or DLP techniques, binder jetting employs a method that directly binds ceramic powders during the printing stage, bypassing the use of a polymer-based matrix. An essential preliminary action is the grinding of ceramic powders to control particle size, which influences the printing and sintering phases, thereby determining the final density and strength of the scaffold [240]. Furthermore, various additives may be introduced during the grinding process to enhance the scaffold's characteristics [117].

Various ceramics, including TCP, calcium sulfate, and HAP, have been adeptly molded into bone scaffolds using this method [102, 103, 117, 241, 242]. Similar to SLA and DLP technologies, the binders applied in this process need to be extracted following completion. The extraction method, however, varies across materials; TCP scaffolds, for example, demand defined curing and sintering temperatures for binder removal [102, 117], in contrast to HAP scaffolds, where a water-based binder necessitates only a drying step is required for its elimination [242].

A particular investigation has enhanced the binder jetting technique for ceramic scaffolds by evaluating factors such as the direction of build, layer thickness, and the level of binder saturation. Of these variables, the direction in which the scaffold is built was identified as the most critical [243]. Consequently, binder jetting stands out as a highly adaptable and effective strategy for customizing ceramic scaffolds.

IV. Robocasting Method

Robocasting introduces a novel methodology for producing ceramic scaffolds, relying on the preliminary formulation of printing ink (e.g., colloidal gel) designed with optimal viscosity for extrusion through a nozzle and the ability to maintain its shape once deposited. This process is intricately linked to the composition of the ceramic printing material, with adjustments likely needed for variations in composition or the introduction of additional additives. A variety of formulations have been explored, such as 45S5 bioactive glass combined with water-soluble carboxymethyl cellulose, a mix of HAP/β-TCP with Pluronic F-127 solution, and a concoction of zirconia/alumina powder with a water-based PF127 solution [244–248]. Robocasting is generally conducted within an oil bath to promote uniform drying. Following the precise deposition of a self-supporting structure, the scaffold is then sintered and goes through any necessary post-processing stages.

### 3.4.2 Methods Used for the Fabrication of Scaffolds from Polymer Materials

I. SLA/DLP Methods

Fabricating polymer-based scaffolds via SLA and DLP is associated with specific challenges, with the most significant being the necessity for a photo-cross-linkable primary polymer that possesses either biocompatibility or biodegradability. This requirement is paramount because, in contrast to ceramic scaffolds, photopolymers are meant to be part of the scaffold's final composition, making their compatibility with biological tissues crucial. Various resins have been effectively employed for bone scaffold fabrication, including poly (trimethylene carbonate) (PTMC) enriched with HAP particles, commercial-grade polyurethane, and poly (propylene fumarate) (PPF) with different molecular weights [249–251]. Furthermore, polyethylene glycol diacrylate (PEGDA) has been combined with decellularized tendon extracellular matrix to serve as a bioink in scaffold production using DLP [252]. Identifying an appropriate photopolymer resin composition represents the main challenge in applying SLA/DLP for scaffold creation. However, once the resin formulation is fine-tuned, the printing process adheres to the standard protocols of SLA/DLP operations.

II. SLS Method

SLS is capable of melting a wide range of polymers for fabrication purposes. However, challenges such as dimensional inaccuracies resulting from excessive bonding of powder, suboptimal density due to partial melting, and inadequate layer adhesion may compromise the integrity of the polymer scaffold [253]. Therefore, the parameters of the printing process play a crucial role in determining the quality of the scaffold. Incorporating ceramic materials into polymer-based SLS scaffolds is a common strategy to improve their bioactivity and mechanical strength.

Numerous combinations of polymers and ceramics, such as PCL/HAP [238], PCL/TCP [254], PCL with HAP [255], polylactide alongside calcium carbonate [256], and PLLA with bioactive glass [257], have been utilized in scaffolds created through SLS techniques. The optimization of sintering parameters in these studies is generally based on the specific materials being printed and the ratio of polymer to ceramics involved.

In a separate study, the influence of various selective laser sintering (SLS) operational parameters, such as the power of the laser, the offset of the beam, and the diameter of the laser beam, on the precision of dimensions and the stiffness of PCL scaffolds was explored [258]. The study identified a significant link between the dimensions of the molten sections inside the scaffold struts, the overall diameter of these struts, and the specific SLS settings employed.

III. FDM Method

Fused deposition modeling (FDM) stands out as a prominent technique in fabricating polymer scaffolds, praised for its precision in adjusting scaffold porosity and structure. The main obstacle is the development of specialized filaments, especially those that embed ceramics within polymer matrices.

Various polymer-ceramic combinations, including PLA with either β-TCP or HAP, have been utilized to create bone scaffolds [259]. The mechanical characteristics and the microstructural details of these scaffolds are significantly influenced by the composition of the filaments, particularly when augmented with ceramic particles.

Specific processing parameters are required by different filaments, which encompass aspects such as the speed of extrusion, applied pressure, and operational temperature [259]. Due to the directional characteristics of heat transfer, the material's ability to conduct heat can influence the accuracy of the produced structures [260]. Moreover, the nozzle diameter in FDM is critical, requiring a close match with the filament diameter to create precise structures [261].

### 3.4.3 Methods Used for the Fabrication of Scaffolds from Metal Materials

I. Selective Laser Melting (SLM) and Electron Beam Melting (EBM)

Selective laser melting (SLM) and electron beam melting (EBM) techniques process metal powders through powder bed fusion (PBF) methods. Through these techniques, objects are constructed layer by layer by fusing powder particles using either a laser in SLM or an electron beam in EBM. SLM utilizes a laser to fully melt the powder, whereas EBM operates with an electron beam in a vacuum environment.

Metal scaffolds are commonly fabricated using bio-inert materials like stainless titanium (Ti), steel (SS), cobalt chromium (CoCr), and tantalum (Ta), as well as biodegradable metals such as magnesium (Mg), zinc (Zn), and iron (Fe).

Although stainless steel is not often the first choice, it has been employed in creating highly porous scaffolds using selective laser melting (SLM) [262]. The focus of numerous studies is on pure titanium (Ti) and Ti-based alloys, appreciated for their bone-like properties [263–265]. Particularly, the Ti-6Al-4V alloy exhibits differing levels of corrosion resistance, depending on its manufacturing process—SLM versus EBM, with EBM showing superior outcomes [263]. The orientation of scaffold construction during fabrication appears to have a minimal effect on osseointegration. Nonetheless, the porosity level, heat treatments, and specific laser adjustments during SLM influence the microstructure and fatigue resistance of the scaffold [266–269].

SLM has been applied to fabricate scaffolds from Ti-TaNb-Zr alloy and pure tantalum (Ta), while cobalt chromium (CoCr) scaffolds have been created using EBM, to promote bone growth [270–272]. Despite facing issues such as low melting points and a tendency toward oxidation, biodegradable metals such as zinc (Zn) and magnesium (Mg) have been effectively processed with SLM. For example, studies on Mg scaffolds have revealed that their biodegradation rates can be modulated via changes in their topological designs [227]. Further investigations have successfully fabricated scaffolds from WE43 magnesium alloy and zinc, exhibiting notable relative density and mechanical strength achievements [149, 273–275]. Additionally, scaffolds made of iron and iron-manganese, created through SLM, have shown varied degradation rates [276].

II. Binder Jetting Method

Metal scaffolds can also be fabricated using binder jetting techniques, which involve the adhesion of metal particles in a powder bed, followed by a sintering process.

Scaffolds composed of Fe–30Mn and Fe-Mn-Ca/Mg alloys were developed using this approach. The process incorporated a water-based organic binder, and the components underwent a dual-stage post-processing treatment. Initially, the components were cured at a lower temperature range (200–230 °C) for 2–3 hours to eliminate the binder and reinforce the structure. Subsequent annealing occurred under a protective atmosphere at elevated temperatures (1200 °C). The resulting 3D-printed Fe-30Mn and Fe-Mn-1Ca scaffolds exhibited open porosity percentages of 36.3% and 52.9%, respectively [277, 278].

In a research effort, binder jetting technology was utilized to fabricate Stainless Steel 316 scaffolds, incorporating various lattice geometries [279]. This approach mirrored the printing and sintering techniques used for the previously mentioned Fe-based scaffolds.

Following the annealing process, a size reduction was observed across all metal components, with the Fe-Mn samples experiencing higher shrinkage and reduced porosity than the Fe-Mn-1Ca samples [278]. However, all the scaffolds produced demonstrated acceptable levels of porosity.

# 4 Prospective Pathways and Challenges

## *4.1 3D Printing In Situ*

3D printing in situ has been gaining traction as an effective method for producing patient-specific bone scaffolds right at the bone defect locations. Leveraging advancements in AM technologies along with the vast array of available biomaterials, this approach presents fresh opportunities for targeted bone regeneration [1].

A technique utilized in situ laser-assisted bioprinting to dispense a collagen bioink blended with nano-hydroxyapatite (nHAP) for addressing calvaria defects in mice, experimenting with various scaffold geometries like disks and rings. The results highlighted how bone healing was affected by the cellular configuration within these scaffolds [280]. Separately, an extrusion-based 3D-printing method deployed methacrylate gelatin bio-ink into a large segmental bone defect in swine, which was subsequently solidified using UV light. Remarkably, this scaffold was printed and set in situ in 12 minutes, leading to notable bone repair after 3 months [281]. Additionally, a handheld melt extrusion printer was employed to create scaffolds from PCL filaments doped with HAP microparticles and zinc oxide nanoparticles, demonstrating superior biocompatibility in a mouse model [282].

These instances emphasize the capabilities of in situ 3D printing, illustrating its ability for quick, site-specific fabrication of bone scaffolds, which holds significant promise for improving personalized bone healing processes.

## *4.2 Customized AM Scaffolds Using Smart Materials*

The use of biomaterials has significantly impacted the field of regenerative medicine, with classifications such as "bioinert," "biocompatible," and "bioactive," reflecting their interaction level within the body. A novel category for biomaterials that adapt to specific cellular signals has been introduced as "smart materials" [283]. These materials are designed to react to external forces such as magnetic fields, electric fields, temperature changes, mechanical pressures, and light exposure, altering their properties in response.

In the realm of customized 3D-printed scaffolds, the choice of bioink determines whether they are considered bioactive, biocompatible, or bioinert. However, these materials traditionally lack the capability to emulate the evolving and dynamic nature of living tissues. To address this limitation, the concept of "4D printing" emerges, defining the development of biomaterials that can independently change over time or react to external stimuli [284]. By integrating the dimension of time with 3D printing technologies, 4D printing enables the creation of scaffolds that undergo change triggered by external inputs like heat, light, or various environmental factors [285].

Materials suitable for 4D printing span across polymers, metals, nanotubes, nanowires, DNA, and proteins, offering a wide spectrum of innovation. Apart from 3D and 4D printing technologies, a variety of smart materials have been manually developed, not utilizing 3D printing. These include materials with piezoelectric properties and the ability to change shape [1].

## 4.3 Scaffold Design Using Deep Learning

Creating personalized 3D-printed scaffolds is a complex and lengthy process that may delay immediate treatment. Therefore, developing sophisticated techniques to analyze digital medical information for scaffold design is crucial. Deep learning, an area of artificial intelligence, offers considerable promise in this regard. It can predict and create bone models from patient-specific data by using convolutional neural networks that have been appropriately trained [286].

One research effort utilized deep learning to enhance pediatric CT scans taken at lower doses, showcasing its capability to manage and process intricate medical imaging data [286]. In a notable advancement, three-dimensional deep learning technology was employed to autonomously produce geometries for cranial implants, indicating its potential for future applications in fabricating personalized bone scaffolds [287]. Further illustrating the capabilities of deep learning, a new technique known as GAN-CIRCLE was utilized to accurately reconstruct high-definition trabecular bone microstructures from low-resolution CT scans [288].

These advancements collectively demonstrate the profound impact that deep learning models can have on fabricating customized scaffolds. This could lead to more efficient, precise clinical applications involving the use of 3D printed scaffolds.

## 4.4 Model of Collaboration for Clinical Use of Customized AM Bone Scaffolds

Implementing customized AM scaffolds in a clinical setting requires a collaborative approach among the patient, clinician, and engineer. This represents a significant shift from the conventional approach, which typically involves the distribution of mass-produced implants to healthcare facilities. Instead, AM bone scaffolds are individually designed, drawing on specific patient medical data for guidance. The process outlined in Fig. 6 demonstrates the sequence of actions for an effective treatment strategy with custom AM bone scaffolds, starting with the clinician's identification of the bone defect. This is followed by a discussion of treatment alternatives with the patient and consultation with engineers regarding the design of the scaffold. The clinician serves as a bridge connecting the patient and engineer,

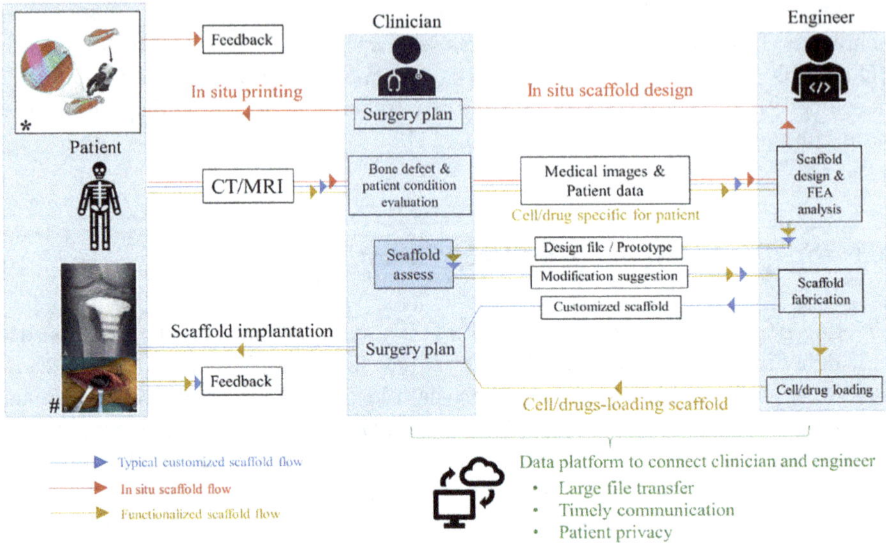


**Fig. 6** The sequence involved in customizing an AM scaffold specific to a patient's requirements. (Images modified from references: [1], * [281], # [30])

ensuring the delivery of the most effective treatment plan. Success in this model is dependent on the strong partnership between the clinician and engineer, with both parties understanding the intricacies of the customization process throughout its phases. In scenarios involving cell or drug-loaded scaffolds, the clinician works directly with the patient for cell harvesting and integrates patient-specific drugs into the scaffold. In contrast, for in situ printing, the clinician carries out the scaffold creation under the guidance of the engineer's recommendations.

A significant challenge to the success of this collaborative approach is the effectiveness of communication. Given the nature of surgical timelines for patients, rapid exchange of information and interaction between clinicians and engineers becomes crucial for the timely customization of scaffolds, a task that can inherently consume considerable time. Consequently, establishing a consolidated data platform is essential for facilitating prompt communication, ensuring the confidentiality of patient information, and handling the transfer of large files due to the extensive amount of medical data required.

An additional barrier lies in the intricacy of the customization process. The optimal specifications for bone scaffolds vary individually, based on the patient's unique medical conditions, requiring detailed pathological examination and customized scaffold design. Such a demanding process could lead to lower efficiency and higher expenses. Nonetheless, the application of artificial intelligence, like the deep-learning approaches discussed in Sect. 4.3, can potentially streamline the analytical and design stages. Meanwhile, thorough quality checks and assessments by clinicians and engineers guarantee the reliability of the end products [1].

## 5 Conclusion

AM bone scaffolds mark a revolutionary step in customized orthopedic solutions, providing clear benefits compared to conventional implants. These scaffolds are tailored to seamlessly align with distinct bone defect geometries and replicate the mechanical and structural properties of the native tissue, according to the specific therapeutic approach and the location of the bone defect. Integral to achieving precise therapeutic outcomes is the customization process, encompassing the imaging of the bone defect, the engineering of the scaffold design, and the careful choice of materials and production techniques. Commitment to these steps ensures that each scaffold meets the distinct requirements of individual patients. Ultimately, the suggested model promotes a partnership between engineers and clinicians, enabling the production of customized AM bone scaffolds for patients. This collaborative effort unlocks the full possibilities of additive manufacturing in orthopedics, marking the beginning of a new phase in customized healthcare solutions.

**Conflicts of Interest** The authors declare no conflicts of interest.

## References


1. Zhou, J., et al., *Customized Additive Manufacturing in Bone Scaffolds—The Gateway to Precise Bone Defect Treatment.* Research, 2023. **6**: p. 0239.
2. Marks Jr, S.C. and S.N.J.A.J.o.A Popoff, *Bone cell biology: the regulation of development, structure, and function in the skeleton.* 1988. **183**(1): p. 1–44.
3. Mohan, S., D.J.J.C.o. Baylink, and r. research, *Bone growth factors.* 1991(263): p. 30–48.
4. Mattavelli, D., et al., *Computer-aided designed 3D-printed polymeric scaffolds for personalized reconstruction of maxillary and mandibular defects: a proof-of-concept study.* European Archives of Oto-Rhino-Laryngology, 2024: p. 1–11.
5. Schindeler, A., et al., *Bone remodeling during fracture repair: The cellular picture.* Semin Cell Dev Biol, 2008. **19**(5): p. 459–66.
6. Khosla, S., J.J. Westendorf, and U.I. Modder, *Concise review: Insights from normal bone remodeling and stem cell-based therapies for bone repair.* Stem Cells, 2010. **28**(12): p. 2124–8.
7. Deng, M., et al., *Modification of PLGA Scaffold by MSC-Derived Extracellular Matrix Combats Macrophage Inflammation to Initiate Bone Regeneration via TGF-beta-Induced Protein.* Adv Healthc Mater, 2020. **9**(13): p. e2000353.
8. Zhao, F., et al., *3D Printing Nanoscale Bioactive Glass Scaffolds Enhance Osteoblast Migration and Extramembranous Osteogenesis through Stimulating Immunomodulation.* Adv Healthc Mater, 2018. **7**(16): p. e1800361.
9. Sun, X., et al., *Three-dimensional bioprinting of multicell-laden scaffolds containing bone morphogenic protein-4 for promoting M2 macrophage polarization and accelerating bone defect repair in diabetes mellitus.* Bioact Mater, 2021. **6**(3): p. 757–769.
10. Wang, Q., et al., *3D-Printed Atsttrin-Incorporated Alginate/Hydroxyapatite Scaffold Promotes Bone Defect Regeneration with TNF/TNFR Signaling Involvement.* Adv Healthc Mater, 2015. **4**(11): p. 1701–8.

11. Cao, Y., et al., *3D printed β-TCP scaffold with sphingosine 1-phosphate coating promotes osteogenesis and inhibits inflammation.* Biochemical and Biophysical Research Communications, 2019. **512**(4): p. 889–895.
12. Ma, L., et al., *Integrating 3D Printing and Biomimetic Mineralization for Personalized Enhanced Osteogenesis, Angiogenesis, and Osteointegration.* ACS Appl Mater Interfaces, 2018. **10**(49): p. 42146–42154.
13. Min, Z., et al., *3D-printed dimethyloxallyl glycine delivery scaffolds to improve angiogenesis and osteogenesis.* Biomater Sci, 2015. **3**(8): p. 1236–44.
14. Zhang, Y., et al., *Mesoporous bioactive glass nanolayer-functionalized 3D-printed scaffolds for accelerating osteogenesis and angiogenesis.* Nanoscale, 2015. **7**(45): p. 19207–21.
15. Bose, S., S. Tarafder, and A. Bandyopadhyay, *Effect of Chemistry on Osteogenesis and Angiogenesis Towards Bone Tissue Engineering Using 3D Printed Scaffolds.* Ann Biomed Eng, 2017. **45**(1): p. 261–272.
16. Barba, A., et al., *Osteogenesis by foamed and 3D-printed nanostructured calcium phosphate scaffolds: Effect of pore architecture.* Acta Biomater, 2018. **79**: p. 135–147.
17. Paris, M., et al., *Scaffold curvature-mediated novel biomineralization process originates a continuous soft tissue-to-bone interface.* Acta Biomater, 2017. **60**: p. 64–80.
18. Lai, Y., et al., *Porous composite scaffold incorporating osteogenic phytomolecule icariin for promoting skeletal regeneration in challenging osteonecrotic bone in rabbits.* Biomaterials, 2018. **153**: p. 1–13.
19. Li, Y., et al., *An effective dual-factor modified 3D-printed PCL scaffold for bone defect repair.* J Biomed Mater Res B Appl Biomater, 2020. **108**(5): p. 2167–2179.
20. Duric, K.S., et al., *Polymethylmethacrylate cranioplasty using low-cost customised 3D printed moulds for cranial defects—a single Centre experience: technical note.* Br J Neurosurg, 2019. **33**(4): p. 376–378.
21. Park, E.K., et al., *Cranioplasty Enhanced by Three-Dimensional Printing: Custom-Made Three-Dimensional-Printed Titanium Implants for Skull Defects.* J Craniofac Surg, 2016. **27**(4): p. 943–9.
22. Kim, S.H., et al., *Staged reconstruction of large skull defects with soft tissue infection after craniectomy using free flap and cranioplasty with a custom-made titanium mesh constructed by 3D-CT-guided 3D printing technology: Two case reports.* Medicine (Baltimore), 2019. **98**(6): p. e13864.
23. Kim, J., et al., *Outcomes of Cranioplasty Using Autologous Bone or 3D-Customized Titanium Mesh Following Decompressive Craniectomy for Traumatic Brain Injury: Differences in Complications.* Journal of Trauma and Injury, 2019. **32**(4): p. 202–209.
24. Hamid, K.S., S.G. Parekh, and S.B. Adams, *Salvage of Severe Foot and Ankle Trauma With a 3D Printed Scaffold.* Foot Ankle Int, 2016. **37**(4): p. 433–9.
25. Perticarini, L., et al., *Clinical and radiological outcomes of acetabular revision surgery with trabecular titanium cups in Paprosky type II and III bone defects.* J Orthop Traumatol, 2021. **22**(1): p. 9.
26. Zou, Y., et al., *Novel exploration of customized 3D printed shoulder prosthesis in revision of total shoulder arthroplasty: A case report.* Medicine (Baltimore), 2018. **97**(47): p. e13282.
27. Suska, F., et al., *Electron Beam Melting Manufacturing Technology for Individually Manufactured Jaw Prosthesis: A Case Report.* J Oral Maxillofac Surg, 2016. **74**(8): p. 1706 e1–1706 e15.
28. Plantz, M.A., et al., *Osteoinductivity and biomechanical assessment of a 3D printed demineralized bone matrix-ceramic composite in a rat spine fusion model.* Acta biomaterialia, 2021. **127**: p. 146–158.
29. Kobbe, P., et al., *Convergence of scaffold-guided bone regeneration and RIA bone grafting for the treatment of a critical-sized bone defect of the femoral shaft.* Eur J Med Res, 2020. **25**(1): p. 70.
30. Lu, M., et al., *A three-dimensional printed porous implant combined with bone grafting following curettage of a subchondral giant cell tumour of the proximal tibia: a case report.* BMC Surg, 2019. **19**(1): p. 29.

31. Lui, H., et al., *Additively Manufactured Multiphasic Bone-Ligament-Bone Scaffold for Scapholunate Interosseous Ligament Reconstruction.* Adv Healthc Mater, 2019. **8**(14): p. e1900133.
32. Park, J.W., et al., *Three-Dimensionally Printed Personalized Implant Design and Reconstructive Surgery for a Bone Tumor of the Calcaneus: A Case Report.* JBJS Case Connect, 2018. **8**(2): p. e25.
33. Rho, J.-Y., et al., *Mechanical properties and the hierarchical structure of bone.* 1998. **20**(2): p. 92–102.
34. Goldstein, S.A.J.J.o.b., *The mechanical properties of trabecular bone: dependence on anatomic location and function.* 1987. **20**(11-12): p. 1055–1061.
35. Pobloth, A.-M., et al., *Mechanobiologically optimized 3D titanium-mesh scaffolds enhance bone regeneration in critical segmental defects in sheep.* Science Translational Medicine, 2018. **10**(423): p. eaam8828.
36. Zheng, Y., et al., *Promotion of tendon growth into implant through pore-size design of a Ti-6Al-4 V porous scaffold prepared by 3D printing.* Materials & Design, 2021. **197**: p. 109219.
37. Lipskas, J., K. Deep, and W. Yao, *Robotic-Assisted 3D Bio-printing for Repairing Bone and Cartilage Defects through a Minimally Invasive Approach.* Sci Rep, 2019. **9**(1): p. 3746.
38. Li, L., et al., *In situ repair of bone and cartilage defects using 3D scanning and 3D printing.* Scientific reports, 2017. **7**(1): p. 9416.
39. Hou, G., et al., *An innovative strategy to treat large metaphyseal segmental femoral bone defect using customized design and 3D printed micro-porous prosthesis: a prospective clinical study.* J Mater Sci Mater Med, 2020. **31**(8): p. 66.
40. Yang, Y., et al., *Dual-functional 3D-printed composite scaffold for inhibiting bacterial infection and promoting bone regeneration in infected bone defect models.* Acta Biomater, 2018. **79**: p. 265–275.
41. Yan, Y., et al., *Vascularized 3D printed scaffolds for promoting bone regeneration.* Biomaterials, 2019. **190-191**: p. 97–110.
42. Alluri, R., et al., *Regional gene therapy with 3D printed scaffolds to heal critical sized bone defects in a rat model.* J Biomed Mater Res A, 2019. **107**(10): p. 2174–2182.
43. Tovar, N., et al., *Form and functional repair of long bone using 3D-printed bioactive scaffolds.* J Tissue Eng Regen Med, 2018. **12**(9): p. 1986–1999.
44. Diao, J., et al., *Bone defect model dependent optimal pore sizes of 3D-plotted beta-tricalcium phosphate scaffolds for bone regeneration.* Small Methods, 2019. **3**(11): p. 1900237.
45. Kim, S.E., et al., *3D printed alendronate-releasing poly(caprolactone) porous scaffolds enhance osteogenic differentiation and bone formation in rat tibial defects.* Biomed Mater, 2016. **11**(5): p. 055005.
46. Han, S.H., et al., *BMP-2 and hMSC dual delivery onto 3D printed PLA-Biogel scaffold for critical-size bone defect regeneration in rabbit tibia.* Biomed Mater, 2020. **16**(1): p. 015019.
47. Zhang, L., et al., *Three-Dimensional Printing-Assisted Masquelet Technique in the Treatment of Calcaneal Defects.* Orthopaedic Surgery, 2021. **13**(3): p. 876–883.
48. Marin, E., et al., *Characterization of grade 2 commercially pure Trabecular Titanium structures.* Materials Science and Engineering: C, 2013. **33**(5): p. 2648–2656.
49. Fu, J., et al., *In Vivo Reconstruction of the Acetabular Bone Defect by the Individualized Three-Dimensional Printed Porous Augment in a Swine Model.* Biomed Res Int, 2020. **2020**: p. 4542302.
50. Basu, B., et al., *Evaluation of implant properties, safety profile and clinical efficacy of patient-specific acrylic prosthesis in cranioplasty using 3D binderjet printed cranium model: A pilot study.* J Clin Neurosci, 2021. **85**: p. 132–142.
51. van de Vijfeijken, S., et al., *The use of cranial resection templates with 3D virtual planning and PEEK patient-specific implants: A 3 year follow-up.* J Craniomaxillofac Surg, 2019. **47**(4): p. 542–547.
52. O'Reilly, E.B., et al., *Computed-tomography modeled polyether ether ketone (PEEK) implants in revision cranioplasty.* J Plast Reconstr Aesthet Surg, 2015. **68**(3): p. 329–38.

53. Kwarcinski, J., et al., *Clinical evaluation of rapid 3D print-formed implants for surgical reconstruction of large cranial defects.* ANZ Journal of Surgery, 2021. **91**(6): p. 1226–1232.
54. Sharma, N., et al., *Quality characteristics and clinical relevance of in-house 3D-printed customized polyetheretherketone (PEEK) implants for craniofacial reconstruction.* Journal of clinical medicine, 2020. **9**(9): p. 2818.
55. Koller, M., et al., *A retrospective descriptive study of cranioplasty failure rates and contributing factors in novel 3D printed calcium phosphate implants compared to traditional materials.* 3D Print Med, 2020. **6**(1): p. 14.
56. Mommaerts, M.Y., P.R. Depauw, and E. Nout, *Ceramic 3D-Printed Titanium Cranioplasty.* Craniomaxillofac Trauma Reconstr, 2020. **13**(4): p. 329–333.
57. Kinsman, M., et al., *Rapid high-fidelity contour shaping of titanium mesh implants for cranioplasty defects using patient-specific molds created with low-cost 3D printing: A case series.* Surg Neurol Int, 2020. **11**: p. 288.
58. Nguyen, B., et al., *Cranioplasty Using Customized 3-Dimensional-Printed Titanium Implants: An International Collaboration Effort to Improve Neurosurgical Care.* World Neurosurg, 2021. **149**: p. 174–180.
59. Shao, H., et al., *Custom Repair of Mandibular Bone Defects with 3D Printed Bioceramic Scaffolds.* J Dent Res, 2018. **97**(1): p. 68–76.
60. Lopez, C.D., et al., *Three dimensionally printed bioactive ceramic scaffold osseoconduction across critical-sized mandibular defects.* J Surg Res, 2018. **223**: p. 115–122.
61. Helal, M.H., et al., *Osteogenesis ability of CAD-CAM biodegradable polylactic acid scaffolds for reconstruction of jaw defects.* J Prosthet Dent, 2019. **121**(1): p. 118–123.
62. Chang, P.C., et al., *Regeneration of critical-sized mandibular defect using a 3D-printed hydroxyapatite-based scaffold: An exploratory study.* J Periodontol, 2021. **92**(3): p. 428–435.
63. Lee, S., et al., *Efficacy of three-dimensionally printed polycaprolactone/beta tricalcium phosphate scaffold on mandibular reconstruction.* Sci Rep, 2020. **10**(1): p. 4979.
64. Sun, H., et al., *3D printing of calcium phosphate scaffolds with controlled release of antibacterial functions for jaw bone repair.* Materials & Design, 2020. **189**: p. 108540.
65. Zhao, Q., et al., *Human periodontal ligament stem cells transplanted with nanohydroxyapatite/chitosan/gelatin 3D porous scaffolds promote jaw bone regeneration in swine.* Stem Cells and Development, 2021. **30**(10): p. 548–559.
66. Hallman, M., et al., *Influence of Geometry and Architecture on the In Vivo Success of 3D-Printed Scaffolds for Spinal Fusion.* Tissue Eng Part A, 2021. **27**(1-2): p. 26–36.
67. Lin, S., et al., *PLGA/beta-TCP composite scaffold incorporating salvianolic acid B promotes bone fusion by angiogenesis and osteogenesis in a rat spinal fusion model.* Biomaterials, 2019. **196**: p. 109–121.
68. Vlad, M.D., et al., *Novel titanium-apatite hybrid scaffolds with spongy bone-like micro architecture intended for spinal application: In vitro and in vivo study.* Materials Science and Engineering: C, 2020. **110**: p. 110658.
69. Otton, J.M., et al., *3D printing from cardiovascular CT: a practical guide and review.* Cardiovasc Diagn Ther, 2017. **7**(5): p. 507–526.
70. Wong, K.C., *3D-printed patient-specific applications in orthopedics.* Orthop Res Rev, 2016. **8**: p. 57–66.
71. Wong, K.C., et al., *Image fusion for computer-assisted bone tumor surgery.* Clin Orthop Relat Res, 2008. **466**(10): p. 2533–41.
72. Dalrymple, N.C., et al., *Introduction to the Language of Three-dimensional Imaging with Multidetector CT.* RadioGraphics, 2005. **25**(5): p. 1409–1428.
73. Rubin, G.D., *3-D imaging with MDCT.* European Journal of Radiology, 2003. **45**: p. S37–S41.
74. Rengier, F., et al., *3D printing based on imaging data: review of medical applications.* Int J Comput Assist Radiol Surg, 2010. **5**(4): p. 335–41.
75. Acar, B. and K. Kamburoğlu, *Use of cone beam computed tomography in periodontology.* World journal of radiology, 2014. **6**(5): p. 139–147.

76. Kumar, M., et al., *Cone beam computed tomography—know its secrets.* Journal of international oral health: JIOH, 2015. **7**(2): p. 64–68.
77. Trace, A.P., et al., *Radiology's Emerging Role in 3-D Printing Applications in Health Care.* J Am Coll Radiol, 2016. **13**(7): p. 856–862.e4.
78. Acar, A.A., et al., *Customized scaffolds for large bone defects using 3D-printed modular blocks from 2D-medical images.* Bio-Design and Manufacturing, 2024. **7**(1): p. 74–87.
79. Buckwalter, K.A., C. Lin, and J.M. Ford, *Managing postoperative artifacts on computed tomography and magnetic resonance imaging.* Semin Musculoskelet Radiol, 2011. **15**(4): p. 309–19.
80. Douglas-Akinwande, A.C., et al., *Multichannel CT: Evaluating the Spine in Postoperative Patients with Orthopedic Hardware.* RadioGraphics, 2006. **26**(suppl_1): p. S97–S110.
81. Stiller, W., *[Principles of multidetector-row computed tomography: part 1. Technical design and physicotechnical principles].* Der Radiologe, 2011. **51**(7): p. 625–37; quiz 638–9.
82. Kotsenas, A.L., et al., *CT Metal Artifact Reduction in the Spine: Can an Iterative Reconstruction Technique Improve Visualization?* AJNR Am J Neuroradiol, 2015. **36**(11): p. 2184–90.
83. Geyer, L.L., et al., *State of the Art: Iterative CT Reconstruction Techniques.* Radiology, 2015. **276**(2): p. 339–357.
84. Coupal, T.M., et al., *Peering through the glare: using dual-energy CT to overcome the problem of metal artefacts in bone radiology.* Skeletal Radiol, 2014. **43**(5): p. 567–75.
85. Graf, H., et al., *RF artifacts caused by metallic implants or instruments which get more prominent at 3 T: an in vitro study.* Magn Reson Imaging, 2005. **23**(3): p. 493–9.
86. Ulbrich, E.J., et al., *STIR sequence with increased receiver bandwidth of the inversion pulse for reduction of metallic artifacts.* AJR Am J Roentgenol, 2012. **199**(6): p. W735–42.
87. Sutter, R., et al., *Reduction of Metal Artifacts in Patients with Total Hip Arthroplasty with Slice-encoding Metal Artifact Correction and View-Angle Tilting MR Imaging.* Radiology, 2012. **265**(1): p. 204–214.
88. Dillenseger, J.P., et al., *An illustrative review to understand and manage metal-induced artifacts in musculoskeletal MRI: a primer and updates.* Skeletal Radiol, 2016. **45**(5): p. 677–88.
89. Ciliberti, F.K., et al., *CT-and MRI-based 3D reconstruction of knee joint to assess cartilage and bone.* Diagnostics, 2022. **12**(2): p. 279.
90. Wong, K., et al., *Use of a patient-specific CAD/CAM surgical jig in extremity bone tumor resection and custom prosthetic reconstruction.* Computer Aided Surgery, 2012. **17**(6): p. 284–293.
91. Boas, F.E. and D. Fleischmann, *CT artifacts: causes and reduction techniques.* Imaging Med, 2012. **4**(2): p. 229–240.
92. Barrett, J.F. and N. Keat, *Artifacts in CT: recognition and avoidance.* Radiographics, 2004. **24**(6): p. 1679–1691.
93. Hargreaves, B.A., et al., *Metal-induced artifacts in MRI.* American journal of roentgenology, 2011. **197**(3): p. 547–555.
94. Winter, L., et al., *MRI-related heating of implants and devices: a review.* Journal of Magnetic Resonance Imaging, 2021. **53**(6): p. 1646–1665.
95. Koch, K., et al., *Magnetic resonance imaging near metal implants.* Journal of Magnetic Resonance Imaging, 2010. **32**(4): p. 773–787.
96. Chong, W. and W. Min, *3D printing of bone tissue engineering scaffolds with osteoconductivity and osteoinductivity.* Front. Bioeng. Biotechnol. Conference Abstract: 10th World Biomaterials Congress., 2016.
97. Vidal, L., et al., *Regeneration of segmental defects in metatarsus of sheep with vascularized and customized 3D-printed calcium phosphate scaffolds.* Scientific Reports, 2020. **10**(1): p. 7068.
98. Fousová, M., et al., *3D printed porous stainless steel for potential use in medicine.* IOP Conference Series: Materials Science and Engineering, 2017. **179**(1): p. 012025.
99. Lantigua, D., et al., *Composite scaffolds from gelatin and bone meal powder for tissue engineering.* Bioengineering, 2021. **8**(11): p. 169.

100. Ni, T., et al., *3D Bioprinting of Bone Marrow Mesenchymal Stem Cell-Laden Silk Fibroin Double Network Scaffolds for Cartilage Tissue Repair.* Bioconjugate Chemistry, 2020. **31**(8): p. 1938–1947.
101. Kaur, G., et al., *A review of bioactive glasses: their structure, properties, fabrication and apatite formation.* Journal of Biomedical Materials Research Part A: An Official Journal of The Society for Biomaterials, The Japanese Society for Biomaterials, and The Australian Society for Biomaterials and the Korean Society for Biomaterials, 2014. **102**(1): p. 254–274.
102. Ke, D. and S. Bose, *Effects of pore distribution and chemistry on physical, mechanical, and biological properties of tricalcium phosphate scaffolds by binder-jet 3D printing.* Additive Manufacturing, 2018. **22**: p. 111–117.
103. Vu, A.A., et al., *Effects of surface area and topography on 3D printed tricalcium phosphate scaffolds for bone grafting applications.* Additive manufacturing, 2021. **39**: p. 101870.
104. Asadi-Eydivand, M., et al., *Structure, Properties, and In Vitro Behavior of Heat-Treated Calcium Sulfate Scaffolds Fabricated by 3D Printing.* PLoS One, 2016. **11**(3): p. e0151216.
105. Qi, X., et al., *Three dimensional printing of calcium sulfate and mesoporous bioactive glass scaffolds for improving bone regeneration in vitro and in vivo.* Sci Rep, 2017. **7**: p. 42556.
106. Chen, S., et al., *3D printed hydroxyapatite composite scaffolds with enhanced mechanical properties.* Ceramics International, 2019. **45**(8): p. 10991–10996.
107. Song, X., et al., *Biomimetic 3D printing of hierarchical and interconnected porous hydroxyapatite structures with high mechanical strength for bone cell culture.* Advanced Engineering Materials, 2019. **21**(1): p. 1800678.
108. Ke, X., et al., *Low-melt bioactive glass-reinforced 3D printing akermanite porous cages with highly improved mechanical properties for lumbar spinal fusion.* J Tissue Eng Regen Med, 2018. **12**(5): p. 1149–1162.
109. Wang, J., et al., *Simultaneous enhancement of vascularization and contact-active antibacterial activity in diopside-based ceramic orbital implants.* Mater Sci Eng C Mater Biol Appl, 2019. **105**: p. 110036.
110. Elsayed, H., et al., *Comparative analysis of wollastonite-diopside glass-ceramic structures fabricated via stereo-lithography.* Advanced Engineering Materials, 2019. **21**(6): p. 1801160.
111. Du, X., et al., *3D printing of mesoporous bioactive glass/silk fibroin composite scaffolds for bone tissue engineering.* Mater Sci Eng C Mater Biol Appl, 2019. **103**: p. 109731.
112. Ghanad, M., et al., *Single-step solution combustion synthesis of porous 1393-B3 glass powders and structural characterization via solid-state NMR spectroscopy.* Ceramics International, 2023. **49**(9): p. 14689–14701.
113. Szurkowska, K., Ł. Szeleszczuk, and J. Kolmas, *Effects of synthesis conditions on the formation of Si-substituted alpha tricalcium phosphates.* International journal of molecular sciences, 2020. **21**(23): p. 9164.
114. Bohner, M., B.L.G. Santoni, and N. Döbelin, *β-tricalcium phosphate for bone substitution: Synthesis and properties.* Acta biomaterialia, 2020. **113**: p. 23–41.
115. Podaropoulos, L., et al., *Bone regeneration using b-tricalcium phosphate in a calcium sulfate matrix.* Journal of Oral Implantology, 2009. **35**(1): p. 28–36.
116. Zhou, Z., et al., *Printability of calcium phosphate: calcium sulfate powders for the application of tissue engineered bone scaffolds using the 3D printing technique.* Materials Science and Engineering: C, 2014. **38**: p. 1–10.
117. Bose, S., et al., *Enhanced In Vivo Bone and Blood Vessel Formation by Iron Oxide and Silica Doped 3D Printed Tricalcium Phosphate Scaffolds.* Ann Biomed Eng, 2018. **46**(9): p. 1241–1253.
118. Nandi, S.K., et al., *3D-printed β-TCP bone tissue engineering scaffolds: Effects of chemistry on in vivo biological properties in a rabbit tibia model.* Journal of Materials Research, 2018. **33**(14): p. 1939–1947.
119. Qi, X., et al., *Three dimensional printing of calcium sulfate and mesoporous bioactive glass scaffolds for improving bone regeneration in vitro and in vivo.* Scientific reports, 2017. **7**(1): p. 1–12.

120. Butscher, A., et al., *New depowdering-friendly designs for three-dimensional printing of calcium phosphate bone substitutes.* Acta biomaterialia, 2013. **9**(11): p. 9149–9158.
121. Zeng, Y., et al., *3D printing of hydroxyapatite scaffolds with good mechanical and biocompatible properties by digital light processing.* Journal of Materials Science, 2018. **53**(9): p. 6291–6301.
122. Shor, L., et al., *Fabrication of three-dimensional polycaprolactone/hydroxyapatite tissue scaffolds and osteoblast-scaffold interactions in vitro.* Biomaterials, 2007. **28**(35): p. 5291–5297.
123. Zhao, S., et al., *Wound dressings composed of copper-doped borate bioactive glass microfibers stimulate angiogenesis and heal full-thickness skin defects in a rodent model.* Biomaterials, 2015. **53**: p. 379–391.
124. Wang, J., et al., *Simultaneous enhancement of vascularization and contact-active antibacterial activity in diopside-based ceramic orbital implants.* Materials Science and Engineering: C, 2019. **105**: p. 110036.
125. Li, X., H. Zhang, and H. Zhang, *Fabrication of β-TCP/Akermanite composite scaffold via DLP and in-situ modification of micro-nano surface morphology for bone repair.* Ceramics International, 2024. **50**(2): p. 2659–2669.
126. El-Ghannam, A., P. Ducheyne, and I.J.J.o.O.R. Shapiro, *Effect of serum proteins on osteoblast adhesion to surface-modified bioactive glass and hydroxyapatite.* 1999. **17**(3): p. 340–345.
127. Singh, S., et al., *Bioglass: A New Era in Modern Dentistry.* European Journal of General Dentistry, 2022. **11**(01): p. 001–006.
128. Chi, C.-Y., et al., *Preparation and in-vitro evaluation of Fe2O3-Doped DP-Bioglass in combination with 3D-printing and selective laser sintering process (3DP-SLS) for alveolar bone augmentation.* Ceramics International, 2021. **47**(9): p. 12725–12734.
129. Lauria, I., et al., *Inkjet printed periodical micropatterns made of inert alumina ceramics induce contact guidance and stimulate osteogenic differentiation of mesenchymal stromal cells.* Acta Biomaterialia, 2016. **44**: p. 85–96.
130. Jin, W. and P.K. Chu, *Orthopedic implants.* Encyclopedia of biomedical engineering, 2019. **1**: p. 3.
131. Nakai, H., et al., *Additively manufactured zirconia for dental applications.* Materials, 2021. **14**(13): p. 3694.
132. Fayazfar, H., et al., *Additive manufacturing of high loading concentration zirconia using high-speed drop-on-demand material jetting.* The International Journal of Advanced Manufacturing Technology, 2020. **109**(9): p. 2733–2746.
133. Liu, F. and X. Wang, *Synthetic polymers for organ 3D printing.* Polymers, 2020. **12**(8): p. 1765.
134. Ligon, S.C., et al., *Polymers for 3D Printing and Customized Additive Manufacturing.* Chem Rev, 2017. **117**(15): p. 10212–10290.
135. Senra, M.N.R. and M.D.V. Marques, *Synthetic Polymeric Materials for Bone Replacement.* Journal of Composites Science, 2020. **4**(4).
136. Goh, B.T., et al., *Novel 3D polycaprolactone scaffold for ridge preservation–a pilot randomised controlled clinical trial.* Clin Oral Implants Res, 2015. **26**(3): p. 271–7.
137. Elawady, D.M., W.I. Ibrahim, and R.B. Osman, *Clinical evaluation of implant overdentures fabricated using 3D-printing technology versus conventional fabrication techniques: a randomized clinical trial.* Int J Comput Dent, 2021. **24**(4): p. 375–384.
138. Shah, F.A., et al., *Long-term osseointegration of 3D printed CoCr constructs with an interconnected open-pore architecture prepared by electron beam melting.* Acta Biomaterialia, 2016. **36**: p. 296–309.
139. Gotman, I.J.J.o.e, *Characteristics of metals used in implants.* 1997. **11**(6): p. 383–389.
140. Valiev, R.Z., E.V. Parfenov, and L.V. Parfenova, *Developing nanostructured metals for manufacturing of medical implants with improved design and biofunctionality.* Materials transactions, 2019. **60**(7): p. 1356–1366.
141. Aviles, T., et al., *Hydroxyapatite formation on coated titanium implants submerged in simulated body fluid.* Materials, 2020. **13**(24): p. 5593.

142. Taniguchi, N., et al., *Effect of pore size on bone ingrowth into porous titanium implants fabricated by additive manufacturing: An in vivo experiment.* Materials Science and Engineering: C, 2016. **59**: p. 690–701.
143. Li, X., et al., *Evaluation of Biological Properties of Electron Beam Melted Ti6Al4V Implant with Biomimetic Coating In Vitro and In Vivo.* PLOS ONE, 2012. **7**(12): p. e52049.
144. Ponader, S., et al., *In vivo performance of selective electron beam-melted Ti-6Al-4V structures.* 2010. **92A**(1): p. 56–62.
145. Wang, X., et al., *Topological design and additive manufacturing of porous metals for bone scaffolds and orthopaedic implants: a review.* 2016. **83**: p. 127–141.
146. Hermawan, H., *Biodegradable metals: state of the art*, in *Biodegradable Metals*. 2012, Springer. p. 13–22.
147. Polo-Corrales, L., et al., *Scaffold design for bone regeneration.* 2014. **14**(1): p. 15–56.
148. Bobe, K., et al., *In vitro and in vivo evaluation of biodegradable, open-porous scaffolds made of sintered magnesium W4 short fibres.* 2013. **9**(10): p. 8611–8623.
149. Zhou, J., et al., *Evolution from Bioinert to Bioresorbable: In Vivo Comparative Study of Additively Manufactured Metal Bone Scaffolds.* Advanced Science, 2023. **n/a**(n/a): p. 2302702.
150. Du, X., et al., *3D printing of pearl/CaSO4 composite scaffolds for bone regeneration.* Journal of Materials Chemistry B, 2018. **6**(3): p. 499–509.
151. Harb, S.V., et al., *3D printed bioabsorbable composite scaffolds of poly (lactic acid)-tricalcium phosphate-ceria with osteogenic property for bone regeneration.* Biomaterials and Biosystems, 2024. **13**: p. 100086.
152. Morra, M., et al., *Surface chemistry and effects on bone regeneration of a novel biomimetic synthetic bone filler.* 2015. **26**(4): p. 159.
153. Rohman, G., et al., *The grafting of a thin layer of poly (sodium styrene sulfonate) onto poly (ε-caprolactone) surface can enhance fibroblast behavior.* 2015. **26**(7): p. 206.
154. Jasmawati, N., et al., *Mg-based porous metals as cancellous bone analogous material: A review.* 2017. **231**(6): p. 544–556.
155. Tellisi, N., et al., *Three Dimensional Printed Bone Implants in the Clinic.* Journal of Craniofacial Surgery, 2018. **29**(8): p. 2363–2367.
156. Oladapo, B.I., et al., *3D printing of PEEK and its composite to increase biointerfaces as a biomedical material- A review.* Colloids Surf B Biointerfaces, 2021. **203**: p. 111726.
157. Oladapo, B.I., et al., *Lattice design and 3D-printing of PEEK with Ca10(OH)(PO4)3 and in-vitro bio-composite for bone implant.* Int J Biol Macromol, 2020. **165**(Pt A): p. 50–62.
158. Chen, S., et al., *Evaluation of BMP-2 and VEGF loaded 3D printed hydroxyapatite composite scaffolds with enhanced osteogenic capacity in vitro and in vivo.* Mater Sci Eng C Mater Biol Appl, 2020. **112**: p. 110893.
159. Monavari, M., et al., *3D printing of alginate dialdehyde-gelatin (ADA-GEL) hydrogels incorporating phytotherapeutic icariin loaded mesoporous SiO2-CaO nanoparticles for bone tissue engineering.* Mater Sci Eng C Mater Biol Appl, 2021. **131**: p. 112470.
160. Wen, B., M. Freilich, and L. Kuhn, *Chapter 55—Bone Tissue Engineering Around Dental Implants*, in *Stem Cell Biology and Tissue Engineering in Dental Sciences*, A. Vishwakarma, et al., Editors. 2015, Academic Press: Boston. p. 749–764.
161. Plantz, M., et al., *Preclinical Safety of a 3D-Printed Hydroxyapatite-Demineralized Bone Matrix Scaffold for Spinal Fusion.* Spine (Phila Pa 1976), 2022. **47**(1): p. 82–89.
162. Driscoll, J.A., et al., *3D-Printed Ceramic-Demineralized Bone Matrix Hyperelastic Bone Composite Scaffolds for Spinal Fusion.* Tissue Eng Part A, 2020. **26**(3-4): p. 157–166.
163. Wang, W., et al., *3D printing of PLA/n-HA composite scaffolds with customized mechanical properties and biological functions for bone tissue engineering.* Composites Part B: Engineering, 2021. **224**: p. 109192.
164. Murphy, S.V. and A. Atala, *3D bioprinting of tissues and organs.* Nature Biotechnology, 2014. **32**(8): p. 773–785.
165. Ventura, R.D., *An Overview of Laser-assisted Bioprinting (LAB) in Tissue Engineering Applications.* Medical Lasers, 2021. **10**(2): p. 76–81.

166. Liu, B., et al., *3D-bioprinted functional and biomimetic hydrogel scaffolds incorporated with nanosilicates to promote bone healing in rat calvarial defect model.* Mater Sci Eng C Mater Biol Appl, 2020. **112**: p. 110905.
167. Li, J., et al., *Bioactive nanoparticle reinforced alginate/gelatin bioink for the maintenance of stem cell stemness.* Mater Sci Eng C Mater Biol Appl, 2021. **126**: p. 112193.
168. Pitacco, P., et al., *3D bioprinting of cartilaginous templates for large bone defect healing.* Acta Biomaterialia, 2023. **156**: p. 61–74.
169. Li, S., et al., *Inhibition of Sympathetic Activation by Delivering Calcium Channel Blockers from a 3D Printed Scaffold to Promote Bone Defect Repair.* Advanced Healthcare Materials, 2022. **11**(16): p. 2200785.
170. Xie, M., et al., *In situ 3D bioprinting with bioconcrete bioink.* Nat Commun, 2022. **13**(1): p. 3597.
171. Marrella, A., et al., *Engineering vascularized and innervated bone biomaterials for improved skeletal tissue regeneration.* Mater Today (Kidlington), 2018. **21**(4): p. 362–376.
172. Nulty, J., et al., *3D bioprinting of prevascularised implants for the repair of critically-sized bone defects.* Acta Biomaterialia, 2021. **126**: p. 154–169.
173. Twohig, C., et al., *A dual-ink 3D printing strategy to engineer pre-vascularized bone scaffolds in-vitro.* Materials Science and Engineering: C, 2021. **123**: p. 111976.
174. Shen, M., et al., *3D bioprinting of in situ vascularized tissue engineered bone for repairing large segmental bone defects.* Mater Today Bio, 2022. **16**: p. 100382.
175. Kim, W. and G. Kim, *Hybrid cell constructs consisting of bioprinted cell-spheroids.* Bioeng Transl Med, 2023. **8**(2): p. e10397.
176. Ma, S., et al., *Functionalized 3D Hydroxyapatite Scaffold by Fusion Peptides-Mediated Small Extracellular Vesicles of Stem Cells for Bone Tissue Regeneration.* ACS Applied Materials & Interfaces, 2024.
177. Ahn, G., et al., *Cleft Alveolus Reconstruction Using a Three-Dimensional Printed Bioresorbable Scaffold With Human Bone Marrow Cells.* J Craniofac Surg, 2018. **29**(7): p. 1880–1883.
178. Vorndran, E., et al., *Simultaneous Immobilization of Bioactives During 3D Powder Printing of Bioceramic Drug-Release Matrices.* Advanced Functional Materials, 2010. **20**(10): p. 1585–1591.
179. Lee, J.H., et al., *Development of a heat labile antibiotic eluting 3D printed scaffold for the treatment of osteomyelitis.* Sci Rep, 2020. **10**(1): p. 7554.
180. Wang, M., et al., *A 3D printed Ga containing scaffold with both anti-infection and bone homeostasis-regulating properties for the treatment of infected bone defects.* J Mater Chem B, 2021. **9**(23): p. 4735–4745.
181. Radhakrishnan, S., et al., *Fabrication of 3D printed antimicrobial polycaprolactone scaffolds for tissue engineering applications.* Mater Sci Eng C Mater Biol Appl, 2021. **118**: p. 111525.
182. Martin, V., et al., *Engineering a multifunctional 3D-printed PLA-collagen-minocycline-nanoHydroxyapatite scaffold with combined antimicrobial and osteogenic effects for bone regeneration.* Mater Sci Eng C Mater Biol Appl, 2019. **101**: p. 15–26.
183. Qiao, Z., et al., *3D-printed composite scaffold with anti-infection and osteogenesis potential against infected bone defects.* RSC Adv, 2022. **12**(18): p. 11008–11020.
184. Tarafder, S. and S. Bose, *Polycaprolactone-coated 3D printed tricalcium phosphate scaffolds for bone tissue engineering: in vitro alendronate release behavior and local delivery effect on in vivo osteogenesis.* ACS Appl Mater Interfaces, 2014. **6**(13): p. 9955–65.
185. Gomez-Cerezo, N., et al., *Mesoporous bioactive glass/varepsilon-polycaprolactone scaffolds promote bone regeneration in osteoporotic sheep.* Acta Biomater, 2019. **90**: p. 393–402.
186. Sun, H., et al., *3D printed calcium phosphate scaffolds with controlled release of osteogenic drugs for bone regeneration.* Chemical Engineering Journal, 2022. **427**.
187. Bai, H., et al., *3D printed porous biomimetic composition sustained release zoledronate to promote osteointegration of osteoporotic defects.* Materials & Design, 2020. **189**: p. 108513.
188. Qiao, S., et al., *3D-printed Ti6Al4V scaffolds coated with freeze-dried platelet-rich plasma as bioactive interface for enhancing osseointegration in osteoporosis.* Materials & Design, 2020. **194**: p. 108825.

189. Sarkar, N. and S. Bose, *Controlled release of soy isoflavones from multifunctional 3D printed bone tissue engineering scaffolds.* Acta Biomater, 2020. **114**: p. 407–420.
190. Wang, C., et al., *Cryogenic 3D printing of porous scaffolds for in situ delivery of 2D black phosphorus nanosheets, doxorubicin hydrochloride and osteogenic peptide for treating tumor resection-induced bone defects.* Biofabrication, 2020. **12**(3): p. 035004.
191. Liu, Y., et al., *3D-printed scaffolds with bioactive elements-induced photothermal effect for bone tumor therapy.* Acta Biomaterialia, 2018. **73**: p. 531–546.
192. Hollister, S.J., *Porous scaffold design for tissue engineering.* Nature materials, 2005. **4**(7): p. 518–524.
193. Zhang, L., et al., *Three-dimensional (3D) printed scaffold and material selection for bone repair.* Acta Biomaterialia, 2019. **84**: p. 16–33.
194. Wang, Z., et al., *Analysis of factors influencing bone ingrowth into three-dimensional printed porous metal scaffolds: A review.* Journal of Alloys and Compounds, 2017. **717**: p. 271–285.
195. Ma, H., et al., *3D-printed bioceramic scaffolds: From bone tissue engineering to tumor therapy.* Acta Biomaterialia, 2018. **79**: p. 37–59.
196. Cheng, A., et al., *Advances in Porous Scaffold Design for Bone and Cartilage Tissue Engineering and Regeneration.* Tissue Eng Part B Rev, 2019. **25**(1): p. 14–29.
197. Cheng, A., et al., *Additively manufactured 3D porous Ti-6Al-4V constructs mimic trabecular bone structure and regulate osteoblast proliferation, differentiation and local factor production in a porosity and surface roughness dependent manner.* Biofabrication, 2014. **6**(4): p. 045007.
198. Sultan, S. and A.P. Mathew, *3D printed scaffolds with gradient porosity based on a cellulose nanocrystal hydrogel.* Nanoscale, 2018. **10**(9): p. 4421–4431.
199. Gupta, D., et al., *Multiscale porosity in compressible cryogenically 3D printed gels for bone tissue engineering.* ACS applied materials & interfaces, 2019. **11**(22): p. 20437–20452.
200. Dang, H.P., et al., *Porous 3D printed scaffolds for guided bone regeneration in a rat calvarial defect model.* Applied Materials Today, 2020. **20**: p. 100706.
201. Diloksumpan, P., et al., *Orthotopic bone regeneration within 3D printed bioceramic scaffolds with region-dependent porosity gradients in an equine model.* Advanced healthcare materials, 2020. **9**(10): p. 1901807.
202. Baptista, R. and M. Guedes, *Morphological and mechanical characterization of 3D printed PLA scaffolds with controlled porosity for trabecular bone tissue replacement.* Materials Science and Engineering: C, 2021. **118**: p. 111528.
203. Ran, Q., et al., *Osteogenesis of 3D printed porous Ti6Al4V implants with different pore sizes.* Journal of the Mechanical Behavior of Biomedical Materials, 2018. **84**: p. 1–11.
204. Bittner, S.M., et al., *Fabrication and mechanical characterization of 3D printed vertical uniform and gradient scaffolds for bone and osteochondral tissue engineering.* Acta biomaterialia, 2019. **90**: p. 37–48.
205. Entezari, A., et al., *Architectural design of 3D printed scaffolds controls the volume and functionality of newly formed bone.* Advanced healthcare materials, 2019. **8**(1): p. 1801353.
206. Fairag, R., et al., *Three-dimensional printed polylactic acid scaffolds promote bone-like matrix deposition in vitro.* ACS applied materials & interfaces, 2019. **11**(17): p. 15306–15315.
207. Rotbaum, Y., et al., *Quasi-static and dynamic in vitro mechanical response of 3D printed scaffolds with tailored pore size and architectures.* Materials Science and Engineering: C, 2019. **96**: p. 176–182.
208. Buenzli, P.R., et al., *Cell proliferation and migration explain pore bridging dynamics in 3D printed scaffolds of different pore size.* Acta Biomaterialia, 2020. **114**: p. 285–295.
209. Shao, H., et al., *Bone regeneration in 3D printing bioactive ceramic scaffolds with improved tissue/material interface pore architecture in thin-wall bone defect.* Biofabrication, 2017. **9**(2): p. 025003.
210. Baptista, R. and M. Guedes, *Porosity and pore design influence on fatigue behavior of 3D printed scaffolds for trabecular bone replacement.* Journal of the mechanical behavior of biomedical materials, 2021. **117**: p. 104378.

211. Deng, F., et al., *3D printed Ti6Al4V bone scaffolds with different pore structure effects on bone ingrowth.* Journal of biological engineering, 2021. **15**: p. 1–13.
212. Wu, R., et al., *Bone tissue regeneration: the role of finely tuned pore architecture of bioactive scaffolds before clinical translation.* Bioactive materials, 2021. **6**(5): p. 1242–1254.
213. Roosa, S.M.M., et al., *The pore size of polycaprolactone scaffolds has limited influence on bone regeneration in an in vivo model.* Journal of Biomedical Materials Research Part A, 2010. **92A**(1): p. 359–368.
214. Abbasi, N., et al., *Porous scaffolds for bone regeneration.* Journal of Science: Advanced Materials and Devices, 2020. **5**(1): p. 1–9.
215. Kaplan, F., et al., *Form and function of bone.* Orthopaedic basic science, 1994. **4**: p. 131–133.
216. Almela, T., et al., *Simulation of cortico-cancellous bone structure by 3D printing of bilayer calcium phosphate-based scaffolds.* Bioprinting, 2017. **6**: p. 1–7.
217. Velioglu, Z.B., et al., *3D-printed poly(lactic acid) scaffolds for trabecular bone repair and regeneration: scaffold and native bone characterization.* Connective Tissue Research, 2019. **60**(3): p. 274–282.
218. Amini, M., A. Reisinger, and D.H. Pahr, *Influence of processing parameters on mechanical properties of a 3D-printed trabecular bone microstructure.* J Biomed Mater Res B Appl Biomater, 2020. **108**(1): p. 38–47.
219. Gomez, S., et al., *Design and properties of 3D scaffolds for bone tissue engineering.* Acta Biomater, 2016. **42**: p. 341–350.
220. Senatov, F.S., et al., *Architected UHMWPE simulating trabecular bone tissue.* Materials Today Communications, 2018. **14**: p. 124–127.
221. Chantarapanich, N., et al., *Scaffold library for tissue engineering: a geometric evaluation.* Comput Math Methods Med, 2012. **2012**: p. 407805.
222. Roohani-Esfahani, S.I., P. Newman, and H. Zreiqat, *Design and Fabrication of 3D printed Scaffolds with a Mechanical Strength Comparable to Cortical Bone to Repair Large Bone Defects.* Sci Rep, 2016. **6**: p. 19468.
223. Qu, H., et al., *Fractal design boosts extrusion-based 3D printing of bone-mimicking radial-gradient scaffolds.* Research, 2021. **2021**.
224. Zhang, M., et al., *3D printing of Haversian bone–mimicking scaffolds for multicellular delivery in bone regeneration.* Science advances, 2020. **6**(12): p. eaaz6725.
225. Gorgin Karaji, Z., et al., *Additively Manufactured and Surface Biofunctionalized Porous Nitinol.* ACS Appl Mater Interfaces, 2017. **9**(2): p. 1293–1304.
226. Wang, H., et al., *Comparison of 3D-printed porous tantalum and titanium scaffolds on osteointegration and osteogenesis.* Mater Sci Eng C Mater Biol Appl, 2019. **104**: p. 109908.
227. Li, Y., et al., *Additively manufactured biodegradable porous magnesium.* Acta Biomater, 2018. **67**: p. 378–392.
228. Li, Y., et al., *Biodegradation-affected fatigue behavior of additively manufactured porous magnesium.* Additive Manufacturing, 2019. **28**: p. 299–311.
229. Kopp, A., et al., *Influence of design and postprocessing parameters on the degradation behavior and mechanical properties of additively manufactured magnesium scaffolds.* Acta Biomater, 2019. **98**: p. 23–35.
230. Konda Gokuldoss, P., S. Kolla, and J. Eckert, *Additive manufacturing processes: Selective laser melting, electron beam melting and binder jetting—Selection guidelines.* materials, 2017. **10**(6): p. 672.
231. Shuai, C., et al., *Optimization of TCP/HAP ratio for better properties of calcium phosphate scaffold via selective laser sintering.* Materials Characterization, 2013. **77**: p. 23–31.
232. Ananth, K.P., N.D. Jayram, and K. Muthusamy, *Novel approach to the 3D printing of biphasic calcium phosphate/molybdenum disulfide composite reinforced with polyamide12.* Smart Materials in Manufacturing, 2024. **2**: p. 100037.
233. Shuai, C., et al., *Structure and properties of nano-hydroxypatite scaffolds for bone tissue engineering with a selective laser sintering system.* Nanotechnology, 2011. **22**(28): p. 285703.

234. Qin, T., et al., *Bioactive tetracalcium phosphate scaffolds fabricated by selective laser sintering for bone regeneration applications.* Materials, 2020. **13**(10): p. 2268.
235. Le Guehennec, L., et al., *In vitro and in vivo biocompatibility of calcium-phosphate scaffolds three-dimensional printed by stereolithography for bone regeneration.* J Biomed Mater Res A, 2020. **108**(3): p. 412–425.
236. Zhang, C., et al., *DLP 3D printed silica-doped HAp ceramic scaffolds inspired by the trabecular bone structure.* Ceramics International, 2022. **48**(19): p. 27765–27773.
237. Zhou, T., et al., *SLA 3D printing of high quality spine shaped β-TCP bioceramics for the hard tissue repair applications.* Ceramics International, 2020. **46**(6): p. 7609–7614.
238. Liu, Z., et al., *Additive manufacturing of hydroxyapatite bone scaffolds via digital light processing and in vitro compatibility.* Ceramics International, 2019. **45**(8): p. 11079–11086.
239. Wang, Z., et al., *Development of a novel aqueous hydroxyapatite suspension for stereolithography applied to bone tissue engineering.* Ceramics International, 2019. **45**(3): p. 3902–3909.
240. Moghadasi, M., et al., *Ceramic binder jetting additive manufacturing: Effects of particle size on feedstock powder and final part properties.* Ceramics International, 2020. **46**(10, Part B): p. 16966–16972.
241. Ahn, J.-H., et al., *3D-printed biodegradable composite scaffolds with significantly enhanced mechanical properties via the combination of binder jetting and capillary rise infiltration process.* Additive Manufacturing, 2021. **41**: p. 101988.
242. Zhou, Z., et al., *Binder jetting additive manufacturing of hydroxyapatite powders: Effects of adhesives on geometrical accuracy and green compressive strength.* Additive Manufacturing, 2020. **36**: p. 101645.
243. Sahu, K.K. and Y.K. Modi, *Multi response optimization for compressive strength, porosity and dimensional accuracy of binder jetting 3D printed ceramic bone scaffolds.* Ceramics International, 2022. **48**(18): p. 26772–26783.
244. Houmard, M., et al., *On the structural, mechanical, and biodegradation properties of HA/β-TCP robocast scaffolds.* Journal of Biomedical Materials Research Part B: Applied Biomaterials, 2013. **101**(7): p. 1233–1242.
245. Eqtesadi, S., et al., *Robocasting of 45S5 bioactive glass scaffolds for bone tissue engineering.* Journal of the European Ceramic Society, 2014. **34**(1): p. 107–118.
246. Lei, L., et al., *Core–Shell Bioactive Ceramic Robocasting: Tuning Component Distribution Beneficial for Highly Efficient Alveolar Bone Regeneration and Repair.* ACS Biomaterials Science & Engineering, 2020. **6**(4): p. 2376–2387.
247. Tabard, L., et al., *Robocasting of highly porous ceramics scaffolds with hierarchized porosity.* Additive Manufacturing, 2021. **38**: p. 101776.
248. Stanciuc, A.-M., et al., *Robocast zirconia-toughened alumina scaffolds: Processing, structural characterisation and interaction with human primary osteoblasts.* Journal of the European Ceramic Society, 2018. **38**(3): p. 845–853.
249. Guillaume, O., et al., *Surface-enrichment with hydroxyapatite nanoparticles in stereolithography-fabricated composite polymer scaffolds promotes bone repair.* Acta Biomater, 2017. **54**: p. 386–398.
250. Feng, Z., et al., *Graphene-Reinforced Biodegradable Resin Composites for Stereolithographic 3D Printing of Bone Structure Scaffolds.* Journal of Nanomaterials, 2019. **2019**: p. 1–13.
251. Nettleton, K., et al., *Molecular Mass-Dependent Resorption and Bone Regeneration of 3D Printed PPF Scaffolds in a Critical-Sized Rat Cranial Defect Model.* Adv Healthc Mater, 2019. **8**(17): p. e1900646.
252. Luo, Y., et al., *Desktop-Stereolithography 3D Printing of a Polyporous Extracellular Matrix Bioink for Bone Defect Regeneration.* Front Bioeng Biotechnol, 2020. **8**: p. 589094.
253. Eshraghi, S. and S. Das, *Micromechanical finite-element modeling and experimental characterization of the compressive mechanical properties of polycaprolactone-hydroxyapatite composite scaffolds prepared by selective laser sintering for bone tissue engineering.* Acta Biomater, 2012. **8**(8): p. 3138–43.

254. Liao, H.T., et al., *Osteogenesis of adipose-derived stem cells on polycaprolactone-beta-tricalcium phosphate scaffold fabricated via selective laser sintering and surface coating with collagen type I.* J Tissue Eng Regen Med, 2016. **10**(10): p. E337–E353.
255. Du, Y., et al., *Selective laser sintering scaffold with hierarchical architecture and gradient composition for osteochondral repair in rabbits.* Biomaterials, 2017. **137**: p. 37–48.
256. Gayer, C., et al., *Development of a solvent-free polylactide/calcium carbonate composite for selective laser sintering of bone tissue engineering scaffolds.* Mater Sci Eng C Mater Biol Appl, 2019. **101**: p. 660–673.
257. Sun, Z., et al., *A Dexamethasone-Eluting Porous Scaffold for Bone Regeneration Fabricated by Selective Laser Sintering.* ACS Appl Bio Mater, 2020. **3**(12): p. 8739–8747.
258. Tortorici, M., et al., *Inner strut morphology is the key parameter in producing highly porous and mechanically stable poly(epsilon-caprolactone) scaffolds via selective laser sintering.* Mater Sci Eng C Mater Biol Appl, 2021. **123**: p. 111986.
259. Mohd Pu'ad, N.A.S., et al., *Review on the fabrication of fused deposition modelling (FDM) composite filament for biomedical applications.* Materials Today: Proceedings, 2020. **29**: p. 228–232.
260. Prajapati, H., et al., *Measurement of anisotropic thermal conductivity and inter-layer thermal contact resistance in polymer fused deposition modeling (FDM).* Additive Manufacturing, 2018. **21**: p. 84–90.
261. Percoco, G., et al., *Mechanobiological approach to design and optimize bone tissue scaffolds 3D printed with fused deposition modeling: A feasibility study.* Materials, 2020. **13**(3): p. 648.
262. Capek, J., et al., *Highly porous, low elastic modulus 316L stainless steel scaffold prepared by selective laser melting.* Mater Sci Eng C Mater Biol Appl, 2016. **69**: p. 631–9.
263. Zhao, B., et al., *Corrosion resistance characteristics of a Ti-6Al-4V alloy scaffold that is fabricated by electron beam melting and selective laser melting for implantation in vivo.* Mater Sci Eng C Mater Biol Appl, 2017. **70**(Pt 1): p. 832–841.
264. Ataee, A., et al., *Ultrahigh-strength titanium gyroid scaffolds manufactured by selective laser melting (SLM) for bone implant applications.* Acta Materialia, 2018. **158**: p. 354–368.
265. Pei, X., et al., *Fabrication of customized Ti6Al4V heterogeneous scaffolds with selective laser melting: Optimization of the architecture for orthopedic implant applications.* Acta Biomater, 2021. **126**: p. 485–495.
266. Ataee, A., Y. Li, and C. Wen, *A comparative study on the nanoindentation behavior, wear resistance and in vitro biocompatibility of SLM manufactured CP-Ti and EBM manufactured Ti64 gyroid scaffolds.* Acta Biomater, 2019. **97**: p. 587–596.
267. Kelly, C.N., et al., *Fatigue behavior of As-built selective laser melted titanium scaffolds with sheet-based gyroid microarchitecture for bone tissue engineering.* Acta Biomater, 2019. **94**: p. 610–626.
268. Zhu, L., et al., *Design and Compressive Fatigue Properties of Irregular Porous Scaffolds for Orthopedics Fabricated Using Selective Laser Melting.* ACS Biomater Sci Eng, 2021. **7**(4): p. 1663–1672.
269. Ginestra, P., et al., *Selective laser melting and electron beam melting of Ti6Al4V for orthopedic applications: a comparative study on the applied building direction.* Materials, 2020. **13**(23): p. 5584.
270. Ma, L., et al., *Immobilizing magnesium ions on 3D printed porous tantalum scaffolds with polydopamine for improved vascularization and osteogenesis.* Mater Sci Eng C Mater Biol Appl, 2020. **117**: p. 111303.
271. Guo, Y., et al., *Study of Bone Regeneration and Osteointegration Effect of a Novel Selective Laser-Melted Titanium-Tantalum-Niobium-Zirconium Alloy Scaffold.* ACS Biomater Sci Eng, 2019. **5**(12): p. 6463–6473.
272. Shah, F.A., et al., *Long-term osseointegration of 3D printed CoCr constructs with an interconnected open-pore architecture prepared by electron beam melting.* Acta Biomater, 2016. **36**: p. 296–309.
273. Li, Y., et al., *Additively manufactured biodegradable porous magnesium.* Acta Biomaterialia, 2018. **67**: p. 378–392.

274. Wen, P., et al., *Densification behavior of pure Zn metal parts produced by selective laser melting for manufacturing biodegradable implants.* Journal of Materials Processing Technology, 2018. **258**: p. 128–137.
275. Wen, P., et al., *Laser additive manufacturing of Zn metal parts for biodegradable applications: Processing, formation quality and mechanical properties.* Materials & Design, 2018. **155**: p. 36–45.
276. Shuai, C., et al., *Selective laser melted Fe-Mn bone scaffold: microstructure, corrosion behavior and cell response.* Materials Research Express, 2020. **7**(1): p. 015404.
277. Chou, D.T., et al., *Novel processing of iron-manganese alloy-based biomaterials by inkjet 3-D printing.* Acta Biomater, 2013. **9**(10): p. 8593–603.
278. Hong, D., et al., *Binder-jetting 3D printing and alloy development of new biodegradable Fe-Mn-Ca/Mg alloys.* Acta Biomater, 2016. **45**: p. 375–386.
279. Vangapally, S., et al., *Effect of Lattice Design and Process Parameters on Dimensional and Mechanical Properties of Binder Jet Additively Manufactured Stainless Steel 316 for Bone Scaffolds.* Procedia Manufacturing, 2017. **10**: p. 750–759.
280. Keriquel, V., et al., *In situ printing of mesenchymal stromal cells, by laser-assisted bioprinting, for in vivo bone regeneration applications.* Sci Rep, 2017. **7**(1): p. 1778.
281. Li, L., et al., *Robotic in situ 3D bio-printing technology for repairing large segmental bone defects.* J Adv Res, 2021. **30**: p. 75–84.
282. Mostafavi, A., et al., *In situ printing of scaffolds for reconstruction of bone defects.* Acta Biomater, 2021. **127**: p. 313–326.
283. Montoya, C., et al., *On the road to smart biomaterials for bone research: definitions, concepts, advances, and outlook.* Bone Res, 2021. **9**(1): p. 12.
284. Amukarimi, S., et al., *Smart biomaterials: From 3D printing to 4D bioprinting.* Methods, 2022. **205**: p. 191–199.
285. Mondal, K. and P.K. Tripathy, *Preparation of smart materials by additive manufacturing technologies: a review.* Materials, 2021. **14**(21): p. 6442.
286. Nagayama, Y., et al., *Deep Learning-based Reconstruction for Lower-Dose Pediatric CT: Technical Principles, Image Characteristics, and Clinical Implementations.* Radiographics, 2021. **41**(7): p. 1936–1953.
287. Wu, C.T., Y.H. Yang, and Y.Z. Chang, *Three-dimensional deep learning to automatically generate cranial implant geometry.* Sci Rep, 2022. **12**(1): p. 2683.
288. Indranil, G., et al. *Deep learning based high-resolution reconstruction of trabecular bone microstructures from low-resolution CT scans using GAN-CIRCLE.* in *Proc.SPIE*. 2020.

# Design and Culture of Biomaterial Scaffolds for Orthopedic Applications

**Jaydon Gibson, Weitong Chen, Alexis V. Graham, Sophie J. McLay, April Guo-Yue, and Lauren B. Priddy**

## 1 Introduction

Biomaterial scaffolds serve as a critical component of tissue engineering. When designing and culturing cell-seeded scaffolds, several key concepts should be considered, including scaffold design, material selection, fabrication method, surface functionalization, cell type and source, cell seeding method, and cell culture environment such as bioreactor systems (Fig. 1). This chapter will explore these key concepts of biomaterial scaffolds, reviewing foundational contributions to the field and providing suggestions for future work.

For orthopedic applications, biomaterial scaffolds can be used in fracture fixation or healing of segmental bone defects. These scaffolds play an important role by

J. Gibson · A. Guo-Yue · L. B. Priddy (✉)
Department of Agricultural and Biological Engineering, Mississippi State University, Mississippi State, MS, USA
e-mail: lbpriddy@abe.msstate.edu

W. Chen
Department of Agricultural and Biological Engineering, Mississippi State University, Mississippi State, MS, USA

Department of Small Animal Medicine & Surgery, University of Georgia, Athens, GA, USA

A. V. Graham
Department of Agricultural and Biological Engineering, Mississippi State University, Mississippi State, MS, USA

Charles River Laboratories, Memphis, TN, USA

S. J. McLay
Department of Agricultural and Biological Engineering, Mississippi State University, Mississippi State, MS, USA

LifeNet Health, Virginia Beach, Virginia, USA



J. Liao, J. Y. Wong (eds.), *Integration and Bridging of Multiscale Bioengineering Designs and Tissue Biomechanics*, https://doi.org/10.1007/978-3-031-81743-4_6

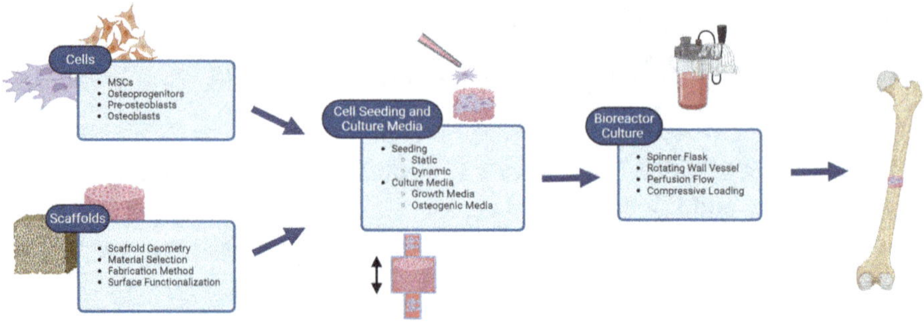


**Fig. 1** Design and culture of cell-seeded biomaterial scaffolds for osteogenesis in orthopedic applications

promoting fracture healing and/or bone regeneration and, ultimately, restoration of skeletal function. To ensure the scaffold has sufficient functionality, scaffold properties such as bioactivity and mechanical strength must be considered. By discussing the principles that comprise scaffold design and fabrication, cell type, and seeding and culture of scaffolds, particularly in physiologically relevant bioreactor systems, this chapter provides insight into the development and use of biomaterial scaffolds in orthopedic applications. Collectively, the information included herein can be leveraged for the long-term goal of enhancing the function of biomaterials for orthopedic medicine.

## 2 Biomaterial Scaffolds

A critical component and foundational element of any tissue engineering approach is the biomaterial scaffold. One of the primary objectives of the scaffold is to mimic biochemical and mechanical functions of the extracellular matrix (ECM) of the tissue which is being regenerated. The ECM is the noncellular part of all tissues, which facilitates cell functions including attachment, proliferation, migration, and differentiation, to name a few. Notably, nearly all cells, with the exception of blood cells, are anchorage-dependent, relying on physical attachment for proliferation and functionality [1]. Scaffolds serve two main purposes in tissue engineering: first, they offer a necessary structure for cellular attachment and proliferation, and second, they provide mechanical support for stimulation that is essential for the growth of tissues.

Cellular adhesion/attachment, wherein a cell binds to a substrate such as another cell, the ECM, or a biomaterial scaffold, is a crucial process in tissue growth and homeostasis, creating and sending signals that regulate cell differentiation, cell cycle, migration, and survival [2]. Cellular attachment to the ECM occurs via the ECM receptors known as integrins, discoidin domain receptors, and syndecans [3]. One study on adrenal cortex and granulosa cell proliferation on petri dishes coated with either collagen types I–IV or natural ECM derived from corneal endothelial cells showed that cells seeded on the native ECM had higher proliferation rates than those seeded on the collagen [4].

Mechanical stimulation plays a critical role in bone growth and development, as the skeleton can sense and respond to mechanical signals, aiding in maintaining tissue homeostasis [5]. These mechanical forces regulate bone development, repair, and regeneration by influencing many cell types which are involved in various phases of bone repair, including neutrophils, macrophages, endothelial cells, osteoblasts, osteoclasts, mesenchymal stem cells, and skeletal stem cells [6]. When it comes to fracture healing, mechanical stimulation is crucial from the onset, helping with normal callus formation and endochondral ossification, the process whereby bone is formed from cartilage [7]. Animal models have demonstrated that mechanical stimulation during the bone remodeling phase significantly enhances callus formation, which promotes effective fracture repair [8]. Mechanical stimulation, applied at appropriate intensities and frequencies during the proliferation and remodeling phases, facilitates bone formation at the fracture site [7]. The mechanotransduction process, which is the process of a cell to sense, integrate, and convert mechanical stimulation into signals, plays a fundamental role in both bone maintenance and bone repair. Osteocytes act as mechanosensors by communicating with each other and responding to stress, strain, and shear fluid stress by transmitting mechanical signals [9]. Various mechanosensitive proteins mediate mechanical sensation in osteoblasts and osteocytes, playing key roles in signal transduction pathways [9]. In the response stage, mechanically stimulated osteocytes produce factors which influence bone formation and adaptation to the mechanical environment [10].

## 3 Scaffold Design

One of the most important parameters in the design of scaffolds is their porosity, as it allows for the exchange of nutrients and oxygen throughout the scaffold and the removal of waste from the scaffold [11]. When describing porosity, four main components should be considered: interconnectivity, pore size, pore shape, and pore distribution. Interconnectivity is the pathway that connects the pores to one another throughout the scaffold, allowing for the transfer of nutrients, migration of cells, and establishment of microvasculature. One study explored the potential for bone ingrowth into scaffolds composed of hydroxyapatite (HA) or β-tricalcium phosphate (β-TCP) with the same mean sizes of pores (100–300 μm) and interconnections between the pores (30–100 μm); while human osteoblasts infiltrated, colonized, and proliferated in interconnections at least 20 μm in size, in both scaffold types, 60 μm was the most frequent interconnection size which contained cells [12]. Indeed, determining the effects of pore size on cell function is critical for enhancing scaffold design and function. In another study, collagen scaffolds with smaller pores (85, 120, 150, and 190 μm) had greater initial cell attachment, presumably due to their larger surface area, than scaffolds with larger (325 μm) pores) [13]. Though the 325 μm pores had a lower initial cell attachment, cells exhibited a higher proliferation rate, likely resulting from increased migration throughout the scaffold. Indeed, pore distribution plays a pivotal role in scaffold design and function, as it can increase the rate of proliferation of bone matrix [14]. Another study on cell growth into two scaffolds with varying porosities—ST1 with porosity of 30% and pore size of ~350 μm and ST2 with porosity of 50% and pore size of ~700 μm—reported higher cell counts at day 14 on ST2 compared to ST1 scaffolds. Of the total pore volume, both scaffolds possessed minimal closed pores (~0.1% in ST1 and ~0.001% in ST2), which was likely advantageous for cell migration through the interconnected pores [15].

## 4 Material Selection

Material selection is an important step in the scaffold design process, to ensure biocompatibility and mechanical function. Classes of materials which have been explored in tissue engineering include polymers, ceramics, polymer-ceramic composites, metals, and decellularized tissues.

### *4.1 Polymers*

Using polymeric materials for biomaterial scaffolds is a relatively modern approach which is popular in the field of tissue engineering. In particular, synthetic polymers afford design versatility not afforded by naturally derived materials. Some of the most commonly studied synthetic polyesters are polylactic acid (PLA), polyglycolic acid (PGA), poly(lactic-co-glycolic) acid (PLGA), and polycaprolactone (PCL) [16]. These polymers are used mainly for their biocompatibility and biodegradability, properties which improve cell survival on the scaffold and allow the scaffold to degrade as the tissue grows to ensure proper remodeling, respectively [17]. PLA is often chosen due to its biodegradability, biocompatibility, and mechanical properties, which can benefit cell proliferation and tissue development [15, 18]. In a 2017 study, Gregor et al. examined three dimensional (3D)-printed PLA scaffolds (ST1 and ST2) with different porosity levels (30% and 50%, respectively) seeded with osteosarcoma cells and reported higher cell proliferation at day 14 on ST2 than ST1 [15]. Another study on PLA scaffolds seeded with human osteoblasts either before or after apatite coating showed good cell adherence and near 100% cell viability on both non-coated and coated scaffolds [18]. Similarly, a study on PLA scaffolds seeded with mouse osteosarcoma cells, either before or after nanohydroxyapatite (nHA) coating, showed that while the PLA-nHA scaffolds had higher initial cell attachment, all scaffolds facilitated cell proliferation through 5 days [19].

PLGA, a copolymer of lactic acid and glycolic acid, is widely used for tissue engineering as it is also biocompatible and biodegradable. Scaffolds composed of PLGA mixed with type I collagen at a 90 to 10 ratio (PLGA to collagen) were seeded with either HeLa cells or NIH-3T3 pre-osteoblast cells and compared with cells seeded on PLGA-only and collagen-only scaffolds [20]. SEM analysis 48 hours after seeding demonstrated good cell attachment for both cell types on all scaffolds. Cell viability on day 7 was highest for HeLa cells on PLGA-collagen scaffolds compared to PLGA-only and collagen-only controls. Interestingly, scaffold type had no effect on the viability of pre-osteoblasts. Another study on PLGA scaffolds, with and without nHA coatings, seeded with MC3T3 cells reported the PLGA scaffolds with no coating facilitated good cell attachment and proliferation at days 1, 3, and 5, though cell number was highest on the combination polydopamine-nHA coating at day 5 [21]. Both studies show PLGA's suitability as an effective base polymer for scaffolds, suggesting that integration of additional bioactive materials may further enhance cell attachment and viability.

### 4.2 *Ceramics*

Inorganic bioceramics such as hydroxyapatite (HA), tricalcium phosphate (TCP), calcium silicates, and bioactive glasses have been widely applied for bone tissue engineering since they possess chemical and structural similarity to the mineral phase of bone, facilitating effective interactions with osteogenic cells (i.e., osteoconductivity) [22–25]. HA can be categorized by particle size of typically micro- or nano-scales (mHA and nHA, respectively). As nHA can effectively create nanotopology, it can improve wettability and cell function [26]. Due to the larger surface area to volume ratio compared to mHA, nHA enables rapid biomineralization [27], can better mechanically strengthen the matrix material [28], and has a higher degradation rate compared to mHA [29]. TCP exhibits various crystalline structures, with β-TCP being widely favored due to its superior structural, mechanical, and biological properties, as well as its lower formation temperature, compared to the α-TCP phase [30].

### 4.3 *Polymer-ceramic Composites*

Despite their advantages, ceramic materials, by nature, are brittle and have limited ductility and tensile strength. As a result, they should be utilized only alongside strong internal fixation in orthopedic procedures until the surrounding bone has sufficiently regenerated [31]. Conversely, as polyesters are typically hydrophobic and show poor osteoconductivity, HA and TCP have been used as support materials (e.g., composite blend or surface coating) to enhance the biological performance of polyesters by improving the hydrophilicity and surface roughness [32, 33] and promoting cell attachment and differentiation [34–38]. Polyester as a polymer binder can reduce the brittleness of HA [39], and nHA blended polyester (e.g., PCL) scaffolds have demonstrated improved mechanical properties compared to polymer alone [25, 40]. Introducing HA and TCP can also buffer the local acidic environment resulting from the acidic byproducts of polyester degradation [35, 41–45]. In the composite material, the presence of nHA reorganizes the polymer chain, increases the crystallinity of the polymer substrate, and inhibits crack growth in the polymer phase, thus enhancing the mechanical properties [34, 46–48]. Expectedly, the amount of HA in the PLGA substrate is a critical variable which affects the mechanical and biological functionality of the composite. 3D-printed PLA scaffolds containing 10% nHA demonstrated a higher compressive modulus than PLGA alone [49]. Interestingly, PLGA with 30% nano-biphasic calcium phosphate showed the highest Young's modulus compared to 10%, 20%, 40%, and 50% [35]. Similarly, PLGA with 20% nHA showed the highest compressive modulus, which decreased with increasing amount of nHA (10%, 30%, and 40%) [50]. Likewise with PCL, 20% β-TCP blended PCL scaffolds exhibited the highest compressive modulus compared to the 0%, 10%, and 30% counterparts [34]. The plateauing of compressive modulus at 20% ceramic content may result from the disruption of the carbon chains and local aggregation of ceramic particles [36, 37]. In addition to enhancing mechanical properties of the polymer matrix, nHA can also enhance the biological properties of polymer matrices. In one study, PLGA with an 85:15 lactic to glycolic acid ratio, along with collagen type I and hydroxyapatite nanorods, were electrospun to form scaffolds [51]. The study evaluated MC3T3-E1 cells, an osteoblastic cell line, seeded on three types of scaffolds: PLGA alone, PLGA-collagen, and PLGA-collagen-HA. Cell viability and alkaline phosphate (ALP) activity, an early indicator of osteoblastic differentiation, were highest in the scaffolds with HA [51]. Interestingly, although both HA-polyester and TCP-polyester composites have been shown to improve cell-substrate interactions compared to polyester alone, cell viability and proliferation on PCL-HA composite scaffolds was superior to that on PCL-TCP, likely due to stronger cell adhesion potentially resulting from more nucleating sites for mineral deposition on PCL-HA composites [25].

Other types of bioceramics, such as calcium silicates and bioglasses (e.g., $CaSiO_3$, β-$Ca_2SiO_4$, and $Na_2O$–$K_2O$–$CaO$–$MgO$–$P_2O_5$–$SiO_2$), also confer promising biological properties. During degradation, the released calcium and silicon ions stimulate cell adhesion, proliferation, and mineral deposition [52, 53]. However, the unsatisfactory mechanical properties due to high porosity and poor densification during the sintering manufacturing process limit the applications to small implants and non-load bearing sites [54]. Similar to HA and TCP, doping these bioceramics and bioglasses into polymers as biocomposite materials is the primary strategy to improve mechanical properties [30, 55].

## 4.4 Metals

Metallic scaffolds have been widely used for bone fracture fixation due to their excellent mechanical properties (e.g., yield strength, hardness, and ultimate tensile strength). One study of rat bone marrow stromal cells (rBMSCs) cultured on titanium scaffolds under various culture conditions showed effective cell growth and ALP activity on the scaffolds [56]. In a recent study, 3D-printed titanium scaffolds with ~92% porosity were tailored for mandibular bone regeneration via their biomechanical strength and sufficient pore size; in a dog model and in clinical trials, good scaffold integration and positive postoperative outcomes were observed [57].

Although metals commonly used for bone implants such as titanium, stainless steel, and cobalt-chromium alloys have high mechanical strength and fracture toughness, they lack similar elastic moduli to bone [58]. This mismatch in elastic moduli can lead to stress shielding and result in reduced stimulation of new bone growth and remodeling and in turn, decreased implant stability [59]. Further, metal alloys can release toxic ions during corrosion, which can elicit inflammation, reducing biocompatibility and further inhibiting osseointegration [58]. If the implant is successful in healing bone tissue sufficiently, it may still cause long-term pain and need to be removed by a second surgical procedure, increasing costs and risk of patient morbidity [58].

Biodegradable alloys are being studied as an alternative to traditional, non-degradable metal implants, as they can temporarily replace the function of the bone for mechanical stabilization until sufficient new bone tissue is formed [60]. These implants could be particularly useful for wrist or ankle fracture fixation, or as a bone void filler [61]. Magnesium alloys show considerable promise as a biodegradable implant given their biocompatibility, ability to induce osteogenesis, and mechanical properties similar to those of bone [62, 63]. Magnesium is the fourth most abundant cation in the human body, and half of the total magnesium is stored in bone tissue [58]. Magnesium has greater fracture toughness than ceramic biomaterials, while having an elastic modulus, density, and compressive yield strength closer to those of bone than do other commonly used metallic implants [58]. With biodegradable implants such as magnesium alloys, stress shielding may be avoided due to the gradual loading on regenerated bone tissue, the need for a secondary surgery would be obviated, and biocompatibility and osteogenesis can be increased [58, 60, 62, 64].

Although magnesium has promising advantages, it tends to degrade rapidly in the highly aqueous and ionic environment of the human body [58]. A byproduct of magnesium degradation is hydrogen gas, which must be resolved efficiently by immune cells. Further, if the implant degrades too rapidly, the mechanical integrity of the implant may not suffice until the bone tissue has become sufficiently load bearing (~weeks to months) [58]. It is therefore necessary to improve the corrosion resistance of magnesium and other degradable metallic biomaterials through methods such as alloying or coating.

### 4.5 Decellularized Matrices

Naturally derived, decellularized materials may function as ideal scaffold materials, particularly for cellular attachment. Indeed, one of the most common sources of biomaterial scaffolds is decellularized extracellular matrix (dECM) [65]. The extracellular matrix (ECM) is one of the fundamental components of all tissues, providing the structural framework for all tissue growth and playing a crucial role in tissue formation and development. The decellularized form of ECM maintains the tissue's structural and (some of its) biological potential, while having the cellular components removed, which reduces the risk of immune rejection. Once the native cellular material has been removed, the desired cells can be seeded (recellularization). There are four main processes of decellularization: perfusion, pressure gradient, supercritical fluid, and immersion with agitation [66]. Each of these processes is complex and requires manual operation and delicate control, making effective decellularization challenging [67]. Researchers are currently exploring automated and semiautomated systems for monitoring and increasing efficiency of the decellularization process [68].

Recently, much research has been done on dECM, focused primarily on understanding the structure and properties of dECM, the techniques to modify dECM [69], and common applications of dECM, including for orthopedic tissue repair [70, 71]. Research on dECM showed these decellularized matrices were composed of collagen, laminin, fibronectin, and glycosaminoglycans (GAGs), which acted as structural, functional, and signaling molecules [70, 72]. One type of naturally derived scaffold used specifically for osteogenesis is demineralized bone matrix (DBM), which is used in bone grafting due to its structure matching that of native bone [69, 73]. Besides orthopedic tissue repair, other examples of successful applications of dECM in tissue engineering have been for cardiac muscle, corneal, and cartilaginous tissue engineering [71].

One consideration which must be assessed with any dECM is the decrease in protein composition between natural and decellularized ECM. For example, in porcine cartilage, the decellularization process led to a large reduction in the collagen and GAG content and a complete loss of elastin, a major component of cartilage [74]. However, another study on the decellularization of human renal cortical tissue showed the preservation of native matrix proteins, including collagen type IV, laminin, and heparin, following decellularization [75].

## 5 Fabrication Methods

When engineering scaffolds for tissue engineering, fabrication methods can be considered either additive manufacturing or subtractive manufacturing. Additive manufacturing is the process of developing a scaffold layer by layer, while subtractive manufacturing is the process of starting with a solid material and then

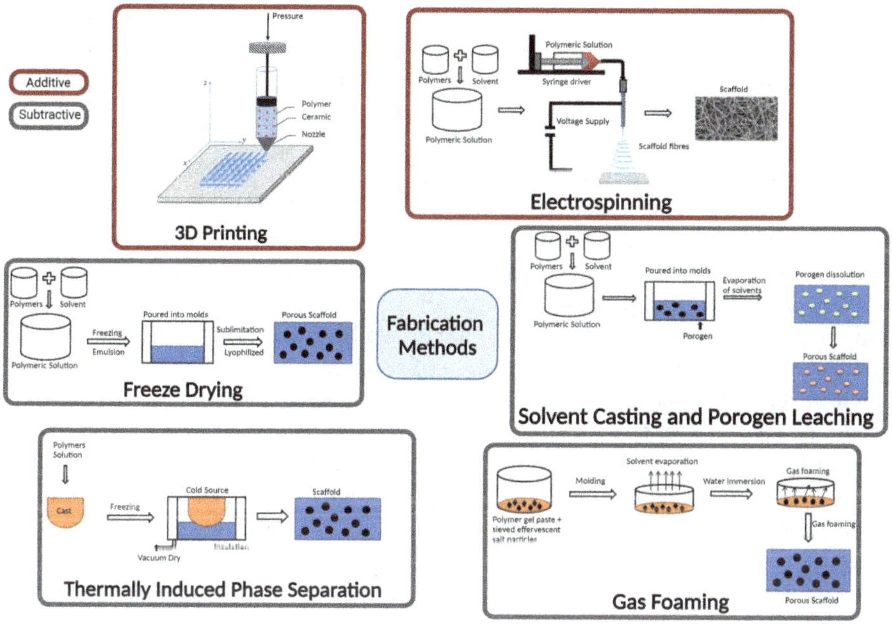


**Fig. 2** Fabrication methods for polymeric scaffolds include 3D printing, electrospinning, freeze-drying, solvent casting, and porogen leaching, thermally induced phase separation (TIPS), and gas foaming. (Adapted from Kumar and Jacob. 2022 and is subject to the terms and conditions of the Creative Commons Attribution (CC BY) license (https://creativecommons.org/licenses/by/4.0/) [84])

removing the material until the desired design is achieved (Fig. 2). Three main methods of additive manufacturing (AM), commonly referred to as 3D printing, are fused deposition modeling (FDM), electrospinning, and powder bed fusion (PBF). Fused deposition modeling is a process that uses filament to print complex scaffolds with intricate designs in a repeated pattern [76]. The most frequently used materials for FDM are polymers, as they are easier to print in a controlled manner. Electrospinning is a method that involves ejecting an electroconductive polymer from a spinneret connected to a high-voltage source to create nanofibrous scaffolds [77]. This method allows for control over parameters such as solution flow rate and distance between the needle and collector, which affect scaffold morphology. The electrospinning method has been widely used with many polymers. Powder bed fusion is a more complex method, as it involves a laser or electron beam that engages with a powdered material from the bottom up and fuses the powder into a solid form [78]. PBF is known for creating intricate designs with better precision than other methods. Materials which can be fabricated via this method are polymers (e.g., in selective laser sintering) as well as metals (e.g., laser powder bed fusion). When introducing porosity, particularly into metal-based biomaterials, AM enables customization of a scaffold to mimic the micro-architecture and properties of human bone [79]. The absorption properties are tunable as well, which allows sufficient loadbearing capacity throughout the bone healing process [79]. Absorption, yield strength, and elastic modulus are tunable chiefly through scaffold geometry, which includes porosity, pore size/shape, pore interconnectivity, and pore distribution. The increase in surface area from AM can facilitate cell adhesion and proliferation when compared to the bulk material counterparts [79].

In subtractive manufacturing, primary methods include porogen leaching/solvent casting, gas foaming, freeze-drying, and thermally induced phase separation. Porogen leaching is a process where a material is placed in a solvent to dissolve it; then, insoluble porogens (like salt or sugar crystals) are added, the solvent is allowed to evaporate, and the porogens are rinsed/leached out, resulting in a porous scaffold [80]. Gas foaming is the process of using a salt crystal that produces gas when mixed with a polymer. By placing the mixture of polymer and salt into hot water, the gas is released which creates pores throughout the polymer scaffold [81]. Freeze-drying involves three main steps: pretreatment, freezing, and drying [82]. The pretreatment stage is when treatments are applied to the material to enhance the structure's stability. The freezing process is when this mixture is cooled past the solvent's triple point, causing the solvent to freeze and form crystals in the structure. Then, the drying process is used to convert these crystals directly from a solid to a gas, leaving pores throughout the scaffold. Thermally induced phase separation (TIPS) is a technique that involves separating a uniform polymer solution into two phases: a polymer-rich phase and a solvent-rich phase [83]. The presence of the two phases allows the formation of pores in the material, as the solvent-rich phase will become pores within the polymer-rich phase.

## 6 Surface Functionalization

Surface functionalization of biomaterials plays a pivotal role in osseointegration and bone tissue engineering by influencing cellular responses at the interface between the biomaterial and the surrounding biological environment. Surface modifications are beneficial for enhancing the biocompatibility, osteoconductivity, and osteoinductivity of biomaterials, thereby promoting cell adhesion, proliferation, differentiation, and, ultimately, bone regeneration [85–87]. By tailoring surface functionalities such as the incorporation of bioactive molecules, chemistry, and topography, surface-functionalized biomaterials can mimic the native bone microenvironment, effectively modulating cell behavior and promoting tissue integration.

Inorganic bioactive coatings (calcium phosphate (CaP)-based materials), such as hydroxyapatite (HA), tricalcium phosphate (TCP), and CaP-bioglass, have garnered significant attention due to their ability to mimic the chemical composition and structure of bone mineral [88, 89]. Since CaP possesses a certain degree of alkalinity, it may serve to neutralize the acidity associated with inflammation and hydrolysis of polyesters when CaP coating is applied to a polyester substrate [90, 91]. Moreover, CaP-based coatings on metallic implants showed improved corrosion resistance, reduced metal ion release, and improved hydrophilicity that benefits cell adhesion [48, 92, 93]. Additionally, CaP-based coatings promote osteoblast attachment, proliferation, differentiation, and osseointegration, which has been demonstrated in several studies [19, 21, 94, 95]. These coatings enhance the bioactivity of biomaterials by providing osteoconductive cues and promoting cell-substrate interactions. Numerous studies have demonstrated the efficacy of inorganic coatings in improving the osseointegration of implants and facilitating new bone formation in vivo [96–98].

Organic bioactive coatings, such as extracellular matrix (ECM), artificial peptides, and growth factors, offer unique advantages in promoting cell–substrate interactions. ECM-based coatings (collagen, fibronectin, gelatin, etc.) create a biomimetic microenvironment that has been shown to support bone regeneration [99–101]. For example, collagen-crosslinked HA coating demonstrated improved cell adhesion, proliferation, and differentiation of MSCs when compared to HA alone [102]. However, it is difficult to control the absorption rate of ECM coatings, and the risks of immunological rejection and microbial infection associated with allogenic and xenogenic ECM coatings can be challenging [103, 104]. Artificial peptide coatings, such as arginine-glycine-aspartic acid (RGD), can be an alternative to avoid the disadvantages of native ECM, while still facilitating cell adhesion and viability/proliferation and ultimately osseointegration [105–107]. Nonetheless, the segment length is still not well controllable due to the complexity of peptide synthesis, limited understanding of structure-activity relationships, and the dynamic nature of integrin-ligand interactions [108]. Additionally, growth factor coatings (e.g., bone morphogenetic proteins (BMPs)) exert potent osteoinductive effects, significantly stimulating bone formation and remodeling, though the controllable release of growth factors to avoid overdosage remains an ongoing area of investigation [45, 109, 110]. Recently, dopamine coating has been identified as an effective modification with great potential that enables the immobilization of bioactive molecules onto biomaterial surfaces, further enhancing their bioactivity and regenerative potential [111, 112].

Surface immobilization with nano-metal particles (nMPs) such as silver (Ag), copper (Cu), copper oxide (CuO), gold (Au), zinc (Zn), zinc oxide (ZnO), and magnesium oxide (MgO) provides additional functionalities beyond promoting osteogenesis [113, 114]. These metal coatings exhibit broad-spectrum antimicrobial properties via the production of reactive oxygen species (ROS), which yields high oxidative stress causing alteration of proteins, inhibition of enzymes, lipid peroxidation, nucleic acid, and membrane damage, and cytolysis of bacteria, ultimately mitigating bacterial adhesion and growth on the biomaterial [115, 116]. Consequently, the risk of implant-associated infections, which are a common complication in orthopedic surgery, can be reduced. Nonetheless, given the cytotoxicity associated with these nano-metal particles, it is crucial to regulate their release from the coating to effectively inhibit bacteria with minimal local concentration and without adverse effects on host cells or tissues [117, 118]. Mokaber et al. developed a silver-containing surface modification (Ag/CaP) via electrochemical coating on titanium substrates to increase the biocompatibility and antibacterial properties of biomaterials for bone regeneration applications [119]. The controlled release of metal ions from coated biomaterials may impart long-lasting antimicrobial activity while maintaining biocompatibility, making metal coatings valuable additions to bone tissue engineering constructs.

Surface topography plays a critical role in cell behavior and tissue response. Surface modification techniques, including the generation of macro-, micro-, and nano-scale surface patterns, can allow for precise control over surface topography, which in turn affects cell adhesion, proliferation, and differentiation [120–122]. Specifically, macro-patterned (surface roughness >50 μm) structures have shown enhanced osseointegration and bone ingrowth, while micro-patterns (surface roughness <2 μm) further increase surface area which can increase protein adsorption on the biomaterial and ion exchange, supporting cell proliferation and formation of an apatite-like layer on the implant surface [120, 121, 123]. Nano-patterns can boost surface hydrophilicity, surface energy, osteoblast proliferation, and mineral deposition [124, 125]. Collectively, the introduction of surface roughness can promote osteogenic differentiation and mineralization, thus facilitating the formation of a stable bone-implant interface and enhancing overall implant performance.

In summary, surface functionalization of biomaterials represents a versatile and effective strategy for enhancing bone tissue engineering outcomes. By tailoring surface properties through inorganic and organic coatings, nano-metal particle incorporation, and surface topography modifications, researchers can engineer biomaterials with tailored bioactivity, antimicrobial properties, osteoconductivity, and osteoinductivity, ultimately advancing the research of bone tissue engineering and improving patient outcomes in orthopedic surgery.

## 7 Selection of Cell Type and Source

When choosing a cell type for studying osteogenesis in biomaterial scaffolds, two critical factors to consider are the stage of differentiation of the cells and whether the cells are primary cells or immortalized cell lines.

### *7.1 Stage of Differentiation*

In bone tissue engineering, osteogenic cells are frequently employed. Common osteogenic cell types include marrow stromal/mesenchymal stem cells (MSCs), osteoprogenitor cells, pre-osteoblasts, and osteoblasts. MSCs are the earliest (least differentiated) stage of the three and allow for examination of how culture conditions may impact osteogenic differentiation. Osteoprogenitor cells and pre-osteoblasts are at a stage between the nascent MSC and the mature osteoblast. Early markers of osteogenesis such as ALP [126] and runt-related transcription factor 2 (Runx2) increase as these cells undergo osteogenic differentiation [127]. These cell types can be useful for studying mid- to late-stage osteogenic differentiation. Osteoblasts are mature bone-forming cells which express late-stage specific markers such as osteopontin (OPN) and osteocalcin (OCN) [127]. When selecting a cell type, one should consider also the study duration and choose the markers of differentiation accordingly. For example, MSCs will require a longer culture duration to express late-stage bone-specific markers, like OPN, than osteoprogenitor cells will. In 3D scaffold culture, compatibility of the assays to measure osteogenic markers should also be considered according to the scaffold type used. For example, OPN, which is often detected by enzyme-linked immunosorbent assay (ELISA) of the culture medium, binds to calcium [128]. Thus, if the scaffold contains calcium, such as those constructed of HA, β-tricalcium phosphate (β-TCP), or bone extracellular matrix (ECM), proteins such as OPN may be difficult to quantitatively distinguish in the culture medium.

## 7.2 *Primary Cells Versus Immortalized Cell Lines*

Cells used in bone tissue engineering research are either primary or immortalized cell lines. Primary cells are isolated from living tissue and behave more like native cells than immortalized cells do. Primary cells can be derived from human or non-human cell sources and are typically a more heterogeneous cell population upon isolation [129–131]. Bone marrow-derived marrow stromal cells (BM-MSCs) are one of the most common cell types used in bone tissue engineering; however, less than 0.01% of the cells in the bone marrow are capable of differentiating into bone-forming cells [132]. Other sources of MSCs, including fat, muscle, and cord blood, as well as embryonic stem cells, have also been considered for use in bone tissue engineering, but BM-MSCs remain the most commonly utilized due to their ease of isolation, typically by bone marrow aspiration [133]. Despite their advantages, primary cells have a limited proliferative potential in vitro and experience senescence, or aging [134, 135]. As a result of this limited proliferative potential, studies commonly report using MSCs between passages 2 and 6, when proliferation and cell morphology are more consistent [33, 136–138]. To overcome the limitation of senescence, immortalized MSC (iMSC) lines have been developed, but currently, iMSCs are not capable of fully maintaining native MSC characteristics [139].

Both primary and immortalized cell lines are available commercially, so the cells have been characterized for quality and purity. Immortalized cell lines, such as MC3T3-E1 mouse pre-osteoblasts and MG63 human osteosarcoma cells, have been modified to replicate indefinitely, providing a more standardized and consistent cell function over the course of an experiment than primary cells. Proliferation and osteogenic differentiation of these cell types are commonly performed in bone tissue engineered scaffolds. For example, viability of MC3T3-E1 and MG63 cells cultured on biodegradable magnesium alloy-based scaffolds was evaluated [140, 141]. Other immortalized cell lines, such as, C2C12 mouse myoblast progenitor cells, can also undergo osteogenic differentiation, especially in the presence of bone morphogenic proteins (BMPs), and have been utilized to study cell attachment, proliferation, and osteogenic differentiation when cultured on materials such as biodegradable PLA and bioactive glass-based composite scaffolds [87].

# 8 Cell Seeding

Effectively seeding the desired cells onto scaffolds is an intricate process that is fundamental to tissue regeneration. The primary types of seeding techniques are referred to as static seeding and perfusion seeding. Two of the most common methods of static seeding are (i) placing the scaffold into a cell suspension, and (ii) dropwise seeding the cell suspension directly onto the scaffold. The seeding efficiency for static seeding has ranged from 10% to 25% [142]. Dynamic, or perfusion, seeding can be broken down into three main categories: oscillation seeding, rotational seeding, and vacuum seeding. Oscillation seeding includes passing the cell suspension back and forth through the scaffold repeatedly; this type of seeding can improve the seeding efficiency between 26% and 63% [143]. Rotational seeding typically consists of a system where the scaffold is rotated in the cell suspension with revolutions per minute (RPM) and time tailored for the intended purpose. This seeding technique has reported seeding efficiencies of 38–90% [142]. Vacuum seeding operates by using either an internal or external vacuum pressure to facilitate cell distribution through the scaffold. This seeding technique has shown a seeding efficiency range from 60% to 90% [142].

# 9 Cell Culture

## 9.1 Growth Media

Cell culture medium is an essential factor in cell culture and has an immediate effect on outcomes by dictating cell survival, proliferation, and differentiation. Basal medium is used to maintain the growth of cells and allow reproducibility of experimental results. Common basal mediums are available with low or high glucose, with or without a pH indicator (phenol red), and with or without L-glutamine [144]. They also typically contain no proteins, lipids, hormones, or growth factors, so they require supplementation with serum, which is commonly sourced from fetal calves (fetal bovine serum (FBS)) and used at concentration of 5–20% [144]. Antibiotics and antimycotics are also commonly added to minimize the risk of biological contamination from bacteria, yeast, fungi, or mycoplasma [144].

Dulbecco's Modified Eagle Medium (DMEM) and Minimum Essential Medium α-modification (α-MEM) are two frequently used basal cell culture media. Harry

Eagle first studied the amino acid requirements of various cells in 1955 and developed the minimum essential medium (MEM) in 1959 [145]. MEM was modified by Dulbecco and Freeman the same year to contain double the amino acids, becoming DMEM [145]. Then, in 1971, MEM was modified to α-MEM by supplementation with non-essential amino acids and vitamins such as Vitamin $B_{12}$, biotin, and ascorbic acid [145]. In studies comparing various culture conditions for MSCs, both DMEM Low Glucose (DMEM LG) and α-MEM were found to consistently support MSC growth [146, 147]. Both DMEM and α-MEM are suitable basal mediums for differentiating MSCs toward the osteogenic lineage, when supplemented with differentiation agents described below [148, 149]. In particular, DMEM was recommended for robust osteogenic differentiation, and α-MEM was found to be more suitable for isolation and expansion of multipotent MSCs [146, 147].

## 9.2 *Osteogenic Medium*

For bone tissue engineering studies, it is important to identify not only the appropriate medium and supplements to use for cell maintenance/proliferation but also which media and supplements affect osteogenic differentiation and at what stage. Commonly, media used for differentiation contains only 1% FBS, rather than the typical 5–20% FBS used in growth media. Osteogenic media also often includes a combination of ascorbic acid (Asc), β-glycerophosphate (β-GP), and/or dexamethasone (Dex) added to the basal medium. When designing a study to evaluate osteogenic differentiation, the stimulants added to the media are important factors to consider (Fig. 3).

Ascorbic acid, otherwise known as vitamin C, facilitates osteogenic differentiation by increasing collagen type I (Col I) secretion, and without Asc, collagen chains cannot form a proper helical structure [150, 151]. Osteoblast differentiation requires contact with a collagen-containing ECM, so the lack of properly formed collagen would hinder the differentiation process and signaling of osteoblasts, including Col I/$\alpha_2\beta_1$ integrin-mediated intracellular signaling [150]. A concentration of 50 μM of Asc was reported as optimal for formation of a mineralized matrix in human MSCs [148]. This is now a commonly used concentration of Asc for osteogenic differentiation studies using MSCs or MC3T3 cells [152, 153].

β-glycerophosphate (β-GP) induces osteogenic differentiation by serving as a source of phosphate needed to produce hydroxyapatite, or bone mineral [150]. The most common concentration of β-GP used in osteogenic media is 10 mM [153–157]. It should be noted that use of β-GP in high concentrations can lead to false-positive results with stem cells. Specifically, β-GP at concentrations over 2 mM in culture with MSCs has been shown to produce dystrophic mineralization or non-apatitic mineralization, which is detected by Alizarin Red S and Von Kossa staining but is not evidence of osteogenic differentiation [150, 158].

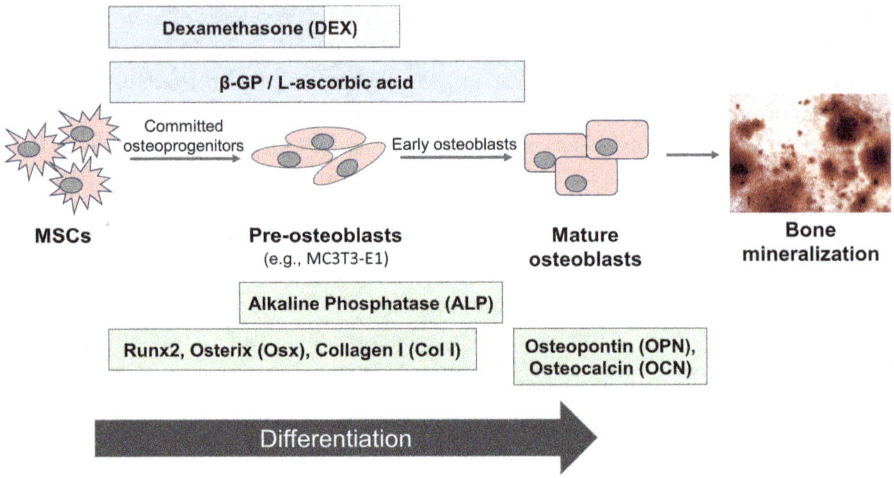


**Fig. 3** Representative timeline of osteogenic differentiation of mesenchymal stem cells (MSCs). Early-stage markers of osteogenesis include runt-related transcription factor 2 (Runx2), osterix (Osx), collagen type I (Col I), and alkaline phosphatase (ALP). Late-stage markers include osteopontin (OPN) and osteocalcin (OCN). Bone mineralization, represented here by alizarin red staining, occurs at the end of osteogenic differentiation. Beta-glycerophosphate (β-GP), ascorbic acid, and dexamethasone (shown at the top in blue) are common osteogenic media supplements. Dexamethasone is reported to have a beneficial impact on early—but not late-stage osteogenesis. (Adapted from Stamnitz and Klimczak (2021) and is subject to the terms and conditions of the Creative Commons Attribution (CC BY) license (https://creativecommons.org/licenses/by/4.0/) [127])

Dexamethasone, a glucocorticoid, is used regularly in addition to Asc and β-GP to elicit an osteogenic response. However, the role of Dex has not been elucidated entirely, and the results of adding Dex are inconsistent. One study reported the transcriptional cofactor four-and-a-half LIM 2 (FHL2) is upregulated by Dex, which potentiates Wnt signaling in MSCs and ultimately results in increased Runx2 expression and osteogenic differentiation in both murine and human MSCs [159]. A study involving flow perfusion culture demonstrated that flow perfusion alone induced osteogenic differentiation in rat MSCs, but when dexamethasone was used in synergy, there was enhanced osteogenic differentiation [56]. Other studies suggest Dex may not be required for development of the osteoblast phenotype [153, 160]. More recently, Dex reportedly promoted bone-lineage commitment in the early stage of differentiation via enhanced Runx2, ALP, and Osterix (Osx) expression but resulted in reduced maturation and terminal differentiation of osteoblasts when used as a long-term (for 10 days in mouse MSCs and 21 days in human MSCs) and high concentration (10 mM) treatment [161]. Long-term treatment with dexamethasone can have increasingly negative effects as early as the pre-osteoblast stage, at day 5 in mouse MSCs and day 14 in human MSCs [161]. Dex was therefore recommended for the early stage of bone-lineage commitment, but not for prolonged or later stages of osteogenesis. In 1985, a dosing study reported the optimal concentration of Dex for osteogenesis was 100 nM, which has been commonly used since [162]. However, in 2001, it was determined that a concentration of 10 nM would be more similar to the physiological level of glucocorticoids [163]. Titanium scaffolds seeded with rat bone marrow MSCs cultured either statically or in a perfusion bioreactor system, with or without 10 nM of Dex, saw opposite effects of Dex based on culture system. In static culture, the presence of Dex resulted in higher cell counts at day 8 compared to culture without Dex. Paradoxically, for the perfusion bioreactor culture, the presence of Dex led to lower cell counts at day 8 compared to culture without Dex [56]. Higher concentrations such as 100 nM have been shown to have positive effects on differentiation, but inhibitory effects on proliferation [163]. Collectively, the impact of Dex on osteogenesis is based on the treatment duration and concentration of Dex, stage of osteoblastic differentiation of the cells, and culture environment [56, 153, 161].

It is recommended that growth medium, a basal media supplemented with FBS and an antibiotic, be used as a negative control for all studies where additional supplements in media are used for differentiation purposes. The purpose of a negative control is to rule out potential noncausal interpretations of results. By evaluating the cell response without osteogenic supplementation, the effects of the scaffold and/or environment can be effectively interpreted. In other words, the assay or study performed can be validated by using a negative growth media control.

## 10 Bioreactor Culture of Cell-Seeded Scaffolds

Static culture of cell-seeded 3D bone scaffolds is generally accepted as disadvantageous, as the explants do not receive fluid flow of cell culture medium, which facilitates nutrient transport, gas exchange, and metabolic waste removal necessary for long-term explant viability [164]. Thus, bioreactor systems in a variety of forms have been developed to apply physiologically relevant stimuli which promote cell growth and osteogenesis (Fig. 4) [165].

### *10.1 Spinner Flasks and Rotating Wall Vessel Bioreactors*

Spinner flasks and rotating wall vessel bioreactors have overcome some of the limitations of static culture [143, 166]. Spinner flasks use a magnetic stir bar to generate convective forces that continually mix the cell culture medium, enhancing nutrient transport and metabolic waste removal as well as eliciting fluid flow-induced shear stress (FFSS) through scaffolds [167]. Rotating wall vessel (RWV) bioreactors simulate a microgravity environment, suspending cells or cell-seeded scaffolds in a laminar flow of culture medium [166, 168]. As a result, RWV bioreactors elicit lower FFSS when compared to spinner flask bioreactors. Due to the low shear stress environment, RWV bioreactors can be effective systems for 2D

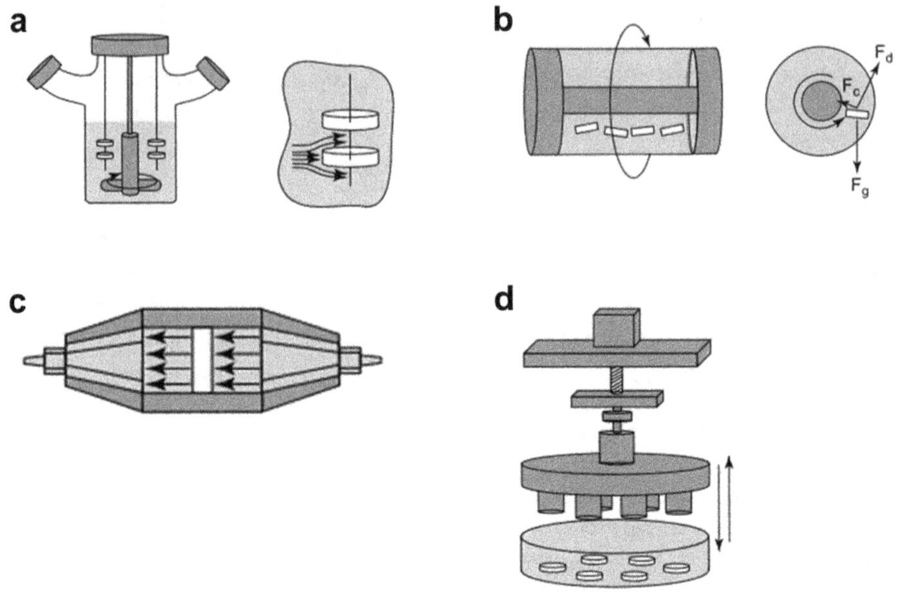


**Fig. 4** Representative bioreactor types used in bone tissue engineering applications. (**a**) Spinner flask bioreactors use a magnetic stir bar to generate convective forces that continually mix cell culture medium, enhancing nutrient transport and metabolic waste removal as well as eliciting fluid flow-induced shear stress (FFSS) through scaffolds. (**b**) Rotating wall vessel (RWV) bioreactors simulate a microgravity environment, suspending cells or cell-seeded scaffolds in a laminar flow of culture medium. (**c**) Perfusion flow bioreactors were introduced to improve long-term scaffold viability, enhance cell seeding efficiency and distribution, and accelerate osteogenesis in 3D tissue-engineered scaffolds. (**d**) Compressive loading bioreactors have been developed to further enhance the functional conditioning of tissue-engineered scaffolds by applying physiologically relevant loading regimens. (Adapted from TRENDS in Biotechnology, Vol. 22 No. 2, Ivan Martin, David Wendt, and Michael Heberer, The role of bioreactors in tissue engineering, Pages 80–86, Copyright (2004), with permission from Elsevier [164])

cell suspensions but are less applicable for stimulating osteogenesis in 3D explant culture [169–171]. Indeed, in a comparative study of spinner flasks and rotating wall vessel bioreactors, spinner flasks enhanced osteoblastic cell proliferation, ALP activity, OCN secretion, and calcium deposition in MSC-seeded scaffolds [172]. However, computational fluid dynamic (CFD) models suggest that spinner flask bioreactors generate inhomogeneous shear stress within 3D scaffolds, since regions of the scaffold closer to the stir bar receive greater shear stress than other regions [167].

## 10.2 Perfusion Flow Bioreactors and Parameter Selection

In the early 2000s, perfusion bioreactors were introduced to enhance cell seeding efficiency and distribution [143, 173, 174], accelerate osteogenesis, and improve long-term scaffold viability in 3D tissue-engineered scaffolds [174] (Fig. 5). For example, Sikavitsas et al. developed a flow perfusion bioreactor system and reported an increase in ALP activity, OPN secretion, and calcium deposition in rat marrow stromal cells at 0.3, 1, and 3 mL/min flow rates compared to statically cultured scaffolds [170]. Bhaskar et al. observed increased DNA content and ALP activity in MSC-seeded tissue-engineered scaffolds in perfusion bioreactor culture when compared to statically cultured scaffolds, highlighting the contribution of FFSS [171]. Harvestine et al. found that hydroxyapatite-poly(lactide-co-glycolide) (HA-PLG) scaffolds combined with mononucleated cells from bone marrow aspirates and preconditioned via perfusion bioreactor culture for 14 days prior to implantation in a rat calvaria defect accelerated bone healing when compared to unconditioned cell-seeded scaffolds [175].

Several studies cite the application of a perfusion flow rate of 1 mL/min in bone tissue engineering bioreactors to promote osteogenesis, metabolic waste removal, and nutrient transport [170, 174, 176]. Other studies have reported improved cell attachment and viability at flow rates less than or equal to 0.1 mL/min compared to higher flow rates and static culture [33, 177–179]. These studies report that higher flow rates, which are favorable for enhanced osteogenesis, can be deleterious to

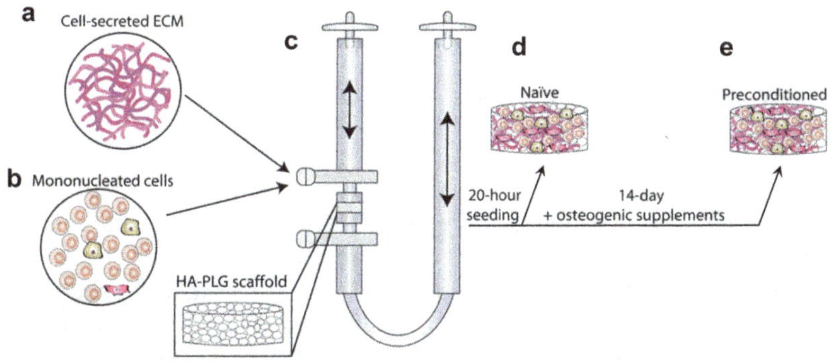


**Fig. 5** Schematic representation of (**a**) cell-secreted extracellular matrix (ECM) being combined with (**b**) mononucleated cells and seeded onto hydroxyapatite-poly(lactide-co-glycolide) (HA-PLG) scaffolds using (**c**) a perfusion flow bioreactor. (**d**) The naïve scaffolds were removed from the perfusion bioreactor after a 20-hour seeding duration, while (**e**) preconditioned scaffolds were continued under perfusion bioreactor culture and supplemented with osteogenic media components for 14 days. The naïve and preconditioned scaffolds were then implanted into bilateral calvarial defects, where the bioreactor preconditioned scaffolds favored vascularization and bone formation in vivo. (Adapted from Harvestine et al. (2020) and is subject to the terms and conditions of the Creative Commons Attribution (CC BY) license (https://creativecommons.org/licenses/by-nc/4.0/) [137])

cell viability [33, 179]. In a study by Cartmell et al., trabecular bone explants were seeded with MC3T3-E1 pre-osteoblast cells and subjected to through-perfusion flow rates of 0.01, 0.1, 0.2, and 1 mL/min [179]. The 0.01 mL/min flow rate favored cell proliferation and viability, while the 0.2 mL/min flow rate enhanced ALP activity, compared to the 0.01 and 0.1 mL/min flow rates, after 7 days of culture. Confocal microscopy revealed that the 1 mL/min flow rate resulted in more cell death after 7 days compared to the lower flow rates; this condition was not included in further analyses (i.e., for proliferation and differentiation) [179]. More recently, Kazimierczak et al. compared osteogenesis of human bone marrow-derived MSCs cultured on HA-based scaffolds for 21 days in one of three conditions: static culture, a flow rate of 0.02 mL/min in a perfusion bioreactor, or a rotating bioreactor with a rotation rate of 1 RPM. The 0.02 mL/min in the perfusion bioreactor demonstrated less expression of osteogenic markers including type I collagen and OPN compared to the rotating system, suggesting the relatively low flow rate in the perfusion bioreactor may have been insufficient to induce significant changes in osteogenesis compared to the rotating bioreactor [165]. A study evaluating the effects of perfusion culture versus static culture reported that when titanium scaffolds were seeded with rat bone marrow MSCs, perfusion-cultured scaffolds (0.3 mL/min for 1 day, 1 mL/min for remainder of study) had cell counts of almost three times the statically cultured scaffolds at day 8 and around double the cell count at day 16 [56]. Taken together, the extent to which flow rate affects cell viability and osteogenesis in bioreactor culture depends on a multitude of factors, including the characteristics of the bioreactor system (e.g., geometry and mechanism of flow) and the scaffold design (e.g., pore size) [180].

## 10.3 Compressive Loading Bioreactors and Parameter Selection

Compressive-loading bioreactors have been developed to further enhance the functional conditioning of tissue-engineered scaffolds [164, 181–183]. For example, Lovecchio et al. reported greater cell viability, proliferation, and enhanced ECM mineralization of human bone marrow mesenchymal stem cells (hBM-MSCs) seeded within a 3D chitosan-graphene scaffold under perfusion-compression conditions (1 mL/minute perfusion flow rate, 1% strain magnitude applied at 1 Hz twice daily for 30 minutes) when compared to static culture [176]. While each mechanical loading regimen applied in bioreactor culture should be tailored to the study objective, previous work, both in vivo and in bioreactors, has identified several physiologically relevant compressive loading parameters which influence osteogenesis in cell-seeded scaffolds, such as strain magnitude, waveform type, strain rate, loading frequency, loading duration, and rest periods.

### 10.3.1 Strain Magnitude

In the 1990s, Frost postulated new bone formation would occur at strain magnitudes between 1000 and 1500 $\mu\varepsilon$, and strain stimuli below 100 $\mu\varepsilon$ or above 3000 $\mu\varepsilon$ were found to induce bone resorption or microdamage [184]. Bioreactor studies involving an array of tissue-engineered bone scaffolds support this. For example, Endres et al. applied a jumping waveform to bovine trabecular bone (300 cycles per day, 1 Hz loading frequency) and found that 2000 $\mu\varepsilon$ was more stimulatory for new bone formation, indicated by an increase in osteoid thickness, compared to 1000, 1500, 3000, and 4000 $\mu\varepsilon$ [185]. Ravichandran et al. applied cyclic, compressive loading to polycaprolactone-β-tricalcium phosphate (PCL-β-TCP) scaffolds seeded with MSCs (4 hours per day, 1 Hz loading frequency) and found that 0.22% strain (2200 $\mu\varepsilon$) was more stimulatory for ALP activity and calcium deposition than 0.88% and 1.1% strain, or 8800 $\mu\varepsilon$ and 11,000 $\mu\varepsilon$, respectively [186].

### 10.3.2 Waveform Type and Strain Rate

Two common types of compressive loading waveforms are sinusoidal and trapezoidal. The sinusoidal waveform may better mimic human locomotion, but it is limited in that the strain rate and loading frequency are dependent on each other: as the strain rate increases, the frequency of the loading waveform also increases. Rubin and Lanyon proposed that the use of a trapezoidal waveform allows for the dissociation of the loading frequency and strain rate [187]. While physiological strain rates corresponding to walking (0.01 Hz) up to sprinting (0.03 Hz) have been quantified in in vivo scenarios, with higher strain rates favoring bone formation [187, 188], optimal strain rates for use in bioreactors require further investigation.

### 10.3.3 Loading Frequency

Often, studies using compressive loading bioreactors implement a loading frequency of approximately 1 Hz to mimic human locomotion [176, 182, 186]. For example, Birmingham et al. compared trabecular bone explants isolated from ovine vertebrae which received physiological, cyclic compressive loading (2000 $\mu\varepsilon$ strain magnitude applied at 1 Hz loading frequency for 10 minutes daily for 15 days) to explants statically cultured and found greater osteogenic activity in the bone explants which received compressive loading [189]. On the other hand, some studies have shown positive effects of using a slower loading frequency (i.e., including rest periods between each cycle).

### 10.3.4 Loading Duration and Rest Periods

The duration of mechanical loading which elicits new bone formation has also been explored using in vivo models [190]. A study applying physiological, cyclic compressive loading with a target 2050 $\mu\varepsilon$ to rooster ulnae showed approximately 36 cycles per bout was sufficient to induce new bone formation, and adding more loading cycles did not result in more bone formation [187]. Another study involving four-point bending of rat tibiae reported that incorporating a "long" (8-hour) rest period between mechanical loading bouts resulted in greater relative bone formation than no rest or 0.5-hour rest periods [191]. Further, when "short" (i.e., seconds) rest periods were introduced between each loading cycle, 14-second rest periods resulted in greater relative bone formation than all shorter rest periods evaluated (0.5, 3.5, and 7 seconds) [191]. Since then, bioreactor studies have incorporated this historical knowledge into compressive loading bioreactor culture. For example, Dua et al. incorporated a 10-second rest period between each compression cycle (42.7 N compressive force, applied for 15 minutes every 8 hours) and saw enhanced cell viability and ECM deposition onto fiber mesh titanium scaffolds in contact with bone explants isolated from bovine metatarsals when compared to statically cultured bone explants, statically cultured bone explants + a fiber mesh metallic scaffold, and bone explants + a fiber mesh metallic scaffold exposed to 150 kPa of constant pressure [192]. Further research into optimal loading duration and rest period is needed for enhancing osteogenesis in bioreactor culture.

## 11 Conclusion

The design, fabrication, and culture of cell-seeded biomaterial scaffolds must be carefully considered for effective use of these scaffolds in orthopedic applications including bone tissue engineering. The main functions of any scaffold are (i) serving as a substrate for cell growth, and (ii) providing support for mechanical stimulation. Critical in the scaffold design process are porosity and other structural parameters, as well as material selection. Depending on the material(s) chosen, the fabrication method is leveraged to enable scaffolds to be tailored to mimic the ECM through their support of cell growth and tissue regeneration. For cell seeding of scaffolds, cell type and method of seeding influence the extent of cell attachment on the scaffold. Finally, appropriate biochemical and mechanical signals which induce proliferation and osteogenic differentiation must be identified. These include culture media, perfusion flow, and/or compressive loading—and the magnitude and timing of each. To this end, many advancements in methods of culturing cell-seeded scaffolds continue to emerge, such as customized bioreactor systems which can provide physiologic spatial and temporal cues to accelerate tissue formation.

## References

1. Chan BP, Leong KW. Scaffolding in tissue engineering: general approaches and tissue-specific considerations. European Spine Journal. 2008 Dec 13;17(S4):467–79.
2. Khalili A, Ahmad M. A Review of Cell Adhesion Studies for Biomedical and Biological Applications. Int J Mol Sci. 2015 Aug 5;16(8):18149–84.
3. Frantz C, Stewart KM, Weaver VM. The extracellular matrix at a glance. J Cell Sci. 2010 Dec 15;123(24):4195–200.
4. Gospodarowicz D, Delgado D, Vlodavsky I. Permissive effect of the extracellular matrix on cell proliferation in vitro (fibroblast growth factor/granulosa cells/adrenal cortex cells/vascular endothelial cells/collagens) [Internet]. Vol. 77, Cell Biology. 1980. Available from: https://www.pnas.org
5. Zhou T, Gao B, Fan Y, Liu Y, Feng S, Cong Q, et al. Piezo1/2 mediate mechanotransduction essential for bone formation through concerted activation of NFAT-YAP1-ß-catenin. Elife. 2020 Mar 18;9.
6. Fazzalari NL. Bone fracture and bone fracture repair. Osteoporosis International. 2011 Jun 27;22(6):2003–6.
7. Song L. Effects of Exercise or Mechanical Stimulation on Bone Development and Bone Repair. Stem Cells Int. 2022;2022:5372229.
8. Isaksson H, Gröngröft I, Wilson W, van Donkelaar CC, van Rietbergen B, Tami A, et al. Remodeling of fracture callus in mice is consistent with mechanical loading and bone remodeling theory. J Orthop Res. 2009 May;27(5):664–72.
9. Li MCM, Chow SKH, Wong RMY, Qin L, Cheung WH. The role of osteocytes-specific molecular mechanism in regulation of mechanotransduction - A systematic review. J Orthop Translat. 2021 Jul;29:1–9.
10. Dobrosak C, Gooi JH. Increased sphingosine-1-phosphate production in response to osteocyte mechanotransduction. Bone Rep. 2017 Dec;7:114–20.
11. Mohammadi H, Sepantafar M, Muhamad N, Bakar Sulong A. How Does Scaffold Porosity Conduct Bone Tissue Regeneration? Adv Eng Mater. 2021 Oct 26;23(10):2100463.
12. Lu JX, Flautre B, Anselme K, Hardouin P, Llur AG, De M, et al. Role of interconnections in porous bioceramics on bone recolonization in vitro and in vivo. Vol. 10, Journal of Materials Science: Materials in Medicine. 1999.
13. Murphy CM, O'Brien FJ. Understanding the effect of mean pore size on cell activity in collagen-glycosaminoglycan scaffolds. Cell Adh Migr. 2010 Jul 28;4(3):377–81.
14. Poh PSP, Valainis D, Bhattacharya K, van Griensven M, Dondl P. Optimization of Bone Scaffold Porosity Distributions. Sci Rep. 2019 Dec 1;9(1).
15. Gregor A, Filová E, Novák M, Kronek J, Chlup H, Buzgo M, et al. Designing of PLA scaffolds for bone tissue replacement fabricated by ordinary commercial 3D printer. J Biol Eng. 2017;11:31.
16. Guo B, Ma PX. Synthetic biodegradable functional polymers for tissue engineering: a brief review. Sci China Chem. 2014 Apr 1;57(4):490–500.
17. Bitar KN, Zakhem E. Design Strategies of Biodegradable Scaffolds for Tissue Regeneration. Biomed Eng Comput Biol. 2014 Jan 8;6:BECB.S10961.
18. Maia-Pinto MOC, Brochado ACB, Teixeira BN, Sartoretto SC, Uzeda MJ, Alves ATNN, et al. Biomimetic Mineralization on 3D Printed PLA Scaffolds: On the Response of Human Primary Osteoblasts Spheroids and In Vivo Implantation. Polymers (Basel). 2020 Dec 27;13(1).
19. Chen W, Nichols L, Brinkley F, Bohna K, Tian W, Priddy MW, et al. Alkali treatment facilitates functional nano-hydroxyapatite coating of 3D printed polylactic acid scaffolds. Materials Science and Engineering: C. 2021 Jan;120:111686.
20. Guzmán-Soria A, Moreno-Serna V, Canales DA, García-Herrera C, Zapata PA, Orihuela PA. Effect of Electrospun PLGA/Collagen Scaffolds on Cell Adhesion, Viability, and Collagen Release: Potential Applications in Tissue Engineering. Polymers (Basel). 2023 Mar 1;15(5).

21. Chen W, Nichols L, Teer L, Clinton K, Priddy LB. A hybrid coating of polydopamine and nano-hydroxyapatite enhances surface properties of 3D printed poly(lactic-co-glycolic acid) scaffolds. J Mater Sci. 2022 Jul 1;57(27):13011–26.
22. Laurencin CT, Attawia M, Borden MD. Advancements in tissue engineered bone substitutes: Current Opinion in Orthopaedics. Curr Opin Orthop [Internet]. 1999;10(6):445–51. Available from: https://journals.lww.com/co-ortho/Abstract/1999/12000/Advancements_in_tissue_engineered_bone_substitutes.5.aspx
23. Kofron MD, Griswold A, Kumbar SG, Martin K, Wen X, Laurencin CT. The implications of polymer selection in regenerative medicine: A comparison of amorphous and semi-crystalline polymer for tissue regeneration. Adv Funct Mater. 2009;19(9):1351–9.
24. Rizzi SC, Heath DJ, Coombes AGA, Bock N, Textor M, Downes S. Biodegradable polymer/hydroxyapatite composites: Surface analysis and initial attachment of human osteoblasts. J Biomed Mater Res [Internet]. 2001 Oct;55(4):475–86. Available from: https://doi.org/10.1002/1097-4636(20010615)55:4<475::aid-jbm1039>3.0.co;2-q
25. Huang B, Caetano G, Vyas C, Blaker J, Diver C, Bártolo P. Polymer-Ceramic Composite Scaffolds: The Effect of Hydroxyapatite and β-tri-Calcium Phosphate. Materials. 2018 Jan 14;11(1):129.
26. Wang X, Jiang M, Zhou Z, Gou J, Hui D. 3D printing of polymer matrix composites: A review and prospective. Compos B Eng [Internet]. 2017;110:442–58. Available from: https://doi.org/10.1016/j.compositesb.2016.11.034
27. Zakaria SM, Zein SHS, Othman MohdR, Jansen JA. Hydroxyapatite nanoparticles: Electrospinning and calcination of hydroxyapatite/polyvinyl butyral nanofibers and growth kinetics. J Biomed Mater Res A [Internet]. 2013 Oct;101A(7):1977–85. Available from: https://doi.org/10.1002/jbm.a.34506
28. Trachtenberg JE, Placone JK, Smith BT, Fisher JP, Mikos AG. Extrusion-based 3D printing of poly(propylene fumarate) scaffolds with hydroxyapatite gradients. J Biomater Sci Polym Ed. 2017;28(6):532–54.
29. Pan Z, Ding J. Poly(lactide-co-glycolide) porous scaffolds for tissue engineering and regenerative medicine. Interface Focus [Internet]. 2012;2(3):366–77. Available from: https://pubmed.ncbi.nlm.nih.gov/23741612/
30. Ribas RG, Schatkoski VM, Montanheiro TL do A, de Menezes BRC, Stegemann C, Leite DMG, et al. Current advances in bone tissue engineering concerning ceramic and bioglass scaffolds: A review. Ceram Int [Internet]. 2019;45(17):21051–61. Available from: https://doi.org/10.1016/j.ceramint.2019.07.096
31. Nickoli MS, Hsu WK. Ceramic-Based Bone Grafts as a Bone Grafts Extender for Lumbar Spine Arthrodesis: A Systematic Review. Global Spine J [Internet]. 2014 Aug [cited 2024 Feb 27];4(3):211–6. Available from: http://dx.doi.org/
32. Xu S, Xu S, Zhou P, Cheng X, Xie Y, Liang C, et al. Selective laser sintering fabrication of nano-hydroxyapatite/poly-epsilon-caprolactone scaffolds for bone tissue engineering applications. Int J Nanomedicine [Internet]. 2013;4197. Available from: http://www.dovepress.com/selective-laser-sintering-fabrication-of-nano-hydroxyapatitepoly-epsil-peer-reviewed-article-IJN
33. Zhao F, Chella R, Ma T. Effects of Shear Stress on 3-D Human Mesenchymal Stem Cell Construct Development in a Perfusion Bioreactor System: Experiments and Hydrodynamic Modeling. Biotechnol Bioeng. 2006;96(Ii):503–5.
34. Habibovic P. In vitro and in vivo bioactivity assessment of a polylactic acid/hydroxyapatite composite for bone regeneration. Biomatter. 2014;4(October 2017):e27664.
35. Chen X, Gao C, Jiang J, Wu Y, Zhu P, Chen G. 3D printed porous PLA/nHA composite scaffolds with enhanced osteogenesis and osteoconductivity in vivo for bone regeneration. Biomedical Materials. 2019 Sep;14(6):065003.
36. Gonçalves EM, Oliveira FJ, Silva RF, Neto MA, Fernandes MH, Amaral M, et al. Three-dimensional printed PCL-hydroxyapatite scaffolds filled with CNTs for bone cell growth stimulation. J Biomed Mater Res B Appl Biomater. 2016 Oct;104(6):1210–9.

37. Oladapo BI, Zahedi SA, Adeoye AOM. 3D printing of bone scaffolds with hybrid biomaterials. Compos B Eng. 2019 Oct;158:428–36.
38. Neufurth M, Wang X, Wang S, Steffen R, Ackermann M, Haep ND, et al. 3D printing of hybrid biomaterials for bone tissue engineering: Calcium-polyphosphate microparticles encapsulated by polycaprolactone. Acta Biomater. 2017 Oct;64:377–88.
39. Khan YM, Katti DS, Laurencin CT. Novel polymer-synthesized ceramic composite-based system for bone repair: An in vitro evaluation. J Biomed Mater Res [Internet]. 2004;69A(4):728–37. Available from: http://doi.wiley.com/10.1002/jbm.a.30051
40. Bruyas A, Lou F, Stahl AM, Gardner M, Maloney W, Goodman S, et al. Systematic characterization of 3D-printed PCL/β-TCP scaffolds for biomedical devices and bone tissue engineering: Influence of composition and porosity. J Mater Res. 2018 Jul 27;33(14):1948–59.
41. Kohn DH, Sarmadi M, Helman JI, Krebsbach PH. Effects of pH on human bone marrow stromal cells in vitro: Implications for tissue engineering of bone. J Biomed Mater Res. 2002;60(2):292–9.
42. Melchels FPW, Feijen J, Grijpma DW. A poly(d,l-lactide) resin for the preparation of tissue engineering scaffolds by stereolithography. Biomaterials. 2009;30(23–24):3801–9.
43. Serra T, Planell JA, Navarro M. High-resolution PLA-based composite scaffolds via 3-D printing technology. Acta Biomater. 2013;9(3):5521–30.
44. Jaidev LR, Chatterjee K. Surface functionalization of 3D printed polymer scaffolds to augment stem cell response. Mater Des. 2019 Jan;161:44–54.
45. Ishack S, Mediero A, Wilder T, Ricci JL, Cronstein BN. Bone regeneration in critical bone defects using three-dimensionally printed β-tricalcium phosphate/hydroxyapatite scaffolds is enhanced by coating scaffolds with either dipyridamole or BMP-2. J Biomed Mater Res B Appl Biomater [Internet]. 2017 Oct;105(2):366–75. Available from: https://pubmed.ncbi.nlm.nih.gov/26513656/
46. Senatov FS, Niaza KV, Zadorozhnyy MYu, Maksimkin AV, Kaloshkin SD, Estrin YZ. Mechanical properties and shape memory effect of 3D-printed PLA-based porous scaffolds. J Mech Behav Biomed Mater. 2016 Apr;57:139–48.
47. Nie L, Chen D, Fu J, Yang S, Hou R, Suo J. Macroporous biphasic calcium phosphate scaffolds reinforced by poly-L-lactic acid/hydroxyapatite nanocomposite coatings for bone regeneration. Biochem Eng J. 2015 Oct;98:29–37.
48. Dávila JL, Freitas MS, Neto PI, Silveira ZC, Silva JVL, d'Ávila MA. Fabrication of PCL/β-TCP scaffolds by 3D mini-screw extrusion printing. J Appl Polym Sci [Internet]. 2016 Oct;133(15):n/a-n/a. Available from: http://doi.wiley.com/10.1002/app.43031
49. Makadia HK, Siegel SJ. Poly Lactic-co-Glycolic Acid (PLGA) as biodegradable controlled drug delivery carrier. Polymers (Basel). 2011;3(3):1377–97.
50. Saleh KJ, Othmani MM El, Tzeng TH, Mihalko WM, Chambers MC, Grupp TM. Acrylic bone cement in total joint arthroplasty: A review. Vol. 34, Journal of Orthopaedic Research. 2016. p. 737–44.
51. Kwak S, Haider A, Chandra Gupta K, Kim S, Kang IK. Micro/Nano Multilayered Scaffolds of PLGA and Collagen by Alternately Electrospinning for Bone Tissue Engineering. Nanoscale Res Lett. 2016;
52. Wu C, Chang J. Degradation, bioactivity, and cytocompatibility of diopside, akermanite, and bredigite ceramics. J Biomed Mater Res B Appl Biomater [Internet]. 2007 Oct [cited 2023 Dec 10];83(1):152–60. Available from: https://pubmed.ncbi.nlm.nih.gov/17318828/
53. Mazón P, De Aza PN. Porous scaffold prepared from α′L-Dicalcium silicate doped with phosphorus for bone grafts. Ceram Int. 2018 Jan 1;44(1):537–45.
54. Baino F. Bioactive glasses—When glass science and technology meet regenerative medicine. Ceram Int. 2018 Sep 1;44(13):14953–66.
55. Kido HW, Gabbai-Armelin PR, Avanzi IR, da Silva AC, Fernandes KR, Fortulan CA, et al. Vacuumed collagen-impregnated bioglass scaffolds: Characterization and influence on proliferation and differentiation of bone marrow stromal cells. J Biomed Mater Res B Appl Biomater [Internet]. 2019 Feb 1 [cited 2023 Dec 10];107(2):211–22. Available from: https://pubmed.ncbi.nlm.nih.gov/29569333/

56. Holtorf HL, Jansen JA, Mikos AG. Flow perfusion culture induces the osteoblastic differentiation of marrow stroma cell-scaffold constructs in the absence of dexamethasone. J Biomed Mater Res A. 2005 Mar 1;72(3):326–34.
57. Li Y, Liu H, Wang C, Yan R, Xiang L, Mu X, et al. 3D printing titanium grid scaffold facilitates osteogenesis in mandibular segmental defects. NPJ Regen Med. 2023 Dec 1;8(1).
58. Staiger MP, Pietak AM, Huadmai J, Dias G. Magnesium and its alloys as orthopedic biomaterials: A review. Vol. 27, Biomaterials. 2006. p. 1728–34.
59. Nagels J, Stokdijk M, Rozing PM. Stress shielding and bone resorption in shoulder arthroplasty. J Shoulder Elbow Surg. 2003;
60. Tsakiris V, Tardei C, Clicinschi FM. Biodegradable Mg alloys for orthopedic implants—A review. Vol. 9, Journal of Magnesium and Alloys. National Engg. Reaserch Center for Magnesium Alloys; 2021. p. 1884–905.
61. Sekar P, S N, Desai V. Recent progress in in vivo studies and clinical applications of magnesium based biodegradable implants—A review. Vol. 9, Journal of Magnesium and Alloys. National Engg. Reaserch Center for Magnesium Alloys; 2021. p. 1147–63.
62. Xie K, Wang N, Guo Y, Zhao S, Tan J, Wang L, et al. Additively manufactured biodegradable porous magnesium implants for elimination of implant-related infections: An in vitro and in vivo study. Bioact Mater. 2022 Feb 1;8:140–52.
63. Hung CC, Chaya A, Liu K, Verdelis K, Sfeir C. The role of magnesium ions in bone regeneration involves the canonical Wnt signaling pathway. Acta Biomater. 2019 Oct 15;98:246–55.
64. Wang N, Ma Y, Shi H, Song Y, Guo S, Yang S. Mg-, Zn-, and Fe-Based Alloys With Antibacterial Properties as Orthopedic Implant Materials. Vol. 10, Frontiers in Bioengineering and Biotechnology. Frontiers Media S.A.; 2022.
65. Amirazad H, Dadashpour M, Zarghami N. Application of decellularized bone matrix as a bioscaffold in bone tissue engineering. Vol. 16, Journal of Biological Engineering. BioMed Central Ltd; 2022.
66. Choudhury D, Tun HW, Wang T, Naing MW. Organ-Derived Decellularized Extracellular Matrix: A Game Changer for Bioink Manufacturing? Trends Biotechnol. 2018 Aug;36(8):787–805.
67. Zhang X, Chen X, Hong H, Hu R, Liu J, Liu C. Decellularized extracellular matrix scaffolds: Recent trends and emerging strategies in tissue engineering. Bioact Mater. 2022 Apr;10:15–31.
68. Price AP, Godin LM, Domek A, Cotter T, D'Cunha J, Taylor DA, et al. Automated Decellularization of Intact, Human-Sized Lungs for Tissue Engineering. Tissue Eng Part C Methods. 2015 Jan;21(1):94–103.
69. Allen AB, Priddy LB, Li MTA, Guldberg RE. Functional Augmentation of Naturally-Derived Materials for Tissue Regeneration. Ann Biomed Eng. 2015 Mar 1;43(3):555–67.
70. Lei J, Priddy LB, Lim JJ, Koob TJ. Dehydrated Human Amnion/Chorion Membrane (dHACM) Allografts as a Therapy for Orthopedic Tissue Repair. Techniques in Orthopaedics. 2017 Sep;32(3):149–57.
71. Isaeva EV, Beketov EE, Arguchinskaya NV, Ivanov SA, Shegay PV, Kaprin AD. Decellularized Extracellular Matrix for Tissue Engineering (Review). Sovremennye tehnologii v medicine. 2022 Jun 30;14(3):57.
72. Ebrahimi Sadrabadi A, Baei P, Hosseini S, Baghaban Eslaminejad M. Decellularized Extracellular Matrix as a Potent Natural Biomaterial for Regenerative Medicine. In 2020. p. 27–43.
73. Ren J, Li Z, Liu W, Fan Y, Qi L, Li S, et al. Demineralized bone matrix for repair and regeneration of maxillofacial defects: A narrative review. J Dent. 2024 Feb;104899.
74. Visscher DO, Lee H, van Zuijlen PPM, Helder MN, Atala A, Yoo JJ, et al. A photo-crosslinkable cartilage-derived extracellular matrix bioink for auricular cartilage tissue engineering. Acta Biomater. 2021 Feb;121:193–203.
75. Nagao RJ, Xu J, Luo P, Xue J, Wang Y, Kotha S, et al. Decellularized Human Kidney Cortex Hydrogels Enhance Kidney Microvascular Endothelial Cell Maturation and Quiescence. Tissue Eng Part A. 2016 Oct;22(19–20):1140–50.

76. Tack P, Victor J, Gemmel P, Annemans L. 3D-printing techniques in a medical setting: a systematic literature review. Biomed Eng Online. 2016 Dec 21;15(1):115.
77. Xue J, Wu T, Dai Y, Xia Y. Electrospinning and Electrospun Nanofibers: Methods, Materials, and Applications. Chem Rev. 2019 Apr 24;119(8):5298–415.
78. Marrey M, Malekipour E, El-Mounayri H, Faierson EJ, Agarwal M. A Novel Framework of Developing a Predictive Model for Powder Bed Fusion Process. 3D Print Addit Manuf. 2024 Feb 1;11(1):179–96.
79. Jahr H, Li Y, Zhou J, Zadpoor AA, Schröder KU. Additively Manufactured Absorbable Porous Metal Implants—Processing, Alloying and Corrosion Behavior. Vol. 8, Frontiers in Materials. Frontiers Media S.A.; 2021.
80. Tran RT, Naseri E, Kolasnikov A, Bai X, Yang J. A new generation of sodium chloride porogen for tissue engineering. Biotechnol Appl Biochem. 2011 Sep 16;58(5):335–44.
81. Harris LD, Kim BS, Mooney DJ. Open pore biodegradable matrices formed with gas foaming. J Biomed Mater Res. 1998 Dec 5;42(3):396–402.
82. Brougham CM, Levingstone TJ, Shen N, Cooney GM, Jockenhoevel S, Flanagan TC, et al. Freeze-Drying as a Novel Biofabrication Method for Achieving a Controlled Microarchitecture within Large, Complex Natural Biomaterial Scaffolds. Adv Healthc Mater. 2017 Nov 31;6(21).
83. Conoscenti G, Schneider T, Stoelzel K, Carfì Pavia F, Brucato V, Goegele C, et al. PLLA scaffolds produced by thermally induced phase separation (TIPS) allow human chondrocyte growth and extracellular matrix formation dependent on pore size. Materials Science and Engineering: C. 2017 Nov;80:449–59.
84. Kumar A, Jacob A. Techniques in scaffold fabrication process for tissue engineering applications: A review. Vol. 10, Journal of Applied Biology and Biotechnology. Open Science Publishers LLP Inc.; 2022. p. 163–76.
85. Dolcimascolo, A.; Calabrese, G.; Conoci, S.; Parenti, R. Innovative biomaterials for tissue engineering. Biomaterial-Supported Tissue Reconstruction or Regeneration. 2019. 1–18 p.
86. Tang Z, Li X, Tan Y, Fan H, Zhang X. The material and biological characteristics of osteoinductive calcium phosphate ceramics. Regen Biomater [Internet]. 2018 Feb 1 [cited 2024 Feb 25];5(1):43. Available from: /pmc/articles/PMC5798025/
87. Houaoui A, Lyyra I, Agniel R, Pauthe E, Massera J, Boissière M. Dissolution, bioactivity and osteogenic properties of composites based on polymer and silicate or borosilicate bioactive glass. Materials Science and Engineering C [Internet]. 2020;107(October 2019):110340. Available from: https://doi.org/10.1016/j.msec.2019.110340
88. Nouri A, Wen C. Introduction to surface coating and modification for metallic biomaterials. Surface Coating and Modification of Metallic Biomaterials [Internet]. 2015 Apr 2 [cited 2024 Feb 25];3–60. Available from: https://www.researchgate.net/publication/282460134_Introduction_to_surface_coating_and_modification_for_metallic_biomaterials
89. Surmenev RA, Surmeneva MA, Ivanova AA. Significance of calcium phosphate coatings for the enhancement of new bone osteogenesis–a review. Acta Biomater [Internet]. 2014 [cited 2024 Feb 25];10(2):557–79. Available from: https://pubmed.ncbi.nlm.nih.gov/24211734/
90. Chen G, Okamura A, Sugiyama K, Wozniak MJ, Kawazoe N, Sato S, et al. Surface modification of porous scaffolds with nanothick collagen layer by centrifugation and freeze-drying. J Biomed Mater Res B Appl Biomater. 2009 Aug 13;90B(2):864–72.
91. Hollinger J, Chaudhari A. Cells and Materials Cells and Materials Bone Regeneration Materials for the Mandibular and Craniofacial Bone Regeneration Materials for the Mandibular and Craniofacial Complex Complex. Cells and Materials [Internet]. 1992 [cited 2024 Feb 26];2. Available from: https://digitalcommons.usu.edu/cellsandmaterials/vol2/iss2/9
92. Dehghanghadıkolaei A, Fotovvati B. Coating Techniques for Functional Enhancement of Metal Implants for Bone Replacement: A Review. Materials (Basel) [Internet]. 2019 Jun 1 [cited 2024 Feb 25];12(11). Available from: https://pubmed.ncbi.nlm.nih.gov/31163586/
93. Rourke AS, Beard MC, Jones SE, Priddy MW, Priddy LB. Hydroxyapatite coating promotes stable physicochemical properties of pure magnesium in a longitudinal degradation study. J Mater Res. 2022 Mar 28;37(6):1231–45.

94. Popkov A V., Gorbach EN, Kononovich NA, Popkov DA, Tverdokhlebov SI, Shesterikov E V. Bioactivity and osteointegration of hydroxyapatite-coated stainless steel and titanium wires used for intramedullary osteosynthesis. Strategies Trauma Limb Reconstr [Internet]. 2017 Aug 1 [cited 2024 Feb 25];12(2):107. Available from: /pmc/articles/PMC5505879/
95. Fu X, Liu P, Zhao D, Yuan B, Xiao Z, Zhou Y, et al. Effects of Nanotopography Regulation and Silicon Doping on Angiogenic and Osteogenic Activities of Hydroxyapatite Coating on Titanium Implant. Int J Nanomedicine [Internet]. 2020 [cited 2024 Feb 25];15:4171–89. Available from: https://pubmed.ncbi.nlm.nih.gov/32606671/
96. Bagherifard A, Joneidi Yekta H, Akbari Aghdam H, Motififard M, Sanatizadeh E, Ghadiri Nejad M, et al. Improvement in osseointegration of tricalcium phosphate-zircon for orthopedic applications: an in vitro and in vivo evaluation. Med Biol Eng Comput [Internet]. 2020 Aug 1 [cited 2024 Feb 26];58(8):1681–93. Available from: https://link.springer.com/article/10.1007/s11517-020-02157-1
97. de Oliveira PGFP, de Melo Soares MS, Silveira e Souza AMM, Taba M, Palioto DB, Messora MR, et al. Influence of nano-hydroxyapatite coating implants on gene expression of osteogenic markers and micro-CT parameters. An in vivo study in diabetic rats. J Biomed Mater Res A [Internet]. 2021 May 1 [cited 2024 Feb 25];109(5):682–94. Available from: https://pubmed.ncbi.nlm.nih.gov/32608088/
98. Makishi S, Watanabe T, Saito K, Ohshima H. Effect of Hydroxyapatite/β-Tricalcium Phosphate on Osseointegration after Implantation into Mouse Maxilla. Int J Mol Sci [Internet]. 2023 Feb 1 [cited 2024 Feb 26];24(4). Available from: /pmc/articles/PMC9965049/
99. Agarwal R, González-García C, Torstrick B, Guldberg RE, Salmerón-Sánchez M, García AJ. Simple coating with fibronectin fragment enhances stainless steel screw osseointegration in healthy and osteoporotic rats. Biomaterials [Internet]. 2015 Sep 1 [cited 2024 Feb 25];63:137–45. Available from: https://pubmed.ncbi.nlm.nih.gov/26100343/
100. von der Mark K, Park J. Engineering biocompatible implant surfaces: Part II: Cellular recognition of biomaterial surfaces: Lessons from cell–matrix interactions. Prog Mater Sci [Internet]. 2013 Apr 1 [cited 2024 Feb 25];58(3):327–81. Available from: https://www.researchgate.net/publication/256922944_Engineering_Biocompatible_Implant_Surfaces_Part_II_Cellular_Recognition_of_Biomaterial_Surfaces_Lessons_from_Cell-Matrix_Interactions
101. Chen W, Xu K, Tao B, Dai L, Yu Y, Mu C, et al. Multilayered coating of titanium implants promotes coupled osteogenesis and angiogenesis in vitro and in vivo. Acta Biomater [Internet]. 2018 Jul 1 [cited 2024 Feb 25];74:489–504. Available from: https://pubmed.ncbi.nlm.nih.gov/29702291/
102. He J, Huang T, Gan L, Zhou Z, Jiang B, Wu Y, et al. Collagen-infiltrated porous hydroxyapatite coating and its osteogenic properties: in vitro and in vivo study. J Biomed Mater Res A [Internet]. 2012 Jul [cited 2024 Feb 25];100(7):1706–15. Available from: https://pubmed.ncbi.nlm.nih.gov/22447757/
103. Meyers SR, Grinstaff MW. Biocompatible and bioactive surface modifications for prolonged in vivo efficacy. Chem Rev [Internet]. 2012 Mar 14 [cited 2024 Feb 25];112(3):1615–32. Available from: https://pubs.acs.org/doi/full/10.1021/cr2000916
104. Zhang BGX, Myers DE, Wallace GG, Brandt M, Choong PFM. Bioactive Coatings for Orthopaedic Implants—Recent Trends in Development of Implant Coatings. Int J Mol Sci [Internet]. 2014 Jul 4 [cited 2024 Feb 25];15(7):11878. Available from: /pmc/articles/PMC4139820/
105. Raphel J, Karlsson J, Galli S, Wennerberg A, Lindsay C, Haugh MG, et al. Engineered protein coatings to improve the osseointegration of dental and orthopaedic implants. Biomaterials. 2016 Mar 1;83:269–82.
106. Kroese-Deutman HC, Van Den Dolder J, Spauwen PHM, Jansen JA. Influence of RGD-loaded titanium implants on bone formation in vivo. Tissue Eng [Internet]. 2005 Nov [cited 2024 Feb 25];11(11–12):1867–75. Available from: https://pubmed.ncbi.nlm.nih.gov/16411833/
107. Elmengaard B, Bechtold JE, Søballe K. In vivo study of the effect of RGD treatment on bone ongrowth on press-fit titanium alloy implants. Biomaterials [Internet]. 2005 Jun [cited 2024 Feb 25];26(17):3521–6. Available from: https://pubmed.ncbi.nlm.nih.gov/15621242/

108. Petrie TA, Raynor JE, Dumbauld DW, Lee TT, Jagtap S, Templeman KL, et al. Multivalent Integrin-Specific Ligands Enhance Tissue Healing and Biomaterial Integration. Sci Transl Med [Internet]. 2010 Aug 8 [cited 2024 Feb 25];2(45):45ra60. Available from: /pmc/articles/PMC3128787/
109. Fung SL, Wu X, Maceren JP, Mao Y, Kohn J. In vitro evaluation of recombinant bone morphogenetic protein-2 bioactivity for regenerative medicine. Tissue Eng Part C Methods. 2019;25(9):553–9.
110. Krishnan L, Priddy LB, Esancy C, Klosterhoff BS, Stevens HY, Tran L, et al. Delivery vehicle effects on bone regeneration and heterotopic ossification induced by high dose BMP-2. Acta Biomater. 2017 Feb;49:101–12.
111. Chen W, Nichols L, Teer L, Clinton K, Priddy LB. A hybrid coating of polydopamine and nano-hydroxyapatite enhances surface properties of 3D printed poly(lactic-co-glycolic acid) scaffolds Materials for life sciences. J Mater Sci [Internet]. 2023 [cited 2023 Apr 4];57. Available from: https://doi.org/10.1007/s10853-022-07442-y
112. Ma H, Luo J, Sun Z, Xia L, Shi M, Liu M, et al. 3D printing of biomaterials with mussel-inspired nanostructures for tumor therapy and tissue regeneration. Biomaterials. 2016 Dec 1;111:138–48.
113. Arjunan A, Robinson J, Al Ani E, Heaselgrave W, Baroutaji A, Wang C. Mechanical performance of additively manufactured pure silver antibacterial bone scaffolds. J Mech Behav Biomed Mater. 2020 Dec 1;112:104090.
114. Gold K, Slay B, Knackstedt M, Gaharwar AK. Antimicrobial Activity of Metal and Metal-Oxide Based Nanoparticles. Adv Ther (Weinh) [Internet]. 2018 Jul 1 [cited 2024 Feb 26];1(3):1700033. Available from: https://onlinelibrary.wiley.com/doi/full/10.1002/adtp.201700033
115. Shuguang Wang, Lawson R, Ray PC, Hongtao Yu. Toxic effects of gold nanoparticles on Salmonella typhimurium bacteria. Toxicol Ind Health [Internet]. 2011 [cited 2024 Feb 26];27(6):547–54. Available from: https://pubmed.ncbi.nlm.nih.gov/21415096/
116. Nathan C, Cunningham-Bussel A. Beyond oxidative stress: an immunologist's guide to reactive oxygen species. Nat Rev Immunol [Internet]. 2013 May [cited 2024 Feb 26];13(5):349–61. Available from: https://pubmed.ncbi.nlm.nih.gov/23618831/
117. Jin G, Qin H, Cao H, Qian S, Zhao Y, Peng X, et al. Synergistic effects of dual Zn/Ag ion implantation in osteogenic activity and antibacterial ability of titanium. Biomaterials. 2014 Sep 1;35(27):7699–713.
118. Shimabukuro M, Ito H, Tsutsumi Y, Nozaki K, Chen P, Yamada R, et al. The Effects of Various Metallic Surfaces on Cellular and Bacterial Adhesion. Metals (Basel) [Internet]. 2019 Oct 25 [cited 2024 Feb 26];9(11):1145. Available from: https://www.mdpi.com/2075-4701/9/11/1145/htm
119. Mokabber T, Cao HT, Norouzi N, Van Rijn P, Pei YT. Antimicrobial Electrodeposited Silver-Containing Calcium Phosphate Coatings. ACS Appl Mater Interfaces [Internet]. 2020 Feb 5 [cited 2024 Feb 29];12(5):5531–41. Available from: https://pubs.acs.org/doi/full/10.1021/acsami.9b20158
120. Annabi, N.; Nichol, J.W.; Zhong, X.; Ji, C.; Koshy, S.; Khademhosseini, A.; Dehghani F. Controlling the porosity and microarchitecture of hydrogels for tissue engineering. Tissue Eng Part B Rev. 2010;16(371–383).
121. Hannink G, Arts JJC. Bioresorbability, porosity and mechanical strength of bone substitutes: What is optimal for bone regeneration? Injury. 2011 Sep 1;42(SUPPL. 2):S22–5.
122. Sachot N, Engel E, Castaño O. Hybrid Organic-Inorganic Scaffolding Biomaterials for Regenerative Therapies. Article Details: Current Organic Chemistry [Internet]. 2014 [cited 2024 Feb 25];18:2299–314. Available from: https://doi.org/10.2174/1385272819666140806200355
123. Mour M, Das D, Winkler T, Hoenig E, Mielke G, Morlock MM, et al. Advances in Porous Biomaterials for Dental and Orthopaedic Applications. Materials [Internet]. 2010 [cited 2024 Feb 25];3(5):2947. Available from: /pmc/articles/PMC5445918/

124. Das K, Bose S, Bandyopadhyay A. TiO2 nanotubes on Ti: Influence of nanoscale morphology on bone cell–materials interaction. J Biomed Mater Res A [Internet]. 2009 Jun 1 [cited 2024 Feb 26];90A(1):225–37. Available from: https://onlinelibrary.wiley.com/doi/full/10.1002/jbm.a.32088
125. Heidari F, Bazargan-Lari R, Razavi M, Fahimipour F, Vashaee D, Tayebi L. Nano-hydroxyapatite and nano-hydroxyapatite/zinc oxide scaffold for bone tissue engineering application. Int J Appl Ceram Technol [Internet]. 2020 Nov 1 [cited 2024 Feb 26];17(6):2752–61. Available from: https://onlinelibrary.wiley.com/doi/full/10.1111/ijac.13596
126. Vimalraj S. Alkaline phosphatase: Structure, expression and its function in bone mineralization. Gene [Internet]. 2020;754(April):144855. Available from: https://doi.org/10.1016/j.gene.2020.144855
127. Stamnitz S, Klimczak A. Mesenchymal stem cells, bioactive factors, and scaffolds in bone repair: From research perspectives to clinical practice. Cells. 2021;10(8).
128. Datta N, Holtorf HL, Sikavitsas VI, Jansen JA, Mikos AG. Effect of bone extracellular matrix synthesized in vitro on the osteoblastic differentiation of marrow stromal cells. Biomaterials. 2005;26(9):971–7.
129. Falcão AM, Palha JA, Ferreira AC, Marques F, Sousa N, Sousa JC. Unbiased Stereological SEZ Proliferation Analysis. Stem Cell Niche [Internet]. 2013;1035(3):121–33. Available from: http://link.springer.com/10.1007/978-1-62703-508-8
130. Mollentze J, Durandt C, Pepper MS. An in Vitro and in Vivo Comparison of Osteogenic Differentiation of Human Mesenchymal Stromal/Stem Cells. Stem Cells Int. 2021;2021.
131. Chan ME, Lu XL, Huo B, Baik AD, Chiang V, Guldberg RE, et al. A trabecular bone explant model of osteocyte-osteoblast co-culture for bone mechanobiology. Cell Mol Bioeng. 2009;2(3):405–15.
132. Pittenger MF, Mackay AM, Beck SC, Jaiswal RK, Douglas R, Mosca JD, et al. Multilineage Potential of Adult Human Mesenchymal Stem Cells. 1999;284(April):143–8.
133. Colnot C. Cell sources for bone tissue engineering: Insights from basic science. Tissue Eng Part B Rev. 2011;17(6):449–57.
134. Zhao X, Han Y, Li J, Cai B, Gao H, Feng W, et al. BMP-2 immobilized PLGA/hydroxyapatite fibrous scaffold via polydopamine stimulates osteoblast growth. Materials Science and Engineering C [Internet]. 2017;78:658–66. Available from: https://doi.org/10.1016/j.msec.2017.03.186
135. Vacanti V, Kong E, Suzuki G, Sato K, Canty JM, Lee T. Phenotypic changes of adult porcine mesenchymal stem cells induced by prolonged passaging in culture. J Cell Physiol. 2005;205(2):194–201.
136. Mitra D, Whitehead J, Yasui OW, Leach JK. Bioreactor culture duration of engineered constructs influences bone formation by mesenchymal stem cells. Biomaterials. 2017 Nov;146:29–39.
137. Whitehead J, Zhang J, Harvestine JN, Kothambawala A, Liu G yu, Leach JK. Tunneling nanotubes mediate the expression of senescence markers in mesenchymal stem/stromal cell spheroids. Stem Cells. 2020 Jan 1;38(1):80–9.
138. Cano-Gamez E, Soskic B, Roumeliotis TI, So E, Smyth DJ, Baldrighi M, et al. Single-cell transcriptomics identifies an effectorness gradient shaping the response of CD4+ T cells to cytokines. Nat Commun. 2020 Apr 14;11(1):1801.
139. Piñeiro-Ramil M, Sanjurjo-Rodríguez C, Castro-Viñuelas R, Rodríguez-Fernández S, Fuentes-Boquete IM, Blanco FJ, et al. Usefulness of mesenchymal cell lines for bone and cartilage regeneration research. Int J Mol Sci. 2019;20(24).
140. Agarwal S, Curtin J, Duffy B, Jaiswal S. Biodegradable magnesium alloys for orthopaedic applications: A review on corrosion, biocompatibility and surface modifications. Materials Science and Engineering C [Internet]. 2016;68:948–63. Available from: https://doi.org/10.1016/j.msec.2016.06.020
141. Rahman M, Li Y, Wen C. HA coating on Mg alloys for biomedical applications: A review. Journal of Magnesium and Alloys [Internet]. 2020;8(3):929–43. Available from: https://doi.org/10.1016/j.jma.2020.05.003

142. Villalona GA, Udelsman B, Duncan DR, McGillicuddy E, Sawh-Martinez RF, Hibino N, et al. Cell-Seeding Techniques in Vascular Tissue Engineering. Tissue Eng Part B Rev. 2010 Jun;16(3):341–50.
143. Wendt D, Marsano A, Jakob M, Heberer M, Martin I. Oscillating perfusion of cell suspensions through three-dimensional scaffolds enhances cell seeding efficiency and uniformity. Biotechnol Bioeng. 2003 Oct 20;84(2):205–14.
144. Weiskirchen S, Schröder SK, Buhl EM, Weiskirchen R. A Beginner's Guide to Cell Culture: Practical Advice for Preventing Needless Problems. Vol. 12, Cells. MDPI; 2023.
145. Yao T, Asayama Y. Animal-cell culture media: History, characteristics, and current issues. Reprod Med Biol. 2017 Apr 21;16(2):99–117.
146. Sotiropoulou PA, Perez SA, Salagianni M, Baxevanis CN, Papamichail M. Characterization of the Optimal Culture Conditions for Clinical Scale Production of Human Mesenchymal Stem Cells. Stem Cells. 2006 Feb 1;24(2):462–71.
147. Grossner T, Haberkorn U, Hofmann J, Gotterbarm T. Effects of Different Basal Cell Culture Media Upon the Osteogenic Response of hMSCs Evaluated by99m Tc-HDP Labeling. Int J Mol Sci. 2022 Jun 1;23(11).
148. Jaiswal N, Haynesworth SE, Caplan AI, Bruder SP. Osteogenic differentiation of purified, culture-expanded human mesenchymal stem cells in vitro. J Cell Biochem. 1997 Feb;64(2):295–312.
149. Manferdini C, Gabusi E, Grassi F, Piacentini A, Cattini L, Zini N, et al. Evidence of specific characteristics and osteogenic potentiality in bone cells from tibia. J Cell Physiol. 2011 Oct 22;226(10):2675–82.
150. Langenbach F, Handschel J. Effects of dexamethasone, ascorbic acid and β-glycerophosphate on the osteogenic differentiation of stem cells in vitro. Stem Cell Res Ther. 2013 Dec 30;4(5):117.
151. Franceschi RT, Iyer BS. Relationship between collagen synthesis and expression of the osteoblast phenotype in MC3T3-E1 cells. Journal of Bone and Mineral Research. 1992 Feb 3;7(2):235–46.
152. Kyllönen L, Haimi S, Mannerström B, Huhtala H, Rajala KM, Skottman H, et al. Effects of different serum conditions on osteogenic differentiation of human adipose stem cells in vitro. Stem Cell Res Ther. 2013;4(1).
153. Sordi MB, Curtarelli RB, da Silva IT, Fongaro G, Benfatti CAM, de Souza Magini R, et al. Effect of dexamethasone as osteogenic supplementation in in vitro osteogenic differentiation of stem cells from human exfoliated deciduous teeth. J Mater Sci Mater Med. 2021 Jan 1;32(1).
154. Yan XZ, Yang W, Yang F, Kersten-Niessen M, Jansen JA, Both SK. Effects of continuous passaging on mineralization of MC3T3-E1 cells with improved osteogenic culture protocol. Tissue Eng Part C Methods. 2014 Mar 1;20(3):198–204.
155. Park JB. The effects of dexamethasone, ascorbic acid, and β-glycerophosphate on osteoblastic differentiation by regulating estrogen receptor and osteopontin expression. Journal of Surgical Research. 2012 Mar;173(1):99–104.
156. Qi H, Ye Z, Ren H, Chen N, Zeng Q, Wu X, et al. Bioactivity assessment of PLLA/PCL/HAP electrospun nanofibrous scaffolds for bone tissue engineering. Life Sci. 2016 Mar 1;148:139–44.
157. Jones SE, Nichols L, Elder SH, Priddy LB. Laser microgrooving and resorbable blast texturing for enhanced surface function of titanium alloy for dental implant applications. Biomedical Engineering Advances. 2023 Jun;5:100090.
158. Bonewald LF, Harris SE, Rosser J, Dallas MR, Dallas SL, Camacho NP, et al. Von Kossa Staining Alone Is Not Sufficient to Confirm that Mineralization In Vitro Represents Bone Formation. Calcif Tissue Int. 2003 May 1;72(5):537–47.
159. Hamidouche Z, Haÿ E, Vaudin P, Charbord P, Schüle R, Marie PJ, et al. FHL2 mediates dexamethasone-induced mesenchymal cell differentiation into osteoblasts by activating Wnt/β-catenin signaling-dependent Runx2 expression. The FASEB Journal. 2008 Nov 24;22(11):3813–22.

160. Beloti MM, Rosa AL, Márcio M, Beloti A, Luiz R. Osteoblast Differentiation of Human Bone Marrow Cells Under Continuous and Discontinuous Treatment with Dexamethasone. Braz Dent J. 2005;16(2).
161. Rimando MG, Wu HH, Liu YA, Lee CW, Kuo SW, Lo YP, et al. Glucocorticoid receptor and Histone deacetylase 6 mediate the differential effect of dexamethasone during osteogenesis of mesenchymal stromal cells (MSCs). Sci Rep. 2016 Nov 30;6.
162. Tenenbaum HC, Heersche JNM. Dexamethasone Stimulates Osteogenesis in Chick Periosteum *in Vitro*. Endocrinology. 1985 Nov;117(5):2211–7.
163. Walsh S, Jordan GR, Jefferiss C, Stewart K, Beresford JN. High concentrations of dexamethasone suppress the proliferation but not the differentiation or further maturation of human osteoblast precursors in vitro: relevance to glucocorticoid-induced osteoporosis. Rheumatology. 2000;
164. Martin I, Wendt D, Heberer M. The role of bioreactors in tissue engineering. Trends Biotechnol. 2004;22(2):80–6.
165. Kazimierczak P, Przekora A. Bioengineered Living Bone Grafts—A Concise Review on Bioreactors and Production Techniques In Vitro. Int J Mol Sci. 2022 Feb 3;23(3):1765.
166. Watson E, Mikos AG. Advances in In Vitro and In Vivo Bioreactor-Based Bone Generation for Craniofacial Tissue Engineering. BME Front. 2023 Jan 31;4.
167. Melke J, Zhao F, Rietbergen B, Ito K, Hofmann S. Localisation of mineralised tissue in a complex spinner flask environment correlates with predicted wall shear stress level localisation. Eur Cell Mater. 2018;36:57–68.
168. Vunjak-Novakovic G, Martin I, Obradovic B, Treppo S, Grodzinsky AJ, Langer R, et al. Bioreactor cultivation conditions modulate the composition and mechanical properties of tissue-engineered cartilage. Journal of Orthopaedic Research. 1999 Jan 18;17(1):130–8.
169. Klein-Nulend J, Bakker AD, Bacabac RG, Vatsa A, Weinbaum S. Mechanosensation and transduction in osteocytes. Bone [Internet]. 2013;54(2):182–90. Available from: https://doi.org/10.1016/j.bone.2012.10.013
170. Sikavitsas VI, Bancroft GN, Van Den Dolder J, Sheffield TL, Jansen JA, Ambrose CG, et al. Fluid flow increases mineralized matrix deposition in three-dimensional perfusion culture of marrow stromal osteoblasts in a dose-dependent manner. Annual International Conference of the IEEE Engineering in Medicine and Biology – Proceedings. 2002;1:884–5.
171. Bhaskar B, Owen R, Bahmaee H, Rao PS, Reilly GC. Design and Assessment of a Dynamic Perfusion Bioreactor for Large Bone Tissue Engineering Scaffolds. Appl Biochem Biotechnol. 2018;185(2):555–63.
172. Sikavitsas VI, Bancroft GN, Mikos AG. Formation of three-dimensional cell/polymer constructs for bone tissue engineering in a spinner flask and a rotating wall vessel bioreactor. J Biomed Mater Res. 2002;62(1):136–48.
173. Engel N, Fechner C, Voges A, Ott R, Stenzel J, Siewert S, et al. An optimized 3D-printed perfusion bioreactor for homogeneous cell seeding in bone substitute scaffolds for future chairside applications. Sci Rep [Internet]. 2021;11(1):1–15. Available from: https://doi.org/10.1038/s41598-021-01516-8
174. Sailon AM, Allori AC, Davidson EH, Reformat DD, Allen RJ, Warren SM. A Novel Flow-Perfusion Bioreactor Supports 3D Dynamic Cell Culture. J Biomed Biotechnol. 2009;873816.
175. Harvestine JN, Gonzalez-Fernandez T, Sebastian A, Hum NR, Genetos DC, Loots GG, et al. Osteogenic preconditioning in perfusion bioreactors improves vascularization and bone formation by human bone marrow aspirates. Sci Adv. 2020;6(7):1–12.
176. Lovecchio J, Gargiulo P, Vargas Luna JL, Giordano E, Sigurjónsson ÓE. A standalone bioreactor system to deliver compressive load under perfusion flow to hBMSC-seeded 3D chitosan-graphene templates. Sci Rep. 2019;9(1):1–11.
177. Bouet G, Cruel M, Laurent C, Vico L, Malaval L, Marchat D. Validation of an in vitro 3D bone culture model with perfused and mechanically stressed ceramic scaffold. Eur Cell Mater. 2015;29:250–67.

178. Mccoy RJ, Jungreuthmayer C, O'Brien FJ. Influence of flow rate and scaffold pore size on cell behavior during mechanical stimulation in a flow perfusion bioreactor. Biotechnol Bioeng. 2012;109(6):1583–94.
179. Cartmell SH, Porter BD, García AJ, Guldberg RE. Effects of Medium Perfusion Rate on Cell-Seeded Three-Dimensional Bone Constructs in Vitro. Tissue Eng. 2003;9(6):1197–203.
180. Bouet G, Marchat D, Cruel M, Malaval L, Vico L. *In Vitro* Three-Dimensional Bone Tissue Models: From Cells to Controlled and Dynamic Environment. Tissue Eng Part B Rev. 2015 Feb;21(1):133–56.
181. Dursun G, Umer M, Markert B, Stoffel M. Designing of an advanced compression bioreactor with an implementation of a low-cost controlling system connected to a mobile application. Processes. 2021;9(6).
182. Schädli GN, Vetsch JR, Baumann RP, de Leeuw AM, Wehrle E, Rubert M, et al. Time-lapsed imaging of nanocomposite scaffolds reveals increased bone formation in dynamic compression bioreactors. Commun Biol [Internet]. 2021;4(1):1–14. Available from: https://doi.org/10.1038/s42003-020-01635-4
183. Bilgen B, Chu D, Stefani R, Aaron RK. Design of a biaxial mechanical loading bioreactor for tissue engineering. J Vis Exp. 2013;(74):e50387.
184. Frost HM. Bone's Mechanostat: A 2003 Update. Anatomical Record - Part A Discoveries in Molecular, Cellular, and Evolutionary Biology. 2003;275(2):1081–101.
185. Armand M, Endres F, MacFarlane DR, Ohno H, Scrosati B. Ionic-liquid materials for the electrochemical challenges of the future. Nat Mater. 2009 Aug 24;8(8):621–9.
186. Ravichandran A, Lim J, Chong MSK, Wen F, Liu Y, Pillay YT, et al. In vitro cyclic compressive loads potentiate early osteogenic events in engineered bone tissue. J Biomed Mater Res B Appl Biomater. 2017;105(8):2366–75.
187. Burr DB, Allen MR. Basic and Applied Bone Biology. Basic and Applied Bone Biology. 2013. 1–373 p.
188. O'Connor JA, Lanyon LE, MacFie H. The influence of strain rate on adaptive bone remodelling. J Biomech. 1982;15(10):767–81.
189. Birmingham E, Kreipke TC, Dolan EB, Coughlin TR, Owens P, McNamara LM, et al. Mechanical Stimulation of Bone Marrow In Situ Induces Bone Formation in Trabecular Explants. Ann Biomed Eng. 2015 Apr 4;43(4):1036–50.
190. Turner CH. Three rules for bone adaptation to mechanical stimuli. Bone. 1998;23(5):399–407.
191. Robling AG, Burr DB, Turner CH. Recovery periods restore mechanosensitivity to dynamically loaded bone. Journal of Experimental Biology. 2001;204(19):3389–99.
192. Dua R, Jones H, Noble PC. Evaluation of bone formation on orthopedic implant surfaces using an ex-vivo bone bioreactor system. Sci Rep [Internet]. 2021;11(1):1–10. Available from: https://doi.org/10.1038/s41598-021-02070-z

# Tendon/Ligament-to-Bone Interface: Biomimetic Strategies for the Structural and Mechanical Integration of the Enthesis

**Xiping Jiang, Wen Xue, Mitchell Kuss, Philipp Streubel, Yaohua He, and Bin Duan**

## 1 Current Clinical Approaches for Enthesis Repair

Tendon bridges muscles to bone, and ligament connects bone to bone. These structures enable load transfer across two widely dissimilar materials, and the soft-to-hard tissue interface that attaches tendon or ligament to bone is also called the enthesis [1, 2]. Orthopedic surgeons frequently encounter the dilemma of ruptured enthesis and torn tendons and ligaments. Rotator cuff tendon tears and anterior cruciate ligament (ACL) ruptures are two of the most common injuries afflicting the shoulder and the knee, respectively, with over 250,000 rotator cuff repairs and

X. Jiang
Department of Orthopedic Surgery, Shanghai Sixth People's Hospital Affiliated to Shanghai Jiao Tong University School of Medicine, Shanghai, China

W. Xue
College of Biological Science and Medical Engineering, Donghua University, Shanghai, China

M. Kuss
Mary & Dick Holland Regenerative Medicine Program, University of Nebraska Medical Center, Omaha, NE, USA

Division of Cardiology, Department of Internal Medicine, University of Nebraska Medical Center, Omaha, NE, USA

P. Streubel
Levitetz Department of Orthopaedic Surgery, Cleveland Clinic Florida, Weston, FL, USA

Y. He (✉)
Department of Orthopedic Surgery, Shanghai Sixth People's Hospital Affiliated to Shanghai Jiao Tong University School of Medicine, Shanghai, China

Department of Orthopedic Surgery, Jinshan District Central Hospital Affiliated to Shanghai University of Medicine & Health Sciences, Jinshan Branch of Shanghai Sixth People's Hospital, Shanghai, China
e-mail: heyaohua@sjtu.edu.cn



J. Liao, J. Y. Wong (eds.), *Integration and Bridging of Multiscale Bioengineering Designs and Tissue Biomechanics*, https://doi.org/10.1007/978-3-031-81743-4_7

about 150,000 ACL reconstructions performed annually in the United States alone [3, 4]. The rotator cuff consists of the supraspinatus, infraspinatus, teres minor and subscapularis, and supports the rotation and stability of the glenohumeral joint of the shoulder [5]. Rotator cuff tears can be divided into four categories based on the tear size: small (<1 cm), medium (1–3 cm), large (3–5 cm), and massive (>5 cm) [6, 7]. Current methods to treat enthesis disease include conservative therapy (e.g. rest, pain control, corticosteroid injection, and physical therapy). When these fail, surgery is indicated, most frequently in the form of arthroscopic repair [8]. Tear size and geometry determine what suture anchor configuration is used during surgical repair (Fig. 1a). Double-row suture bridging repairs have been shown to provide increased footprint contact area and enhanced ultimate force to failure [9, 10]. Despite the improvement in repair techniques, the retear rate of rotator cuff tears with small and medium size is still about 20% [11]. In some instances, rotator cuff tears are not repairable and require the use of other surgical alternatives such as superior capsular reconstruction (SCR), subacromial balloon spacer placement, biologic tuberoplasty, tendon transfers, or even reverse total shoulder arthroplasty. Common sources for autograft include fascia lata, and long head of the biceps tendon (Fig. 1b). Tendon transfers include latissimus dorsi tendon transfer and lower trapezius tendon transfer, while a dermal allograft patch is a viable alternative for SCR (Fig. 1c) [12–15]. The failure rate of SCR treatment for large and massive rotator cuff tears ranges from 20% to 40% with 1-year follow-up [16, 17]. The recovery of muscle strength takes at least 1 year after the SCR surgery, with potential complications of graft detachment and infection [18]. A subacromial balloon spacer that can be placed arthroscopically for massive rotator cuff repair has been gaining popularity in recent years (Fig. 1d). This biodegradable implant reduces the friction gliding between humeral head and acromion and provides vertical stability of the humeral head within the glenoid during arm elevation [19]. The implant is composed of a copolymer of poly-L-lactide-eco-ε-caprolactone that degrades in vivo within 12 months after surgery [20]. The application of this method in clinical trials shows conflicting results. Atoun et al. demonstrated that the balloon spacer reduced pain and improved shoulder function for at minimum 5 years after implantation [21]. A double-blinded multicenter randomized control trial indicated that the balloon spacer implantation was not recommended with little

B. Duan (✉)
Mary & Dick Holland Regenerative Medicine Program, University of Nebraska Medical Center, Omaha, NE, USA

Division of Cardiology, Department of Internal Medicine, University of Nebraska Medical Center, Omaha, NE, USA

Department of Surgery, College of Medicine, University of Nebraska Medical Center, Omaha, NE, USA

Department of Mechanical and Materials Engineering, University of Nebraska-Lincoln, Lincoln, NE, USA
e-mail: bin.duan@unmc.edu

benefit and potentially harmful outcome compared with the debridement-only group [22]. The ultimate option for massive irreparable rotator cuff tears is reverse total shoulder arthroplasty (RTSA) (Fig. 1e). RTSA has the ability to achieve better shoulder function in the setting of significant loss of active shoulder elevation and pseudoparalysis [23, 24]. For the knee, ACL controls anterior translation of the tibia relative to the femur and prevents extreme movement of tibia rotation [4]. ACL injury is commonly associated with involvement of other structures such as meniscus, articular cartilage, subchondral bone, or other ligaments [25, 26]. ACL reconstruction with autologous hamstring tendons is the most popular procedure in several national registries. Hamstrings consist of both gracilis and semitendinosus tendons, which are harvested from the ipsilateral knee during surgery. Several fixation techniques exist. One common alternative is the fixation of the autograft of the femur using a suspensory cortical button and stabilization at the tibia via an interference screw (Fig. 1f) [27, 28]. There are also clinical artificial ligaments for ACL reconstruction, such as ligament advanced reinforcement system (LARS), which is made of terephthalic polyethylene polyester [29]. Although the short-term follow-up of the artificial ligament shows favorable results, robust long-term evaluation is lacking and ambiguous, with several associated complications reported [30]. The long-term complications of ACL reconstruction involve knee osteoarthritis, failure of the graft, and arthrofibrosis. The developing of osteoarthritis may not be prevented by ACL reconstruction. The rates of graft failure and arthrofibrosis are both around 6% [31–33].

Current treatment for tendon/ligament-to-bone interface injuries still needs to be improved due to the high retear rate and complications after surgery. The complex transitional zone of the enthesis fails to regenerate by using current clinical strategies, resulting in unsatisfactory outcomes. The comprehensive understanding of the biology of the enthesis is the essential foundation for inspiring innovative regenerative therapies. The biomimetic strategies based on current understanding of the enthesis have demonstrated promising outcomes. In the following parts of the review, we summarize the advances in development and maturation of the enthesis, followed by a description of the structural and mechanical properties of the enthesis. Next, the biomimetic strategies are discussed in various sections, including utilizing the diverse extracellular matrix, encapsulating different levels of mineralization, recapitulating different cell types, mimicking the collagen fiber orientation at the enthesis, and promoting in situ stem cell differentiation and regeneration using various growth factors and metallic ions. Finally, we highlight the present difficulties and future directions of the tendon/ligament-to-bone interface tissue engineering.

## 2 Advances in the Development and Maturation of the Enthesis

The embryonic enthesis is derived from a distinct pool of Sox9- and Scx-positive cells [34]. During this timeframe, enthesis cells also express growth and differentiation factors (Gdf5), which are distinct from the primary cartilage [35]. These cells

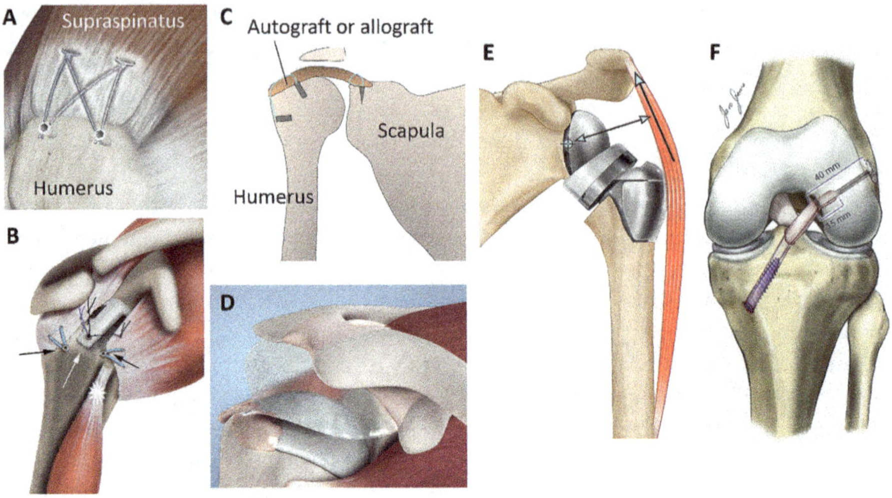


**Fig. 1** Current approaches for enthesis repair. (**a**) Linked and double-row suture repair for small and medium rotator cuff tears (Reproduced with permission from Ref. [9]). (**b**) Long head of biceps tendon as the autograft for irreparable rotator cuff tear (Reproduced with permission from Ref. [12]). A partial repair of the infraspinatus and subscapularis tendons is achieved with suture anchors (black arrows). The lateral end of the cut long head of biceps tendon (LHBT) was fixed with an anchor while incorporating the infraspinatus and subscapularis tendons via a side-to-side marginal repair (white arrow). The distal long head of the biceps can be fixed with a separate anchor (asterisk). (**c**) Irreparable rotator cuff repair managed with superior capsular reconstruction (Ref. [13]). (**d**) Subacromial balloon spacer to treat massive irreparable rotator cuff tear (Reproduced with permission from Ref. [19]). (**e**) Reverse shoulder total arthroplasty for the treatment of massive irreparable rotator cuff tear (Reproduced with permission from Ref. [24]). (**f**) Hamstrings autograft for ACL reconstruction (Reproduced with permission from Ref. [27])

are regulated by fibroblast growth factor (FGF), transforming growth factor beta (TGFβ), and bone morphogenetic protein (BMP) signaling pathways to determine their fate and differentiation [36, 37]. The embryonically Sox9+ progenitors are replaced with Gli1+ lineage cells in migrating enthesis (postnatally will develop into fibrocartilaginous enthesis). The expression of Gli1 is controlled by sonic hedgehog (Hh) signaling in utero and regulated by Indian Hh signaling after birth [38]. Indian Hh/parathyroid hormone (PTrH)-related protein feedback loop also manipulates the mineralization of the enthesis [36]. Inactivating Hh signaling reduces the area of mineralized fibrocartilage by 32%, and leads to coalesce collagen fibers at the insertion site, resulting in more than 100% increased failure insertion strain [39]. At later embryonic stages, Gli1 lineage cells, in response to Hh signaling, are distinct from tenocytes and chondrocytes, which eventually occupy the entire mature enthesis [40]. Both the Gli1+ cells and activated Hh signaling are modulated by the muscle loading and are necessary for the fibrocartilage mineralization and biomechanical strength of the enthesis [40]. Unmineralized Gli1+ cells are mineralized via the activated Hh signaling to generate the mineralized fibrocartilage at the enthesis [41]. Gli1+ cells also have the ability to promote the healing of

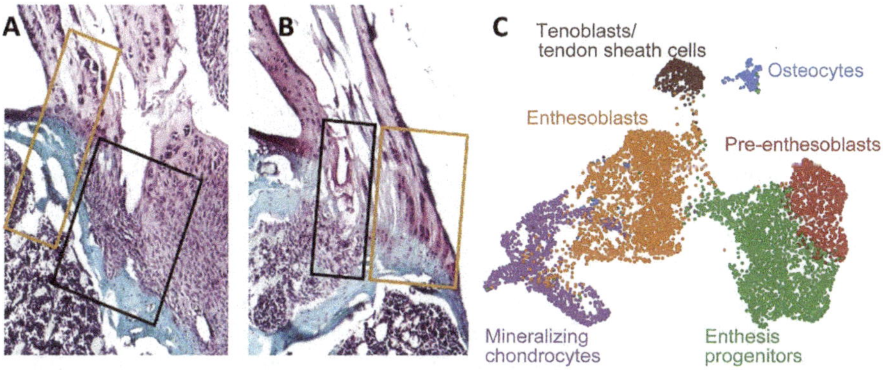


**Fig. 2** Enthesis development and maturation. (**a** and **b**) Histological staining of injured entheses at post-operative day 28 ($n = 6$–8/group). The left image is the control group with only collagen gel delivered. The right image is the treatment group with Gli1+ cells delivered via collagen gel. Safranin O is applied for histological section staining. The black and yellow rectangles in the images represent injured and intact enthesis areas, respectively (Reproduced with permission from Ref. [43]). (**c**) Clustering the six resident cell types of enthesis mesenchymal cells (Reproduced with permission from Ref. [43])

immature entheses after laceration, but the regenerative capacity is absent after these cells differentiated into mature mineralized fibrochondrocytes. Colocalized expression of Gli1-Cre and Ki67, indicating proliferating cells, was observed in injured immature entheses but not in mature entheses [42]. Transplantation of Gli1+ cells into injured enthesis also promoted healing process (Fig. 2a and b) [43]. The cell surface transmembrane proteins patched (Ptch) and smoothened (Smo) are correlated with the Hh signaling. Smo translocates to the primary cilia stimulated by the binding of Hh to Ptch, and activates downstream genes such as Gli1 that are Gli-dependent [44, 45]. Primary cilium is a non-motile organelle enclosed by a membrane that is connected to the plasma membrane but can detect and convoy extracellular signaling [46]. Enthesis development and mineralization depend on the primary cilia. Ciliary gene deletion, specifically on tendon tissue, resulted in a 67% decreased expression of collagen X, a marker for hypertrophic chondrocytes. For adaptation to enthesis loading, it relied on Hh signaling through primary cilia [47]. Single-cell RNA sequencing of the entheses mesenchymal cells over postnatal stages identified six resident cell types: tenoblasts/tendon sheath cells, enthesoblasts, mineralizing chondrocytes, osteocytes, pre-enthesoblasts, and enthesis progenitors (Fig. 2c). Transcription factor regulons of these cells were identified. With a specific focus on Gli1+ progenitors, their clonogenicity and multipotency were demonstrated [43]. There are other transcription factors and signaling pathways that are related to enthesis development and maturation, for instance, Scleraxis (Scx), secreted protein acidic and rich in cysteine (SPARC), Connexin 43 (Cx43), and nuclear factor kappa-light-chain-enhancer of activated B cells (NF-κB) pathway [48–50]. Specifically, Cx43 was linked to the hedgehog signaling and participated in the enthesis response to loading [49]. With all these signaling pathways and molecules that regulates the enthesis formation and maturation, simply expression of these molecules are inadequate to regenerate a functional enthesis after injury [51].

Muscle loading is required for the enthesis maturation and mineralization [36]. Three loading animal models of rotator cuff tendinopathy are frequently applied: underuse loading via botulinum toxin-A (BtxA) injection into the supraspinatus muscle; overuse loading through treadmill running; destabilization loading by generating infraspinatus tendon defect surgically [52]. BtxA-unloading reduced the ultimate load, strength, and toughness of the enthesis without affecting the cross-sectional area of the insertion site, which was associated with significantly disorganized collagen fiber alignment and the changes of micrometer-scale hydroxyapatite crystals [53]. Supraspinatus muscle unloading also caused severe defects in bone formation and trabecular bone architecture. The bony malformation was partially due to the sustained high osteoclast activity, which should be suppressed with maturation after birth [54]. Destabilization of the shoulder had similar effects as underuse models, causing decreased enthesis strength, reduced mineralization levels at the enthesis, and significantly increased expression of inflammation and extracellular matrix remodeling genes [52]. For overuse activity with treadmill animal models, strengthened enthesis toughness and decreased expression of genes associated with wound healing were surprisingly observed [52]. However, the treadmill model has its drawbacks for overuse loading studies. The examination of enthesis maturation before 7–8 weeks of age is limited due to inability for treadmill running at such a young age [55]. The loading dose for this approach is variable and hard to control [56]. The systemic effects of inflammation and metabolic conditioning changes induced by cardiovascular load also affect the results [57, 58]. Electrical nerve stimulation is an alternative option for repeated muscle stimulation, but it will also cause injury and infection due to the invasive requirement of subdermal needle electrode placement [59]. Recently, optogenetics is employed to noninvasively induce specific muscle contraction for in vivo overuse loading in mice from neonatal to mature. The pulsatile stimulation of precise muscle is achieved by channelrhodopsin-2 (ChR2)-expressing skeletal muscle cells exposed to collimated blue light-emitting diode (LED) light [60]. ChR2 functions as a light-sensitive cation channel, located on the cell membrane, to activate depolarization of the membrane [61]. For overuse model using optogenetic-induced loading, decreased strength and increased collagen damage were observed in young enthesis, with little alteration in adult mice [60].

Therefore, a comprehensive understanding of enthesis development and maturation is fundamental for generating an innovative enthesis regeneration strategy. Appropriate application of Gli1 + cell, precise regulation of Hh signaling and enthesis mineralization, and proper utility of muscle loading at the enthesis demonstrate promising outlook for functional enthesis regeneration.

## 3 Structural and Mechanical Properties of the Enthesis

Based on the histological structure, the enthesis can be divided into direct and indirect insertion. Direct enthesis bridges the tendon/ligament to bone via a fibrocartilaginous enthesis, such as the attachment sites of rotator cuff, ACL, and Achilles tendon. Indirect enthesis connects the tendon/ligament indirectly to the periosteum rather than directly to the bone through a fibrous tissue [62]. Since current clinical problems lie on the unsatisfactory reconstruction of direct enthesis, we will focus on the fibrocartilage tendon/ligament to bone interface in this review.

The fibrocartilage enthesis has four different areas in general: tendon/ligament, nonmineralized fibrocartilage (NMF), mineralized fibrocartilage (MF), and bone. The fibrocartilage enthesis enables stress dispersion and proper load transfer between two dissimilar tissues, tendon/ligament (200 MPa tensile modulus) and bone (20 GPa tensile modulus) [63, 64]. The length of the fibrocartilaginous enthesis averages from 100 μm to 1 mm depending on different species and age [65, 66]. The four parts of the enthesis have gradient variations in four categories: cells, collagen fiber orientation, extracellular matrix (ECM), and mineralization level (Table 1) [11, 67–69]. For the cell phenotypes, the tendon/ligament zone is populated by tenocytes and fibroblasts arranged in linear arrays [70]. The NMF zone is populated by fibrochondrocytes organized in rows [67]. The MF zone is occupied by hypertrophic fibrochondrocytes arranged in columns [68]. The bone zone is resided with osteoblasts, osteocytes, and osteoclasts [64]. The collagen fibers splay outwards when approaching the insertion site, featuring the transition from aligned orientation at the tendon part to disorganized arrangement at the enthesis [2]. At the attachment footprint of the enthesis, a smaller and denser insertion area is hidden in the naked-eye observed interface footprint where the tendon fibers directly insert into bone [71]. The diameter of the collagen fibers at the tendon part also changes into thinner interface fibers with a symmetrical splaying angle up to 15°

**Table 1** Gradient structures of the four sections at the enthesis

| | Tendon | Nonmineralized Fibrocartilage | Mineralized fibrocartilage | Bone |
|---|---|---|---|---|
| Cells | Tenocyte<br>Fibroblast | Fibrochondrocyte | Hypertrophic fibrochondrocyte | Osteoblast<br>Osteocyte<br>Osteoclast |
| Fiber orientation | Aligned | ⟶ | ⟶ | Disorganized |
| ECM | Collagen I | Collagen II, III | Collagen II, X | Collagen I, HAp |
| Mineralization | Low | ⟶ | ⟶ | High |

[72]. The orientation and design of the collagen fibers at the enthesis is controversial for their contribution to reduce stress concentration and increasing toughness and strength. Rossetti et al. indicated the heterogeneous structure was beneficial for stress dispersion [72], whereas Sartori et al. demonstrated the stress was expected to peak at the enthesis due to the high fiber curvatures and reduced cross-sectional area [73]. The subset of collagen fibers that are differentially recruited enables toughness maintenance with difference angle of force directions [71]. In terms of the ECM, tendon/ligament consists mostly type I collagen [68]. NMF zone is mainly composed of type II and III collagen [69]. MF zone consists of type II collagen, with significant amounts of type X collagen [67]. Bone zone is composed of mineralized type I collagen [69]. The mineralization level increases from the tendon to the attachment site, which has a major stress dispersion effect at the insertion area [67]. Both tensile and compressive forces are applied to the enthesis area, and the MF zone exhibits greater compressive mechanical properties than the NMF zone [74]. Mineral defects of the MF zone will result in decreased strength at the enthesis [75]. The loading environment also affects the mineralization distribution at the enthesis. Unloading will reduce bioapatite crystal size and disorganize the crystallographic structure of the enthesis, decreasing the ability for energy absorption [76]. As for the mineralization formation at the enthesis during the early development, stresses around cells in the MF zone can reach a similar level to the early adulthood, because of the cell volume fraction effects at the cell level to mitigate the low muscle stresses during development [77]. With maturation, the humeral head enlarges geometrically with animal mass, whereas the enthesis isometrically increases with muscle cross section growth. The increase of the attachment area according to the muscle forces strength maintains the constant stress at the enthesis [78]. The mineral composition and microstructure of the enthesis also adapts to the loading environment to retain the strength at the enthesis [71]. Furthermore, different loading to failure modes at the enthesis is associated with different toughening mechanisms. Little energy dissipation is observed for the monotonic loading to failure mode, while in cyclical loading mode, damage was first seen within the tendon and enthesis with energy absorbed, eventually resulting in failure at the NMF zone [71].

## 4 Biomimetic Strategies for Enthesis Regeneration

With the gradient structures and mechanical properties at the enthesis, five major biomimetic strategies have been developed: mimicking different types of cells at the enthesis, applying various growth factors and metallic ions to stimulate different cell differentiations, and generating gradient variation of fiber alignment, depositing different ECM across the fabricated scaffold, and developing gradient mineralization. In this section, we will summarize current biomimetic strategies based on these categories.

## 4.1 Cell-Based Strategies

Stem cells play a vital role in tissue engineering and regenerative medicine. Various types of stem cells have been applied to enthesis regeneration. The most widely used stem cells are bone marrow-derived mesenchymal stem cells (BMSCs) and adipose-derived mesenchymal stem cells (ADMSCs) [79–81]. However, the disadvantages of these stem cells are the invasive techniques required during harvesting and donor site morbidities [82, 83]. Stem cells from other sources without invasive procedures are reported, such as umbilical cord-derived mesenchymal stem cells (UCSCs) and urine-derived stem cells (USCs) [84–86]. Both UCSCs and USCs can be noninvasively harvested, possess low immunogenicity, and have the potential for multilineage differentiation [87, 88]. Recently, Zuo et al. identified that costal-cartilage-derived stem cells have the potential for enthesis regeneration, exhibiting high clonogenic capacity, great adaptation to post-transplantation microenvironments, and strong abilities of tenogenic and osteogenic differentiation [89]. Tendon-derived stem cells (TDSCs), fibroblasts, osteoblasts, and terminally differentiated cells like chondrocytes are also employed to regenerate different regions of the enthesis [90–92]. These cell-based biomimetic strategies are summarized in Table 2. Most strategies mimicked the tendon and bone zones using either co-culture of different types of cells or pre-differentiating the same stem cells. TDSCs, fibroblasts, and pre-differentiated stem cells are the commonly applied options for tendon zone. Osteoblasts, pre-differentiated stem cells, and stem cells with hydroxyapatite (HAp) or tricalcium phosphate (TCP) addition are the potential options for bone zones. As for the fibrocartilage zone, not all of the biomimetic strategies attempt to recapitulate this region, but various options are still available, such as terminally differentiated chondrocytes, pre-differentiated stem cells, stem cells supplemented with differentiation medium, and BMSCs cultured in between fibroblasts and osteoblasts. 3D bioprinting technique and co-culture system are implemented to generate zonal regions with different types of cells. Most studies demonstrate favorable outcomes in terms of ECM deposition, collagen fiber organization, and heterogenous structure regeneration. Cells alone cannot provide the necessary toughness for the enthesis. Accompanied with appropriate scaffolds are needed for the cell-based regenerative strategies. However, the results of biomechanical properties vary among studies, with some showing favorable or near-native mechanical strength and some indicating no improvement or benefit to strengthen the enthesis. Further studies need to elucidate the ultimate role of implanted cells in enthesis regeneration and find potential mechanisms to explain their results.

## 4.2 Growth Factor and Metallic Ion-Based Strategies

Growth factors play a significant role in cell differentiation and ECM deposition [97]. With regional applications of different growth factors, multilineage differentiation of the same type of stem cells can be achieved, promoting the regeneration

**Table 2** The application of different types of cells for enthesis regeneration

| Cell type | Biomimetic tendon zone | Biomimetic fibrocartilage or bone zone | Major outcomes | References |
|---|---|---|---|---|
| BMSCs | TGFβ for tenogenic differentiation of BMSCs | TGFβ and dexamethasone for osteogenic differentiation of BMSCs. | In vivo: a rapid enthesis is formed with native bone, which becomes vascularized and innervated; organized collagen fascicles are developed in some samples. | [79] |
| BMSCs<br>Osteoblasts<br>Fibroblasts | Fibroblasts for tendon zone | BMSCs cultured in between fibroblasts and osteoblasts for fibrocartilage zone; osteoblasts and HAp coating for bone zone. | In vitro on a hybrid silk scaffold: BMSCs cultured in between fibroblasts and osteoclasts differentiate into fibrocartilage lineage; BMSCs seeded adjacent to osteoclasts have calcium deposition, while no calcium deposition of BMSCs is found adjacent to the fibroblasts. | [93] |
| Fibroblasts<br>Osteoblasts | Fibroblasts for tendon zone | Osteoblasts for bone zone. | In vitro on a PLA woven fabric: No significant ECM deposition is found in different zones of the scaffold. | [90] |
| BMSCs<br>Chondrocytes<br>Osteoblasts | BMSCs for tendon zone | Chondrocytes with chondroitin sulfate and hyaluronic acid sodium salt for fibrocartilage zone; osteoclasts and HAp for bone zone. | In vitro on a silk scaffold: differentiation at different regions of the scaffold is observed; in vivo: bone-ligament insertion is regenerated with favorable mechanical property and stratified structure. | [92] |
| TDSCs<br>Pre-OB from ADMSCs | TDSCs for tendon zone | Pre-OB from ADMSCs for bone zone. | In vitro: Coculturing different lineage cells can express different zone-related markers through cell-cell interaction and modulating medium conditions. | [80, 94] |

| | | | | |
|---|---|---|---|---|
| TDSCs<br>BMSCs | TDSCs for tendon zone | BMSCs for bone zone | In vivo: TDSCs and BMSCs encapsulate in Mo-containing silicate bioceramics for 3D bioprinting, and the scaffold can induce integrative regeneration of the enthesis. | [91] |
| TDSCs<br>BMSCs | TDSCs for tendon zone | BMSCs for bone zone. | In vivo: TDSCs and BMSCs co-cultured on a decellularized tendon scaffold with gradient osteogenic induction medium demonstrate favorable outcomes for enthesis regeneration. | [95] |
| Fibroblasts<br>BMSCs<br>Osteoblasts | Fibroblasts and TGFβ3 for tendon zone | BMSCs and TGFβ3 for fibrocartilage zone; osteoblasts and TCP for bone zone. | In vivo: 3D-printed multilineages of cell encapsulated scaffold can stimulate chondrogenesis and be effective for heterogenous tissue regeneration. | [96] |
| Pre-differentiated<br>ADMSCs | ADMSCs with pre-tenogenic differentiation | ADMSCs with pre-chondrogenic and osteogenic differentiation for fibrocartilage and bone zone, respectively. | In vivo: the 3D-bioprinted scaffold promotes enthesis regeneration with a superior histological score and improved collagen fiber arrangement. However, the differentiation patterns result in no significant improvement for structure and mechanical property. | [81] |

*BMSCs* bone marrow-derived mesenchymal stem cells, *TGFβ* transforming growth factor beta, *HAp* hydroxyapatite, *PLA* poly-L-lactic acid, *TDSCs* tendon-derived stem cells, *pre-OB* pre-differentiated osteoblasts, *ADMSCs* adipose-derived mesenchymal stem cells, *TCP* tricalcium phosphate of the heterogenous enthesis structure. Three major challenges for growth factor utilization are the choice of scaffold and stem cells, the appropriate selectivity of growth factors, and the delivery method of the growth factor onto or into the scaffold. Table 3 summarizes the current strategies to overcome these challenges. Synthesized materials combined with electrospinning or 3D printing technique are frequently employed for scaffold fabrication. Polycaprolactone (PCL), poly (L-lactide-co-glycolide) acid (PLGA), and poly-lactide acid (PLA) are the commonly used synthesized materials because of their good biocompatibility and mechanical properties. The utilization of silk fibroin, gelatin methacrylate (GelMA), alginate-fibrin construct, and decellularized ECM are also reported. Most of the approaches seed or add autologous stem cells on the scaffold to promote differentiation of these additional stem cells, while some studies recruit endogenous stem cells to the scaffold to facilitate different differentiations of these recruited cells. The commonly employed autologous stem cells are ADMSCs and BMSCs. The utilization of USCs and UCSCs are also viable options. As for the choice of growth factors, platelet-derived growth factor-β (PDGF-BB) and connective tissue growth factor (CTGF) are the preferences for tenogenic differentiation. Transforming growth factor β3 (TGFβ3) is the option for chondrogenic differentiation. Bone morphogenetic protein (BMP) is the choice for osteogenic differentiation, especially BMP2 and BMP4. The selected growth factors can be delivered to the scaffold through several strategies, like heparin immobilization, directly add or mix these growth factors with materials when fabricating, encapsulating these growth factors with PLGA microspheres, and coating the surface of materials with polydopamine (PD). Fusing the collagen-binding peptide (CBP) with the N-terminal of growth factors can be applied when the scaffolds consist of ECM [86]. Metal ions also have the capacity to regulate metabolism and facilitate tissue repair [98]. Different active ions can promote stem cells to differentiate into different lineages for tissue regeneration [99]. Copper ions play a role in bone repair, and zinc oxide nanoparticles can be used for tendon repair [100, 101]. Based on these findings, a thiol-gelatin-based hydrogel crosslinking with gradient Cu and Zn ions through sulfhydryl groups has been developed to reconstruct the enthesis [102]. Both growth factors and metallic ions have shown good outcomes to promote multilineage differentiation of stem cells in vitro and in vivo, facilitating enthesis regeneration. However, the combination of materials and adequate cells with growth factors or metal ions limits their application in clinic. Temporally and spatially controlled release of growth factors or metal ions is needed to enable precise regulation of the enthesis regeneration. The optimal selectivity of growth factors, metal ions, and delivery methods still requires further investigation, since the current literature claims their choice as effective without comparison using other potential options.

**Table 3** The application of different growth factors and metallic ions for enthesis regeneration

| Scaffold and cells | Growth factors | Delivery method | Major outcomes | References |
|---|---|---|---|---|
| PCL/Pluronic F127; ADMSCs | PDGF-BB for tenogenesis; BMP-2 for osteogenesis | Heparin immobilization | In vitro: higher PDGF-BB promotes tenogenesis of ADMSCs; higher BMP-2 promotes osteogenesis of ADMSCs. | [103] |
| Alginate-fibrin construct; BMSCs | CTGF for tenogenesis; TGFβ3 for chondrogenesis | Directly add growth factors into the scaffold when fabricating | In vitro: Spatially distinct arrangement of cell phenotypes and matrix deposition is observed when cocultures cells in a dual chamber system. | [104] |
| PLGA, PLA | BMP-2 for osteogenesis | Directly add BMP-2 into the PLGA solution | In vivo: electrospinning microfiber-reinforced scaffold improves the mechanical property and facilitates the fibrocartilage formation at the enthesis. | [105] |
| PLA; ADMSCs | PDGF-BB for tenogenesis | PD coating the nanofibers | In vitro: gradient biochemical cues are necessary for tenogenesis of ADMSCs, and gradient PD coating of PDGF-BB is beneficial for enthesis regeneration. | [106] |
| Silk fibroin; ADMSCs | TGFβ2 for tenogenesis; TGFβ2 and GDF5 for chondrogenesis | Heparin immobilization | In vitro: TGFβ2 promotes tenogenesis and collagen I protein expression, and the combined delivery of TGFβ2 and GDF5 promotes chondrogenesis and collagen II protein expression. | [107] |
| PCL, PLGA | CTGF for tenogenesis; CTGF and TGFβ3 for chondrogenesis; BMP2 for osteogenesis | Encapsulated in PLGA microsphere | In vivo: Released growth factors from 3D printed scaffold facilitate endogenous stem/progenitor cells recruitment and promote regional differentiation. | [108] |
| GelMA; BMSCs | TGFβ3 and BMP4 for teno-, chondro-, osteo-genesis | Directly added into GelMA solution when fabricating | In vitro: different protein deposition patterns are observed within this multiphasic hydrogel. | [109] |

(continued)

**Table 3** (continued)

| Scaffold and cells | Growth factors | Delivery method | Major outcomes | References |
|---|---|---|---|---|
| PLGA; BMSCs | TGFβ3 and BMP2 for tenogenesis and chondrogenesis | Directly added | In vitro: zonal differentiation of BMSCs on the electrospinning fibers is achieved through gradient distribution of biochemical cues. | [110] |
| Decellularized enthesis matrix; USCs | GDF7 for tenogenesis; TGFβ3 for chondrogenesis; BMP2 for osteogenesis | Fusing a CBP into the N-terminal of these growth factors | In vivo: the regenerated enthesis shows similarity with native enthesis in histology, composition, and mechanical strength in a canine model. | [86] |
| RGD-alginate and fibrinogen, PLGA; UCSCs | BMP2 for osteogenesis; TGFβ for chondrogenesis; bFGF for tenogenesis | Encapsulated in PLGA microsphere | In vivo: the regenerated enthesis demonstrates native-like properties in terms of the collagen alignment, fibrocartilage formation, and gradient structure regeneration. | [85] |
| Thiol-gelatin | Copper ions for osteogenesis; zinc ions for tenogenesis | Mix copper or zinc sulfate solution with gelatin solution | In vivo: Gelatin and ions are crosslinked with sulfhydryl groups, and the hydrogel shows good mechanical property, biocompatibility, and anti-bacterial properties. | [102] |

*PCL* polycaprolactone, *PDGF-BB* platelet-derived growth factor-β, *BMP* bone morphogenetic protein, *TGFβ* transforming growth factor β, *CTGF* connective tissue growth factor, *PLGA* poly (L-lactide-co-glycolide) acid, *PLA* poly-lactide acid, *PD* polydopamine, *GDF* growth/differentiation factor, *GelMA* gelatin methacrylate, *CBP* collagen binding peptide, *RGD* arginine-glycine-aspartic acid, *bFGF* basic fibroblast growth factor

## 4.3 Align-to-Random Scaffold

The collagen fiber orientation is also a key factor for cellular differentiation and tissue regeneration [67]. The fibers change from aligned at the tendon midstance to random as approaching the insertion site. To mimic the topographic cue of the enthesis, a variety of strategies have been developed to recapitulate the collagen fiber orientation. Electrospinning is the most popular approach because of the ECM-like microarchitecture, favorable mechanical strength, controllable morphology, alignment, and electrical and chemical properties of the fabricated nanofibers [111]. PCL, PLGA, and PLA are the most commonly used materials for enthesis repair [105, 106, 112]. Xie et al. manipulate the electric field with a unique two-stapler-shaped metal frame and use these metal frames as the collector. Nanofibers with random orientations will be deposited on the metal, while aligned nanofibers will lie across the air gap between the two metal frames [112]. Meshes with different materials or multiple layers are also generated by this similar technique using a dual-drum collector [113, 114]. The alignment of electrospun nanofibers can also alter the status of the collector. Random oriented nanofibers will be produced with stationary collectors, while rotating collectors will generate aligned nanofibers [105, 110, 115, 116]. Similarly, collector rotating at a low speed of 200 rpm will create random fibers, and aligned fibers will be developed when the rotating speed increases to 1400 rpm [106]. The graded features of fiber orientation can also be manipulated by altering the exposure time to ethanol vapor. Immersed in 70% v/v ethanol solution, it takes a long period of time to wield nanofibers, while exposure to ethanol vapor significantly shortens the wielding time to 3 min. A gradient topographical cue will be generated by adding an ethanol solution into a vial that contains the nanofiber mesh [117]. The electrospinning mesh with gradient fiber orientations can convert from 2D structure to 3D gradient scaffold with either gas foaming technique [118], increasing the porosity of the scaffold to facilitate cell penetration, or a weaving technique, significantly strengthening the mechanical properties of the scaffold [119].

Several other strategies have also been developed to mimic the aligned-to-random feature of the enthesis. Silk fibroin scaffolds with aligned oriented structure is fabricated through directional freezing followed by freeze-drying, while salt leaching and freeze-drying techniques result in silk fibroin scaffolds with randomly oriented structure. Combining these two approaches together can generate scaffolds with anisotropic and isotropic pores [120]. The biphasic silk fibroin scaffold can further mimic the enthesis with gradient growth factors mobilization with heparin functionalization [107]. Combining directional solidification with liquid phase co-synthesis followed by freeze-drying also fabricates biphasic scaffolds with aligned-to-random porosity distribution. The scaffold platform is based on collagen and glycosaminoglycan (GAG) [121]. Decellularized tendon ECM partially disorganized with ultrasonic treatment also generates aligned-to-random fibers to mimic the enthesis. The collagen fibers of tendon ECM without ultrasonic treatment maintain the aligned orientation, while the aligned microstructure will be disorganized with ultrasonic involvement [122]. Rod-shaped cellulose nanocrystals (CNC), obtained from the crystalline regions of the cellulose microfibrils, have diamagnetic capacity. Mixing CNC within hydrogel enables the manipulation of CNC orientation via magnetic fields, which will fabricate anisotropic hydrogels. The application of magnetic field at part of hydrogel will result in anisotropic hydrogels at one end and isotropic hydrogels at the other end [123].

Collectively, various strategies have been developed to generate the aligned-to-random scaffolds. Most of the studies are designed in vitro with promoting stem cells to differentiate into tenogenic and osteogenic lineages as effective outcomes. However, few studies have investigated the mechanical strength that will be impaired with the transition from aligned orientation to disorganization. Previous studies have raised concerns about the highly curved structure of the enthesis [73]. The stress concentration at the enthesis may concentrate in this area. Thus, more in vivo experiments are needed to confirm the benefits of the aligned-to-random scaffold, especially the biomechanical properties.

### 4.4 Mimicking the ECM of the Enthesis

It is extremely challenging to generate a scaffold with completely the same components of the native enthesis tissue structure. The application of decellularized ECM provides the opportunity to overcome this difficulty. Utilization of decellularized allogenic and xenogeneic ECM for enthesis repair is also the most promising approach to be approved for clinical operation because clinical products derived from mammalian ECM, such as human dermis and porcine small intestinal submucosa, have been widely employed for the rotator cuff repair [64, 124]. However, current available decellularized ECM patches fail to recapitulate the heterogeneous structure of the enthesis. To improve the enthesis healing, Liu et al. developed a decellularized tendon-fibrocartilage-bone (TFB) composite with multiple layers (Fig. 3a). The advance of this strategy is that when harvesting the patellar tendon graft, it preserves the attached bone for decellularization, and the enthesis between the harvested tendon and bone is retained as well. BMSCs fabricated cell sheets are integrated into the decellularized scaffold, and the scaffold with cell sheets is sutured back to the detached tendon insertion site for rotator cuff repair (Fig. 3b) [125]. Similar approaches are also reported by Bronsein et al. [126], Farnebo et al. [127], and Su et al. [128], investigating decellularized composite harvested from different sites or using different animal models. Tang et al. further improve the TFB composite with a book-type triphasic scaffold for enthesis repair (Fig. 3c). The book-shaped decellularized tendon, fibrocartilage, and bone scaffold are stacked together, and the heterogeneous structure of tendon-fibrocartilage-bone composite is fabricated (Fig. 3d and e) [129]. The decellularized ECM components can also act to functionalize the electrospun fibers. Through physical adsorption and covalent conjugation, the decellularized ligament-ECM and cartilage-ECM are regionally immobilized to the surface of the PCL electrospun fibers [130]. Besides,

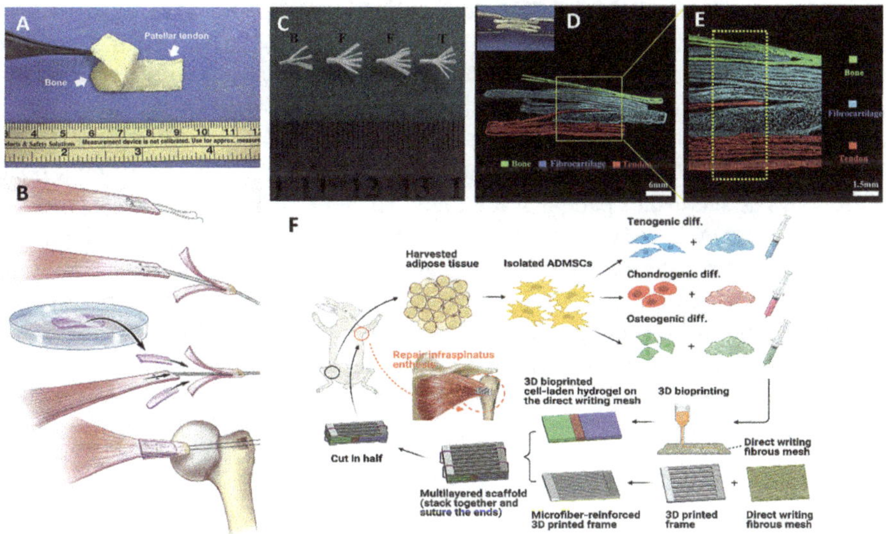


**Fig. 3** Mimicking the ECM of the enthesis. (**a**) An image of harvested TFB composite with multilayers (Reproduced with permission from Ref. [125]). (**b**) Procedures of rotator cuff repair using TFB composite with stem cell sheets in a canine model. Modified Mason-Allen stitch is applied to suture the detached tendon. The detached tendon with sutures is sandwiched between the two layers of TFB composite. Stem cell sheets are placed on the upper and lower surface of the detached tendon. The TFB composite is sutured back to the greater tuberosity through two transosseous tunnels and the bony section of the TFBC composite can fit into a bone trough (Reproduced with permission from Ref. [125]). (**c**) The book-type TFB composite has four parts. B bone, F fibrocartilage, T tendon (Reproduced with permission from Ref. [129]). (**d** and **e**) A decellularized bone scaffold, two decellularized fibrocartilage scaffolds, and one decellularized tendon scaffold are stack together to fabricate the book-type TFB composite. The composite is scanned with synchrotron radiation microcomputed tomography (SR-μCT) (Reproduced with permission from Ref. [129]). (**f**) Schematic outline of the 3D-bioprinted scaffold for enthesis regeneration using pre-differentiated stem cells and polymers (Reproduced with permission from Ref. [81])

the decellularized ECM can be converted into bioinks for 3D bioprinting. The bioinks with different ECM formulations and stem cells are 3D-bioprinted into multilayer structures to regenerate the enthesis complexity (Fig. 3f). The bioinks of decellularized ECM alone cannot provide sufficient mechanical strength to support the whole structure. Polyurethane (PU) and PCL are prepared for the 3D printing as the backbone of the whole structure [131, 132]. Thus, the decellularized TFB composite is the most promising strategy to apply in clinic because the enthesis complex microstructure is well preserved and the repair outcome demonstrates increased neo-fibrocartilage formation and improved mechanical strength of the enthesis. However, the mechanical property of the regenerated enthesis is still far from the native enthesis with the ultimate stress for the repaired one as $4.50 \pm 1.11$ MPa, and the native enthesis as $55.64 \pm 7.97$ MPa [125]. Strategies to enhance the strength of the TFB composite are needed in future experiments. 3D bioprinting strategy using bioinks derived from decellularized ECM is another encouraging approach to employ in clinic without the use of cells and FDA-unapproved biomaterials. The gradient alteration of the enthesis structure can be mimicked by 3D bioprinting different bioinks with various decellularized ECM. 3D-bioprinted personalized active bone implants have been utilized to treat bone defects in patients using patient's autologous platelet-rich plasma (PRP) with the combination of PCL and β-tricalcium phosphate (β-TCP) [133]. The 3D-bioprinted enthesis may have the opportunity to treat the massive rotator cuff tear as well. However, there are still some critical problems to solve for the 3D-bioprinting technique. The mechanical properties of currently 3D-bioprinted scaffold are still unsatisfactory. The attachment of the scaffold into the bone at the tendon insertion site is hard to achieve. More precise printing pattern or structural design is needed since current strategies fail to mimic the exact ECM deposition and structure of the enthesis.

## 4.5 Different Levels of Mineralization

Given the gradient mineralization at the enthesis, studies have developed various strategies to mimic the different levels of mineralization across the microscale region. Li et al. use tens times concentrated simulated body fluid (SBF), which consists of concentrated calcium and phosphate ions ten times higher than those in human plasma or SBF. The concentrated SBF is added at constant rates into a glass vial with electrospun nanofiber meshes (Fig. 4a). The gradient profile can be adjusted by altering the concentration of SBF, the tilting angle of the scaffold in the vial, and the SBF adding rate to the vial [134]. Their follow-up study shows that the gradient mineralization levels of the electrospun nanofibers enable graded osteogenic differentiation of seeded ADMSCs [135]. Similar approaches using SBF for gradient mineralization are also applied in other scaffold designs [106, 130, 136–138]. Collagen-based scaffolds can be mineralized through sequentially immersing in calcium chloride dehydrate solution ($CaCl_2 \cdot 2H_2O$) and disodium hydrogen phosphate solution ($Na_2HPO_4$) [84, 139]. Combining the unmineralized and mineralized scaffold will result in the composite having different levels of mineralization. Harley et al. also apply this strategy for mineralization of collagen and GAG-based scaffolds [121, 140]. Qu et al. demonstrate a method for tendon mineralization by combining $Na_2HPO_4$ extraction and fetuin treatment [141]. Another widely applied strategy is incorporating nanoscale HAp into the electrospinning polymer solution for co-electrospinning [110, 116, 142]. The co-electrospun nanofibers can be further converted to 3D scaffolds by rolling [105], crocheting [143], or weaving [144]. Adding HAp into hydrogel system can also generate different mineral distribution across the scaffold [123].

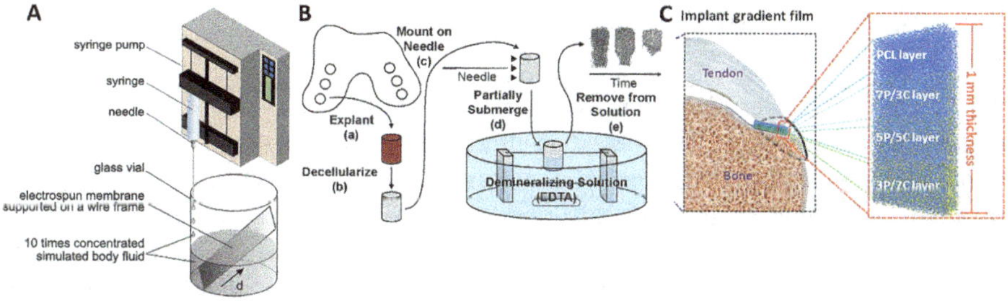


**Fig. 4** Scaffolds with different levels of mineralization to mimic the enthesis structure. (**a**) Schematic illustration of fabricating a gradient coating of calcium phosphate on the surface of electrospun nanofibers. Ten times concentrated simulated body fluid (SBF) is dipped into a vial at a constant rate. (Reprinted with permission from Ref. [134]. Copyright 2023 American Chemical Society). (**b**) Schematic illustration of producing bone scaffolds with gradient mineral (Reprinted with permission from Ref. [145]. Copyright 2023 American Chemical Society). (**c**) A multilayer scaffold is generated by combining PCL with gradient calcium phosphate silicate (CPS) content through tape casting technique, and the multilayered scaffolds are applied in vivo to investigate its performance for enthesis healing (Reproduced with permission from Ref. [147])

Partially and spatially demineralizing the bone biopsies is another approach to create gradient mineralization across the scaffold. The bone biopsies are first explanted using a coring bit, decellularized, and mounted on a needle for demineralization. The bone biopsies are partially submerged in the EDTA demineralizing solution. Finally, the bone scaffolds with gradient mineral are generated and removed from the EDTA solution (Fig. 4b) [145, 146]. Besides, tape casting polymer solution with different proportion of calcium phosphate silicate (CPS) also produces a composite film with gradient mineral distribution (Fig. 4c) [147]. Swelling a composite film consisting of PCL and HAp nanoparticles with a PCL solution can create tunable thickness of the mineral gradient via changing the extent of swelling [148]. A hierarchically structured scaffold for enthesis regeneration can be fabricated by a layer-by-layer coating strategy of PLGA/HAp and combined with interconnected gelatin beads at the bottom. An array of channels created using $CO_2$ laser creates aligned features at the top of the scaffold to mimic the tendon structure. Gradient mineral distribution in the middle part of the scaffold mimics the fibrocartilage zone. Dissolving the gelatin beads generates the porous structure at the bottom along with high mineral contents to mimic the bone properties [149]. Overall, the advances to fabricate gradient mineral distribution of the scaffold are remarkable, with promising potential in promoting enthesis healing. However, most of the scaffolds are still investigated in vitro studies. The robust evaluation of in vivo validation is still lacking, especially the mechanical strength of these scaffolds after implantation. Further comprehensive studies to testify whether the gradient mineral distribution is beneficial for stress dissipation at the enthesis are still needed.

# 5 Summaries and Perspectives

During the past decades, the regeneration of a functional and integrative tendon/ligament-to-bone interface has obtained tremendous attention across the world due to the large number of patients who required surgery annually and the great economic burden. In this review, we focus on various aspects of the enthesis disease. To gain an understanding of current clinical treatment for this disease, we introduce both the traditional and newly developed strategies that are applied in the clinic to reconstruct the torn enthesis, especially of the rotator cuff and ACL. However, the persistent retear rates after surgical intervention present and opportunity to improve the outcomes of current approaches to treat these conditions by developing innovative biological and tissue engineering strategies. The understanding of the development and maturation of the enthesis is fundamental to further advancing the design of original biomaterials.

We summarize the current research of the enthesis biology, emphasizing the important role of Gli1+ cells, Hh signaling pathway, the onset of mineral gradient at the enthesis, and the muscle loading during the development and maturation. Rotator cuff tears can be classified into four categories based on the tear size. A significant number of tears might be treated with innovative biological treatment generated from developmental biology, especially in the setting of large tears with multi-tendon involvement. Furthermore, in the absence of repairable tissue, innovative biomaterials could be of great benefit. Biomimetic strategies have gained much attention in recent years because of the promising results in regenerating the enthesis. We first outline the structural and mechanical properties of the enthesis and then summarize the current biomimetic approaches into five categories: mimicking the enthesis through different cell types, growth factors, topographical cues, ECM deposition, and mineral distribution. All of the strategies have shown favorable results for enthesis repair from different aspects. However, in our opinion, one critical problem to address is the relatively weak mechanical strength of the regenerated enthesis compared with the native enthesis. In other words, these biomimetic strategies seem to reconstruct the enthesis at the structural level, but the mimics of heterogenous structures fail to convert into mechanical strengthening of the enthesis.

To further augment the mechanical property of the enthesis, there are trends to mimic the mechanical gradient at the enthesis without the concern of biological cues. Mozdzen et al. fabricate a collagen-based scaffold with PLA fiber reinforcement. The scaffold possesses mechanical resilience because there are gradient profiles of PLA fiber arrays [150]. Adhesive film or hydrogel is another approach for enthesis reattachment to the insertion site [151–153]. Choi et al. demonstrate a strong, stiff, and adhesive hydrogel that possesses high tensile modulus and strength while reserving robust adhesion to bone without chemical modification of the bone surface [153]. Fang et al. show a mechanically competent chitosan-based adhesive hydrogel that exhibits the strength to improve the enthesis repair (about 1 MPa, 20% of current suture repair strength) [152]. To precisely mimic bone-tendon-like properties, Ker et al. fabricate a UV-cross-linkable and phototunable PU-based scaffold with good biocompatibility that can mimic both the range of tendon-like tensile and compressive properties to bone-like mechanical properties by adjusting the crosslinking time of UV exposure [154]. Chang et al. generates a continuous polyethylene glycol (PEG) hydrogel interface between the osseous and tendinous tissue compartments that can reduce the strain concentration at the interface area [155].

Collectively, mimicking the biological structure of the enthesis enables better integration of the fabricated scaffolds with the tendon/ligament and bone tissue, while mimicking and improving the mechanical properties of the repaired enthesis improves the strength of the tendon/ligament-to-bone interface. Incorporating the two strategies together in future scaffold design might produce encouraging strategies for enthesis regeneration.

## References


1. M. Benjamin, H. Toumi, J.R. Ralphs, G. Bydder, T.M. Best, S. Milz, Where tendons and ligaments meet bone: attachment sites ('entheses') in relation to exercise and/or mechanical load, Journal of Anatomy 208(4) (2006) 471–490.
2. T. Lei, T. Zhang, W. Ju, X. Chen, B.C. Heng, W. Shen, Z. Yin, Biomimetic strategies for tendon/ligament-to-bone interface regeneration, Bioactive Materials 6(8) (2021) 2491–2510.
3. L.V. Gulotta, D. Kovacevic, J.R. Ehteshami, E. Dagher, J.D. Packer, S.A. Rodeo, Application of Bone Marrow-Derived Mesenchymal Stem Cells in a Rotator Cuff Repair Model, The American Journal of Sports Medicine 37(11) (2009) 2126–2133.
4. F. Rayan, S.K. Nanjayan, C. Quah, D. Ramoutar, S. Konan, F.S. Haddad, Review of evolution of tunnel position in anterior cruciate ligament reconstruction, World Journal of Orthopedics 6(2) (2015) 252–262.
5. G. Deprés-tremblay, A. Chevrier, M. Snow, M.B. Hurtig, S. Rodeo, M.D. Buschmann, Rotator cuff repair: a review of surgical techniques, animal models, and new technologies under development, Journal of Shoulder and Elbow Surgery 25(12) (2016) 2078–2085.
6. X. Jiang, S. Wu, M. Kuss, Y. Kong, W. Shi, P.N. Streubel, T. Li, B. Duan, 3D printing of multilayered scaffolds for rotator cuff tendon regeneration, Bioactive Materials 5(3) (2020) 636–643.
7. C.C. Schmidt, C.D. Jarrett, B.T. Brown, Management of Rotator Cuff Tears, Journal of Hand Surgery 40(2) (2015) 399–408.
8. L.S. Oh, B.R. Wolf, M.P. Hall, B.A. Levy, R.G. Marx, Indications for Rotator Cuff Repair: A Systematic Review, Clinical Orthopaedics and Related Research 455 (2007).
9. P.J. Denard, S.S. Burkhart, The evolution of suture anchors in arthroscopic rotator cuff repair, Arthroscopy 29(9) (2013) 1589–1595.
10. A.J. Sheean, R.U. Hartzler, S.S. Burkhart, Arthroscopic Rotator Cuff Repair in 2019: Linked, Double Row Repair for Achieving Higher Healing Rates and Optimal Clinical Outcomes, Arthroscopy 35(9) (2019) 2749–2755.
11. N. Saveh-Shemshaki, L. S.Nair, C.T. Laurencin, Nanofiber-based matrices for rotator cuff regenerative engineering, Acta Biomaterialia 94 (2019) 64–81.
12. A. Boutsiadis, S. Chen, C. Jiang, H. Lenoir, P. Delsol, J. Barth, Long Head of the Biceps as a Suitable Available Local Tissue Autograft for Superior Capsular Reconstruction: "The Chinese Way", Arthroscopy Techniques 6(5) (2017) e1559-e1566.

13. N. Yamamoto, E. Itoi, Treatment of irreparable rotator cuff tears with superior capsular reconstruction, Journal of Experimental Orthopaedics 8(1) (2021) 23.
14. S. Namdari, P. Voleti, K. Baldwin, D. Glaser, G.R. Huffman, Latissimus Dorsi Tendon Transfer for Irreparable Rotator Cuff Tears: A Systematic Review, JBJS 94(10) (2012).
15. L.E. Stoll, J.L. Codding, Lower Trapezius Tendon Transfer for Massive Irreparable Rotator Cuff Tears, Orthopedic Clinics of North America 50(3) (2019) 375–382.
16. Y.-K. Kim, K.-H. Jung, K.-Y. Choi, S.-W. Kang, R.-J. Choi, Arthroscopic Superior Capsular Reconstruction With Tensor Fascia Lata Allograft for Irreparable Rotator Cuff Tears: Clinical and Radiologic Outcomes for a Minimum 1-Year Follow-Up, Arthroscopy 39(6) (2023) 1415–1424.
17. T.D. Pashuck, A.M. Hirahara, J.L. Cook, C.R. Cook, W.J. Andersen, M.J. Smith, Superior Capsular Reconstruction Using Dermal Allograft Is a Safe and Effective Treatment for Massive Irreparable Rotator Cuff Tears: 2-Year Clinical Outcomes, Arthroscopy 37(2) (2021) 489–496.e1.
18. S.-J. Shin, S. Lee, J.Y. Hwang, W. Lee, K.H. Koh, Superior Capsular Reconstruction Using Acellular Dermal Allograft Combined With Remaining Rotator Cuff Augmentation Improved Shoulder Pain and Function at 1 Year After The Surgery, Arthroscopy 38(4) (2022) 1089–1098.
19. A. Castagna, R. Garofalo, E. Maman, A.C. Gray, E.A. Brooks, Comparative cost-effectiveness analysis of the subacromial spacer for irreparable and massive rotator cuff tears, International Orthopaedics 43(2) (2019) 395–403.
20. U. Srikumaran, R. Russo, F. Familiari, Subacromial Balloon Spacer for Massive Irreparable Rotator Cuff Tears, Arthroscopy 39(3) (2023) 576–577.
21. E. Atoun, M. Oulianski, I. Bachar-Avnieli, A. Artamonov, R. Gilat, O. Lubovsky, P.J. Rosinsky, Subacromial Balloon Spacer for Irreparable Rotator Cuff Tear Treatment Shows Improved Pain and Strength at Five-Year Follow-Up, Arthroscopy (S0749-8063(23)00655-2) (2023) Online.
22. A. Metcalfe, H. Parsons, N. Parsons, J. Brown, J. Fox, E. Gemperlé Mannion, A. Haque, C. Hutchinson, R. Kearney, I. Khan, T. Lawrence, J. Mason, N. Stallard, M. Underwood, S. Drew, S.R. team, Subacromial balloon spacer for irreparable rotator cuff tears of the shoulder (START:REACTS): a group-sequential, double-blind, multicentre randomised controlled trial, The Lancet 399(10339) (2022) 1954–1963.
23. S.W. Young, M. Zhu, C.G. Walker, P.C. Poon, Comparison of functional outcomes of reverse shoulder arthroplasty with those of hemiarthroplasty in the treatment of cuff-tear arthropathy: a matched-pair analysis, Journal of Bone and Joint Surgery 95(10) (2013) 910–915.
24. C.P. Roche, Reverse Shoulder Arthroplasty Biomechanics, Journal of Functional Morphology and Kinesiology, 2022.
25. D.L. Bernholt, N.N. DePhillipo, M.D. Crawford, Z.S. Aman, W.J. Grantham, R.F. LaPrade, Incidence of Displaced Posterolateral Tibial Plateau and Lateral Femoral Condyle Impaction Fractures in the Setting of Primary Anterior Cruciate Ligament Tear, The American Journal of Sports Medicine 48(3) (2020) 545–553.
26. D.C. Fithian, L.W. Paxton, D.H. Goltz, Fate of the anterior cruciate ligament-injured knee, Orthopedic Clinics of North America 33(4) (2002) 621–636.
27. J.P. Martinez-Cano, L.M. Zamudio-Castilla, J.A. Cuadros-Potes, L.C. Ibarra-Mosquera, F.M. Mejia-Lopez, Quadrupled Semitendinosus ACL Reconstruction Combining Cortical Button in Femur and Interference Screw in Tibia, Arthroscopy Techniques 9(1) (2020) e9-e14.
28. L. Siegel, C. Vandenakker-Albanese, D. Siegel, Anterior Cruciate Ligament Injuries: Anatomy, Physiology, Biomechanics, and Management, Clinical Journal of Sport Medicine 22(4) (2012).
29. J. Chen, J. Xu, A. Wang, M. Zheng, Scaffolds for tendon and ligament repair: review of the efficacy of commercial products, Expert Review of Medical Devices 6(1) (2009) 61–73.
30. M.T. Rodrigues, R.L. Reis, M.E. Gomes, Engineering tendon and ligament tissues: present developments towards successful clinical products, Journal of Tissue Engineering and Regenerative Medicine 7(9) (2013) 673–686.

31. N.S. Crawford, B.R. Waterman, J.H. Lubowitz, Long-term failure of anterior cruciate ligament reconstruction, Arthroscopy 29(9) (2013) 1566–1571.
32. R. Rousseau, C. Labruyere, C. Kajetanek, O. Deschamps, K.G. Makridis, P. Djian, Complications After Anterior Cruciate Ligament Reconstruction and Their Relation to the Type of Graft: A Prospective Study of 958 Cases, The American Journal of Sports Medicine 47(11) (2019) 2543–2549.
33. T.L. Sanders, H.M. Kremers, A.J. Bryan, W.K. Kremers, M.J. Stuart, A.J. Krych, Procedural intervention for arthrofibrosis after ACL reconstruction: trends over two decades, Knee Surgery, Sports Traumatology, Arthroscopy 25(2) (2017) 532–537.
34. E. Blitz, A. Sharir, H. Akiyama, E. Zelzer, Tendon-bone attachment unit is formed modularly by a distinct pool of Scx- and Sox9-positive progenitors, Development 140(13) (2013) 2680–2690.
35. N.A. Dyment, Y. Hagiwara, B.G. Matthews, Y. Li, I. Kalajzic, D.W. Rowe, Lineage tracing of resident tendon progenitor cells during growth and natural healing, PLoS ONE 9(4) (2014) e96113.
36. E. Zelzer, E. Blitz, M.L. Killian, S. Thomopoulos, Tendon-to-bone attachment: From development to maturity, Birth Defects Research Part C: Embryo Today: Reviews 102(1) (2014) 101–112.
37. R.R. Roberts, L. Bobzin, C.S. Teng, D. Pal, C.T. Tuzon, R. Schweitzer, A.E. Merrill, FGF signaling patterns cell fate at the interface between tendon and bone, Development 146(15) (2019) dev170241.
38. N. Felsenthal, S. Rubin, T. Stern, S. Krief, D. Pal, B.A. Pryce, R. Schweitzer, E. Zelzer, Development of migrating tendon-bone attachments involves replacement of progenitor populations, Development 145(24) (2018) dev165381.
39. A.P. Breidenbach, L. Aschbacher-Smith, Y. Lu, N.A. Dyment, C.-F. Liu, H. Liu, C. Wylie, M. Rao, J.T. Shearn, D.W. Rowe, K.E. Kadler, R. Jiang, D.L. Butler, Ablating hedgehog signaling in tenocytes during development impairs biomechanics and matrix organization of the adult murine patellar tendon enthesis, Journal of Orthopaedic Research 33(8) (2015) 1142–1151.
40. A.G. Schwartz, F. Long, S. Thomopoulos, Enthesis fibrocartilage cells originate from a population of Hedgehog-responsive cells modulated by the loading environment, Development 142(1) (2015) 196–206.
41. N.A. Dyment, A.P. Breidenbach, A.G. Schwartz, R.P. Russell, L. Aschbacher-Smith, H. Liu, Y. Hagiwara, R. Jiang, S. Thomopoulos, D.L. Butler, D.W. Rowe, Gdf5 progenitors give rise to fibrocartilage cells that mineralize via hedgehog signaling to form the zonal enthesis, Developmental Biology 405(1) (2015) 96–107.
42. A.G. Schwartz, L.M. Galatz, S. Thomopoulos, Enthesis regeneration: a role for Gli1+ progenitor cells, Development 144(7) (2017) 1159–1164.
43. F. Fang, Y. Xiao, E. Zelzer, K.W. Leong, S. Thomopoulos, A mineralizing pool of Gli1-expressing progenitors builds the tendon enthesis and demonstrates therapeutic potential, Cell Stem Cell 29(12) (2022) 1669–1684.e6.
44. L.P. Lai, J. Mitchell, Indian hedgehog: Its roles and regulation in endochondral bone development, Journal of Cellular Biochemistry 96(6) (2005) 1163–1173.
45. Y. Wang, Z. Zhou, C.T. Walsh, A.P. McMahon, Selective translocation of intracellular Smoothened to the primary cilium in response to Hedgehog pathway modulation, Proceedings of the National Academy of Sciences 106(8) (2009) 2623–2628.
46. Z. Anvarian, K. Mykytyn, S. Mukhopadhyay, L.B. Pedersen, S.T. Christensen, Cellular signalling by primary cilia in development, organ function and disease, Nature Reviews Nephrology 15(4) (2019) 199–219.
47. F. Fang, A.G. Schwartz, E.R. Moore, M.E. Sup, S. Thomopoulos, Primary cilia as the nexus of biophysical and hedgehog signaling at the tendon enthesis, Science Advances 6(44) (2020) eabc1799.
48. A.C. Abraham, S.A. Shah, M. Golman, L. Song, X. Li, I. Kurtaliaj, M. Akbar, N.L. Millar, Y. Abu-Amer, L.M. Galatz, S. Thomopoulos, Targeting the NF-κB signaling pathway in chronic tendon disease, Science Translational Medicine 11(481) (2019) eaav4319.

49. H. Shen, A.G. Schwartz, R. Civitelli, S. Thomopoulos, Connexin 43 Is Necessary for Murine Tendon Enthesis Formation and Response to Loading, Journal of Bone and Mineral Research 35(8) (2020) 1494–1503.
50. T. Wang, A. Wagner, R. Gehwolf, W. Yan, F.S. Passini, C. Thien, N. Weissenbacher, Z. Lin, C. Lehner, H. Teng, C. Wittner, Q. Zheng, J. Dai, M. Ni, A. Wang, J. Papadimitriou, T. Leys, R.S. Tuan, S. Senck, J.G. Snedeker, H. Tempfer, Q. Jiang, M.H. Zheng, A. Traweger, Load-induced regulation of tendon homeostasis by SPARC, a genetic predisposition factor for tendon and ligament injuries, Science Translational Medicine 13(582) (2021) eabe5738.
51. X.-H. Deng, A. Lebaschi, C.L. Camp, C.B. Carballo, N.W. Coleman, J. Zong, B.M. Grawe, S.A. Rodeo, Expression of Signaling Molecules Involved in Embryonic Development of the Insertion Site Is Inadequate for Reformation of the Native Enthesis: Evaluation in a Novel Murine ACL Reconstruction Model, JBJS 100(15) (2018).
52. A.C. Abraham, F. Fang, M. Golman, P. Oikonomou, S. Thomopoulos, The role of loading in murine models of rotator cuff disease, Journal of Orthopaedic Research 40(4) (2022) 977–986.
53. A.G. Schwartz, J.H. Lipner, J.D. Pasteris, G.M. Genin, S. Thomopoulos, Muscle loading is necessary for the formation of a functional tendon enthesis, Bone 55(1) (2013) 44–51.
54. A.M. Tatara, J.H. Lipner, R. Das, H.M. Kim, N. Patel, E. Ntouvali, M.J. Silva, S. Thomopoulos, The role of muscle loading on bone (Re)modeling at the developing enthesis, PLoS ONE 9(5) (2014).
55. Y.J. Kim, H.J. Kim, W.J. Lee, J.K. Seong, A comparison of the metabolic effects of treadmill and wheel running exercise in mouse model, Laboratory Animal Research 36(1) (2020) 3.
56. I. Lerman, B.C. Harrison, K. Freeman, T.E. Hewett, D.L. Allen, J. Robbins, L.A. Leinwand, Genetic variability in forced and voluntary endurance exercise performance in seven inbred mouse strains, Journal of Applied Physiology 92(6) (2002) 2245–2255.
57. L.G. Koch, O.J. Kemi, N. Qi, S.X. Leng, P. Bijma, L.J. Gilligan, J.E. Wilkinson, H. Wisløff, M.A. Høydal, N. Rolim, P.M. Abadir, E.M. van Grevenhof, G.L. Smith, C.F. Burant, Ø. Ellingsen, S.L. Britton, U. Wisløff, Intrinsic Aerobic Capacity Sets a Divide for Aging and Longevity, Circulation Research 109(10) (2011) 1162–1172.
58. M. Svensson, P. Rosvall, A. Boza-Serrano, E. Andersson, J. Lexell, T. Deierborg, Forced treadmill exercise can induce stress and increase neuronal damage in a mouse model of global cerebral ischemia, Neurobiology of Stress 5 (2016) 8–18.
59. S.N. Rezvani, A.E.C. Nichols, R.W. Grange, L.A. Dahlgren, P.G. Brolinson, V.M. Wang, A novel murine muscle loading model to investigate Achilles musculotendinous adaptation, Journal of Applied Physiology 130(4) (2021) 1043–1051.
60. E. Ganji, S.N. Lamia, M. Stepanovich, N. Whyte, R.W. Goulet, A.C. Abraham, M.L. Killian, Optogenetic-induced muscle loading leads to mechanical adaptation of the Achilles tendon enthesis in mice, Science Advances 9(25) (2023) eadf4683.
61. G. Nagel, T. Szellas, W. Huhn, S. Kateriya, N. Adeishvili, P. Berthold, D. Ollig, P. Hegemann, E. Bamberg, Channelrhodopsin-2, a directly light-gated cation-selective membrane channel, Proceedings of the National Academy of Sciences 100(24) (2003) 13940–13945.
62. P.P.-Y. Lui, P. Zhang, K.-M. Chan, L. Qin, Biology and augmentation of tendon-bone insertion repair, Journal of Orthopaedic Surgery and Research 5(1) (2010) 59.
63. X. Jiang, M. Wojtkiewicz, C. Patwardhan, S. Greer, Y. Kong, M. Kuss, X. Huang, J. Liao, Y. Lu, A. Dudley, R.L. Gundry, M. Fuchs, P. Streubel, B. Duan, The effects of maturation and aging on the rotator cuff tendon-to-bone interface, The FASEB Journal 35(12) (2021) e22066.
64. X. Zhang, D. Bogdanowicz, C. Erisken, N.M. Lee, H.H. Lu, Biomimetic scaffold design for functional and integrative tendon repair, Journal of Shoulder and Elbow Surgery 21(2) (2012) 266–277.
65. R.R. Cooper, S. Misol, Tendon and ligament insertion. A light and electron microscopic study, JBJS 52(1) (1970) 1–20.
66. I.N.E. Wang, S. Mitroo, F.H. Chen, H.H. Lu, S.B. Doty, Age-dependent changes in matrix composition and organization at the ligament-to-bone insertion, Journal of Orthopaedic Research 24(8) (2006) 1745–1755.

67. S. Font Tellado, E.R. Balmayor, M. Van Griensven, Strategies to engineer tendon/ligament-to-bone interface: Biomaterials, cells and growth factors, Advanced Drug Delivery Reviews 94 (2015) 126–140.
68. S. Patel, J.-M. Caldwell, S.B. Doty, W.N. Levine, S. Rodeo, L.J. Soslowsky, S. Thomopoulos, H.H. Lu, Integrating soft and hard tissues via interface tissue engineering, Journal of Orthopaedic Research 36(4) (2018) 1069–1077.
69. I. Calejo, R. Costa Almeida, R.L. Reis, M.E. Gomes, Enthesis Tissue Engineering: Biological Requirements Meet at the Interface, Tissue Engineering Part B: Reviews 25(4) (2019) 330–356.
70. H.H. Lu, S. Thomopoulos, Functional Attachment of Soft Tissues to Bone: Development, Healing, and Tissue Engineering, Annual Review of Biomedical Engineering 15(1) (2013) 201–226.
71. M. Golman, A.C. Abraham, I. Kurtaliaj, B.P. Marshall, Y.J. Hu, A.G. Schwartz, X.E. Guo, V. Birman, P.J. Thurner, G.M. Genin, S. Thomopoulos, Toughening mechanisms for the attachment of architectured materials: The mechanics of the tendon enthesis, Science Advances 7(48) (2021) eabi5584.
72. L. Rossetti, L.A. Kuntz, E. Kunold, J. Schock, K.W. Müller, H. Grabmayr, J. Stolberg-Stolberg, F. Pfeiffer, S.A. Sieber, R. Burgkart, A.R. Bausch, The microstructure and micromechanics of the tendon–bone insertion, Nature Materials 16(6) (2017) 664–670.
73. J. Sartori, H. Stark, Tracking tendon fibers to their insertion—a 3D analysis of the Achilles tendon enthesis in mice, Acta Biomaterialia 120 (2021) 146–155.
74. K.L. Moffat, W.-H.S. Sun, P.E. Pena, N.O. Chahine, S.B. Doty, G.A. Ateshian, C.T. Hung, H.H. Lu, Characterization of the structure–function relationship at the ligament-to-bone interface, Proceedings of the National Academy of Sciences 105(23) (2008) 7947–7952.
75. A.C. Deymier, Y. An, J.J. Boyle, A.G. Schwartz, V. Birman, G.M. Genin, S. Thomopoulos, A.H. Barber, Micro-mechanical properties of the tendon-to-bone attachment, Acta Biomaterialia 56 (2017) 25–35.
76. A.C. Deymier, A.G. Schwartz, Z. Cai, T.L. Daulton, J.D. Pasteris, G.M. Genin, S. Thomopoulos, The multiscale structural and mechanical effects of mouse supraspinatus muscle unloading on the mature enthesis, Acta Biomaterialia 83 (2019) 302–313.
77. Y. Liu, A.G. Schwartz, V. Birman, S. Thomopoulos, G.M. Genin, Stress amplification during development of the tendon-to-bone attachment, Biomechanics and Modeling in Mechanobiology 13(5) (2014) 973–983.
78. A.C. Deymier-Black, J.D. Pasteris, G.M. Genin, S. Thomopoulos, Allometry of the Tendon Enthesis: Mechanisms of Load Transfer Between Tendon and Bone, Journal of Biomechanical Engineering 137(11) (2015).
79. J. Ma, M.J. Smietana, T.Y. Kostrominova, E.M. Wojtys, L.M. Larkin, E.M. Arruda, Three-Dimensional Engineered Bone–Ligament–Bone Constructs for Anterior Cruciate Ligament Replacement, Tissue Engineering Part A 18(1–2) (2011) 103–116.
80. I. Calejo, R. Costa-Almeida, A.I. Gonçalves, D. Berdecka, R.L. Reis, M.E. Gomes, Bi-directional modulation of cellular interactions in an in vitro co-culture model of tendon-to-bone interface, Cell Proliferation 51(6) (2018) e12493.
81. X. Jiang, Y. Kong, M. Kuss, J. Weisenburger, H. Haider, R. Harms, W. Shi, B. Liu, W. Xue, J. Dong, J. Xie, P. Streubel, B. Duan, 3D bioprinting of multilayered scaffolds with spatially differentiated ADMSCs for rotator cuff tendon-to-bone interface regeneration, Applied Materials Today 27 (2022) 101510.
82. H. Uthoff, A. Spenner, W. Reckelkamm, B. Ahrens, G. Wölk, R. Hackler, F. Hardung, J.r. Schaefer, A. Scheffold, H. Renz, U. Herz, Critical Role of Preconceptional Immunization for Protective and Nonpathological Specific Immunity in Murine Neonates, The Journal of Immunology 171(7) (2003) 3485–3492.
83. C.-F. Hsieh, P. Alberton, E. Loffredo-Verde, E. Volkmer, M. Pietschmann, P. Müller, M. Schieker, D. Docheva, Scaffold-free Scleraxis-programmed tendon progenitors aid in significantly enhanced repair of full-size Achilles tendon rupture, Nanomedicine 11(9) (2016) 1153–1167.

84. J.-H. Yea, T.S. Bae, B.J. Kim, Y.W. Cho, C.H. Jo, Regeneration of the rotator cuff tendon-to-bone interface using umbilical cord-derived mesenchymal stem cells and gradient extracellular matrix scaffolds from adipose tissue in a rat model, Acta Biomaterialia 114 (2020) 104–116.
85. L. Bai, Q. Han, Z. Meng, B. Chen, X. Qu, M. Xu, Y. Su, Z. Qiu, Y. Xue, J. He, J. Zhang, Z. Yin, Bioprinted living tissue constructs with layer-specific, growth factor-loaded microspheres for improved enthesis healing of a rotator cuff, Acta Biomaterialia 154 (2022) 275–289.
86. C. Chen, Q. Shi, M. Li, Y. Chen, T. Zhang, Y. Xu, Y. Liao, S. Ding, Z. Wang, X. Li, C. Zhao, L. Sun, J. Hu, H. Lu, Engineering an enthesis-like graft for rotator cuff repair: An approach to fabricate highly biomimetic scaffold capable of zone-specifically releasing stem cell differentiation inducers, Bioactive Materials 16 (2022) 451–471.
87. C.H. Jo, O.-S. Kim, E.-Y. Park, B.J. Kim, J.-H. Lee, S.-B. Kang, J.H. Lee, H.S. Han, S.H. Rhee, K.-S. Yoon, Fetal mesenchymal stem cells derived from human umbilical cord sustain primitive characteristics during extensive expansion, Cell and Tissue Research 334(3) (2008) 423–433.
88. X. Ji, M. Wang, F. Chen, J. Zhou, Urine-Derived Stem Cells: The Present and the Future, Stem Cells International 2017 (2017) 4378947.
89. R. Zuo, J. Liu, Y. Zhang, H. Zhang, J. Li, J. Wu, Y. Ji, S. Mao, C. Li, Y. Zhou, Y. Wu, D. Cai, Y. Sun, C. Zhang, In situ regeneration of bone-to-tendon structures: Comparisons between costal-cartilage derived stem cells and BMSCs in the rat model, Acta Biomaterialia 145 (2022) 62–76.
90. J.O. Cooper, J.D. Bumgardner, J.A. Cole, R.A. Smith, W.O. Haggard, Co-cultured tissue-specific scaffolds for tendon/bone interface engineering, Journal of Tissue Engineering 5 (2014) 2041731414542294.
91. L. Du, C. Qin, H. Zhang, F. Han, J. Xue, Y. Wang, J. Wu, Y. Xiao, Z. Huan, C. Wu, Multicellular Bioprinting of Biomimetic Inks for Tendon-to-Bone Regeneration, Advanced science 10(21) (2023) 2301309.
92. H. Li, J. Fan, L. Sun, X. Liu, P. Cheng, H. Fan, Functional regeneration of ligament-bone interface using a triphasic silk-based graft, Biomaterials 106 (2016) 180–192.
93. P. He, K.S. Ng, S.L. Toh, J.C.H. Goh, In Vitro Ligament–Bone Interface Regeneration Using a Trilineage Coculture System on a Hybrid Silk Scaffold, Biomacromolecules 13(9) (2012) 2692–2703.
94. E.D. Silva, P.S. Babo, R. Costa-Almeida, R.M.A. Domingues, B.B. Mendes, E. Paz, P. Freitas, M.T. Rodrigues, P.L. Granja, M.E. Gomes, Multifunctional magnetic-responsive hydrogels to engineer tendon-to-bone interface, Nanomedicine: Nanotechnology, Biology and Medicine 14(7) (2018) 2375–2385.
95. J. Lyu, L. Chen, J. Zhang, X. Kang, Y. Wang, W. Wu, H. Tang, J. Wu, Z. He, K. Tang, A microfluidics-derived growth factor gradient in a scaffold regulates stem cell activities for tendon-to-bone interface healing, Biomaterials Science 8(13) (2020) 3649–3663.
96. Y. Cao, S. Yang, D. Zhao, Y. Li, S.S. Cheong, D. Han, Q. Li, Three-dimensional printed multiphasic scaffolds with stratified cell-laden gelatin methacrylate hydrogels for biomimetic tendon-to-bone interface engineering, Journal of Orthopaedic Translation 23 (2020) 89–100.
97. I.H.A. Ali, D.P. Brazil, Bone morphogenetic proteins and their antagonists: current and emerging clinical uses, British Journal of Pharmacology 171(15) (2014) 3620–3632.
98. J.H.F. de Baaij, J.G.J. Hoenderop, R.J.M. Bindels, Magnesium in Man: Implications for Health and Disease, Physiological Reviews 95(1) (2015) 1–46.
99. C. Wu, Y. Zhou, M. Xu, P. Han, L. Chen, J. Chang, Y. Xiao, Copper-containing mesoporous bioactive glass scaffolds with multifunctional properties of angiogenesis capacity, osteostimulation and antibacterial activity, Biomaterials 34(2) (2013) 422–433.
100. A. Ewald, C. Käppel, E. Vorndran, C. Moseke, M. Gelinsky, U. Gbureck, The effect of Cu(II)-loaded brushite scaffolds on growth and activity of osteoblastic cells, Journal of Biomedical Materials Research Part A 100A(9) (2012) 2392–2400.
101. A. Yousefi, F. Sarrafzadeh-Rezaei, S. Asri-Rezaei, A.A. Farshid, M. Behfar, Fabrication of novel tubular scaffold for tendon repair from chitosan in combination with zinc oxide nanoparticles, Veterinary Research Forum 9(2) (2018) 105–111.

102. R. Yang, G. Li, C. Zhuang, P. Yu, T. Ye, Y. Zhang, P. Shang, J. Huang, M. Cai, L. Wang, W. Cui, L. Deng, Gradient bimetallic ion–based hydrogels for tissue microstructure reconstruction of tendon-to-bone insertion, Science Advances 7(26) (2021) eabg3816.
103. H.K. Min, S.H. Oh, J.M. Lee, G.I. Im, J.H. Lee, Porous membrane with reverse gradients of PDGF-BB and BMP-2 for tendon-to-bone repair: In vitro evaluation on adipose-derived stem cell differentiation, Acta Biomaterialia 10(3) (2014) 1272–1279.
104. E. Harris, Y. Liu, G. Cunniffe, D. Morrissey, S. Carroll, K. Mulhall, D.J. Kelly, Biofabrication of soft tissue templates for engineering the bone–ligament interface, Biotechnology and Bioengineering 114(10) (2017) 2400–2411.
105. J. He, N. Jiang, T. Qin, W. Zhang, Z. Liu, Y. Liu, D. Li, Microfiber-reinforced nanofibrous scaffolds with structural and material gradients to mimic ligament-to-bone interface, Journal of Materials Chemistry B 5(43) (2017) 8579–8590.
106. S.K. Madhurakkat Perikamana, J. Lee, T. Ahmad, E.M. Kim, H. Byun, S. Lee, H. Shin, Harnessing biochemical and structural cues for tenogenic differentiation of adipose derived stem cells (ADSCs) and development of an in vitro tissue interface mimicking tendon-bone insertion graft, Biomaterials 165 (2018) 79–93.
107. S. Font Tellado, S. Chiera, W. Bonani, P.S.P. Poh, C. Migliaresi, A. Motta, E.R. Balmayor, M. van Griensven, Heparin functionalization increases retention of TGF-β2 and GDF5 on biphasic silk fibroin scaffolds for tendon/ligament-to-bone tissue engineering, Acta Biomaterialia 72 (2018) 150–166.
108. S. Tarafder, J.A. Brito, S. Minhas, L. Effiong, S. Thomopoulos, C.H. Lee, In situ tissue engineering of the tendon-to-bone interface by endogenous stem/progenitor cells, Biofabrication 12(1) (2020) 015008.
109. A. Hurley-Novatny, N. Arumugasaamy, M. Kimicata, H. Baker, A.G. Mikos, J.P. Fisher, Concurrent multi-lineage differentiation of mesenchymal stem cells through spatial presentation of growth factors, Biomedical Materials 15(5) (2020) 055035.
110. N. Jiang, J. He, W. Zhang, D. Li, Y. Lv, Directed differentiation of BMSCs on structural/compositional gradient nanofibrous scaffolds for ligament-bone osteointegration, Materials Science and Engineering: C 110 (2020) 110711.
111. Z. Lin, X. Zhao, S. Chen, C. Du, Osteogenic and tenogenic induction of hBMSCs by an integrated nanofibrous scaffold with chemical and structural mimicry of the bone–ligament connection, Journal of Materials Chemistry B 5(5) (2017) 1015–1027.
112. J. Xie, X. Li, J. Lipner, C.N. Manning, A.G. Schwartz, S. Thomopoulos, Y. Xia, "Aligned-to-random" nanofiber scaffolds for mimicking the structure of the tendon-to-bone insertion site, Nanoscale 2(6) (2010) 923–926.
113. S. Samavedi, P. Vaidya, P. Gaddam, A.R. Whittington, A.S. Goldstein, Electrospun meshes possessing region-wise differences in fiber orientation, diameter, chemistry and mechanical properties for engineering bone-ligament-bone tissues, Biotechnology and Bioengineering 111(12) (2014) 2549–2559.
114. J. Cai, J. Wang, K. Ye, D. Li, C. Ai, D. Sheng, W. Jin, X. Liu, Y. Zhi, J. Jiang, J. Chen, X. Mo, S. Chen, Dual-layer aligned-random nanofibrous scaffolds for improving gradient microstructure of tendon-to-bone healing in a rabbit extra-articular model, Int J Nanomedicine 2018(13) (2018) 3481—3492.
115. A.P. Kishan, A.B. Robbins, S.F. Mohiuddin, M. Jiang, M.R. Moreno, E.M. Cosgriff-Hernandez, Fabrication of macromolecular gradients in aligned fiber scaffolds using a combination of in-line blending and air-gap electrospinning, Acta Biomaterialia 56 (2017) 118–128.
116. S. Cong, Y. Sun, J. Lin, S. Liu, J. Chen, A Synthetic Graft With Multilayered Co-Electrospinning Nanoscaffolds for Bridging Massive Rotator Cuff Tear in a Rat Model, The American Journal of Sports Medicine 48(8) (2020) 1826–1836.
117. H. Li, T. Wu, J. Xue, Q. Ke, Y. Xia, Transforming Nanofiber Mats into Hierarchical Scaffolds with Graded Changes in Porosity and/or Nanofiber Alignment, Macromolecular Rapid Communications 41(3) (2020) 1900579.

118. S. Chen, A. McCarthy, J.V. John, Y. Su, J. Xie, Converting 2D Nanofiber Membranes to 3D Hierarchical Assemblies with Structural and Compositional Gradients Regulates Cell Behavior, Advanced Materials 32(43) (2020) 2003754.
119. I. Calejo, R.L. Reis, R.M.A. Domingues, M.E. Gomes, Texturing Hierarchical Tissues by Gradient Assembling of Microengineered Platelet-Lysates Activated Fibers, Advanced Healthcare Materials 11(8) (2022) 2102076.
120. S. Font Tellado, W. Bonani, E.R. Balmayor, P. Foehr, A. Motta, C. Migliaresi, M. van Griensven, Fabrication and Characterization of Biphasic Silk Fibroin Scaffolds for Tendon/Ligament-to-Bone Tissue Engineering, Tissue Engineering Part A 23(15–16) (2017) 859–872.
121. S.R. Caliari, D.W. Weisgerber, W.K. Grier, Z. Mahmassani, M.D. Boppart, B.A.C. Harley, Collagen Scaffolds Incorporating Coincident Gradations of Instructive Structural and Biochemical Cues for Osteotendinous Junction Engineering, Advanced Healthcare Materials 4(6) (2015) 831–837.
122. H. Liu, L. Yang, E. Zhang, R. Zhang, D. Cai, S. Zhu, J. Ran, V. Bunpetch, Y. Cai, B.C. Heng, Y. Hu, X. Dai, X. Chen, H. Ouyang, Biomimetic tendon extracellular matrix composite gradient scaffold enhances ligament-to-bone junction reconstruction, Acta Biomaterialia 56 (2017) 129–140.
123. M.C. Echave, R.M.A. Domingues, M. Gómez-Florit, J.L. Pedraz, R.L. Reis, G. Orive, M.E. Gomes, Biphasic Hydrogels Integrating Mineralized and Anisotropic Features for Interfacial Tissue Engineering, ACS Applied Materials & Interfaces 11(51) (2019) 47771–47784.
124. O. Hakimi, P.-A. Mouthuy, A. Carr, Synthetic and degradable patches: an emerging solution for rotator cuff repair, International Journal of Experimental Pathology 94(4) (2013) 287–292.
125. Q. Liu, Y. Yu, R.L. Reisdorf, J. Qi, C.-K. Lu, L.J. Berglund, P.C. Amadio, S.L. Moran, S.P. Steinmann, K.-N. An, A. Gingery, C. Zhao, Engineered tendon-fibrocartilage-bone composite and bone marrow-derived mesenchymal stem cell sheet augmentation promotes rotator cuff healing in a non-weight-bearing canine model, Biomaterials 192 (2019) 189–198.
126. J.A. Bronstein, C.Y.L. Woon, S. Farnebo, A.W. Behn, T. Schmitt, H. Pham, A.B. Castillo, J. Chang, Physicochemical Decellularization of Composite Flexor Tendon–Bone Interface Grafts, Plastic and Reconstructive Surgery 132(1) (2013).
127. S. Farnebo, C.Y.L. Woon, J.A. Bronstein, T. Schmitt, D.P. Lindsey, H. Pham, A.B. Castillo, J. Chang, Decellularized Tendon-Bone Composite Grafts for Extremity Reconstruction: An Experimental Study, Plastic and Reconstructive Surgery 133(1) (2014).
128. M. Su, Q. Zhang, Y. Zhu, S. Wang, J. Lv, J. Sun, P. Qiu, S. Fan, K. Jin, L. Chen, X. Lin, Preparation of Decellularized Triphasic Hierarchical Bone-Fibrocartilage-Tendon Composite Extracellular Matrix for Enthesis Regeneration, Advanced Healthcare Materials 8(19) (2019) 1900831.
129. Y. Tang, C. Chen, F. Liu, S. Xie, J. Qu, M. Li, Z. Li, X. Li, Q. Shi, S. Li, X. Li, J. Hu, H. Lu, Structure and ingredient-based biomimetic scaffolds combining with autologous bone marrow-derived mesenchymal stem cell sheets for bone-tendon healing, Biomaterials 241 (2020) 119837.
130. D. Olvera, B.N. Sathy, D.J. Kelly, Spatial Presentation of Tissue-Specific Extracellular Matrix Components along Electrospun Scaffolds for Tissue Engineering the Bone–Ligament Interface, ACS Biomaterials Science & Engineering 6(9) (2020) 5145–5161.
131. S. Chae, Y. Sun, Y.-J. Choi, D.-H. Ha, I. Jeon, D.-W. Cho, 3D cell-printing of tendon-bone interface using tissue-derived extracellular matrix bioinks for chronic rotator cuff repair, Biofabrication 13(3) (2021) 035005.
132. S. Chae, U. Yong, W. Park, Y.-m. Choi, I.-H. Jeon, H. Kang, J. Jang, H.S. Choi, D.-W. Cho, 3D cell-printing of gradient multi-tissue interfaces for rotator cuff regeneration, Bioactive Materials 19 (2023) 611–625.
133. Y. Hao, B. Cao, L. Deng, J. Li, Z. Ran, J. Wu, B. Pang, J. Tan, D. Luo, W. Wu, The first 3D-bioprinted personalized active bone to repair bone defects: A case report, Int J Bioprint 9(2) (2022) 654.

134. X. Li, J. Xie, J. Lipner, X. Yuan, S. Thomopoulos, Y. Xia, Nanofiber Scaffolds with Gradations in Mineral Content for Mimicking the Tendon-to-Bone Insertion Site, Nano Letters 9(7) (2009) 2763–2768.
135. W. Liu, J. Lipner, J. Xie, C.N. Manning, S. Thomopoulos, Y. Xia, Nanofiber Scaffolds with Gradients in Mineral Content for Spatial Control of Osteogenesis, ACS Applied Materials & Interfaces 6(4) (2014) 2842–2849.
136. B. Zou, Y. Liu, X. Luo, F. Chen, X. Guo, X. Li, Electrospun fibrous scaffolds with continuous gradations in mineral contents and biological cues for manipulating cellular behaviors, Acta Biomaterialia 8(4) (2012) 1576–1585.
137. X. Li, R. Cheng, Z. Sun, W. Su, G. Pan, S. Zhao, J. Zhao, W. Cui, Flexible bipolar nanofibrous membranes for improving gradient microstructure in tendon-to-bone healing, Acta Biomaterialia 61 (2017) 204–216.
138. P. Chen, L. Li, L. Dong, S. Wang, Z. Huang, Y. Qian, C. Wang, W. Liu, L. Yang, Gradient Biomineralized Silk Fibroin Nanofibrous Scaffold with Osteochondral Inductivity for Integration of Tendon to Bone, ACS Biomaterials Science & Engineering 7(3) (2021) 841–851.
139. B. Soo Kim, E. Ji Kim, J. Suk Choi, J. Hoon Jeong, C. Hyunchul Jo, Y. Woo Cho, Human collagen-based multilayer scaffolds for tendon-to-bone interface tissue engineering, Journal of Biomedical Materials Research Part A 102(11) (2014) 4044–4054.
140. D.W. Weisgerber, D.O. Kelkhoff, S.R. Caliari, B.A.C. Harley, The impact of discrete compartments of a multi-compartment collagen–GAG scaffold on overall construct biophysical properties, Journal of the Mechanical Behavior of Biomedical Materials 28 (2013) 26–36.
141. J. Qu, A.R. Thoreson, Q. Chen, K.-N. An, P.C. Amadio, C. Zhao, Tendon gradient mineralization for tendon to bone interface integration, Journal of Orthopaedic Research 31(11) (2013) 1713–1719.
142. S. Samavedi, S.A. Guelcher, A.S. Goldstein, A.R. Whittington, Response of bone marrow stromal cells to graded co-electrospun scaffolds and its implications for engineering the ligament-bone interface, Biomaterials 33(31) (2012) 7727–7735.
143. I. Calejo, R. Costa-Almeida, R.L. Reis, M.E. Gomes, A Textile Platform Using Continuous Aligned and Textured Composite Microfibers to Engineer Tendon-to-Bone Interface Gradient Scaffolds, Advanced Healthcare Materials 8(15) (2019) 1900200.
144. X. Xie, J. Cai, Y. Yao, Y. Chen, A.u.R. Khan, J. Wu, X. Mo, A woven scaffold with continuous mineral gradients for tendon-to-bone tissue engineering, Composites Part B: Engineering 212 (2021) 108679.
145. A.J. Boys, H. Zhou, J.B. Harrod, M.C. McCorry, L.A. Estroff, L.J. Bonassar, Top-Down Fabrication of Spatially Controlled Mineral-Gradient Scaffolds for Interfacial Tissue Engineering, ACS Biomaterials Science & Engineering 5(6) (2019) 2988–2997.
146. S.-K. He, L.-J. Ning, X. Yao, R.-N. Hu, J. Cui, Y. Zhang, W. Ding, J.-C. Luo, T.-W. Qin, Hierarchically Demineralized Cortical Bone Combined With Stem Cell–Derived Extracellular Matrix for Regeneration of the Tendon-Bone Interface, The American Journal of Sports Medicine 49(5) (2021) 1323–1332.
147. W. Su, J. Guo, J. Xu, K. Huang, J. Chen, J. Jiang, G. Xie, J. Zhao, S. Zhao, C. Ning, Gradient composite film with calcium phosphate silicate for improved tendon -to-Bone intergration, Chemical Engineering Journal 404 (2021) 126473.
148. J. Qiu, J. Ahn, D. Qin, S. Thomopoulos, Y. Xia, Biomimetic Scaffolds with a Mineral Gradient and Funnel Shaped Channels for Spatially Controllable Osteogenesis, Advanced Healthcare Materials 11(9) (2022) 2100828.
149. C. Zhu, S. Pongkitwitoon, J. Qiu, S. Thomopoulos, Y. Xia, Design and Fabrication of a Hierarchically Structured Scaffold for Tendon-to-Bone Repair, Advanced Materials 30(16) (2018) 1707306.
150. L.C. Mozdzen, A. Vucetic, B.A.C. Harley, Modifying the strength and strain concentration profile within collagen scaffolds using customizable arrays of poly-lactic acid fibers, Journal of the Mechanical Behavior of Biomedical Materials 66 (2017) 28–36.

151. S.W. Linderman, M. Golman, T.R. Gardner, V. Birman, W.N. Levine, G.M. Genin, S. Thomopoulos, Enhanced tendon-to-bone repair through adhesive films, Acta Biomaterialia 70 (2018) 165–176.
152. F. Fang, R.T.H. Linstadt, G.M. Genin, K. Ahn, S. Thomopoulos, Mechanically Competent Chitosan-Based Bioadhesive for Tendon-to-Bone Repair, Advanced Healthcare Materials 11(10) (2022) 2102344.
153. S. Choi, J.R. Moon, N. Park, J. Im, Y.E. Kim, J.-H. Kim, J. Kim, Bone-Adhesive Anisotropic Tough Hydrogel Mimicking Tendon Enthesis, Advanced Materials 35(3) (2023) 2206207.
154. D.F.E. Ker, D. Wang, A.W. Behn, E.T.H. Wang, X. Zhang, B.Y. Zhou, Á.E. Mercado-Pagán, S. Kim, J. Kleimeyer, B. Gharaibeh, Y. Shanjani, D. Nelson, M. Safran, E. Cheung, P. Campbell, Y.P. Yang, Functionally Graded, Bone- and Tendon-Like Polyurethane for Rotator Cuff Repair, Advanced Functional Materials 28(20) (2018) 1707107.
155. R.A. Sun Han Chang, J.F. Shanley, M.E. Kersh, B.A.C. Harley, Tough and tunable scaffold-hydrogel composite biomaterial for soft-to-hard musculoskeletal tissue interfaces, Science Advances 6(34) eabb6763.

# Bioengineering of Tendon Grafts for Tendon Defect Treatment

**Prasad Sawadkar, Jason Wong, and Vivek Mudera**

## 1 Introduction

Currently, worldwide, there are an estimated 1.71 people who have musculoskeletal disorders; the majority of them are from tendons and joints (WHO, 2022). Prolonged cyclic load on these tendon-related injuries results in a tendon rupture. The cause of tendon ruptures is complex, involving external and internal factors. Extrinsic factors refer to traumatic events that cause direct tissue laceration or the application of forces surpassing the tendon structure's physiological tolerance. However, intrinsic factors are defined by internal variables, such as genetic predispositions, that cause degenerative changes in the tendon matrix [1].

The identification and diagnosis of tendon damage are made more accessible by using specialised orthopaedic evaluation procedures. The Simmonds test is a diagnostic manoeuvre specifically developed to evaluate the integrity of the Achilles tendon. This assessment entails performing plantar flexion motions in the lower extremities to assess the functional capability and integrity of the tendon. The O'Brien test is used to determine the existence of lesions or disturbances in the tendons linked to the glenohumeral joint. This diagnostic process involves bending

P. Sawadkar (✉)
Division of Surgery and Interventional Science, University College London, London, UK

The Griffin Institute, Northwick Park Institute for Medical Research, Northwick Park Hospital, Harrow, UK
e-mail: prasad.sawadkar@ucl.ac.uk

J. Wong
Blond McIndoe Laboratories, Division of Cell Matrix and Regenerative Medicine, MAHSC, University of Manchester, Manchester, UK

V. Mudera
Division of Surgery and Interventional Science, University College London, London, UK

 
J. Liao, J. Y. Wong (eds.), *Integration and Bridging of Multiscale Bioengineering Designs and Tissue Biomechanics*, https://doi.org/10.1007/978-3-031-81743-4_8

the shoulder to a 90-degree angle and slightly moving the arm away from the body by 10 degrees in relation to the sagittal plane. The purpose is to isolate and identify possible tendon rips or ruptures.

Furthermore, the introduction and incorporation of sophisticated diagnostic imaging methods have significantly improved the accuracy and dependability of tendon pathology diagnoses and the aforementioned physical examination procedures. Ultrasonography is a crucial tool in this situation, as it provides immediate visualisation of soft tissue structures and allows for the dynamic evaluation of tendon shape and performance. Magnetic resonance imaging (MRI) enhances diagnostic tools by producing detailed images of both soft and hard tissues, allowing for a thorough assessment of tendon health and detecting related musculoskeletal abnormalities. Radiography, also known as conventional X-ray imaging, is primarily used to evaluate bone structures. However, it can also indirectly aid in the diagnostic process by revealing changes in joint alignment or the existence of calcific deposits, which may indicate underlying tendon issues [2].

## 2 Understanding Tendon Defects

### *2.1 Anatomy and Function of Tendons*

A tendon is a complex structure of collagen fibrils responsible for providing the tendon with its primary source of tensile strength. The average sizes of these fibrils range from 10 to 500 nm, varying depending on the species and age. They have a consistent periodicity of 67 nm and are a few millimetres in length. During the first stages of development, fibrils exhibit an unimodal diameter distribution [3, 4]. However, as they age, fibrils undergo maturation and acquire a distribution characterised by two distinct modes. The collagen fibrils in the tendon combine laterally to create parallel-aligned structures known as tendon fibres. These fibres are the fundamental microscopic units of the tendon and can be subjected to mechanical testing [5–7]. The rat tail collagen fibres have a diameter ranging from 5 $\mu$m to 30 $\mu$m, while human tendon fibres have a diameter of approximately 300 $\mu$m and are organised in fasciculi. The collagen concentration is directly related to the diameter of the fibres [8] (Fig. 1).

Tendon fascicles, consisting of collagen fibrils, are enveloped by the epitenon, also known as the tendon sheath. The sheath encompasses the tendon and provides it with nerve, lymphatic, and vascular support [10, 11]. The tendon comprises around 60% collagen by dry weight, with type I collagen accounting for around 95% of this amount. The collagen formation in the tendon results from the heterotrimeric nature of two similar chains [11–14]. The chains referred to are the $\alpha$1 (I) chain and $\alpha$2 (I)-chain, which combine to form collagen type III and/or collagen type V, along with the remaining 5% of collagen [15]. Collagen types II, IX, X, XI, and XII are present in small quantities, while collagen type X is located in the mineralised

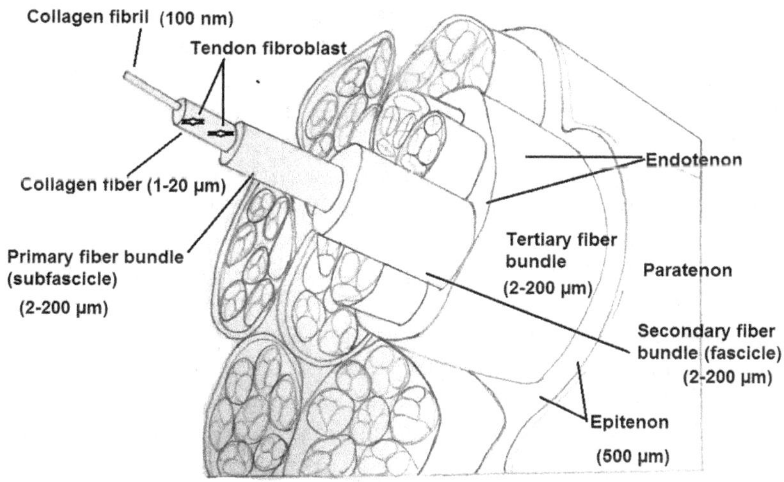


**Fig. 1** Schematic multiunit hierarchical structure of the tendon, such as collegen fibril, composed to form collegen fibre whose architecture bundle up to form primary, secondary and tertiary fascicles enclosed by epiteneon. (Image adapted and modified from [9])

region of the tendon-calcaneus junction. During the initial phase of development, collagen III is present in the fascicles and contributes to their growth. However, as the tendon matures, there is a reduction in the expression of collagen III, which is accompanied by an increase in the diameter of the fibrils. Conversely, collagen V does not contribute to the regulation of fibril diameter throughout the developmental phase [16]. Collagen III is not a prominent constituent in the typical tendon, but it rapidly creates crosslinks and stabilises the site of repair during the healing process [17]. Collagen XII functions as a lubricant for collagen I fibrils during movement [18].

Tendon fibroblasts (TFs) are specialised fibroblasts that reside in tendons. Their primary function is to regulate the extracellular matrix (ECM) turnover, enhancing the tendon's tensile strength and cross-linked areas, thus maintaining homeostasis [11, 19]. The tendon has a lower overall cell count compared to other tissues. However, the contractions of these cells, which involve the cytoskeletal system and impose strain on the extracellular matrix (ECM), are crucial for the healing process of the tendon [20, 21].

## 2.2 *Tendon Injuries and Defects*

Tendons are intricate fibrous structures that possess viscoelastic qualities, which play a crucial role in their mechanical durability and usefulness. Tendons possess several characteristics which enable them to efficiently transfer forces generated by muscles to bones, hence aiding movement. Surprisingly, among highly skilled athletes, the Achilles tendon, with a cross-sectional area of 1 $cm^2$, may endure stresses ranging from 500 to 1000 kg before it breaks. The tendon's crucial function in high-performance sports and physical activities highlights its intricate biological design [22].

The stress-strain curve of tendons can be used to analyse their mechanical behaviour when subjected to external force, allowing for the identification of different phases of reaction to stretching. At strains of up to 2%, tendons exhibit a nonlinear reaction called the toe region [23]. This phase is marked by the straightening of the collagen fibres in the tendon, resulting in increased flexibility as the fibres adapt to the imposed tension. Once the strain is beyond this threshold, the tendon demonstrates a progressively linear reaction. During the 2% to 4% strain range, the collagen fibres undergo alignment and lose their crimped form. This results in the tendon's elasticity, as reflected by Young's modulus, becoming noticeable [24]. The linear section emphasises the tendon's ability to endure and react to strain within its physiological thresholds.

Once the strain surpasses 4%, the tendon's characteristics shift from elastic to more rigid and plastic-like. This indicates that the tendon has beyond its elastic limit and is experiencing irreversible deformation. This crucial shift is a phenomenon that occurs over time, during which tendons become less effective at absorbing energy when subjected to high rates of loading. As a result, they become more fragile and prone to injury. The individual collagen fibres within the tendon undergo rupture, resulting in a decrease in mechanical integrity [25, 26].

When subjected to greater levels of strain, specifically at 8–10%, tendons experience a failure threshold, resulting in visible rupture [27]. The moment at which failure occurs differs across tendons and is affected by factors such as the individual's age, degree of physical activity, and overall health. It is worth mentioning that avian flexor tendons have the potential to endure strains of up to 14% [28]. This emphasises the differences in mechanical characteristics that exist among various species and types of tendons. The remarkable ability of birds in this regard can be ascribed to their distinct biomechanical needs, which include the need for both flexibility and strength during flight activities, therefore requiring a greater tolerance for strain.

Comprehending these mechanical characteristics and stages is essential for preventing and treating tendon injuries and advancing synthetic or bioengineered tendon replacements in regenerative medicine. By customising rehabilitation programmes and training regimens to align with the inherent biomechanics of tendons, it is possible to optimise athletic performance while reducing the likelihood of injury. This approach ensures that the stress placed on tendons remains within their natural limits of flexibility and power (Fig. 2).

The mechanical characteristics and stages are essential for preventing and treating tendon injuries and advancing the creation of synthetic or bioengineered tendon replacements in regenerative medicine. Customising rehabilitation programmes and training regimens to align with the inherent biomechanics of tendons can optimise

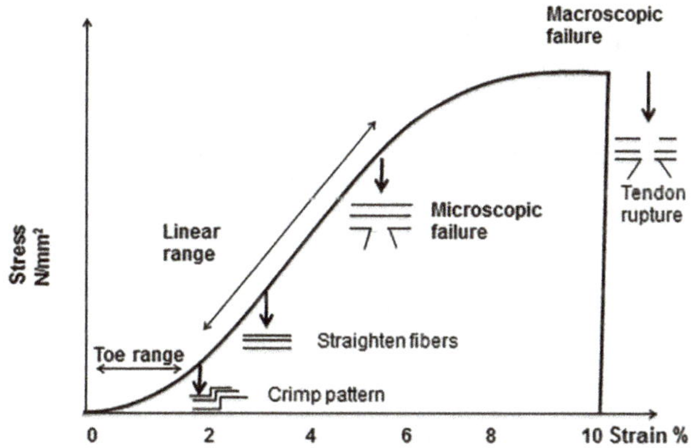


**Fig. 2** Mechanical stress and strain pattern in a tendon under applied force. (Adapted and reproduced from [11])

athletic performance while reducing the likelihood of injury. This approach ensures that the stress placed on tendons remains within their physiological limits for flexibility and power.

## 3 Pathophysiology of Tendon Injuries and the Healing Process

The tendon healing process is complex, consisting of highly coordinated stages marked by distinct cellular activity and metabolic processes. Gaining a deeper understanding of these stages offers valuable insights into the difficulties and constraints of tendon repair and possible therapeutic strategies to improve recovery [29].

The inflammatory phase is the earliest reaction to a tendon injury, which commences promptly after the injury and lasts for around one week. This stage is crucial for initiating the healing process. The process entails enlisting immune cells, such as neutrophils, monocytes, and macrophages, to the site of injury. These cells play a crucial role in removing waste, combating infection, and secreting cytokines and growth factors that facilitate the later phases of the healing process. Chemotactic factors are crucial in this phase as they attract immune cells to the injury site, preparing for the repair process [30]. The proliferative phase, which takes place from the first to the third week after the damage, follows the inflammatory phase. At this stage, the main activities consist of fibroblasts synthesising new collagen, the production of new blood vessels (angiogenesis), and creating granulation tissue. Fibroblasts play a crucial role in this phase by synthesising neo-collagen, mostly type III, creating the initial tissue regeneration framework. The proliferative phase is characterised by a significant increase in cellular activity and a new extracellular matrix forming, which serves as the foundation for tendon repair [31] (Fig. 3).

The ultimate phase of tendon healing is the remodelling stage, which typically spans a duration of 8–10 weeks following the injury. The collagen fibres undergo

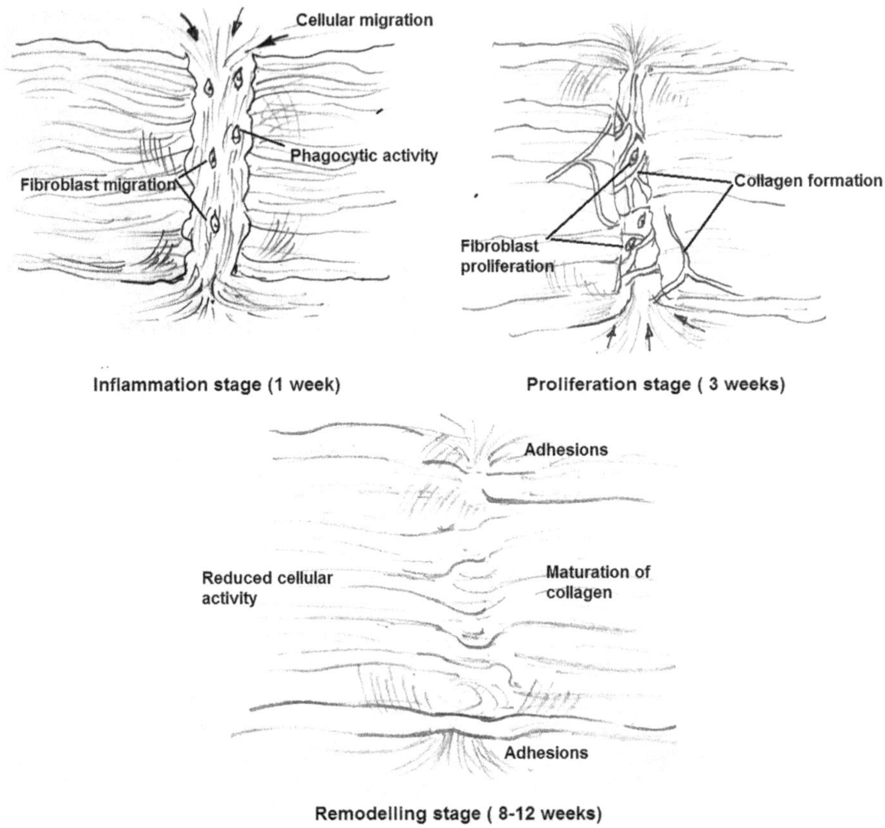


**Fig. 3** Mechanism of tendon healing at 1 week, where phagocytic activity with fibroblast migration increases, called an inflammation stage. 3 week collagen synthesis occurs with fibroblast proliferation called the proliferation stage and by 12 weeks, the remodelling stage, with collagen maturation and reduced fibroblast activity is completed. (Image adapted and modified from [32])

maturation during this stage, transforming from type III to type I collagen. This change results in a tendon structure that is stronger and biomechanical characteristics. The difference can be explained by various variables, such as inadequate more closely resembles its original form. However, the remodelling phase also underscores the constraints of tendon healing, as the newly generated tendon frequently lacks the initial structure and blood flow to tendons, which hampers the process of restructuring and supplying nutrients [29, 33, 34].

The process of tendon recovery is hindered by the development of scar tissue, which possesses reduced flexibility and strength compared to the original tendon tissue. This can result in diminished mobility and a heightened susceptibility to recurring injuries. To tackle these difficulties, the research concentrates on developing sophisticated therapeutic methods such as growth factor therapy, stem cell therapy, and tissue engineering. These tactics are designed to optimise the innate healing process, stimulate the growth of tendon tissue, and eventually optimise recovery results.

# 4 Current Treatment Strategies

## 4.1 Overview of Conventional Tendon Repair Techniques

The current treatment options for tendon rupture include both non-surgical and surgical approaches. In 1918, Bunnell suggested that rather than repairing a torn tendon, it should be removed and substituted with a tendon graft [35]. During the 1960s, it was documented that tendon repair should be performed using primary procedures [36]. Subsequently, a range of techniques have been employed to handle tendon rupture injuries. Non-surgical interventions are recommended for individuals who are less physically active or have a persistent medical condition, hence avoiding the need for surgical procedures [37] (Fig. 4).

The surgical treatment of a ruptured tendon is difficult due to the complex process of tendon healing, which involves alterations in the structural, nutritional, and biochemical properties of the tendon. Additionally, there is a risk of adhesion formation in the tendon, which can substantially impact clinical outcomes [38]. Multiple suture techniques have been employed for tendon repair, as the mechanical strength of the repaired tendon graft relies solely on the suture technique and material used, rather than the tendon or tendon graft itself [39, 40].

Suture procedures need to have strong and easily placed suture knots in the tendon, with minimum separation at the repair site. The repaired tendons should have a smooth connection minimal damage to their blood supply, and the sutures should be strong enough to support the repair until healing takes place at the repair site [32, 38].

Prior to the 1980s, the common method for repairing tendon rupture was using two strands. However, more current surgical literature suggests the use of four strands, six strands, or even eight strands [32, 41]. In 1985, Savage introduced a method of suturing using six strands and six anchor points, which was found to be considerably more robust than the Bunnell suture technique that employs only four strands [42]. Furthermore, repairs using 8–strand techniques exhibit a 43% increase in strength compared to repairs using other strands in both cadavers and in situ models [43–46]. However, these repairs are not commonly performed. Consequently, a greater number of core suture strands leads to an augmented improvement in the biomechanical properties of the repaired tendon. Nevertheless, this can impact the density of cells surrounding the suture locations and cause significant harm to the tendon, compromising the tendon repair process [47].

When a tendon ruptures, it is not possible to directly repair it because tendons are often under normal stress. After a rupture, the ends of the tendon pull back, creating

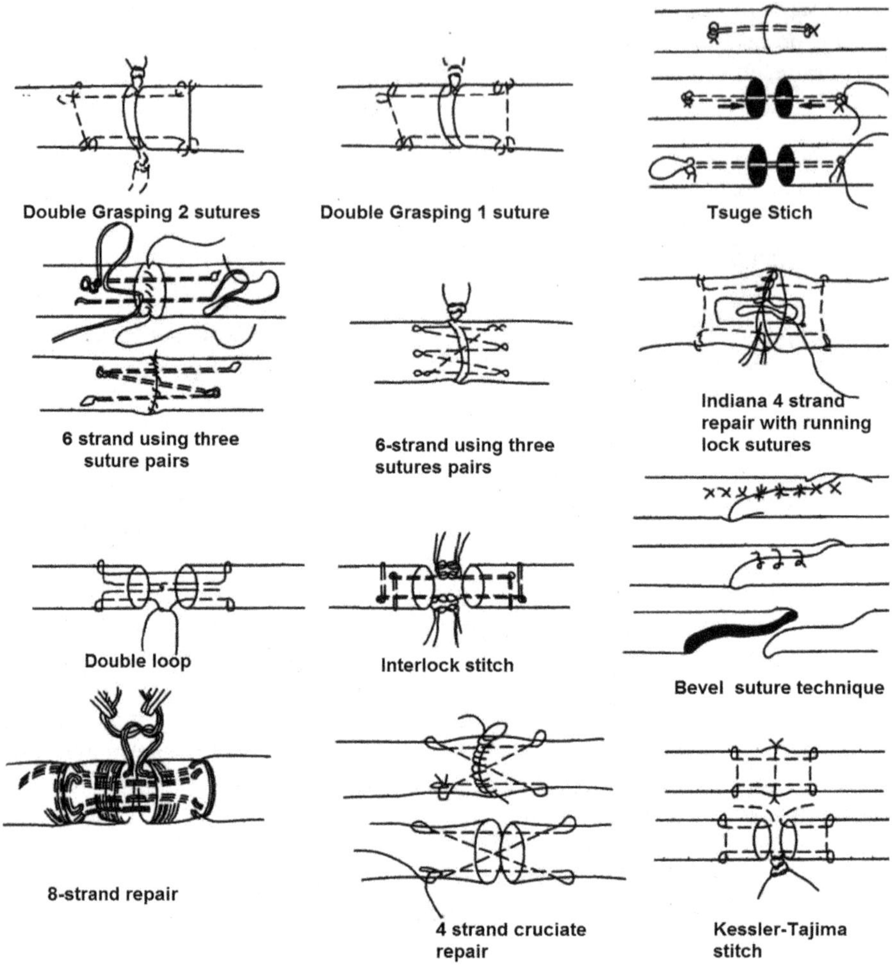


**Fig. 4** Some standard core suture techniques used for tendon rupture. (Images adapted and modified from [32])

a gap between the torn tendon ends or between the tendon and its bony attachment. Thus, to address this shortage, tendon grafts are presently employed.

## 4.2 *Current Grafting Treatment for Tendon Rupture*

**Autograft** Autografts from the quads, hamstring, patellar, and quadriceps tendons are typically considered the most effective method for repairing the anterior cruciate ligament (ACL) in tendon repair. The choice for autografts stems from their capacity to provide a strong and flexible alternative for ACL restoration, closely resembling the qualities of the original ligament. Studies have emphasised the efficacy of utilising quadriceps tendon and hamstring tendon autografts in ACL reconstruction. These studies have demonstrated that this procedure can achieve successful outcomes while minimising bone loss [48, 49]. The main factors influencing the use of autografts for ACL restoration are their safety, ability to be replicated, and compatibility with the immune system. These qualities make autografts a desirable alternative for surgeons and patients [50]. The graft's autologous nature reduces the likelihood of immunological rejection and enhances the integration with the host tissue, therefore facilitating the healing process [51]. However, the assessment of ACL restoration's long-term functional results utilising autografts is insufficient. The lack of research in this area highlights the need for longitudinal studies to determine autografts' long-term effectiveness and durability in restoring knee stability and functionality.

Furthermore, although autografts offer a feasible approach to ACL restoration, the method of harvesting them is not devoid of problems. Harvesting tissue from the patient's body might result in several detrimental consequences, such as nerve impairment, haemorrhage, and an elevated susceptibility to infection. The initial hurdles after surgery are made more difficult by the possibility of extended immobilisation, which is essential for the process of healing and rehabilitation. The concept of donor site morbidity refers to a range of problems that arise from the area where the graft is taken [51]. The morbidity associated with this condition might present itself in different ways, including the experience of discomfort, weakness, and a decrease in the functionality of the area from which the tissue was taken. These effects can diminish the overall advantages of the treatment.

Given these factors, the selection of autografts for ACL restoration requires a meticulous evaluation of their benefits compared to the possible disadvantages. This highlights the significance of advancing surgical methodologies and post-operative care procedures that reduce the negative effects on the donor site and improve patient results. Moreover, exploring alternate graft materials and bioengineering strategies shows great potential for improving ACL repair procedures. This could potentially provide solutions that address the drawbacks of current autograft techniques. Continued evaluation of autografts' long-term functional outcomes is essential in establishing their efficacy and enhancing ACL restoration procedures as the field of orthopaedic surgery progresses. These study efforts will not only enhance our comprehension of autograft effectiveness throughout time but also aid in the creation of inventive therapeutic alternatives that tackle the intricacies of ACL injuries.

This extended passage offers a thorough examination of the use of autografts in ACL restoration, emphasising the advantages and difficulties and underscoring the necessity for additional investigation into the enduring consequences and substitute options.

**Allograft** Allografts, which are tissues transferred from one individual to another of the same species, offer a notable benefit in tendon repair and reconstruction by avoiding the negative effects on the donor site that sometimes occur with autografts, where tissue is taken from the patient's own body. Although tendon allografts have the potential to be beneficial, their use is limited in the United Kingdom due to difficulties in finding appropriate donors and worries about the spread of infectious illnesses [52].

Tendon allografts, obtained from cadavers or organ donors, consist of several varieties, such as the Achilles, tibialis anterior, tibialis posterior, rotator cuff, patellar, semitendinosus, and gracilis tendons. These allografts have the benefit of shortening surgical procedures, but they require thorough screening for impurities like viruses, bacteria, and fungi before transplantation to assure patient safety.

Pre-transplant sterilisation of allografts is essential in order to minimise the potential for disease spread. Two main techniques are ethylene Oxide (EO) sterilisation and gamma radiation. EO sterilisation, frequently employed for medical devices, exhibits efficacy against both bacteria and viruses [53] Nonetheless, clinical studies have revealed that EO and its residues can contribute to tendon allograft failure in roughly 22% of patients, attributable to a heightened generation of interleukin-1 (IL-1), which increases inflammation, particularly in the bone-patellar tendon-bone complex [54].

Gamma radiation is known for its effectiveness in killing bacteria and viruses. In UK tissue banks, doses of 1–3.5 Megarad (Mrad) are commonly used to decontaminate germs, while doses of up to 4.0 Mrad are used to inactivate viruses such as HIV [55]. Nevertheless, this kind of sterilisation may undermine the biomechanical soundness of the tendon, potentially diminishing its strength by as much as 15% [56, 57].

Furthermore, allografts are vulnerable to contamination with Clostridium septicum after being stored, which presents a potential danger of severe infection. Despite comprehensive screening procedures, it is impossible to completely prevent the possibility of spreading viral illnesses like HIV and Hepatitis C Virus (HCV). This is partly because serological testing may not be able to identify infections during the "window period," which is the initial phase of infection when conventional tests cannot yet detect the viruses. The window period for HIV ranges from 28 to 90 days, whereas for HCV, it ranges from 38 to 94 days [58]. During this period, infected donors may inadvertently transfer these viruses through their donated tissues [59].

To summarise, tendon allografts have various benefits in surgical procedures, such as shorter operation time and avoidance of complications at the donor site. However, their utilisation is limited due to difficulties with finding suitable donors, risks of disease transmission, and potential effects on graft strength after sterilisation. These considerations require a careful approach to their use, highlighting the significance of rigorous screening and sterilisation procedures to guarantee the safety and effectiveness of allograft transplantation.

Xenograft: Xenografts, obtained from species other than humans, offer a promising approach in tendon repair and reconstruction. This is due to the advantageous utilisation of the mechanical resemblances between collagen derived from animals and human tendons. These grafts, made mostly from collagen derived from cows, horses, or pigs, are designed to serve as a practical substitute for grafts taken from the patient's own body or from another person, by closely matching the mechanical qualities necessary for normal tendon function.

Several commercially accessible xenogeneic tendon transplants have evolved, each with distinct processing procedures and sources of origin. An example of this is BioBlanket, a porous matrix created by Kensey Nash Corporation in the United States. It is made from a combination of acid-soluble and fibrous collagen obtained from bovine sources. It has been utilised in soft tissue and rotator cuff repairs, with the benefit of naturally breaking down within a year when inserted into a living organism. However, the use of this has been negatively affected by the triggering of notable inflammatory reactions in certain instances, which reveals a possible drawback in its ability to interact well with living organisms [60].

TissueMend, manufactured by TEI Biosciences in Massachusetts, USA, is derived from bovine dermis and generated using chemical and enzymatic methods specifically tailored to remove host DNA. This scaffold has a greater genetic element than previous grafts, which raises concerns about its acceptability for human implantation due to the possibility of triggering immunological responses [61].

OrthoADAPT, developed by Pegasus Biologics in California, USA, is a biocompatible structure made from the equine pericardium, which has been stripped of its cellular components. The main components of this substance are primarily type I collagen (90%) and type II collagen (10%), resulting in a composition that closely resembles human tendon tissue. The product is offered in three different versions, namely FX, PX, and MX, each varying in cross-linking level to meet certain mechanical requirements [62]. Nevertheless, despite its favourable composition, there is still little information regarding its clinical applicability or validation through clinical trials.

The field of tendon repair alternatives, including autografts, allografts, and xenografts, is filled with difficulties. Each approach has inherent limitations: autografts, although advantageous in terms of eliminating the risk of immune rejection, can result in donor site morbidity and are constrained by the availability of appropriate tissue; allografts, while addressing donor site concerns, are costly, necessitate specialised storage conditions, and may elicit immune responses or encounter rejection; xenografts, intended to replicate human tendon mechanics, also entail the potential for immune rejection due to their non-human source [63].

These issues highlight the intricate nature of creating optimal materials for tendon regeneration that mimic the biomechanical characteristics of human tendons, guarantee biocompatibility, minimise inflammatory reactions, and prevent immunological rejection. The continuous pursuit of the 'gold standard' in tendon restoration highlights the need for additional research and development in this area to maximise the equilibrium between mechanical characteristics, biocompatibility, and clinical effectiveness.

## 5 Principles of Tendon Tissue Engineering

Tissue engineering is a multidisciplinary field that combines biology, engineering, and material science. Its main goal is to imitate natural biological processes in order to regenerate, repair, or replace damaged tissues and organs. In the last 30 years, advancements in material science have revealed that artificial scaffolds, made from a combination of natural and synthetic polymers, may imitate the complex structure of natural tissues. This allows for the restoration of their function.

The scaffolds function as three-dimensional structures that provide support for cell attachment and growth and direct tissue development by imitating the extracellular matrix (ECM)—an intricate network of proteins and polysaccharides that offer structural and biochemical support to neighbouring cells. By creating scaffolds that mimic the extracellular matrix's physical, chemical, and biological signals, scientists may provide conditions that promote the transformation of precursor cells into specific cell types. This facilitates the growth of functional tissue structures.

Natural polymers, including collagen, gelatin, chitosan, and alginate, are highly desirable because they interact with cells and provide attachment sites, closely resembling the original extracellular matrix (ECM). These polymers are known for their biocompatibility and bioactivity. Synthetic polymers, such as polyglycolic acid (PGA), polylactic acid (PLA), and their copolymers (PLGA), offer the ability to tailor mechanical properties, degradation rates, and structural stability. This enables precise manipulation of the scaffold's physical characteristics and degradation timeline to align with the tissue regeneration process.

An essential element of scaffold design is the integration of self-degradation capabilities. Biodegradable materials are designed to break down slowly over time through hydrolysis or enzymatic activity, eliminating the need for surgical extraction and allowing for new tissue growth. The process of disintegration is precisely calibrated to align with the pace of tissue regeneration, guaranteeing that the scaffold offers temporary assistance throughout the crucial initial phases of healing and tissue development, prior to completely dissolving and leaving behind a fully merged and operational tissue. By combining these scaffolds with cells and bioactive substances, such as growth factors, the regenerative capacity of synthetic tissues is significantly augmented. The combination of these components acts in synergy to promote vital biological activity necessary for tissue repair, such as cell migration, cell proliferation, and extracellular matrix deposition.

The progress in fabrication technologies, including electrospinning, 3D printing, and self-assembly, has broadened the potential for producing scaffolds with intricate designs and particular capabilities tailored to different tissues. These techniques allow for the creation of scaffolds that have exact manipulation of pore size, shape, and arrangement of bioactive signals, thereby closely resembling the hierarchical structure of actual tissues.

Tissue engineering is advancing and has the potential to transform regenerative medicine. It aims to develop bioengineered constructions that repair and replace damaged tissues and organs, restoring their function and smoothly integrating with the body's tissues. This advancement highlights a future in which the restoration of intricate tissues and organs becomes a concrete actuality, greatly influencing patient care and the medical industry.

The successful production of tendon scaffolds for use inside the body depends on several crucial features that guarantee their functionality, integration, and eventual substitution by the host's own extracellular matrix (ECM). These qualities are crucial in directing the design and manufacturing of scaffolds that can efficiently facilitate tendon repair and regeneration.

## 5.1 Biocompatibility

Ensuring biocompatibility is of utmost importance when designing scaffolds for tissue engineering purposes. The scaffold must create a favourable environment for cell adhesion and growth in the immediate period and seamlessly merge with the surrounding tissue in the long run. Having a high porosity is crucial for enabling cell adhesion and promoting the effective migration of cells into the scaffold. In addition, the implanted scaffold should induce a regulated immunological response, characterised by the activation of macrophages. This response is advantageous because it releases growth factors, cytokines, and proteolytic enzymes that are crucial for the repair mechanisms [64]. Nevertheless, it is imperative to prevent profound immune responses, as they might impede the healing process and perhaps result in the rejection of the graft [65].

## 5.2 Biodegradability

The notion of tissue engineering focuses on creating scaffolds using materials that can be easily replaced by the host's extracellular matrix (ECM). This approach involves the total disintegration and removal of the scaffold from the body, without inflicting any harm to the surrounding tissues or organs. To prevent any negative impact on the host organism, it is imperative to completely eliminate all byproducts of the scaffold [64, 65].

## 5.3 Mechanical Property

Ensuring that tendon scaffolds have the same mechanical characteristics as the surrounding tissue is a difficult but necessary condition. The scaffold must accurately reproduce the natural tendon's tensile strength, elasticity, and mechanical load response. It is crucial to consider that the mechanical requirements placed on tendons differ among individuals and are affected by factors such as age and amount of physical activity. Hence, scaffolds necessitate the incorporation of adjustable mechanical characteristics and degradation rates to adapt to these fluctuations while preserving their structural integrity and porosity, which are crucial for cellular adhesion and tissue integration.

### 5.4 Architecture of the Scaffold

The structural design of the scaffold is crucial for its proper functioning. An optimal scaffold should exhibit a significantly porous architecture to facilitate cell adhesion, movement, and multiplication. Furthermore, increased permeability should enhance the dispersion of essential substances and the elimination of metabolic byproducts, while also stimulating the formation of new blood vessels. The scaffold must be structured in a way that replicates the natural alignment of tendon fibres, in order to facilitate the development of organised tissue that is capable of withstanding physiological loads [66].

### 5.5 Fabrication Technology

The fabrication techniques employed for tendon scaffolds must strictly conform to Good Manufacturing Practice (GMP) standards and adhere to applicable regulatory requirements, including those mandated by the European Union. These criteria guarantee the safety, quality, and effectiveness of the scaffolds. Moreover, the manufacturing process should be economically efficient in comparison to conventional procedures, enabling the scaffolds to be easily accessible and ready for use. Immediate access to these scaffolds is essential for doctors, eliminating the need for extra surgical operations to obtain autografts or prepare allografts [65].

To successfully create tendon scaffolds that can operate effectively in living organisms, it is crucial to achieve a delicate equilibrium between factors like biocompatibility, biodegradability, mechanical qualities, architectural design, and appropriate production techniques. By considering these essential attributes, scientists and engineers may create scaffolds that not only facilitate the regrowth of tendon tissue but also smoothly merge with the biological systems of the recipient, thus enabling progress in the fields of regenerative medicine and tissue engineering.

## 6 Bioengineering Techniques for Tendon Grafts

### 6.1 Natural Polymers

Natural polymers are known for their biocompatibility and their ability to imitate the original biological extracellular matrix (ECM). Examples of such materials include collagen, silk, chitosan-hyaluronan, and alginate-chitosan hybrid polymers. Chitosan scaffolds have shown a remarkable ability to produce collagen type I, resembling the characteristics of natural tendons, including cellular affinity and extracellular matrix (ECM) formation [67–70] In addition, tendon grafts that have been bioengineered using porcine small intestine submucosa (SIS) have demonstrated potential in mending and reinforcing flexor tendon lesions in rabbit models. This is achieved through the deposition of collagen and an elevation in transforming growth factor-β (TGF-β) levels. Hence, natural polymers are favoured in tendon tissue engineering because of their intrinsic benefits compared to other scaffold materials [71].

Collagen, by itself, does not have robust mechanical characteristics. However, combining its biomechanical capabilities with other polymers and/or mesenchymal stem cells (MSCs) can improve its biomechanical capabilities. This combination creates an optimal habitat for tendon fibroblasts within the scaffold matrix. We prioritise type I collagen grafts because they are highly preserved and prevalent in tendons, making up around 95% of the total collagen in tendons. Collagen is well-suited for reproducing the characteristics of tendons within a living organism [72–75].

We used type I collagen derived from rat tail tendons to produce tendon grafts. The collagen was dissolved in acetic acid at a 2.15 mg/ml concentration. The polymerisation method entailed the in vitro neutralisation of collagen using sodium hydroxide, forming non-cross-linked anisotropic collagen fibrils. The hydrogel that is produced contains 98% fluid and can encapsulate tendon fibroblasts, thereby mimicking living tissue in vivo within a time frame of 20 min. Nevertheless, this hydrogel's mechanical strength is lower than the original tendon tissue, renowned for its exceptional durability [76–80]. Brown et al. proposed a method called 'plastic compression' to compress collagen hydrogels to overcome this constraint quickly. This approach is based on the direct relationship between the weight applied and the level of compression, which helps release fluid from the surface where the fluid exits. The utilisation of spiral rolling of the graft greatly improves its mechanical characteristics, bringing it closer in alignment with the features of natural tendons [81].

## 6.2 *Decellularised Scaffold*

Decellularised tendon scaffolds provide a new method for tissue engineering to maintain the mechanical strength of natural polymers while utilising their natural compatibility with living tissue and structural benefits. This method is especially appealing since it preserves the natural structure of the tendon, which is essential for recreating the mechanical conditions and facilitating effective integration and healing in the body.

The procedure entails extracting cellular constituents from tendon tissues, resulting in a mostly extracellular matrix (ECM) protein structure. The ECM scaffold preserves the intricate hierarchical organisation of the tendon, encompassing its collagen fibres that play a crucial role in determining the tendon's mechanical characteristics. The elimination of biological components reduces the likelihood of immunological rejection, enhancing the scaffold's compatibility with the host tissue.

Conventional therapies, such as autografts, allografts, and synthetic materials, have certain drawbacks, such as complications at the donor site, the potential for disease transmission, and insufficient bonding with the recipient's tissue. Decellularised tendon scaffolds provide a novel solution to these difficulties by serving as a framework that can facilitate the infiltration of cells, the development of blood vessels, and the remodelling of tissues after being implanted.

Nevertheless, the decellularisation process is intricate and necessitates meticulous optimisation to eliminate cellular material without affecting the scaffold's structural and mechanical characteristics. Different methodologies, encompassing physical, chemical, and enzymatic approaches, have been investigated, each with distinct benefits and constraints. Freeze-thaw cycles can disturb cell membranes, while detergents can dissolve cellular components. Enzymatic interventions can additionally assist in eliminating residual cellular components. However, it is crucial to maintain a delicate equilibrium in these activities in order to save the essential extracellular matrix (ECM) components that provide the tendon with its robustness and flexibility [82, 83].

Although there have been hopeful advancements, there are still considerable obstacles in the process of standardising decellularisation techniques to guarantee consistent, effective, and safe use in clinical applications. Further work is needed to examine the effects of decellularisation on the biomechanical qualities, bioactivity, and capacity of the scaffold to support recellularisation and integration into host tissue. Furthermore, it is imperative to conduct extended research to comprehend the in vivo characteristics of these scaffolds, such as their capacity for degradation, inflammation, and integration with adjacent tissues.

## *6.3 Cell-Free Scaffold*

The emergence of scaffold-free tendon tissue engineering is a progressive option to overcome the constraints of traditional scaffold-based techniques in tendon healing. This novel approach centres on harnessing the inherent ability of cells to produce and arrange their own extracellular matrix (ECM), resulting in the formation of a biomimetic framework for tendon grafting, without the requirement of an external scaffold.

Scaffold-free techniques rely on the idea that cells, when presented with the appropriate environmental signals and physical conditions, can create and build a dense and organised ECM typical of native tendon tissue. The ECM plays a crucial role in tissue formation by offering both structural support and essential biological signals required for tissue development and successful integration after implantation.

The main benefit of scaffold-free tendon engineering is its capacity for developing a healing tissue that is totally biological and closely replicates the native tendon's structure and chemistry. Contrary to conventional scaffolds, scaffold-free structures consist exclusively of biological components, reducing the likelihood of the body's immunological responses, inflammation, and rejection. This strategy harnesses the body's natural healing capabilities, promoting a more harmonic integration with the host tissue and potentially improving the repair process [84].

Nevertheless, the progress in creating scaffold-free tendon grafts encounters notable obstacles, specifically in reproducing the mechanical characteristics of natural tendon tissues. Tendons possess a specialised structure that enables them to endure significant pulling forces. Their exceptional mechanical qualities are a result of their well-organised collagen fibre arrangement, as well as other components of the extracellular matrix and certain cell types. Attaining such a degree of structural organisation and mechanical robustness in a laboratory setting, solely relying on extracellular matrix generated from cells, is an intricate and demanding endeavour.

The present study endeavours are concentrated on comprehending the determinants that impact the formation and arrangement of extracellular matrix (ECM) by cells in a controlled environment, encompassing the influence of mechanical stimulation, culture circumstances, and cell kinds. Researchers are currently investigating techniques including three-dimensional bioprinting, cell sheet engineering, and dynamic culture systems to improve the deposition and alignment of extracellular matrix (ECM). The goal is to create scaffold-free structures that closely mimic natural tendons' mechanical and structural characteristics.

However, there are still notable deficiencies in our capacity to regulate the composition, shape, and mechanical characteristics of the extracellular matrix (ECM) in scaffold-free systems. Additional research is required to enhance the conditions that facilitate the development of a functional tendon-like tissue, capable of enduring the mechanical stresses experienced by tendons in a living organism. This includes investigations into the biological signals that direct tissue formation and the creation of innovative bioreactor systems capable of replicating physiological conditions. The notion of scaffold-free tendon tissue engineering arises as an innovative alternative intended to overcome the constraints linked to conventional scaffold-based methods in tendon restoration. This novel approach centres on harnessing the inherent capacities of cells to produce and arrange their extracellular matrix (ECM) to generate a tendon graft. As a result, a biomimetic structure is formed without requiring an external scaffold.

Scaffold-free methods depend on the concept that when given suitable environmental signals and physical circumstances, cells can generate and organise a compact and well-structured extracellular matrix (ECM) similar to the natural tissue of tendons. The ECM plays a crucial role in tissue formation by providing both structural support and required biological signals for tissue maturation and integration after implantation.

## 6.4 Synthetic Scaffolds

Synthetic polymers like poly(lactide-co-glycolide) (PLGA), polyglycolide (PGA), and polylactide are leading materials in tendon tissue engineering due to their FDA approval and the ability to customise them for specific biomedical uses [85]. These polymers provide notable benefits in the field, such as the ability to be precisely engineered, predictable biomechanical properties, and the capability to imitate the structure of the natural tendon using electrospun fibre alignment. When these polymers are combined with biological components such as bone marrow stromal cells (BMSCs), tendon fibroblasts, and growth factors, they have been demonstrated to enhance tendon repair significantly, both in vivo and in vitro, by promoting cell attachment, proliferation, and differentiation, which are crucial for effective tissue regeneration [86–89].

Although synthetic polymers possess favourable characteristics, their application in tendon tissue engineering is accompanied by various obstacles and restrictions. A significant issue arises from the absence of comprehensive, extended-duration studies that assess the efficacy and compatibility of these materials. This study gap raises concerns over the resilience of the repairs and the long-term consequences for people undergoing these procedures [90].

Another notable concern pertains to the deterioration process of artificial polymers. Although intended to break down naturally in the body, the byproducts of this process may cause harmful consequences such as toxicity and immunological responses. These reactions have the potential to undermine the structural integrity of the repair site, resulting in inflammation, fibrosis, or reduced ability to mend. Controlling the pace of degradation and the characteristics of the byproducts is crucial to guarantee that they are compatible with the body's ability to absorb or eliminate these substances without causing any harm.

Cell adherence to the surface of synthetic scaffolds is another crucial component impacting the effectiveness of tendon restoration. The surface features of these polymers must be carefully tailored to facilitate robust cell attachment, which is essential for integrating the scaffold with the host tissue and the following regeneration process. Inadequate cellular adhesion can result in compromised tissue healing, marked by inadequate integration of the scaffold, insufficient tissue development, and potentially the breakdown of the repair process.

In order to tackle these difficulties, current research is concentrated on altering the surface characteristics of artificial scaffolds to improve the adherence and growth of cells. Researchers are now investigating many methods to enhance the performance of a surface. These methods include applying bioactive chemicals as a coating, using plasma treatment to modify the surface chemistry, and incorporating nanotopographical characteristics such as biocompatibility and functioning of these materials. Moreover, progress in scaffold design is focused on developing more intricate composite materials that integrate the advantageous characteristics of artificial polymers with natural substances or bioactive molecules to replicate the biochemical and mechanical conditions of native tendon tissue closely.

Developing scaffolds that progressively release growth factors or other bioactive compounds as they degrade is another way to facilitate healing and tissue integration. The controlled release of substances can assist in directing the regeneration process, minimising the likelihood of negative responses to the breakdown products, and promoting the necessary cellular functions for tissue restoration. To tackle the problem of degrading byproducts and their possible toxicity, it is necessary to employ novel techniques in the field of polymer chemistry. Researchers are studying the production of new polymers that breakdown into non-toxic, easily metabolisable byproducts. These materials would significantly decrease the likelihood of negative immune responses and create a more favourable environment for tissue healing and integration.

Incorporating artificial frameworks with cellular therapy, such as using BMSCs and tendon fibroblasts, presents a mutually beneficial strategy for tendon restoration. This concept encompasses a comprehensive approach to healing by combining a scaffold that facilitates cell attachment and proliferation with cells that can directly contribute to tissue regeneration. Nevertheless, the crucial objective still lies in guaranteeing these cells' efficient transportation and viability, as well as their successful incorporation into the recipient tissue.

While synthetic polymers present a promising path for tendon tissue engineering, their successful application needs a multidimensional approach that addresses the issues of long-term biocompatibility, degradation byproducts, and cell-scaffold interactions. Advancements in materials science, surface engineering, and cellular biology are expected to lead to superior synthetic scaffolds for tendon healing. These scaffolds will offer enhanced safety, effectiveness, and functional restoration.

## *6.5 Clinical Applications*

In regenerative medicine and orthopaedic surgery, researchers investigate different bio-scaffolding materials to find viable treatment methods for tendon restoration. This is especially important for instances involving significant rotator cuff tendon abnormalities. Among these options, a substantial and notable strategy is using Porcine Small Intestinal Submucosa (SIS) and bio-scaffolds made of collagen types I and III, along with autologous tendon fibroblasts. Chen et al. used these materials to treat large rotator cuff tendon defects in a rabbit model in their investigation. Their results showed that, although the initial examination of tissue samples four weeks after the implantation did not show as positive outcomes as the control groups, a later evaluation at eight weeks revealed a significant decrease in both inflammation and cellular activity, closely resembling the properties of natural tendon tissue [91]. The available information indicates that including tendon fibroblasts in the scaffold significantly improves the repair process, surpassing scaffolds that lack cellular components.

However, Iannotti et al. have raised doubts about the effectiveness of using porcine SIS in human patients. They conducted a randomised control trial with 30 patients with rotator cuff injuries of varying severity. The study determined that using swine small intestinal submucosa (SIS) did not result in better clinical results or enhanced healing of tendons [92].

Another method entails utilising GraftJacket™, a product generated from human cadaver skin that undergoes chemical processing to remove unnecessary components, making it suitable for tendon augmentation. The utilisation of GraftJacket™ in multiple research trials has shown its capacity to improve significantly healing results without negatively impacting muscle growth and decreasing inflammation. Although the mechanical performance of the treatment and the resulting improvements in orthopaedic scores are evident, many patients still experience recurring tears. This emphasises the necessity for additional research and refinement of follow-up protocols [93–98].

In recent years, investigations have been into synthetic options for tendon grafts, including poly-L-lactic acid (PLA), polylactide-co-glycolide (PLGA), and polyurethane. Nevertheless, the limited knowledge of the mechanisms by which they degrade and their possible harmful effects on cells have dampened the enthusiasm for their broad use in therapeutic settings.

Decellularisation is a process that shows promise in utilising the mechanical strength of natural tissue for grafting applications. Farnebo et al. and Pan et al. have shown that decellularised tendon-bone composite grafts and tendon slices have the ability to enhance biomechanics and promote the creation of tendon structure greatly [82, 99]. Although there have been improvements in this area, the importance of the decellularisation process and the scarcity of appropriate grafts highlight the need for further improvement and investigation in order to utilise this technology in clinical environments effectively.

There has been notable progress in advancing bio-scaffolds and synthetic polymers for tendon restoration, but the search for an optimal graft material persists. The complex nature of tissue engineering, combined with the specific difficulties encountered in tendon restoration, requires a comprehensive strategy that includes the fields of material science, cellular biology, and clinical surgery. As the field progresses, researchers and clinicians must work closely together to improve current procedures and investigate new approaches to improve tendon restoration technologies' effectiveness, safety, and availability.

## 6.6 Future Perspectives

The field of tendon graft bioengineering is experiencing notable progress due to the incorporation of state-of-the-art biofabrication techniques, the introduction of hybrid scaffolds, the creation of "smart" scaffolds, and a transition towards cell-free and minimalist methods. The advancements in 3D printing and electrospinning are transforming the production of tendon scaffolds, allowing for the development of constructs that accurately mimic genuine tendons' intricate hierarchical structure and mechanical properties. Simultaneously, introducing hybrid scaffolds, which combine natural and synthetic materials, provides a potential approach for using the compatibility and activity of organic substances with the strength of synthetic polymers, enabling a more effective tendon restoration. Furthermore, integrating bioactive compounds into scaffolds, resulting in "smart" platforms, signifies the advent of a new era of scaffolds that may actively participate in the healing process by regulating the biological environment to enhance tissue regeneration. Finally, the growing fascination in cell-free and minimalist approaches, which take advantage of the body's natural regenerative capacities, represents a significant change in simplifying tendon restoration methods, potentially making the journey to clinical use more efficient. These new developments indicate a substantial change in the field of tendon graft bioengineering. They have the potential to rethink how tendon restoration is approached, making it more effective and compatible with patients.

The use of Artificial Intelligence (AI) and Machine Learning (ML) in the domain of tendon graft bioengineering is poised to transform the industry by utilising predictive modelling, optimising scaffold qualities, expediting biomaterial discovery, and improving imaging and diagnosis. AI and ML algorithms can analyse large datasets from patient records, imaging studies, and treatment outcomes. This enables the development of personalised treatment methods, improving surgical planning, scaffold design, and rehabilitation protocols to meet each patient's specific needs. Moreover, these technologies play a crucial role in adjusting scaffolds' composition and production parameters, guaranteeing that they closely mimic the mechanical and biological properties of genuine tendons. The utilisation of AI in the expedited recognition and evaluation of novel biomaterials holds the potential to substantially diminish the time and resources necessary for their creation, hence expediting the transition of promising materials from the laboratory to clinical trials. Furthermore, using AI-enhanced imaging techniques can improve the identification and ongoing assessment of tendon injuries and graft integration. Machine learning models have the ability to detect subtle changes over time, hence boosting the accuracy of therapeutic interventions and the monitoring of progress. AI and ML together represent a new age in tendon repair and regeneration, offering precise and efficient methods that can adapt to individual needs. This advancement in personalised medicine and optimising clinical outcomes in tendon graft bioengineering is a tremendous leap forward.

## 7 Conclusion

In conclusion, the area of tendon graft bioengineering is on the verge of a revolutionary period, propelled by a combination of cutting-edge biofabrication techniques, hybrid scaffolding, intelligent materials, and cell-free methods. This chapter has examined the subtle and complex aspects of these developing patterns, clarifying their ability to reshape the tendon repair and regeneration field. The introduction of 3D printing and electrospinning technologies has brought exceptional accuracy to the construction of scaffolds. These scaffolds not only imitate the natural structure of tendons but also offer improved integration and healing capabilities. Simultaneously, the investigation of hybrid scaffolds, which combine the advantages of natural and synthetic materials, provides a refined method for enhancing the biomechanical and bioactive characteristics necessary for successful tendon restoration.

Moreover, the implementation of scaffolds infused with bioactive compounds signifies a substantial advancement towards scaffolds that can actively impact the healing process, providing novel opportunities for focused therapy and tissue regeneration. At the same time, the move towards cell-free and minimalist methods highlights a larger trend of lowering the complexity of regenerative treatments. This can potentially improve their capacity to be applied in practice and decrease the likelihood of causing an immune response.

The significance of AI and ML in this field cannot be exaggerated. Predictive modelling and data-driven insights can potentially customise treatment plans to an unparalleled extent, optimising all aspects of tendon repair, ranging from surgical planning to post-operative rehabilitation. Furthermore, AI and ML are poised to completely transform the process of optimising scaffold qualities, uncovering new biomaterials, and enhancing diagnostic imaging. This will result in a faster rate of innovation and improved patient outcomes.

Nevertheless, the path ahead is not devoid of obstacles. To fully harness the potential of these advances, we must overcome challenges related to biocompatibility, matching mechanical properties, and navigating regulatory procedures. Moreover, the expense and availability of sophisticated therapies continue to be crucial factors, emphasising the requirement for scalable solutions that may be universally implemented. Although facing these difficulties, the potential for enhancing tendon restoration using bioengineering is substantial. Integrating knowledge and skills from several fields like engineering, biology, clinical practice, and data science is crucial for advancing beyond the existing limits of possibility. The incorporation of regenerative rehabilitation, the examination of worldwide health consequences, and the quest for customised treatments emphasise the comprehensive approach necessary to address the intricate issue of tendon injuries.

To summarise, the progress made in the field of tendon graft bioengineering, as discussed in this chapter, is the result of a combination of innovative ideas, technological developments, and scientific investigation. This progress shows significant potential for the future of regenerative medicine. As we further investigate these emerging patterns, the possibility of creating more efficient, individualised, and easily reachable therapies for tendon injuries is vast. The fusion of state-of-the-art bioengineering methodologies with the capabilities of artificial intelligence and machine learning presents novel opportunities for the discipline, potentially improving patients' well-being on a global scale. Despite obstacles and possibilities, the future of tendon repair and regeneration appears promising. By persistently working together and introducing new ideas, we can navigate the way ahead successfully.

## References


1. Koob TJ, Summers AP. Tendon—bridging the gap. Comparative biochemistry and physiology Part A, Molecular & integrative physiology. 2002; 133(4):905–9.
2. Starkey C, Brown SD, Ryan J. Examination of orthopedic and athletic injuries: FA Davis; 2009.
3. Canty EG, Kadler KE. Procollagen trafficking, processing and fibrillogenesis. J Cell Sci. 2005; 118(Pt 7):1341–53.
4. Moore MJ, De Beaux A. A quantitative ultrastructural study of rat tendon from birth to maturity. Journal of Anatomy. 1987; 153:163–9.
5. Franchi M, Trire A, Quaranta M, Orsini E, Ottani V. Collagen structure of tendon relates to function. Scientific World Journal. 2007; 7:404–20.
6. Curwin S. Biomechanics of tendon and the effects of immobilization. Foot Ankle Clin. 1997; 2:371–89.
7. O'Brien M. Structure and metabolism of tendons. Scand J Med Sci Sports. 1997; 7(2):55–61.
8. Elliott DH. Structure and function of mammalian tendon. Biological reviews of the Cambridge Philosophical Society. 1965; 40:392–421.
9. eOrthopod.com. Ankle Anatomy. Medical Multimedia Group, LLC. 2015.
10. Kastelic J, Galeski A, Baer E. The multicomposite structure of tendon. Connect Tissue Res. 1978; 6(1):11–23
11. Wang JH. Mechanobiology of tendon. J Biomech. 2006; 39(9):1563–82.
12. Fleischmajer R, MacDonald ED, Perlish JS, Burgeson RE, Fisher LW. Dermal collagen fibrils are hybrids of type I and type III collagen molecules. J Struct Biol. 1990; 105(1–3):162–9.
13. Elliott DH. Structure and function of mammalian tendon. Biological Reviews. 1965; 40(3):392–421.
14. Gelse K, Poschl E, Aigner T. Collagens—structure, function, and biosynthesis. Adv Drug Deliv Rev. 2003; 55(12):1531–46.
15. Fukuta S, Oyama M, Kavalkovich K, Fu FH, Niyibizi C. Identification of types II, IX and X collagens at the insertion site of the bovine achilles tendon. Matrix Biol. 1998; 17(1):65–73.
16. Birk DE, Mayne R. Localization of collagen types I, III and V during tendon development. Changes in collagen types I and III are correlated with changes in fibril diameter. Eur J Cell Biol. 1997; 72(4):352–61.
17. Liu SH, Yang RS, al-Shaikh R, Lane JM. Collagen in tendon, ligament, and bone healing. A current review. Clin Orthop Relat Res. 1995(318):265–78.
18. Walchli C, Koch M, Chiquet M, Odermatt BF, Trueb B. Tissue-specific expression of the fibril-associated collagens XII and XIV. J Cell Sci. 1994; 107 (Pt 2):669–81.
19. Killian ML, Cavinatto L, Galatz LM, Thomopoulos S. The role of mechanobiology in tendon healing. J Shoulder Elbow Surg. 2012; 21(2):228–37.
20. Liu Y, Ramanath HS, Wang DA. Tendon tissue engineering using scaffold enhancing strategies. Trends Biotechnol. 2008; 26(4):201–9.
21. Dahlmann-Noor AH, Martin-Martin B, Eastwood M, Khaw PT, Bailly M. Dynamic protrusive cell behaviour generates force and drives early matrix contraction by fibroblasts. Exp Cell Res. 2007; 313(20):4158–69.
22. Kongsgaard M, Aagaard P, Kjaer M, Magnusson SP. Structural Achilles tendon properties in athletes subjected to different exercise modes and in Achilles tendon rupture patients. J Appl Physiol (1985). 2005; 99(5):1965–71.
23. Wilmink J, Wilson AM, Goodship AE. Functional significance of the morphology and micromechanics of collagen fibres in relation to partial rupture of the superficial digital flexor tendon in racehorses. Res Vet Sci. 1992; 53(3):354–9.
24. Butler DL, Goldstein SA, Guilak F. Functional tissue engineering: the role of biomechanics. J Biomech Eng. 2000; 122(6):570–5.
25. Fyfe I, Stanish WD. The use of eccentric training and stretching in the treatment and prevention of tendon injuries. Clin Sports Med. 1992; 11(3):601–24.

26. Maquirriain J. Achilles Tendon Rupture: Avoiding Tendon Lengthening during Surgical Repair and Rehabilitation. The Yale Journal of Biology and Medicine. 2011; 84(3):289–300.
27. Butler DL, Grood ES, Noyes FR, Zernicke RF. Biomechanics of ligaments and tendons. Exerc Sport Sci Rev. 1978; 6:125–81.
28. Devkota AC, Weinhold PS. Mechanical response of tendon subsequent to ramp loading to varying strain limits. Clin Biomech (Bristol, Avon). 2003; 18(10):969–74.
29. Sharma P, Maffulli N. Tendon injury and tendinopathy: healing and repair. J Bone Joint Surg Am. 2005; 87(1):187–202.
30. James R, Kesturu G, Balian G, Chhabra AB. Tendon: Biology, Biomechanics, Repair, Growth Factors, and Evolving Treatment Options. The Journal of Hand Surgery. 2008; 33(1):102–12.
31. Garner WL, McDonald JA, Koo M, Kuhn C, 3rd, Weeks PM. Identification of the collagen-producing cells in healing flexor tendons. Plast Reconstr Surg. 1989; 83(5):875–9.
32. Strickland JW. Development of flexor tendon surgery: twenty-five years of progress. J Hand Surg Am. 2000; 25(2):214–35.
33. Almekinders LC, Almekinders SV. Outcome in the treatment of chronic overuse sports injuries: a retrospective study. J Orthop Sports Phys Ther. 1994; 19(3):157–61.
34. Sharma P, Maffulli N. Biology of tendon injury: healing, modeling and remodeling. J Musculoskelet Neuronal Interact. 2006; 6(2):181–90.
35. Bunnell S. Repair of tendons in the fingers and description of two new instruments. Surg Gynecol Obstet. 1918; 26:103–10.
36. Verdan CE. Primary repair of flexor tendons. The Journal of Bone & Joint Surgery. 1960; 42(4):647–57.
37. Soroceanu A, Sidhwa F, Aarabi S, Kaufman A, Glazebrook M. Surgical Versus Nonsurgical Treatment of Acute Achilles Tendon Rupture 2012 2012–12–05 00:00:00. 2136–43 p.
38. Strickland JW. Flexor Tendon Injuries: II. Operative Technique. J Am Acad Orthop Surg. 1995; 3(1):55–62.
39. Ketchum LD. Suture materials and suture techniques used in tendon repair. Hand Clin. 1985; 1(1):43–53.
40. Lawrence TM, Davis TR. A biomechanical analysis of suture materials and their influence on a four-strand flexor tendon repair. J Hand Surg Am. 2005; 30(4):836–41.
41. Moriya K, Yoshizu T, Maki Y, Tsubokawa N, Narisawa H, Endo N. Clinical outcomes of early active mobilization following flexor tendon repair using the six-strand technique: short- and long-term evaluations. J Hand Surg Eur Vol. 2014.
42. Savage R. In vitro studies of a new method of flexor tendon repair. J Hand Surg Br. 1985; 10(2):135–41.
43. Uslu M, Isik C, Ozsahin M, Ozkan A, Yasar M, Orhan Z, et al. Flexor tendons repair: effect of core sutures caliber with increased number of suture strands and peripheral sutures. A sheep model. Orthop Traumatol Surg Res. 2014; 100(6):611–6.
44. Osei DA, Stepan JG, Calfee RP, Thomopoulos S, Boyer MI, Potter R, et al. The Effect of Suture Caliber and Number of Core Suture Strands on Zone II Flexor Tendon Repair: A Study in Human Cadavers. Journal of Hand Surgery. 2013; 39(2):262–8.
45. Dovan TT, Ditsios KT, Boyer MI. Eight-strand core suture technique for repair of intrasynovial flexor tendon lacerations. Tech Hand Up Extrem Surg. 2003; 7(2):70–4.
46. Silva MJ, Hollstien SB, Fayazi AH, Adler P, Gelberman RH, Boyer MI. The effects of multiple-strand suture techniques on the tensile properties of repair of the flexor digitorum profundus tendon to bone. J Bone Joint Surg Am. 1998; 80(10):1507–14.
47. Wong JK, Alyouha S, Kadler KE, Ferguson MW, McGrouther DA. The cell biology of suturing tendons. Matrix Biol. 2010; 29(6):525–36.
48. Paessler HH, Mastrokalos DS. Anterior cruciate ligament reconstruction using semitendinosus and gracilis tendons, bone patellar tendon, or quadriceps tendon-graft with press-fit fixation without hardware. A new and innovative procedure. Orthop Clin North Am. 2003; 34(1):49–64.

49. Williams RJ, 3rd, Hyman J, Petrigliano F, Rozental T, Wickiewicz TL. Anterior cruciate ligament reconstruction with a four-strand hamstring tendon autograft. Surgical technique. J Bone Joint Surg Am. 2005; 87 Suppl 1(Pt 1):51–66.
50. Slone HS, Romine SE, Premkumar A, Xerogeanes JW. Quadriceps Tendon Autograft for Anterior Cruciate Ligament Reconstruction: A Comprehensive Review of Current Literature and Systematic Review of Clinical Results. Arthroscopy: The Journal of Arthroscopic & Related Surgery. 2015; 31(3):541–54.
51. Niklason LE. Engineering of bone grafts. Nat Biotech. 2000; 18(9):929–30.
52. Robertson A, Nutton RW, Keating JF. Current trends in the use of tendon allografts in orthopaedic surgery. J Bone Joint Surg Br. 2006; 88(8):988–92.
53. Phillips CR, Kaye S. The sterilizing action of gaseous ethylene oxide.1. Review. American Journal of Hygiene. 1949; 50(3):270–9.
54. Silvaggio VJ, Fu FH, Georgescu HI, Evans CH. The induction of IL-1 by freeze-dried ethylene oxide-treated bone-patellar tendon-bone allograft wear particles: an in vitro study. Arthroscopy: the journal of arthroscopic & related surgery: official publication of the Arthroscopy Association of North America and the International Arthroscopy Association. 1993; 9(1):82–6.
55. Vangsness CT, Jr., Triffon MJ, Joyce MJ, Moore TM. Soft tissue for allograft reconstruction of the human knee: a survey of the American Association of Tissue Banks. Am J Sports Med. 1996; 24(2):230–4.
56. Fideler BM, Vangsness CT, Jr., Lu B, Orlando C, Moore T. Gamma irradiation: effects on biomechanical properties of human bone-patellar tendon-bone allografts. Am J Sports Med. 1995; 23(5):643–6.
57. McGilvray KC, Santoni BG, Turner AS, Bogdansky S, Wheeler DL, Puttlitz CM. Effects of (60)Co gamma radiation dose on initial structural biomechanical properties of ovine bone—patellar tendon—bone allografts. Cell Tissue Bank. 2011; 12(2):89–98.
58. Barbour SA, King W. The safe and effective use of allograft tissue—an update. Am J Sports Med. 2003; 31(5):791–7.
59. Fishman JA, Greenwald MA, Grossi PA. Transmission of infection with human allografts: essential considerations in donor screening. Clin Infect Dis. 2012; 55(5):720–7.
60. Sullivan EK, Kamstock DA, Turner AS, Goldman SM, Kronengold RT. Evaluation of a flexible collagen surgical patch for reinforcement of a fascial defect: Experimental study in a sheep model. Journal of Biomedical Materials Research Part B: Applied Biomaterials. 2008; 87B(1):88–94.
61. Seldes RM, Abramchayev I. Arthroscopic Insertion of a Biologic Rotator Cuff Tissue Augmentation After Rotator Cuff Repair. Arthroscopy: the journal of arthroscopic & related surgery: official publication of the Arthroscopy Association of North America and the International Arthroscopy Association. 2006; 22(1):113–6.
62. Johnson W, Inamasu J, Yantzer B, Papangelou C, Guiot B. Comparative in vitro biomechanical evaluation of two soft tissue defect products. J Biomed Mater Res B Appl Biomater. 2007.
63. Tasaki M, Wamala I, Tena A, Villani V, Sekijima M, Pathiraja V, et al. High Incidence of Xenogenic Bone Marrow Engraftment in Pig-to-Baboon Intra-Bone Bone Marrow Transplantation. American Journal of Transplantation. 2015.
64. Dong Z, Li Y, Zou Q. Degradation and biocompatibility of porous nano-hydroxyapatite/polyurethane composite scaffold for bone tissue engineering. Applied Surface Science. 2009; 255(12):6087–91.
65. O’Brien FJ. Biomaterials scaffolds for tissue engineering. Materials Today. 2011; 14(3):88–95.
66. Phelps EA, Landázuri N, Thulé PM, Taylor WR, García AJ. Bioartificial matrices for therapeutic vascularization. Proceedings of the National Academy of Sciences of the United States of America. 2010; 107(8):3323–8.
67. Bagnaninchi PO, Yang Y, Zghoul N, Maffulli N, Wang RK, Haj AJ. Chitosan microchannel scaffolds for tendon tissue engineering characterized using optical coherence tomography. Tissue Eng. 2007; 13(2):323–31.

68. Majima T, Funakosi T, Iwasaki N, Yamane ST, Harada K, Nonaka S, et al. Alginate and chitosan polyion complex hybrid fibers for scaffolds in ligament and tendon tissue engineering. J Orthop Sci. 2005; 10(3):302–7.
69. Xiong Y, Wan L. Preparation and characterization of an extracellular matrix of artificial tendon tissue from natural macromolecules. Sheng Wu Gong Cheng Xue Bao. 2008; 24(11):1907–11.
70. Sawaguchi N, Majima T, Funakoshi T, Shimode K, Harada K, Minami A, et al. Effect of cyclic three-dimensional strain on cell proliferation and collagen synthesis of fibroblast-seeded chitosan-hyaluronan hybrid polymer fiber. J Orthop Sci. 2010; 15(4):569–77.
71. Murphy KD, Mushkudiani IA, Kao D, Levesque AY, Hawkins HK, Gould LJ. Successful incorporation of tissue-engineered porcine small-intestinal submucosa as substitute flexor tendon graft is mediated by elevated TGF-beta1 expression in the rabbit. J Hand Surg Am. 2008; 33(7):1168–78.
72. Kraus A, Woon C, Raghavan S, Megerle K, Pham H, Chang J. Co-culture of human adipose-derived stem cells with tenocytes increases proliferation and induces differentiation into a tenogenic lineage. Plast Reconstr Surg. 2013; 132(5):754e-66e.
73. Rajan N, Habermehl J, Cote MF, Doillon CJ, Mantovani D. Preparation of ready-to-use, storable and reconstituted type I collagen from rat tail tendon for tissue engineering applications. Nat Protoc. 2006; 1(6):2753–8.
74. Evans JH, Barbenel JC. Structural and mechanical properties of tendon related to function. Equine Vet J. 1975; 7(1):1–8.
75. Riley GP, Harrall RL, Constant CR, Chard MD, Cawston TE, Hazleman BL. Glycosaminoglycans of human rotator cuff tendons: changes with age and in chronic rotator cuff tendinitis. Ann Rheum Dis. 1994; 53(6):367–76.
76. Sawadkar P, Alexander S, Tolk M, Wong J, McGrouther D, Bozec L, et al. Development of a surgically optimized graft insertion suture technique to accommodate a tissue-engineered tendon in vivo. Biores Open Access. 2013; 2(5):327–35.
77. Sawadkar P, Alexander S, Mudera V. Tissue-engineered collagen grafts to treat large tendon defects. 2014.
78. Sawadkar P, Sibbons P, Ahmed T, Bozec L, Mudera V. Engineering of a Functional Tendon Using Collagen As a Natural Polymer. ACS Biomaterials Science & Engineering. 2019; 5(10):5218–28.
79. Sawadkar P, Wong J, Mudera V. A Novel Tenorrhaphy Suture Technique with Tissue Engineered Collagen Graft to Repair Large Tendon Defects. JoVE. 2021(178):e57696.
80. Sawadkar P, Alexander S, Mudera V. Tissue-engineered collagen grafts to treat large tendon defects. Regenerative medicine. 2014; 9(3):249–51.
81. Brown RA, Wiseman M, Chuo CB, Cheema U, Nazhat SN. Ultrarapid Engineering of Biomimetic Materials and Tissues: Fabrication of Nano- and Microstructures by Plastic Compression. Advanced Functional Materials. 2005; 15(11):1762–70.
82. Pan J, Liu GM, Ning LJ, Zhang Y, Luo JC, Huang FG, et al. Rotator cuff repair using a decellularized tendon slices graft: an in vivo study in a rabbit model. Knee Surg Sports Traumatol Arthrosc. 2014.
83. Yin Z, Chen X, Zhu T, Hu JJ, Song HX, Shen WL, et al. The effect of decellularized matrices on human tendon stem/progenitor cell differentiation and tendon repair. Acta Biomater. 2013; 9(12):9317–29.
84. Ni M, Rui YF, Tan Q, Liu Y, Xu LL, Chan KM, et al. Engineered scaffold-free tendon tissue produced by tendon-derived stem cells. Biomaterials. 2013; 34(8):2024–37.
85. Caplan AI, Bruder SP. Mesenchymal stem cells: building blocks for molecular medicine in the 21st century. Trends Mol Med. 2001; 7(6):259–64.
86. Manning CN, Schwartz AG, Liu W, Xie J, Havlioglu N, Sakiyama-Elbert SE, et al. Controlled delivery of mesenchymal stem cells and growth factors using a nanofiber scaffold for tendon repair. Acta Biomater. 2013; 9(6):6905–14.
87. Sahoo S, Toh SL, Goh JC. A bFGF-releasing silk/PLGA-based biohybrid scaffold for ligament/tendon tissue engineering using mesenchymal progenitor cells. Biomaterials. 2010; 31 (11):2990–8.

88. Stoll C, John T, Endres M, Rosen C, Kaps C, Kohl B, et al. Extracellular matrix expression of human tenocytes in three-dimensional air-liquid and PLGA cultures compared with tendon tissue: implications for tendon tissue engineering. J Orthop Res. 2010; 28(9):1170–7.
89. Cao D, Liu W, Wei X, Xu F, Cui L, Cao Y. In vitro tendon engineering with avian tenocytes and polyglycolic acids: a preliminary report. Tissue Eng. 2006; 12(5):1369–77.
90. Henson F, Getgood A. The use of scaffolds in musculoskeletal tissue engineering. Open Orthop J. 2011; 5 Suppl 2:261–6.
91. Chen JM, Willers C, Xu J, Wang A, Zheng MH. Autologous tenocyte therapy using porcine-derived bioscaffolds for massive rotator cuff defect in rabbits. Tissue Eng. 2007; 13(7):1479–91.
92. Iannotti JP, Codsi MJ, Kwon YW, Derwin K, Ciccone J, Brems JJ. Porcine small intestine submucosa augmentation of surgical repair of chronic two-tendon rotator cuff tears. A randomized, controlled trial. J Bone Joint Surg Am. 2006; 88(6):1238–44.
93. Adams JE, Zobitz ME, Reach JS, Jr., An KN, Steinmann SP. Rotator cuff repair using an acellular dermal matrix graft: an in vivo study in a canine model. Arthroscopy: the journal of arthroscopic & related surgery: official publication of the Arthroscopy Association of North America and the International Arthroscopy Association. 2006; 22(7):700–9.
94. Burkhead Jr WZ, Schiffern SC, Krishnan SG. Use of Graft Jacket as an Augmentation for Massive Rotator Cuff Tears. Seminars in Arthroplasty. 2007; 18(1):11–8.
95. Lee DK. Achilles tendon repair with acellular tissue graft augmentation in neglected ruptures. J Foot Ankle Surg. 2007; 46(6):451–5.
96. Dopirak R, Bond J, Snyder S. Arthroscopic total rotator cuff replacement with an acellular human dermal allograft matrix 2007 January 1, 2007. 7–15 p.
97. Bond JL, Dopirak RM, Higgins J, Burns J, Snyder SJ. Arthroscopic replacement of massive, irreparable rotator cuff tears using a GraftJacket allograft: technique and preliminary results. Arthroscopy: the journal of arthroscopic & related surgery: official publication of the Arthroscopy Association of North America and the International Arthroscopy Association. 2008; 24(4):403–9 e1.
98. Coons DA, Alan Barber F. Tendon graft substitutes-rotator cuff patches. Sports Med Arthrosc. 2006; 14(3):185–90.
99. Farnebo S, Woon CY, Bronstein JA, Schmitt T, Lindsey DP, Pham H, et al. Decellularized tendon-bone composite grafts for extremity reconstruction: an experimental study. Plast Reconstr Surg. 2014; 133(1):79–89.

# Part III
# Applications in Cardiac and Heat Valve Bioengineering

# Cardiomyocyte Proliferation and Cardiac Cell Therapy

**Fan Li and Wuqiang Zhu**

Heart failure has been a leading cause of morbidity and mortality in the world. Myocardial infarction is the most common cause of heart failure which is frequently a consequence of coronary artery disease [1]. During myocardial infarction, the blood flow decreases or stops supply to a part of the heart, which leads to severe damage to cardiac muscle. Unlike lower vertebrate and neonatal mammals, adult cardiomyocytes are not capable of proliferation to regenerate the damaged myocardium. Instead, the damaged heart undergoes extensive remodeling to replenish the dead cardiomyocytes with fibrotic scars and pathological hypertrophy, which significantly impairs normal cardiac function and, in some cases, results in the development of chronic heart failure [2]. Studies on some lower vertebrates [3–5] and during the early neonatal period in mammals [6, 7] revealed the regulatory mechanism of cardiomyocyte proliferation and the functions of non-cardiomyocyte cells in cardiac regeneration. A number of promising strategies for promoting cardiac regeneration and repair are currently being developed [8].

## 1 Heart Regeneration

### *1.1 Heart Regeneration in Lower Vertebrate*

Unlike humans, many lower vertebrates exhibit remarkable regenerative abilities in limbs, appendages, and many other internal organs following injury. Among these animals, amphibians have been extensively studied for their ability to regenerate

F. Li · W. Zhu (✉)
Department of Cardiovascular Diseases, Physiology and Biomedical Engineering, Center for Regenerative Biotherapeutics, Mayo Clinic Arizona, Scottsdale, AZ, USA
e-mail: Zhu.Wuqiang@mayo.edu

 
J. Liao, J. Y. Wong (eds.), *Integration and Bridging of Multiscale Bioengineering Designs and Tissue Biomechanics*, https://doi.org/10.1007/978-3-031-81743-4_9

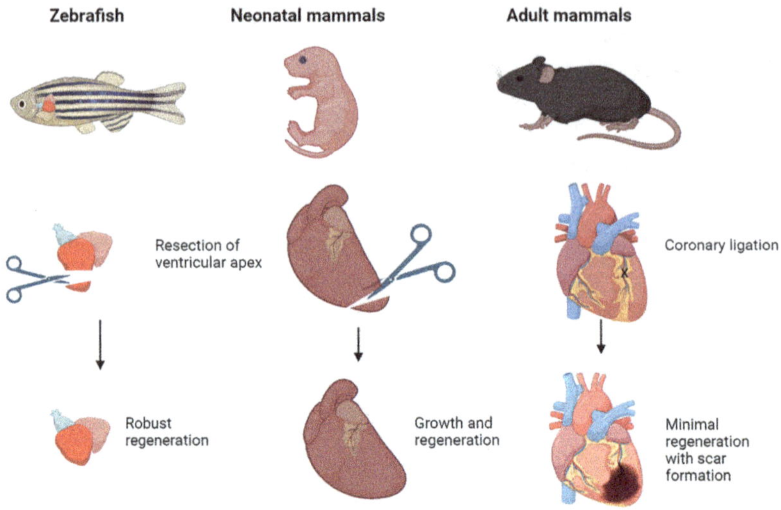


**Fig. 1** Heart regeneration in different species. Zebrafish and newborn mice exhibit a strong ability to regenerate, while adult mammals generally have limited or no ability to regenerate after injury. The mouse is used to represent mammals, including rats, pigs, and sheep

their hearts and other organs. Research on cardiac regeneration in eastern newts, particularly the adult newt (*Notophthalmus viridescens*), has demonstrated that cardiomyocytes play an important role in heart regeneration. Their hearts respond to injury by initiating cardiomyocyte proliferation at the injury site, as evidenced by increased DNA synthesis and mitosis. This process involves the dedifferentiation of cardiomyocytes near the injury, followed by their proliferation and subsequent tissue remodeling [4, 9]. Studies conducted in 2011 provided insights into the full regeneration of the newt heart within approximately 70 days after ventricular resection and mechanical injury [10]. Adult salamanders, such as newts, also exhibit impressive cardiac regeneration abilities [3].

More recently, the zebrafish has emerged as an excellent model for studying cardiac regeneration due to its substantial regenerative capacity and genetic tractability. In 2002, Poss and colleagues demonstrated that the adult zebrafish heart, around 1–2 years old, could fully regenerate following surgical amputation of the cardiac apex, which corresponds to a loss of roughly 20% of the total ventricular mass (Fig. 1). This regeneration process is characterized by robust proliferation of cardiomyocytes, primarily localized at the leading epicardial edge of the new myocardium [5]. Kikuchi et al. used genetic lineage in which the cardiomyocyte-specific *cmcl2* promoter drives the expression of tamoxifen-inducible Cre recombinase tracing to confirm the primary contribution of preexisting cardiomyocytes, especially those expressing the *gata4* (Gata transcription factors encode zinc-finger DNA-binding protein 4) gene, in the regenerative process [11].

## 1.2 Heart Regeneration in Neonatal and Adult Mammals

The concept of cardiomyocyte cell cycle activity has been explored in rodents since the 1960s[12]. Various techniques, including metabolic assays for deoxyribonucleotide analog incorporation into DNA, microfluorometric or flow cytometric analyses for quantifying nuclear DNA content, tritiated thymidine incorporation, immune-based assays monitoring DNA synthesis through bromodeoxyuridine (BrdU), and proliferating cell nuclear antigen (PCNA) analyses, have been used to assess cellular proliferation [12]. Research by Field's team employed a transgenic mouse model where the nLAC reporter gene is regulated by the cardiac α-*MHC* promoter. This allowed for the efficient and objective identification of cardiomyocyte nuclei through X-GAL staining, and DNA synthesis was determined by thymidine incorporation. Their findings confirmed that cardiomyocyte synthesis is exceptionally rare in adult mouse hearts. In intact and postinfarction adult ventricles, only 0.0006% and 0.0083% of muscle cells, respectively, were actively proliferating [13]. Studies involving surgical resection of the ventricular apex in neonatal mice at different ages revealed that 1-day-old (postnatal day 1, P1) mice displayed a regenerative response that restored the damaged heart to normal anatomy and function within 21 days. In contrast, P7 mice failed to regenerate their myocardium and developed significant fibrosis (Fig. 1) [14]. Similar observations have also been reported in P1 rats after apical resection [15]. Neonatal rabbits demonstrated natural heart regeneration after myocardial infarction at P1, resulting in recovery of normal cardiac geometry and function within 3 weeks. However, this regenerative potential is significantly reduced by age P4, with minimal regeneration observed after P7 [16].

In neonatal pigs, acute myocardial infarction was induced on P1, P2, P3, P7, and P14. The results showed that the hearts of P1 animals displayed little evidence of scar formation or wall thinning on day 30 post-myocardial infarction, and cell cycle activity was increased at P7. Systolic function at day 30 post-myocardial infarction was largely restored in P1 animals and partially restored in P2 animals but significantly impaired when myocardial infarction was induced on P3 or later. It was demonstrated that the neonatal pig heart is capable of regeneration after myocardial infarction during the first 2 days of life [7]. Genome-wide analyses of the infarct zone revealed a strong transcriptional signature of fibrosis in 14-day-old animals that was absent in 2-day-old pigs, which instead had enrichment for cytokinesis genes [17]. In fetal Dorset sheep (between 65 and 75 days of gestation), the response to myocardial infarction resulted in robust cardiomyocyte proliferation [18]. However, adult sheep showed a decline in ejection fraction and increased akinetic myocardial segment size 4 weeks following myocardial infarction. In contrast, the fetal sheep can preserve ejection fraction and have no akinetic myocardial segment 4 weeks postinfarction [19]. Interestingly, findings from clinical studies showed that the neonatal human heart also seems to possess the capacity for regeneration given that rapid cardiac functional recovery was observed in newborn infants with myocardial infarction due to congenital anomalies of coronary vessels or blood clotting factor mutations [6, 20].

# 2 Molecules that Regulate Cardiomyocyte Cell Cycle

Many regulatory molecules that control the cardiomyocyte cell cycle have recently been identified, including cyclins, transcription factors, miRNAs, and metabolism (Fig. 2).

## 2.1 *Cyclins*

In general, positive cell cycle regulators (for example, cyclins, cyclin-dependent kinases [CDKs], and proto-oncogenes) are highly expressed in embryonic hearts where cardiomyocyte cell cycle activity is relatively high. Most of these positive cell cycle regulators are downregulated in the adult hearts. In general, expression patterns of negative cell cycle regulatory genes (for example, cyclin-dependent kinase inhibitors [CKIs]) are frequently increased in adult hearts, where cardiomyocyte cell

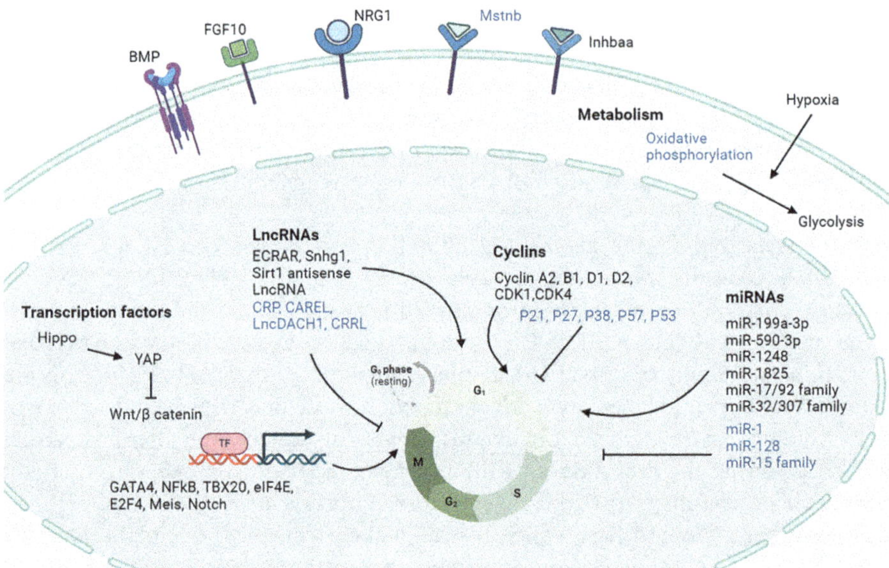


**Fig. 2** Molecules that regulate cardiomyocyte cell cycle. Overexpressing cyclins or inhibiting cyclin-dependent kinase inhibitors (CKIs) promotes cardiomyocyte proliferation. Growth factors such as Neuregulin1 (NRG1) and Fibroblast Growth Factors (FGF) stimulate pro-proliferation pathways, including YAP, Notch, and Wnt/β-catenin, and affect the expression of transcription factors that enhance cardiomyocyte proliferation. Furthermore, certain miRNAs and lncRNAs can control the expression or function of cyclins, CKIs, or transcription factors, thus modulating cardiomyocyte proliferation. Hypoxia induces glycolysis, which further supports cardiomyocyte proliferation. Pro-proliferative molecules are denoted in black, while inhibitory molecules are marked in blue

cycle activity is largely absent [21]. Activation of p38 in vivo by MKK3bE reduces BrdU incorporation in fetal cardiomyocytes by 17.6%. In contrast, cardiac-specific *p38α* knockout mice show a 92.3% increase in neonatal cardiomyocyte mitosis. Furthermore, inhibition of p38 in adult cardiomyocytes promotes cytokinesis [22]. P38 regulates the expression of genes required for mitosis in cardiomyocytes, including cyclin A and cyclin B[22]. The overexpression of cyclin A2 [23], cyclin D1 [24, 25], and cyclin D2 [26] can stimulate cardiomyocyte DNA synthesis or mitosis, respectively, in adult mice. Overexpression of cyclin D2 activates the cell cycle and enhances proliferation in cultured human-induced pluripotent stem cell-derived cardiomyocytes (hiPSC-CMs). Furthermore, it results in an increased number of engrafted hiPSC-CMs after transplantation into infarcted mouse hearts, primarily by promoting cell proliferation [27–29]. Combined overexpression of CDK1, CDK4, cyclin B1, and cyclin D1 efficiently induces cell division in postmitotic mouse, rat, and human cardiomyocytes [30]. Overexpression of cyclin G1 promotes DNA synthesis but inhibits cytokinesis in neonatal cardiomyocytes, resulting in an increase in the number of binuclear myocytes. In contrast, inactivation of the cyclin G1 gene in mice reduced the degree of polyploidy and multinucleation of cardiomyocytes [31]. Triple transfection of siRNAs targeting CKIs, specifically *p21*(Waf1/Cip1), *p27*(Kip1), and *p57*(Kip2), induces both neonatal and adult cardiomyocytes to enter the S phase, leading to an increased nuclei-to-cardiomyocyte ratio, and also increases cytokinesis [32].

P53 exerts its function by activating the transcription of genes that induce cell cycle arrest at either the G1/S or G2 phase. Additionally, p53 activates genes responsible for initiating DNA repair and triggering apoptosis. Its activity is finely regulated through ubiquitination by the E3 ubiquitin ligase and murine double minute 2 (MDM2), which targets p53 for proteasomal degradation. In zebrafish, following heart injury, there is an increase in *mdm2* transcripts, coupled with a transient decrease in p53 protein levels. Zebrafish lacking functional p53 exhibit heightened indicators of cardiomyocyte proliferation during regeneration. Conversely, the inhibition of injury-induced cardiomyocyte proliferation is observed in zebrafish with transgenic Mdm2 blockade [33]. Surprisingly, the absence of p53 alone did not lead to DNA synthesis in adult mouse cardiomyocytes. Intriguingly, the concurrent deletion of p53 and MDM2 effectively spurs proliferation in adult cardiomyocytes in vivo. The loss of p53/MDM2 results in the downregulation of p21 and p27 while simultaneously inducing Cyclin E expression [34].

## 2.2 *Transcription Factors*

Transcription factors involved in early cardiomyocyte differentiation also promote cardiomyocyte proliferation, including GATA4 [35], nuclear factor kappa B (NF-κβ) [36], T-box transcription factor 20 (TBX20) [37], notch-activated CSL [38], and Wnt/β-catenin-activated TCF/LEF [39]. Cyclin D2 and Cdk4 are direct transcriptional targets of GATA4 [35]. Another key transcription factor is Meis homeobox 1 (MEIS1), which has been identified as a critical regulator of the cardiomyocyte cell cycle. Deletion of *Meis1* in mouse cardiomyocytes extends the postnatal proliferative window of cardiomyocytes and reactivates cardiomyocyte mitosis in the adult heart without negative effects on cardiac function. Conversely, overexpression of *Meis1* inhibits neonatal myocyte proliferation and neonatal heart regeneration by activating CKIs such as p15, p16, and p21 [40]. The deletion of *Hoxb13*, a MEIS1 cofactor, in cardiomyocytes, extends the postnatal window of cardiomyocyte proliferation and reactivates the cardiomyocyte cell cycle in the adult heart. Notably, the double knockout of *Meis1* and *Hoxb13* in adult hearts results in widespread cardiomyocyte mitosis, sarcomere disassembly, and improved left ventricular systolic function following myocardial infarction [41]. Using a hydrogel-based delivery method for small interfering RNA-mediated silencing of *Rb1* and *Meis2* in vivo following myocardial infarction improves the proliferation of adult cardiomyocytes, cell survivability, and enhances peri-infarct angiogenesis [42]. The eIF4E family of translation initiation factors, which binds to 5′ methylated caps, is a critical determinant of mRNA translation, serving as the rate-limiting step in this process. In addition to the well-known eIF4E factors, there is another group called eIF4E1c that plays a crucial role in the normal production of cardiomyocytes in zebrafish hearts during both animal growth and injury-induced regeneration. Ribosome profiling studies in mutant hearts have revealed changes in the translation efficiency of mRNAs known to regulate cardiomyocyte proliferation [43]. E2F4 is another important transcription factor which is an essential factor for cardiomyocyte mitosis [44]. Moreover, forkhead box M1(FoxM1) is known to regulate cardiomyocyte proliferation in adult zebrafish following cardiac injury [45]. In addition, overexpression of reprogramming factors, octamer-binding transcription factor 4 (*Oct4*), SRY-box transcription factor 2 (*Sox2*), Kruppel-like factor 4 (*Klf4*), and *c-Myc*, which were originally used to generate the first induced pluripotent stem cells from fibroblasts [46], has the capacity to induce de-differentiation and proliferation of adult cardiomyocytes [47].

The Hippo kinase cascade is a pivotal regulator of organ growth, acting to inhibit Wnt signaling and restrain cardiomyocyte proliferation. Hippo-deficient embryos exhibit overgrown hearts due to increased cardiomyocyte proliferation. Specifically, Hippo signaling restrains the pro-growth Wnt/β-catenin-YAP interaction during the transition from rapid progenitor cell growth to measured growth in the maturing heart [39]. Low-density lipoprotein receptor-related protein 6 (LRP6) is a single-span transmembrane protein that plays a key role in the transduction of the canonical Wnt/β-catenin signaling pathway. Deficiency of Lrp6 has been associated with increased cardiomyocyte proliferation at various stages in both mouse hearts and iPSC-CMs. In mouse hearts, *Lrp6* deficiency leads to a reduction in the size of myocardial infarction and an improvement in heart function. The ING5/P21 signaling pathway has been identified as responsible for mediating the effects of LRP6 on cardiomyocyte proliferation [48]. The core components of the mammalian Hippo signaling pathway include MST1 and MST2, members of the Ste20 family of kinases, which are homologous to Drosophila HIPPO. These kinases form active complexes with the WW repeat scaffolding protein Salvador (SALV), also known as WW45, which phosphorylates the large tumor suppressor homolog (LATS) kinase. Mammals have two *Lats* genes, *Lats1* and *Lats2*, homologous to Drosophila warts. LATS kinases complex with MOB to phosphorylate YAP and TAZ, two related transcriptional coactivators. Upon phosphorylation, YAP and TAZ, the most downstream components of Hippo signaling, are excluded from the nucleus and rendered transcriptionally inactive. Cardiomyocyte-specific *Sulv* KO mice showed evidence of regenerated myocardium with reduced scar size after apex resection at P8, and, following left anterior descending occlusion, there was functional recovery and reduced scar size when analyzed at 21 days after occlusion in P8 and adult hearts [49]. A major target of this kinase cascade is YAP1, a transcriptional coactivator that is inactivated by Hippo kinase activity. Activation of YAP1 has been shown to stimulate the proliferation of postnatal cardiomyocytes, both in culture and the intact heart. YAP1 exerts its regulatory effects by binding to DNA-binding transcription factors, which is essential for YAP1-induced cell proliferation [50]. Conversely, forced expression of a constitutively active form of YAP in the adult heart stimulates cardiac regeneration and improves contractility after myocardial infarction [51].

## 2.3 Growth Factors

Some studies showed that the Neuregulin1 (NRG1)/Erb-B2 receptor tyrosine kinase 4 (ErbB4) signaling pathway is involved in cardiac regeneration. NRG1, a growth factor, binds to its receptor, ErbB4, stimulating cardiomyocyte proliferation and enhancing heart repair in response to injury. Genetic inactivation of ErbB4 reduces cardiomyocyte proliferation, while its upregulation promotes it. NRG1 injection in adult mice induces cardiomyocyte cell-cycle activity, enhancing myocardial regeneration and improving heart function post-myocardial infarction [52]. ERBB2, an NRG1 co-receptor, also plays a crucial role; its inhibition disrupts cardiomyocyte proliferation. In zebrafish, NRG1 reactivation induces dedifferentiation, muscle hyperplasia, and cardiac regeneration [53]. ERBB2 activation and its downstream pathways, such as ERK, AKT, and GSK3β/β-catenin signaling, mediate cardiomyocyte dedifferentiation and proliferation, ultimately facilitating cardiac regeneration in neonatal, juvenile, and adult cardiomyocytes [54]. ERBB2 drives YAP activation and EMT-like processes during cardiac regeneration, manifested by cytoskeletal remodeling, junction dissolution, migration, and extracellular matrix (ECM) turnover [55]. Two TGF-β family ligands, mstnb and inhbaa, have opposing effects. After cryoinjury, mstnb is consistently downregulated, while inhbaa is upregulated in the adult zebrafish heart. Loss of mstnb function and activation of inhbaa both promote cardiomyocyte proliferation and enhance cardiac regeneration. Importantly, overexpression of inhbaa alone is sufficient to induce cardiomyocyte proliferation independently of the NRG–ErbB signaling pathway [56]. However, it was recently reported that recombinant Neuregulin 1 does not activate cardiomyocyte DNA synthesis in normal or infarcted adult mice [57]. In this study, adult mice were subjected to nine consecutive daily injections of recombinant NRG1β1 (comprising Neuregulin 1 amino acid residues 176–256) or vehicle. Interestingly, the NRG1β1 treatment increased DNA synthesis in non-cardiomyocytes but inhibited baseline rates of cardiomyocyte DNA synthesis in normal mice. Myeloid-derived growth factor (*Mydgf*) which is predominantly expressed by endothelial cells promotes cardiomyocyte proliferation by activating c-MYC/FOXM1 pathway and improves heart regeneration both in neonatal and adult mice after cardiac injury [58]. FGF10 enhances cardiac regeneration by elevating cardiomyocyte renewal and limiting fibrosis, affecting *Meis1* expression, Hippo signaling, and metabolic pathways [59]. BMP signaling is activated in cardiomyocytes and promotes their proliferation [60].

## 2.4 miRNAs

Eulalio et al. performed a high-content microscopy and high-throughput functional screening for human miRNAs that promoted neonatal cardiomyocyte proliferation using a whole-genome miRNA library. They identified 40 miRNAs that significantly increased both DNA synthesis and cytokinesis in neonatal mice and rat cardiomyocytes. Two of these miRNAs, miR-590 and miR-199a, were selected for further investigation and demonstrated the ability to induce cell cycle re-entry in adult cardiomyocytes ex vivo. These miRNAs also promoted cardiomyocyte proliferation in both neonatal and adult animals. In a mouse model of myocardial infarction, these miRNAs led to substantial cardiac regeneration and nearly complete restoration of cardiac functional parameters [61]. Among these miRNAs, miR-199a-3p, miR-590-3p, miR-1248, and miR-1825, as well as two extensive miRNA families, the miR-17/92 and the miR-302/367 clusters, played pivotal roles in the regulation of cardiac cell proliferation during development. Knockout experiments revealed that their absence led to cardiac hypoplasia, while overexpression expanded the cardiomyocyte pool during embryogenesis [62, 63]. Notably, miR-19a/19b, members of the miR-17-92 cluster, were found to be induced in heart failure patients. Intracardiac injection of miR-19a/19b mimics enhanced cardiomyocyte proliferation and stimulated cardiac regeneration in response to myocardial infarction injury [64]. Another study demonstrated that a single-dose intracardiac injection of pro-regenerative microRNAs, miR-199a-3p and miR-590-3p, improved cardiac function after myocardial infarction [65]. Additionally, transient treatment of mice with miR-302/367 mimics promoted cardiac regeneration without adverse effects on physiological function [63]. The most effective of these miRNAs activate nuclear localization of the master transcriptional cofactor YAP and induce expression of YAP-responsive genes. In particular, miR-199a-3p directly targets two mRNAs coding for proteins impinging on the Hippo pathway, the upstream YAP inhibitory kinase TAOK1, and the E3 ubiquitin ligase β-TrCP, which leads to YAP degradation. Several of the pro-proliferative miRNAs (including miR-199a-3p) also inhibit filamentous actin depolymerization by targeting *Cofilin2*, a process that by itself activates YAP nuclear translocation. Thus, activation of YAP and modulation of the actin cytoskeleton are major components of the pro-proliferative action of miR-199a-3p and other miRNAs that induce cardiomyocyte proliferation [66]. Besides these pro-cell cycle miRNAs, some miRNAs act to inhibit cardiomyocyte proliferation. Notable examples include miR-1, the miR-15 family, and miR-128. For instance, miR-1 represses cyclin D1 expression and, in turn, inhibits cardiomyocyte proliferation in neonatal mouse hearts [67]. Members of the miR-15 family, such as miR-195, exert modulatory effects on neonatal heart regeneration by restraining postnatal cardiomyocyte proliferation. Overexpression of miR-195 in cardiomyocytes disrupts the regenerative response of the 1-day-old mouse heart, resulting in adult-like remodeling after myocardial infarction. Conversely, inhibiting the miR-15 family from an early postnatal age until adulthood increases myocyte proliferation in the adult heart and enhances left ventricular systolic function following adult myocardial infarction [68]. Moreover, miR-128 is upregulated in cardiomyocytes during the postnatal transition from proliferation to terminal differentiation. In neonatal mice, cardiac-specific overexpression of miR-128 impairs cardiomyocyte proliferation and cardiac function. Conversely, miR-128 deletion extends the proliferation of postnatal cardiomyocytes by enhancing the expression of the chromatin modifier SUZ12 (SUZ12 polycomb repressive complex 2 subunit) which suppresses *p27* expression and activates the positive cell cycle regulators cyclin E and CDK2. Additionally, deletion of miR-128 promotes cell cycle re-entry in adult cardiomyocytes, thereby reducing fibrosis levels and attenuating cardiac dysfunction in response to myocardial infarction [69].

## 2.5 *Long Noncoding RNAs and Circular RNAs*

Long noncoding RNAs (LncRNAs) have emerged as important regulators of cardiac development and aging [70]. lncRNAs are nonprotein coding transcripts longer than 200 nucleotides. LncRNAs could act as competing endogenous RNAs (ceRNAs) that sponge miRNAs, which in turn reduce the regulatory effect of miRNAs on their target mRNAs. Recent studies have shown that LncRNAs play important roles in cardiac regeneration. Chen et al. identified a novel upregulated fetal lncRNA and named endogenous cardiac regeneration-associated regulator (ECRAR) which promotes cardiomyocyte proliferation by forming a positive feedback loop with the E2F1-ERK1/2 signaling pathway, initiating cell cycle progression [71]. *Sirt1* antisense long noncoding RNA enhances the stability of SIRT1 [72], and Snhg1 forms a positive feedback loop with c-MYC to sustain PI3K/AKT signaling activation [73] to promote cardiomyocyte proliferation and improve cardiac function post-myocardial infarction. Instead, long noncoding RNA, CPR, repression of *Mcm3* expression [74], CAREL acts as a competing endogenous RNA of miR-296 which targets *Trp53inp1* and *Itm2a* [75]. LncDACH1 directly binds to protein phosphatase 1 catalytic subunit alpha (PP1A) and, in turn, limits its dephosphorylation activity, thus enhancing YAP1 phosphorylation and reducing its nuclear translocation [76]. CRRL regulates cardiomyocyte regeneration by acting as an endogenous miR-199a-

3p sponge that decreases the activity of its target gene (*Hopx*) [77], inhibiting the regenerative capability of neonatal hearts and impairing heart regeneration in postnatal and ischemic hearts.

Circular RNAs (circRNAs) are a new class of noncoding RNAs that adopt a covalently closed loop structure, lack a 3′ terminus, and are resistant to RNA exonucleases [78]. CircNfix hinders cardiomyocyte proliferation by enhancing the interaction between YBX1 (Y-box binding protein 1) and NEDD4L (an E3 ubiquitin ligase), leading to YBX1 degradation through ubiquitination and inhibition of cyclin A2 and cyclin B1 expression. Moreover, it acts as a sponge for miR-214, promoting *Gsk3β* expression and suppressing β-catenin activity [79]. CircHipk3 promotes cardiomyocyte proliferation by increasing Notch1 intracellular domain (N1ICD) acetylation, thereby increasing N1ICD stability and preventing its degradation. In addition, circHipk3 acts as a sponge for miR-133a, promoting connective tissue growth factor (CTGF) expression, thereby activating endothelial cells and promoting coronary vascular endothelial cell proliferation, migration, tube formation ability, and subsequent angiogenesis [80]. Furthermore, hypoxia-induced circRNA-Whsc1 in endothelial cells, CircWhsc1, is packaged into cardiac endothelial cell-derived extracellular vesicles and transported to cardiomyocyte, activating the pTRIM59/pSTAT3/cyclin B2 pathway, thereby enhancing post-myocardial infarction cardiac regeneration [81].

## 2.6 Metabolism

A recent study by Huang's team showed that the inactivation of thyroid hormone signaling reduces cardiomyocyte polyploidization, delays cell cycle exit, and retains cardiac regenerative potential in adult mice. In hearts with inactivated thyroid hormone, there is downregulation in genes related to oxidative phosphorylation, the citrate cycle, cardiac muscle contraction, and numerous mitochondrial genes. Thyroid hormone treatment inhibits zebrafish heart regeneration, which implies that loss of heart regenerative capacity in adult mammals might be triggered by increasing thyroid hormones [82]. Studies by Puente et al. reveal a critical shift in mitochondrial metabolism and the role of reactive oxygen species (ROS) in cardiomyocyte proliferation. Initially, postnatal mouse and zebrafish hearts have low mitochondrial complexity and lack DNA damage response markers, but as they transition to oxidative metabolism, increased mitochondrial ROS production triggers cardiomyocyte cell cycle arrest through DNA damage response activation. Their findings demonstrate that inhibiting ROS scavenging or the DNA damage response pathway can delay this cell cycle arrest, emphasizing the significance of ROS and DNA damage response in regulating cardiomyocyte proliferation [83]. In a subsequent study, they found that gradual exposure to systemic hypoxemia in mice, achieved by gradually reducing inspired oxygen, leads to inhibition of oxidative metabolism and decreased oxidative DNA damage. This promotes cardiomyocyte mitosis and improves myocardial regeneration. Exposure to hypoxemia after myocardial infarction induces a robust regenerative response in mice [84]. In zebrafish, glycolysis is shown to be essential for cell cycle reentry, as inhibiting glycolysis impairs cardiomyocyte proliferation [85]. Furthermore, improved mitochondrial oxidation by enhanced pyruvate conversion to acetyl-CoA resulted in a decreased number of proliferating cardiomyocytes following cardiac injury; in contrast, *Pdk3b* which promotes the conversion of pyruvate to lactate overexpression increased the number of proliferating cardiomyocytes following cardiac injury [86]. Inhibition of succinate dehydrogenase by malonate treatment promotes cardiomyocyte cell proliferation [87]. Adeno-associated virus (AAV9) mediated knock down of Acyl CoA synthase long-chain family member 1 (ACSL1) which is the key enzyme that regulates lipid metabolism enhances the capacity for myocardial regeneration both in mice and primary cardiomyocytes [88]. A muscle-specific isoform of carnitine palmitoyltransferase *Cpt1b* which is required for mitochondrial uptake of fatty acids and subsequent fatty acid oxidation was upregulated during cardiomyocyte maturation in the first week after birth. Inactivation of CPT1B induces hyperplastic and hypertrophic growth of cardiomyocytes. Deletion of *Cpt1b* in adult cardiomyocytes reverses cell cycle arrest. Metabolic studies reveal profound changes in energy metabolism and accumulation of α-ketoglutarate in *Cpt1b*-mutant cardiomyocytes, leading to activation of the α-ketoglutarate-dependent lysine demethylase KDM5. Activated KDM5 demethylates broad H3K4me3 domains in genes that drive cardiomyocyte maturation, lowering their transcription levels and shifting cardiomyocytes into a less mature state, thereby promoting proliferation [89].

## 3 Cells that Contribute to Myocardial Regeneration

To elucidate the source of newly formed cardiomyocytes during heart regeneration, two independent research groups generated transgenic lines expressing tamoxifen-inducible Cre recombinase ($CreER^{T2}$) under the regulatory sequences of the cardiomyocyte-specific promoter *cmlc2* (*myl7*). Upon crossing this line with a Cre-responsive strain, these research teams observed that cardiomyocytes in the regenerate originated from the proliferation of preexisting cardiomyocytes, rather than progenitor cells [90, 91]. Additionally, proliferating cardiomyocytes underwent limited dedifferentiation, characterized by the disassembly of their sarcomeric structure, detachment from one another, and expression of regulators of cell cycle progression [11, 91].

In noninfarcted adult hearts, the largest population of nonmyocyte cells comprises endothelial cells (51.8 ± 4.7%), followed by CD90+ fibroblasts (26.5 ± 4.3%) and leukocytes (19.9 ± 0.7%) [92]. The inevitable consequence of prolonged ischemia is the demise of all cardiomyocytes and endothelial cells in the infarct core. Irreversible cardiomyocyte injury, indicated by sarcolemma disruption and the presence of small amorphous densities in the mitochondria, manifests after 20–40 minutes of sustained severe ischemia [93]. Dying cells trigger

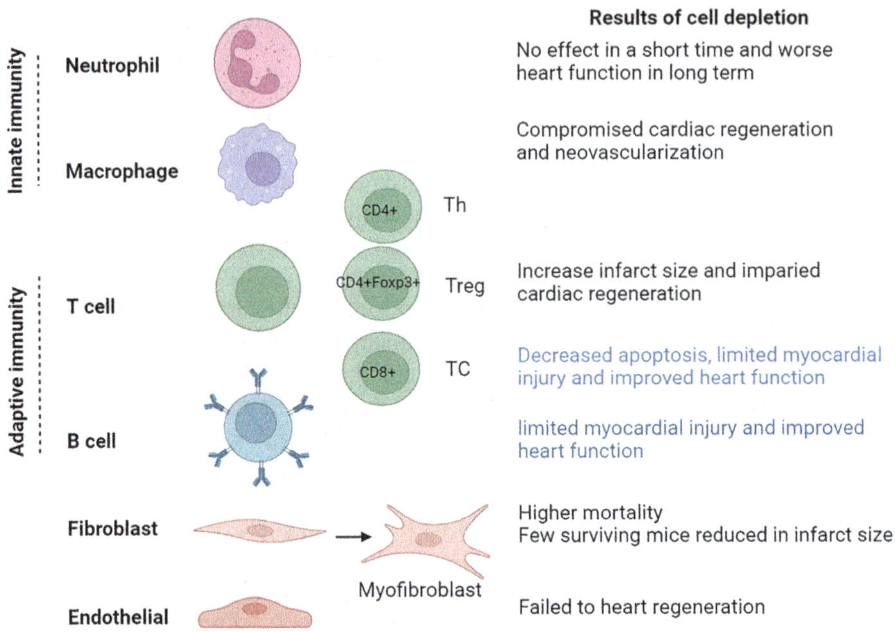


**Fig. 3** The impact of noncardiomyocytes on heart regeneration

an inflammatory reaction, activating reparative pathways that ultimately culminate in the formation of a scar. The process of infarct healing can be delineated into three overlapping phases: the inflammatory phase, the proliferative phase, and the maturation phase [94]. We have summarized the major cell types that are important for heart regeneration (Fig. 3).

## 3.1 *Innate Immunity*

The cellular components of innate immunity are myeloid cells, including monocytes, macrophages, dendritic cells, and natural killer cells, as well as neutrophilic, basophilic, and eosinophilic granulocytes [95].

In response to myocardial infarction, the cardiac tissue undergoes a complex and dynamic immune response involving various immune cell types [92]. Ischemia alone for 90 minutes induces mild morphological changes to the coronary microcirculation, with increased vascular permeability. Ischemia for 30 minutes, followed by 60 minutes of reperfusion, induces massive microvascular injury with endothelial cell death, loss of vascular integrity, edema formation, and intramyocardial hemorrhage [96]. The induction of endothelial adhesion molecules, such as P-selectin and E-selectin, facilitates the binding of these molecules to their leukocyte ligands. This interaction captures neutrophils and monocytes, enabling them to roll along the venular endothelium. Consequently, this process attracts neutrophils, aiding in the removal of necrotic tissue in infarcted zones [97–101]. Single-cell RNA sequencing on neonatal hearts at various time points following myocardial infarction showed an increase in the percentage of immune-related cell types, including macrophages, monocytes, lymphocytes, and granulocytes (from 7.90% to 32.85% at P1 + 1 dpi (days post-injury) and from 4.35% to 11.28% at P8 + 1 dpi) [102]. Neutrophil infiltration occurs within several hours, peaks at 1–3 days, and starts to decline from 5 days after myocardial infarction [103, 104], which has high expression of proinflammatory markers at day 1 and anti-inflammatory markers at days 5 and 7 post-myocardial infarction [105]. In adult mice, macrophages are numerically the predominant immune cells infiltrating the infarcted myocardium and increase in number over the first week post-myocardial infarction. In the early phase, M1 macrophages dominate (1–3 days post-myocardial infarction), while M2 macrophages become predominant after 5 days. The M2 macrophages expressed high levels of reparative genes in addition to proinflammatory genes to the same levels as in M1 macrophages [104]. In zebrafish, macrophage cluster 1 (MC1) cells exhibited the highest level of *il1b* expression and specifically expressed proinflammatory factors *tnfa* and *csf3b* with enriched GO terms related to inflammatory response and neutrophil chemotaxis. Largely due to an expansion of the number of MC1 cells, the overall expression of *tnfa, csf3b,* and other proinflammatory cytokines *ccl35.1, ccl34.4,* and *cxcl11.1* was transiently upregulated at 2 and 7 dpi. In contrast, MC2 cells highly expressed genes involved in antigen presentation, such as *cd74a, cd74b,* and *mhc2dab,* and related to GO terms such as antigen processing and presentation, suggestive of a role in immune surveillance. MC2's relative frequency within the MC population decreased initially (2d) but subsequently increased over time, and MC2 became the dominant MC subpopulation at 14 dpi [106]. The predominant subset of dendritic cells was myeloid DC, which peaked in number on day 7. Th1 and regulatory T-cells were the predominant subsets of $CD4^+$ T-cells, whereas Th2 and Th17 cells were minor populations. $CD8^+$ T-cells, $\gamma\delta$T cells, B cells, natural killer cells, and NKT cells peaked on day 7 post-myocardial infarction [104].

### 3.1.1 Neutrophils

Neutrophil accumulation occurs rapidly, progressing between 2 and 30 minutes after reperfusion in response to ischemia-reperfusion (I/R) injury [107]. Monoclonal antibodies against L-selectin and P-selectin reduce myocardial necrosis, preserve coronary endothelial function, and attenuate neutrophil accumulation in the ischemic and reperfused feline myocardium [97, 108]. P-selectin deficient mice showed decreased infarct size after 30 min of coronary occlusion and 2 h of reperfusion [109]. In contrast, no difference in infarct size was noted after a 60-minute ischemic period [110]. Horckmans et al. found the same result that after prolonged ischemia, there is no difference in infarct size between control and neutrophil-depleted mice 24 h postinfarction [111]. However, neutrophil depletion leads to worsening of cardiac function 7–14 days after infarction. It manifests as an increase in left ventricular size at end-systole and a significant decrease in left ventricular ejection fraction and cardiac output [111]. The neutrophil secretome, neutrophil gelatinase-associated lipocalin, promotes macrophage polarization towards M2c; M2c polarization is required for apoptotic cell clearance [111]. This mechanism could elucidate why neutrophil depletion has no immediate impact after ischemia but proves detrimental in the long-term.

#### 3.1.2 Macrophages

Macrophages play a crucial role in the regenerative capacity of the heart in organisms such as zebrafish and salamanders. In macrophage-depleted salamanders, the ability to regenerate their hearts is impaired, leading to a disruption of the regeneration program, increased fibroblast activation, altered ECM profile, poor cardiac function, and decreased survival [112]. However, macrophage depletion in these organisms doesn't directly affect cardiomyocyte proliferation. Instead, it severely impacts vessel density [112]. In zebrafish, macrophage pre-depletion at 1 day prior to cardiac injury delayed macrophage recruitment and compromised heart regeneration even though the overall macrophage numbers were restored within the first week [113]. Delayed macrophage recruitment results in prolonged expression of genes related to inflammation and disrupted energy metabolism during cardiac repair [113]. Resident macrophage-deficient zebrafish exhibited defects in revascularization, cardiomyocyte survival, debris clearance, and ECM remodeling/scar resolution without functional compensation from the circulating/monocyte-derived macrophages [114]. Medaka, another teleost-like zebrafish, does not have the regeneration ability of the heart as zebrafish. Compared with zebrafish, there is a delayed and reduced macrophage recruitment in medaka, along with delayed neutrophil clearance. Suggesting that delayed macrophage recruitment in zebrafish impairs neovascularization, neutrophil clearance, cardiomyocyte proliferation, and scar resolution. In contrast, stimulating Toll-like receptor signaling in medaka enhanced immune cell dynamics and promoted neovascularization, neutrophil clearance, cardiomyocyte proliferation, and scar resolution [113]. Compared with the P8 heart which cannot regenerate after myocardial infarction, the P1 heart showed higher percentages of macrophages and monocytes at 1 day post-myocardial infarction, whereas P8 hearts showed higher percentages of macrophages and monocytes at 3 days post-myocardial infarction, reflecting different dynamics of the neonatal immune response to cardiac injury [102].

### 3.2 *Adaptive Immunity*

The central cellular components of adaptive immunity are T- and B-cells that arise from lymphoid progenitor cells in the bone marrow [115]. Both B- and T-cell levels peak around day 7 after myocardial infarction [104]. Mature T-cells are divided into three major types: $CD4^+$ T-helper (Th), $CD8^+$ T-cytotoxic (Tc), and $CD4^+$ $Foxp3^+$ T-regulatory (Treg) cells [107].

### 3.2.1 T Lymphocytes

$CD3^+CD8^+$ and $CD3^+CD4^+$ T lymphocytes accumulated in the injured myocardium as early as day 1 and peaked at day 3 after myocardial infarction. Infiltration of $CD4^+$ T cells was twofold higher than that of $CD8^+$ T-cells [116]. Depletion of $CD8^+$ T lymphocytes decreases apoptosis within the ischemic myocardium, hampers inflammatory response, limits myocardial injury, and improves heart function in the permanent coronary artery ligation model. These effects are recapitulated in mice with *granzyme B*-deficient $CD8^+$ T-cells. The protective effect of CD8 depletion on heart function is confirmed by using a model of I/R in pigs [116]. However, in another I/R model, the infarct size did not differ between mice without $CD8^+$ T lymphocytes and control mice [107].

Following myocardial infarction, the absolute cell numbers of Treg cells increase, peaking at around day 7 after myocardial infarction in adult mice [117]. These Treg cells are primarily derived from the circulating compartment [118]. Ablation of Treg cells has been shown to lead to increased infarct size, cardiac deterioration, and increased numbers of both inflammatory myeloid cells and lymphocytes, which are associated with M1-like macrophage polarization [117, 119]. Treg cells modulate the expression of M2-associated genes in monocytic cells by producing TGFβ1, IL-13, and IL-10 [117]. Increased number of Treg cells in the infarcted heart inducing improved cardiac function and attenuated interstitial fibrosis and cardiac apoptosis in the rats after myocardial infarction [120]. When Treg cells were ablated, it impaired the endogenous regenerative capability of P3 heart after cryoinfarction [121]. In contrast, adoptive transfer of $FOXP3^+$ Treg promotes neonatal heart regeneration of NOD/SCID mice which are deficient in T-cells, but harbor macrophages that fail to regenerate their injured myocardium as early as P3 [121]. Treg directly promotes the proliferation of both mouse and human cardiomyocytes in a paracrine manner [119, 121]; and their secreted factors such as C-C motif chemokine ligand 24 (CCL24), growth arrest-specific gene-6 (GAS6), or AREG potentiate neonatal cardiomyocyte proliferation [121]; the overexpression of six factors secreted by Tregs (*Cst7*, tumor necrosis factor ligand superfamily member 11 [*Tnfsf11*], *Il33*, fibrinogen-like protein 2 [*Fgl2*], *Matn2*, and insulin-like growth factor 2[*Igf2*]) reproduces the therapeutic effect of Treg injection [119]. Besides, Tregs mediate the protective effects of *Ccl17* deletion on myocardial [122].

In addition, γδ T lymphocytes, a type of unconventional T cells, are a major source of IL-17A which is increased after myocardial I/R injury. Anti-IL-17A mAb treatment or IL-17A knockout markedly ameliorated I/R injury, as demonstrated by reduced infarct size, reduced cardiac troponin T levels, and improved cardiac function. This improvement is associated with a reduction in cardiomyocyte apoptosis and neutrophil infiltration through regulating ICAM-1 and E-selectin expression and stimulates endothelial activation and neutrophil-endothelial cell adhesion. On the contrary, repletion of exogenous IL-17A induced the opposite effect [123]. Compared with the regeneratable P1 mice heart, the P8 mice heart has more $CD4^{+}FOXP3^{-}$ conventional T-cells per mg tissue within the first week after injury [124]. $CD4^{+}$ T-cell–deficient mouse strains show accentuated cardiac inflammation, impaired wound healing, aggravated left ventricular remodeling, and impaired survival [125]. However, ablation of $CD4^{+}$ T-cell reactivates neonatal heart regeneration after cryoinjury, and this effect was not recapitulated in the adult heart [124]. Further studies are needed to differentiate between Tregs and CD4+ cells to determine the function of $CD4^{+}FOXP3^{-}$ conventional T-cells in cardiac regeneration. Overall, Treg cells might rather play a protective role in myocardial infarction.

#### 3.2.2 B Lymphocytes

Studies using animal models of myocardial infarction have shown that B lymphocytes infiltrate the infarcted myocardium as early as 1-day post-myocardial infarction and persist until the healing phase at day 7 [104]. Depletion of mature B lymphocytes using genetic (Baff receptor deficiency) or antibody-mediated methods (CD20- or Baff-specific antibody) has been found to limit myocardial injury and improve heart function. These effects were also observed in mice with B-cell selective *Ccl7* deficiency [126]. Additionally, the adoptive transfer of regulatory B-cells (Bregs) has been shown to ameliorate ventricular remodeling in a murine myocardial infarction model. Mechanistically, Bregs reduce the expression of C-C motif chemokine receptor 2 (CCR2) in monocytes, which inhibits proinflammatory monocyte recruitment to the heart from peripheral blood and their mobilization from the bone marrow [127].

### *3.3 Fibroblasts*

Repair of the infarcted myocardium is dependent on the mobilization and activation of reparative fibroblasts and myofibroblasts, which are responsible for depositing ECM proteins. These proteins play a crucial role in preventing the catastrophic rupture of the ventricle. Activated fibroblasts in the heart following a myocardial infarction can originate from two distinct sources: the epicardium and the endocardium [128]. The most reliable markers for fibroblasts in the myocardium are the transcription factor TCF21 [129] and the growth factor receptor platelet-derived growth factor receptor α (PDGFRα) [130]. Fibroblast transformation to myofibroblast expresses fibrogenic (e.g., Periostin; POSTN) and/or contractile (e.g., alpha smooth muscle actin (αSMA)) proteins [131].

In a zebrafish cryoinjury model, the damaged area was about 18.5% at 4 days post-injury and remained at 15.9% at 7 days. However, by 14 days post-injury, it was reduced by around 50%. This reduction continued at 21 days, 30 days, and eventually, by 60 days post-injury, the damage was nearly absent [132]. Initially, there is thrombosis, fibroblast accumulation, and collagen deposition, but later, cardiac cells re-enter the cell cycle, invade the infarct area, and regenerate new myocardium, resulting in scarless recovery [132]. Hu et al. used single-cell transcriptomics and spatiotemporal analysis and found there are three clusters of fibroblasts in zebrafish, characterized by expression of *col11a1a, col12a1a,* and *nppc*, respectively. They were present at the peak of regeneration (3 and 7 days post-injury) but absent before and after. Depleting *col12a1a*-expressing cells impaired heart regeneration, highlighting their pro-regeneration role [128]. Strikingly, genetically limiting the fibrotic response by ablating *col1a2*-expressing cells (specific to cardiac fibroblasts) hindered cardiomyocyte proliferation [133].

To achieve full cardiac regeneration, early fibrosis must be resolved to make room for new functional heart tissue. In species with limited regenerative capacity, like adult mammals, persistent early fibrosis leads to long-term cardiac deterioration. Utilizing mouse GFP reporter lines, Pinto et al. revealed that quiescent TCF21 and PDGFRα expressing fibroblasts constitute approximately 15% of the non-myocyte cardiac cell population in the uninjured myocardium, while activated fibroblasts and myofibroblasts are rare or nonexistent [134]. Following injury, fibroblasts transition into a highly proliferative state, reaching their peak activity between 2 and 4 days post-injury [135, 136]. Subsequently, these activated fibroblasts undergo differentiation into myofibroblasts from myocardial infarction day 3 to myocardial infarction day 7 [135, 136], which generate ECM proteins to provide structural support to the injured region. As the recovery progresses, typically by 7–10 days post-injury, myofibroblasts cease their proliferation, and there is a decrease in the expression of αSMA as the scar matures [135]. Mice subjected to myocardial infarction injury with ablation of Periostin$^+$ myofibroblast cells or lacking MBNL1, an RNA-binding protein essential for fibroblast-to-myofibroblast transformation, experienced higher mortality rates due to ventricular wall rupture, which is consistent with lethality in *Postn*$^{-/-}$ mice subject to the same injury due to a defect in the formation of a protective scar, or in a subset of developing embryos when collagen-dependent structural regions are formed in the cardiovascular system in and around the heart, suggest that periostin-expressing myofibroblasts are required to generate a protective scar after myocardial infarction injury [131, 137]. The few surviving mice showed reduced collagen levels in the infarct area, emphasizing the role of myofibroblasts in cardiac regeneration [131, 137] and indicating the importance of these myofibroblast-related processes in cardiac regeneration. Nevertheless, in vitro studies demonstrated that co-culturing immature cardiomyocytes from neonatal hearts with fibroblasts from adult hearts or human stem cell-derived cardiomyocytes with adult human cardiac fibroblasts impaired their proliferation [138]. Periostin induces the proliferation of differentiated cardiomyocytes and promotes cardiac repair [139]. In the strain of neonatal mice subjected to myocardial infarction, periostin deficiency resulted in post-myocardial infarction cardiac fibrosis and dysfunction caused by attenuated myocardial regeneration and angiogenesis effects mediated by the PI3K/GSK3β/cyclin D1 signaling [140]. However, these results were not readily reproducible by other groups [141]. These findings underscore the crucial role of fibroblasts and myofibroblasts in cardiac regeneration.

## *3.4 Endothelial Cells*

In the zebrafish cryoinjury model, angiogenic sprouting into the injured area starts as early as 15 h after injury [142]. Regenerating coronary endothelial cells upregulate *vegfc* upon cardiac injury in adult zebrafish, VEGFc signaling promotes coronary endothelial cell proliferation, and VEGFc also stimulates the emilin2a/cxcl8a expression in epicardium-derived cells and coronary endothelial to promote coronary revascularization during cardiac regeneration in zebrafish [143]. The new coronary plexus is guided by the endocardium at least in part via VEGFa signaling. Cardiomyocytes use the regenerated vascular scaffold to repopulate the injured area [144]. Fish expressing inducible dominant negative *vegfa* displayed minimal revascularization of the damaged area. In the absence of fast angiogenic revascularization, cardiomyocyte proliferation did not occur, and the heart failed to regenerate, retaining a fibrotic scar. Suggested a fast endothelial invasion allows efficient revascularization of the injured area, which is necessary to support the replenishment of new tissue and achieve efficient heart regeneration [142]. Regenerating coronary endothelial cells upregulate *vegfc,* and blocking VEGFc signaling leads to reduced cardiomyocyte dedifferentiation and proliferation after cardiac injury in zebrafish [143]. Gene cellular communication network factor 2a (*ccn2a*), previously known as *ctgfa*, is also induced in endocardial cells in the injured tissue and regulates cardiomyocyte proliferation and repopulation of the damaged tissue. In *ccn2a* mutants, cardiomyocyte proliferation and repopulation are disrupted despite apparently unaffected revascularization; instead, *ccn2a* overexpression enhances cardiomyocyte proliferation and improves the resolution of transient collagen deposition [145].

In mice, after myocardial infarction, new microvessels form in the infarct border zone, the subendocardial space, and the epicardium. Between 2 and 4 days after coronary artery ligation, the capillary network in the border zone starts expanding, with extensive branching and vessel sprouting into the infarct core. After 7 days, most capillary endothelial cells in the border zone cease to proliferate, and some newly formed capillaries enlarge [146]. Utilizing various lineage-tracing mouse models for endothelial cells, it has been found that the majority of coronary vessels in the injured mouse hearts are derived from preexisting coronary vessel endothelial cells [147, 148]. After injury, cardiomyocytes stimulate angiogenesis by upregulating the expression of the transcription factor, Zinc finger E-box-binding homeobox 2 (ZEB2). ZEB2, in turn, induces the release of thymosin β4 (TMSB4) and prothymosin α (PTMA), paracrine factors, that enhance endothelial cell migration and stimulate angiogenesis [149]. In vitro studies have reported that preformed endothelial networks improve the survival and spatial organization of cardiomyocytes in a three-dimensional configuration [150].

## 4 Characterization of Regenerative Myocardium

Cardiac regeneration primarily involves the proliferation of existing cardiomyocytes. While cardiomyocytes display a proliferative capacity in the fetus, this potential rapidly diminishes after birth, coinciding with characteristic changes in cardiomyocyte maturation. Key biological processes in cardiomyocyte maturation encompass myofibril maturation, maturation of electrophysiology and $Ca2^{+}$ handling, metabolic maturation, proliferation-to-hypertrophy transition, and cardiomyocyte integration into mature tissue (Fig. 4) [151]. Transmission electron microscopy reveals distinct features in mature sarcomeres, such as clearer banding

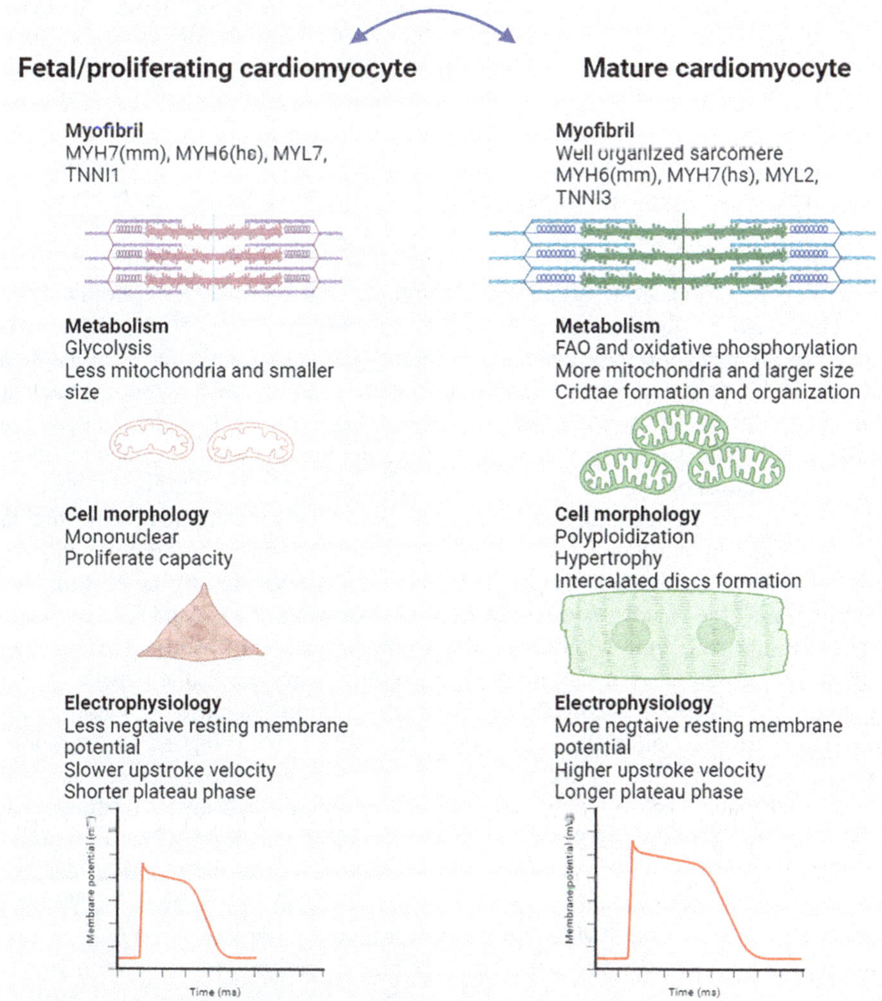


**Fig. 4** Major differences between fetal and mature cardiomyocytes

indicating improved alignment of sarcomere filaments. Z-lines widen and align more effectively, and the distance between Z-lines (sarcomere length) increases to approximately 2.2 μm in diastole in mature cardiomyocytes. Myofibril maturation involves sarcomeric isoform switching, where several sarcomere components transition from a fetal to an adult isoform due to transcriptional changes or alternative splicing. In rodents, among the most well-known is the myosin heavy chain (MYH) switch from fetal MYH7 to adult MYH6 [152, 153]. The gene *MYL7*, often known as *MLC-2a*, primarily governs the expression of the regulatory light chain of myosin in all early fetal cardiomyocytes. However, as ventricular cardiomyocytes mature, there is a shift to the expression of *MYL2*, also known as *MLC-2v*, and *MYL7* expression becomes confined to atrial cardiomyocytes [154]. Fetal cardiomyocytes primarily express slow skeletal troponin I (*TNNI1*), and this is replaced by cardiac troponin I (*TNNI3*) in mature cardiomyocytes [155]. Immature cardiomyocytes differ in important ways from mature cardiomyocytes in electrophysiology. First, the resting membrane potential of immature cardiomyocytes is less negative [156]. Second, the upstroke velocity of immature cardiomyocytes (~15–30 V/s) is slower. Third, the plateau phase of the action potential is longer in mature cardiomyocytes [151, 157]. Whereas mature ventricular cardiomyocytes exhibit low automaticity, immature cardiomyocytes and pluripotent stem cell-derived cardiomyocytes (PSC-CMs) spontaneously beat a phenotype that likely contributes to arrhythmia when PSC-CMs are transplanted in myocardial infarction models [158]. The metabolic transition from immature cardiomyocytes to mature cardiomyocytes is orchestrated by the activation of metabolic transcriptional regulators, including peroxisome proliferator-activated receptor gamma, coactivator 1 alpha/beta (*Ppargc1a/b*), peroxisome proliferator-activated receptor alpha (*Ppara*), nuclear respiratory factor 1/2(*Nrf1/2*), and estrogen-related receptor alpha/beta/gamma (*Esrra/b/g*) [159]. These regulators orchestrate the upregulation of genes involved in fatty acid metabolism, oxidative phosphorylation, and mitochondrial biogenesis, while concurrently downregulating glycolytic genes. This metabolic transformation is fundamental in facilitating the shift from fetal to adult energy requirements [14]. In humans, the cardiomyocyte proliferation rate declines rapidly postnatally but does not reach the steady-state rate of <1% per year until the second decade of life [160]. Despite cell cycle withdrawal, the postnatal heart increases in size through a proportional increase in cardiomyocyte volume, a process called maturational hypertrophy, with no new myocytes contributing to the increase in overall heart size [161]. Another hallmark of cardiomyocyte maturation during the proliferation-to-hypertrophy transition is polyploidization. In murine cardiomyocytes, the final round of the cell cycle involves karyokinesis without cytokinesis, leading most mature cardiomyocytes (~90%) to contain two diploid nuclei (binucleation) [162–164]. By contrast, in adult humans, ~75% of cardiomyocytes are mononuclear, but the majority of these nuclei are polyploid due to DNA endoreplication without karyokinesis [165, 166]. Maturational integration of cardiomyocytes into cardiac tissues requires the formation of specialized cardiomyocyte-cardiomyocyte junctions called intercalated discs (ICDs), which occur 2–3 weeks after birth in mice. ICDs are hybrid junctions comprising three major types of cell adhesions including desmosomes, adherens junctions, and gap junctions [167]. Immature cardiomyocytes lack ICDs, and ICD components are either not expressed, localized to the interior of cells, or throughout the cell surface. As cardiomyocytes mature, these molecules redistribute to cell termini to form ICDs [151, 168].

Proliferating cardiomyocytes and immature cardiomyocytes share the same characteristics. Cardiomyocyte proliferation is characterized by sarcomere disassembly, particularly the marginalization of sarcomere structures to the cell periphery during mitosis [14, 91]. The zebrafish heart regenerates very efficiently and will regrow cardiac muscle without scarring [5]. Lineage tracing experiments revealed a model in which the proliferation of preexisting cardiomyocytes replaces the myocardium that was lost during the injury [11, 91]. After the injury, cardiomyocytes near the injury area (so-called border zone) start to reenter the cell cycle [11, 91, 169]. Wu et al. [60] utilized the tomo-seq method and observed that 3 days after cryoinjury, cardiomyocytes near the injury area exhibited enrichment of genes highly expressed in the embryonic myocardium (e.g., *nppb*, *vmhc* [the *MYH6* ortholog], *actc1*, and *desma*). Using a single-cell RNA-sequencing approach, Honkoop et al. [85] found cardiomyocytes in the border zone disassemble sarcomeric structures and re-express markers of embryonic cardiomyocytes suggesting their dedifferentiation. Moreover, these cells have reduced expression of mitochondrial genes and reduced mitochondrial activity, while glycolysis gene expression and glucose uptake are increased, suggesting metabolic reprogramming. Furthermore, blocking glycolysis impaired the ability of the heart muscle cells to divide, indicating that this switch is necessary for the heart to regenerate. In the uninjured heart, cardiomyocytes showed ordered arrangements of sarcomeres and mitochondria with clearly defined z-lines under transmission electron microscopy. At 5 dpa (days after amputation), many of the cardiomyocytes display a disorganized sarcomeric structure along with the appearance of intercellular spaces and lost z lines [91].

Using optical mapping, Kikuchi et al. [11] found that at 7 dpa near the apex of the ventricle, conduction was markedly slower compared to the uninjured control. By 14 dpa, conduction improved. Additionally, electrical activity consistently failed to propagate into the regenerating apex at 7 dpa, while impulses were again conducted throughout the ventricle by 14 dpa when a substantial number of new *gata4*$^+$ cardiomyocytes appeared. By 30 dpa, conduction at the apex appeared normal. Furthermore, there was a significant reduction in the maximum depolarization rate at 7 dpa, which was fully restored by 14 dpa. This indicates that electrical coupling of new apical cardiomyocytes begins to occur around 2 weeks post-injury, with full coupling in the restored wall by 30 dpa.

## 5 Cell Therapy for the Treatment of Heart Failure

In addition to manipulating genes that affect the cardiomyocyte cell cycle to promote endogenous cardiomyocyte proliferation, both cell-based and non-cell-based therapies have emerged as new methods for heart remuscularization (Fig. 5).

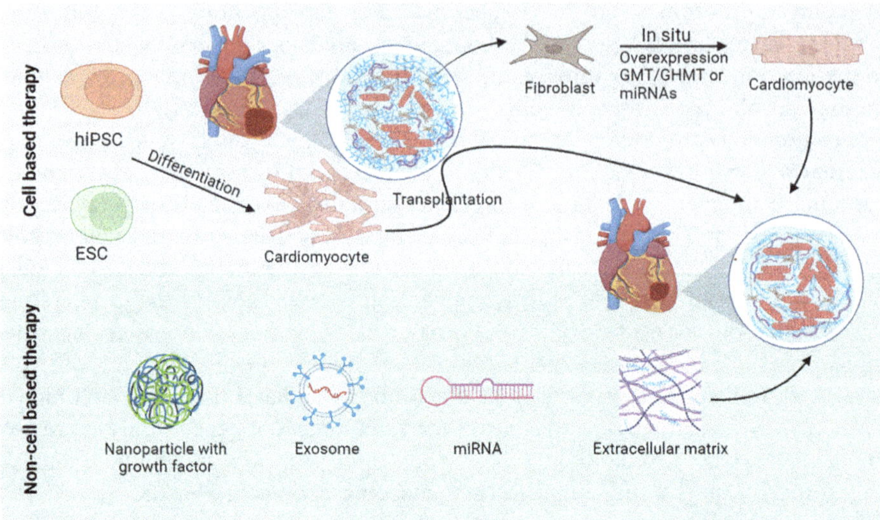


**Fig. 5** Cardiac cell therapy. Human-induced pluripotent stem cells (hiPSCs)—or embryonic stem cells (ESCs)—derived cardiomyocytes are transplanted into the injured area to regenerate injured myocardium. Overexpression of transcription factors (GMT: *Gata4, Mef2c, Tbx5*; or GHMT: *Gata4, Hand2, Mef2c, Tbx5*) or specific miRNAs is employed to reprogram fibroblasts into cardiomyocytes. Additionally, exosomes derived from stem cells and miRNAs, growth factors, and extracellular matrix proteins have been used to promote cardiomyocyte proliferation and heart regeneration

The directed differentiation of mouse and human embryonic stem cells (ESCs) and induced pluripotent stem cells into cardiomyocytes has become a highly promising approach for heart regeneration, enabling the production and expansion of cardiomyocytes in vitro [170].

## 5.1 *Cell-Based Therapy for Heart Repair*

Several studies have demonstrated that mESC-derived cardiomyocytes [171, 172] and more recently hiPSC-derived cardiomyocytes(hiPSC-CMs) [173, 174] can successfully engraft into the hearts of mice or rats posttransplantation. Although current technology allows for efficient differentiation of human PSCs into cardiomyocytes, these PSC-CMs exhibit immature phenotypes that resemble fetal cardiomyocytes [175]. Large animal studies revealed a risk of potentially life-threatening ventricular arrhythmias following injection of PSC-CMs into nonhuman primate or pig models [158, 176, 177].

However, a significant challenge in cell transplantation is the limited retention and survival of transplanted cells, often constituting only a small percentage, regardless of the cell type or study setting, whether in animal models or clinical scenarios. By using *CCND2* (cyclin D2) overexpressing cardiomyocytes (hiPSC-CCND2$^{OE}$CMs), the grafted size is tripled compared with transplanted with wildtype cardiomyocytes (hiPSC-CCND2$^{WT}$CMs) at 1 month after treatment [27]. Six months later, engrafted cells occupied >50% of the scarred region in CCND2$^{OE}$ animals and exceeded the number of engrafted cells in CCND2$^{WT}$ animals by ~eightfold. Optical mapping of action potential propagation and optogenetics studies indicated that the engrafted hiPSC-CCND2$^{OE}$CMs were electrically coupled to each other and the cells of the native myocardium [28]. hiPSC-CCND2$^{OE}$CMs also enhance the efficacy for myocardial repair in a swine model of myocardial infarction [29, 178]. Small molecules were used to improve hiPSC-CMs adhesion [179], proliferation [179], and survival [180] to improve the transplantation efficacy.

Reprogramming of cardiac fibroblasts to cardiomyocytes with defined transcription factors, GMT (*Gata4, Mef2c, and Tbx5*) or GHMT (*Gata4, Hand2, Mef2c, and Tbx5*), and microRNAs (miRNA-1, -133, -208, and -499) [181] represents an attractive alternative for repair of the adult heart. Remarkably, inhibition of pro-fibrotic signaling using small molecules that target the TGFβ or Rho-associated kinase pathways converts embryonic fibroblasts into functional cardiomyocyte-like cells, with an efficiency up to 60% [182]. Viral delivery of combinations of cardiogenic transcription factors in mice post-myocardial infarction has shown significant generation of new cardiomyocytes with improved cardiac systolic function [182, 183].

## 5.2 *Exosomes and Other Cell-Derived Products for Heart Repair*

The stem cell-based approach did provide modest cardioprotective benefits, paradoxically, the transplanted stem cells rarely differentiated into cardiomyocytes [8]. The researchers theorize that the paracrine effects of growth factors, miRNAs, and stem cell-secreted exosomes are responsible for, and may be critical for, cardiac repair and regeneration. The therapeutic effect of exosomes derived from different stem cells has been extensively studied in animal models. In general, exosomes isolated from mouse ESCs [184], BM-MSCs [185–187], iPSCs [188, 189], and endothelial progenitor cells (EPC) [190] have proven their ability to improve heart regeneration in terms of ejection fraction, fractional shortening, reduction of infarcted size, increased angiogenesis, and decrease in oxidative stress.

The ECM is the noncellular component of all the tissues. Recently, ECM has been shown to play an important role in cardiac regeneration and repair after cardiac injury [191]. Embryonic cardiac fibroblasts have been reported to promote cardiomyocyte proliferation via paracrine signaling, embryonic, but not adult, cardiac fibroblasts secrete high levels of fibronectin, collagen, and heparin-binding EGF-like growth factor (HBEGF), which collaboratively interact and regulate mitotic activity in cardiomyocytes through β1 integrin signaling, while adult cardiac fibroblasts induce cardiomyocyte hypertrophy [192]. Periostin has been shown to stimulate cardiomyocyte proliferation and cardiac regeneration after myocardial infarction in neonates and adults [139, 140]. Agrin, another component of neonatal ECM, is required for the full regenerative capacity of neonatal mouse hearts [193]. In vivo, a single administration of Agrin promotes cardiac regeneration through improved angiogenesis, inflammatory suppression, and cell cycle in adult mice [193] and pig [194]. In Wu et al. study, they found that SLIT2 and nephronectin (NPNT), these two ECM proteins secreted by neonatal cardiac fibroblasts promote mitotic rounding and cytokinesis but not cell cycle entry of postnatal cardiomyocytes [195]. Single administration of ECM from zebrafish (zECM), particularly the healing variety, enables cardiac functional recovery and regeneration of adult mouse heart tissues after acute myocardial infarction. Inhibition of ErbB2 activity abrogates the beneficial effects of zECM administration, indicating the possible involvement of ErbB2 signaling in zECM-mediated regeneration [196]. Inject ECM microparticles including collagen I, fibronectin, laminin, and fibrillin promote heart regeneration in mice with post-ischemic heart injury [197]. Intramyocardial injection of platelet gel promotes endogenous repair and augments cardiac function in rats with myocardial infarction [198].

## 6 Future Directions

Pigs are popular large mammals for cardiac preclinical testing due to their similarities with humans in terms of organ size and physiology. Studying heart regeneration in pig models could offer insight into human postnatal cardiac regenerative potential and facilitate the development of effective therapeutics for human cardiac regenerative repair. The neonatal pig heart is capable of regeneration during the first 2 days of life. However, this regenerative ability diminishes as cardiomyocytes exit the cell cycle shortly after birth [7], and precisely evaluating the generation of new cardiomyocytes in pigs is challenging due to the lack of genetic reporters and lineage-tracing tools, unlike smaller model organisms such as mice and zebrafish. Techniques such as EdU or BrdU incorporation and staining for markers such as Ki67 and phosphorylated histone H3 are commonly used to detect cardiomyocyte cell cycling, but these methods primarily detect DNA synthesis and don't provide a clear distinction between multinucleation, polyploidization of existing nuclei, and de novo generation of cardiomyocytes [199]. Addressing these issues and gaining a better understanding of the role of noncardiomyocytes in cardiac regeneration in large animals will be essential.

It is also worth noting that gene therapy methods aimed at stimulating cardiomyocyte proliferation or transdifferentiation into cardiomyocytes can result in an immature fetal-like phenotype [151, 175]. Large animal studies revealed a risk of potentially life-threatening ventricular arrhythmias following injection of PSC-CMs into nonhuman primate or pig models [158, 176, 177]. Future research will need to focus on finding this balance and optimizing methods to promote both proliferation and maturation simultaneously for successful cardiac regeneration.

## References

1. Savarese, G. *et al.* Global burden of heart failure: a comprehensive and updated review of epidemiology. *Cardiovasc Res* **118**, 3272-3287 (2023). https://doi.org/10.1093/cvr/cvac013
2. Laflamme, M. A. & Murry, C. E. Heart regeneration. *Nature* **473**, 326-335 (2011). https://doi.org/10.1038/nature10147
3. Becker, R. O., Chapin, S. & Sherry, R. Regeneration of the ventricular myocardium in amphibians. *Nature* **248**, 145-147 (1974). https://doi.org/10.1038/248145a0
4. Oberpriller, J. O. & Oberpriller, J. C. Response of the adult newt ventricle to injury. *J Exp Zool* **187**, 249-253 (1974). https://doi.org/10.1002/jez.1401870208
5. Poss, K. D., Wilson, L. G. & Keating, M. T. Heart regeneration in zebrafish. *Science* **298**, 2188-2190 (2002). https://doi.org/10.1126/science.1077857
6. Haubner, B. J. *et al.* Functional Recovery of a Human Neonatal Heart After Severe Myocardial Infarction. *Circ Res* **118**, 216-221 (2016). https://doi.org/10.1161/CIRCRESAHA.115.307017
7. Zhu, W. *et al.* Regenerative Potential of Neonatal Porcine Hearts. *Circulation* **138**, 2809-2816 (2018). https://doi.org/10.1161/CIRCULATIONAHA.118.034886
8. Hashimoto, H., Olson, E. N. & Bassel-Duby, R. Therapeutic approaches for cardiac regeneration and repair. *Nat Rev Cardiol* **15**, 585-600 (2018). https://doi.org/10.1038/s41569-018-0036-6
9. Bader, D. & Oberpriller, J. O. Repair and reorganization of minced cardiac muscle in the adult newt (Notophthalmus viridescens). *J Morphol* **155**, 349-357 (1978). https://doi.org/10.1002/jmor.1051550307
10. Witman, Nevin et al. Recapitulation of developmental cardiogenesis governs the morphological and functional regeneration of adult newt hearts following injury. Developmental biology vol. 354,1 (2011): 67–76. https://doi:10.1016/j.ydbio.2011.03.021
11. Kikuchi, K. *et al.* Primary contribution to zebrafish heart regeneration by gata4(+) cardiomyocytes. *Nature* **464**, 601-605 (2010). https://doi.org/10.1038/nature08804
12. Soonpaa, M. H. & Field, L. J. Survey of studies examining mammalian cardiomyocyte DNA synthesis. *Circ Res* **83**, 15-26 (1998). https://doi.org/10.1161/01.res.83.1.15
13. Soonpaa, M. H. & Field, L. J. Assessment of cardiomyocyte DNA synthesis in normal and injured adult mouse hearts. *Am J Physiol* **272**, H220-226 (1997). https://doi.org/10.1152/ajpheart.1997.272.1.H220
14. Porrello, E. R. *et al.* Transient regenerative potential of the neonatal mouse heart. *Science* **331**, 1078-1080 (2011). https://doi.org/10.1126/science.1200708
15. Zogbi, C. *et al.* Early postnatal rat ventricle resection leads to long-term preserved cardiac function despite tissue hypoperfusion. *Physiol Rep* **2** (2014). https://doi.org/10.14814/phy2.12115
16. Wang, H. *et al.* A neonatal leporine model of age-dependent natural heart regeneration after myocardial infarction. *J Thorac Cardiovasc Surg* **164**, e389-e405 (2022). https://doi.org/10.1016/j.jtcvs.2021.08.013
17. Ye, L. *et al.* Early Regenerative Capacity in the Porcine Heart. *Circulation* **138**, 2798-2808 (2018). https://doi.org/10.1161/CIRCULATIONAHA.117.031542
18. Hodges, M. M., Zgheib, C. & Liechty, K. W. A Large Mammalian Model of Myocardial Regeneration After Myocardial Infarction in Fetal Sheep. *Adv Wound Care (New Rochelle)* **10**, 174-190 (2021). https://doi.org/10.1089/wound.2018.0894
19. Herdrich, B. J. *et al.* Regenerative healing following foetal myocardial infarction. *Eur J Cardiothorac Surg* **38**, 691-698 (2010). https://doi.org/10.1016/j.ejcts.2010.03.049
20. Cesna, S., Eicken, A., Juenger, H. & Hess, J. Successful treatment of a newborn with acute myocardial infarction on the first day of life. *Pediatr Cardiol* **34**, 1868-1870 (2013). https://doi.org/10.1007/s00246-012-0417-2
21. Pasumarthi, K. B. & Field, L. J. Cardiomyocyte cell cycle regulation. *Circ Res* **90**, 1044-1054 (2002). https://doi.org/10.1161/01.res.0000020201.44772.67

22. Engel, F. B. *et al.* p38 MAP kinase inhibition enables proliferation of adult mammalian cardiomyocytes. *Genes Dev* **19**, 1175-1187 (2005). https://doi.org/10.1101/gad.1306705
23. Chaudhry, H. W. *et al.* Cyclin A2 mediates cardiomyocyte mitosis in the post-mitotic myocardium. *J Biol Chem* **279**, 35858-35866 (2004). https://doi.org/10.1074/jbc.M404975200
24. Soonpaa, M. H. *et al.* Cyclin D1 overexpression promotes cardiomyocyte DNA synthesis and multinucleation in transgenic mice. *J Clin Invest* **99**, 2644-2654 (1997). https://doi.org/10.1172/JCI119453
25. Tane, S. *et al.* Repression of cyclin D1 expression is necessary for the maintenance of cell cycle exit in adult mammalian cardiomyocytes. *J Biol Chem* **289**, 18033-18044 (2014). https:/ /doi.org/10.1074/jbc.M113.541953
26. Pasumarthi, K. B., Nakajima, H., Nakajima, H. O., Soonpaa, M. H. & Field, L. J. Targeted expression of cyclin D2 results in cardiomyocyte DNA synthesis and infarct regression in transgenic mice. *Circ Res* **96**, 110-118 (2005). https://doi.org/10.1161/01.RES.0000152326.91223.4F
27. Zhu, W., Zhao, M., Mattapally, S., Chen, S. & Zhang, J. CCND2 Overexpression Enhances the Regenerative Potency of Human Induced Pluripotent Stem Cell-Derived Cardiomyocytes: Remuscularization of Injured Ventricle. *Circ Res* **122**, 88-96 (2018). https://doi.org/10.1161/CIRCRESAHA.117.311504
28. Fan, C. *et al.* Cardiomyocytes from CCND2-overexpressing human induced-pluripotent stem cells repopulate the myocardial scar in mice: A 6-month study. *J Mol Cell Cardiol* **137**, 25-33 (2019). https://doi.org/10.1016/j.yjmcc.2019.09.011
29. Zhao, M. *et al.* Cyclin D2 Overexpression Enhances the Efficacy of Human Induced Pluripotent Stem Cell-Derived Cardiomyocytes for Myocardial Repair in a Swine Model of Myocardial Infarction. *Circulation* **144**, 210-228 (2021). https://doi.org/10.1161/CIRCULATIONAHA.120.049497
30. Mohamed, T. M. A. *et al.* Regulation of Cell Cycle to Stimulate Adult Cardiomyocyte Proliferation and Cardiac Regeneration. *Cell* **173**, 104-116 e112 (2018). https://doi.org/10.1016/j.cell.2018.02.014
31. Liu, Z., Yue, S., Chen, X., Kubin, T. & Braun, T. Regulation of cardiomyocyte polyploidy and multinucleation by CyclinG1. *Circ Res* **106**, 1498-1506 (2010). https://doi.org/10.1161/CIRCRESAHA.109.211888
32. Di Stefano, V., Giacca, M., Capogrossi, M. C., Crescenzi, M. & Martelli, F. Knockdown of cyclin-dependent kinase inhibitors induces cardiomyocyte re-entry in the cell cycle. *J Biol Chem* **286**, 8644-8654 (2011). https://doi.org/10.1074/jbc.M110.184549
33. Shoffner, A., Cigliola, V., Lee, N., Ou, J. & Poss, K. D. Tp53 Suppression Promotes Cardiomyocyte Proliferation during Zebrafish Heart Regeneration. *Cell Rep* **32**, 108089 (2020). https://doi.org/10.1016/j.celrep.2020.108089
34. Stanley-Hasnain, S. *et al.* p53 and Mdm2 act synergistically to maintain cardiac homeostasis and mediate cardiomyocyte cell cycle arrest through a network of microRNAs. *Cell Cycle* **16**, 1585-1600 (2017). https://doi.org/10.1080/15384101.2017.1346758
35. Rojas, A. *et al.* GATA4 is a direct transcriptional activator of cyclin D2 and Cdk4 and is required for cardiomyocyte proliferation in anterior heart field-derived myocardium. *Mol Cell Biol* **28**, 5420-5431 (2008). https://doi.org/10.1128/MCB.00717-08
36. Karra, R., Knecht, A. K., Kikuchi, K. & Poss, K. D. Myocardial NF-kappaB activation is essential for zebrafish heart regeneration. *Proc Natl Acad Sci U S A* **112**, 13255-13260 (2015). https://doi.org/10.1073/pnas.1511209112
37. Chakraborty, S., Sengupta, A. & Yutzey, K. E. Tbx20 promotes cardiomyocyte proliferation and persistence of fetal characteristics in adult mouse hearts. *J Mol Cell Cardiol* **62**, 203-213 (2013). https://doi.org/10.1016/j.yjmcc.2013.05.018
38. Collesi, C., Zentilin, L., Sinagra, G. & Giacca, M. Notch1 signaling stimulates proliferation of immature cardiomyocytes. *J Cell Biol* **183**, 117-128 (2008). https://doi.org/10.1083/jcb.200806091

39. Heallen, T. *et al.* Hippo pathway inhibits Wnt signaling to restrain cardiomyocyte proliferation and heart size. *Science* **332**, 458-461 (2011). https://doi.org/10.1126/science.1199010
40. Mahmoud, A. I. *et al.* Meis1 regulates postnatal cardiomyocyte cell cycle arrest. *Nature* **497**, 249-253 (2013). https://doi.org/10.1038/nature12054
41. Nguyen, N. U. N. *et al.* A calcineurin-Hoxb13 axis regulates growth mode of mammalian cardiomyocytes. *Nature* **582**, 271-276 (2020). https://doi.org/10.1038/s41586-020-2228-6
42. Alam, P. *et al.* Inhibition of Senescence Associated Genes Rb1 and Meis2 in Adult Cardiomyocytes Results in Cell Cycle Reentry and Cardiac Repair Post-Myocardial Infarction. *J Am Heart Assoc* **8**, e012089 (2019). https://doi.org/10.1161/JAHA.119.012089
43. Rao, A. *et al.* The translation initiation factor homolog eif4e1c regulates cardiomyocyte metabolism and proliferation during heart regeneration. *Development* **150** (2023). https://doi.org/10.1242/dev.201376
44. van Amerongen, M. J., Diehl, F., Novoyatleva, T., Patra, C. & Engel, F. B. E2F4 is required for cardiomyocyte proliferation. *Cardiovasc Res* **86**, 92-102 (2010). https://doi.org/10.1093/cvr/cvp383
45. Zuppo, D. A. *et al.* Foxm1 regulates cardiomyocyte proliferation in adult zebrafish after cardiac injury. *Development* **150** (2023). https://doi.org/10.1242/dev.201163
46. Takahashi, K. & Yamanaka, S. Induction of pluripotent stem cells from mouse embryonic and adult fibroblast cultures by defined factors. *Cell* **126**, 663-676 (2006). https://doi.org/10.1016/j.cell.2006.07.024
47. Chen, Y. *et al.* Reversible reprogramming of cardiomyocytes to a fetal state drives heart regeneration in mice. *Science* **373**, 1537-1540 (2021). https://doi.org/10.1126/science.abg5159
48. Wu, Y. *et al.* LRP6 downregulation promotes cardiomyocyte proliferation and heart regeneration. *Cell Res* **31**, 450-462 (2021). https://doi.org/10.1038/s41422-020-00411-7
49. Heallen, T. *et al.* Hippo signaling impedes adult heart regeneration. *Development* **140**, 4683-4690 (2013). https://doi.org/10.1242/dev.102798
50. von Gise, A. *et al.* YAP1, the nuclear target of Hippo signaling, stimulates heart growth through cardiomyocyte proliferation but not hypertrophy. *Proc Natl Acad Sci U S A* **109**, 2394-2399 (2012). https://doi.org/10.1073/pnas.1116136109
51. Xin, M. *et al.* Hippo pathway effector Yap promotes cardiac regeneration. *Proc Natl Acad Sci U S A* **110**, 13839-13844 (2013). https://doi.org/10.1073/pnas.1313192110
52. Bersell, K., Arab, S., Haring, B. & Kuhn, B. Neuregulin1/ErbB4 signaling induces cardiomyocyte proliferation and repair of heart injury. *Cell* **138**, 257-270 (2009). https://doi.org/10.1016/j.cell.2009.04.060
53. Gemberling, M., Karra, R., Dickson, A. L. & Poss, K. D. Nrg1 is an injury-induced cardiomyocyte mitogen for the endogenous heart regeneration program in zebrafish. *Elife* **4** (2015). https://doi.org/10.7554/eLife.05871
54. D'Uva, G. *et al.* ERBB2 triggers mammalian heart regeneration by promoting cardiomyocyte dedifferentiation and proliferation. *Nat Cell Biol* **17**, 627-638 (2015). https://doi.org/10.1038/ncb3149
55. Aharonov, A. *et al.* ERBB2 drives YAP activation and EMT-like processes during cardiac regeneration. *Nat Cell Biol* **22**, 1346-1356 (2020). https://doi.org/10.1038/s41556-020-00588-4
56. Dogra, D. *et al.* Opposite effects of Activin type 2 receptor ligands on cardiomyocyte proliferation during development and repair. *Nat Commun* **8**, 1902 (2017). https://doi.org/10.1038/s41467-017-01950-1
57. Reuter, S., Soonpaa, M. H., Firulli, A. B., Chang, A. N. & Field, L. J. Recombinant neuregulin 1 does not activate cardiomyocyte DNA synthesis in normal or infarcted adult mice. *PLoS One* **9**, e115871 (2014). https://doi.org/10.1371/journal.pone.0115871
58. Wang, Y. *et al.* Mydgf promotes Cardiomyocyte proliferation and Neonatal Heart regeneration. *Theranostics* **10**, 9100-9112 (2020). https://doi.org/10.7150/thno.44281
59. Hubert, F. *et al.* FGF10 promotes cardiac repair through a dual cellular mechanism increasing cardiomyocyte renewal and inhibiting fibrosis. *Cardiovasc Res* **118**, 2625-2637 (2022). https://doi.org/10.1093/cvr/cvab340

60. Wu, C. C. *et al.* Spatially Resolved Genome-wide Transcriptional Profiling Identifies BMP Signaling as Essential Regulator of Zebrafish Cardiomyocyte Regeneration. *Dev Cell* **36**, 36-49 (2016). https://doi.org/10.1016/j.devcel.2015.12.010
61. Eulalio, A. *et al.* Functional screening identifies miRNAs inducing cardiac regeneration. *Nature* **492**, 376-381 (2012). https://doi.org/10.1038/nature11739
62. Chen, J. *et al.* mir-17-92 cluster is required for and sufficient to induce cardiomyocyte proliferation in postnatal and adult hearts. *Circ Res* **112**, 1557-1566 (2013). https://doi.org/10.1161/CIRCRESAHA.112.300658
63. Tian, Y. *et al.* A microRNA-Hippo pathway that promotes cardiomyocyte proliferation and cardiac regeneration in mice. *Sci Transl Med* **7**, 279ra238 (2015). https://doi.org/10.1126/scitranslmed.3010841
64. Gao, F. *et al.* Therapeutic role of miR-19a/19b in cardiac regeneration and protection from myocardial infarction. *Nat Commun* **10**, 1802 (2019). https://doi.org/10.1038/s41467-019-09530-1
65. Lesizza, P. *et al.* Single-Dose Intracardiac Injection of Pro-Regenerative MicroRNAs Improves Cardiac Function After Myocardial Infarction. *Circ Res* **120**, 1298-1304 (2017). https://doi.org/10.1161/CIRCRESAHA.116.309589
66. Torrini, C. *et al.* Common Regulatory Pathways Mediate Activity of MicroRNAs Inducing Cardiomyocyte Proliferation. *Cell Rep* **27**, 2759-2771 e2755 (2019). https://doi.org/10.1016/j.celrep.2019.05.005
67. Gan, J. *et al.* microRNA-1 inhibits cardiomyocyte proliferation in mouse neonatal hearts by repressing CCND1 expression. *Ann Transl Med* **7**, 455 (2019). https://doi.org/10.21037/atm.2019.08.68
68. Porrello, E. R. *et al.* Regulation of neonatal and adult mammalian heart regeneration by the miR-15 family. *Proc Natl Acad Sci U S A* **110**, 187-192 (2013). https://doi.org/10.1073/pnas.1208863110
69. Huang, W. *et al.* Loss of microRNA-128 promotes cardiomyocyte proliferation and heart regeneration. *Nat Commun* **9**, 700 (2018). https://doi.org/10.1038/s41467-018-03019-z
70. Devaux, Y. *et al.* Long noncoding RNAs in cardiac development and ageing. *Nat Rev Cardiol* **12**, 415-425 (2015). https://doi.org/10.1038/nrcardio.2015.55
71. Chen, Y. *et al.* Long Non-coding RNA ECRAR Triggers Post-natal Myocardial Regeneration by Activating ERK1/2 Signaling. *Mol Ther* **27**, 29-45 (2019). https://doi.org/10.1016/j.ymthe.2018.10.021
72. Li, B. *et al.* Sirt1 Antisense Long Noncoding RNA Promotes Cardiomyocyte Proliferation by Enhancing the Stability of Sirt1. *J Am Heart Assoc* **7**, e009700 (2018). https://doi.org/10.1161/JAHA.118.009700
73. Li, M. *et al.* LncRNA Snhg1-driven self-reinforcing regulatory network promoted cardiac regeneration and repair after myocardial infarction. *Theranostics* **11**, 9397-9414 (2021). https://doi.org/10.7150/thno.57037
74. Ponnusamy, M. *et al.* Long Noncoding RNA CPR (Cardiomyocyte Proliferation Regulator) Regulates Cardiomyocyte Proliferation and Cardiac Repair. *Circulation* **139**, 2668-2684 (2019). https://doi.org/10.1161/CIRCULATIONAHA.118.035832
75. Cai, B. *et al.* The Long Noncoding RNA CAREL Controls Cardiac Regeneration. *J Am Coll Cardiol* **72**, 534-550 (2018). https://doi.org/10.1016/j.jacc.2018.04.085
76. Cai, B. *et al.* Targeting LncDACH1 promotes cardiac repair and regeneration after myocardium infarction. *Cell Death Differ* **27**, 2158-2175 (2020). https://doi.org/10.1038/s41418-020-0492-5
77. Chen, G. *et al.* Loss of long non-coding RNA CRRL promotes cardiomyocyte regeneration and improves cardiac repair by functioning as a competing endogenous RNA. *J Mol Cell Cardiol* **122**, 152-164 (2018). https://doi.org/10.1016/j.yjmcc.2018.08.013
78. Tang, L., Li, P., Jang, M. & Zhu, W. Circular RNAs and Cardiovascular Regeneration. *Front Cardiovasc Med* **8**, 672600 (2021). https://doi.org/10.3389/fcvm.2021.672600
79. Huang, S. *et al.* Loss of Super-Enhancer-Regulated circRNA Nfix Induces Cardiac Regeneration After Myocardial Infarction in Adult Mice. *Circulation* **139**, 2857-2876 (2019). https://doi.org/10.1161/CIRCULATIONAHA.118.038361

80. Si, X. *et al.* circRNA Hipk3 Induces Cardiac Regeneration after Myocardial Infarction in Mice by Binding to Notch1 and miR-133a. *Mol Ther Nucleic Acids* **21**, 636-655 (2020). https://doi.org/10.1016/j.omtn.2020.06.024
81. Wei, G. *et al.* Extracellular vesicle-derived CircWhsc1 promotes cardiomyocyte proliferation and heart repair by activating TRIM59/STAT3/Cyclin B2 pathway. *J Adv Res* (2022). https://doi.org/10.1016/j.jare.2022.12.014
82. Hirose, K. *et al.* Evidence for hormonal control of heart regenerative capacity during endothermy acquisition. *Science* **364**, 184-188 (2019). https://doi.org/10.1126/science.aar2038
83. Puente, B. N. *et al.* The oxygen-rich postnatal environment induces cardiomyocyte cell-cycle arrest through DNA damage response. *Cell* **157**, 565-579 (2014). https://doi.org/10.1016/j.cell.2014.03.032
84. Nakada, Y. *et al.* Hypoxia induces heart regeneration in adult mice. *Nature* **541**, 222-227 (2017). https://doi.org/10.1038/nature20173
85. Honkoop, H. *et al.* Single-cell analysis uncovers that metabolic reprogramming by ErbB2 signaling is essential for cardiomyocyte proliferation in the regenerating heart. *Elife* **8** (2019). https://doi.org/10.7554/eLife.50163
86. Fukuda, R. *et al.* Stimulation of glycolysis promotes cardiomyocyte proliferation after injury in adult zebrafish. *EMBO Rep* **21**, e49752 (2020). https://doi.org/10.15252/embr.201949752
87. Bae, J. *et al.* Malonate Promotes Adult Cardiomyocyte Proliferation and Heart Regeneration. *Circulation* **143**, 1973-1986 (2021). https://doi.org/10.1161/CIRCULATIONAHA.120.049952
88. Li, Y. *et al.* Targeting ACSL1 promotes cardiomyocyte proliferation and cardiac regeneration. *Life Sci* **294**, 120371 (2022). https://doi.org/10.1016/j.lfs.2022.120371
89. Li, X. *et al.* Inhibition of fatty acid oxidation enables heart regeneration in adult mice. *Nature* **622**, 619-626 (2023). https://doi.org/10.1038/s41586-023-06585-5
90. Lepilina, A. *et al.* A dynamic epicardial injury response supports progenitor cell activity during zebrafish heart regeneration. *Cell* **127**, 607-619 (2006). https://doi.org/10.1016/j.cell.2006.08.052
91. Jopling, C. *et al.* Zebrafish heart regeneration occurs by cardiomyocyte dedifferentiation and proliferation. *Nature* **464**, 606-609 (2010). https://doi.org/10.1038/nature08899
92. Quaife-Ryan, G. A. *et al.* Multicellular Transcriptional Analysis of Mammalian Heart Regeneration. *Circulation* **136**, 1123-1139 (2017). https://doi.org/10.1161/CIRCULATIONAHA.117.028252
93. Frangogiannis, N. G. The immune system and cardiac repair. *Pharmacol Res* **58**, 88-111 (2008). https://doi.org/10.1016/j.phrs.2008.06.007
94. Frangogiannis, N. G. The mechanistic basis of infarct healing. *Antioxid Redox Signal* **8**, 1907-1939 (2006). https://doi.org/10.1089/ars.2006.8.1907
95. Chaplin, D. D. Overview of the immune response. *J Allergy Clin Immunol* **125**, S3-23 (2010). https://doi.org/10.1016/j.jaci.2009.12.980
96. Hollander, M. R. *et al.* Correction: Dissecting the Effects of Ischemia and Reperfusion on the Coronary Microcirculation in a Rat Model of Acute Myocardial Infarction. *PLoS One* **11**, e0166809 (2016). https://doi.org/10.1371/journal.pone.0166809
97. Weyrich, A. S., Ma, X. Y., Lefer, D. J., Albertine, K. H. & Lefer, A. M. In vivo neutralization of P-selectin protects feline heart and endothelium in myocardial ischemia and reperfusion injury. *J Clin Invest* **91**, 2620-2629 (1993). https://doi.org/10.1172/JCI116501
98. Burns, A. R. *et al.* P-selectin mediates neutrophil adhesion to endothelial cell borders. *J Leukoc Biol* **65**, 299-306 (1999). https://doi.org/10.1002/jlb.65.3.299
99. Ley, K., Allietta, M., Bullard, D. C. & Morgan, S. Importance of E-selectin for firm leukocyte adhesion in vivo. *Circ Res* **83**, 287-294 (1998). https://doi.org/10.1161/01.res.83.3.287
100. Granger, D. N., Benoit, J. N., Suzuki, M. & Grisham, M. B. Leukocyte adherence to venular endothelium during ischemia-reperfusion. *Am J Physiol* **257**, G683-688 (1989). https://doi.org/10.1152/ajpgi.1989.257.5.G683

101. Osborn, L. *et al.* Direct expression cloning of vascular cell adhesion molecule 1, a cytokine-induced endothelial protein that binds to lymphocytes. *Cell* **59**, 1203-1211 (1989). https://doi.org/10.1016/0092-8674(89)90775-7
102. Wang, Z. *et al.* Cell-Type-Specific Gene Regulatory Networks Underlying Murine Neonatal Heart Regeneration at Single-Cell Resolution. *Cell Rep* **35**, 109211 (2021). https://doi.org/10.1016/j.celrep.2021.109211
103. Christia, P. *et al.* Systematic characterization of myocardial inflammation, repair, and remodeling in a mouse model of reperfused myocardial infarction. *J Histochem Cytochem* **61**, 555-570 (2013). https://doi.org/10.1369/0022155413493912
104. Yan, X. *et al.* Temporal dynamics of cardiac immune cell accumulation following acute myocardial infarction. *J Mol Cell Cardiol* **62**, 24-35 (2013). https://doi.org/10.1016/j.yjmcc.2013.04.023
105. Ma, Y. *et al.* Temporal neutrophil polarization following myocardial infarction. *Cardiovasc Res* **110**, 51-61 (2016). https://doi.org/10.1093/cvr/cvw024
106. Ma, H. *et al.* Functional coordination of non-myocytes plays a key role in adult zebrafish heart regeneration. *EMBO Rep* **22**, e52901 (2021). https://doi.org/10.15252/embr.202152901
107. Yang, Z. *et al.* Myocardial infarct-sparing effect of adenosine A2A receptor activation is due to its action on CD4+ T lymphocytes. *Circulation* **114**, 2056-2064 (2006). https://doi.org/10.1161/CIRCULATIONAHA.106.649244
108. Ma, X. L. *et al.* Monoclonal antibody to L-selectin attenuates neutrophil accumulation and protects ischemic reperfused cat myocardium. *Circulation* **88**, 649-658 (1993). https://doi.org/10.1161/01.cir.88.2.649
109. Palazzo, A. J., Jones, S. P., Anderson, D. C., Granger, D. N. & Lefer, D. J. Coronary endothelial P-selectin in pathogenesis of myocardial ischemia-reperfusion injury. *Am J Physiol* **275**, H1865-1872 (1998). https://doi.org/10.1152/ajpheart.1998.275.5.H1865
110. Briaud, S. A. *et al.* Leukocyte trafficking and myocardial reperfusion injury in ICAM-1/P-selectin-knockout mice. *Am J Physiol Heart Circ Physiol* **280**, H60-67 (2001). https://doi.org/10.1152/ajpheart.2001.280.1.H60
111. Horckmans, M. *et al.* Neutrophils orchestrate post-myocardial infarction healing by polarizing macrophages towards a reparative phenotype. *Eur Heart J* **38**, 187-197 (2017). https://doi.org/10.1093/eurheartj/ehw002
112. Godwin, J. W., Debuque, R., Salimova, E. & Rosenthal, N. A. Heart regeneration in the salamander relies on macrophage-mediated control of fibroblast activation and the extracellular landscape. *NPJ Regen Med* **2** (2017). https://doi.org/10.1038/s41536-017-0027-y
113. Lai, S. L. *et al.* Reciprocal analyses in zebrafish and medaka reveal that harnessing the immune response promotes cardiac regeneration. *Elife* **6** (2017). https://doi.org/10.7554/eLife.25605
114. Wei, K. H. *et al.* Comparative single-cell profiling reveals distinct cardiac resident macrophages essential for zebrafish heart regeneration. *Elife* **12** (2023). https://doi.org/10.7554/eLife.84679
115. Hofmann, U. & Frantz, S. Role of lymphocytes in myocardial injury, healing, and remodeling after myocardial infarction. *Circ Res* **116**, 354-367 (2015). https://doi.org/10.1161/CIRCRESAHA.116.304072
116. Santos-Zas, I. *et al.* Cytotoxic CD8(+) T cells promote granzyme B-dependent adverse post-ischemic cardiac remodeling. *Nat Commun* **12**, 1483 (2021). https://doi.org/10.1038/s41467-021-21737-9
117. Weirather, J. *et al.* Foxp3+ CD4+ T cells improve healing after myocardial infarction by modulating monocyte/macrophage differentiation. *Circ Res* **115**, 55-67 (2014). https://doi.org/10.1161/CIRCRESAHA.115.303895
118. Xia, N. *et al.* A Unique Population of Regulatory T Cells in Heart Potentiates Cardiac Protection From Myocardial Infarction. *Circulation* **142**, 1956-1973 (2020). https://doi.org/10.1161/CIRCULATIONAHA.120.046789

119. Zacchigna, S. *et al.* Paracrine effect of regulatory T cells promotes cardiomyocyte proliferation during pregnancy and after myocardial infarction. *Nat Commun***9**, 2432 (2018). https://doi.org/10.1038/s41467-018-04908-z
120. Tang, T. T. *et al.* Regulatory T cells ameliorate cardiac remodeling after myocardial infarction. *Basic Res Cardiol***107**, 232 (2012). https://doi.org/10.1007/s00395-011-0232-6
121. Li, J. *et al.* Regulatory T-cells regulate neonatal heart regeneration by potentiating cardiomyocyte proliferation in a paracrine manner. *Theranostics***9**, 4324-4341 (2019). https://doi.org/10.7150/thno.32734
122. Feng, G. *et al.* CCL17 Aggravates Myocardial Injury by Suppressing Recruitment of Regulatory T Cells. *Circulation***145**, 765-782 (2022). https://doi.org/10.1161/CIRCULATIONAHA.121.055888
123. Liao, Y. H. *et al.* Interleukin-17A contributes to myocardial ischemia/reperfusion injury by regulating cardiomyocyte apoptosis and neutrophil infiltration. *J Am Coll Cardiol***59**, 420-429 (2012). https://doi.org/10.1016/j.jacc.2011.10.863
124. Li, J. *et al.* Specific ablation of CD4(+) T-cells promotes heart regeneration in juvenile mice. *Theranostics***10**, 8018-8035 (2020). https://doi.org/10.7150/thno.42943
125. Hofmann, U. *et al.* Activation of CD4+ T lymphocytes improves wound healing and survival after experimental myocardial infarction in mice. *Circulation***125**, 1652-1663 (2012). https://doi.org/10.1161/CIRCULATIONAHA.111.044164
126. Zouggari, Y. *et al.* B lymphocytes trigger monocyte mobilization and impair heart function after acute myocardial infarction. *Nat Med***19**, 1273-1280 (2013). https://doi.org/10.1038/nm.3284
127. Jiao, J. *et al.* Regulatory B cells improve ventricular remodeling after myocardial infarction by modulating monocyte migration. *Basic Res Cardiol***116**, 46 (2021). https://doi.org/10.1007/s00395-021-00886-4
128. Hu, B. *et al.* Origin and function of activated fibroblast states during zebrafish heart regeneration. *Nat Genet***54**, 1227-1237 (2022). https://doi.org/10.1038/s41588-022-01129-5
129. Acharya, A. *et al.* The bHLH transcription factor Tcf21 is required for lineage-specific EMT of cardiac fibroblast progenitors. *Development***139**, 2139-2149 (2012). https://doi.org/10.1242/dev.079970
130. Chen, L., Acciani, T., Le Cras, T., Lutzko, C. & Perl, A. K. Dynamic regulation of platelet-derived growth factor receptor alpha expression in alveolar fibroblasts during realveolarization. *Am J Respir Cell Mol Biol***47**, 517-527 (2012). https://doi.org/10.1165/rcmb.2012-0030OC
131. Kanisicak, O. *et al.* Genetic lineage tracing defines myofibroblast origin and function in the injured heart. *Nat Commun***7**, 12260 (2016). https://doi.org/10.1038/ncomms12260
132. Chablais, F., Veit, J., Rainer, G. & Jazwinska, A. The zebrafish heart regenerates after cryoinjury-induced myocardial infarction. *BMC Dev Biol***11**, 21 (2011). https://doi.org/10.1186/1471-213X-11-21
133. Sanchez-Iranzo, H. *et al.* Transient fibrosis resolves via fibroblast inactivation in the regenerating zebrafish heart. *Proc Natl Acad Sci U S A***115**, 4188-4193 (2018). https://doi.org/10.1073/pnas.1716713115
134. Pinto, A. R. *et al.* Revisiting Cardiac Cellular Composition. *Circ Res***118**, 400-409 (2016). https://doi.org/10.1161/CIRCRESAHA.115.307778
135. Fu, X. *et al.* Specialized fibroblast differentiated states underlie scar formation in the infarcted mouse heart. *J Clin Invest***128**, 2127-2143 (2018). https://doi.org/10.1172/JCI98215
136. Farbehi, N. *et al.* Single-cell expression profiling reveals dynamic flux of cardiac stromal, vascular and immune cells in health and injury. *Elife***8** (2019). https://doi.org/10.7554/eLife.43882
137. Davis, J. *et al.* MBNL1-mediated regulation of differentiation RNAs promotes myofibroblast transformation and the fibrotic response. *Nat Commun***6**, 10084 (2015). https://doi.org/10.1038/ncomms10084
138. Wang, Y. *et al.* Single-cell analysis of murine fibroblasts identifies neonatal to adult switching that regulates cardiomyocyte maturation. *Nat Commun***11**, 2585 (2020). https://doi.org/10.1038/s41467-020-16204-w

139. Kuhn, B. *et al.* Periostin induces proliferation of differentiated cardiomyocytes and promotes cardiac repair. *Nat Med***13**, 962-969 (2007). https://doi.org/10.1038/nm1619
140. Chen, Z. *et al.* Ablation of periostin inhibits post-infarction myocardial regeneration in neonatal mice mediated by the phosphatidylinositol 3 kinase/glycogen synthase kinase 3beta/cyclin D1 signalling pathway. *Cardiovasc Res***113**, 620-632 (2017). https://doi.org/10.1093/cvr/cvx001
141. Lorts, A., Schwanekamp, J. A., Elrod, J. W., Sargent, M. A. & Molkentin, J. D. Genetic manipulation of periostin expression in the heart does not affect myocyte content, cell cycle activity, or cardiac repair. *Circ Res***104**, e1-7 (2009). https://doi.org/10.1161/CIRCRESAHA.108.188649
142. Marin-Juez, R. *et al.* Fast revascularization of the injured area is essential to support zebrafish heart regeneration. *Proc Natl Acad Sci U S A***113**, 11237-11242 (2016). https://doi.org/10.1073/pnas.1605431113
143. El-Sammak, H. *et al.* A Vegfc-Emilin2a-Cxcl8a Signaling Axis Required for Zebrafish Cardiac Regeneration. *Circ Res***130**, 1014-1029 (2022). https://doi.org/10.1161/CIRCRESAHA.121.319929
144. Marin-Juez, R. *et al.* Coronary Revascularization During Heart Regeneration Is Regulated by Epicardial and Endocardial Cues and Forms a Scaffold for Cardiomyocyte Repopulation. *Dev Cell***51**, 503-515 e504 (2019). https://doi.org/10.1016/j.devcel.2019.10.019
145. Mukherjee, D. *et al.* Ccn2a is an injury-induced matricellular factor that promotes cardiac regeneration in zebrafish. *Development***148** (2021). https://doi.org/10.1242/dev.193219
146. Dube, K. N. *et al.* Recapitulation of developmental mechanisms to revascularize the ischemic heart. *JCI Insight***2** (2017). https://doi.org/10.1172/jci.insight.96800
147. He, L. *et al.* Preexisting endothelial cells mediate cardiac neovascularization after injury. *J Clin Invest***127**, 2968-2981 (2017). https://doi.org/10.1172/JCI93868
148. Li, Z. *et al.* Single-cell transcriptome analyses reveal novel targets modulating cardiac neovascularization by resident endothelial cells following myocardial infarction. *Eur Heart J***40**, 2507-2520 (2019). https://doi.org/10.1093/eurheartj/ehz305
149. Gladka, M. M. *et al.* Cardiomyocytes stimulate angiogenesis after ischemic injury in a ZEB2-dependent manner. *Nat Commun***12**, 84 (2021). https://doi.org/10.1038/s41467-020-20361-3
150. Narmoneva, D. A., Vukmirovic, R., Davis, M. E., Kamm, R. D. & Lee, R. T. Endothelial cells promote cardiac myocyte survival and spatial reorganization: implications for cardiac regeneration. *Circulation***110**, 962-968 (2004). https://doi.org/10.1161/01.CIR.0000140667.37070.07
151. Guo, Y. & Pu, W. T. Cardiomyocyte Maturation: New Phase in Development. *Circ Res***126**, 1086-1106 (2020). https://doi.org/10.1161/CIRCRESAHA.119.315862
152. Selewa, A. *et al.* Systematic Comparison of High-throughput Single-Cell and Single-Nucleus Transcriptomes during Cardiomyocyte Differentiation. *Sci Rep***10**, 1535 (2020). https://doi.org/10.1038/s41598-020-58327-6
153. Reiser, P. J., Portman, M. A., Ning, X. H. & Schomisch Moravec, C. Human cardiac myosin heavy chain isoforms in fetal and failing adult atria and ventricles. *Am J Physiol Heart Circ Physiol***280**, H1814-1820 (2001). https://doi.org/10.1152/ajpheart.2001.280.4.H1814
154. Kubalak, S. W., Miller-Hance, W. C., O'Brien, T. X., Dyson, E. & Chien, K. R. Chamber specification of atrial myosin light chain-2 expression precedes septation during murine cardiogenesis. *J Biol Chem***269**, 16961-16970 (1994).
155. Bedada, F. B. *et al.* Acquisition of a quantitative, stoichiometrically conserved ratiometric marker of maturation status in stem cell-derived cardiac myocytes. *Stem Cell Reports***3**, 594-605 (2014). https://doi.org/10.1016/j.stemcr.2014.07.012
156. Horvath, A. *et al.* Low Resting Membrane Potential and Low Inward Rectifier Potassium Currents Are Not Inherent Features of hiPSC-Derived Cardiomyocytes. *Stem Cell Reports***10**, 822-833 (2018). https://doi.org/10.1016/j.stemcr.2018.01.012
157. Karbassi, E. *et al.* Cardiomyocyte maturation: advances in knowledge and implications for regenerative medicine. *Nat Rev Cardiol***17**, 341-359 (2020). https://doi.org/10.1038/s41569-019-0331-x

158. Chong, J. J. *et al.* Human embryonic-stem-cell-derived cardiomyocytes regenerate non-human primate hearts. *Nature* **510**, 273-277 (2014). https://doi.org/10.1038/nature13233
159. Dorn, G. W., 2nd, Vega, R. B. & Kelly, D. P. Mitochondrial biogenesis and dynamics in the developing and diseased heart. *Genes Dev* **29**, 1981-1991 (2015). https://doi.org/10.1101/gad.269894.115
160. Bergmann, O. *et al.* Dynamics of Cell Generation and Turnover in the Human Heart. *Cell* **161**, 1566-1575 (2015). https://doi.org/10.1016/j.cell.2015.05.026
161. Velayutham, N., Agnew, E. J. & Yutzey, K. E. Postnatal Cardiac Development and Regenerative Potential in Large Mammals. *Pediatr Cardiol* **40**, 1345-1358 (2019). https://doi.org/10.1007/s00246-019-02163-7
162. Li, F., Wang, X., Capasso, J. M. & Gerdes, A. M. Rapid transition of cardiac myocytes from hyperplasia to hypertrophy during postnatal development. *J Mol Cell Cardiol* **28**, 1737-1746 (1996). https://doi.org/10.1006/jmcc.1996.0163
163. Fang, J. *et al.* Cellular polyploidy in organ homeostasis and regeneration. *Protein Cell* **14**, 560-578 (2023). https://doi.org/10.1093/procel/pwac064
164. Li, F., Wang, X. & Gerdes, A. M. Formation of binucleated cardiac myocytes in rat heart: II. Cytoskeletal organisation. *J Mol Cell Cardiol* **29**, 1553-1565 (1997). https://doi.org/10.1006/jmcc.1997.0403
165. Brodsky, V., Sarkisov, D. S., Arefyeva, A. M., Panova, N. W. & Gvasava, I. G. Polyploidy in cardiac myocytes of normal and hypertrophic human hearts; range of values. *Virchows Arch* **424**, 429-435 (1994). https://doi.org/10.1007/BF00190566
166. Olivetti, G. *et al.* Aging, cardiac hypertrophy and ischemic cardiomyopathy do not affect the proportion of mononucleated and multinucleated myocytes in the human heart. *J Mol Cell Cardiol* **28**, 1463-1477 (1996). https://doi.org/10.1006/jmcc.1996.0137
167. Vermij, S. H., Abriel, H. & van Veen, T. A. Refining the molecular organization of the cardiac intercalated disc. *Cardiovasc Res* **113**, 259-275 (2017). https://doi.org/10.1093/cvr/cvw259
168. Zhu, R. *et al.* Physical developmental cues for the maturation of human pluripotent stem cell-derived cardiomyocytes. *Stem Cell Res Ther* **5**, 117 (2014). https://doi.org/10.1186/scrt507
169. Itou, J. *et al.* Migration of cardiomyocytes is essential for heart regeneration in zebrafish. *Development* **139**, 4133-4142 (2012). https://doi.org/10.1242/dev.079756
170. Sadek, H. & Olson, E. N. Toward the Goal of Human Heart Regeneration. *Cell Stem Cell* **26**, 7-16 (2020). https://doi.org/10.1016/j.stem.2019.12.004
171. Cai, J. *et al.* Transplantation of embryonic stem cell-derived cardiomyocytes improves cardiac function in infarcted rat hearts. *Cytotherapy* **9**, 283-291 (2007). https://doi.org/10.1080/14653240701247838
172. Moon, S. H. *et al.* The use of aggregates of purified cardiomyocytes derived from human ESCs for functional engraftment after myocardial infarction. *Biomaterials* **34**, 4013-4026 (2013). https://doi.org/10.1016/j.biomaterials.2013.02.022
173. Caspi, O. *et al.* Transplantation of human embryonic stem cell-derived cardiomyocytes improves myocardial performance in infarcted rat hearts. *J Am Coll Cardiol* **50**, 1884-1893 (2007). https://doi.org/10.1016/j.jacc.2007.07.054
174. Laflamme, M. A. *et al.* Cardiomyocytes derived from human embryonic stem cells in pro-survival factors enhance function of infarcted rat hearts. *Nat Biotechnol* **25**, 1015-1024 (2007). https://doi.org/10.1038/nbt1327
175. Ghiroldi, A., Piccoli, M., Ciconte, G., Pappone, C. & Anastasia, L. Regenerating the human heart: direct reprogramming strategies and their current limitations. *Basic Res Cardiol* **112**, 68 (2017). https://doi.org/10.1007/s00395-017-0655-9
176. Liu, Y. W. *et al.* Human embryonic stem cell-derived cardiomyocytes restore function in infarcted hearts of non-human primates. *Nat Biotechnol* **36**, 597-605 (2018). https://doi.org/10.1038/nbt.4162
177. Romagnuolo, R. *et al.* Human Embryonic Stem Cell-Derived Cardiomyocytes Regenerate the Infarcted Pig Heart but Induce Ventricular Tachyarrhythmias. *Stem Cell Reports* **12**, 967-981 (2019). https://doi.org/10.1016/j.stemcr.2019.04.005

178. Joshi, J. *et al.* Optogenetic Control of Engrafted Human Induced Pluripotent Stem Cell-Derived Cardiomyocytes in Live Mice: A Proof-of-Concept Study. *Cells***11** (2022). https://doi.org/10.3390/cells11060951
179. Zhao, M. *et al.* Y-27632 preconditioning enhances transplantation of human-induced pluripotent stem cell-derived cardiomyocytes in myocardial infarction mice. *Cardiovasc Res***115**, 343-356 (2019). https://doi.org/10.1093/cvr/cvy207
180. Fan, C. *et al.* Myocardial protection by nanomaterials formulated with CHIR99021 and FGF1. *JCI Insight***5** (2020). https://doi.org/10.1172/jci.insight.132796
181. Jayawardena, T. M. *et al.* MicroRNA-mediated in vitro and in vivo direct reprogramming of cardiac fibroblasts to cardiomyocytes. *Circ Res***110**, 1465-1473 (2012). https://doi.org/10.1161/CIRCRESAHA.112.269035
182. Zhao, Y. *et al.* High-efficiency reprogramming of fibroblasts into cardiomyocytes requires suppression of pro-fibrotic signalling. *Nat Commun***6**, 8243 (2015). https://doi.org/10.1038/ncomms9243
183. Jayawardena, T. M. *et al.* MicroRNA induced cardiac reprogramming in vivo: evidence for mature cardiac myocytes and improved cardiac function. *Circ Res***116**, 418-424 (2015). https://doi.org/10.1161/CIRCRESAHA.116.304510
184. Khan, M. *et al.* Embryonic stem cell-derived exosomes promote endogenous repair mechanisms and enhance cardiac function following myocardial infarction. *Circ Res***117**, 52-64 (2015). https://doi.org/10.1161/CIRCRESAHA.117.305990
185. Zhang, Z. *et al.* Pretreatment of Cardiac Stem Cells With Exosomes Derived From Mesenchymal Stem Cells Enhances Myocardial Repair. *J Am Heart Assoc***5** (2016). https://doi.org/10.1161/JAHA.115.002856
186. Zhu, L. P. *et al.* Hypoxia-elicited mesenchymal stem cell-derived exosomes facilitates cardiac repair through miR-125b-mediated prevention of cell death in myocardial infarction. *Theranostics***8**, 6163-6177 (2018). https://doi.org/10.7150/thno.28021
187. Zhao, J. *et al.* Mesenchymal stromal cell-derived exosomes attenuate myocardial ischaemia-reperfusion injury through miR-182-regulated macrophage polarization. *Cardiovasc Res***115**, 1205-1216 (2019). https://doi.org/10.1093/cvr/cvz040
188. Dougherty, J. A. *et al.* Extracellular Vesicles Released by Human Induced-Pluripotent Stem Cell-Derived Cardiomyocytes Promote Angiogenesis. *Front Physiol***9**, 1794 (2018). https://doi.org/10.3389/fphys.2018.01794
189. Ong, S. G., Lee, W. H., Zhou, Y. & Wu, J. C. Mining Exosomal MicroRNAs from Human-Induced Pluripotent Stem Cells-Derived Cardiomyocytes for Cardiac Regeneration. *Methods Mol Biol***1733**, 127-136 (2018). https://doi.org/10.1007/978-1-4939-7601-0_10
190. Yue, Y. *et al.* Interleukin-10 Deficiency Alters Endothelial Progenitor Cell-Derived Exosome Reparative Effect on Myocardial Repair via Integrin-Linked Kinase Enrichment. *Circ Res***126**, 315-329 (2020). https://doi.org/10.1161/CIRCRESAHA.119.315829
191. Vu, T. V. A. *et al.* Extracellular Matrix-Based Approaches in Cardiac Regeneration: Challenges and Opportunities. *Int J Mol Sci***23** (2022). https://doi.org/10.3390/ijms232415783
192. Ieda, M. *et al.* Cardiac fibroblasts regulate myocardial proliferation through beta1 integrin signaling. *Dev Cell***16**, 233-244 (2009). https://doi.org/10.1016/j.devcel.2008.12.007
193. Bassat, E. *et al.* The extracellular matrix protein agrin promotes heart regeneration in mice. *Nature***547**, 179-184 (2017). https://doi.org/10.1038/nature22978
194. Baehr, A. *et al.* Agrin Promotes Coordinated Therapeutic Processes Leading to Improved Cardiac Repair in Pigs. *Circulation***142**, 868-881 (2020). https://doi.org/10.1161/CIRCULATIONAHA.119.045116
195. Wu, C. C., Jeratsch, S., Graumann, J. & Stainier, D. Y. R. Modulation of Mammalian Cardiomyocyte Cytokinesis by the Extracellular Matrix. *Circ Res***127**, 896-907 (2020). https://doi.org/10.1161/CIRCRESAHA.119.316303
196. Chen, W. C. *et al.* Decellularized zebrafish cardiac extracellular matrix induces mammalian heart regeneration. *Sci Adv***2**, e1600844 (2016). https://doi.org/10.1126/sciadv.1600844

197. Wang, X. *et al.* Injectable Extracellular Matrix Microparticles Promote Heart Regeneration in Mice with Post-ischemic Heart Injury. *Adv He althc Mater* **11**, e2102265 (2022). https://doi.org/10.1002/adhm.202102265
198. Cheng, K. *et al.* Intramyocardial injection of platelet gel promotes endogenous repair and augments cardiac function in rats with myocardial infarction. *J Am Coll Cardiol* **59**, 256-264 (2012). https://doi.org/10.1016/j.jacc.2011.10.858
199. Marchiano, S. *et al.* Gene editing to prevent ventricular arrhythmias associated with cardiomyocyte cell therapy. *Cell Stem Cell* **30**, 396-414 e399 (2023). https://doi.org/10.1016/j.stem.2023.03.010

# Revolutionizing Pediatric Valve Replacement: The Future of Elastin-Rich Tissue-Engineered Solutions

**Claudia Ponce Aportela, Daniela Alvarado, Keya Pandya, Ariadna Herrera, Asad Mirza, and Sharan Ramaswamy**

## Abbreviations

| | |
|---|---|
| IL-1β | Interleukin-1 beta |
| IGF-1 | Insulin-like growth factor 1 |
| MIP-1α | Macrophage inflammatory protein-1 alpha |
| MCP-1 | Monocyte chemoattractant protein-1 |
| PDGF-BB | Platelet-derived growth factor-BB |
| VEGF | Vascular endothelial growth factor |
| M-CSF | Human macrophages |
| PSIS | Porcine small intestinal submucosa |
| EERV | Engineered elastin-rich valves |
| VICs | Valvular interstitial cells |
| VECs | Valvular endothelial cells |
| ECM | Extracellular matrix |
| TAVR | Transcatheter valve replacements |
| MVP | Mitral valve prolapse |
| VHD | Valvular heart diseases |
| CHDs | Congenital heart defects |

C. Ponce Aportela · D. Alvarado · K. Pandya · A. Herrera · A. Mirza · S. Ramaswamy (✉)
Department of Biomedical Engineering, Florida International University, Miami, FL, USA
e-mail: Cponc021@fiu.edu; dalva214@fiu.edu; kpand017@fiu.edu; aherr154@fiu.edu; amirz013@fiu.edu; sramaswa@fiu.edu



J. Liao, J. Y. Wong (eds.), *Integration and Bridging of Multiscale Bioengineering Designs and Tissue Biomechanics*, https://doi.org/10.1007/978-3-031-81743-4_10

# 1 Introduction

## 1.1 Problem Statement

Congenital heart defects (CHDs) represent the most common type of birth anomaly in the United States, affecting approximately 40,000 newborns annually [1]. Of these cases, around 25% prove fatal before the child reaches their first year, often due to severe cardiac malformations [1, 2]. Among the spectrum of CHDs, valvular heart diseases (VHD) are particularly significant due to their contributions to cardiovascular morbidity [3, 4]. Their primary inefficiency is represented by either valvular stenosis or insufficiency. While valvular stenosis is known as a reduced elasticity of the leaflets, leading to narrowed openings and reduced blood flow, valvular regurgitation on the opposite end is characterized by an incomplete closing resulting in leakage and regurgitant flow [5].

In treating VHD, various conditions may be encountered such as aortic valve stenosis [6], pulmonary valve stenosis [7], and tricuspid valve stenosis/atresia (malformed or absent valve), which all affect blood flow between the left heart ventricle and the atria [7]. Transitioning to the focal point of this chapter, we delve into mitral valve disease, a critical variation of VHD. This specific condition, encompassing both mitral valve prolapse and stenosis [8], pertains to the valve situated between the left atrium and ventricle. Mitral valve prolapse (MVP) involves the abnormal bulging of one or both mitral valve flaps into the left atrium during systolic contraction [8, 9]. Normally, these flaps close tightly to prevent backward blood flow, but in MVP, they are overly flexible or large, causing potential backflow or mitral regurgitation [9]. In contrast to MVP, mitral stenosis involves the narrowing (stenosis) of the mitral valve. This narrowing restricts the flow of blood from the left atrium to the left ventricle causing ischemia and ultimately can lead to death [10].

Heart valves consist of three distinct layers: the spongiosa, fibrosa, and ventricularis. The fibrosa, the innermost layer, is rich in collagen and valvular interstitial cells, providing structural strength [11]. The middle spongiosa layer primarily contains proteoglycans, contributing to the valve's cushioning and flexibility [12]. The ventricularis, adjacent to the ventricle, comprises elastin sheets and is lined with valvular endothelial cells [13]. This layer is crucial for the valve's elasticity, enabling the proper biomechanics for valve opening and closing [13]. In a properly functioning heart valve, the leaflets are firmly anchored to the ventricular wall through chordae tendineae and papillary muscles, guaranteeing the smooth and synchronized operation of the valve [14]. The typical mitral valve facilitates the passage of blood from the left atrium to the left ventricle during diastole and effectively seals off the left atrium from the left ventricle during systole. This crucial role hinges upon the unblemished interplay and integrity of its key components: the mitral annulus, leaflets, chordae tendineae, and papillary muscles [15].

In congenital heart valve defects, structural abnormalities can manifest in various aspects of the valve mechanism. For instance, in the case of the leaflet, it may exhibit excessive size, thickening, or irregular shapes due to myxomatous degeneration [16]. Conversely, when considering the chordae tendineae, they may become elongated [17] or ruptured [18]. Any such aberrations within these anatomical components can lead to the development of mitral regurgitation [19], a condition characterized by the backward leakage of blood into the left atrium and subsequently into the left ventricle [20]. Over time, this can result in left ventricular enlargement and, in severe cases, culminate in heart failure or even sudden cardiac death.

In cases in which repair is not possible for valves, valves may be replaced. Some replacements in adults are made by mechanical valves or even bioprosthetic valves which require open-heart surgery [21]. Currently, the most common procedure is the use of bioprosthetic valves with transcatheter valve replacements (TAVR), which instead of open-heart surgery, deliver a new bioprosthetic valve via a catheter [22]. A catheter is threaded through a large blood vessel, usually the femoral artery in the leg, but alternative access routes can also include the subclavian artery, direct aortic access, and even transapical access through the apex of the left ventricle [23]. For infants, this procedure is not viable due to the need for larger arteries, leaving open-heart surgery as the sole alternative. Open-heart surgery poses significant risks for these young patients because the valves must be replaced as the child's heart grows [24]. Since recovering from open-heart surgery is particularly challenging, especially for immature immune systems, and frequent replacements are often necessary, it presents a life-threatening situation for these children, depriving them of the possibility of full recovery [25].

## *1.2 Allogenic Elastin Heart Valve Replacements*

Addressing this issue calls for the development of a valve capable of adapting to an infant's growth. In the pursuit of designing such a heart valve, it has become evident that the valve must not only seamlessly integrate with the surrounding tissues and support natural growth post-implantation, but it must also exhibit compatibility with the immune system. Utilizing biocompatible scaffolds that mimic the natural protein structures of heart valves offers a promising alternative for dysfunctional heart valves. This approach becomes feasible as long as the scaffold permits cell integration and the gradual formation of surrounding tissues, potentially eliminating the need for valve replacement altogether. In previous studies [26], it has been shown that if the tissue does not completely surround the surface of the valve after implantation in a period of 3 months, there is going to be an immune reaction as a consequence. Consequently, if the scaffold to be used had self-secreted proteins that are allogenic to the species of implantation, then the immune reaction could be avoided altogether [26]. Since one of the most important components in a heart valve is elastin [27, 28], it would be optimal to have such protein attached to the scaffold in question [28]; however, it is particularly challenging to create elastin as the body takes a long time for replacement. This is due to collagen being more easily produced in the body with elastin being mostly produced with a low turnover rate and complex synthesis process [29]. The primary formation of elastin, known as elastogenesis, predominantly occurs during fetal development and early neonatal stages being produced by mother stem cells in organs such as blood vessels, lungs, and skin [30]; however, post-adolescence, the synthesis of elastin significantly diminishes, resulting in mature elastic fibers having to maintain their biomechanical functions almost throughout the lifespan of an individual [31]. The Cardiovascular Therapeutics Laboratory at Florida International University has innovatively applied this concept to develop allogenic elastin for heart valve applications [32]. Introducing a specialized layer of human stem cells is a pioneered technique for generating elastin layer sheets. This process is facilitated by cellular signaling and the strategic incorporation of specific growth factors. The resultant production of allogenic elastin-rich valves, as detailed in the research by Gonzalez et al. [32], marks a significant advancement in the field. The use of allogenic elastin is particularly noteworthy for its critical role in somatic growth, attributed to its unique chemotactic properties [29]. This development not only represents a breakthrough in heart valve therapy but also underscores the potential of allogenic elastin in regenerative medicine.

Chemotaxis is especially important since it works by activating nearby cells between the vascular and lymphatic systems to go toward the site of implantation [33, 34]. In other words, it is used as a signaling mechanism that helps with wound healing, tissue morphogenesis, and axon guidance [35].

### *1.3 Aims*

To ensure the sustainability of the elastin-rich regenerative valves following implantation, it is imperative to conduct a series of in vitro immune studies. We suggest that, within a milieu composed of various natural valvular cells, a discernible enhancement in extracellular matrix (ECM) synthesis will occur. Quantitative and qualitative evaluations through the examination of proteomic growth could prospectively demonstrate such development with two specific aims:

I. Examine the in vitro expression and functional performance of engineered elastin-rich heart valves in the presence of native valvular cells, comparing their efficacy to non-elastin enriched scaffolds.
II. Examine the in vitro growth potential and functionality of the engineered elastin-rich heart valve by exposing it to native valvular and immune cells and compare its performance with raw scaffolds that lack the elastin-rich component.

Expanding on the initial explanation, the designed elastin component plays a pivotal role in cardiac valve regeneration by mimicking the natural extracellular matrix, thereby providing a scaffold that not only supports structural integrity but also promotes the biological functions essential for valve repair and regeneration. This biomimetic approach is crucial for integrating with the native cellular architecture of the cardiac valve environment, ensuring compatibility and enhancing regenerative outcomes.

For a comprehensive and robust evaluation of the elastin component's efficacy, it is imperative to involve a diverse range of cells that are inherently present within the valvular ecosystem. This includes not only the valvular interstitial cells (VICs) and valvular endothelial cells (VECs), which play significant roles in maintaining valve function and integrity, but also a broader spectrum of cellular participants. VICs are known for their critical role in producing and remodeling the extracellular matrix, thereby contributing to the mechanical strength and flexibility of the heart valves. VECs, on the other hand, form a protective lining that modulates blood flow and valve permeability, and their health is essential for preventing thrombotic events.

Initial testing phases should focus on observing proteomic changes induced by the interaction between the engineered elastin and these valvular cells. An increase in the production of specific proteins associated with repair and regeneration, as indicated by interactions with VICs and VECs, would signal a successful integration of the elastin component with the native valve environment [36]. This includes monitoring the release of proteins critical for extracellular matrix formation, cellular adhesion, and the modulation of inflammatory responses.

Furthermore, the involvement of immune cells such as macrophages in the evaluation process is crucial. Macrophages are versatile cells capable of adopting pro-inflammatory or pro-regenerative roles in response to environmental cues. Their response to the implanted elastin component can provide invaluable insights into whether the engineered tissue is perceived as a foreign body eliciting an inflammatory reaction or as a compatible scaffold that facilitates tissue regeneration. By analyzing macrophage behavior, researchers can assess the biomaterial's potential to modulate the immune environment in favor of regeneration. This duality in macrophage function underscores the importance of designing biomaterials that not only avoid triggering adverse immune reactions but also actively promote a regenerative immune response.

## 2 Materials and Methodologies

### *2.1 Individual Expansion of Human Valvular Endothelial Cells and Human Interstitial Endothelial Cells*

Human valvular endothelial cells (hVECs, Lonza, Walkersville, MD, USA; Catalog #: 00225974) and human valvular interstitial cells (hVICs, Innoprot, Bizkaia, Spain, Catalog #: P10462) were platted and expanded following the manufacturing company's protocol. Basically, cells were conditioned with their perspective media supplemented by the company of origin; the media used for hVECs was EGM-2 MV (Lonza, Walkersville, MD, USA; Catalog #: CC-3202) and, for hVICs was Fibroblast Medium II (Plus) (Innoprot, Bizkaia, Spain, Catalog #: P60166). Cells were grown into confluency under an incubator in a humidified 5% $CO_2$ at 37 °C. The cells were expanded until cell density reached $8 \times 10^5$ cells/cm$^2$.

## 2.2 *In Vitro Co-culture of hVECs and hVICs in Engineered Elastin-Rich Valves vs. Unseeded Raw PSIS Scaffolds*

ECM-engineered elastin-rich valves (EERV) were created by seeding human stem cells onto porcine small intestinal submucosa (PSIS) scaffolds employing the protocol designed by Gonzalez et al. [32]. An experiment was conducted to evaluate the phenotypical response that would be produced when the EERV with the elastin-rich component on the surface of the scaffolds was exposed to hVICs and hVECs, which are common cell types that can be found in the area of implantation inside the heart. Raw PSIS (RPSIS, scaffold by itself) was used as a comparison control group. These two different groups were compared by combining $n = 2$ EERV and $n = 2$ RPSIS scaffolds in a co-culture of hVECs and hVICs for 8 days in a rotisserie in an incubator at 5% $CO_2$ at 37 °C. The media for co-culture was Dulbecco's Modified Eagle's Medium (DMEM; Corning, Fisher Scientific, Pittsburgh, PA; Catalog #: MT10013CV), Fetal Bovine Serum (FBS, Corning, Fisher Scientific, Pittsburgh, PA; Catalog #: MT35010CV) at 10%, and Penicillin/Streptomycin at 2% (Cytiva, HyClone™, Fisher Scientific, Pittsburgh, PA; Catalog #: SV300010). Media was further supplemented with 2 ng/ml basic fibroblast growth factor (bFGF, Corning™; Fisher Scientific, Pittsburgh, PA; Catalog #: CB-40060) as well as 82 μg/mL L-Ascorbic acid 2-phosphate (AA2P; Fisher Scientific, Catalog #: NC0602549), which stimulates the formation of collagen production.

## 2.3 *Histological Assessment of Engineered Elastin-Rich Valves vs. Previously Unseeded PSIS Scaffolds*

After 8 days, both groups were placed in formalin at 10% (w/v) for 24 hours and embedded in optimal cutting temperature compound (Fisher HealthCare, Houston, TX; Catalog #: 4585) and stored at −80 °C. Constructs were cryostat sectioned at 16 μm thickness, positioned on glass slides (TruBond 380; Newcomer Supply, Middleton, WI), and allowed to air dry at room temperature. The specimens were stained with MOVATs Pentachrome (American Stat Lab, CA, Catalog #: NC9321143) to determine the different constituents secreted by the co-culture, specifically collagen (yellow/orange), mucins (blue/green), and elastin (purple/black) (Fig. 1).

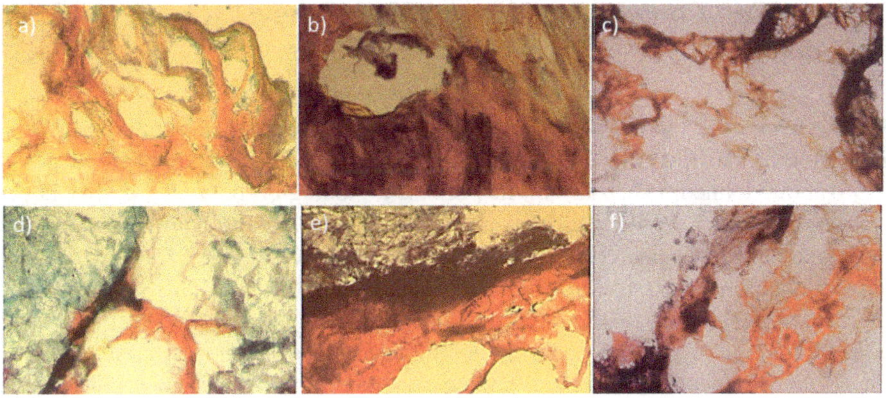


**Fig. 1** Raw PSIS scaffolds (**a–c**) ECM secretion quantification in comparison to engineered elastin-rich valves (**d–f**) after 8 days static rotisserie co-culture with hVICs and hVECs ($n = 2$ specimens/group, sectioned at 16 μm from the outer surface). Collagen (yellow/orange), fibrin or muscle (red), mucins (blue/green), and elastin (purple/black), magnification 10x

## *2.4 Spatial Intensity Heat Mapping for Assessment of Percent Extracellular Matrix Content*

Samples were imaged by using a Nikon TE300 Inverted Microscope Phase Contrast at 10×, 20×, and 40× magnification. Spatial intensity was quantified using a heat-mapping intensity code that denoised normalized images (MATLAB, MathWorks, Inc., Natick, MA, USA). Heat code intensities were distinguished by the intensity of color at each intersection point reflecting the expression level of the particular protein expressed in the given sample. The proteins identified by the MOVATs stain were collagen (yellow), elastin (black), proteoglycans_glycosaminoglycans/mucins (green), and fibrin/muscle (red), respectively, in which there were sometimes combinations of each of these expressed proteins. The intensity average was reported as average ± SD. Student t-test was used in Minitab (Minitab Inc.) to compare the two groups. Equal variances were assumed.

## *2.5 Monocyte Differentiation into Macrophage Platting*

Human macrophages (M-CSF, PromoCell, Heidelberg, Germany, catalog #: C-28056), monocyte-derived, single donor (hMDM-MCSF (−)) were acquired and platted at M0 without polarized activation at 1.5 million cells. Cells were further supplemented with macrophage generation medium XF (PromoCell, Heidelberg, Germany, catalog #: C-12915). The serum AB off-the-clot was added to the media to maintain the viability of the cells during the differentiation phase.

### 2.6 In Vitro Co-culture of Macrophages with hVICs and hVECs

Further testing was done by adding immune cells to the co-coculture. EERV and RPSIS scaffolds were co-cultured with hVICs, hVECs, and human macrophages at M0 without defined polarization. The purpose of adding un-polarized macrophages was to examine what polarization would arise from the co-culture with the other two cell types. The latter procedure was done by using hVICs and hVECs as previously explained before. The two cell types were added to the EERV/RPSIS scaffolds for 8 days allowing the cells to seed onto the scaffolds. Afterward, seeded scaffolds were added for another 8 days in media containing macrophages as it is demonstrated that the tissue M2 peak regeneration cytokine expression happens at day 7 [37]. At the 8-day mark in the rotisserie, the two types of scaffolds were removed, and media was centrifuged at 1500 RPM for 15 min with the purpose of removing the debris. Subsequently, the resulting media was stored at −80 °C. The tissue samples were embedded in OCT and stored at −80 °C.

### 2.7 Cytokine Assessment of In Vitro Co-culture Media Containing Macrophages, hVICs, and hVECs

Cytokine assessment of the culture media that had been in triple co-culture with hVICs, hVECs, and macrophages was assessed by using a customized human cytokine ELSA array (Signosis, Santa Clara, California, US, catalog # Cus-EA-1001_006297). The protocol was followed as recommended by the company, and absorbances were measured in a spectrometer at 450 nm. The panel of cytokines quantified included a range of pro-inflammatory and anti-inflammatory markers, starting with tumor necrosis factor-alpha (TNFα), a pivotal mediator of acute inflammation [38]. Interleukin-1 alpha (IL-1α) is also noteworthy for its role in systemic responses to infection and injury, propagating inflammatory signals [39]. Moreover, vascular endothelial growth factor (VEGF) serves as a key regulator of angiogenesis and tissue repair [40], while IL-6, a multifunctional cytokine, is essential for both inflammation and B-cell maturation [41]. On the other hand, IL-8 serves as a chemotactic cytokine attracting migration toward the zone of inflammation such as immune cells, namely, neutrophils and granulocytes [42]. IL-12 is crucial for the differentiation of naive T-cells into Th1 cells for antibody production [43], and IL-13 plays a significant role in immune response regulation and tissue remodeling [44].

Platelet-derived growth factor-BB (PDGF-BB) is recognized for its mitogenic effects during wound healing being an activator of mesenchymal cells [45]. Similarly, monocyte chemoattractant protein-1 (MCP-1) is instrumental in recruiting monocytes to injury and infection sites [46]. Macrophage inflammatory protein-1 alpha (MIP-1α) contributes to inflammatory responses and leukocyte trafficking [47]. IL-4 is key in guiding naive helper T-cells to Th2 cells [48], while IL-10,

**Table 1** Measurement of cytokine absorbance intensity in two samples each of EERV and RPSIS culture media, following a 16-day incubation period under static rotisserie conditions for three cell types of co-cultures, hVICs, hVECs, and macrophages

| Cytokine expressed | Elastin-rich | Elastin-rich | Raw PSIS | Raw PSIS | Sample media |
|---|---|---|---|---|---|
| TNFa | 0.260 | 1.370 | 0.652 | 0.447 | 0.415 |
| IL-1a | 0.342 | 1.257 | 0.379 | 0.746 | 0.310 |
| IL-8 | 1.382 | 1.292 | 0.652 | 0.583 | 0.414 |
| VEGF | 0.695 | 1.272 | 0.584 | 0.562 | 0.461 |
| IL-6 | 1.329 | 1.172 | 1.158 | 2.023 | 0.349 |
| IL-12 | 0.841 | 1.285 | 0.492 | 0.453 | 0.308 |
| IL-13 | 0.918 | 1.480 | 0.585 | 0.338 | 0.366 |
| PDGF-BB | 0.851 | 0.738 | 0.348 | 0.558 | 0.559 |
| MCP-1 | 0.506 | 0.661 | 0.506 | 0.583 | 0.746 |
| MIP-1a | 1.865 | 0.883 | 0.391 | 0.346 | 0.363 |
| IL-4 | 1.183 | 0.600 | 0.372 | 0.368 | 0.348 |
| IL-10 | 1.841 | 0.912 | 0.509 | 1.020 | 0.428 |
| Leptin | 0.703 | 0.779 | 0.511 | 0.708 | 0.401 |
| IGF-1 | 0.979 | 1.241 | 0.480 | 0.409 | 0.315 |
| Adiponectin | 1.494 | 1.030 | 0.641 | 0.873 | 0.223 |
| IL-1B | 0.885 | 0.458 | 0.300 | 0.682 | 0.364 |

an anti-inflammatory cytokine, is vital in moderating the immune response to pathogens [49].

Leptin, beyond its known role in energy regulation, influences immune function [50]. Insulin-like growth factor 1 (IGF-1) promotes growth across various cell types [51]. Adiponectin is engaged in metabolic processes and imparts anti-inflammatory effects on the vascular system [52]. Lastly, interleukin-1 beta (IL-1β) stands out as a potent pro-inflammatory cytokine integral to the body's infection response mechanisms [53] (Table 1). Sample media was measured as a control group between these two groups, and results were analyzed in Minitab Solutions Analytics™ as T-tests between groups.

# 3 Experimental Results and Conclusions

## 3.1 *Comparative Analysis of Extracellular Matrix in Elastin-Enriched vs. Unseeded PSIS Bio-Scaffold Valves After Co-culturing hVICs and hVECs, Utilizing MOVATS's Penta Chrome Staining for Spatial Intensity Quantification*

In our study, the ECM of two groups—EERV and RPSIS—was evaluated after an 8-day rotisserie co-culture with human valvular interstitial cells (hVICs) and human valvular endothelial cells (hVECs) (Fig. 1). The analysis revealed a significantly

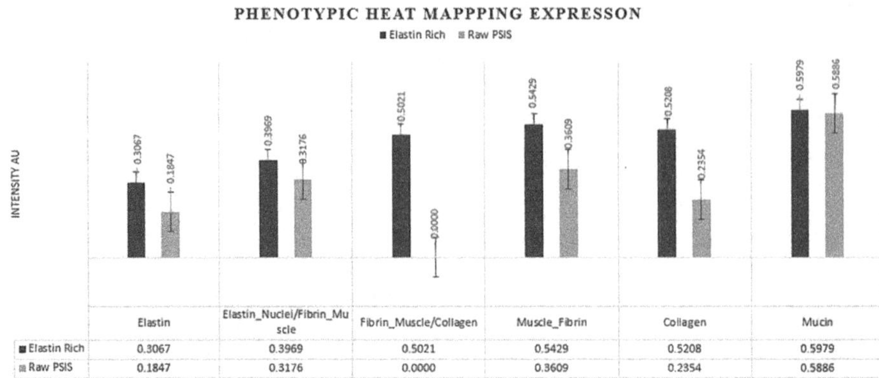


**Chart 1** Average ± SD of ECM secretion by hVICs + hVECs seeded on the EERV and RPSIS bio-scaffolds, magnification 10x

higher elastin expression in EERV, with a mean difference of 0.1220 arbitrary units (AU) ($p \leq 0.05$) (Chart 1) and an increase in the elastin/fibrin_muscle ratio (ER) by 0.0793 AU ($p \leq 0.05$) (Chart 1). Notably, a unique combination of fibrin_muscle/collagen was found only in the EERV group, averaging 0.5021 AU (Chart 1). Muscle integration, indicative of hVICs recruitment, showed a higher value in the EERV group with a statistical difference of 0.1820 AU ($p \leq 0.05$) (Chart 1). While the difference in collagen concentration between the groups was not statistically significant ($p \geq 0.05$), a trend toward increased collagen levels was observed in the EERV group. Additionally, mucins showed a difference of 0.0093 AU (Chart 1) although this did not reach statistical significance.

### 3.2 *Advanced Evaluation of De Novo Extracellular Matrix in Elastin-Enriched Valves and Raw Unseeded PSIS Bio-Scaffolds Using MOVATS Pentachrome Staining: Spatial Intensity Analysis Post Co-culture with hVICs, hVECs, and Macrophages*

The developed heart valves underwent further examination to assess their interaction with immune cells. We assessed their response to immune cells using histological methods after an 8-day rotisserie co-culture with hVICs and hVECs, followed by an 8-day period with added macrophages (Fig. 2). The results, presented in Chart 2, show a significant difference in the expression of specific proteins between the two valve types. The EERV exhibited a higher expression of elastin, with an intensity value of 0.1361 AU, compared to 0.0666 AU in the RPSIS ($p \leq 0.05$) (Chart 2). Similarly, for the combination of fibrin_muscle/nuclei phenotype, EERV recorded a

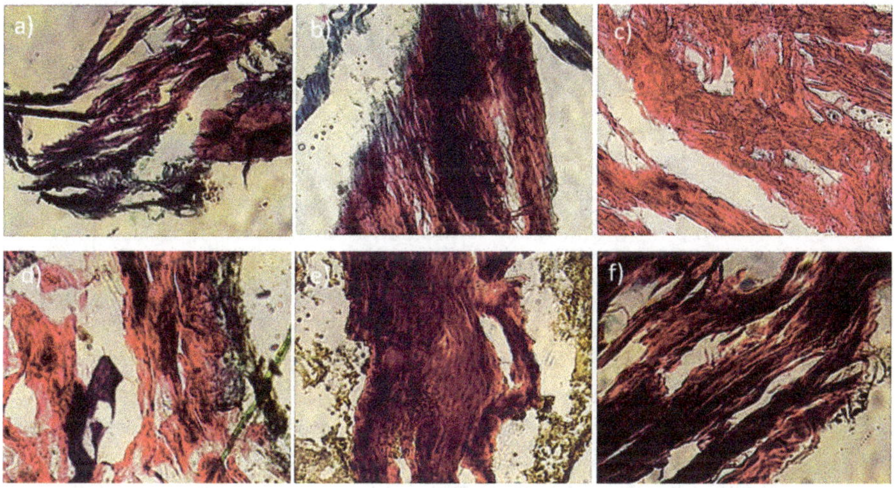


**Fig. 2** Raw PSIS scaffolds (**a–c**) ECM secretion quantification in comparison to engineered elastin-rich valves (**d–f**) after 16 days static rotisserie co-culture with hVICs, hVECs, and, macrophages ($n = 2$ specimens/group, sectioned at 16 μm from the outer surface). Collagen (yellow/orange), fibrin or muscle (red), mucins (blue/green), and elastin (purple/black), magnification 10x

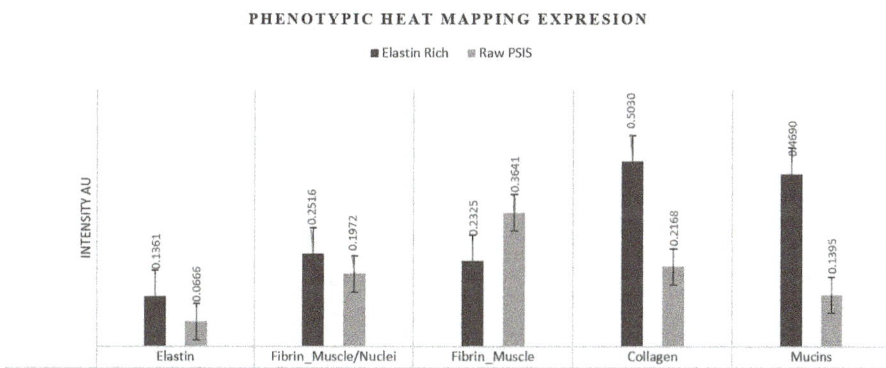


| | Elastin | Fibrin_Muscle/Nuclei | Fibrin_Muscle | Collagen | Mucins |
|---|---|---|---|---|---|
| Elastin Rich | 0.1361 | 0.2516 | 0.2325 | 0.5030 | 0.4690 |
| Raw PSIS | 0.0666 | 0.1972 | 0.3641 | 0.2168 | 0.1395 |

**Chart 2** Average ± SD of ECM secretion by hVICs + hVECs+ macrophages seeded on the EERV and RPSIS bio-scaffolds after 8 days rotisserie, magnification 10x

higher expression at 0.2516 AU compared to 0.1972 AU in RPSIS ($p \leq 0.05$) (Chart 2). The trend continued with collagen, where EERV showed an expression level of 0.5030 AU, markedly higher than the 0.2168 AU observed in RPSIS ($p \leq 0.05$), and Mucins with an expression of 0.4690 AU in EERV against 0.1395 AU in RPSIS ($p \leq 0.05$) (Chart 2). In contrast, the RPSIS group demonstrated a significantly higher expression of fibrin_muscle in isolation, at 0.3641 AU, compared to 0.2325 AU in EERV ($p \leq 0.05$) (Chart 2).

### 3.3 Cytokine Comparison Assessment of Culture Media Derived from Co-culture of Elastin-Enriched Valves and Raw Unseeded PSIS Bio-Scaffolds in hVICs, hVECs, and Macrophages

Upon the conclusion of an 8-day rotisserie culture involving macrophages, a comprehensive cytokine profile was delineated for all samples under scrutiny. The subsequent assay involved the recovery of the media, which was then subjected to spectrophotometric analysis at a wavelength of $\lambda = 450$ nm, as delineated in Fig. 3. The cytokines quantified included TNFα, IL-1α, VEGF, IL-12, IL-13, PDGF-BB, MCP-1, MIP-1α, IL-4, IL-10, leptin, IGF-1, adiponectin, and IL-1β. The results evidenced a consistently higher absorbance for most measured cytokines in the EERV samples relative to the RPSIS (Graph 1). An exception to this trend was noted for IL-6, which exhibited a superior absorbance in the RPSIS samples (Fig. 3). Notably, statistical analysis revealed significant differences in the expression levels of IL-8 and IGF-1 ($p \leq 0.05$) (Fig. 3).

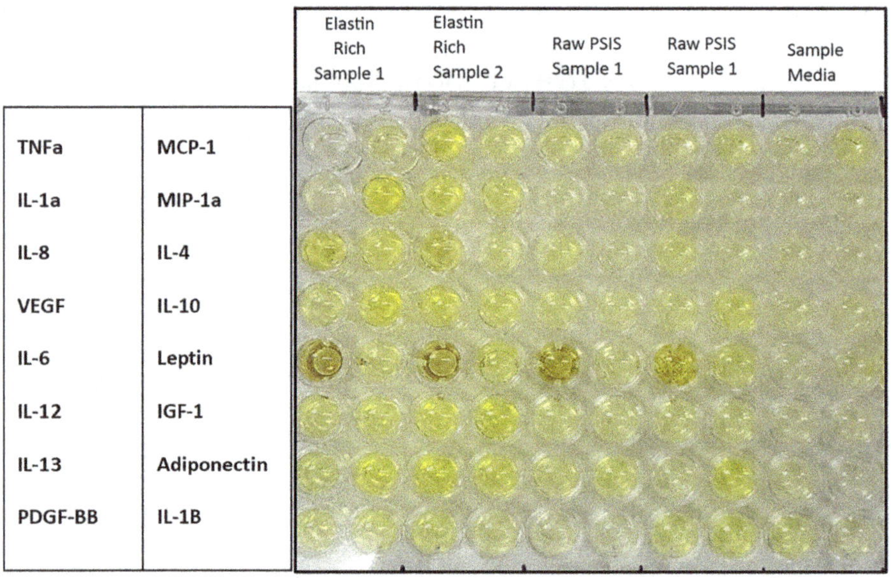


**Fig. 3** Cytokine expression plate of culture media samples recuperated from samples of 8-day static rotisserie co-culture of EERV/RPSIS in hVIC, hVEC, and macrophages

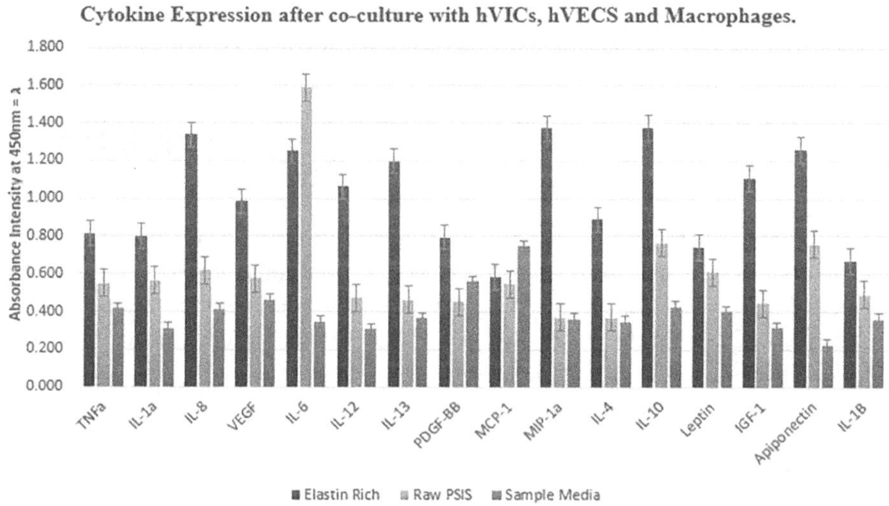


**Graph 1** Comparison of cytokine absorption intensity in independent samples of EERV and RPSIS culture media after a 16-day period of static rotisserie incubation

## *3.4 Discussion and Conclusions*

In the comparative analysis between the EERV and the RPSIS, the EERV not only exhibited higher proteomic secretion when in the presence of native valvular cells but also maintained this elevated secretion upon exposure to immune cells. This suggests that the EERV's ECM has a superior composition and functionality, which is particularly evident in its elastin content, the balance between elastin and fibrin/muscle appearance, and the integration with muscular structures by itself. The distinctive combination of fibrin, muscle, and collagen in the EERV may confer a structural advantage, potentially enhancing the valve's mechanical properties and durability. In addition, an identified trend in EERV is that glycosaminoglycans which are stained with a green hue denoted as mucins (Charts 1 and 2) were expressed with a higher phenotypic intensity, which is of high importance given its role in natural heart valves as functional and structural component [54]. In contrast, while fibrin's presence often signals inflammatory processes, its histological staining characteristic is very similar to those of muscle cells. The redder hues in stained images correlate with increased recruitment of valvular cells, indicating more active cell recruitment in the EERV incorporation of repair processes. In contrast, the RPSIS predominantly expressed higher fibrin in association with muscle cells when exposed to macrophages (Chart 2) which would mean that there is possible new tissue inflammation as this category of red parts without nuclei present was seen more profoundly in the RPSIS. This differential expression could be consequential for the valve's functionality and its immune response after 2 weeks of cytokine expression with the union and co-culture of all these different cell types.

Furthermore, exploring cytokine expression when the two groups were placed under co-culture provides deeper insight into the signaling pathways and phenotypic responses observed for a more nuanced understanding. A broad spectrum of cytokines showed higher absorbances in the EERV group, pointing to an elevated presence or stability of these signaling molecules within the engineered tissue environment for most studied cytokines (Graph 1). This was not uniform across all cytokines, however, as IL-6 was found in greater concentrations in the RPSIS samples. As we delve more into this cytokine, it mostly expresses during acute phase inflammation and fibrinogen production which could correspond with the red hues found before for phenotype expression of fibrin when macrophages were added (Chart 2). With this, noteworthy is the statistical significance of IL-8 and IGF-1 expression in the EERV group, hinting at the possibility that the elastin-rich matrix could be selectively modulating the secretion or degradation of these particular cytokines. These cytokines are critical mediators of inflammatory and growth responses, respectively, and their altered levels suggest that the elastin-rich matrix may be affecting the tissue's microenvironment in a specific and targeted manner with chemotaxis and growth tissue repair.

## References

1. S. M. Reller MD, Riehle-Colarusso T, Mahle WT, Correa A., "Prevalence of congenital heart defects in Atlanta," *J Pediatr,* vol. 153:807–13, 1998–2005.
2. K. N. Lopez, S. A. Morris, S. K. Sexson Tejtel, A. Espaillat, and J. L. Salemi, "US Mortality Attributable to Congenital Heart Disease Across the Lifespan From 1999 Through 2017 Exposes Persistent Racial/Ethnic Disparities," *Circulation,* vol. 142, no. 12, pp. 1132–1147, 2020, https://doi.org/10.1161/circulationaha.120.046822.
3. K. S. Hoffman JL, "The incidence of congenital heart disease.," *J Am Coll Cardiol.,* vol. 39(12):1890–1900., 2002.
4. L. B. Beerman. Overview of Congenital Cardiovascular Anomalies
5. A. Adere, M. Bedru, and M. Afework, "Etiologies and Patterns of Valvular Heart Disease Among Cardiac Patients at the Cardiac Center of Ethiopia During February 2000 to April 2022," *International Journal of General Medicine,* vol. 16, pp. 2849–2856, 2023, https://doi.org/10.2147/ijgm.s420591.
6. H. S. L. Adams, S. Ashokkumar, A. Newcomb, A. I. Macisaac, R. J. Whitbourn, and S. Palmer, "Contemporary review of severe aortic stenosis," *Internal Medicine Journal,* vol. 49, no. 3, pp. 297–305, 2019, https://doi.org/10.1111/imj.14071.
7. P. M. Shah, "Tricuspid and pulmonary valve disease evaluation and management," *Rev Esp Cardiol,* vol. 63, no. 11, pp. 1349–65, Nov 2010, https://doi.org/10.1016/s1885-5857(10)70259-0.
8. S. C. Harb and B. P. Griffin, "Mitral Valve Disease: A Comprehensive Review," *Curr Cardiol Rep,* vol. 19, no. 8, p. 73, Aug 2017, https://doi.org/10.1007/s11886-017-0883-5.
9. B. Del Forno *et al.*, "Mitral valve regurgitation: a disease with a wide spectrum of therapeutic options," *Nature Reviews Cardiology,* vol. 17, no. 12, pp. 807–827, 2020, https://doi.org/10.1038/s41569-020-0395-7.
10. V. Galusko *et al.*, "Management of mitral stenosis: a systematic review of clinical practice guidelines and recommendations," *European Heart Journal—Quality of Care and Clinical Outcomes,* 2021, https://doi.org/10.1093/ehjqcco/qcab083.

11. S. Jana, A. Lerman, and R. D. Simari, "In Vitro Model of a Fibrosa Layer of a Heart Valve," *ACS Appl Mater Interfaces,* vol. 7, no. 36, pp. 20012–20, Sep 16 2015, https://doi.org/10.1021/acsami.5b04805.
12. E. H. Stephens, C. K. Chu, and K. J. Grande-Allen, "Valve proteoglycan content and glycosaminoglycan fine structure are unique to microstructure, mechanical load and age: Relevance to an age-specific tissue-engineered heart valve," *Acta Biomater,* vol. 4, no. 5, pp. 1148–60, Sep 2008, https://doi.org/10.1016/j.actbio.2008.03.014.
13. R. M. Buchanan and M. S. Sacks, "Interlayer micromechanics of the aortic heart valve leaflet," *Biomechanics and Modeling in Mechanobiology,* vol. 13, no. 4, pp. 813–826, 2014, https://doi.org/10.1007/s10237-013-0536-6.
14. Y. Topilsky, "Mitral Regurgitation: Anatomy, Physiology, and Pathophysiology—Lessons Learned From Surgery and Cardiac Imaging," *Frontiers in Cardiovascular Medicine,* vol. 7, 2020, https://doi.org/10.3389/fcvm.2020.00084.
15. C. J. Ross, J. Zheng, L. Ma, Y. Wu, and C.-H. Lee, "Mechanics and Microstructure of the Atrioventricular Heart Valve Chordae Tendineae: A Review," *Bioengineering,* vol. 7, no. 1, p. 25, 2020, https://doi.org/10.3390/bioengineering7010025.
16. A. O'Donnell and K. E. Yutzey, "Mechanisms of heart valve development and disease," *Development,* vol. 147, no. 13, p. dev183020, 2020, https://doi.org/10.1242/dev.183020.
17. K. Obase *et al.*, "Elongation of chordae tendineae as an adaptive process to reduce mitral regurgitation in functional mitral regurgitation," *European Heart Journal—Cardiovascular Imaging,* vol. 17, no. 5, pp. 500–509, 2016, https://doi.org/10.1093/ehjci/jev314.
18. Y. Koda, T. Tsukube, M. Hoshino, N. Yagi, H. Ishibashi-Ueda, and K. Okada, "Structural properties in ruptured mitral chordae tendineae measured by synchrotron-based X-ray phase computed tomography," *Journal of Synchrotron Radiation,* vol. 30, no. 5, pp. 995–1002, 2023, https://doi.org/10.1107/s1600577523005167.
19. R. Aguilar-Lopez *et al.*, "Acute Mitral Regurgitation Due to Chordae Tendineae Rupture: A Rare Presentation of Cardiac Amyloidosis," *Am J Case Rep,* vol. 23, p. e936545, Jul 4 2022, https://doi.org/10.12659/AJCR.936545.
20. E. A. Ahmed *et al.*, "Optimal Management of Mitral Regurgitation Due to Ruptured Mitral Chordae Tendineae in Patients With Hypertrophic Cardiomyopathy," *Seminars in Thoracic and Cardiovascular Surgery,* vol. 35, no. 3, pp. 476–482, 2023, https://doi.org/10.1053/j.semtcvs.2022.05.008.
21. F. Sawaya, T. H. Jørgensen, L. Søndergaard, and O. De Backer, "Transcatheter Bioprosthetic Aortic Valve Dysfunction: What We Know So Far," *Frontiers in Cardiovascular Medicine,* vol. 6, 2019, https://doi.org/10.3389/fcvm.2019.00145.
22. J. G. Webb and D. Dvir, "Transcatheter Aortic Valve Replacement for Bioprosthetic Aortic Valve Failure," *Circulation,* vol. 127, no. 25, pp. 2542–2550, 2013, https://doi.org/10.1161/circulationaha.113.000631.
23. R. R. Makkar *et al.*, "Five-Year Outcomes of Transcatheter or Surgical Aortic-Valve Replacement," *New England Journal of Medicine,* vol. 382, no. 9, pp. 799–809, 2020, https://doi.org/10.1056/nejmoa1910555.
24. R. Henaine, F. Roubertie, M. Vergnat, and J. Ninet, "Valve replacement in children: a challenge for a whole life," *Arch Cardiovasc Dis,* vol. 105, no. 10, pp. 517–28, Oct 2012, https://doi.org/10.1016/j.acvd.2012.02.013.
25. A. F. Elmahrouk *et al.*, "Mitral valve replacement in infants and younger children," *Scientific Reports,* vol. 11, no. 1, 2021, https://doi.org/10.1038/s41598-021-94779-0.
26. Brittany A. Gonzalez, "Bioscaffold Valve with and without Mechanically Conditioned Stem Cells for the Treatment of Critical Mitral Valve Diseases in the Young," no. FIDC009539, 11-13-2020, 10.25148/etd.FIDC009539.
27. C. Deborde, D. T. Simionescu, C. Wright, J. Liao, L. N. Sierad, and A. Simionescu, "Stabilized Collagen and Elastin-Based Scaffolds for Mitral Valve Tissue Engineering," *Tissue Engineering Part A,* vol. 22, no. 21–22, pp. 1241–1251, 2016, https://doi.org/10.1089/ten.tea.2016.0032.

28. I. Gonzalez De Torre, M. Alonso, and J.-C. Rodriguez-Cabello, “Elastin-Based Materials: Promising Candidates for Cardiac Tissue Regeneration,” *Frontiers in Bioengineering and Biotechnology,* vol. 8, 2020, https://doi.org/10.3389/fbioe.2020.00657.
29. A. Heinz, “Elastases and elastokines: elastin degradation and its significance in health and disease,” *Critical Reviews in Biochemistry and Molecular Biology,* vol. 55, no. 3, pp. 252–273, 2020, https://doi.org/10.1080/10409238.2020.1768208.
30. L. Baumann *et al.*, “Clinical Relevance of Elastin in the Structure and Function of Skin,” *Aesthet Surg J Open Forum,* vol. 3, no. 3, p. ojab019, Sep 2021, https://doi.org/10.1093/asjof/ojab019.
31. H. Trębacz and A. Barzycka, “Mechanical Properties and Functions of Elastin: An Overview,” *Biomolecules,* vol. 13, no. 3, p. 574, 2023, https://doi.org/10.3390/biom13030574.
32. B. A. Gonzalez *et al.*, “Stem Cell-Secreted Allogeneic Elastin-Rich Matrix with Subsequent Decellularization for the Treatment of Critical Valve Diseases in the Young,” *Bioengineering,* vol. 9, no. 10, p. 587, 2022, https://doi.org/10.3390/bioengineering9100587.
33. H. Mensing, “[Importance of fibroblast chemotaxis in wound healing and tumor cell evasion],” *Klin Wochenschr,* vol. 63, no. 4, pp. 145–51, Feb 15 1985, https://doi.org/10.1007/BF01732167. Bedeutung der Fibroblasten-Chemotaxis fur Wundheilung und Tumorzelleva-sion.
34. S.-S. Wang *et al.*, “Interaction Between Blood Vasculatures and Lymphatic Vasculatures During Inflammation,” *Journal of Inflammation Research,* vol. Volume 16, pp. 3271–3281, 2023, https://doi.org/10.2147/jir.s414891.
35. L. Hinck, “The Versatile Roles of “Axon Guidance” Cues in Tissue Morphogenesis,” *Developmental Cell,* vol. 7, no. 6, pp. 783–793, 2004, https://doi.org/10.1016/j.devcel.2004.11.002.
36. L. Cheng *et al.*, “Generation and characterization of cardiac valve endothelial-like cells from human pluripotent stem cells,” *Communications Biology,* vol. 4, no. 1, 2021, https://doi.org/10.1038/s42003-021-02571-7.
37. Y. Oishi and I. Manabe, “Macrophages in inflammation, repair and regeneration,” *International Immunology,* vol. 30, no. 11, pp. 511–528, 2018, https://doi.org/10.1093/intimm/dxy054.
38. H. Zelová and J. Hošek, “TNF-α signalling and inflammation: interactions between old acquaintances,” *Inflammation Research,* vol. 62, no. 7, pp. 641–651, 2013, https://doi.org/10.1007/s00011-013-0633-0.
39. N. C. Di Paolo and D. M. Shayakhmetov, “Interleukin 1α and the inflammatory process,” *Nature Immunology,* vol. 17, no. 8, pp. 906–913, 2016, https://doi.org/10.1038/ni.3503.
40. L. Pérez-Gutiérrez and N. Ferrara, “Biology and therapeutic targeting of vascular endothelial growth factor A,” *Nature Reviews Molecular Cell Biology,* vol. 24, no. 11, pp. 816–834, 2023/11/01 2023, https://doi.org/10.1038/s41580-023-00631-w.
41. S. Kang and T. Kishimoto, “Interplay between interleukin-6 signaling and the vascular endothelium in cytokine storms,” *Experimental & Molecular Medicine,* vol. 53, no. 7, pp. 1116–1123, 2021, https://doi.org/10.1038/s12276-021-00649-0.
42. M. Baggiolini and I. Clark-Lewis, “Interleukin-8, a chemotactic and inflammatory cytokine,” *FEBS Letters,* vol. 307, no. 1, pp. 97–101, 1992, https://doi.org/10.1016/0014-5793(92)80909-z.
43. M. D. Powell, K. A. Read, B. K. Sreekumar, D. M. Jones, and K. J. Oestreich, “IL-12 signaling drives the differentiation and function of a TH1-derived TFH1-like cell population,” *Scientific Reports,* vol. 9, no. 1, 2019, https://doi.org/10.1038/s41598-019-50614-1.
44. R. L. Gieseck, M. S. Wilson, and T. A. Wynn, “Type 2 immunity in tissue repair and fibrosis,” *Nature Reviews Immunology,* vol. 18, no. 1, pp. 62–76, 2018, https://doi.org/10.1038/nri.2017.90.
45. G. F. Pierce, T. A. Mustoe, B. W. Altrock, T. F. Deuel, and A. Thomason, “Role of platelet-derived growth factor in wound healing,” (in eng), *J Cell Biochem,* vol. 45, no. 4, pp. 319–26, Apr 1991, https://doi.org/10.1002/jcb.240450403.
46. S. Singh, D. Anshita, and V. Ravichandiran, “MCP-1: Function, regulation, and involvement in disease,” (in eng), *Int Immunopharmacol,* vol. 101, no. Pt B, p. 107598, Dec 2021, https://doi.org/10.1016/j.intimp.2021.107598.

47. E. Marciniak *et al.*, "The Chemokine MIP-1α/CCL3 impairs mouse hippocampal synaptic transmission, plasticity and memory," *Scientific Reports,* vol. 5, no. 1, p. 15862, 2015, https://doi.org/10.1038/srep15862.
48. J. A. Walker and A. N. J. McKenzie, "TH2 cell development and function," *Nature Reviews Immunology,* vol. 18, no. 2, pp. 121–133, 2018/02/01 2018, https://doi.org/10.1038/nri.2017.118.
49. K. N. Couper, D. G. Blount, and E. M. Riley, "IL-10: The Master Regulator of Immunity to Infection," *The Journal of Immunology,* vol. 180, no. 9, pp. 5771–5777, 2008, https://doi.org/10.4049/jimmunol.180.9.5771.
50. V. Abella *et al.*, "Leptin in the interplay of inflammation, metabolism and immune system disorders," *Nature Reviews Rheumatology,* vol. 13, no. 2, pp. 100-109, 2017/02/01 2017, https://doi.org/10.1038/nrrheum.2016.209.
51. S. Al-Samerria and S. Radovick, "The Role of Insulin-like Growth Factor-1 (IGF-1) in the Control of Neuroendocrine Regulation of Growth," *Cells,* vol. 10, no. 10, p. 2664, 2021, https://doi.org/10.3390/cells10102664.
52. X. Hui, K. S. Lam, P. M. Vanhoutte, and A. Xu, "Adiponectin and cardiovascular health: an update," *British Journal of Pharmacology,* vol. 165, no. 3, pp. 574–590, 2012, https://doi.org/10.1111/j.1476-5381.2011.01395.x.
53. N. Kaneko, M. Kurata, T. Yamamoto, S. Morikawa, and J. Masumoto, "The role of interleukin-1 in general pathology," *Inflammation and Regeneration,* vol. 39, no. 1, 2019, https://doi.org/10.1186/s41232-019-0101-5.
54. J. J. Lovekamp, D. T. Simionescu, J. J. Mercuri, B. Zubiate, M. S. Sacks, and N. R. Vyavahare, "Stability and function of glycosaminoglycans in porcine bioprosthetic heart valves," *Biomaterials,* vol. 27, no. 8, pp. 1507–1518, 2006, https://doi.org/10.1016/j.biomaterials.2005.08.003.

# Molecular-Level Modifications for Improving Immune Tolerance and Calcification Resistance of Bioprosthetic Heart Valves

**Zhongshi Wu, Yuhong Liu, Zhenjie Tang, Zhenjin Jiang, Mingzhe Song, and Liang Yi**

## 1 Introduction

Valvular heart disease (VHD) is a serious threat to human health worldwide, especially for the elderly. Due to the lack of effective preventive and therapeutic drugs, prosthetic valve replacement has become the most effective treatment strategy for VHD. With the increasing aging population, the number of valve replacements has steadily increased. Since bioprosthetic heart valves (BHV) have significant advantages over mechanical heart valves (MHV) in hemodynamics and blood compatibility, BHV is a better choice for older patients. In addition to older patients, young patients can also choose the BHV. In recent years, the invention of interventional valves has also expanded the clinical application of the BHVs.

Poor durability clearly limits the clinical application of BHV, and improving the durability of biological valves is a key issue that urgently needs to be solved. BHVs are prone to structural valve deterioration (SVD) after implantation, which eventually leads to valve stenosis or regurgitation and reduces valve durability [1, 2]. Currently, it is believed that valve failure is a complex, immune-mediated process with calcification as the main manifestation, which integrates atherosclerosis and inflammatory response, insufficient endothelialization, and acute thrombosis against xenografts [2, 3]. Therefore, to improve the anti-calcification and anti-thrombosis performance of the BHVs, reduce immune rejection, and promote endothelialization, thereby prolonging the service life of the BHVs and expanding the indications for clinical application of the BHVs, has become cutting-edge research.

Z. Wu (✉) · Y. Liu · Z. Tang · Z. Jiang · M. Song · L. Yi
Engineering Laboratory of Human Province for Cardiovascular Biomaterials, The Second Xiangya Hospital of Central South University Changsha, Hunan, China
e-mail: Owenzswu@csu.edu.cn; yuhong5423@csu.edu.cn; tangzhenjie@csu.edu.cn; songmingzhe@hnca.org.cn; 238202130@csu.eu.cn

 
J. Liao, J. Y. Wong (eds.), *Integration and Bridging of Multiscale Bioengineering Designs and Tissue Biomechanics*, https://doi.org/10.1007/978-3-031-81743-4_11

## 2 Characteristics of BHV Materials from Different Species

### 2.1 Mammals

In the current study of BHVs materials, glutaraldehyde cross-linked bovine and porcine bioprostheses are the main species sources of classical xenogeneic valve materials, but the calcification and degeneration of BHVs have not been completely solved. Different species of biomaterials have their own advantages; therefore, the selection of appropriate species of biomaterials is also a research aspect to improve the durability performance of BHVs. Many researchers have explored biomaterials from different species and provided new strategies for the development of novel BHV materials. For example, Naimark WA investigated the pericardium of calves, dogs, pigs, and sheep using structural/mechanical analysis methods. The results showed that thinner dog and pig pericardium had the same viscoelasticity as thicker bovine and sheep pericardium; however, they were stiffer and contained more abundant type III collagen. This means that cross-linking with glutaraldehyde (GA) had a better effect [4].

Aisa Rassoli [5] compared the histology and mechanics of the pericardium of porcine, equine, and bovine and found that the equine pericardium has a wavy collagen bundle structure similar to that of bovine pericardium, while the collagen bundles in porcine pericardium are thinner and regular in structure. Compared with bovine pericardium, equine pericardium tissue is mechanically and histologically more suitable for BHVs. Compared with bovine pericardium, equine pericardium tissue is more suitable for the manufacturing of cardiac valve prostheses mechanically and histologically.

WM Neethling [6] performed subcutaneous embedding and in vitro experimental evaluation of glutaraldehyde-crosslinked kangaroo pericardium, bovine pericardium, and porcine aortic valve in rats and found that kangaroo pericardium has a densely packed collagen matrix with higher ductility and significantly lower calcification potential than bovine pericardium and porcine aortic valve, and the kangaroo pericardium is considered to have low calcification potential and can be a candidate for cardiovascular biomaterials.

Jifu Mao [7] found that the average thickness of donkey pericardium was significantly lower than bovine pericardium and equine pericardium but slightly higher than porcine pericardium. It has a similar wavy collagen bundle structure to bovine pericardium, with sufficient maximum tensile stress and tensile strain.

Our research group also compared the characteristics of bovine pericardium in different species (cattle, yak, and Australian cattle) [8], and the experimental results showed that the average thickness of yak pericardium was significantly lower than that of Australian bovine pericardium and cattle pericardium, and the mechanical properties and anti-calcification property were better than those of pericardium in other species, which was an great potential biomaterial for Transcatheter Aortic Valve Implantation (TAVI).

## 2.2 *Birds or Fish*

In addition to mammals, research teams have studied materials derived from other species, such as birds or fish. For example, M. Martín Maestro compared glutaraldehyde cross-linked ostrich pericardium with bovine pericardium [9] and showed that ostrich pericardium was superior in water content and biomechanical properties, and the degree of calcification was also significantly lower than bovine pericardium. Li Binhan et al [10] used glutaraldehyde cross-linked acellular fish swim bladder as valve material. Compared with bovine pericardium, fish swim bladder-derived biomaterials are rich in elastin and have a higher elastic modulus. The results demonstrated that the swim bladder has better anti-calcification performance, cytocompatibility, hemocompatibility, and enzyme stability. However, due to the poor homology between fish and birds and humans, the targeted area of materials is small, not suitable for large-scale production, immunogenicity is difficult to remove, and there are still great challenges.

# 3 Immunogenicity of Biological Valve Materials and Removal Techniques

## 3.1 *Definition and Mechanism of Immunogenicity*

Immunogenicity refers to a feature or ability of the biological valve materials implanted in the body to stimulate an immune response. The immunogenicity of biological valve materials depends on their inherent antigenicity and the immune response of the host immune system. The proximity of an antigen's species origin to humans affects its immunogenicity. The more distant the species from humans, the less homologous its antigenic components (e.g., proteins or peptides) are to humans and the more immunogenic it is. Conversely, the closer the species is to humans, the more homologous the primate proteins are to humans and the less immunogenic they are. The immunogenicity of an antigen is related to its physicochemical properties. Antigens with complex structures, such as proteins containing aromatic amino acids (especially tyrosine), tend to have stronger immunogenicity. The most abundant collagen in the extracellular matrix of biological valve materials is an interspecies conserved sequence protein. It is generally believed that there is little variation in collagen between different species, resulting in weak antigenicity [11]. Gorham et al. [12] conducted antibody cross-reactivity on collagen from both autologous and allogeneic sources and confirmed that collagen has weak immunogenicity. Collagen has almost no immunogenicity when its structure is intact, but after primary structural destruction, it exposes antigenic epitopes, which lead to immune reactions. This may also be one of the reasons for the chronic graft immune response caused by the slow degradation of biological valve materials during the late stage of implantation after cross-linking. The cellular and extracellular matrix protein components in native biological valve materials are considered to be the main sources of immunogenicity in biological valve materials, with proteins and lipids on the cell membrane being the most important sources eliciting in vivo immune responses [11].

## 3.2 Immunogenicity of Cellular Components

The cells or residual cellular components in BHVs are important factors leading to immunogenicity. Failure to fully remove these cellular substances allows them to bind to immune receptors and trigger a strong immune response. Transplantation-related proteins, such as α-1,3 galactose (Gal) antigens, and major histocompatibility complexes (MHCs) are important sources of immunogenicity. For xenotransplantation, α-1,3-Gal antigen is the main antigen that stimulates transplant failure and can stimulate xenograft rejection reactions in primates/humans to porcine or bovine porcine tissues and organs [13]. α-1,3-galactose (Gal) antigen is a cell surface antigen containing a polylactamine core terminal residue, widely present in non-primate mammals. The alpha-Gal antigen is not synthesized due to a base mismatch in the galactosyltransferase gene in humans and primates. Instead, the bodies of humans and primates synthesized a large number of anti-Gal antibodies that recognize Gal antigens on the cell surface [14]. When a xenograft containing α-Gal antigen is transplanted, anti-Gal immunoglobulins in the human body will bind to Gal antigen epitopes, activating complementary reactions and antibody-dependent cell-mediated cytotoxicity, which will lead to the failure of xenotransplantation [15]. The major histocompatibility complex (MHC) is a type of species-specific antigenic protein, widely distributed on the surface of all nucleated cells, and is also an important source of immunogenicity in BHV materials [16, 17]. The MHC protein consists of two subtypes: MHC I and MHC II. MHC I is expressed on the surface of all nucleated cells and recognizes endogenous intracellular antigens and induces CD8 + T cell responses, which can directly invade cells with antigen targets. MHC II is mainly expressed on the surface of antigen-presenting cells (APCs), inducing CD4 + T cell responses, which clear antigens mainly by recruiting other inflammatory cells and producing cytokines.

Additionally, an increasing number of antigens have been found to be associated with posttransplant immune responses. In 2008, Kim et al. first detected the expression of a novel non-Gal carbohydrate antigen, N-glycolylneuraminic acid (Neu5Gcin porcine endothelial cells) [18]. NeuGc is widely distributed in mammals such as monkeys and apes but has not been found in humans and birds. The absence of NeuGc in human glycoconjugates is due to a partial deletion in the gene that encodes CMP-NeuAc hydroxylase [19, 20]. The expression of Neu5Gc is controlled by the activity of cytidine monophosphate-N-acetylneuraminate hydroxylase (CMAH). When implanting Neu5Gc-containing xenogeneic tissue material into the human body, the anti-Neu5Gc IgG will recognize the antigen and exhibit cell and/or complement-dependent cytotoxicity, leading to endothelial cell activation, complement deposition, E-selectin expression, increase secretion of pro-inflammatory cytokines, and changes in monocyte function [21, 22]. One study found that the Neu5Gc content in porcine pericardium is significantly higher than that of bovine pericardium [18]. This may correlate with a more severe immune response after implantation of porcine pericardial valves in vivo.

Carbohydrate antigen (SDA) produced by porcine β1,4 N-acetylgalactosaminyl transferase-2 (B4GALNT2) is a glycan antigen that is thought to be closely related to immunogenicity. The analysis of non-galactose antibodies induced after porcine-to-baboon xenograft heart transplantation indicates that porcine B4GALNT2-derived polysaccharide is an immunogenic xenograft antigen [23]. Porcine B4GALNT2 is homologous to humans and synthesizes the human SDA blood group antigen. Most people produce low levels of anti-SDA IgM, which can agglutinate with red blood cells. This suggests that there may be an early antibody-mediated immune injury to SDA expressed by endothelial cells and organs in porcine. However, clinical products originating from bovine tissues, such as bovine pericardial valves, do not express B4GALNT2 gene-associated SDA antigen.

## 3.3 Immunogenicity of Extracellular Matrix

Decellularized BHVs still do not completely eliminate immunogenicity, and immunogenicity caused by the extracellular matrix remains an important factor in valve failure. In addition to its cellular components, the ECM is also an important source of immunogenicity in BHVs. The interaction between ECM and an immune system is mainly mediated by damage-associated molecular patterns (DAMPs). DAMPs are molecules that may trigger immune responses when cells and ECM are damaged during the preparation or implantation of BHVs. Nuclear and mitochondrial DNA, reactive oxygen species (ROS), phospholipids on cell membranes, and fragmented ECM components (such as hyaluronic acid or fibronectin) are the most famous DAMPs [24]. DAMP can be recognized by macromolecules primarily expressed on innate immune cells, namely transmembrane and cytoplasmic receptors (PRRs). After being recognized by PRRs, DAMP promotes the expression of inflammation-related genes and induces the release of pro-inflammatory cytokines, including interleukins (IL1, IL-6), tumor necrosis factor (TNF), and chemokines, thereby recruiting immune cells [25]. The innate immune response induced by DAMP has a certain destructive effect on BHVs, but the induced adaptive immune response is the important factor affecting the lifespan of BHVs [26]. As the main mediator of adaptive immune response, T-cells require the simultaneous presence of two signals for activation, including antigenic substances and co-stimulatory signals [27]. The decellularization-induced DAMP increases the expression of MHC II and co-stimulatory molecules on receptors of APC and promotes T-cell recognition of antigens and MHC molecules in biological valve tissues [25]. In addition, DAMP activates the complementary cascade in APC to release C3a, C3b, and C5a components. C3a and C5a are effective chemical attractants and induce polarization of helper T-cell 1 (Th1and Th17) in CD4+ T-cells. Th1 and Th17 induce macrophage M0 polarization toward M1, increasing the production of pro-inflammatory cytokines and exacerbating the inflammatory response. The C3b component binds to the corresponding receptors on M1 macrophages and triggers the release of matrix metalloproteinases (MMPs), degrading EMC and causing a vicious cycle [28].

## *3.4 Removal of Immunogenic Substances*

### 3.4.1 Decellularization Techniques

Cells or residual cellular components are important factors leading to immunogenicity. Decellularization is a major method to eliminate immunogenicity in biomaterials. Since disruption of the constituent protein ultrastructure of the ECM leading to changes in the secondary conformation can cause an immune response, and residual cellular debris or nucleic acid fragments may also cause antigenicity leading to a low-grade immune response, the main goal of the decellularized process is to remove cells and nucleic acids while maintaining the intrinsic ultrastructure [29]. There have been research [29] teams who have set minimum standards for decellularization: (a) lack of visible histological staining in nuclear material; (b) the content of DNA should be less than 50 ng/mg of the dry weight of the ECM; (c) the length of DNA fragments should be <200 bp; and (d) lack of damage to the ECM structure (collagen, Glycosaminoglycans (GAGs), fibronectin, etc.) and the lack of cytotoxicity in decellularized ECM.

Traditional decellularization methods can be divided into physical, chemical, and biological methods [30, 31] and can be used alone or combined to disrupt the cell membrane in BHVs. These methods, combined with perfusion, shaking, and washing, can remove cellular components and leave the recellulatable ECM. Most physical decellularization methods are used before or in combination with chemical or enzymatic methods to improve the efficiency of reagents and tissue permeability. Freeze-thaw cycles and altered osmotic pressure can be used to destroy cells prior to chemical treatment. Freeze-thaw cycling produces ice crystals that disrupt cells but also cause damage to the ECM. Hypotonic and hypertonic solutions produce a solute concentration gradient that leads to cell membrane cleavage by osmotic pressure [32]. Osmotic decellularization has little effect on the morphology of the valve tissue. Hypotonic solutions cause less damage to the extracellular matrix of porcine valve leaflets than other physical or chemical methods, and collagen fibers are well preserved, but they do not completely remove the cellular components, and there is a greater loss for glycosaminoglycans [33]. Adding chemical reagents or enzymes to the hypotonic fluid can significantly improve the effect of decellularization, but it can also cause certain damage to the ECM. Ultrasound uses ultrasonic transducers to cause physical vibrations on the cell membrane, leading to cell apoptosis. It has a better decellularization effect but can be damaging to the ECM [34]. Ultrasound is typically used in conjunction with detergents such as sodium dodecyl sulfate (SDS) to enhance their ability to remove residual DNA. Their combination does not cause superimposed damage to the ECM. Ultrasound treatment is more effective than shaking and rinsing in helping to remove residual DNA from tissues [35]. Vacuum-assisted treatment has been applied as a physical method for decellularization, which reduces the processing time by accelerating the diffusion of chemical reagents or enzymes into the tissue [36]. Ma Lie et al. compared the decellularization efficiency of heart valves treated with vacuum and atmospheric pressure, demonstrating that vacuum treatment can improve the efficiency of decellularization without damaging mechanical properties [37]. Supercritical carbon dioxide (scCO2) has been proposed as a novel decellularization technology, which removes cells through supercritical extraction. Studies that have used scCO2 as a decellularization aid have reported that collagen and elastin content was largely unaffected after treatment. However, even if collagen content appears to be preserved after scCO2 treatment, the organization of collagen fiber is altered as a side effect of tissue dehydration or due to the reduction of other major ECM components such as GAGs [38]. The addition of water or ethanol modifiers to reduce tissue dehydration and the use of higher scCO2 pressures can be more effective in reducing the DNA content of tissues and improving decellularized efficiency [39].

The chemical method is currently the most widely used decellularization method, which exposes tissues to chemicals through soaking or perfusion, leading to cell damage. Ethanol (alcohol) derivatives such as methanol, ethanol, isopropanol, and glycerol have been used for tissue decellularization. They induce cell lysis by replacing intracellular fluid through osmotic pressure, leading to dehydration and destruction of cell membranes [40, 41]. Acids and bases are also common chemicals that can achieve the effect of decellularization by dissolving cellular components and denaturing DNA [41].

Detergents, including ionic, nonionic, and zwitterionic detergents, are the most commonly used chemical reagents for cell removal. Sodium dodecyl sulfate is the most common ionic detergent. Its amphiphilic nature makes it a highly effective surfactant capable of disrupting the phospholipid bilayer of exposed cells to achieve decellularization [42]. Sodium deoxycholate (SDC) is also a common ionic detergent that induces cleavage of cell membrane components by partially integrating cholate into the cell membrane, leading to cell lysis [42].

Polyethylene glycol octylphenyl ether (Triton X-100) is a representative of nonionic detergents, which provides gentler treatment by disrupting lipid-lipid, lipid-protein, and DNA-protein interactions to dissolve cell membranes and dissociate DNA from proteins but leave protein-protein interactions intact. Because nonionic detergents do not denature proteins, they can remove cellular contents without damaging the structure and orientation of collagen [43].

Alcohol ether sulfate (AES) is a novel zwitterionic detergent with an excellent protein solubilization effect, which can achieve the effect of removing cells by disrupting the cell membrane [44].

Enzymes are the most widely used biological methods that target protein bonds, disrupt normal cellular processes, and thus eliminate cells. Enzymes can provide high specificity for removing cellular residues or undesirable ECM components. However, complete cell removal by enzyme treatment alone is difficult, and enzyme residues may impair cell recellularization or cause adverse immune responses.

Nucleases (such as DNases and RNases) can cleave nucleic acid sequences, thus aiding in the clearance of nucleotides after cell lysis in tissues. Endonucleases such as benzonase may be more effective than exonucleases because they can cleave nucleotides in the middle of the sequence, thus more effectively segmenting DNA. Similarly, nonrestrictive endonucleases are more effective in segmenting DNA compared to sequence-dependent endonucleases [45].

The main endonuclease used for heart valve decellularization is trypsin, a serine protease commonly used as an enzymatic decellularization agent. This enzyme cleaves the entire peptide chain at the C-terminus of arginine or lysine. The use of a pancreatin enzyme for the decellularization of the porcine aorta resulted in severe damage to the ECM, particularly the loss of collagen and glycosaminoglycans, despite the complete removal of cellular components [46].

Researchers use cytotoxic agents to decellularize cells. Gilles et al. used latrunculin (Latrunculin B) and DNase to decellularize the anterior tibial muscles, preserving intact glycosaminoglycans while removing DNA [47], but it cannot remove some immunogenic components such as phospholipids [57]. Manuela et al. used Latruculin B to decellularize the bovine saphenous vein, which showed a better decellularization effect and lower inflammatory response compared to SDS treatment after implantation [48].

### 3.4.2 Gene Knockout Technology

The application of decellularization technology in the removal of immunogenicity of biomaterials is mainly based on two important hypotheses: (1) heterologous antigens are only found within cells and (2) the removal of cells under light microscopy is equivalent to the removal of immunogenicity of biomaterials. However, an increasing number of studies have confirmed that even after complete cell removal, the immunogenicity of biomaterials can still be detected [49]. Importantly, the persistent antigenicity of the removal of cells' biomaterial has been shown to stimulate recipient graft-specific adaptive immune responses, leading to structural damage of the biomaterial [50]. Due to the lack of correlation between cellular components of biomaterials and transplant-associated antigens or overall antigenicity, it is insufficient to rely solely on tissue decellularization when determining the degree of tissue immunogenicity. The use of gene knockout technology and targeted removal of specific known heterologous antigens can effectively remove the immunogenicity of heterologous tissue materials. Kuwaki et al. transplanted hearts from $\alpha$1, 3-galactosyltransferase-knockout pigs (GalT-KO) into baboons with no rejection [51]. Recently, a team from the University of Maryland edited ten genomes of pigs and transplanted pig hearts into a 57-year-old patient with nonischemic cardiomyopathy, reducing immune rejection and enabling patient survival for 60 days [52].

### 3.4.3 Protein Solubilization

Researchers have developed a set of antigen removal methods based on antigen solubility in order to remove unknown antigen components. Its principle is an adaptation of protein extraction techniques in proteomics [53]. Simple and effective separation of high solubility (hydrophilic) cytoplasmic proteins and low solubility (hydrophobic) membranes are associated proteins through protein compatibilization [54]. Researchers have added reducing agents and salts to reduce the residual antigen in bovine pericardium by 80% compared to decellularizing with a low osmotic solution of 60% and compared to decellularizing with 0.1% sodium dodecyl sulfate, without affecting the structural and functional characteristics of the ECM [55]. The graft-specific humoral adaptive immune response is more sensitive to residual antigens in the ECM than the cell-mediated immune response, providing direction for future research on xenobiotic materials.

### 3.4.4 Cross-Linking Technology

The cross-linking treatment of BHVs is also another common way to reduce immunogenicity. Cross-linking treatment can stabilize and mask the immunogenic components of biological materials, thereby improving the durability of valves. In 1971, the first biologic heart valves made from pericardial materials cross-linked with glutaraldehyde (GA) were put into clinical use with success by Marianne-Jean Ionesco [56]. Due to the strong cross-linking effect of glutaraldehyde on collagen fibers, it can stabilize and mask immune sites on collagen fibers, reducing tissue immunogenicity [57]. However, increasing evidence suggests that glutaraldehyde cross-linking can only partially mask immunogenicity with many related adverse reactions. Residual glutaraldehyde exhibits cytotoxic effects that prevent host cell attachment, migration, and proliferation, thereby increasing the risk of calcification and tissue fatigue [58]. Some studies demonstrated that antigenic masking after glutaraldehyde fixation is incomplete or reversible, and leads to chronic graft-specific immune responses and limited valve lifespan [59]. Other crosslinking agents such as EDC/NHS, epoxy compounds, polyphenols, and epoxides can also reduce the immunogenicity of tissues but are characterized by incomplete ECM cross-linking and slow degradation, leading to poor performance in vivo (Table 1).

## 4 Calcification Mechanism and Anti-calcification Strategy for BHV Materials

Porcine aortic valve leaflet and bovine pericardium are the main xenogeneic tissue biomaterials for the fabrication of BHVs currently, but they have some differences in composition and the cell types of porcine aortic valve leaflet primarily include

**Table 1** Methods and mechanisms of immunogenicity removal

| | | Mechanism | Pros | Cons | Ref |
|---|---|---|---|---|---|
| Decellularization | | | | | |
| Physical methods | Freeze and thaw cycles | Formation of ice crystals in intracellular spaces during rapid freezing and thawing, leading to rupture of cell membranes and cell lysis | No adverse effects on the ECM components and biomechanical properties | Remnant DNA fragments | [60, 61] |
| | Hypotonic and hypertonic solutions | Solute concentration gradient leads to cell membrane cleavage | Effective removal of cellular material<br>Maintains ECM protein | Remnant DNA fragments | [62] |
| | Vacuum | Facilitating penetration of enzymes or chemical detergents into tissues | Improves decellularization efficiency without compromising mechanical properties | Insufficiency and need for chemical and enzymatic co-treatment | [37] |
| | Supercritical carbon Dioxide | The pressure and supercritical environment bursts cells and depressurization removes cellular material from the extracellular matrix<br>The removal of cell components occurs via super-critical extraction | (1) Maintains ECM proteins and mechanical properties<br>(2) low cytotoxicity | Cellular or phospholipid residues | [38] |

| | | | | | |
|---|---|---|---|---|---|
| Chemical methods | Acids and bases:NaOH, HCl | Solubilize cytoplasmatic components of cells; disruption of nucleic acids and denature proteins | Reduced pathogen-associated immunogenicity due to simultaneous sterilization | Insufficient cell removal<br>Decreases salt- and acid-soluble<br>ECM proteins and damage mechanical performance | [29, 63] |
| | Ionic detergents:SDS, SDC, etc. | Dissolve cell membranes and nuclear membranes<br>Protein denaturation | High potency in removing protein antigens, especially in dense tissues | Disrupting the structure of ECM and releasing DAMP exposes hidden antigenic sites<br>Damaged GAG and growth factor<br>Cytotoxicity | [29, 64, 65] |
| | Nonionic detergents:Triton X-100, etc. | Solubilize membrane proteins; disrupts lipid–lipid<br>and protein–lipid interactions | Removal of DNA and membrane immunogens as well as lipids | Damaging ECM<br>Low efficiency in dense tissues<br>Cytotoxicity<br>Endocrine-disrupting properties | [44, 64] |
| | Zwitterionic detergents:CHAPS, Sulfobetaines, etc. | Solubilize membrane proteins; disrupts lipid–lipid<br>and protein–lipid interactions | Effective agents for solubilizing both Hydrophobic and hydrophilic immunogens | Damaging GAG and elastin<br>Low efficiency in dense tissues | [65, 66] |

(continued)

**Table 1** (continued)

| | | Mechanism | Pros | Cons | Ref |
|---|---|---|---|---|---|
| Biological agents | Nucleases | Catalyze the hydrolytic cleavage of ribonucleotide and deoxyribonucleotide chains | Eliminating nuclear antigens | Disrupting the structure of ECM<br>Releasing DAMP exposes hidden antigenic sites<br>Retention of enzymes reduces decellularization | [67] |
| | Trypsin | Cleaves various peptide bonds | Promoting cell-ECM dissociation | Disrupting the structure of ECM<br>Retention of enzymes reduces decellularization | [38, 68] |
| | Chelating agents:EDTA | | Promoting cell-ECM dissociation and nuclear material removal | Weak cell and antigen removal efficacy | [29, 69] |
| Gene knockout | | Targeted removal of xenoantigens known to elicit an immune response, e.g., αGal, Neu 5Gc, Sda | Substantial reduction of xenogeneic tissue-induced immune responses | Genetic modification may adversely affect the properties of biomaterials<br>Costs are high | [50, 70] |
| Protein solubilization | | Maintaining the solubility of heterogeneous proteins by means of reducing agents and salts that allow them to diffuse into aqueous solutions | Significantly reducing residual antigens without compromising biomaterial functional properties | Residual cell nuclei | [55] |

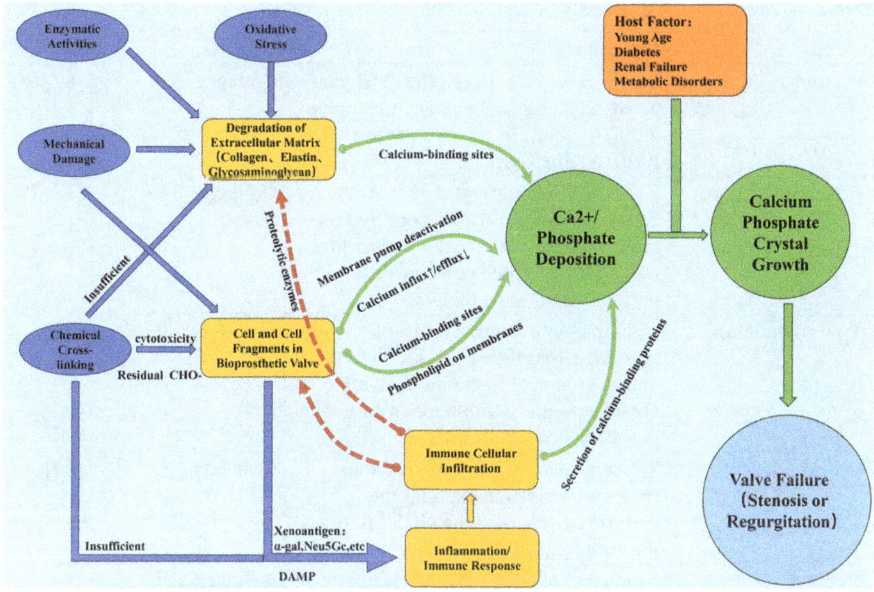


**Fig. 1** Mechanisms of calcification in bioprosthetic heart valves

endothelial cells and interstitial cells, and bovine pericardium primarily includes epithelial cells and interstitial cells. Both of the extracellular matrix components are mainly composed of collagen, elastin, and glycosaminoglycans, which provide mechanical strength and resilience, and reduce shear stress for valve materials, respectively [56]. However, the content of elastic fibers and glycosaminoglycans in porcine aortic valve leaflets was more abundant than in bovine pericardium. Calcification is the primary cause of the failure of BHVs in vivo. At present, the exact mechanism underlying the calcification of BHVs remains unclear although potential theories include immunological rejection, inflammatory cell aggregation, aldehyde toxicity, extracellular matrix degradation, hemodynamic alterations, and disorders involving the metabolism of calcium and phosphorus (Fig. 1). Optimized treatment or modification according to calcification mechanisms may improve the durability of the BHVs in vivo.

The main factors involved in the calcification of BHV are mechanical stress, chemical treatment, enzymatic hydrolysis, oxidative stress, residual cellular components, and extracellular matrix degradation, causing immune response and calcium phosphate deposition in the body. Ultimately, valve failure occurs due to increased calcium phosphate crystals and can be further accelerated by multiple host factors (DAMP: damage-associated molecular patterns) (Table 2).

**Table 2** Tissue cross-linking methods and their mechanism

| Methods | Mechanism | Location | Ref |
|---|---|---|---|
| Glutaraldehyde | The interaction between the aldehyde group of glutaraldehyde and the amino group of lysine or hydroxyproline forms a Schiff base | Collagen | [67, 68] |
| Hydroxy chromium | HC complexes through Cr3 + as the centrosome, followed by coordination cross-linking in molecular form with free carboxyl groups on matrix proteoglycans and collagen | Collagen | [77–79] |
| Epoxy compounds | Combine with amino groups of lysine/hydroxylysine and secondary amino groups of histidine in alkaline conditions, and combine with carboxyl groups in acidic conditions | Collagen | [63] |
| Carbodiimides | The adjacent primary amino group forms an amide tendon with the carboxyl group, and the NHS improves the coupling efficiency | Collagen | [64] |
| Genipin | Binds free amino groups on lysine, hydroxylysine, and arginine and polymerizes them together to form intermolecular and intramolecular cross-links | Collagen and elastin | [84] |
| Photoreactive reagents | Photosensitizer-mediated tryptophan and tyrosine absorb light wave radiation at specific wavelengths and produce intermolecular cross-links | Collagen and elastin | [85, 86] |
| Ribose | Nucleophilic addition reaction to amino group, amadorian rearrangement, and tautomerization | Collagen and elastin | [60] |
| Polyphenolic compounds | Phenolic hydroxyl groups are functional groups that act on hydrophobic regions to form intermolecular hydrogen bonds | Collagen and elastin | [88, 89] |
| Metal ions | Irreversibly bind to elastin in vivo and alter the spatial structure of elastin | Elastin | [96, 97] |
| Polyphenols combined with metal ions | Polyphenol compounds are rich in negatively charged phenolic hydroxyl groups and calcium ions may bind to pericardium-bound polyphenols resulting in calcification; the introduction of metal ions can competitively inhibit the binding of calcium ions to polyphenols | Collagen and elastin | [109] |

(continued)

**Table 2** (continued)

| | | | |
|---|---|---|---|
| Sodium metaphosphate | Sodium metaphosphate cross-linked glycosaminoglycans, sodium metaphosphate esterification, or etherification of hydroxyl groups in glycosaminoglycans and other reactions to improve mechanical properties and stability | Glycosaminoglycan | [104] |

## 4.1 Cellular Component

Regarding pericardial or leaflet tissue-derived biomaterials, their cells are mainly endothelial cells and interstitial cells. The immunogenicity of residual donor cells and biomaterial cellular debris makes it easy for recipients to trigger an immune response upon implantation, accelerating the xenografts' deterioration [71]. Calcium salt crystals tend to nucleate and concentrate in remaining cells, and the residual cellular components are essential contributors to tissue calcification [72]. Significant dendritic calcium deposition has been seen in the early phases of bovine pericardial tissue treated with glutaraldehyde and transplanted subcutaneously in rats [73]. This may be due to residual cellular components that are cleaved into cellular debris by initiating the host cell apoptotic signaling pathway after implantation. The cellular debris and phospholipids trigger an inflammatory response that leads to calcification.

**Anti-calcification Strategy** Removing endothelial cells, interstitial cells, and other cellular components further minimizes the immunogenicity of xenogeneic tissue materials and reduces calcification. Specific methods are described in Sect. 3.3 of this chapter.

## 4.2 Collagen

Collagen fibers are the most abundant extracellular matrix components of heart valves and pericardium, accounting for ~70% of tissue dry weight. Collagen fibers and elastic fibers are intertwined to form the fundamental skeleton of an extracellular matrix. Any damage to the skeleton's stability and integrity might increase the calcification of BHVs. Generally, collagen fibers with intact structure have the characteristics of collagen fibers that often exhibit minimal immunogenicity and high biocompatibility and are least liable to cause calcification. Nevertheless, limited mechanical strength, poor thermal stability, and weak resistance to enzymatic hydrolysis are the drawbacks of native collagen. In vivo, blood flow stress accelerates the leaflet's destruction of the structure, exposing antigen and calcium nucleodeposition sites, causing calcium nucleation and inflammatory immune response, and ultimately leading to calcification.

**Anti-calcification Strategy Based on Collagen** To maintain the structural integrity of collagen fibers, various physical, chemical, and emerging biological cross-linking technologies have been used to alter or modify the molecular connectivity of collagen fibers and increase the stability of internal structures. Chemical cross-linking techniques—represented by glutaraldehyde, and also hydroxychromium, epoxy compounds, and carbodiimides (EDC/NHS)—are the most frequently employed. These techniques can create stable chemical linkages between biomacromolecules [74, 75]. In addition, research is taking place on innovative biological and physical cross-linking methods, such as enzymatic oxidative cross-linking, dye-mediated photooxidation, and dehydration heat treatment [76, 77]. The cross-linking of collagen fibers is described as follows:

### 4.2.1 Glutaraldehyde

Currently, glutaraldehyde cross-linking treatment is the primary technique used to manufacture commercial BHVs; it is characterized by its ease of utilization, low cost, and high degree of collagen cross-linking. The cross-linking by the glutaraldehyde aldehyde group forms covalent bonds with the free amino groups of hydroxyproline or lysine. Furthermore, glutaraldehyde cross-links antigenic components in biomaterials, including peptides, glycolipids, lipoproteins, and lipopolysaccharides, to form insoluble macromolecules that minimize material antigenicity and effectively hide antigenic determinants [78]. Biological materials that have been cross-linked with glutaraldehyde exhibit more compressed tissue, enhanced resistance to enzymatic hydrolysis, and higher biological stability. However, glutaraldehyde cross-linked residual aldehyde groups are cytotoxic and tend to cause inactivation of cells in vivo in the persistent host [79]. Valve calcification, on the other hand, is often thought to begin in devitalized connective tissue cells. Cell inactivation allows membrane phospholipids to react with free calcium, resulting in nucleation of calcium and phosphorus deposits [72]. The initial calcified precipitates eventually enlarge and merge, producing severely calcified nodules that harden and weaken the tissue, resulting in debilitating loss of function of the bioprosthesis calcification.

Based on the side effects of the glutaraldehyde residual aldehyde group, eliminating or blocking a residual aldehyde group is an important way to improve the anti-calcification performance of glutaraldehyde cross-linked biomaterials. Modification methods are described below:

1. Adequate rinsing significantly reduces the free aldehyde content [80]. The amine groups of collagen fibers can be covalently bound to free aldehyde groups which can also bind to collagen fibers by physisorption. Free aldehydes on the surface can be reduced if repeatedly rinsed using a phosphate buffer.

2. To neutralize residual aldehyde groups, the groups are blocked using amino compounds such as oleic acid, binary amines, and amino acids (glycine, taurine, arginine, glutamic acid, etc.) [81–83]. Moreover, it has been confirmed that glutathione can efficiently eliminate the remaining aldehyde group of bovine pericardium that has been cross-linked with glutaraldehyde and can enhance the anti-calcification performance and biocompatibility of biomaterials. This can all be achieved without affecting the mechanical properties of cross-linked biomaterials.
3. Chemical reactions, such as reduction reaction, acetal reaction, and addition reaction, are used to convert the free aldehyde group into a hydroxyl group and carboxyl group and eliminate the aldehyde group in tissue. For example, sodium borohydride can reduce aldehyde groups to alcohols, neutralize them using citric acid, and remove the groups with sodium bisulfite, further reducing the cytotoxicity of glutaraldehyde and reducing tissue calcification [80, 84, 85]. However, these reaction processes may lead to weakened performance of cross-linked tissues and the generation of new toxic substances and, therefore, not widely recommended for use in clinical applications.
4. The no-touch technique is another method where more aldehyde groups and free glutaraldehyde may persist in the tissue and valve surface when glutaraldehyde-cross-linked BHVs are preserved in low-dose glutaraldehyde solution. The current generation of commercial valves employs preservation technologies such as capping and no-glutaraldehyde touch, capping of ethanolamine and sodium borohydride, and capping of glycerol and ethanol dehydration to preserve bioprostheses [86, 87]. These techniques have been demonstrated to significantly decrease the content of acidic and aldehyde groups in tissue, reduce tissue toxicity, and enhance valve anti-calcification performance.

### 4.2.2 Hydroxy Chromium

Jin et al. proposed that among the extracellular matrix components of bovine pericardium, the free carboxyl groups carried by collagen polypeptide chain molecules and a large number of free carboxyl groups contained in tissues are the main causes of bovine pericardial valve calcification [88]. Lei used coordination compounds to cross-link the free carboxyl groups of bovine pericardium and combined glutaraldehyde and hydroxychromium to cross-link the amino and carboxyl groups to improve the anti-calcification effect of bioprosthetic valve materials [89, 90]. The anti-calcification effect of hydroxychromium is the formation of HC complexes through Cr3 + as the centrosome, followed by coordination cross-linking in molecular form with free carboxyl groups on matrix proteoglycans and collagen in valve tissue. This cross-linking is a multi-point binding in three-dimensional space, which not only seals the calcium and phosphorus binding sites and stabilizes proteoglycans but also reduces the water-soluble space required for hydroxyapatite formation by cross-linked collagen fibers, thus reducing the calcification of biomaterials.

### 4.2.3 Polyepoxy Compound

Epoxy compound (EC) can combine with amino and carboxyl groups of collagens to produce chemical modification and cross-linking. EC has good hydrophilicity and low cytotoxicity. EC cross-linked biomaterial has fine biomechanical properties and anti-calcification properties and is considered an alternative to glutaraldehyde. In addition, EC can react with biomaterials under both acidic and alkaline conditions. In alkaline conditions (pH > 8), EC mainly reacts with amino groups (optimum pH = 8.5 ~ 9.0) and has high cross-linking efficiency for collagen fibers; in acidic conditions (pH < 6), EC mainly reacts with carboxyl groups (optimum pH = 4.5 ~ 5.0) but does not react with amino groups at the same time. Therefore, the cross-linking reaction conditions can be adjusted according to actual needs, such as soft tissue can be obtained by cross-linking under acidic conditions and hard spongy tissue can be obtained under alkaline conditions. Ethylene glycol diglycidyl ether and 1,4-butanediol diglycidyl ether are used more frequently for cross-linking biological valve materials and have carbon chain lengths and spatial conformations similar to glutaraldehyde [75]. Structurally, they all have at least two or more active functional groups capable of binding to amino groups of collagen to form stable covalent cross-links [91]. The advantages of epoxy compounds are low toxicity, good solubility, and their good binding ability to hydroxyl and carboxyl groups of collagen fibers [92].

### 4.2.4 EDC/NHS

EDC is a water-soluble carbodiimide and is commonly used as an activator of carboxyl groups. EDC can precipitate collagen cross-linking, which reacts with carboxyl groups on aspartate and glutamate residues in collagen to form a precipitant of unstable urea derivatives. NHS, on the other hand, is an amino group activator that enhances the stability of carbodiimide cross-linked products by forming more stable esters. EDC/NHS do not enter the collagen matrix during cross-linking but instead turn into water-soluble urea derivatives and are therefore called zero-length crosslinkers [93]. Raghavan D combined with EDC/NHS and glutaraldehyde cross-linked porcine aortic valves showed that the two combinations had better anti-calcification properties than glutaraldehyde alone [94]. However, the protein by-products produced by EDC/NHS cross-linking have some biological toxicity, thus limiting their clinical application.

### 4.2.5 Other Types of Crosslinkers

*Genipin* is a natural compound with low toxicity and belongs to the iridoid class, which binds free amino groups on lysine, hydroxylysine, and arginine and polymerizes them to form intermolecular and intramolecular cross-links [95]. However, genipin is expensive and has limited clinical application. Under UV irradiation, a nonspecific photoinduced reaction group initiates a cross-linking reaction with nearby molecules. The majority of photoreactive crosslinkers are heterobifunctional agents that have reactive groups that are triggered by light as well as groups that react with thiol or amino groups.

*Photoreactive reagents* are mainly based on azbin and carbene chemistry, and their precursors include arylazides, bisacridone, and diphenylmethanone, among others [96]. Photoreactive reagents, tryptophan and tyrosine, form intermolecular cross-links when they absorb light wave radiation at specific wavelengths through the action of photosensitizers. Studies have shown that pericardial biomaterial prepared by riboflavin 5′-monophosphate (FMN) or methylene blue initiated photo cross-linking has stable collagen and good anti-calcification properties [97].

*Ribose* is a pentasaccharide that is highly reducing and cross-linked to biomaterials. Findings have shown that ribose has a definite cross-linking effect on collagen in bovine pericardium in which the possible reaction process is a nucleophilic addition reaction between a ribose ring opening and the amino group of collagen fibers to generate enamines. Enamines at different sites undergo amadorian rearrangement reaction by themselves, tautomerization to form olefins, and repeat nucleophilic addition reaction, thus forming stable chemical bond cross-linking between two collagen fibers and improving the mechanical properties and anti-calcification properties of bovine pericardium materials [71].

## 4.3 *Elastin*

Elastin is another major component of the ECM of native pericardium or valves. Elastin is mainly interdigitally linked by peptide chains composed of glycine (30%), proline (10%), hydroxyproline (1%), and alanine to form an extendable three-dimensional network that coexists with collagen fibers and confers elasticity and tensile resistance to tissues [98]. Because glutaraldehyde does not cross-link elastin, elastin that has not been cross-linked tends to be further degraded and absorbed by host cells after implantation in vivo. The loss of elastin may lead to reduced elastic return and tissue elasticity [99, 100]. The BHV could calcify and suffer structural damage as a result of elastin degradation [101]. The elastin degradation location is the aggregation point where calcification occurs. Fragmented elastin is more likely to cause calcification, and the inflammatory response induced after biomaterial implantation in vivo can regulate the expression of matrix metalloproteinases [72]. This results in elastic fiber fragmentation and degradation, and the production of elastin-derived peptides triggers a variety of cellular responses, such as increased osteoblast expression. Fragmented elastin has an increased affinity for calcium due to enhanced polarity, promoting free calcium ion enrichment and deposition and becoming the site of mineral nucleation and growth.

**Anti-calcification Strategies Based on Elastin** Currently, exogenous elastin introduction, metal ion cross-linking, polyphenol compound cross-linking, elastase activity suppression in vivo, and other methods often employ elastin protection strategies.

### 4.3.1 Polyphenolic Compounds

Polyphenol compounds are secondary metabolites in plants, widely distributed in fruits and vegetables, and are commonly used as natural antioxidants, free radical scavengers, and cardiovascular protective agents. The protective agents contain a large number of hydrophobic aromatic moieties and stabilize elastin by hydrogen bonding with hydrophobic amino acids in elastic fibers [99, 100]. At present, it is mainly reported that tannic acid, Pentagalloyl glucose (PGG), proanthocyanidins, quercetin, tea polyphenols, curcumin, etc. [102–106] are naturally safe and nontoxic polyphenols and can crosslink with collagen fibers in addition to crosslinked elastin. The principle is to specifically bind the phenolic hydroxyl group to the hydroxyproline hydroxyl group, e-amino group, amide carbonyl group in collagen, and hydrophobic structural region of elastin (such as proline) and form multiple hydrogen bonds and protect them from degradation. Naringenin and dihydromyricetin have been proposed as new polyphenol crosslinkers for elastin cross-linking. This combination with glutaraldehyde has good anti-calcification effects and biocompatibility.

### 4.3.2 Metal Ions

Metal ions ($Al^{3+}$, $Mg^{2+}$, $Fe^{3+}$, etc.) irreversibly bind to elastin in vivo and alter the spatial structure of elastin. Narendra Vyavahare found that aluminum ions can enhance elastin anti-calcification ability, and in addition, aluminum ions combined with elastin also prevent matrix metalloproteinases and alkaline phosphatase-mediated elastin dissolution, further inhibiting elastin calcification [107]. It has also been shown that $Mg^{2+}$ can interact with elastin and promote its conformational change, with at least two conformational distinct metal ion binding sites on elastin molecules [108].

### 4.3.3 Polyphenols Combined with Metal Ions

The oxygen atoms of polyphenols have strong coordination ability and easily form cyclic chelates with central ions such as $Ni^{2+}$, $Bi^{3+}$, $Zn^{2+}$, $Cr^{6+}$, $Ag^{+}$, $Sr^{2+}$, $Hg^{2+}$, $Fe^{3+}$, $Al^{3+}$, $Fe^{2+}$, $Cu^{2+}$, and $Mn^{2+}$. These chelates also have good biological activity [109]. Because polyphenol compounds are rich in negatively charged phenolic hydroxyl groups, and calcium ions may bind to pericardium-bound polyphenols resulting in calcification, the introduction of metal ions can competitively inhibit the binding of calcium ions to polyphenols, while further stabilizing the polyphenol structure, with a positive synergistic anti-calcification effect. For example, Wang et al. found that the use of polyphenols combined with trivalent metal ions (aluminum ions and iron ions) can effectively resist elastin calcification. Using the hydroxyl groups rich in polyphenols to form hydrogen bonds with elastin components in the porcine pericardium plays a role in stabilizing elastin. Then, introducing metal ions in combination with polyphenols is expected to inhibit BHV calcification [110].

In addition to protecting elastin by cross-linking fixation with crosslinkers and increasing elastin content and stability using exogenous protoelastin, this approach is also one of the strategies to inhibit biomaterials. Soluble elastin molecules are mainly cross-linked through a chain-locking element structure to form insoluble elastin in nature; for example, Wang Yunbing's group used exogenous protoelastin (TE)/lysine oxidase (LOXL) and tea polyphenols to increase the content of additional elastin and stabilize elastin, which in turn inhibits calcification [109]. It has been shown that epichlorohydrin-cross-linked BHV can inhibit the expression of MMP-9 and MMP-2 in tissues, and both enzymes have the ability to degrade elastin [111].

## 4.4 *Glycosaminoglycans*

Glycosaminoglycans are high molecular weight, hydrophilic, and highly negatively charged polysaccharides that are a general term for glycans in proteoglycan macromolecules, are integrated into the extracellular matrix in a highly ordered structure, and are the main components in the extracellular matrix of the pericardium or heart valves in addition to collagen and elastin [112]. Glycosaminoglycans are present in the cavernous layer of the heart valve cusps and prevent local compression buckling during valve opening and closing. The glycosaminoglycan species present in heart valves are mainly hyaluronic acid and chondroitin sulfate, and their main functions are to maintain the morphology of heart valves, inhibit platelet adhesion, interact with growth factors and receptor factors, and play a role in the conduction of cell signals and angiogenesis [112].

Schoen FJ concluded that glycosaminoglycans decrease shear stress and increase water content in the tissue of the BHV, thereby slowing the process of calcification and improving valve durability [113]. Loss of glycosaminoglycans leads to the destruction of pericardial tissue to form voids, which allow the deposition of hydroxyapatite crystals. For example, Herrero et al. found that pericardial calcification worsened after the removal of glycosaminoglycans from pericardial tissue [114].

**Anti-calcification Strategy Based on Glycosaminoglycans** Glycosaminoglycans do not contain active amino groups, and glutaraldehyde cross-linking technology cannot achieve stabilization of glycosaminoglycans in tissues. In the current study, there are few reports on how to efficiently inhibit the degradation of glycosaminoglycans. The research methods mainly focus on increasing the introduction of exogenous glycosaminoglycans and cross-linking fixation and protection against enzymatic hydrolysis to increase the glycosaminoglycan content in tissues and reduce the loss of components.

### 4.4.1 Exogenous Polysaccharide Coating

Hyaluronic acid and chondroitin sulfate are two major types of glycosaminoglycan in heart valves and have repeating units of polysaccharides that are very similar structures to starch. Wang Yunbing et al. used a combination of exogenous hyaluronic acid/chondroitin sulfate and sodium metaphosphate to increase the content and stability of glycosaminoglycans [115]. Some studies have also shown that increasing hyaluronic acid (HA) and its derivatives in tissue prevented the formation of calcium phosphate deposits, reduced the shear stress distribution of the BHV leaflets, and further improved the anti-calcification properties of BHVs [116].

### 4.4.2 Protection Against Enzymatic Hydrolysis and Cross-linking Fixation

Neomycin is an aminoglycoside antibiotic that is an inhibitor of glycosaminoglycans. Neomycin binds to glycosaminoglycans and attaches to the extracellular matrix, stabilizing glycosaminoglycans by reducing the site of action of glycosaminoglycans by binding to the carboxyl group of glycosaminoglycans. Raghavan D et al. found that the degradation of glycosaminoglycans in bovine pericardial valves was inhibited by treating with neomycin before glutaraldehyde cross-linking. [117]. Cross-linking immobilization against glycosaminoglycan components has also been investigated, such as Mercuri JJ using carbodiimide (EDC) and N-hydroxysuccimide (NHS) to increase resistance to enzymes by cross-linking the carboxyl groups of glycosaminoglycans [118]. Alternatively, changing the spatial conformation of glycosaminoglycans to increase resistance to enzymes, such as using periodate to inhibit the degradation of glycosaminoglycans by cross-linking with double-stranded diols in the uronic acid moiety of glycosaminoglycans. Sodium metaphosphate can improve the mechanical properties and stability of biomaterials through esterification or etherification reactions with hydroxyl groups in glycosaminoglycans.

The ideal treatment of BHV materials is not satisfied by just one or two treatment strategies. The main focus of current research is how to optimize cross-linking technology to enhance the mechanical properties, biocompatibility, and stability of biomaterials, and reduce SVD. According to the characteristics of different components in biomaterials and the advantages of different cross-linking reagents, composite cross-linking and modification may have better potential.

## 5 Endothelialization Reduces Calcification of BHV

Endothelialization of the BHV contributes to valve durability due to prolonged contact with blood. The native valve tissue is mainly composed of collagen, elastin, glycoprotein, endothelial cells, and valve mesenchymal cells. When the endothelial cells are covered completely, they exert their barrier function in vivo. At present, commercial BHVs are mainly derived from bovine pericardium and porcine aortic valve biomaterials, and there is a lack of coverage of normal endothelial cells on the leaflets. This results in a lack of normal cell shielding effect, antithrombotic effect, and endocrine function of BHV in vivo. Glutaraldehyde-crosslinked tissue, due to residual toxicity, is not conducive to endothelial cell regeneration and proliferation. Therefore, rapid endothelialization, or surface modification techniques that effectively capture endothelial cells or endothelial progenitor cells in the blood, can significantly improve the hemocompatibility of valves and improve the service life of BHVs.

Endothelial cells and mesenchymal cells in valve tissue maintain the active function of valve tissue. Endothelial cells play a very important role in preventing thrombosis and calcium deposition. Therefore, the study of endothelialization of BHVs has been increasing in recent years [119, 120]. The surface modification technology of BHV is the main area of current research, in order to improve the biocompatibility of the interface and accelerate endothelialization. At present, the main technical strategies include: (1) modification of the material interface with affinity/hydrophobic, zwitterionic, and specific functional group modification techniques by physical modification or chemical modification on the biomaterial interface [121]; (2) loading endothelialization bioactive factors at the biomaterial interface by chemical grafting, including cytokines that promote endothelial cell adhesion, proliferation, or migration, such as VEGF, REDV, SDF-α, and CD34 antibodies to capture endothelial cells or endothelial cells in the blood [122–126]; and (3) through the cultivation, proliferation, and differentiation of endothelial cells in vitro (such as umbilical vein endothelial cells, mesenchymal stem cells, etc.), they are replanted into the host to achieve endothelial cell coverage of the BHV [127, 128].

The majority of BHVs' endothelialization procedures are now in the research stage, and there are still numerous obstacles to overcome before they can be used in clinical applications. The obstacles mainly include [129, 130] sustained release control of active factors; source and expansion of exogenous endothelial cells; maintenance of in vivo activity; control of immune-inflammatory response; biosafety of in vitro culture; and how to sterilize and preserve them. These are important technological issues that will be resolved in the future.

## 6 Summary

The clinical demand for BHVs is steadily increasing, while the current BHVs still have the problem of poor durability, mainly due to immune response, calcification, and structural degeneration. This section summarizes the source, processing, and modification techniques of bioprosthetic valve materials, the mechanism of calcification and immunogenicity in vivo, the prevention of failure techniques, and the current research status and focus areas of BHVs.

## References


1. Schoen F J, Levy R J. Calcification of tissue heart valve substitutes: progress toward understanding and prevention [J]. The Annals of thoracic surgery, 2005, 79(3): 1072–1080.
2. Paradis J M, Del Trigo M, Puri R, et al. Transcatheter Valve-in-Valve and Valve-in-Ring for Treating Aortic and Mitral Surgical Prosthetic Dysfunction [J]. J Am Coll Cardiol, 2015, 66(18): 2019–2037.
3. Pibarot P, Dumesnil J G. Prosthetic heart valves: selection of the optimal prosthesis and long-term management [J]. Circulation, 2009, 119(7): 1034–1048.
4. Naimark W A, Lee J M, Limeback H, et al. Correlation of structure and viscoelastic properties in the pericardia of four mammalian species [J]. The American journal of physiology, 1992, 263(4 Pt 2): H1095–1106.
5. Rassoli A, Fatouraee N, Guidoin R, et al. Comparison of tensile properties of xenopericardium from three animal species and finite element analysis for bioprosthetic heart valve tissue [J]. Artif Organs, 2020, 44(3): 278–287.
6. Neethling W M, Papadimitriou J M, Swarts E, et al. Kangaroo versus porcine aortic valve tissue—valve geometry morphology, tensile strength and calcification potential [J]. The Journal of cardiovascular surgery, 2000, 41(3): 341–348.
7. Mao J, Rassoli A, Tong Y, et al. Donkey pericardium compares favorably with commercial xenopericardia used in the manufacture of transcatheter heart valves [J]. Artif Organs, 2019, 43(10): 976–987.
8. Song M, Tang Z, Liu Y, et al. Yak Pericardium as an Alternative Biomaterial for Transcatheter Heart Valves [J]. Frontiers in bioengineering and biotechnology, 2021, 9: 766991.
9. Maestro M M, Turnay J, Olmo N, et al. Biochemical and mechanical behavior of ostrich pericardium as a new biomaterial [J]. Acta biomaterialia, 2006, 2(2): 213–219.
10. Libinhan. Study on swim bladder derived cardiovascular biomaterialswith anti-calcification properties [D], 2020.
11. Kawecki M, Łabuś W, Klama-Baryla A, et al. A review of decellurization methods caused by an urgent need for quality control of cell-free extracellular matrix' scaffolds and their role in regenerative medicine [J]. Journal of biomedical materials research Part B, Applied biomaterials, 2018, 106(2): 909–923.
12. Gorham S D, Hyland T P, French D A, et al. Cellular invasion and breakdown of three different collagen films in the lumbar muscle of the rat [J]. Biomaterials, 1990, 11(2): 113–118.
13. Galili U. Discovery of the natural anti-Gal antibody and its past and future relevance to medicine [J]. Xenotransplantation, 2013, 20(3): 138–147.
14. Kobayashi T, Cooper D K. Anti-Gal, alpha-Gal epitopes, and xenotransplantation [J]. Subcellular biochemistry, 1999, 32: 229–257.
15. Galili U. The alpha-gal epitope and the anti-Gal antibody in xenotransplantation and in cancer immunotherapy [J]. Immunology and cell biology, 2005, 83(6): 674–686.
16. Chong A S, Habal M V. From bench to bedside: reversing established antibody responses and desensitization [J]. Current opinion in organ transplantation, 2022, 27(5): 376–384.
17. Heidt S, Haasnoot G W, van der Linden-Van oevelen M J H, et al. Highly Sensitized Patients Are Well Served by Receiving a Compatible Organ Offer Based on Acceptable Mismatches [J]. Frontiers in immunology, 2021, 12: 687254.
18. Kim Y G, Gil G C, Harvey D J, et al. Structural analysis of alpha-Gal and new non-Gal carbohydrate epitopes from specific pathogen-free miniature pig kidney [J]. Proteomics, 2008, 8(13): 2596–2610.

19. Irie A, Koyama S, Kozutsumi Y, et al. The molecular basis for the absence of N-glycolylneuraminic acid in humans [J]. J Biol Chem, 1998, 273(25): 15866–15871.
20. Martin P T, Camboni M, Xu R, et al. N-Glycolylneuraminic acid deficiency worsens cardiac and skeletal muscle pathophysiology in α-sarcoglycan-deficient mice [J]. Glycobiology, 2013, 23(7): 833–843.
21. Pham T, Gregg C J, Karp F, et al. Evidence for a novel human-specific xeno-auto-antibody response against vascular endothelium [J]. Blood, 2009, 114(25): 5225–5235.
22. Saethre M, Baumann B C, Fung M, et al. Characterization of natural human anti-non-gal antibodies and their effect on activation of porcine gal-deficient endothelial cells [J]. Transplantation, 2007, 84(2): 244–250.
23. Tseng Y L, Moran K, Dor F J, et al. Elicited antibodies in baboons exposed to tissues from alpha1,3-galactosyltransferase gene-knockout pigs [J]. Transplantation, 2006, 81(7): 1058–1062.
24. Kostrzewa-Nowak D, Ciechanowicz A, Clark J S C, et al. Damage-Associated Molecular Patterns and Th-Cell-Related Cytokines Released after Progressive Effort [J]. Journal of clinical medicine, 2020, 9(3).
25. Gong T, Liu L, Jiang W, et al. DAMP-sensing receptors in sterile inflammation and inflammatory diseases [J]. Nature reviews Immunology, 2020, 20(2): 95–112.
26. Wong M L, Wong J L, Vapniarsky N, et al. In vivo xenogeneic scaffold fate is determined by residual antigenicity and extracellular matrix preservation [J]. Biomaterials, 2016, 92: 1–12.
27. Callemeyn J, Lamarthée B, Koenig A, et al. Allorecognition and the spectrum of kidney transplant rejection [J]. Kidney international, 2022, 101(4): 692–710.
28. Franzin R, Stasi A, Fiorentino M, et al. Corrigendum: Inflammaging and Complement System: A Link Between Acute Kidney Injury and Chronic Graft Damage [J]. Frontiers in immunology, 2020, 11: 630855.
29. Chakraborty J, Roy S, Ghosh S. Regulation of decellularized matrix mediated immune response [J]. Biomater Sci, 2020, 8(5): 1194–1215.
30. Salti H, Kramer L, Nelz S C, et al. Decellularization of PRECISION-cut kidney slices-application of physical and chemical methods [J]. Biomed Mater, 2023, 18(2).
31. Paulo Zambon J, Atala A, Yoo J J. Methods to generate tissue-derived constructs for regenerative medicine applications [J]. Methods (San Diego, Calif), 2020, 171: 3–10.
32. Meyer S R, Chiu B, Churchill T A, et al. Comparison of aortic valve allograft decellularization techniques in the rat [J]. Journal of biomedical materials research Part A, 2006, 79(2): 254–262.
33. Tudorache I, Cebotari S, Sturz G, et al. Tissue engineering of heart valves: biomechanical and morphological properties of decellularized heart valves [J]. The Journal of heart valve disease, 2007, 16(5): 567–573; discussion 574.
34. Cebotari S, Tudorache I, Jaekel T, et al. Detergent decellularization of heart valves for tissue engineering: toxicological effects of residual detergents on human endothelial cells [J]. Artif Organs, 2010, 34(3): 206–210.
35. Azhim A, Yamagami K, Muramatsu K, et al. The use of sonication treatment to completely decellularize blood arteries: a pilot study [J]. Annual International Conference of the IEEE Engineering in Medicine and Biology Society IEEE Engineering in Medicine and Biology Society Annual International Conference, 2011, 2011: 2468–2471.
36. Butler C R, Hynds R E, Crowley C, et al. Vacuum-assisted decellularization: an accelerated protocol to generate tissue-engineered human tracheal scaffolds [J]. Biomaterials, 2017, 124: 95–105.
37. Luo Y, Ma L. Bioprosthetic heart valves with reduced immunogenic residuals using vacuum-assisted decellularization treatment [J]. Biomed Mater, 2020, 15(6): 065012.
38. Duarte M M, Silva I V, Eisenhut A R, et al. Contributions of supercritical fluid technology for advancing decellularization and postprocessing of viable biological materials [J]. Materials horizons, 2022, 9(3): 864–891.
39. Guler S, Aslan B, Hosseinian P, et al. Supercritical Carbon Dioxide-Assisted Decellularization of Aorta and Cornea [J]. Tissue engineering Part C, Methods, 2017, 23(9): 540–547.

40. Levy R J, Vyavahare N, Ogle M, et al. Inhibition of cusp and aortic wall calcification in ethanol- and aluminum-treated bioprosthetic heart valves in sheep: background, mechanisms, and synergism [J]. The Journal of heart valve disease, 2003, 12(2): 209–216; discussion 216.
41. Montoya C V, Mcfetridge P S. Preparation of ex vivo-based biomaterials using convective flow decellularization [J]. Tissue engineering Part C, Methods, 2009, 15(2): 191–200.
42. Fang N T, Xie S Z, Wang S M, et al. Construction of tissue-engineered heart valves by using decellularized scaffolds and endothelial progenitor cells [J]. Chinese medical journal, 2007, 120(8): 696–702.
43. Yu B T, Li W T, Song B Q, et al. Comparative study of the Triton X-100-sodium deoxycholate method and detergent-enzymatic digestion method for decellularization of porcine aortic valves [J]. European review for medical and pharmacological sciences, 2013, 17(16): 2179–2184.
44. Song M, Yi L, Tang Z, et al. Preliminary study of improving immune tolerance in vivo of bioprosthetic heart valves through a novel antigenic removal method [J]. Frontiers in bioengineering and biotechnology, 2023, 11: 1141247.
45. Petersen T H, Calle E A, Zhao L, et al. Tissue-engineered lungs for in vivo implantation [J]. Science, 2010, 329(5991): 538–541.
46. Yang M, Chen C Z, Wang X N, et al. Favorable effects of the detergent and enzyme extraction method for preparing decellularized bovine pericardium scaffold for tissue engineered heart valves [J]. Journal of biomedical materials research Part B, Applied biomaterials, 2009, 91(1): 354–361.
47. Gillies A R, Smith L R, Lieber R L, et al. Method for decellularizing skeletal muscle without detergents or proteolytic enzymes [J]. Tissue engineering Part C, Methods, 2011, 17(4): 383–389.
48. Gui L, Chan S A, Breuer C K, et al. Novel utilization of serum in tissue decellularization [J]. Tissue engineering Part C, Methods, 2010, 16(2): 173–184.
49. Hülsmann J, Grün K, El Amouri S, et al. Transplantation material bovine pericardium: biomechanical and immunogenic characteristics after decellularization vs. glutaraldehyde-fixing [J]. Xenotransplantation, 2012, 19(5): 286–297.
50. Senage T, Paul A, Le Tourneau T, et al. The role of antibody responses against glycans in bioprosthetic heart valve calcification and deterioration [J]. Nature medicine, 2022, 28(2): 283–294.
51. Kuwaki K, Tseng Y L, Dor F J, et al. Heart transplantation in baboons using alpha1,3-galactosyltransferase gene-knockout pigs as donors: initial experience [J]. Nature medicine, 2005, 11(1): 29–31.
52. Mohiuddin M M, Singh A K, Scobie L, et al. Graft dysfunction in compassionate use of genetically engineered pig-to-human cardiac xenotransplantation: a case report [J]. Lancet, 2023, 402(10399): 397–410.
53. Arai S, Lacerda C, Orton E C. Tissue-gel electrophoresis enhances antigen removal from porcine aortic valve and bovine pericardium [J]. The Journal of heart valve disease, 2010, 19(6): 753–758.
54. Privé G G. Detergents for the stabilization and crystallization of membrane proteins [J]. Methods (San Diego, Calif), 2007, 41(4): 388–397.
55. Wong M L, Leach J K, Athanasiou K A, et al. The role of protein solubilization in antigen removal from xenogeneic tissue for heart valve tissue engineering [J]. Biomaterials, 2011, 32(32): 8129–8138.
56. Umashankar P R, Arun T, Kumari T V. Short duration gluteraldehyde cross linking of decellularized bovine pericardium improves biological response [J]. Journal of biomedical materials research Part A, 2011, 97(3): 311–320.
57. Gu L, Shan T, Ma Y X, et al. Novel Biomedical Applications of Crosslinked Collagen [J]. Trends in biotechnology, 2019, 37(5): 464–491.
58. Gu Y, Wang F, Wang R, et al. Preparation and evaluation of decellularized porcine carotid arteries cross-linked by genipin: the preliminary results [J]. Cell and tissue banking, 2018, 19(3): 311–321.

59. Dalgliesh A J, Parvizi M, Lopera-Higuita M, et al. Graft-specific immune tolerance is determined by residual antigenicity of xenogeneic extracellular matrix scaffolds [J]. Acta biomaterialia, 2018, 79: 253–264.
60. Li N, Li Y, Gong D, et al. Efficient decellularization for bovine pericardium with extracellular matrix preservation and good biocompatibility [J]. Interactive cardiovascular and thoracic surgery, 2018, 26(5): 768–776.
61. Xing Q, Yates K, Tahtinen M, et al. Decellularization of fibroblast cell sheets for natural extracellular matrix scaffold preparation [J]. Tissue engineering Part C, Methods, 2015, 21(1): 77–87.
62. Greco K V, Francis L, Somasundaram M, et al. Characterisation of porcine dermis scaffolds decellularised using a novel non-enzymatic method for biomedical applications [J]. J Biomater Appl, 2015, 30(2): 239–253.
63. Syed O, Walters N J, Day R M, et al. Evaluation of decellularization protocols for production of tubular small intestine submucosa scaffolds for use in oesophageal tissue engineering [J]. Acta biomaterialia, 2014, 10(12): 5043–5054.
64. Keane T J, Swinehart I T, Badylak S F. Methods of tissue decellularization used for preparation of biologic scaffolds and in vivo relevance [J]. Methods (San Diego, Calif), 2015, 84: 25–34.
65. Kasravi M, Ahmadi A, Babajani A, et al. Immunogenicity of decellularized extracellular matrix scaffolds: a bottleneck in tissue engineering and regenerative medicine [J]. Biomaterials research, 2023, 27(1): 10.
66. Tsuchiya T, Balestrini J L, Mendez J, et al. Influence of pH on extracellular matrix preservation during lung decellularization [J]. Tissue engineering Part C, Methods, 2014, 20(12): 1028–1036.
67. Yu C, Bianco J, Brown C, et al. Porous decellularized adipose tissue foams for soft tissue regeneration [J]. Biomaterials, 2013, 34(13): 3290–3302.
68. Rieder E, Kasimir M T, Silberhumer G, et al. Decellularization protocols of porcine heart valves differ importantly in efficiency of cell removal and susceptibility of the matrix to recellularization with human vascular cells [J]. The Journal of thoracic and cardiovascular surgery, 2004, 127(2): 399–405.
69. Emami A, Talaei-Khozani T, Vojdani Z, et al. Comparative assessment of the efficiency of various decellularization agents for bone tissue engineering [J]. Journal of biomedical materials research Part B, Applied biomaterials, 2021, 109(1): 19–32.
70. Casós K, Llatjós R, Blasco-Lucas A, et al. Differential Immune Response to Bioprosthetic Heart Valve Tissues in the α1,3Galactosyltransferase-Knockout Mouse Model [J]. Bioengineering (Basel, Switzerland), 2023, 10(7).
71. Qi X, Jiang Z, Song M, et al. A Novel Crosslinking Method for Improving the Anti-Calcification Ability and Extracellular Matrix Stability in Transcatheter Heart Valves [J]. Frontiers in bioengineering and biotechnology, 2022, 10: 909771.
72. Bonetti A, Marchini M, Ortolani F. Ectopic mineralization in heart valves: new insights from in vivo and in vitro procalcific models and promising perspectives on noncalcifiable bioengineered valves [J]. Journal of thoracic disease, 2019, 11(5): 2126–2143.
73. Liu J, Zhong S, Lan H, et al. Mapping the calcification of bovine pericardium in rat model by enhanced micro-computed tomography [J]. Biomaterials, 2014, 35(29): 8305–8311.
74. Tan H, Wu B, Li C, et al. Collagen cryogel cross-linked by naturally derived dialdehyde carboxymethyl cellulose [J]. Carbohydrate polymers, 2015, 129: 17–24.
75. Zhuravleva I Y, Karpova E V, Oparina L A, et al. Cross-linking method using pentaepoxide for improving bovine and porcine bioprosthetic pericardia: A multiparametric assessment study [J]. Mater Sci Eng C Mater Biol Appl, 2021, 118: 111473.
76. Li S, Hu S, Zhou J, et al. In-vitro endothelialization of patches of bovine jugular vein stabilized by dye-mediated photooxidation [J]. The Journal of heart valve disease, 2003, 12(5): 625–627.
77. Haugh M G, Jaasma M J, O'brien F J. The effect of dehydrothermal treatment on the mechanical and structural properties of collagen-GAG scaffolds [J]. Journal of biomedical materials research Part A, 2009, 89(2): 363–369.

78. Manji R A, Lee W, Cooper D K C. Xenograft bioprosthetic heart valves: Past, present and future [J]. International journal of surgery (London, England), 2015, 23(Pt B): 280–284.
79. Guldner N W, Jasmund I, Zimmermann H, et al. Detoxification and endothelialization of glutaraldehyde-fixed bovine pericardium with titanium coating: a new technology for cardiovascular tissue engineering [J]. Circulation, 2009, 119(12): 1653–1660.
80. Lopez-Moya M, Melgar-Lesmes P, Kolandaivelu K, et al. Optimizing Glutaraldehyde-Fixed Tissue Heart Valves with Chondroitin Sulfate Hydrogel for Endothelialization and Shielding against Deterioration [J]. Biomacromolecules, 2018, 19(4): 1234–1244.
81. Zilla P, Weissenstein C, Human P, et al. High glutaraldehyde concentrations mitigate bioprosthetic root calcification in the sheep model [J]. The Annals of thoracic surgery, 2000, 70(6): 2091–2095.
82. Lee C, Kim S H, Choi S H, et al. High-concentration glutaraldehyde fixation of bovine pericardium in organic solvent and post-fixation glycine treatment: in vitro material assessment and in vivo anticalcification effect [J]. European journal of cardio-thoracic surgery: official journal of the European Association for Cardio-thoracic Surgery, 2011, 39(3): 381–387.
83. Braile M C, Carnevalli N C, Goissis G, et al. In vitro properties and performance of glutaraldehyde-crosslinked bovine pericardial bioprostheses treated with glutamic acid [J]. Artif Organs, 2011, 35(5): 497-501.
84. Kim K C, Kim S H, Kim Y J. Detoxification of Glutaraldehyde Treated Porcine Pericardium Using L-arginine & NABH(4) [J]. The Korean journal of thoracic and cardiovascular surgery, 2011, 44(2): 99–107.
85. Kim S S, Lim S H, Cho S W, et al. Tissue engineering of heart valves by recellularization of glutaraldehyde-fixed porcine valves using bone marrow-derived cells [J]. Experimental & molecular medicine, 2006, 38(3): 273–283.
86. Tamagnini G, Bourguignon T, Rega F, et al. Device profile of the Inspiris Resilia valve for aortic valve replacement: overview of its safety and efficacy [J]. Expert review of medical devices, 2021, 18(3): 239–244.
87. Bavaria J E, Griffith B, Heimansohn D A, et al. Five-year Outcomes of the COMMENCE Trial Investigating Aortic Valve Replacement With RESILIA Tissue [J]. The Annals of thoracic surgery, 2023, 115(6): 1429–1436.
88. Changping, Haiqing X, Qiqing D, et al. Application of HC-processed bovine pericardium in the surgery of congenital heart disease [J]. Journal of Cardiovascular Surgery (Electronic Edition), 2017, 6(02): 81–82.
89. Jinlei, Zhuxiaodong, Zhujunming, Lihongtao, Liuwei. Experimental study of hydroxy-chromium to slow down calcification of biological flaps I: Anti-calcification effect of hydroxychromium-modified bovine pericardial test pieces in a rat subcutaneous implantation model [J]. Chinese Journal of Thoracic and Cardiovascular Surgery, 1996, (03): 176–178.
90. Jinlei, Zhuxiaodong, Chenlanying, Zhujunming, Lihongtao, Liuwei. Experimental study of hydroxychromium to slow down calcification of biological flaps II: Effect of external ligands on the anticalcification effect of hydroxychromium [J]. Chinese Journal of Thoracic and Cardiovascular Surgery, 1996, (04): 238–240.
91. Birnie A J, English J S. Ethylene oxide allergy may have been confirmed by patch testing to a similar epoxy compound [J]. Contact dermatitis, 2006, 55(2): 126.
92. Kamiński K, Kujdowicz M, Kajta M, et al. Enhanced delivery of daidzein into fibroblasts and neuronal cells with cationic derivatives of gamma-cyclodextrin for the control of cellular glycosaminoglycans [J]. European journal of pharmaceutics and biopharmaceutics: official journal of Arbeitsgemeinschaft fur Pharmazeutische Verfahrenstechnik eV, 2015, 91: 111–119.
93. Tam H, Zhang W, Feaver K R, et al. A novel crosslinking method for improved tear resistance and biocompatibility of tissue based biomaterials [J]. Biomaterials, 2015, 66: 83–91.
94. Leong J, Munnelly A, Liberio B, et al. Neomycin and carbodiimide crosslinking as an alternative to glutaraldehyde for enhanced durability of bioprosthetic heart valves [J]. J Biomater Appl, 2013, 27(8): 948–960.

95. Inostroza-Brito K E, Collin E C, Majkowska A, et al. Cross-linking of a biopolymer-peptide co-assembling system [J]. Acta biomaterialia, 2017, 58: 80–89.
96. Paramelle D, Miralles G, Subra G, et al. Chemical cross-linkers for protein structure studies by mass spectrometry [J]. Proteomics, 2013, 13(3–4): 438–456.
97. Liu C, Qiao W, Cao H, et al. A riboflavin-ultraviolet light A-crosslinked decellularized heart valve for improved biomechanical properties, stability, and biocompatibility [J]. Biomater Sci, 2020, 8(9): 2549–2563.
98. Weiqingrong. Research on Bovine Pericardium as Controlled Degradation Material of Tissue Engineering [D], 2003.
99. Daamen W F, Veerkamp J H, Van Hest J C, et al. Elastin as a biomaterial for tissue engineering [J]. Biomaterials, 2007, 28(30): 4378–4398.
100. Debelle L, Tamburro A M. Elastin: molecular description and function [J]. The international journal of biochemistry & cell biology, 1999, 31(2): 261–272.
101. Daamen W F, Nillesen S T, Hafmans T, et al. Tissue response of defined collagen-elastin scaffolds in young and adult rats with special attention to calcification [J]. Biomaterials, 2005, 26(1): 81–92.
102. Wang X, Ma B, Chang J. Preparation of decellularized vascular matrix by co-crosslinking of procyanidins and glutaraldehyde [J]. Biomed Mater Eng, 2015, 26(1-2): 19–30.
103. Isenburg J C, Simionescu D T, Vyavahare N R. Tannic acid treatment enhances biostability and reduces calcification of glutaraldehyde fixed aortic wall [J]. Biomaterials, 2005, 26(11): 1237–1245.
104. Lei Y, Yang L, Guo G, et al. EGCG and enzymatic cross-linking combined treatments for improving elastin stability and reducing calcification in bioprosthetic heart valves [J]. Journal of biomedical materials research Part B, Applied biomaterials, 2019, 107(5): 1551–1559.
105. Wang Y, Ma B, Yin A, et al. Polycaprolactone vascular graft with epigallocatechin gallate embedded sandwiched layer-by-layer functionalization for enhanced antithrombogenicity and anti-inflammation [J]. Journal of controlled release: official journal of the Controlled Release Society, 2020, 320: 226–238.
106. Lei Y, Xia Y, Wang Y. The tropoelastin and lysyl oxidase treatments increased the content of insoluble elastin in bioprosthetic heart valves [J]. J Biomater Appl, 2018, 33(5): 637–646.
107. Zhu X, Ma K, Zhou K, et al. Reversal of phosphate-induced ORAI1 expression, store-operated Ca(2+) entry and osteogenic signaling by MgCl(2) in human aortic smooth muscle cells [J]. Biochemical and biophysical research communications, 2020, 523(1): 18–24.
108. Starcher B C, Urry D W. Specificity and sites of calcium ion interactions with elastin [J]. Bioinorganic chemistry, 1974, 3(2): 107–114.
109. Chen J, Pan S, Zhou J, et al. Assembly of Bioactive Nanoparticles via Metal-Phenolic Complexation [J]. Advanced materials (Deerfield Beach, Fla), 2022, 34(10): e2108624.
110. Jin W, Guo G, Chen L, et al. Elastin Stabilization Through Polyphenol and Ferric Chloride Combined Treatment for the Enhancement of Bioprosthetic Heart Valve Anticalcification [J]. Artif Organs, 2018, 42(11): 1062–1069.
111. Hornebeck W, Emonard H, Monboisse J C, et al. Matrix-directed regulation of pericellular proteolysis and tumor progression [J]. Seminars in cancer biology, 2002, 12(3): 231–241.
112. Cigliano A, Gandaglia A, Lepedda A J, et al. Fine structure of glycosaminoglycans from fresh and decellularized porcine cardiac valves and pericardium [J]. Biochemistry research international, 2012, 2012: 979351.
113. Raghavan D, Starcher B C, Vyavahare N R, Neomycin binding preserves extracellular matrix in bioprosthetic heart valves during in vitro cyclic fatigue and storage [J]. Acta biomaterialia, 2009, 5(4): 983–992.
114. Herrero-Mendez A, Palomares T, Castro B, et al. Generation of tunable glycosaminoglycan hydrogels to mimic extracellular matrices [J]. Journal of tissue engineering and regenerative medicine, 2016, 10(12): 1000–1011.
115. Lei Y, Ning Q, Tang Y, et al. Exogenous hyaluronic acid and chondroitin sulfate crosslinking treatment for increasing the amount and stability of glycosaminoglycans in bioprosthetic heart valves [J]. Journal of materials science Materials in medicine, 2019, 30(3): 38.

116. Marinho E, JR., Carbonari R C, Malmonge S M, et al. Mechanical behavior of bovine pericardium treated with hyaluronic acid derivative for bioprosthetic aortic valves [J]. Journal of biomedical materials research Part B, Applied biomaterials, 2019, 107(7): 2273–2280.
117. Raghavan D, Simionescu D T, Vyavahare N R. Neomycin prevents enzyme-mediated glycosaminoglycan degradation in bioprosthetic heart valves [J]. Biomaterials, 2007, 28(18): 2861–2868.
118. Mendoza-Novelo B, Avila E E, Cauich-Rodríguez J V, et al. Decellularization of pericardial tissue and its impact on tensile viscoelasticity and glycosaminoglycan content [J]. Acta biomaterialia, 2011, 7(3): 1241–1248.
119. Dainese L, Guarino A, Burba I, et al. Heart valve engineering: decellularized aortic homograft seeded with human cardiac stromal cells [J]. The Journal of heart valve disease, 2012, 21(1): 125-134.
120. Dai J, Qiao W, Shi J, et al. Modifying decellularized aortic valve scaffolds with stromal cell-derived factor-1α loaded proteolytically degradable hydrogel for recellularization and remodeling [J]. Acta biomaterialia, 2019, 88: 280–292.
121. Wei Q, Becherer T, Angioletti-Uberti S, et al. Protein interactions with polymer coatings and biomaterials [J]. Angewandte Chemie (International ed in English), 2014, 53(31): 8004–8031.
122. Zhou J, Nie B, Zhu Z, et al. Promoting endothelialization on decellularized porcine aortic valve by immobilizing branched polyethylene glycolmodified with cyclic-RGD peptide: an in vitro study [J]. Biomed Mater, 2015, 10(6): 065014.
123. Marinval N, Morenc M, Labour M N, et al. Fucoidan/VEGF-based surface modification of decellularized pulmonary heart valve improves the antithrombotic and re-endothelialization potential of bioprostheses [J]. Biomaterials, 2018, 172: 14–29.
124. Assmann A, Delfs C, Munakata H, et al. Acceleration of autologous in vivo recellularization of decellularized aortic conduits by fibronectin surface coating [J]. Biomaterials, 2013, 34(25): 6015–6026.
125. Williams J K, Miller E S, Lane M R, et al. Characterization of CD133 Antibody-Directed Recellularized Heart Valves [J]. J Cardiovasc Transl Res, 2015, 8(7): 411–420.
126. Yang L, Huang X, Deng L, et al. Pre-mounted dry TAVI valve with improved endothelialization potential using REDV-loaded PEGMA hydrogel hybrid pericardium [J]. Journal of materials chemistry B, 2020, 8(13): 2689–2701.
127. Cebotari S, Lichtenberg A, Tudorache I, et al. Clinical application of tissue engineered human heart valves using autologous progenitor cells [J]. Circulation, 2006, 114(1 Suppl): I132–137.
128. Kajbafzadeh A M, Ahmadi Tafti S H, Mokhber-Dezfooli M R, et al. Aortic valve conduit implantation in the descending thoracic aorta in a sheep model: The outcomes of pre-seeded scaffold [J]. International journal of surgery (London, England), 2016, 28: 97–105.
129. Vincentelli A, Wautot F, Juthier F, et al. In vivo autologous recellularization of a tissue-engineered heart valve: are bone marrow mesenchymal stem cells the best candidates? [J]. The Journal of thoracic and cardiovascular surgery, 2007, 134(2): 424–432.
130. Jover E, Fagnano M, Angelini G, et al. Cell Sources for Tissue Engineering Strategies to Treat Calcific Valve Disease [J]. Frontiers in cardiovascular medicine, 2018, 5: 155.

# Part IV
# Multiscale Biomechanics and Modeling

# Virtual Heart Therapy: Multiscale Models of Cardiac Disease Treatment and Recovery

**Johane H. Bracamonte and Jeffrey W. Holmes**

## 1 Introduction

Computational modeling, a well-established tool in biomedical engineering, holds immense potential for advancing personalized medicine [1]. This potential is acknowledged by numerous public and private agencies worldwide, which provide support and funding for computer model-based research. For instance, the Interagency Modeling and Analysis Group (IMAG) is a collaborative program involving multiple government agencies in the United States and Canada that provides financial support for research on the modeling of biological, biomedical, and behavioral systems [2]. The Virtual Physiological Human, a European initiative, is dedicated to developing an integrated computer model that comprehensively describes the human body as a complex system. Its goal is to provide a descriptive and predictive framework to support personalized healthcare solutions and reduce the reliance on animal models in research [3].

Computational models have the potential to impact the management of cardiovascular disease by providing noninvasive tools to supplement patient imaging and clinical measurements [4]. More significantly, they can generate predictions of the acute and long-term outcomes of medical devices and treatments. The heart, a complex dynamic organ, functions through the coordinated contraction and relaxation of protein assemblies, which are regulated by cascades of biomolecular processes and influenced by the vascular and neurohormonal systems (Fig. 1). The focus of cardiovascular medicine is to deal with organ and system-scale manifestations of impaired biological processes that can occur at any biological scale, from genetic defects (as in familial cardiomyopathies) to large-scale injuries (such as

J. H. Bracamonte · J. W. Holmes (✉)
Department of Biomedical Engineering, University of Alabama at Birmingham, Birmingham, AL, USA
e-mail: jbracamo@uab.edu; holmesjw@uab.edu

 
J. Liao, J. Y. Wong (eds.), *Integration and Bridging of Multiscale Bioengineering Designs and Tissue Biomechanics*, https://doi.org/10.1007/978-3-031-81743-4_12

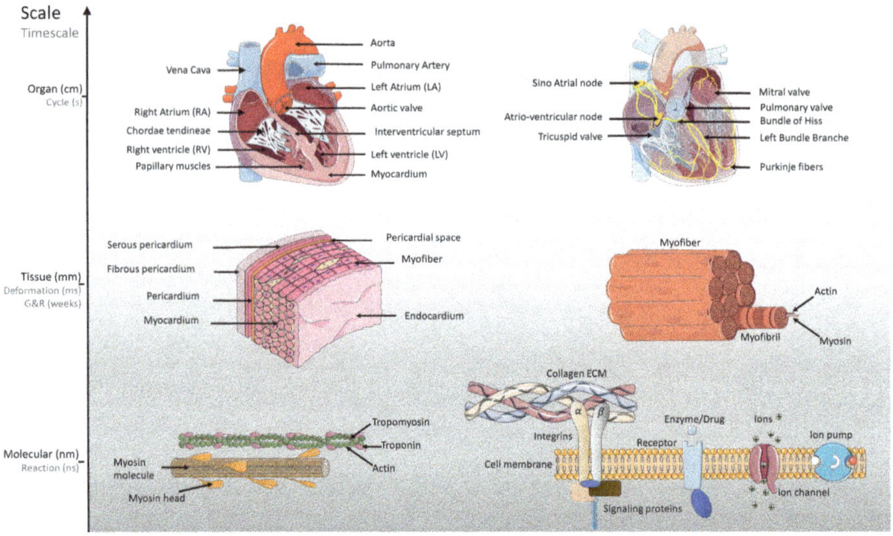


**Fig. 1** Anatomic and functional structures of the heart at organ, tissue, and molecular scales relevant to multiscale modeling of heart disease and treatment. Images adapted from Servier Medical Art, open access licensed under CC BY 4.0

in myocardial infarction). Management of heart disease usually involves multiple therapeutic approaches, from drug treatments targeting key signaling pathways at molecular scales to surgery targeting tissue-scale structures. In some cases, medical treatment aims to directly treat the underlying cause of disease, but in others, it aims to offset the effects of disease and palliate symptoms by intentionally modulating other aspects of cardiovascular physiology. The physiological processes most relevant to describing cardiac function, its ailments, and treatments in computational models are electrophysiology, tissue mechanics, hemodynamics, and growth and remodeling [5].

Creating a detailed computational model that encompasses all of the biology and physics involved in heart function, from the molecular to the organ scale, is impractical given the complexity of the system, and personalizing a model of such complexity would be even more daunting. Consequently, the design of personalized multiphysics-multiscale models of heart disease and treatment necessitates strategic allocation of resources to specific physiological processes and biological scales, depending on the nature of the heart disease and therapeutic intervention. The limitations of available computational resources and clinical or experimental data are crucial in the selection of models of different levels of complexity and in shaping the multiscale modeling approach (Fig. 2) [6].

The application of multiscale models of the heart has followed the logical evolution of simulating first healthy conditions in animal models and humans. Later, they were applied to simulate pathological conditions and, only recently, applied to the study of therapeutic interventions [1, 6]. Multiple review articles

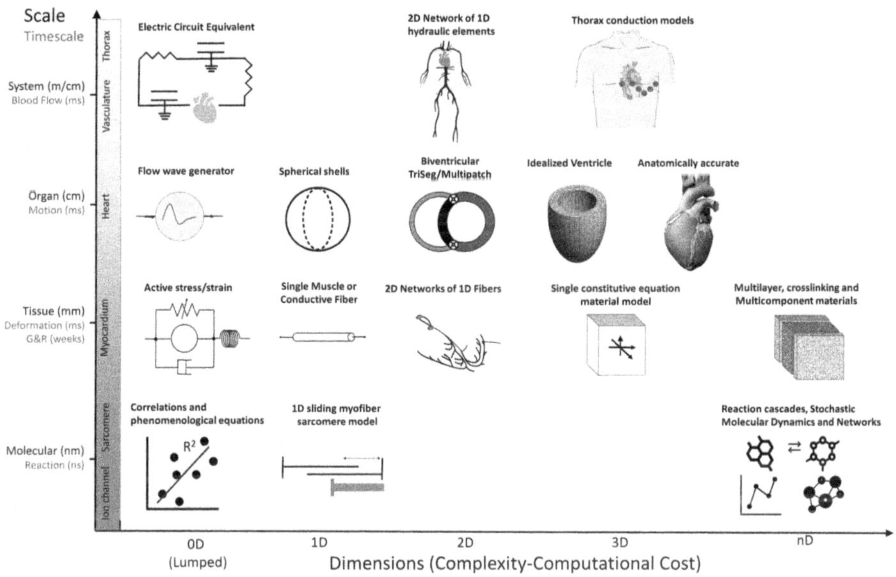


**Fig. 2** Map of most common submodules employed in multiscale models of heart disease and treatment according to the scale of physiological process and their complexity. From tissue scale and up, the geometrical dimensions resolved by the model determine the number of equations to be solved simultaneously, and thus, its complexity and computational cost. Aside from some models of sarcomere contraction, most molecular processes are modeled with no geometrical detail but as chains of chemical reactions with a level of complexity depending on the degrees of freedom (dimensions in a nongeometrical sense) of the resulting system of equations. The image of the anatomically accurate heart is reprinted/adapted with permission from Peirlinck et al. 2021 [1], Springer Nature

summarize the developments of multiscale models for healthy and diseased hearts, while fewer publications focus on multiscale models of cardiac treatment. To fill this gap, we present in this chapter a comprehensive review of the most recent developments in multiphysics-multiscale modeling of treatment for heart disease, placing special emphasis on the level of detail and complexity assigned to the modeled physics and scales. We aim to provide the reader with an overview of the state-of-the-art multiscale modeling of treatment for heart disease and to highlight the current limitations and future directions toward the implementation of this analysis technique into everyday clinical practice.

To achieve this, we structure this chapter into five sections. Section 2 briefly discusses heart anatomy, structure, and function at organ, tissue, and cellular biological scales, with a focus on specific features relevant to the models discussed thereafter. In Sect. 3, we present a description of the mathematical models most relevant for the multiphysics-multiscale formulations presented later, and we discuss the strengths and limitations of models with different degrees of complexity for electrophysiology, cardiac mechanics, and growth and remodeling at organ, tissue, and subcellular scales. In Sect. 4, we present a comprehensive review of applications of multiscale models in simulating treatments for six different types of heart disease; in each case, we also briefly review each disease's etiology, manifestations, and treatment and prior models of disease that are relevant for understanding the treatment models. Finally, in Sect. 5, we discuss the current limitations and challenges of integrating heart therapy models into personalized medicine and the future steps needed to bring these technologies into clinical practice.

# 2 Heart Anatomy, Structure, and Function

## *2.1 Heart as a Pumping Organ*

The heart comprises four muscular chambers that alternately relax and contract with each heartbeat. Deoxygenated blood returns from the body to the right atrium, flows through the tricuspid valve into the right ventricle (RV) and is pumped through the pulmonary valve into the pulmonary circulation, where it exchanges oxygen and carbon dioxide with air in the lungs. The oxygenated blood enters the left atrium, flows through the mitral valve into the left ventricle (LV), and is pumped through the aortic valve into the systemic circulation [7].

Due to its critical role in supplying blood to the body, the LV is the primary focus of many treatments and models. However, concepts developed for the LV are also often adapted for the other chambers. At the simplest level, with each heartbeat, the LV must fill with blood, generate sufficient pressure to push blood into the systemic circulation, eject some fraction of the blood it contains, and then relax and begin to fill again. Filling occurs during a period termed diastole and depends on the upstream pressure in the left atrium, the passive stiffness of the myocardium, the geometry of the LV, and the resistance to flow through the mitral valve; each of these determinants of diastolic function is altered in at least one clinical condition discussed in this review. Ejection of blood into the aorta occurs during systole, and it depends on the ability of the myocardium to generate active force, the geometry of the LV, and the resistance to ejection imposed by the aortic valve and systemic circulation; diseases can also impact each of these determinants of systolic function.

## *2.2 Tissue-Scale Structure and Function*

The heart contains multiple integrated subsystems, each with a specialized composition and function: the extracellular fibrous matrix, the cardiac muscle, an electrical conduction system, a coronary vascular system, and a series of valves (Fig. 1). The main functions of the extracellular fibrous matrix are providing structural support for the cardiac muscle, contributing to the anisotropic stiffness of the cardiac tissue, providing firm anchoring for the valves, and isolating the electrical activity of the atria from the ventricles [7]. Furthermore, recent evidence points to the additional roles of the fibrous extracellular matrix as a carrier of intercellular signals [8].

The cardiac tissue is composed of three layers: the endocardium, myocardium, and epicardium (Fig. 1). The inner and outermost layers, epicardium and endocardium, respectively, function as semipermeable barriers isolating the cardiac tissue and are thought to modulate cardiac development and function through biochemical signaling. The myocardium is the middle and thickest layer containing the cardiac muscle responsible for the pumping function of the heart. For this reason, most, if not all, of multiscale models of heart function only consider the myocardial layer. The myocardium comprises primarily cardiomyocytes, a specialized tubular cell type that contains fiber-like structures called myofibrils (Fig. 1). Neighboring cardiomyocytes link to each other to make up an intricate system of contractile muscle fibers called myofibers [9]. The orientation of the myofibers gradually changes across the thickness of the myocardium; at the epicardium, myofibers follow a left-handed helical path from apex to top, and with increasing depth below the surface, they transition gradually to circumferential at the mid-wall and then a right-handed helical path at the endocardium. This helical array is responsible for the characteristic twisting motion of the heart during active contraction [10].

The active contraction of the myofibers is triggered by the propagation of an electrical potential and distributed by a conduction system. The electrical impulses are generated by specialized pacemaker cells at the sinoatrial node near the superior vena cava in the right atrium. The electrical impulse is first conducted through the atria until it reaches the atrioventricular (AV) node between the interatrial and interventricular septum walls, where it is delayed, allowing full contraction of the atrial chambers before the onset of ventricular contraction. From the AV node, the electrical impulse is propagated into the bundle of His, which then splits into the left and right bundle branches. Next, a network of specialized Purkinje fibers embedded in the endocardium distributes the electrical potential through the ventricles. Then, the potential is transmitted from the specialized fibers into the myocardium, propagating through cell-to-cell gap junctions to reach all of the myocytes [7, 9].

The coronary vasculature distributes oxygenated blood throughout the myocardium to meet the high oxygen demands required to support contraction approximately once per second for a lifetime. The right and left coronary arteries originate at the aortic root and subdivide into a series of surface vessels before penetrating deeper into the myocardium. The right coronary artery typically supplies the right atrium, septum, and about one-third of the septal wall, while the left coronary artery supplies the rest of the heart [7, 11]. In addition to its primary function of supplying oxygen to the myocardium, the pressurized coronary vasculature influences ventricular mechanics, and those mechanics in turn influence the timing and volume of coronary blood flow [12].

The cardiac valvular apparatus, composed of the aortic, mitral, pulmonary, and tricuspid valves, is responsible for maintaining the unidirectionality of blood flow through the heart chambers and keeping the ventricular chambers closed during isovolumic contraction, allowing the ventricles to develop sufficient pressure to eject blood into the circulation. The main valve elements are fibrous structures called leaflets or cusps that open and close to allow or impede blood flow. The mitral and tricuspid valves separating the atrial and ventricular spaces are anchored to papillary muscles originating from the ventricular wall by a set of chordae tendineae [13]. These papillary muscles play important roles in both valvular and ventricular function [14].

## 2.3 Subcellular Scale Structures and Function

Cardiomyocytes are tubular branching cells responsible for cardiac muscle contraction. The contractile function of cardiomyocytes is carried by chains of myofibrils contained within the cell and modulated by ion channels and gap junctions on the cell's surface and intracellular calcium storage within the sarcoplasmic reticulum [9].

The fundamental contractile units of cardiac myocytes are called sarcomeres. Sarcomeres are composed of multiple proteins that organize into thick and thin filaments called myofilaments. Thick myofilaments contain a protein called myosin, while thin filaments contain the protein actin. During contraction, actin and myosin bind to form "cross-bridges," which cyclically attach, change conformation, and then detach, producing force and/or myocyte shortening via sliding of the myofilaments relative to one another [9, 15].

The propagating electrical signal (action potential) that triggers contraction reflects changes in the gradient of electric potential across the membrane of cardiac cells originating from the transport and flow of ions. The sarcolemma is a plasma membrane surrounding the outside of the cardiomyocyte with the purpose of separating extracellular and intracellular contents. Invaginations of the sarcolemma into the cytoplasm of the cardiomyocyte are populated by calcium channels, sodium-calcium exchangers, calcium ATPases, and beta-adrenergic receptors that allow for the exchange of ions with the extracellular medium. Gap junctions join neighboring cardiomyocytes together, facilitating the flow of ions between neighboring cells. The low resistance of the gap junctions allows depolarization to spread quickly, facilitating the rapid transmission of action potentials to produce a synchronized contraction of the cardiomyocytes [15].

The contraction-relaxation cycle in a cardiomyocyte begins when the action potential arrives via the inflow of Na+ ions. Once the membrane depolarizes beyond a threshold, the membrane sodium and calcium channels open, allowing the influx of Na+ and Ca+ into the cell and rapid membrane depolarization. Once the intracellular potential reaches its maximum, the Na+ channels close. Next, the relatively small increase in Ca+ concentration within the cell leads to the activation of the ryanodine receptors in the sarcoplasmic reticulum and the bulk release of Ca+ within the cardiomyocyte (a process called calcium-triggered-calcium-release). The intracellular calcium binds to the troponin C proteins on the thin myofilaments to trigger the contraction process in the sarcomere units. During this phase, potassium channels and chloride channels are activated, allowing the outflow of K+ and inflow of Cl- ions. These ionic flows counter the increasing concentration of positive Ca+, maintaining the transmembrane potential at a seemingly stable level, and are responsible for the relatively long plateau observed in the cardiac action potential. Next, the cell experiences a rapid repolarization during which calcium channels close and the potassium channels remain open, leading to a net outflow of positive K+ ions until the cardiomyocyte reaches the resting potential. Ion pumps help restore concentrations of Ca+, K+, and Cl- to their original values while maintaining the transmembrane potential at the resting level. Calcium ions are returned to the sarcoplasmic reticulum by specialized ion pumps. As the cytoplasmic concentration of Ca+ drops to resting levels, troponin C and the cross-bridges revert to their resting conformations and the muscle relaxes [16].

## 2.4 *Cardiac Growth and Remodeling*

One of the most remarkable features of the cardiac muscle is its capacity to adapt its size, shape, and structure in response to a combination of mechanical and chemical stimuli. This process is observed in physiological conditions such as development from embryo to adulthood, pregnancy, and exercise. Of particular interest in this chapter is the adaptive response to pathological conditions such as pressure or volume overload of the cardiac chambers, myocardial infarction, and alterations in the subcellular signaling pathways governing the regular function of the heart.

The predominating scientific consensus is that mammalian ventricular myocytes lose their capacity for cell division during the perinatal period, so cardiac hypertrophy in developed mammals is mainly due to an increase in the size of already existing cardiomyocytes [17]. Other manifestations of cardiac remodeling include changes in mass and composition of the extracellular matrix, translating into macroscopic changes in tissue compliance [18].

Global cardiac remodeling is typically classified into two major patterns at the organ scale: eccentric and concentric remodeling. Concentric remodeling occurs in response to increases in ventricular pressure, as in aortic stenosis and hypertension, and consists of growth in cardiomyocyte thickness by deposition of sarcomeres in parallel, leading to the thickening of the heart wall; it is often accompanied by fibrosis [19, 20]. Eccentric remodeling occurs in response to increases in the volume the heart is pumping, for example, during mitral regurgitation. It consists of dilation of the chamber driven by the longitudinal growth of cardiomyocytes through deposition of sarcomeres in series and is accompanied in some cases by loss of interstitial collagen [18, 21, 22]. Some pathologies, such as aortic valve regurgitation, will induce a combination of pressure and volume overload, yielding a mixed remodeling pattern. Other pathologies such as myocardial infarction are highly localized and induce a heterogeneous remodeling pattern. In myocardial infarction, the interruption of local blood flow leads to the apoptosis of cardiomyocytes within the infarct region. Dead cardiomyocytes are replaced by fibrotic scar tissue, which is less compliant than the healthy myocardium and is incapable of active contraction. Remaining myocardial tissue unaffected by the injury adapts to maintain the cardiac output with a mixed remodeling pattern that depends on the location and extent of the infarct [19, 23].

Notably, in some physiological situations (such as pregnancy or exercise), cardiac growth and remodeling appear to be wholly or mostly reversible once the mechanical and neurohormonal environment returns to normal. Similarly, cardiac remodeling associated with various pathologies may be reversed in some situations if medical treatment resolves the inciting stimuli [24–26]. However, if the pathological condition is severe enough, or if it remains untreated for long enough, cardiac remodeling may become maladaptive and lead to heart failure. Heart failure is characterized by irreversible growth and remodeling of cardiac tissue and profound impairment of cardiac function, with a significant impact on quality of life, and is ultimately lethal [27].

There is consensus that mechanical stimulation is a potent modulator of cardiac remodeling in vitro and in vivo; however, there is still considerable debate regarding what features of the mechanical environment are sensed by cardiomyocytes and how they are translated to a molecular-scale stimulus [19]. In addition, multiple in vitro and in vivo studies have identified biochemical stimuli for cardiac remodeling, including α-adrenergic agonists, peptide growth factors, cytokines, and vasoactive peptides, to name a few [17]. These extracellular stimuli stimulate a plethora of intracellular signaling reactions, with pathways that activate the mitogen-activated protein kinases (MAPK) and the c-Jun N-terminal kinases (JNK) being two of the most commonly implicated drivers of cardiomyocyte hypertrophy [17, 28]. A comprehensive review of the intracellular signaling pathways related to cardiac hypertrophy and fibrosis falls outside the scope of this chapter.

## 3 Representing Key Aspects of Heart Function in Models

Multiscale models of heart disease and treatment aim to integrate our collective knowledge of physics, anatomy, physiology, and biology. The main advantage of the multiscale modeling approach is the capacity to capture properties and behaviors emerging from the interactions between scales that isolated components or scales of the system cannot reproduce. Macroscopic biomechanics consists of the formulation of balances of mass, momentum, and energy applied to tissues and organs. However, most biological systems are open and continuously interact with their surroundings; thus, the conservation principles must be handled carefully. Models of molecular biology processes such as signaling cascades or metabolic reactions often consist of large systems of differential equations or mass balance equations. At the molecular level, models may incorporate statistical methods for reproducing the stochastic nature of molecular binding and protein folding (Fig. 2).

The selection of the sub-models that constitute the multiscale approach is critical for successfully reproducing the phenomena of interest. At first glance, it might seem that incorporating the most sophisticated models of each subsystem would produce the best multiscale formulation; however, two main limitations exist. The first limitation is the computational expense, which may render the solution or routine use of very complex models impractical. The computational expense is mainly associated with the number of equations to be solved simultaneously and to a lesser extent the mathematical complexity of the model describing the phenomena of interest. For example, the computational expense of a finite element method-based model will be more sensitive to changes in the size of the finite element mesh, which increases the number of equations by multiples of 3 per each added element, than to the complexity of the material constitutive equation. The second limitation is that the ability to confidently parameterize the model drops as the number of parameters and interactions between them increases, so that very complex models may be numerically solvable yet biologically meaningless. Furthermore, increasing the number of undetermined parameters also increases the amount of data required to properly calibrate the model, increases uncertainty around the model predictions, and can introduce numerical instabilities. Thus, when building a multiscale model, it is important to select components that are sufficiently detailed to describe the most relevant observable features at the scale of interest while being simple enough to remain computationally tractable and enable rational interpretation of the parameters and results.

In cardiovascular modeling of disease and treatment, more resources are typically allocated to the scale at which the treatment of interest will act. For example, models of drug therapies often represent details of their binding and effects on receptors, signaling pathways, and other molecular processes. By contrast, models of annuloplasty treatments for secondary mitral regurgitation generally focus more attention on a detailed anatomic description of the left ventricle and valvular apparatus, potentially coupled with representations of organ-scale mechanics and fluid flow.

Due to its computational expense, the solution of the Navier-Stokes equations (mass and momentum conservation in fluids) in a computational fluid dynamics (CFD) framework is only employed in heart models when details of intraventricular blood flow patterns are relevant. A proper description of intraventricular hemodynamics requires the consideration of the effect of the ventricular walls as the source of blood flow momentum. This effect can be implemented by prescribing a moving boundary condition on the fluid domain, where the boundary displacement can be obtained from the solution of the ventricular wall mechanics or prescribed directly from image-based measurements. A more computationally exhaustive alternative is the complete solution of a fluid-structure-interaction (FSI) problem, which implies allocating most resources to the solution of the mechanical phenomena at a macroscopic scale.

The heart is coupled to the pulmonary and systemic circulations, and changes in vascular resistance and compliance due to reflex responses, hormonal changes, and drug treatment can be important determinants of how a given treatment impacts the heart. For models focused on the heart, these effects can be captured well through relatively simple 0D or lumped formulations, either with a closed-loop circulation model for the whole arterial and venous vasculature or a three-element Windkessel model for portions of the vasculature. This modeling approach effectively reproduces the effect of reflexes and drugs on cardiac output and blood pressure by adjusting parameters for vascular resistance and compliance. Balances of blood flow and momentum, practically incorporated as congruent pressure gradients, couple the models of circulation hemodynamics and heart mechanics. In the summary tables we specify which model of circulatory hemodynamics is employed in each work reviewed.

In this section, we review some foundational models of cardiac function and remodeling at organ, tissue, and molecular scales that serve as the basis of many multiscale models of heart disease and treatment. In the next section, we review and discuss multiscale models of specific diseases and treatments (Fig. 2).

## *3.1 Models of Cardiac Anatomy, Structure, and Function*

### 3.1.1 Models at Organ and Tissue Scale

Remarkably simple models of the heart can capture the changes in LV pressure and volume during the cardiac cycle and the interactions with the pulmonary and systemic circulations that help govern filling and ejection. For example, Sagawa and colleagues developed a time-varying elastance model in which they represented the LV as a thin-walled sphere (1D anatomic model) with a prescribed time-varying function E(t) that specified the ratio of LV pressure to volume throughout the cardiac cycle [29]. When coupled to a circuit model of the resistance and compliance of the systemic circulation, this time-varying elastance model provided tremendous insight into how changes in vascular properties affect pumping and blood pressure (ventricular-vascular coupling), how drugs that increase or decrease force generation by myocytes affect overall pump performance, and even the functional impacts of features of disease such as acute myocardial infarction (loss of contraction in a region of the heart) or LV dilation [14].

These simple 1D models are computationally inexpensive, so they are convenient when the effects of the pathology and treatment are sufficiently described by global metrics of heart morphology and dynamics or when computational resources are better allocated elsewhere in a multiscale model. A relevant limitation of this model is that it assumes each chamber is homogenous and contracts uniformly, which is inappropriate when spatial gradients or material or anatomical irregularities are central to the problem of interest. This limitation can be partially addressed by using multiple 1D compartments that are adequately constrained, an approach that has been used successfully to model biventricular interaction and myocardial infarction [30–32].

The application of 2D anatomical models can significantly improve the precision of cardiac models at a relatively low additional computational expense. For example, the propagation of action potentials through the system of specialized conducting fibers can be described surprisingly well by a 2D network of unidimensional electrical conductors [33]. The TriSeg 2D biventricular model by Lumens et al. assumes the ventricular walls and septum to be thick-walled spherical shells joined together at mid-wall [34]. This approach captures relevant features of biventricular coupling and septal wall mechanics, such as septal flash and rebound and left branch block dyssynchrony [35–37].

Fully 3D models of the heart range from idealized paraboloid representations of a single ventricle to image-based patient-specific reconstructions of the whole heart or its structures. Advances in medical imaging and machine learning techniques are enabling the delivery of 3D anatomical representations of the heart with increasing precision and efficiency, and the level of anatomical detail in published cardiac models is expanding in parallel. For example, the Living Heart Project (LHP) by Dassault Systems in collaboration with researchers at the University of Berkeley and Stanford produced an anatomically detailed model of an adult human heart, including substructures such as coronary arteries and valvular apparatus, built from CT scans and MRI images [38]. This anatomical precision, however, comes with a tremendous computational cost, particularly when incorporated into a multiscale model. For this reason, highly detailed 3D models are usually implemented when the details of the geometry and anatomy are particularly relevant to the analysis, such as for patient-specific surgical planning or the design of catheter-deployable prosthetic valves.

Any 2D or 3D model of cardiac mechanics and electrophysiology requires discretizing the anatomical domain into subdomains or elements. Balance equations are formulated for each element and its surroundings, including the domain's boundaries, which requires including a material model, or constitutive equation, to solve the boundary value problem. In other words, the solution of electrophysiology and mechanics of the heart at an organ level with a multidimensional anatomical model requires including a tissue-scale model of electrical and/or mechanical material properties. Cardiac function at the organ scale is then computed by integrating the simultaneous solution of the governing equations over all the subdomains. Numerical methods, such as the finite element method, are employed for the numerical solution of the resulting system of equations.

The primary goal of most organ-scale electrophysiology models is to predict the propagation of the action potential that triggers muscular contraction. The bidomain model is one of the most comprehensive approaches for numerical simulation of cardiac tissue electrophysiology at the tissue scale. The model can be directly derived from Maxwell's equations of electromagnetism and consists of a system of two partial differential equations; the first is a diffusion-reaction equation relating the transmembrane potential and the transmembrane ionic currents, and the second differential equation represents the transport of the active potential in the extracellular domain. The solution of the bidomain model requires the specification of conductivity tensors in the intracellular and extracellular domains and a model of transmembrane ion currents. The monodomain model is a popular simplification of the bidomain model, derived from the assumption that both the intracellular and extracellular domains share the same anisotropy ratios. The components of the conductivity tensor can be derived from a tissue-scale structural model with preferential conductivity in the direction of specialized fibers. Some models employ this continuum model of action potential diffusion for myocardial tissue coupled to a 2D network of 1D elements representing Purkinje fibers. The models for transmembrane ion currents can either be phenomenological or incorporate subcellular models of ion dynamics, as will be described in detail in Sect. 3.1.2.

The mechanics of the heart at an organ scale are described in multidimensional models by the momentum balance between all discrete elements and the boundary, with active contraction as a source of momentum. The required constitutive equation of material behavior is a tissue-scale structural model. The constitutive model is often formulated in the framework of continuum mechanics and hyperelasticity as a functional relation between the scalar stored energy density function and the deformation gradient tensor or the strain tensor. The scalar energy density function is the amount of deformational energy stored per unit volume defined such that its derivatives with respect to strain yield the stress tensors.

The energy density equation is often formulated as the sum of the stored energy by volume changes and the stored energy by isochoric deformation. Incompressibility is a common assumption in the modeling of cardiovascular tissue, so energy storage due to volume changes is often ignored. However, the validity of this assumption has been challenged [39, 40]. Tissue-scale models of the myocardium must account for its nonlinear anisotropic behavior. Initially, this was addressed by phenomenological equations that used exponential functional forms to fit experimental data. The formulation by Guccione et al., a modification of Fung's orthotropic equation, is one of the most common phenomenological models of the myocardium [6]. Humphrey and Yin used biaxial testing of the myocardium to develop a constitutive model that comprises polynomial functions of invariants of the deformation gradient tensor. Fiber-family models such as the Holzapfel-Ogden models assume that families of 1D fibers, each with specific mechanical behavior, orientation distribution, and volume fraction, are embedded within an isotropic continuum matrix.

The generation of momentum by active contraction can be incorporated into the material model through active stress or active strain models. The former assumes the stress tensor can be formulated as the sum of a passive and active component; the latter is formulated as a multiplicative decomposition of the deformation tensor into an elastic tensor and an inelastic active deformation tensor. Most of the multidimensional models reviewed herein implement the active stress model and assume the active stress is aligned with the myofiber direction. The models of active tension can be phenomenological or molecular-scale representations of sarcomere dynamics. Most phenomenological equations of active stress are simplifications of the chemo-mechanical model of Hunter et al. [46] that link the magnitude of active stress to the fiber length and the activation status. The activation status can be defined as a function of time or can be coupled to the solution of an electrophysiology model. For readers interested in more details on this topic, we recommend the review by Avazmohammadi et al. offering valuable insights into conceptual, functional, and empirical considerations for the use and design of myocardial constitutive equations [41].

### 3.1.2 Models at Cellular and Molecular Scale

Active stress is the tissue-scale manifestation of the coordinated contraction of sarcomere units at a molecular scale. Mechanistic models of sarcomere dynamics are typically based either on the sliding filament theory or the cross-bridge theory. The former assumes the active shortening of sarcomeres is a consequence of the relative sliding of actin filaments over myosin filaments. The cross-bridge theory assumes the sliding of actin filaments is driven by the rotation of cross-bridges. Both simplified theories reproduce the observed proportionality between generated active force and the size of the overlapping region of actin and myosin filaments where cross-bridges can be formed [42, 43]. The model proposed by Rice et al. assumes the tropomyosin-troponin units and their interaction with neighboring units to be represented by discrete magnetic dipoles (also known as the Ising approach) to model thin myofilament activation [44]. The model is formulated as a system of ordinary differential equations describing cross-bridge cycling, which is computationally efficient but lacks spatial detail and interactions with neighboring filaments. The model was later expanded by Washio et al. in a 1D geometrical model of a sarcomere, including the interaction with its nearest neighbor. This sarcomere model successfully reproduced experimental data on the contractile response of sarcomere bundles to different calcium transients [45]. Calcium modulates sarcomere activity by inducing conformational changes in the troponin-tropomyosin complex, revealing the binding sites and enabling the formation of cross-bridges [43]. Molecular ion transport models can therefore be coupled to models of sarcomere dynamics through the intracellular Ca + concentration to provide a full molecular-scale description of active contraction [46].

Single-cell, single-channel experiments on animal myocytes inspired many early models of the cellular action potential. The seminal work of Luo and Rudy in 1991 and 1994 on modeling ion transport in guinea pig ventricular cells established the foundation for many modern models of action potential [47, 48]. The Luo-Rudy model of single-cell electrophysiology assumes the myocyte to be a cylindrical structure with the rate of change of action potential depending on the balance of ionic currents and the capacitance of the membrane. The model accounts for ionic currents through voltage-gated channels, diffusion mechanisms, and intracellular ion concentration dynamics. Gated channels are formulated as Hodgkin-Huxley elements. The dynamic Luo-Rudy model (an expansion of the original formulation) balances the ionic currents through the sodium and potassium channels, sodium-potassium pumps, calcium pumps, L-type calcium channels, nonspecific calcium-activated channels, and sodium-calcium exchanger on the plasma membrane, as well as the calcium-induced calcium release channel and calcium pump on the membrane of sarcoplasmic reticulum with calcium buffers in the myoplasm.

The dynamic Luo-Rudy model has been frequently revised and modified to improve its accuracy for other cell types including human myocytes. However, human cardiomyocytes differ from other mammals in some key features, such as the relative weight of ionic currents on action potential generation, action potential wave shape and duration, and potential restitution, which may influence mechanisms of arrhythmia dynamics [49]. The increasing availability of human cardiomyocyte lines and the cloning of human ion channels led to the formulation of specialized models of cellular electrophysiology for humans. One of the most comprehensive yet computationally tractable models was formulated by Ten Tusscher et al. (2004) [48]. In Ten Tusscher models and their variations, the cell membrane is modeled as a capacitor connected in parallel with variable resistances and batteries representing the ionic currents. It assumes that layers of cardiac cells behave as a continuum so that the rate of change of local action potential results from the balance of ionic currents and current diffusion. The model accounts for 12 different ionic currents and pumps and the intracellular ion dynamics of potassium, sodium, and calcium, including calcium-induced-calcium release from the sarcoplasmic reticulum. The model requires the simultaneous solution of 16 variables and the definition of 40 model parameters calibrated for specific cell types. Due to the convenient balance between physiological detail and computational effectiveness, the Ten Tusscher model is a popular option for integration into multiscale cardiac mechanics models where precise descriptions of membrane ion currents are relevant, such as modeling drug therapies with channel blockers such as amlodipine, a commonly prescribed calcium channel blocker.

When incorporating electrophysiology into a multiscale model focused on the tissue scale, it is often convenient to implement a simpler model of cellular ion transport, such as the model by Bernus et al., which reduces the number of unknown variables to six by reformulating the equations for ion currents and fixing some intracellular ion concentrations [49]. Alternatively, phenomenological equations of cellular active potential, such as the 2-equation 6-parameter Fitzhugh-Nagumo model of tissue excitation, are a computationally inexpensive option when no detail of the molecular electrophysiology is required for an adequate description of the problem of interest [50].

These mechanistic molecular models are mostly calibrated to cell and tissue scale measurements from in vitro experiments, with model parameters chosen to reproduce the most salient features emerging from the simultaneous contribution of many molecular processes operating simultaneously. Statistical models are often better suited to describing the stochastic nature of molecular processes; for example, Markovian models relying on Monte Carlo simulations may provide advantages for predicting sarcomere dynamics [51–53] and ion transmembrane transport [53–55]. However, statistical approaches relying on Monte Carlo simulations may become computationally expensive because of the number of samples required to yield smooth and statistically significant results [46].

## 3.2 Models of Growth and Remodeling

The effect of chemo-mechanical stimulation on the growth and remodeling of soft tissues is an active field of study bridging biomechanics and mechanobiology. A detailed review of growth and remodeling models applied to cardiac tissue can be found elsewhere [56–59]. In this section, we summarize the principles of models of growth and remodeling applied to the study of heart disease and treatment.

Most models of cardiovascular growth and remodeling are formulated within the kinematic growth framework developed by Rodriguez et al. (1994) [60]. This theoretical framework represents growth and remodeling at the tissue scale using continuum mechanics concepts like those employed in plasticity. In continuum mechanics, the deformation tensor quantifies a body's departure from its reference (baseline) configuration. The kinematic growth theory assumes that the deformation tensor can be decomposed into the product of an inelastic growth tensor and an elastic deformation gradient tensor. The growth tensor represents all changes in shape and size due to growth occurring in a hypothetical unloaded configuration, while the elastic deformation tensor reconciles the new "grown" unloaded state with the actual material configuration. This approach requires the definition of two constitutive equations: the strain energy function (see Sect. 3.1.1) and an evolution equation for the growth tensor.

There is substantial evidence that mechanical stimulation directly triggers cardiac growth and remodeling, from in vitro experiments on cardiomyocytes to the in vivo response of animal models and humans to injuries and pathologies that are mechanical in nature (such as pressure and volume overload). However, the details of the underlying molecular mechanisms linking mechanical stimulation to cardiac growth and remodeling are still not fully understood. Consequently, there are many phenomenological models of cardiovascular tissue growth as a function of mechanical stimulation, typically quantified as either tissue-scale stress or strain. Witzenburg et al. (2017) implemented and directly compared the most common phenomenological growth models for cardiac hypertrophy [61], while Yoshida and Holmes (2021) reviewed their application to a variety of physiologic and pathologic conditions [62]. These authors found that the phenomenological model by Kerckhoffs et al. [63] was the most successful in predicting a broad range of physiologic and pathologic responses. Kerckhoffs' growth model consists of a pair of sigmoidal equations that separately govern fiber and cross-fiber growth. The predicted rate of growth is computed from a combination of fiber, cross-fiber, radial, and cross-fiber radial strains using equations containing six independent parameters [63].

The phenomenological growth equations developed by Kerckhoffs and others were intended to describe hypertrophy—an increase in mass—and were usually tested against data from animal models where the growth was deliberately triggered by an intervention such as damaging a heart valve (experimental volume overload) or constricting the aorta (experimental pressure overload).

In the clinical setting, many treatments aim to slow or reverse hypertrophy that has already occurred, and phenomenological models have been less successful in capturing this regression of hypertrophy ("reverse remodeling") following treatment. The standard Kerckhoffs models assume that the rate of growth depends on the departure of mechanical strains from a homeostatic setpoint value, meaning that growth reversal can only be achieved if strains fall below the setpoint. Yoshida et al. (2020) found that the removal of mechanical overload does not necessarily bring strains down to or below their homeostatic value once growth and remodeling have occurred, so the standard formulation of Kerckhoffs models could not predict the reverse remodeling following aortic unbanding in animal models of PO. To address this issue, the authors proposed updating the reference configuration employed to calculate the growth stimuli, reflecting the gradual evolution of the homeostatic state of the tissue as might be expected due to the turnover of its constituents (see below) [64]. In Yoshida's model, the homeostatic strain values gradually increased during simulated pressure overload, so that strains fell below the new setpoints after removing the overload, resulting in a reversal of growth as seen in experiments. This time-evolving homeostatic state approach was also used by Oomen et al. to predict ventricular recovery in a model of cardiac resynchronization therapy [37].

Humphrey and Rajagopal proposed a constrained mixture theory [65, 66] that explicitly represents the continuing turnover of various tissue components and the effects of this turnover on subsequent remodeling. A growth and remodeling model formulated through the constrained mixture theory will account for the production and degradation rate of relevant tissue constituents that are constrained to deform together. Typically, the model assumes that each constituent possesses its own stress-free reference configuration, which evolves as new material is deposited at the current (loaded) configuration, and material deposited at previous reference configurations is degraded. This model requires the solution of an additional differential transport equation for each tissue constituent, plus constraint conditions to maintain thermodynamic consistency [67, 68]. Constrained mixture theory has been employed to study a range of vascular diseases and therapies and the development of tissue-engineered vascular grafts [66]. However, this growth and remodeling framework is not employed by any of the cardiac models reviewed herein.

In recent years, the field of systems biology has developed methods for modeling and analyzing large networks of interconnected reactions, such as the metabolic pathways that underlie energy generation, storage, and use by cells or the intracellular cascades of signaling molecules and reactions that transduce mechanical and chemical inputs into gene expression and other fundamental cellular responses. Metabolic and signaling networks clearly play a critical role in growth and remodeling of any tissue, and Ryall et al. (2012) showed that a model of the network of signaling pathways most relevant to cardiomyocyte hypertrophy [69] could reproduce a wealth of data on stimulation of hypertrophy by both stretch and hormones. These biomolecular models can replace phenomenological equations in kinematic growth frameworks or be utilized as source terms for a specific chemical species in a constrained mixture theory. A notable example is the research by Saucerman and coworkers, who developed network models that integrate signaling pathways through which mechanical and hormonal stimuli interact to yield changes in myocyte size and protein synthesis [28, 70, 71]. These models have been incorporated into multiscale modeling frameworks of ventricular hypertrophy under physiological (pregnancy) and pathological conditions (pressure overload) [70, 71]. The main advantage of such models is the incorporation of biochemical modulators of cellular behavior, such as hormones and drugs, that play critical roles in both pathophysiology and treatment in most clinical conditions.

# 4 Multiscale Models of Heart Disease, Therapy, and Recovery

## *4.1 Hypertension and Aortic Stenosis*

### 4.1.1 Disease Manifestation and Treatment

Hypertension is a sustained increase in systemic blood pressure, which is often asymptomatic. Clinicians typically use a threshold of 140/90 or above for systolic/diastolic blood pressure to diagnose hypertension, and it has multiple underlying causes. Hypertension is generally associated with stiffening and obstruction of blood vessels, and thus, is correlated to lifestyle risk factors such as sedentarism, tobacco and alcohol use, and high fat or salt diet, as well as to weight, age, and congenital conditions [72].

Untreated hypertension can cause kidney disease, heart disease, and stroke. The heart responds to the persistent pressure overload by remodeling, with thickening of the ventricular walls and interstitial fibrosis in the myocardium, leading to a structural reinforcement of the cardiac muscle. One popular hypothesis is that a biologic feedback loop that regulates wall stress triggers these structural changes in order to restore stresses in the myocardium to homeostatic levels. However, this reinforcement and stiffening of the ventricles eventually restricts the passive filling of the ventricles and leads to diastolic dysfunction and heart failure [73].

Treatment of hypertension typically aims to address its primary cause, such as recommending changes in lifestyle or prescribing drugs that promote vasodilation (calcium channel blockers or inhibitors of the renin-angiotensin system and signaling) or prevent fluid retention (diuretics) to lower arterial pressures.

Pressure overloading of the heart can also be secondary to focused pathologies, such as aortic stenosis, in which the aortic valve becomes fibrotic and the valvular opening narrows, thus increasing vascular resistance and cardiac workload and inducing patterns of ventricular growth and remodeling similar to those observed in hypertension. Treatment for aortic stenosis aims to facilitate forward cardiac output by replacement of the aortic valve through open-chest surgery, transcatheter valve replacement, or balloon valvuloplasty, a palliative treatment option when valve replacement is not possible or too risky [74, 75].

### 4.1.2 Multiscale Models of the Disease

Cardiac models of chronic hypertension typically focus on the maladaptive remodeling of the LV in response to sustained pressure overload, which can eventually lead to diastolic dysfunction and heart failure. Most of our knowledge on how acute and chronic hypertension affects heart function, growth, and remodeling comes from animal models of aortic constriction. In these experiments, the diameter of the aorta is reduced at a single location by placing a band around the aorta [76, 77]. The data collected from these animal models have been exhaustively analyzed and used to design and develop computer models of the cardiac response to pressure overload.

A comprehensive review of computational multiscale models of pressure overload and its effects on cardiac function, growth, and remodeling can be found elsewhere, and the interested reader is recommended to consult the work by Niestrawska et al. [57]. When the model focuses on the heart, hypertension is modeled by either a prescribed high pressure as a boundary condition or an increase in vascular resistance in a coupled circulation model [4, 78]. Witzenburg and Holmes (2018) proposed one of the simplest and most effective models to predict the organ-scale effects of chronic pressure overload; in this model, a 1D organ-scale spherical representation of the ventricles was coupled to a 0D system-scale lumped model of hemodynamic circulation and a tissue-scale Kerckhoffs model of hypertrophy. Aiming to reproduce the experimental data from aortic banding experiments, hypertension was modeled as a sudden increase in systemic vascular resistance to flow. Despite the relative simplicity of this model and low computational cost, the model showed excellent accuracy in predicting changes in left ventricular mass and free wall thickness [32].

### 4.1.3 Multiscale Models of Therapy

A main challenge for the development of multiscale models of pressure overload treatment is the diversity of etiologies and therapeutic options. For example, the surgical solution for aortic stenosis will produce an acute and sustained relief of pressure overload on the left ventricle, while drug therapies often partially alleviate pressure overload while also directly modulating hypertrophic and fibrotic signaling pathways in cardiomyocytes, producing a wide range of possible outcomes in different patients receiving different treatments (Table 1).

A multiscale study on the acute reversal of hypertension and its effects on the heart was proposed by Yoshida et al. (2020). They utilized data from studies of canine aortic banding and subsequent removal to calibrate a multiscale model of ventricular growth and remodeling capable of reproducing ventricular recovery after acute relief of pressure overload. These authors employed a biventricular 3D electromechanical organ-scale model of the heart incorporating transverse isotropy and a Hill-type active contraction model, a closed-loop lumped model of the circulation, and a Kerckhoffs-like growth model [63]. The electromechanical model was coupled to a closed-loop lumped model of the circulation through

**Table 1** Multiscale models of treatments for pressure overload

| Reference | Condition | Treatment | Physics | Scales | | | |
|---|---|---|---|---|---|---|---|
| | | | | System | Organ | Tissue | Molecular |
| Yoshida et al., 2020 [62] | Aortic constriction | Constriction release | *Mechanics* | | 3D biventricular | Fung-type with active tension | |
| | | | *Hemodynamics* | 0D closed circulation | | | |
| | | | *Growth* | | | Kerckhoffs model | |
| Estrada et al., 2021 [70] | Aortic constriction | Drug therapies | *Mechanics* | Pressure BC | 3D single ventricle | Isotropic Mooney-Rivlin with active stress | |
| | | | *Growth* | | | | Saucerman-type network |

momentum and flow balance. To model ventricular hypertrophy and recovery, the authors implemented a kinematic growth framework and a modified Kerckhoffs growth model. The authors aimed to reproduce canine experimental data of cardiac hypertrophy during 54 days of aortic constriction and reversal of hypertrophy during 54 days after the release of the aortic band. The authors observed that the standard Kerckhoffs growth model formulation was unable to predict the reversal of growth following aortic unbanding because mechanical strains never fell below the baseline homeostatic setpoint, a condition required by the model to reverse growth. The authors concluded that Kerckhoffs-like kinematic growth models must include an evolving reference (homeostatic) configuration in order to reproduce the observed reversal of hypertrophy (see Sect. 3.2). The authors warned that other similar models that consider the departure of stress or strain from a homeostatic setpoint as the trigger for growth may face similar limitations. Finally, the authors highlighted the impact of the hemodynamic adaptive response following aortic unbanding on the modulation of mechanical strains.

Estrada et al. (2021) proposed a noteworthy model of the therapeutic effects of drug treatments for hypertension incorporating both mechanical and intracellular signaling effects. They employed a cell-scale network model of hypertrophic signaling pathways in the myocyte to drive growth, rather than a phenomenological growth equation. The network consisted of 107 species and 207 reactions representing established intracellular signaling pathways underlying changes in cardiomyocyte size and protein synthesis, with inputs for levels of mechanical stimuli and concentrations of Angiotensin II, norepinephrine, isoproterenol, endothelin 1, and natriuretic peptides ANP and BNP. The mechanical stimuli for the network model were linearly mapped to the strain output of an organ-scale 3D anatomical model of the canine LV incorporating MRI-derived geometry, realistic transmural variation of fiber orientations across the wall, and an isotropic Mooney Rivlin hyperelastic constitutive model with a time-dependent active stress component. The 3D mechanical model was coupled to a closed-loop lumped circulation model through mass and momentum balance. The growth model was implemented within a kinematic growth framework by linearly mapping the cell-scale myocyte area change predictions to components of the growth deformation tensor. The organ-scale model was solved using a FEBio solver with a custom-made plugin for growth, which called the solution of the network model using Netflux (Fig. 3). Drug effects were implemented in the network model by nulling the activation level for the receptor blocked by each drug. The model was calibrated to experimental data on concentric hypertrophy driven by aortic constriction and by isoproterenol infusion. The calibrated model was validated against independent experimental data on both drug treatments and genetic interventions in the setting of aortic constriction. The model adequately predicted ventricular mass growth differences between mice with untreated aortic constriction and those receiving β blockers and angiotensin receptor blockers [70].

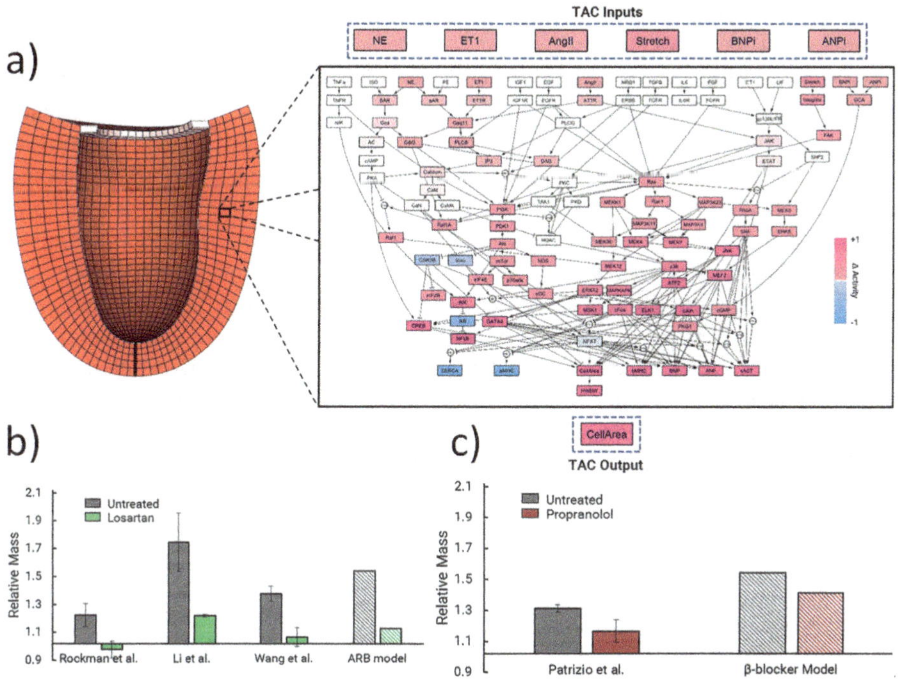


**Fig. 3** Modeling the effect of drug therapies on pressure-overload-induced ventricular hypertrophy. (**a**) Schematic representation of a multiscale model consisting of a 3D organ-scale model of the left ventricle with uniform single-equation tissue constitutive equation coupled to a network model of cardiomyocyte hypertrophy pathways. Simulations of the effect of drug therapies with (**b**) losartan and (**c**) propranolol on the LV mass of animal models of thoracic artery constriction. (Reprinted/adapted with permission from Estrada et al. 2021 [70], Springer Nature)

## *4.2 Valvular Disease*

### 4.2.1 Disease Manifestation and Treatment

Each one of the chambers of the heart is equipped with a discharge valve to ensure unidirectional blood flow, and the operation of these valves is controlled primarily by surrounding pressures and flows. The main elements of heart valves are fibrous structures called leaflets or cusps that flap to allow or impede blood flow. Primary heart valve disease may involve regurgitation (backflow due to deficient closing), stenosis (hardening and thickening of the leaflets producing elevated resistance to flow), or atresia (a congenital disease where the heart valve is partially or entirely absent). Secondary or functional valve disease occurs when valvular dysfunction is the result of a primary injury or pathology, such as myocardial infarction [13].

In the presence of valvular regurgitation, the affected chamber must receive and eject more blood with each cardiac cycle in order to maintain the same net forward flow. This volume overload increases contractile work and energy consumption and leads to eccentric remodeling characterized primarily by dilation of the affected chamber [79]. Dilation imposes an additional burden on the myocardium, which must increase contractility to generate normal pressures in the presence of dilation. When compensatory mechanisms are eventually depleted, patients experience systolic dysfunction and heart failure [22, 80]. Valvular stenosis also increases the contractile work and energy required to maintain normal cardiac output, by forcing the affected chamber to generate higher pressures to push blood through the stenotic valve. This "pressure overload" leads to concentric remodeling, where thickening of the walls of the chamber may help stabilize it mechanically but cannot fully alleviate elevated energy requirements and other detrimental impacts of valve dysfunction.

Risk factors for the occurrence of valvular disease include hypertension, age, myocardial infarction, and congenital disorders. The only definitive treatment for valvular disease is to repair or replace the dysfunctional valve. Palliative treatments include beta-blockers, angiotensin II receptor blockers, angiotensin converting enzyme inhibitors, and diuretics that will either promote the increase of forward flow by reducing systemic vascular resistance or alleviate volume overload by reducing blood volume [79].

### 4.2.2 Multiscale Models of the Disease

Detailed finite element models of the valvular apparatus are computationally expensive, so many multiscale models of the cardiac impacts of valvular disease simulate the condition by adding backflow paths to 0D lumped models of the circulation [81] that determine loading conditions on the heart. Witzenburg and Holmes added a mitral regurgitation resistance bypassing the pressure-dependent diode representing the mitral valve and coupled them to a 1D spherical model of the LV. The authors calibrated the model to data from canine experimental mitral regurgitation, adjusting the mitral regurgitation resistance to reproduce the observed regurgitant fraction. Using the tissue-scale Kerckhoffs equation of growth, this inexpensive approach predicted eccentric hypertrophy of the LV as observed in experimental volume overload [32]. Rondanina et al. (2020a, 2020b) employed a similar multiscale model to predict left ventricular hypertrophy following aortic stenosis and regurgitation with a stress-driven phenomenological growth law [82, 83]. Likewise, Lee et al. (2015) studied the effect of mitral regurgitation on an organ-scale 3D model of biventricular electrophysiology by coupling it to a system-scale 0D lumped model of circulation. With this multiscale model, the authors evaluated the increased stroke work required to maintain observed hemodynamics during experimental mitral regurgitation [84]. Overall, these examples suggest that a relatively simple approach to modeling the hemodynamics of primary mitral regurgitation may be sufficient to represent many of the most important impacts on cardiac function and remodeling.

By contrast, reproducing the key features of secondary valve regurgitation may require a detailed model to capture the functional relation between the primary pathology and the consequential valve dysfunction. Peirlinck et al. (2021) used the detailed Living Heart Project (LHP) anatomic model to simulate mitral regurgitation secondary to myocardial infarction; this model included a detailed representation of mitral valve leaflets, chordae tendineae, and papillary muscles. Myocardial infarction was simulated as stiffer noncontractile material in the infarct zone and resulted in an asymmetric valve operation, leading to partial valve opening and regurgitation during systole [1]. Although computationally expensive, this detailed approach might one day enable customized surgical treatment based on the severity of myocardial infarction and its secondary valvular dysfunction, as discussed below [85].

### 4.2.3 Multiscale Models of the Therapy

As discussed in Sect. 4.2.1, the most effective therapies for valvular disease involve tissue-scale surgical or transcatheter interventions for the restoration of valvular function, sometimes involving the implantation of a medical device. In this section, we review multiscale models of treatments for valvular disease and their effects on cardiac function, some of which focus on the optimization of medical devices (Table 2).

Annuloplasty remains one of the most effective treatments for both primary and secondary valvular insufficiency. However, more than half of patients will show recurrent regurgitation within 3–5 years after intervention. Multiscale models have been proposed as a supporting tool to improve the understanding of the factors influencing valvular dysfunction and the success of repairs and to provide quantitative guidelines in the design and sizing of prosthetic annuli [86]. Baillargeon et al. (2015) proposed a multiscale model of the annuloplasty procedure as treatment for mitral valve regurgitation secondary to myocardial infarction. They combined the LHP organ-scale 4-chamber anatomic model with a Holzapfel-type passive material and an active tension model dependent on local electrical potential, with an action potential propagation model consisting of a Purkinje fiber network embedded in a homogeneous conductive material. The model also included the mitral valvular apparatus, which consists of the chordae tendineae, papillary muscles, leaflets, and a lumped model of circulation. Myocardial infarction was modeled by suppressing active contractile stress in a subregion of the left ventricle, including the posterior papillary muscle. Simulated infarction was sufficient to induce mitral valve dysfunction in the absence of any other changes, resulting in partial systolic opening and regurgitant flow. Simulating annuloplasty by creating tight frictionless contact spots (representing sutures) between the mitral valve annulus and a geometrical model of a commercial device resolved mitral valve regurgitation. Peirlinck et al. 2021 [1] built on this work, minimizing predicted valve leakage by optimizing three dimensions of the prosthetic ring (curvature length and height of the sub-valvular element).

Cerclage annuloplasty is a novel alternative to conventional annuloplasty that has shown promise in animal models and humans and is currently being tested in a randomized clinical trial (NHLBI Transcatheter Mitral Cerclage Ventriculoplasty Early Feasibility Study NCT03929913) [87, 88]. This treatment consists of a

**Table 2** Multiscale models of treatments for mitral valve disease

| Reference | Condition | Treatment | Physics | Scales | | | |
|---|---|---|---|---|---|---|---|
| | | | | System | Organ | Tissue | Molecular |
| Baillargeon et al. 2015 [86]<br>Peirlinck et al. 2021 [1] | Mitral valve regurgitation secondary to MI | Annuloplasty | *Mechanics* | | 3D 4 chamber LHP | Holzapfel-like material with active tension | |
| | | | *Valve mechanics* | | 3D papillary muscles<br>2D leaflets<br>1D chordae | Hyperelastic leaflets<br>Rigid device | |
| | | | *Hemodynamics* | 0D open Windkessel x2 | | | |
| | | | *Electrophysiology* | | 3D 4 chamber LHP<br>1D Purkinje network | Monodomain | |
| Lee et al. 2019 [90] | Mitral valve disease | Cerclage annuloplasty | *Mechanics* | Pressure BC | 3D biventricular | Hyperelastic with active tension | Rice sarcomere model |
| | | | *Valve mechanics* | 0D lumped valves | | | |
| | | | *Hemodynamics* | 0D closed circulation | | | |
| | | | *Electrophysiology* | | 3D biventricular | Monodomain | Luo-Rudy-like model |

| Masuda et al. 2021 [92] | Mitral valve disease | Mitral valve repair | *Mechanics* | | 3D biventricular 1D atria TVE | | |
|---|---|---|---|---|---|---|---|
| | | | *Valve mechanics* | 0D lumped valve | | | |
| | | | *Hemodynamics* | 0D open Windkessel x2 | | | |
| | | | *Electrophysiology* | | 3D biventricular. 3D Purkinje layer | Bidomain | Ten Tusscher × 3 cell types |
| Peirlinck et al. 2021 [1] | Mitral valve stenosis | Edge-to-edge valve repair | *Mechanics* | | 3D 4 chamber LHP | Holzapfel-like material with active tension | |
| | | | *Valve mechanics* | | 3D papillary muscles 2D leaflets 1D chordae | Hyperelastic leaflets elastic device | |
| | | | *Hemodynamics* | 0D Open Windkessel x2 | | | |
| | | | *Electrophysiology* | | 3D 4 chamber LHP 1D Purkinje network | Monodomain | |

Notes: *LHP* living heart project
Ten Tusscher × 3 cell types: the Ten Tusscher model was formulated for endocardial, mid-wall, and epicardial myocytes with different sets of parameters
*TVE* time-varying elastance
Windkessel ×2: two independent Windkessel models for pulmonary and arterial vasculature with grounded end

circumferential cerclage suture wrapping the coronary sinus and the basal septal myocardium transversally and secured within the right atrium; the suture is tightened to approximate the free and septum walls, effectively reducing the regurgitant orifice. Lee et al. (2019) developed a multiscale model to evaluate the acute outcome of mitral valve cerclage annuloplasty on patient-specific anatomical models. They explored the correlation between simulated changes in stress distribution following the procedure and clinical outcomes at 6 months after intervention [89, 90]. An organ-scale model of biventricular anatomy was built from patient-specific CT scans assuming a homogeneous orthotropic hyperelastic material with an active stress component and a mono-domain equation coupled to a Luo-Rudy-like ion current model for action potential propagation. The Rice active tension model linked the estimation of calcium transients to the active stress via the integration of actively generated cellular forces. The procedure was modeled by prescribing a distribution of external forces acting normally on the base of the LV. The authors observed that simulated placement of the suture decreased LV wall stresses. They also found that cases that presented larger drops in ventricular wall stress following intervention corresponded to patients who showed better signs of reverse remodeling in a follow-up visit. The authors hypothesized that further tightening of the cerclage annuloplasty may benefit patients who did not show ventricular recovery and suggested that such multiscale models could support preoperative virtual rehearsal of the procedure.

When the ventricle-to-atrium leakage is caused by the deterioration or damage of the valvular apparatus, an effective treatment is either the replacement or repair of the dysfunctional valve which oftentimes is complemented with annuloplasty. Mitral valve replacement is associated with elevated morbidity and mortality when compared to valve repair, so it is only performed when no other therapeutic options are feasible or available [91]. Masuda et al. (2021) explored the acute effect of valvular replacement on cardiac function using a multiscale model formulated through the UT-Heart platform. CT scan-derived organ-scale 3D models of biventricular anatomy were coupled to 1D organ-scale time-dependent elastance models of the atria and two system-scale circuit models for arterial and pulmonary vasculature hemodynamics. The biventricular model included orthotropic hyperelastic passive mechanics with an active stress component linked to a molecular sarcomere model. The bidomain model was implemented with a thin highly conductive endocardial layer representing the Purkinje network and coupled to the Ten Tusscher model of ion currents differentiating endocardial, mid-wall, and epicardial myocytes. The sarcomere and ion transport models were coupled via calcium transients. The sarcomere model allows the estimation of contractile work as ATP consumption by the cross-bridges. Mitral regurgitation and valve replacement were modeled in the system by adding a backflow path to the lumped model of pulmonary vasculature, while valve replacement was simulated by shutting it down. The authors report that eliminating the regurgitant flow significantly reduced the contractile energy demand and increased the mechanical efficiency of the cardiovascular system by 4%, while the model reproduced expected trends for changes in ventricular dimensions and ejection fraction [92].

Edge-to-edge leaflet approximation is a mitral valve repair technique also known as the Alfieri procedure, which consists of stitching or clipping together the anterior and posterior leaflet at a single point, effectively narrowing the mitral orifice and reducing regurgitant flow. This technique has gained special attention in the past decade due to the development and successful incorporation of transcatheter approaches. The success of the technique depends on the ability of the valve leaflets and the implant to withstand the mechanical loads during the cardiac cycle. Peirlinck et al. (2021) [1] repurposed the model by Baillargeon et al. (2015) to estimate the approximation force necessary to maintain the edge-to-edge repair in the case of degenerative valve disease. Degenerative mitral valve disease was modeled by disengaging several marginal chordae from the anterior leaflet. The mitral valve repair was modeled at a tissue scale by attaching a stiff connector element at a single point of the anterior and posterior leaflet (Fig. 4). The authors report that the peak approximation force occurred at diastole and strongly depended on the stiffness of the connector element. A second, lesser peak in approximation force was identified in systole during isovolumetric contraction. Furthermore, the estimated force magnitudes and timing showed reasonable agreement with measurements performed in animal models. The authors argue that the simulated force depended on anatomy, device properties, and location, making the case for further studies on patient-specific anatomies [1].

## *4.3 Arrhythmia*

### 4.3.1 Disease Manifestation and Treatment

Arrhythmia is a sustained alteration of the rhythm of the heartbeat including beating at a slow rate (bradycardia), fast rate (tachycardia), or irregularly. Atrial fibrillation (AF) is one of the most common types of arrhythmias, in which additional activation pulses are generated in the atria, preventing coordinated contraction of these two chambers and affecting the filling of the ventricles. Patients with sustained arrhythmia can manifest anxiety, chest pain, confusion, difficulty breathing, dizziness and fainting, fluid build-up, and general weakness [93].

The specific mechanisms that trigger the onset of arrhythmias are still poorly understood. However, the condition is more common in older adults, survivors of heart attacks, tobacco and alcohol users, and patients with coronary artery disease, heart failure, and other conditions compromising the integrity of cardiac tissue. Arrhythmia is often treated with drugs that affect transmembrane ion dynamics, such as calcium, sodium, and potassium channel blockers. These drugs may be complemented by therapies that promote forward cardiac output and treat heart failure, such as beta-blockers, angiotensin-converting enzyme inhibitors, angiotensin receptor blockers, diuretics, and digoxin [94].

Patients with atrial fibrillation are initially treated with drugs and/or electrical cardioversion (an external electric shock to restore the normal rhythm). In many

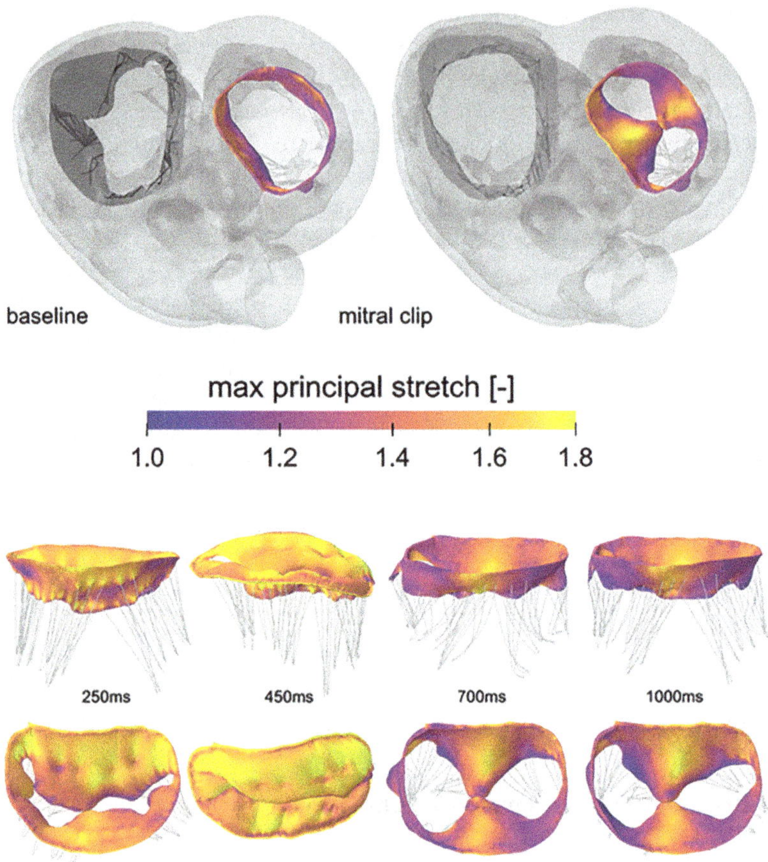


**Fig. 4** Modeling the Alfieri valvuloplasty as a therapy for mitral valve regurgitation. The multiscale model consists of an anatomically accurate representation of the four heart chambers, mitral valve leaflets, chordae, and LV papillary muscles, uniform mechanical, and conduction material models with a network of Purkinje fibers and coupled to a lumped model of the circulation. The model is employed to estimate the mechanical loading induced in the mitral leaflets by clipping them together to reduce regurgitation flow. (Reprinted/adapted with permission from Peirlinck et al. 2021 [1], Springer Nature)

cases, fibrillation recurs and is treated with ablation, a process that intentionally damages muscle, resulting in the formation of scar tissue that can block errant electrical conduction. Implanted pacemakers (devices that electrically stimulate the heart) are often employed to treat chronic and persistent arrhythmias, and implantable defibrillators may also be used for conditions that may provoke cardiac arrest [93].

### 4.3.2 Multiscale Models of Disease

Arrhythmias are a clear example of a physiological phenomenon where feedback interactions across hierarchical levels determine the organ-scale integrative behavior. Multiscale modeling of cardiac arrhythmia consists of linking our increasing knowledge of single-cell action potential dynamics and modeling to organ-scale continuum field equations. The organ-scale model, typically consisting of a form of the bidomain or monodomain equations, predicts the propagation of the action potential through the tissue; the local active potential is fed to molecular-scale models of ion transport dynamics, which determine the duration and intensity of the activation cycle that is fed back to the organ-scale model. Arrhythmia is induced in such models either by setting multiple points of active potential generation at the organ scale or by altering ion transport dynamics at the molecular scale, depending on the focus of the study. A detailed review of the historical development of multiscale models of atrial fibrillation and ventricular arrhythmia can be found in the works of Heijman et al. (2021) and Trayanova and Boyle (2014), respectively [95, 96].

Many multiscale arrhythmia models focus on the electrophysiology of the pathology, neglecting tissue deformation during the cardiac cycle and related mechanical effects. This decoupled electrophysiologic model is adequate to reproduce many relevant features of arrhythmia, such as the role of anatomic structures on reentrant arrhythmia dynamics. A notable example is the work by Bishop and Plank (2012), which employed a detailed organ-scale anatomical model of the rabbit's heart coupled to a bidomain propagation model and a molecular model of rabbit myocyte activation. The study concluded that fine-scale anatomic structures, such as coronary vasculature and endocardial structures, play only a marginal role in arrhythmic patterns and reentrant waveforms, suggesting a simplified model capturing the shape and size of atrial anatomy is enough to simulate the maintenance of cardiac arrhythmias [97].

Despite the success of electrical-only models, in vivo, the electrical and mechanical functions of the heart are tightly coupled, each capable of affecting the other. A large body of work on mechano-electric feedback is intentionally omitted here because related clinical syndromes such as commotio cordis are relatively rare. However, this is a fascinating area of cardiac physiology, and an excellent review by Quinn and Kohl can provide a good introduction to some of the relevant computational models [98]. This section focuses instead on models of how electrical dysfunction or therapies that modify electrical function affect mechanics. For example, Moyer et al. (Annals of BME 2015) coupled a finite element model of the left atrium to a lumped-parameter model of blood flow and simulated known features of remodeling due to paroxysmal (intermittent) atrial fibrillation (AF) [99]. They reported that physical boundary conditions such as movement of the mitral annulus, compression from adjacent structures, and boundary conditions on fluid flow into and out of the atrium were critical to reproducing pressure-volume behavior and wall motion measured in normal atria. When simulated individually, no one feature of atrial remodeling associated with paroxysmal AF (dilation, fibrosis, and elevated pressures) could reproduce observed mechanical dysfunction. However, these three features, in combination, could explain 84% of the differences in pressure-volume loops between normal and AF atria. Including impaired relaxation of the myocardium—a hypothesized but unproven feature of atrial remodeling—further improved the match between model and data.

### 4.3.3 Multiscale Models of Therapy

Improvements in multiscale modeling of cardiac electrophysiology over the last two decades have allowed researchers to address increasingly realistic, clinically driven problems. Trayanova and Chang (2016) reviewed electrophysiologic models of therapies for the arrhythmic human heart, focusing on patient-specific electrophysiological models [100]. In this section, we focus on multiscale modeling developments that offer insights into the adequate administration of anti-arrhythmic therapies and their limitations, complementing the above-cited review by adding multiphysics-multiscale models and more recent developments (Table 3).

Some of the earliest multiscale models of therapies incorporating molecular phenomena focused on electrical pacing as a therapeutic alternative for atrial fibrillation (AF). Uldry et al. (2010, 2012) sought to understand the reported variability in clinical response to pacing during AF. They coupled an organ-level model of electrical conduction in an anatomically accurate bi-atrial geometry to a Luo-Rudy molecular model of transmembrane ion dynamics. They simulated the effect of rapid pacing at different locations of the atria on the capture and termination of AF. Initial studies found that atrial tissue properties, pacing location, and cycle length could all affect the predicted treatment outcome [101]. The same model was later used to test a septal pacing algorithm, which provided a more consistent suppression of reentrant waves and termination of simulated AF compared to other pacing locations [102]. This model was further expanded by Russu et al. (2014) to explore atrial substrate heterogeneity as a potential cause for patient-to-patient variability in response to pacing therapy. Russu and coworkers incorporated a more sophisticated Courtemanche-Ramirez-Nattel model of transmembrane ion dynamics accounting for atrial substrate heterogeneity. The authors concluded that substrate heterogeneity can significantly affect the outcome of septal fast pacing as a therapy for atrial fibrillation, which might partly explain the disparity of clinical outcomes for this treatment option [103].

Although drug therapies for the suppression of arrhythmias are an appealing clinical approach, the alteration of ion channel dynamics can have undesirable side effects, which paradoxically include dosage-related risks of arrhythmia and cardiac arrest. Thus, pro-arrhythmic risk evaluation is a critical step in developing new drugs. The reported failure of single-cell in vitro models to predict the pro-arrhythmic effect of drugs and the high cost of preclinical and clinical trials have brought attention to using multiscale models as a support tool for drug screening [1, 104]. A notable example of an in silico anti-arrhythmic drug test was developed by Moreno et al. (2011). This work combined a 40-parameter Ten Tusscher model

**Table 3** Multiscale models of treatments for arrhythmia

| Reference | Condition | Treatment | Physics | Scales | | | |
|---|---|---|---|---|---|---|---|
| | | | | System | Organ | Tissue | Molecular |
| Uldry et al. 2012 [102] | Arrhythmia | Pacing | *Electrophysiology* | | 3D atria | Monodomain homogeneous conductivity | Luo-Rudy |
| Rusu et al. 2014 [103] | Arrhythmia | Pacing | *Electrophysiology* | | 3D biatrial | Monodomain | Luo-Rudy membrane and Courtemanche–Ramirez–Nattel human atrial cell model |
| Moreno et al. 2017 [54] | Arrhythmia | Drug therapy | *Electrophysiology* | | 3 layered 3D biventricular | 3 layered monodomain | Ten Tusscher with Markovian model of sodium channel |
| Sahli-Costabal et al. 2018, 2019 [105, 106] | Mitral valve stenosis | Edge-to-edge valve repair | *Mechanics* | | 3D 4 chamber LHP | Holzapfel-like material with active tension | |
| | | | *Hemodynamics* | 0D open Windkessel ×2 | | | |
| | | | *Electrophysiology* | | 3D 4 chamber LHP 1D Purkinje network | Monodomain fibrous material | O'Hara Rudy × 4 cell-types |

(continued)

**Table 3** (continued)

| Reference | Condition | Treatment | Physics | Scales | | | |
|---|---|---|---|---|---|---|---|
| | | | | System | Organ | Tissue | Molecular |
| Hörmann et al. 2017 [109] | Atrial fibrillation | Ablation | *Mechanics* | | 3D biatrial | Isotropic hyperelastic with active stress (noncontractile scar) | |
| | | | *Hemodynamics* | 0D Open Windkessel | | | |
| | | | *Electrophysiology* | | 3D biatrial | Monodomain Nonconductive scar | Bueno-Ovorio |
| Phung et al. 2017 [110] | Atrial fibrillation | Ablation | *Mechanics* | | 3D atria | Orthotropic hyperplastic with active stress (noncontractile scar) | |
| | | | *Hemodynamics* | 0D Open Windkessel ×2. | | | |

Notes: *LHP* living heart project
O'Hara Rudy × 3 cell types: the O'Hara-Rudy model was formulated for endocardial, mid-wall, and epicardial myocytes with different sets of parameters
Windkessel ×2: two independent Windkessel models for pulmonary and arterial vasculature with grounded end

of transmembrane ion dynamics with a stochastic 8-state Markovian model of drug binding to ion channel receptors. They coupled this molecular model to a 3D organ-scale conduction model of the human LV, including a Purkinje fiber network, and calibrated model parameters to estimate the rate of diffusion and receptor affinity for the anti-arrhythmic drugs flecainide and lidocaine. Finally, they used the calibrated model to predict the effect of drugs in a normal heart in the setting of spontaneous ventricular ectopy (a common trigger for arrhythmia). The results showed good correspondence to experiments on rabbit hearts and correctly reproduced the clinically relevant concentrations at which flecainide and lidocaine exacerbate arrhythmia in human cell constructs [54]. Similarly, Sahli-Costabal et al. (2018, 2019) proposed a multiscale model to estimate the effect of ion channel blockers on cardiac QT intervals as a major risk factor for ventricular arrhythmia. These investigators employed an O'Hara-Rudy of ion dynamics (a modification of the Ten Tusscher model) for four types of cell types (Purkinje, endocardium, mid wall, and epicardium myocytes), each with its own set of transmembrane ion currents and corresponding internal parameters. The transmembrane molecular model was coupled to a monodomain single-layered conduction model at the organ scale on a 3D anatomically accurate bi-atrial geometry. Drug effects were modeled by blocking specific ion channels to assess the resulting alterations of regional action potential and wave propagation patterns (Fig. 5). The model was first used to test a low-risk drug, ranolazine, and a high-risk drug, quinidine, and predicted a QT interval prolongation of 19.4% for ranolazine against a 78.4% prolongation by quinidine [105]. The model was then embedded in a machine-learning framework to screen the effect of 30 common drugs on prolonging QT intervals. The outcomes of this study showed good agreement with changes in QT interval durations observed in randomized clinical trials for dofetilide, quinidine, ranolazine, and verapamil. In addition, the authors suggested that compounds that block the rapid delayed rectifier potassium current have the greatest prolonging effect on the QT interval and that blocking the L-type calcium current and late sodium current shortens the QT interval [106]. These findings inspired an algorithm for arrhythmogenic risk based on multiscale modeling and machine learning tools [107].

The therapeutic concept behind catheter ablation is to replace regions of the arrhythmic atria with nonconductive scar tissue to isolate the points of origin of arrhythmogenesis and/or to interrupt reentrant waves promoting fibrillation. Many different scar patterns are employed clinically, alone and in combination, making it difficult to directly compare how different scar patterns might impact function in one patient. The ability to simulate the replacement of muscle with a scar in any desired amount and location provides an opportunity to explore the impact of different potential choices and the tradeoff between improved electrical function and reduced mechanical function. Hörmann et al. (2017) used a multiscale model to compare the effect of different ablation techniques on the systolic mechanical performance of the atria. They simulated two 3D anatomically accurate biatrial geometries, one from a healthy individual and one from a patient with hypertrophied atria. The electrophysiology model consisted of a monodomain conduction problem at an organ level, coupled with a simplified 4 variable-8 equation model of transmem-

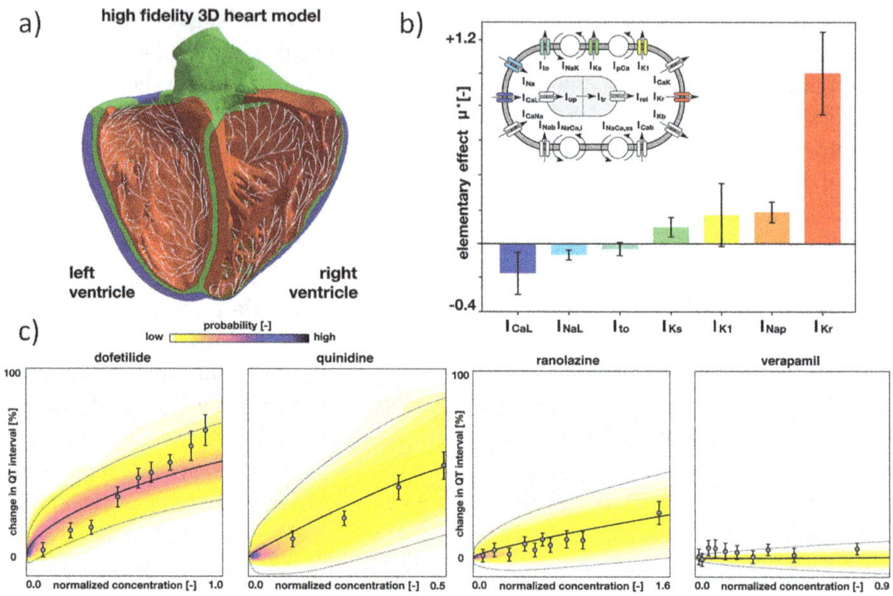


**Fig. 5** Modeling the effect of anti-arrhythmia drug therapies on the ECG QT interval using a multiscale model and machine learning tools. The multiscale model consists of (**a**) a 3D anatomically detailed biventricular model with uniform conductive properties and 1D Purkinje fibers coupled to (**b**) a Ten Tusscher model of ion transport. (**c**) Predicted dose-dependent probability maps of the effect of dofetilide, quinidine, ranolazine, and verapamil on QT interval duration. (Reprinted/adapted with permission from Sahli et al. 2019 [106], Elsevier)

brane action potentials by Bueno-Ovorio [108]. Boundary conditions on fluid flow included a 3-parameter Windkessel model of the LV and a no-backflow condition at the pulmonary veins. Ablation injuries were modeled as subregions with no conductivity and no contractility [109]. In a paper published the same month, Phung et al. (2017) conducted a similar series of simulations, including passive and active atrial mechanics, using a model with fewer electrophysiologic details but more realistic boundary conditions [110]. Both groups predicted a progressive decline in active atrial function with increasing total scar volume, suggesting that more aggressive ablation strategies to address the electrical disruption from AF may come at the cost of diminished atrial function. In addition, Phung et al. found that ablation in some locations caused less mechanical disruption than others, suggesting some potential for customizing procedures to account for the tradeoff between ablation scar volume and disruption of mechanical function (Fig. 6). However, neither of these models included sufficient electrical detail to predict the likelihood that any given ablation pattern would terminate or prevent AF, a capability that would be essential for personalized preoperative planning.

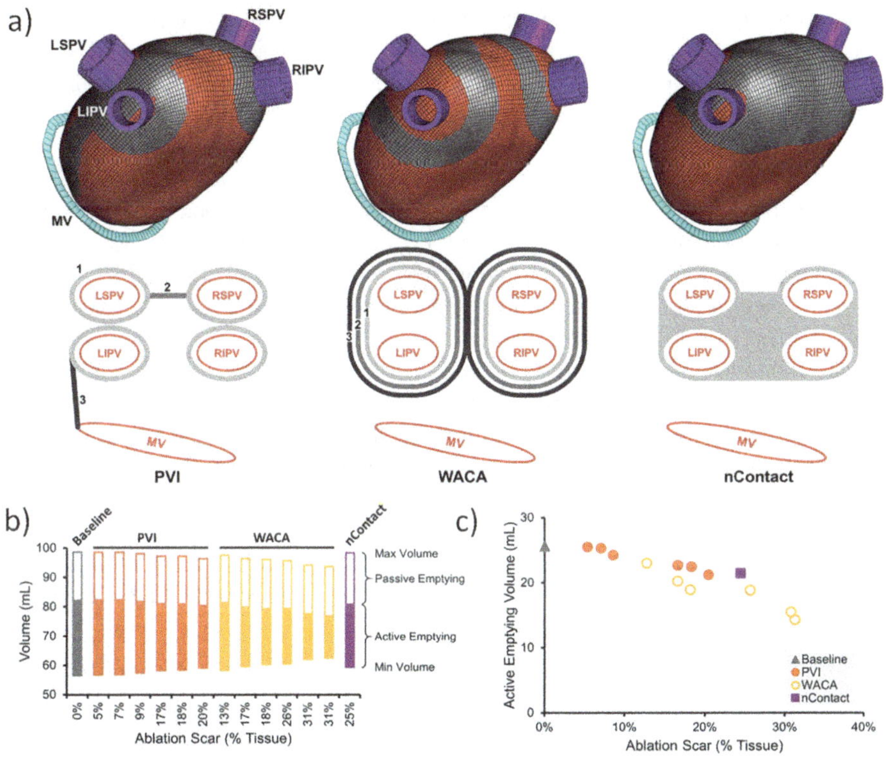


**Fig. 6** Study of the mechanical performance of the left atrium after different ablation patterns as a treatment for atrial fibrillation. The multiscale model consists of an anatomically accurate 3D model of the atrium with uniform single-equation mechanical behavior and contraction, coupled to a 1D spherical model of LV mechanics and lumped model of the circulation. (**a**) Ablation patterns in 3D anatomy and 2D maps. (**b**) Effect of ablation patterns and scar size on atrial volumes, indicating contributions from passive (blank portion of the bars) and active (colored portion) emptying. (c) Effect of ablation pattern and scar size on active emptying volume of the atrium. (Reprinted/adapted with permission from Phung et al. 2017 [110], Wiley)

## *4.4 Bundle Branch Block and Dyssynchrony*

### 4.4.1 Disease Manifestation and Treatment

In cardiac dyssynchrony, the propagation of the active potential wave is spatially uneven, making the heart chambers beat out of sync. This condition can severely affect pump function and locally elevate peak stretches and stresses. Among the most salient symptoms of cardiac dyssynchrony include chest pain, confusion, anxiety, shortness of breath, dizziness and fainting, fluid build-up, and weakness [111].

Injuries to specialized conducting fibers in the interventricular septum (bundle branches) following a heart attack are one of the leading causes of cardiac dyssynchrony. However, it has also been associated with congenital conditions, aging, tobacco and alcohol use, heart remodeling following a primary injury or disease, and heart failure. In normal cardiac conduction, the action potential wave is propagated simultaneously through the left and right branch bundles, with the septum being activated from left to right. Left bundle branch block (LBBB) consists of the disruption of the normal distribution of electrical impulses, such that the left ventricle activation is delayed and less well synchronized, which impairs cardiac function and promotes mitral regurgitation, LV dilation, and asymmetric hypertrophy [112, 113].

The treatment of cardiac dyssynchrony includes using pacemakers and cardiac resynchronization therapy. Cardiac resynchronization therapy (CRT) represents a significant leap in treating dyssynchronous heart failure [114]. CRT uses modified pacemakers to electrically stimulate the LV in ways that reduce regional differences in the timing of active contraction. At its simplest, it might involve implanting a lead in the latest-activated region of the LV and stimulating that region simultaneously with the earliest-activated region. When successful, CRT can not only halt dilation and progressive deterioration of cardiac function but even reverse them, decreasing mortality and improving quality of life [115, 116].

### 4.4.2 Multiscale Models of Disease

Modeling dyssynchronous hearts requires implementing a multidimensional anatomic model that can be discretized in sections with independent, active contraction functions. The time dependence of the active contraction of each anatomic segment can be directly imposed or can be obtained from the solution of an action potential propagation model. When dyssynchrony is the product of injury-induced rerouting of the action potential propagation wavefront, this can be implemented by representing the scar tissue as a nonconductive region in the organ-scale model of electrophysiology. Walmsley et al. (2015) proposed a modification of the 2D organ-scale TriSeg biventricular model by Lumens et al. wherein each chamber and septum wall is subdivided into spherical-shell patches mechanically linked in series. Each patch corresponds to one of the standardized AHA anatomical segments of the heart and was assigned its own mechanical properties and time-dependent active contraction curves. The activation curves were directly prescribed based on electrocardiographic measurements from a human patient with a left bundle branch block. The heart model was coupled to a lumped model of the circulation, and simulation parameters were calibrated to reproduce patient-specific vitals and hemodynamic records. The estimations of regional strain by the model showed good qualitative and quantitative correspondence to MRI-derived measurements. The authors also found that estimated regional strain was more sensitive to the timing of activation than to its spatial location, meaning that anatomical detail is secondary to electrical parameters such as QRS duration in modeling LBBB adequately. This observation has promising implications for the development of low-cost, simplified models of cardiac dyssynchrony [36]. Similarly, Owashi et al. (2020, 2022) employed a 2D organ-scale model of the LV discretized in the standard AHA segments coupled to a 0D lumped model of circulation and 1D spherical compartment models of the right ventricle and atria. This work, however, incorporated a simple model of organ-scale electrophysiology. The electrophysiology model comprised 26 interconnected cellular automata that could assume four different activation states. The interconnection of the electrophysiology network was defined to reproduce the activation patterns observed in the clinical records of three patients of LBBB. The authors validated the model by comparing estimations of regional strain to patient-specific measurements based on echocardiographic images that resulted in root mean squared errors below 4% [117, 118].

In 2009, Kerckhoffs et al. proposed a detailed organ-scale 3D model of a canine heart to assess the effect of scar size on the performance of failing hearts with ischemic LBBB. The organ-scale model consisted of the biventricular anatomy with realistic myofiber orientation and a Purkinje conduction network coupled to two 0D lumped circuits for systemic and pulmonary circulation. The myocardium was assumed to be an isotropic hyperelastic material, and infarct regions of 7%, 14%, 21%, and 31% were simulated, assuming the infarct region is not conducting, noncontractile, and about 15 times stiffer than healthy myocardium. The propagation of active potential was calculated with a modified Fitzhugh–Nagumo kinetics. This early work demonstrated the capabilities of a detailed 3D multiscale model for predicting activation and contraction patterns in dyssynchronous hearts and demonstrated how the scar size determines the impairment in cardiac function [119]. In 2018, Finsberg et al. compared the ventricular contractility between healthy and dyssynchronous human hearts using patient-specific organ-scale 3D anatomical models derived from 4D echocardiography. The comparison was carried out on the simulation outputs from 14 patient-specific models, seven healthy and seven with ischemic LBBB. The patient-specific parameters for the model were calibrated in two steps. First, the passive material properties and unloaded configuration were estimated using the geometric and pressure information at early and late diastole. Second, the active and anisotropic material properties were obtained by minimizing an error function defined as a weighted average of ventricular volume and strain error. The authors observed that patient-specific customization produced more compliance for the healthy subgroup, while active contraction was significantly lower for the LBBB group [120]. This work is an example of the use of medical imaging as the main data source for developing customized, detailed computer models of cardiac dyssynchrony. However, customization required the solution of two different optimization processes, each implying the iterative solution of a computationally expensive multiscale problem. Further considerations about coupling multiscale electromechanical models for the study of dyssynchronous and failing hearts can be found in the review by Kuijpers et al. [121]

### 4.4.3 Multiscale Models of Therapy

In theory, the freedom to customize pacemaker lead locations and adjust electrical settings on the devices should allow physicians to account for individual variability in electrical conduction patterns, scar from prior myocardial infarctions, and other features. Nevertheless, in practice, there are far too many options to explore during an implantation procedure, and approximately half of patients fail to respond to CRT. This motivated the formulation of multiscale models that could predict clinical outcomes from various CRT strategies and potentially support the personalization of therapy in individual patients. Multiscale models for CRT require both electrophysiological and mechanical components (Table 4). The model proposed by Okada et al. (2017) provides a detailed description of electrophysiology by incorporating separate Ten Tusscher models of transmembrane ion dynamics for endocardial, mid wall, and epicardial myocytes, coupled with a bidomain model solved on an MRI-derived patient-specific 3D biventricular model and torso. Intracellular calcium concentrations from the electrophysiology simulation modulated a model of sarcomere contraction and active stress, while passive mechanical behavior was assumed to be orthotropic hyperelastic. The mechanical biventricular model was coupled to a 1D time-varying elastance model of the atria and a lumped model of the vasculature. The multiscale model was customized to reproduce the preoperatory electrocardiogram and hemodynamic parameters of each patient under study. Then, CRT was simulated, and model outputs were compared to echocardiographic follow-up data collected at an average of 4 months after CRT. The authors explored possible correlations between simulation outputs and clinical indexes of cardiac function at the 4-month follow-up and found that the simulated rate of ventricular pressure rise (dP/dt) significantly correlated directly to the patient's ejection fraction, suggesting the value of simulation outputs as a surrogate for invasive catheter manometry [122]. In a follow-up study, Isotani et al. (2020) sought to evaluate the predictive capabilities of the multiscale model, so they performed a prospective blind comparison of simulation outputs to the clinical outcomes in a 3-month follow-up of a cohort of eight patients with cardiac dyssynchrony and New York Heart Association functional class II/III classification treated with CRT. The model was customized to each patient's preoperative data, and CRT was iteratively simulated while changing the position of two leads to identify the optimal lead placement. The study was purely observational, and the simulations did not influence the clinical procedure. The authors report that the simulated rate of ventricular pressure at the optimal lead position correlated to patient-specific responsiveness to CRT at the 3-month follow-up [123].

The models discussed so far focus on therapies' acute or short-term effects. Nevertheless, heart failure and other cardiovascular diseases often develop and progress over decades, and treatments may work partly or even primarily by modifying this long-term remodeling process. Thus, the multiscale model proposed by Oomen et al. (2022) stands out as one of the few to consider the long-term effects of CRT on cardiac remodeling and reversal of hypertrophy. This was achieved by incorporating a modified Kerckhoffs model for strain-dependent growth into a

**Table 4** Multiscale models of treatments for left bundle branch block

| Reference | Condition | Treatment | Physics | Scales | | | |
|---|---|---|---|---|---|---|---|
| | | | | System | Organ | Tissue | Molecular |
| Okada et al. 2017 [122] Isotani et al. 2020 [123] | Myocardial infarction and left branch block | Cardiac resynchronization therapy | *Mechanics* | | 3D biventricular 1D atria TVE | Humphrey and Lin with active stress | Washio-Okada Sarcomere |
| | | | *Hemodynamics* | 0D open Windkessel × 2 | | | |
| | | | *Electrophysiology* | 3D torso | 3D bilayer biventricular | Bidomain | Ten Tusscher |
| Oomen et al. 2022 [37] | Myocardial infarction and left branch block | Cardiac resynchronization therapy | *Mechanics* | | 2D triSeg multipatch | Exponential passive with Hill-type active stress | |
| | | | *Hemodynamics* | 0D closed circulation | | | |
| | | | *Electrophysiology* | | 3D biventricular | | |
| | | | *Growth* | | | Kerckhoffs | |

Notes: Windkessel x2: two independent Windkessel models for pulmonary and arterial vasculature with grounded end multiscale electromechanical physiology model. Following Yoshida et al.'s (2020) approach, a time-evolving reference configuration for the homeostatic level of strain was included to facilitate the simulation of both hypertrophy and its reversal [64]. The electrophysiological model consisted of a 3D organ-scale monodomain model of active potential propagation; the resulting time-resolved distribution of active potential was mapped to a 2D layout of 16 AHA segments for the LV and five segments for the RV. The mechanical model consisted of a 2D organ-scale TriSeg biventricular model with orthotropic exponential passive behavior at the tissue scale and an active stress term dependent on the segment-wise active potential distribution. The coupling of the electromechanical model to the growth model assumed the timescale of a heartbeat to be several orders of magnitude smaller than the timescale of tissue growth and remodeling. Thus, the solution of the electromechanical model assumed constant cardiac mass over a cardiac cycle to produce estimations of ventricular strain, which were fed to the growth model to produce the change of mass during each growth step. The mass change was applied to the unloaded configuration, and a new simulation of the electromechanical beat was performed, closing the coupling loop (Fig. 7). The model parameters were calibrated to data from canine experimental left branch block. The model was then validated with CRT experimental canine data, which consisted of time-resolved measurements of ventricular volume, pressure, and LV wall mass over 4 weeks of untreated LBBB and 4 weeks following CRT therapy. The model was applied to explore the effect of lead placement on long-term ventricular adaptative response. The authors report that the predicted remodeling outcome varied with lead location and ischemia location and did not necessarily correlate with acute improvement in QRS duration [32]. These findings challenge the practice of targeting the optimization of the QRS complex during CRT as a guide for lead placement and invite further study on the identification of new clinical indexes for risk stratification of patients with dyssynchronous cardiopathies.

## *4.5 Familial Cardiomyopathy*

### 4.5.1 Disease Manifestation and Treatment

Familial cardiomyopathy is a congenital disease that affects the heart muscle. The most common phenotype is dilated cardiomyopathy, a condition in which the volume of the chambers increases without a proportionate increase in wall thickness, increasing the contractile force required to maintain a normal stroke volume. Compensation mechanisms such as reflex activation of beta-adrenergic hormones and signaling pathways may provide the additional contractional force required to maintain heart function for some period but are ultimately exhausted, leading to heart failure. Another common phenotype is familial hypertrophic cardiomyopathy (FHC), which is characterized by the thickening of ventricular walls unexplained

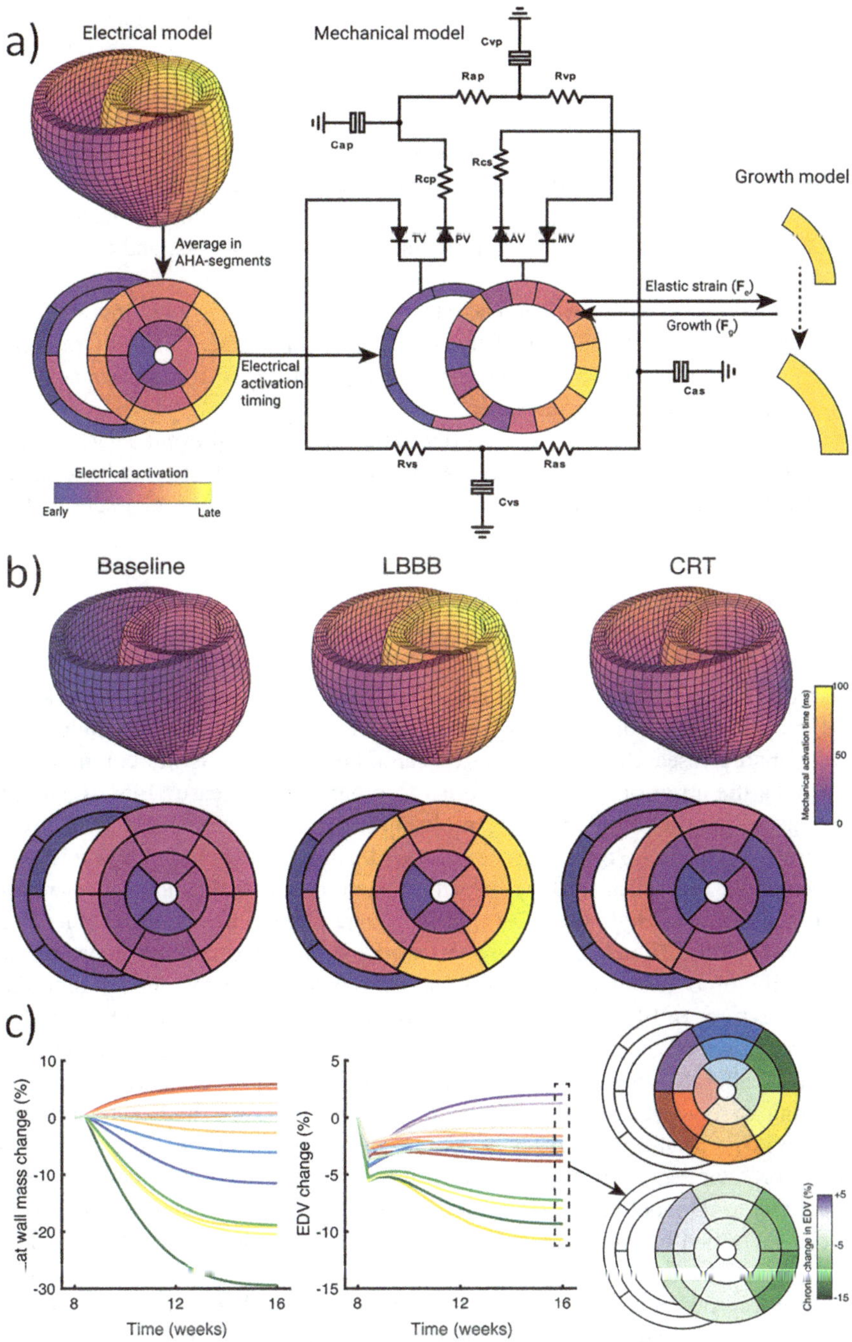


**Fig. 7** (**a**) Diagram of a multiscale model of left bundle branch block (LBBB) and cardiac resynchronization therapy (CRT) consisting of a 3D biventricular model of action potential propagation with uniform conductive material properties, a 2D multipatch model of ventricular mechanics with uniform tissue single-equation constitutive equation, and tissue-scale phenomenological model of growth. (**b**) Comparison of the distribution of muscle activation on a healthy heart, during LBBB, and after CRT from the electrophysiology model and mappings into a 2D segmentation for coupling to the mechanical model. (**c**) Simulation of the effect of CRT on left ventricular mass and volume on a 2D segmentation. (Reprinted/adapted with permission from Oomen et al. 2021 [37], Springer Nature)

by secondary causes, leading to obstruction of the left ventricular outflow tract and diastolic dysfunction.

Genetic linkage studies have associated familial cardiomyopathy with mutations to a number of genes encoding for sarcomere and cardiomyocyte cytoskeleton proteins. The disease often presents in mid-adulthood with the new onset of arrhythmia, shortness of breath, fatigue, faintness and dizziness, and fluid build-up [124].

No cure exists for the condition, and treatments to alleviate symptoms are limited. Drug therapies usually consist of the same drugs prescribed for the treatment of heart failure: beta-blockers, angiotensin II receptor blockers, angiotensin converting enzyme inhibitors, and diuretics.

The use of implantable cardioverter–defibrillators to prevent sudden cardiac death has successfully reduced mortality among FHC patients, as have surgical procedures that remove excess myocardium from the intraventricular septum [125].

### 4.5.2 Multiscale Models of Disease

The fact that so many different mutations can lead to cardiomyopathy has posed a major challenge in developing comprehensive predictive models for the onset and progression of the pathological phenotype. The consensus in familial cardiomyopathy research is that impaired sarcomere function is the common thread that links the array of known mutations to organ-scale disease [125]. Testing this hypothesis requires implementing molecular models of sarcomere structure and functions detailed enough to reproduce the effects of conformational irregularities. For example, Wu et al. (2011) presented a molecular-scale model of sarcomere dysfunction caused by a cardiac troponin I gene mutation associated with familial cardiomyopathy. The effect of the gene mutation was incorporated as an increased sensitivity of troponin to calcium concentrations. While the model reproduced typical features of FHC such as slower myocardial relaxation, the authors highlight the prediction of a net production of lactate by the myocardial metabolism in scenarios of high pumping energy demand. The production of lactate results in poor coupling of glucose oxidation to glycolysis, a condition found in hypertrophied hearts [126]. Khalilimeybodi et al. (2023) developed a systems biology network model of established intracellular signaling pathways relating sarcomere calcium sensitivity to remodeling phenotype based chiefly on experimental murine data. The model identified calcium-mediated myofilament force development and calcium-calmodulin kinase signaling reactions as key players in manifesting hypertrophic and dilated cardiomyopathy [127]. Few attempts have been made to incorporate these molecular-scale models into a multiscale framework that includes the heart at an organ scale. Syomyn et al. (2020) explored the hemodynamic effects of mutations in the regulatory sarcomere protein tropomyosin, specifically Ile284Val, associated with hypertrophic cardiomyopathy (HCM) and Asp230Asn, associated with dilated cardiomyopathy (DCM). The model consisted of an organ-scale 3D idealized elliptical model of the LV with passive and active stress components, isotropic hyperelastic passive behavior, and an active stress function depending on two molecular parameters representing actin-myosin interactions. The organ-scale model was coupled to a system-scale lumped model of hemodynamics and a molecular-scale model of calcium kinetics, and the ventricle was assumed to contract synchronously (i.e., no model of electrical propagation). The mutations were assumed to affect only the binding of calcium to tropomyosin, strain dependence of the binding, and the number of myosin filaments per cross-sectional area in the myocardial model. Cardiac remodeling was imposed as a quasistatic change in ventricular dimensions. The study suggested that ventricular remodeling tends to preserve LV function, countering the impaired contractile function of the sarcomere [128]. Further considerations for developing multiscale models of familial cardiomyopathy, challenges, and future directions can be found in the insightful work of Campbell and MacCulloch [125].

### 4.5.3 Multiscale Models of Therapy

Modeling the effect of drug therapies for familial cardiomyopathies logically requires a similar level of detail of sarcomere dynamics as in the pathology models. The project SILICOFCM (In silico trials for drug tracing the effects of sarcomere protein mutations leading to familial cardiomyopathy) agglomerates an international consortium of academic, clinical, and private institutions aiming to develop a computational platform for in silico clinical trials of FCM. The SILICOFCM platform incorporates the muscle simulation code (MUSICO), a molecule-to-fiber multiscale model of cardiac and skeletal muscle, and finite element models of biventricular electromechanics. The molecular model incorporates stochastic dynamics of cross-bridges, myosin binding, and calcium ion regulation of thin-filament switches.

As a proof of concept for the SILICOFCM platform, Filipovic et al. (2022) simulated the effect of Entresto and Digoxin drug therapies on cardiac function in the setting of familial dilated cardiomyopathy [129]. Drug effects were simulated by imposing reported changes in ion currents and myosin cross-bridge cycling at the molecular scale and relaxing vasoconstriction at the system scale. The authors found the changes in the molecular model scaled up from modulating calcium transients to determining organ-scale ventricular pressure-volume loops. The simulated effect of both drugs on a patient-specific model showed reasonable agreement with reports of blood pressure and cardiac function from clinical trials (Table 5).

## *4.6 Heart Failure*

### 4.6.1 Disease Manifestation and Treatment

Heart failure is a general term used for the late stages of heart disease when the organ can no longer supply the cardiac output needed to deliver oxygen- and nutrient-rich

**Table 5** Multiscale models of treatments for familial cardiomyopathy

| Reference | Condition | Treatment | Physics | Scales | | | |
|---|---|---|---|---|---|---|---|
| | | | | System | Organ | Tissue | Molecular |
| Filipovic et al. 2022 [129] | Dilated cardiomyopathy | Drug therapy | *Mechanics* | | 3D biventricular | Holzapfel and Hunter with active stress | |
| | | | *Hemodynamics* | 0D open Windkessel x2 | 3D Navier-Stokes | | |
| | | | *Electrophysiology* | | 3D biventricular | Monodomain | Mijailovich-Prodanovic surrogate model |

Notes: Windkessel x2: two independent Windkessel models for pulmonary and arterial vasculature with grounded end blood to the body. This stage is usually reached following a long series of reflex, hormonal, and cell signaling alterations, as well as physical growth and remodeling in response to mutation, injury, or disease. Some of the most common symptoms include arrhythmia, chest pain, fatigue, shortness of breath, dizziness, fainting, chronic cough, and fluid build-up. Heart failure may lead to liver and kidney disease, pulmonary hypertension, and sudden cardiac arrest [130].

Heart failure is an incurable condition, with a heart transplant being the only definitive treatment. Palliative therapies meant to improve symptoms typically include a combination of beta-blockers, angiotensin II receptor blockers, converting enzyme inhibitors, and diuretics. Surgical intervention to treat the primary cause of heart failure, such as valve repair, can potentially improve symptoms and increase life expectancy.

Ventricular assist devices (VAD) are a temporary treatment for heart failure consisting of an external pump that drives blood from the heart's lower chambers to the rest of the body. Given the shortage of heart donors relative to the number of patients with heart failure and the constantly improving survival rate with new LVAD technologies (2-year survival rates over 80%), it is foreseeable that patients will be supported by LVADs for longer periods of time [131]. Implantation of LVAD poses an acute risk of infections and adverse immune responses, while the most common causes for rehospitalization include arrhythmia, bleeding, neurological events, and pump thrombosis [131, 132].

### 4.6.2 Multiscale Models of Disease

The transition from early adaptations that often appear to compensate for an initial injury or dysfunction to decompensated heart failure remains poorly understood, and the timing of this transition in an individual patient is very difficult to predict. It is known that this transition not only consists of anatomical changes at the organ scale and compositional changes at the tissue scale but also remodeling at the molecular scale, such as downregulation and modulation of adrenergic signaling pathways that regulate myocardial contractility. Thus, a comprehensive multiscale model of the transition to heart failure is yet to be formulated and depends on further understanding of the mechanisms underlying maladaptive remodeling.

While the specific mechanisms triggering the transition to heart failure remain to be elucidated, phenomenological equations for growth and remodeling can potentially describe this process. For example, Peirlinck et al. (2019) used a strain-based phenomenological growth equation implemented on an organ scale accurate 3D model of pig hearts to predict cardiac remodeling in response to chronic pressure and volume overload. The myocardium was assumed to be a passive material with a fiber-family constitutive equation, and the overload stimulus was assumed to be static. A machine learning algorithm was implemented to calibrate the parameters of the growth law to experimental measurements on pig animal models of heart failure, which the authors consider an important step toward the multiscale characterization of heart failure [133]. Modeling of hearts already in failure is an increasingly straightforward process, where the hypertrophic anatomy can be built from available image-based data and passive and active properties calibrated to reproduce the impaired diastolic and systolic function.

### 4.6.3 Multiscale Models of Therapy

The LVAD technology has undergone multiple design and conceptual changes since its first successful clinical trial in 2001. The commercial technology shifted from the heavy, prone-to-failure pulsatile-flow membrane devices to the lighter and more reliable centrifugal and axial continuous flow pump technology [134]. Computational modeling plays an important role in the design and improvement of the LVAD pumps [135]; in this section, however, we focus on multiscale models of the assisted heart (Table 6).

One of the main concerns during this design paradigm shift was how introducing a continuous flow generator may affect systemic blood flow distribution and pressure waveforms. This concern was addressed by Bonnemain et al. (2013) through a multiscale model that coupled a 1D time-varying elastance compartmental model of the ventricles to a 1D Navier-Stokes model of intraventricular flow and a lumped model of systemic circulation with the LVAD as a flow source. The study finds that pump speed determines the shape of flow and pressure waveforms in the heart and arterial system, while the pump cannulation site (ascending vs. descending aorta) has little to no effect on model outputs (Fig. 8) [136]. However, it has been argued that the classical time-varying elastance model is inadequate to describe heart function in the presence of an LVAD, as the bovine experiments by Vandenberghe et al. (2006) suggest that maximum elastance depends on pump-induced flow [137].

Aiming to study the effect of LVAD on pressure pulsatility, Cox et al. (2009) implemented a 1D single-sarcomere fiber model with an active stress component to describe ventricular mechanics, overcoming the limitations of the time-varying elastance. The ventricular model was coupled to a lumped parameter circulation loop and a baroreflex model for heart rate. The LVAD was modeled as either a continuous or pulsatile flow source following the pump characteristic curve. Notably, the authors report that the pulsatile flow assistance did not solve the decreased systemic pressure pulsatility, although it significantly improved coronary flow and diastolic unloading [138]. Lim et al. (2012) also compared the continuous and pulsatile operation of LVADs, focusing on its effect on ventricular pressure and contractile work with a more sophisticated model. With this purpose, the authors integrated a 3D organ-level monodomain model of active potential propagation to a Luo-Rudy-like ion current model, with 3D ventricular hyperelastic passive mechanics and the Rice sarcomere model for active contraction. The LVAD was modeled as a continuous or pulsatile 0D flow source and coupled to a closed-loop lumped circulation and the ventricular model. The authors found that pulsatile-assisted flow significantly reduced ventricular peak pressure but had minor effects on ATP consumption [139]. Both of these works suggest that the benefit of pulsatile-

**Table 6** Multiscale models of left ventricular assist devices for the treatment of heart failure

| Reference | Condition | Treatment | Physics | Scales | | | |
|---|---|---|---|---|---|---|---|
| | | | | System | Organ | Tissue | Molecular |
| Bonnemain et al. 2013 [136] | Heart failure | Continuous LVAD | *Mechanics* | | 1D ventricle TVE | | |
| | | | *LVAD* | 0D flow generator | | | |
| | | | *Hemodynamics* | 0D closed circulation | | | |
| Cox et al. 2009 [138] | Heart failure | Continuous vs. pulsatile LVAD | *Mechanics* | | 1D single fiber | Exponential passive with active stress | |
| | | | *LVAD* | 0D flow generator | | | |
| | | | *Hemodynamics* | 0D closed circulation | | | |
| Lim et al. 2012 [139] | Heart failure | Continuous vs. pulsatile LVAD | *Mechanics* | | 3D biventricular | Hyperelastic with active tension | Rice sarcomere model |
| | | | *LVAD* | 0D flow generator | | | |
| | | | *Hemodynamics* | 0D closed circulation | | | |
| | | | *Electrophysiology* | | 3D biventricular | Monodomain | Luo-Rudy-like model |

(continued)

**Table 6** (continued)

| Reference | Condition | Treatment | Physics | Scales | | | |
|---|---|---|---|---|---|---|---|
| | | | | System | Organ | Tissue | Molecular |
| Sack et al. 2016, 2018 [140, 141] | Heart failure | Continuous LVAD | *Mechanics* | | 3D biventricular | Fiber-family passive with active stress | |
| | | | *LVAD* | 0D flow generator | | | |
| | | | *Hemodynamics* | 0D closed circulation | | | |
| | | | *Electrophysiology* | | | 0D calcium-dependent uniform activation | |
| Bakir et al. 2018 [142] | Heart failure | Continuous LVAD | *Mechanics* | | 3D biventricular | Doll and Schweizerhof passive with active stress | |
| | | | *LVAD* | 0D flow generator | 3D Cannulation | | |
| | | | *Hemodynamics* | 0D closed circulation | 3D Navier-Stokes | | |
| | | | *Electrophysiology* | | 3D Biventricular | Nash and Panfilov | |

Notes: *TVE* time-varying elastance

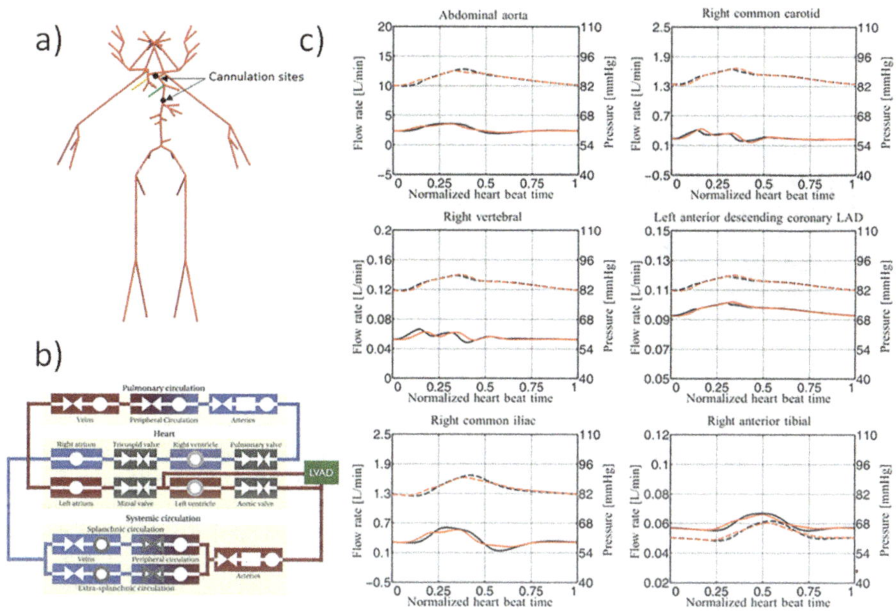


**Fig. 8** Study of the systemic effects of left ventricular assist devices (LVAD), rotational speed, and canulation site. The multiscale model consists of 1D models of ventricular mechanics and hemodynamics coupled to a lumped model of hemodynamics and a 0D model of the LVAD as a flow source. The LVAD discharge is connected to the ascending aorta (AA) and descending aorta (DA). (**a**) Arterial circulation lumped network with an indication of the canulation sites. (**b**) Diagram of circulation and lumped flow elements. (**c**) Flow (solid lines) and pressure (dashed lines) comparison between AA (black lines) and DA (red lines) cannulation in six arteries. (Reprinted/adapted with permission from Bonnemain et al. 2013 [136], Elsevier)

assisted flow is marginal, consistent with the commercial success of continuous-flow LVADs.

3D organ-level models of anatomically accurate ventricular anatomy were also employed to study mechanical loading and deformation of the heart chambers by LVAD operation as possible sources of device-induced morbidity. Sack et al. (2016) compared the ventricular wall stress distribution between a normal heart, a failing heart, and an assisted failing heart. The analysis was performed on anatomically accurate 3D organ-scale models of the whole heart. The passive mechanical properties were modeled at a tissue scale with a fiber-family constitutive equation with a time dependent active stress component. Finally, the LVAD was modeled as a continuous flow generator on the lumped model circuit (Fig. 9). The authors report that introducing LVADs reduces stress magnitude on the failing LV but has little effect on RV stress [140]. In a subsequent publication, the same model was employed to study the effect of LVADs on interventricular mechanics. The simulations suggest that increasing LVAD operating speeds may induce leftward septal shift and unnatural septum bending, leading to stress concentration. Notably,

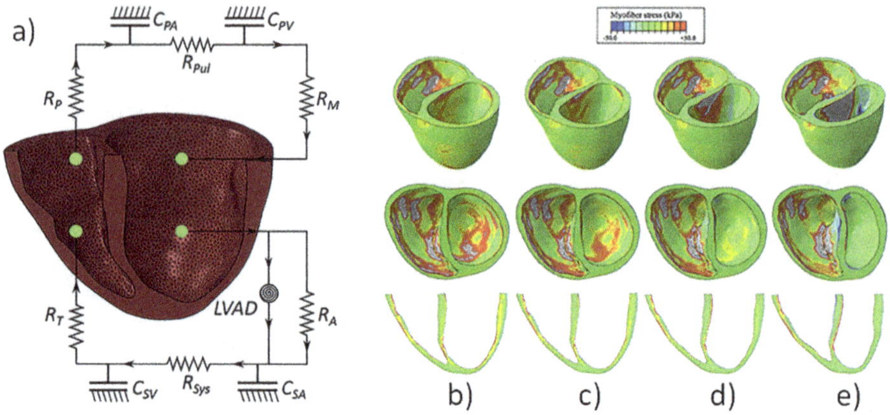


**Fig. 9** Modeling the effect of LVAD pump speed on biventricular mechanics. (**a**) Diagram of the multiscale model consisting of an anatomically accurate 3D biventricular model with uniform single-equation material behavior and uniform active contraction, coupled to a 0D model of the circulation and LVAD as a flow source. Myofiber stress distribution on short and long axis cuts at LVAD rotational speeds of (**b**) 0 rpm, (**c**) 8 k rpm, (**d**) 10 k rpm, and (**e**) 12 k rpm. Note how increasing LVAD pump speed induces interventricular septum shift. (Reprinted/adapted with permission from Sack et al. 2018 [141], Frontiers Media)

LVADs significantly improve diastolic overload in the LV although they also induce shifts in the RV pressure-volume loops toward larger volumes and pressures, which may induce RV remodeling and other complications [141].

Bakir et al. (2018) proposed a multiscale model to study how LVAD-heart interaction affected pump power demand. The authors produced a 3D organ-level model of an idealized biventricular anatomy with embedded Purkinje fibers [142]. Electrical activation was initiated at the Purkinje network with one-way coupling to the surrounding myocardium. Active tension was modeled with a 6-equation phenomenological model, while passive mechanics followed an orthotropic hyperelastic constitutive equation. Intraventricular blood flow was modeled with full Navier-Stokes CFD simulation assuming laminar flow and coupled to a lumped parameter model of a closed loop circulation as boundary conditions. The CFD solution was strongly coupled to the wall mechanics solution by equating blood and tissue velocity at the endocardium surface formulated as a numerical fluid-structure interaction (FSI) problem. A 0D ODE model related the LVAD hydraulic state to the instantaneous power demand of the pump to maintain the flow speed fixed. The model was applied assuming impaired systolic function as in the failing heart and tested for several LVAD flow speeds. The authors observed that at sufficiently high rotor speeds the aortic ejection was interrupted, and the LV pressure and volume were significantly decreased. If the condition was sustained for enough cardiac cycles, the LV cavity collapsed, which was accompanied by a drop in pump power demand. The authors suggest that these models could be used for the development of automatic safety and control mechanisms for LVADs.

# 5 Current Limitations and Future Developments

The studies reviewed here showcase the versatility of multiscale modeling approaches for the simulation of heart disease and therapeutic alternatives. However, their widespread application to personalized medicine still requires overcoming several limitations. One essential limitation of many of these models is the use of phenomenological equations to describe molecular processes such as sarcomere contraction, mechanotransduction, and growth and remodeling. Although phenomenological models are computationally inexpensive, they must be re-calibrated against data for each different situation (for example, in the presence or absence of growth-modulating drug treatments). This limitation is being actively addressed by incorporating models of cellular and subcellular dynamics representing our growing understanding of myocyte molecular biology. In the case of active stress and active tension, the incorporation of sarcomere models coupled to ion transport models provides broader potential applicability, but the simplification of activation depending solely on cytoplasmic calcium concentration requires further attention [5].

It is generally accepted that mechanical stimulation modulates cellular signaling, communication, and tissue-scale growth and remodeling. However, what mechanical stimuli are sensed by cardiomyocytes and how they relate to tissue-scale mechanics remains largely unknown. All existing models of mechanics-driven growth and remodeling at tissue and cellular scales thus rely on ad-hoc assumptions and phenomenological laws derived from empirical observations. This is further obscured in animal and clinical research by the fact that the true mechanical state of the cardiac tissue cannot be resolved by any in vivo measurement. Rigorously, the mechanical state should be estimated from a zero-strain and zero-load reference configuration, which, given that the heart chambers are permanently pressurized, does not occur in vivo, and cannot be measured by in vivo imaging [143]. Moreover, even when unloaded, the myocardial tissue is pre-stressed owing to the configuration of the extracellular matrix [144]. Many studies show that in vivo estimations of material properties and stress assuming the diastolic configuration as a zero-strain reference led to significant discrepancies to measurements on excised tissue. To overcome this, some studies impose the diastolic strain as reported from healthy animal models [32], a questionable approach for the study of diseased hearts. Alternatively, some researchers propose iterative methods to resolve the unloaded configuration on finite element models, such as the backward displacement method by Rajagopal and coworkers [145].

While most of the models presented herein focus on the acute effects of therapy, it is important to note that these metrics do not necessarily correlate with the long-term effects that ultimately determine a patient's health and clinical outcomes. Predicting these long-term outcomes requires the simulation of extended periods of time and the incorporation of models of growth and remodeling, which is inherently computationally expensive. The models reviewed above that do address the long-term effects of therapy typically rely on low-dimensional models for mechanics

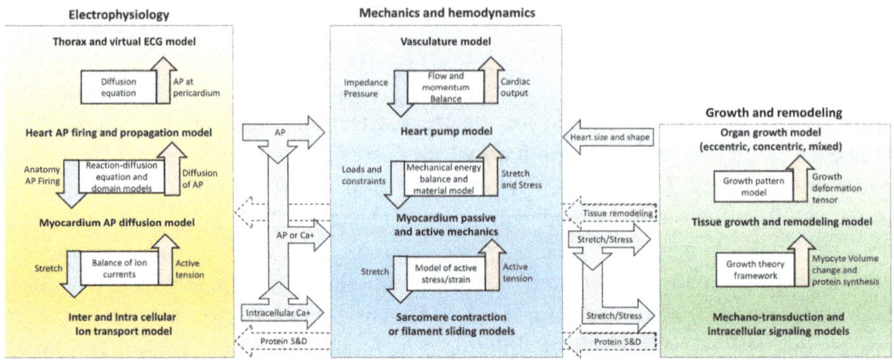


**Fig. 10** Diagram of dataflow in typical multiphysics-multiscale models of heart disease and treatment. White boxes indicate physical principles applied for inter-scale coupling; vertical arrows indicate data communication across scales, while horizontal arrows indicate data communication across physical models. The electrophysiology model produces time-resolved estimations of action potential (AP) distributions at the tissue scale or intracellular ion concentration at the molecular scale. The output of the electrophysiology model is fed as an input for the mechanical model to simulate active contraction and pumping action. The mechanical model produces an estimation of myocardial stress and strain that can be fed to models of growth and remodeling. The growth and remodeling models produce an estimation of changes in tissue mass and composition that are fed back to the mechanical model. Notably, most models of heart disease and treatment neglect any multiphysics feedback to the electrophysiology model. The development of models based on constrained mixture theory may be used to establish feedback from tissue-scale G&R models to adjust mechanical and electrophysiological material properties. Alternatively, G&R molecular models can feed protein synthesis/degradation rates to molecular models of sarcomere contraction and cellular ion transport

and electrophysiology, and most apply phenomenological laws for growth. Even the work by Estrada et al. (2021), the only work reviewed here incorporating a molecular model for growth, applies a kinematic growth framework that only considers changes in myocardial mass while ignoring changes in myocyte shape and material properties [70]. It is thought, however, that remodeling is associated with changes in ECM and myocyte structure and function, affecting not only passive properties at the tissue scale but also changes in intracellular signaling, ion dynamics, excitation-contraction coupling, and other molecular effects. The incorporation of constrained-mixture models can potentially address some of these limitations by introducing means to couple the growth and remodeling to electrophysiology and mechanical models at tissue and molecular scales [66]. Notably, the works reviewed here only consider the effects of growth on heart mechanics but not on electrophysiology (Fig. 10). A first step toward studying the effect of growth on the conduction of the active potential is proposed by Arumugan et al. (2019), who solved the electrophysiology models on each new configuration generated by the growth model [50].

Another obstacle in the path to incorporating multiscale models for personalized medicine is related to the availability of clinical data for the customization of the model. Even accounting for the most recent developments of new technology for medical imaging and tests, the time, cost, and risk associated with clinical exams prevent the collection of information critical for adjusting the parameters of the models. Data scarcity can be partially solved by complementing available patient-specific tests with data drawn from population-based studies and animal models at the cost of introducing further uncertainty into the model predictions.

Machine learning tools are valuable for managing uncertainty introduced by disparate data sources and statistically based molecular models. Complementing multiscale modeling with computational statistics and other physics-informed machine learning algorithms is gaining traction in computational biomedical engineering to improve the reliability of computer-assisted predictions. Recent efforts have successfully integrated machine learning analysis into a multiscale model of patient-specific healthy and pathological heart mechanics [133, 146–148]. Physics-informed machine learning also opens the opportunity to test medical treatments under different conditions and different model settings, a process that has been coined as "virtual clinical trials." Virtual clinical trials are already streamlined for some drug treatments [1, 104, 106, 129], showing the potential to reduce the cost and risk exposure associated with conventional trials by predicting drug efficacy, secondary effects, and potential risks.

## 6 Conclusions

We reviewed the most recent developments in multiscale simulation of therapies for heart disease here, showing the power of this technique to predict their acute and long-term effects. We discussed how the design of multiscale models requires balancing resource economy with model completeness while showing how different research groups preserve functional relations between the physics and scales relevant to each type of disease and treatment.

One of the most exciting features of multiscale modeling is its ability to replicate cross-scale synergetic effects, such as drug-induced arrhythmogenesis. These effects, observed in animal models and clinical practice, have proven to be elusive for isolated cell experiments and single-scale computational models. This underscores the importance of establishing functional relations between physiological processes at different biological scales. It also highlights the advantage of multiscale modeling over black-box models developed for machine learning and artificial intelligence algorithms. While multiscale models have limitations, such as parameter and prediction uncertainty, these can be effectively managed through statistical methods, a strength of machine learning algorithms. The potential of physics-informed machine learning to complement the strengths of both techniques is a promising avenue for future developments in the modeling of medical therapies, providing statistically reliable predictions and identifying the physiological processes relevant to observed outcomes. This comprehensive approach can bridge the gap between biomedical computational research, clinical trials, and personalized medicine.

**Acknowledgments** This work is supported by the American Heart Association Grant 23POST1026645 (JHB) (https://doi.org/10.58275/AHA.23POST1026645.pc.gr.161204) and NIH/NHLBI Grant R01 HL159945 (JWH).

## References

1. Peirlinck, M., Costabal, F. S., Yao, J., Guccione, J. M., Tripathy, S., Wang, Y., Ozturk, D., Segars, P., Morrison, T. M., Levine, S., and Kuhl, E., 2021, "Precision Medicine in Human Heart Modeling," Biomech Model Mechanobiol, **20**(3), pp. 803–831.
2. Peng, G. C. Y., 2021, "The Structure and Function of the Interagency Modeling and Analysis Group" [Online]. Available: https://www.imagwiki.nibib.nih.gov/sites/default/files/MSM%20Viral%20PandemicsWG_011421_Peng.pdf. [Accessed: 13-Mar-2024].
3. Viceconti, M., and Hunter, P., 2016, "The Virtual Physiological Human: Ten Years After," Annual Review of Biomedical Engineering, **18**(1), 103–123.
4. Marsden, A. L., and Esmaily-Moghadam, M., 2015, "Multiscale Modeling of Cardiovascular Flows for Clinical Decision Support," Appl Mech Rev, **67**(3).
5. Aboelkassem, Y., Powers, J. D., McCabe, K. J., and McCulloch, A. D., 2019, "Multiscale Models of Cardiac Muscle Biophysics and Tissue Remodeling in Hypertrophic Cardiomyopathies," Curr Opin Biomed Eng, **11**, pp. 35–44.
6. Bracamonte, J. H., Saunders, S. K., Wilson, J. S., Truong, U. T., and Soares, J. S., 2022, "Patient-Specific Inverse Modeling of In Vivo Cardiovascular Mechanics with Medical Image-Derived Kinematics as Input Data: Concepts, Methods, and Applications," Applied Sciences, **12**(8), p. 3954.
7. Arackal, A., and Alsayouri, K., 2023, "Histology, Heart," StatPearls [Online]. Available: https://www.ncbi.nlm.nih.gov/books/NBK545143/. [Accessed: 07-Dec-2023].
8. Bowers, S. L. K., Meng, Q., and Molkentin, J. D., 2022, "Fibroblasts Orchestrate Cellular Crosstalk in the Heart through the ECM," Nature Cardiovascular Research 2022 1:4, **1**(4), pp. 312–321.
9. Saxton, A., Tariq, M. A., and Bordoni, B., 2023, "Anatomy, Thorax, Cardiac Muscle," StatPearls [Online]. Available: https://www.ncbi.nlm.nih.gov/books/NBK535355/. [Accessed: 15-Nov-2023].
10. Nishitani, S., Torii, N., Imai, H., Haraguchi, R., Yamada, S., and Takakuwa, T., 2020, "Development of Helical Myofiber Tracts in the Human Fetal Heart: Analysis of Myocardial Fiber Formation in the Left Ventricle From the Late Human Embryonic Period Using Diffusion Tensor Magnetic Resonance Imaging," Journal of the American Heart Association: Cardiovascular and Cerebrovascular Disease, **9**(19).
11. Ogobuiro, I., Wehrle, C. J., and Tuma, F., 2023, "Anatomy, Thorax, Heart Coronary Arteries," StatPearls [Online]. Available: https://www.ncbi.nlm.nih.gov/books/NBK534790/. [Accessed: 07-Dec-2023].
12. Nordsletten, D. A., Niederer, S. A., Nash, M. P., Hunter, P. J., and Smith, N. P., 2011, "Coupling Multi-Physics Models to Cardiac Mechanics," Prog Biophys Mol Biol, **104**(1–3), pp. 77–88.
13. Misfeld, M., and Sievers, H.-H., 2007, "Heart Valve Macro- and Microstructure," Philosophical Transactions of the Royal Society B: Biological Sciences, **362**(1484), pp. 1421–1436.
14. Holmes, J. W., Takayama, Y., LeGrice, I., and Covell, J. W., 1995, "Depressed Regional Deformation near Anterior Papillary Muscle," Am J Physiol Heart Circ Physiol, **269**(1).
15. Martin, A. A., Thompson, B. R., Hahn, D., Angulski, A. B. B., Hosny, N., Cohen, H., and Metzger, J. M., 2022, "Cardiac Sarcomere Signaling in Health and Disease," Int J Mol Sci, **23**(24).
16. Grant, A. O., 2009, "Cardiac Ion Channels," Circ Arrhythm Electrophysiol, **2**(2), pp. 185–194.

17. Sugden, P. H., 1999, “Signaling in Myocardial Hypertrophy,” Circ Res, **84**(6), pp. 633–646.
18. Melenovsky, V., 2013, “Cardiac Adaptation to Volume Overload,” Cardiac Adaptations: Molecular Mechanisms, **4**, pp. 167–199.
19. Opie, L. H., Commerford, P. J., Gersh, B. J., and Pfeffer, M. A., 2006, “Controversies in Ventricular Remodelling,” The Lancet, **367**(9507), pp. 356–367.
20. Lorell, B. H., and Carabello, B. A., 2000, “Left Ventricular Hypertrophy,” Circulation, **102**(4), pp. 470–479.
21. Ahmed, M. I., Andrikopoulou, E., Zheng, J., Ulasova, E., Pat, B., Kelley, E. E., Powell, P. C., Denney, T. S., Lewis, C., Davies, J. E., Darley-Usmar, V., and Dell’Italia, L. J., 2022, “Interstitial Collagen Loss, Myocardial Remodeling, and Function in Primary Mitral Regurgitation,” Basic to Translational Science, **7**(10), pp. 973–981.
22. Dillon, A. R., Dell’Italia, L. J., Tillson, M., Killingsworth, C., Denney, T., Hathcock, J., and Botzman, L., 2012, “Left Ventricular Remodeling in Preclinical Experimental Mitral Regurgitation of Dogs,” Journal of Veterinary Cardiology, **14**(1), pp. 73–92.
23. Witzenburg, C. M., and Holmes, J. W., 2019, “The Impact of Hemodynamic Reflex Compensation Following Myocardial Infarction on Subsequent Ventricular Remodeling,” J Biomech Eng, **141**(9).
24. Liu, B., Neil, D. A. H., Bhabra, M., Patel, R., Barker, T. A., Nikolaidis, N., Billing, J. S., Hayer, M., Baig, S., Price, A. M., Vijapurapu, R., Treibel, T. A., Edwards, N. C., and Steeds, R. P., 2022, “Reverse Myocardial Remodeling Following Valve Repair in Patients With Chronic Severe Primary Degenerative Mitral Regurgitation,” Cardiovascular Imaging, **15**(2), pp. 224–236.
25. Hwang, J., Kim, S. M., Park, S.-J., Cho, E. J., Kim, E. K., Chang, S.-A., Lee, S.-C., Choe, Y. H., and Park, S. W., 2017, “Assessment of Reverse Remodeling Predicted by Myocardial Deformation on Tissue Tracking in Patients with Severe Aortic Stenosis: A Cardiovascular Magnetic Resonance Imaging Study,” Journal of Cardiovascular Magnetic Resonance, **19**(1), p. 80.
26. Chipeta, P., Young Shim, C., Hong, G. R., Kim, D., Cho, I. J., Lee, S., Chang, H. J., Chang, B. C., Ha, J. W., and Chung, N., 2016, “Time Course of Left Atrial Reverse Remodelling after Mitral Valve Surgery and the Impact of Left Ventricular Global Longitudinal Strain in Patients with Chronic Severe Mitral Regurgitation,” Interact Cardiovasc Thorac Surg, **23**(6), pp. 876–882.
27. Berry, J. M., Naseem, R. H., Rothermel, B. A., and Hill, J. A., 2007, “Models of Cardiac Hypertrophy and Transition to Heart Failure,” Drug Discov Today Dis Models, **4**(4), pp. 197–206.
28. Frank, D. U., Sutcliffe, M. D., and Saucerman, J. J., 2018, “Network-Based Predictions of in Vivo Cardiac Hypertrophy.,” J Mol Cell Cardiol, **121**, pp. 180–189.
29. Suga, H., and Sagawa, K., 1974, “Instantaneous Pressure-Volume Relationships and Their Ratio in the Excised, Supported Canine Left Ventricle,” Circ Res, **35**(1), pp. 117–126.
30. Sunagawa, K., Maughan, W. L., and Sagawa, K., 1983, “Effect of Regional Ischemia on the Left Ventricular End-Systolic Pressure-Volume Relationship of Isolated Canine Hearts.,” Circ Res, **52**(2), pp. 170–178.
31. Santamore, W. P., and Burkhoff, D., 1991, “Hemodynamic Consequences of Ventricular Interaction as Assessed by Model Analysis,” American Journal of Physiology-Heart and Circulatory Physiology, **260**(1), pp. H146–H157.
32. Witzenburg, C. M., and Holmes, J. W., 2018, “Predicting the Time Course of Ventricular Dilation and Thickening Using a Rapid Compartmental Model,” J Cardiovasc Transl Res, **11**(2), pp. 109–122.
33. Sahli Costabal, F., Hurtado, D. E., and Kuhl, E., 2016, “Generating Purkinje Networks in the Human Heart,” J Biomech, **49**(12), pp. 2455–2465.
34. Lumens, J., Delhaas, T., Kirn, B., and Arts, T., 2009, “Three-Wall Segment (TriSeg) Model Describing Mechanics and Hemodynamics of Ventricular Interaction,” Ann Biomed Eng, **37**(11), p. 2234.

35. Walmsley, J., Huntjens, P. R., Prinzen, F. W., Delhaas, T., and Lumens, J., 2016, "Septal Flash and Septal Rebound Stretch Have Different Underlying Mechanisms," Am J Physiol Heart Circ Physiol, **310**(3), pp. H394–H403.
36. Walmsley, J., Arts, T., Derval, N., Bordachar, P., Cochet, H., Ploux, S., Prinzen, F. W., Delhaas, T., and Lumens, J., 2015, "Fast Simulation of Mechanical Heterogeneity in the Electrically Asynchronous Heart Using the MultiPatch Module," PLoS Comput Biol, **11**(7), p. e1004284.
37. Oomen, P. J. A., Phung, T. K. N., Weinberg, S. H., Bilchick, K. C., and Holmes, J. W., 2022, "A Rapid Electromechanical Model to Predict Reverse Remodeling Following Cardiac Resynchronization Therapy," Biomech Model Mechanobiol, **21**(1), pp. 231–247.
38. Baillargeon, B., Rebelo, N., Fox, D. D., Taylor, R. L., and Kuhl, E., 2014, "The Living Heart Project: A Robust and Integrative Simulator for Human Heart Function," European Journal of Mechanics - A/Solids, **48**(1), pp. 38–47.
39. Yin, F. C., Chan, C. C., and Judd, R. M., 1996, "Compressibility of Perfused Passive Myocardium," American Journal of Physiology-Heart and Circulatory Physiology, **271**(5 40-5), pp. 1864–1870.
40. Avazmohammadi, R., Soares, J. S., Li, D. S., Eperjesi, T., Pilla, J., Gorman, R. C., and Sacks, M. S., 2020, "On the in Vivo Systolic Compressibility of Left Ventricular Free Wall Myocardium in the Normal and Infarcted Heart," J Biomech, **107**, p. 109767.
41. Avazmohammadi, R., Soares, J. S., Li, D. S., Raut, S. S., Gorman, R. C., and Sacks, M. S., 2019, "A Contemporary Look at Biomechanical Models of Myocardium," Annu Rev Biomed Eng, **21**(1), pp. 417–442.
42. Rassier, D. E., 2017, "Sarcomere Mechanics in Striated Muscles: From Molecules to Sarcomeres to Cells," Am J Physiol Cell Physiol, **313**(2), pp. C134–C145.
43. Powers, J. D., Malingen, S. A., Regnier, M., and Daniel, T. L., 2021, "The Sliding Filament Theory Since Andrew Huxley: Multiscale and Multidisciplinary Muscle Research," Annu Rev Biophys, **50**(1), pp. 373–400.
44. Rice, J. J., Wang, F., Bers, D. M., and De Tombe, P. P., 2008, "Approximate Model of Cooperative Activation and Crossbridge Cycling in Cardiac Muscle Using Ordinary Differential Equations," Biophysical Journal, **95**(5), 2368.
45. Washio, T., Okada, J. I., Sugiura, S., and Hisada, T., 2012, "Approximation for Cooperative Interactions of a Spatially-Detailed Cardiac Sarcomere Model," Cell Mol Bioeng, **5**(1), pp. 113–126.
46. Aboelkassem, Y., Powers, J. D., McCabe, K. J., and McCulloch, A. D., 2019, "Multiscale Models of Cardiac Muscle Biophysics and Tissue Remodeling in Hypertrophic Cardiomyopathies," Curr Opin Biomed Eng, **11**, pp. 35–44.
47. Luo, C. H., and Rudy, Y., 1994, "A Dynamic Model of the Cardiac Ventricular Action Potential. I. Simulations of Ionic Currents and Concentration Changes.," Circ Res, **74**(6), pp. 1071–1096.
48. Ten Tusscher, K. H. W. J., Noble, D., Noble, P. J., Panfilov, A. V, Tusscher, T., and Univ, U., 2004, "A Model for Human Ventricular Tissue," Am J Physiol Heart Circ Physiol, **286**, pp. 1573–1589.
49. Ten Tusscher, K. H. W. J., Bernus, O., Hren, R., and Panfilov, A. V., 2006, "Comparison of Electrophysiological Models for Human Ventricular Cells and Tissues," Prog Biophys Mol Biol, **90**(1–3), pp. 326–345.
50. Arumugam, J., Mojumder, J., Kassab, G., and Lee, L. C., 2019, "Model of Anisotropic Reverse Cardiac Growth in Mechanical Dyssynchrony," Scientific Reports 2019 9:1, **9**(1), pp. 1–12.
51. Campbell, S. G., Lionetti, F. V., Campbell, K. S., and McCulloch, A. D., 2010, "Coupling of Adjacent Tropomyosins Enhances Cross-Bridge-Mediated Cooperative Activation in a Markov Model of the Cardiac Thin Filament," Biophys J, **98**(10), pp. 2254–2264.
52. Aboelkassem, Y., Bonilla, J. A., McCabe, K. J., and Campbell, S. G., 2015, "Contributions of Ca2+-Independent Thin Filament Activation to Cardiac Muscle Function," Biophys J, **109**(10), pp. 2101–2112.

53. Mijailovich, S. M., Kayser-Herold, O., Stojanovic, B., Nedic, D., Irving, T. C., and Geeves, M. A., 2016, "Three-Dimensional Stochastic Model of Actin–Myosin Binding in the Sarcomere Lattice," Journal of General Physiology, **148**(6), pp. 459–488.
54. Moreno, J. D., Zhu, Z. I., Yang, P. C., Bankston, J. R., Jeng, M. T., Kang, C., Wang, L., Bayer, J. D., Christini, D. J., Trayanova, N. A., Ripplinger, C. M., Kass, R. S., and Clancy, C. E., 2011, "A Computational Model to Predict the Effects of Class I Anti-Arrhythmic Drugs on Ventricular Rhythms," Sci Transl Med, **3**(98).
55. Clancy, C. E., and Rudy, Y., 1999, "Linking a Genetic Defect to Its Cellular Phenotype in a Cardiac Arrhythmia," Nature, **400**(6744), pp. 566–569.
56. Sharifi, H., Mann, C. K., Rockward, A. L., Mehri, M., Mojumder, J., Lee, L. C., Campbell, K. S., and Wenk, J. F., 2021, "Multiscale Simulations of Left Ventricular Growth and Remodeling," Biophysical Reviews, **13**(5), pp. 729–746.
57. Niestrawska, J. A., Augustin, C. M., and Plank, G., 2020, "Computational Modeling of Cardiac Growth and Remodeling in Pressure Overloaded Hearts—Linking Microstructure to Organ Phenotype," Acta Biomater, **106**, pp. 34–53.
58. Kroon, W., Delhaas, T., Arts, T., and Bovendeerd, P., 2009, "Computational Modeling of Volumetric Soft Tissue Growth: Application to the Cardiac Left Ventricle," Biomech Model Mechanobiol, **8**(4), pp. 301–309.
59. Arts, T., Lumens, J., Kroon, W., Donker, D., Prinzen, F., and Delhaas, T., 2010, "Patient-Specific Modeling of Cardiovascular Dynamics with a Major Role for Adaptation," Patient-Specific Modeling of the Cardiovascular System: Technology-Driven Personalized Medicine, pp. 21–41.
60. Rodriguez, E. K., Hoger, A., and McCulloch, A. D., 1994, "Stress-Dependent Finite Growth in Soft Elastic Tissues," J Biomech, **27**(4), pp. 455–467.
61. Witzenburg, C. M., and Holmes, J. W., 2017, "A Comparison of Phenomenologic Growth Laws for Myocardial Hypertrophy," J Elast, **129**(1–2), pp. 257–281.
62. Yoshida, K., and Holmes, J. W., 2021, "Computational Models of Cardiac Hypertrophy," Prog Biophys Mol Biol, **159**, pp. 75–85.
63. Kerckhoffs, R. C. P., Omens, J. H., and McCulloch, A. D., 2012, "A Single Strain-Based Growth Law Predicts Concentric and Eccentric Cardiac Growth during Pressure and Volume Overload," Mech Res Commun, **42**, pp. 40–50.
64. Yoshida, K., McCulloch, A. D., Omens, J. H., and Holmes, J. W., 2020, "Predictions of Hypertrophy and Its Regression in Response to Pressure Overload," Biomech Model Mechanobiol, **19**(3), pp. 1079–1089.
65. Baek, S., Gleason, R. L., Rajagopal, K. R., and Humphrey, J. D., 2007, "Theory of Small on Large: Potential Utility in Computations of Fluid-Solid Interactions in Arteries," Comput Methods Appl Mech Eng, **196**(31–32), pp. 3070–3078.
66. Humphrey, J. D., 2021, "Constrained Mixture Models of Soft Tissue Growth and Remodeling – Twenty Years After," J Elast, **145**, pp. 49–75.
67. Gebauer, A. M., Pfaller, M. R., Braeu, F. A., Cyron, C. J., and Wall, W. A., 2023, "A Homogenized Constrained Mixture Model of Cardiac Growth and Remodeling: Analyzing Mechanobiological Stability and Reversal," Biomech Model Mechanobiol, **22**(6), pp. 1983–2002.
68. Cyron, C. J., Aydin, R. C., and Humphrey, J. D., 2016, "A Homogenized Constrained Mixture (and Mechanical Analog) Model for Growth and Remodeling of Soft Tissue," Biomech Model Mechanobiol, **15**(6), pp. 1389–1403.
69. Ryall, K. A., Holland, D. O., Delaney, K. A., Kraeutler, M. J., Parker, A. J., and Saucerman, J. J., 2012, "Network Reconstruction and Systems Analysis of Cardiac Myocyte Hypertrophy Signaling," Journal of Biological Chemistry, **287**(50), pp. 42259–42268.
70. Estrada, A. C., Yoshida, K., Saucerman, J. J., and Holmes, J. W., 2021, "A Multiscale Model of Cardiac Concentric Hypertrophy Incorporating Both Mechanical and Hormonal Drivers of Growth," Biomech Model Mechanobiol, **20**(1), pp. 293–307.

71. Yoshida, K., Saucerman, J. J., and Holmes, J. W., 2022, "Multiscale Model of Heart Growth during Pregnancy: Integrating Mechanical and Hormonal Signaling," Biomech Model Mechanobiol, **21**(4), pp. 1267–1283.
72. Mills, K. T., Stefanescu, A., and He, J., 2020, "The Global Epidemiology of Hypertension," Nature Reviews Nephrology 2020 16:4, **16**(4), pp. 223–237.
73. Carey, R. M., Moran, A. E., and Whelton, P. K., 2022, "Treatment of Hypertension: A Review," JAMA, **328**(18), pp. 1849–1861.
74. Kvaslerud, A. B., Santic, K., Hussain, A. I., Auensen, A., Fiane, A., Skulstad, H., Aaberge, L., Gullestad, L., and Broch, K., 2021, "Outcomes in Asymptomatic, Severe Aortic Stenosis," PLoS One, **16**(4), p. e0249610.
75. Kang, D.-H., Park, S.-J., Lee, S.-A., Lee, S., Kim, D.-H., Kim, H.-K., Yun, S.-C., Hong, G.-R., Song, J.-M., Chung, C.-H., Song, J.-K., Lee, J.-W., and Park, S.-W., 2020, "Early Surgery or Conservative Care for Asymptomatic Aortic Stenosis," New England Journal of Medicine, **382**(2), pp. 111–119.
76. Sasayama, S., Ross, J., Franklin, D., Bloor, C. M., Bishop, S., and Dilley, R. B., 1976, "Adaptations of the Left Ventricle to Chronic Pressure Overload.," Circ Res, **38**(3), pp. 172–178.
77. Carabello, B. A., Zile, M. R., Tanaka, R., and Cooper, G., 1992, "Left Ventricular Hypertrophy Due to Volume Overload versus Pressure Overload," American Journal of Physiology-Heart and Circulatory Physiology, **263**(4), pp. H1137–H1144.
78. Marsden, A. L., 2014, "Optimization in Cardiovascular Modeling," Annu Rev Fluid Mech, **46**(1), pp. 519–546.
79. Otto, C. M., Nishimura, R. A., Bonow, R. O., Carabello, B. A., Erwin, J. P., Gentile, F., Jneid, H., Krieger, E. V., Mack, M., McLeod, C., O'Gara, P. T., Rigolin, V. H., Sundt, T. M., Thompson, A., and Toly, C., 2021, "2020 ACC/AHA Guideline for the Management of Patients With Valvular Heart Disease: A Report of the American College of Cardiology/American Heart Association Joint Committee on Clinical Practice Guidelines," Circulation, **143**(5), pp. E72–E227.
80. Nagatsu, M., Zile, M. R., Tsutsui, H., Schmid, P. G., DeFreyte, G., Cooper IV, G., and Carabello, B. A., 1994, "Native Beta-Adrenergic Support for Left Ventricular Dysfunction in Experimental Mitral Regurgitation Normalizes Indexes of Pump and Contractile Function," Circulation, **89**(2), pp. 818–826.
81. Rausch, M. K., Famaey, N., Shultz, T. O., Bothe, W., Miller, D. C., and Kuhl, E., 2012, "Mechanics of the Mitral Valve," Biomechanics and Modeling in Mechanobiology 2012 12:5, **12**(5), pp. 1053–1071.
82. Rondanina, E., and Bovendeerd, P. H. M., 2020, "Stimulus–Effect Relations for Left Ventricular Growth Obtained with a Simple Multi-Scale Model: The Influence of Hemodynamic Feedback," Biomech Model Mechanobiol, **19**(6), pp. 2111–2126.
83. Rondanina, E., and Bovendeerd, P. H. M., 2020, "Evaluation of Stimulus-Effect Relations in Left Ventricular Growth Using a Simple Multiscale Model," Biomech Model Mechanobiol, **19**(1), pp. 263–273.
84. Lim, K. M., Hong, S. B., Lee, B. K., Shim, E. B., and Trayanova, N., 2015, "Computational Analysis of the Effect of Valvular Regurgitation on Ventricular Mechanics Using a 3D Electromechanics Model," Journal of Physiological Sciences, **65**(2), pp. 159–164.
85. Rausch, M. K., Genet, M., and Humphrey, J. D., 2017, "An Augmented Iterative Method for Identifying a Stress-Free Reference Configuration in Image-Based Biomechanical Modeling," J Biomech, **58**, pp. 227–231.
86. Baillargeon, B., Costa, I., Leach, J. R., Lee, L. C., Genet, M., Toutain, A., Wenk, J. F., Rausch, M. K., Rebelo, N., Acevedo-Bolton, G., Kuhl, E., Navia, J. L., and Guccione, J. M., 2015, "Human Cardiac Function Simulator for the Optimal Design of a Novel Annuloplasty Ring with a Sub-Valvular Element for Correction of Ischemic Mitral Regurgitation," Cardiovasc Eng Technol, **6**(2), pp. 105–116.
87. Park, Y. H., Chon, M. K., Lederman, R. J., Sung, S. C., Je, H. G., Choo, K. S., Lee, S. H., Shin, E. S., Kim, J. S., Hwang, K. W., Lee, S. Y., Chun, K. J., Kim, C. M., and Kim, J. H.,

2017, "Mitral Loop Cerclage Annuloplasty for Secondary Mitral Regurgitation: First Human Results," JACC Cardiovasc Interv, **10**(6), pp. 597–610.

88. Rogers, T., Greenbaum, A. B., Babaliaros, V. C., Foerst, J. R., Khan, J. M., Bruce, C. G., Stine, A. M., Satler, L. F., Perdoncin, E., Gleason, P. T., Lisko, J. C., Tian, X., Miao, R., Sachdev, V., Chen, M. Y., and Lederman, R. J., 2022, "Transcatheter Mitral Cerclage Ventriculoplasty: From Bench To," JACC Cardiovasc Interv, **15**(12), p. 1249.
89. Lee, C.-H., Laurence, D. W., Ross, C. J., Kramer, K. E., Babu, A. R., Johnson, E. L., Hsu, M.-C., Aggarwal, A., Mir, A., Burkhart, H. M., Towner, R. A., Baumwart, R., and Wu, Y., 2019, "Mechanics of the Tricuspid Valve—From Clinical Diagnosis/Treatment, In-Vivo and In-Vitro Investigations, to Patient-Specific Biomechanical Modeling," Bioengineering, **6**(2), p. 47.
90. Lee, K. E., Kim, K. T., Lee, J. H., Jung, S., Kim, J.-H., and Shim, E. B., 2019, "Computational Analysis of the Electromechanical Performance of Mitral Valve Cerclage Annuloplasty Using a Patient-Specific Ventricular Model," The Korean Journal of Physiology & Pharmacology, **23**(1), p. 63.
91. Carabello, B. A., 2008, "The Current Therapy for Mitral Regurgitation," J Am Coll Cardiol, **52**(5), pp. 319–326.
92. Masuda, H., Miyagawa, S., Sugiura, S., Washio, T., Okada, J. ichi, Ueno, T., Toda, K., Kuratani, T., Hisada, T., and Sawa, Y., 2021, "An Application of a Patient-Specific Cardiac Simulator for the Prediction of Outcomes after Mitral Valve Replacement: A Pilot Study," Journal of Artificial Organs, **24**(3), pp. 351–357.
93. Kingma, J., Simard, C., and Drolet, B., 2023, "Overview of Cardiac Arrhythmias and Treatment Strategies," Pharmaceuticals, **16**(6), p. 844.
94. Elliott, A. D., Middeldorp, M. E., Van Gelder, I. C., Albert, C. M., and Sanders, P., 2023, "Epidemiology and Modifiable Risk Factors for Atrial Fibrillation," Nature Reviews Cardiology, **20**(6), pp. 404–417.
95. Heijman, J., Sutanto, H., Crijns, H. J. G. M., Nattel, S., and Trayanova, N. A., 2021, "Computational Models of Atrial Fibrillation: Achievements, Challenges, and Perspectives for Improving Clinical Care," Cardiovasc Res, **117**(7), pp. 1682–1699.
96. Trayanova, N. A., and Boyle, P. M., 2014, "Advances in Modeling Ventricular Arrhythmias: From Mechanisms to the Clinic," WIREs Systems Biology and Medicine, **6**(2), pp. 209–224.
97. Bishop, M. J., and Plank, G., 2012, "The Role of Fine-Scale Anatomical Structure in the Dynamics of Reentry in Computational Models of the Rabbit Ventricles," J Physiol, **590**(18), pp. 4515–4535.
98. Alexander Quinn, T., and Kohl, P., 2021, "Cardiac Mechano-Electric Coupling: Acute Effects of Mechanical Stimulation on Heart Rate and Rhythm," Physiol Rev, **101**(1), pp. 37–92.
99. Moyer, C. B., Norton, P. T., Ferguson, J. D., and Holmes, J. W., 2015, "Changes in Global and Regional Mechanics Due to Atrial Fibrillation: Insights from a Coupled Finite-Element and Circulation Model," Ann Biomed Eng, **43**(7), pp. 1600–1613.
100. Trayanova, N. A., Chang, K. C., Trayanova, N. A., and Chang, K. C., 2016, "How Computer Simulations of the Human Heart Can Improve Anti-Arrhythmia Therapy," J Physiol, **594**(9), pp. 2483–2502.
101. Uldry, L., Virag, N., Jacquemet, V., Vesin, J.-M., and Kappenberger, L., 2010, "Optimizing Local Capture of Atrial Fibrillation by Rapid Pacing: Study of the Influence of Tissue Dynamics.," Ann Biomed Eng, **38**(12), pp. 3664–73.
102. Uldry, L., Virag, N., Lindemans, F., Vesin, J. M., and Kappenberger, L., 2012, "Atrial Septal Pacing for the Termination of Atrial Fibrillation: Study in a Biophysical Model of Human Atria," EP Europace, **14**(5), pp. v112–v120.
103. Rusu, A., Jacquemet, V., Vesin, J. M., and Virag, N., 2014, "Influence of Atrial Substrate on Local Capture Induced by Rapid Pacing of Atrial Fibrillation," EP Europace, **16**(5), pp. 766–773.
104. Zeigler, A. C., Chandrabhatla, A. S., Christiansen, S. L., Nelson, A. R., Holmes, J. W., and Saucerman, J. J., 2021, "Network Model-Based Screen for FDA-Approved Drugs Affecting Cardiac Fibrosis," CPT Pharmacometrics Syst Pharmacol, **10**(4), pp. 377–388.

105. Sahli Costabal, F., Yao, J., and Kuhl, E., 2018, "Predicting Drug-Induced Arrhythmias by Multiscale Modeling," Int J Numer Method Biomed Eng, **34**(5), p. e2964.
106. Sahli Costabal, F., Matsuno, K., Yao, J., Perdikaris, P., and Kuhl, E., 2019, "Machine Learning in Drug Development: Characterizing the Effect of 30 Drugs on the QT Interval Using Gaussian Process Regression, Sensitivity Analysis, and Uncertainty Quantification," Comput Methods Appl Mech Eng, **348**, pp. 313–333.
107. Sahli-Costabal, F., Seo, K., Ashley, E., and Kuhl, E., 2020, "Classifying Drugs by Their Arrhythmogenic Risk Using Machine Learning," Biophys J, **118**(5), pp. 1165–1176.
108. Bueno-Orovio, A., Cherry, E. M., and Fenton, F. H., 2008, "Minimal Model for Human Ventricular Action Potentials in Tissue," J Theor Biol, **253**(3), pp. 544–560.
109. Hörmann, J. M., Bertoglio, C., Nagler, A., Pfaller, M. R., Bourier, F., Hadamitzky, M., Deisenhofer, I., and Wall, W. A., 2017, "Multiphysics Modeling of the Atrial Systole under Standard Ablation Strategies," Cardiovasc Eng Technol, **8**(2), pp. 205–218.
110. Phung, T. N., Moyer, C. B., Norton, P. T., Ferguson, J. D., and Holmes, J. W., 2017, "Effect of Ablation Pattern on Mechanical Function in the Atrium," Pacing and Clinical Electrophysiology, **40**(6), pp. 648–654.
111. Francia, P., Balla, C., Paneni, F., and Volpe, M., 2007, "Left Bundle-Branch Block—Pathophysiology, Prognosis, and Clinical Management," Clin Cardiol, **30**(3), pp. 110–115.
112. Vernooy, K., Verbeek, X. A. A. M., Peschar, M., Crijns, H. J. G. M., Arts, T., Cornelussen, R. N. M., and Prinzen, F. W., 2005, "Left Bundle Branch Block Induces Ventricular Remodelling and Functional Septal Hypoperfusion," Eur Heart J, **26**(1), pp. 91–98.
113. van Oosterhout, M. F. M., Prinzen, F. W., Arts, T., Schreuder, J. J., Vanagt, W. Y. R., Cleutjens, J. P. M., and Reneman, R. S., 1998, "Asynchronous Electrical Activation Induces Asymmetrical Hypertrophy of the Left Ventricular Wall," Circulation, **98**(6), pp. 588–595.
114. Bristow, M. R., Saxon, L. A., Boehmer, J., Krueger, S., Kass, D. A., de Marco, T., Carson, P., DiCarlo, L., DeMets, D., White, B. G., DeVries, D. W., and Feldman, A. M., 2004, "Cardiac-Resynchronization Therapy with or without an Implantable Defibrillator in Advanced Chronic Heart Failure," N Engl J Med, **350**(21), pp. 2140–2150.
115. Brignole, M., Auricchio, A., Baron-Esquivias, G., Bordachar, P., Boriani, G., Breithardt, O.-A., Cleland, J., Deharo, J.-C., Delgado, V., Elliott, P. M., Gorenek, B., Israel, C. W., Leclercq, C., Linde, C., Mont, L., Padeletti, L., Sutton, R., Vardas, P. E., Zamorano, J. L., Achenbach, S., Baumgartner, H., Bax, J. J., Bueno, H., Dean, V., Deaton, C., Erol, C., Fagard, R., Ferrari, R., Hasdai, D., Hoes, A. W., Kirchhof, P., Knuuti, J., Kolh, P., Lancellotti, P., Linhart, A., Nihoyannopoulos, P., Piepoli, M. F., Ponikowski, P., Sirnes, P. A., Tamargo, J. L., Tendera, M., Torbicki, A., Wijns, W., Windecker, S., Kirchhof, P., Blomstrom-Lundqvist, C., Badano, L. P., Aliyev, F., Bansch, D., Baumgartner, H., Bsata, W., Buser, P., Charron, P., Daubert, J.-C., Dobreanu, D., Faerestrand, S., Hasdai, D., Hoes, A. W., le Heuzey, J.-Y., Mavrakis, H., McDonagh, T., Merino, J. L., Nawar, M. M., Nielsen, J. C., Pieske, B., Poposka, L., Ruschitzka, F., Tendera, M., van Gelder, I. C., and Wilson, C. M., 2013, "2013 ESC Guidelines on Cardiac Pacing and Cardiac Resynchronization Therapy: The Task Force on Cardiac Pacing and Resynchronization Therapy of the European Society of Cardiology (ESC). Developed in Collaboration with the European Heart Rhythm Association (EHRA)," Europace, **15**(8), pp. 1070–1118.
116. Tracy, C. M., Epstein, A. E., Darbar, D., Dimarco, J. P., Dunbar, S. B., Estes, N. A. M., Ferguson, T. B., Hammill, S. C., Karasik, P. E., Link, M. S., Marine, J. E., Schoenfeld, M. H., Shanker, A. J., Silka, M. J., Stevenson, L. W., Stevenson, W. G., and Varosy, P. D., 2012, "2012 ACCF/AHA/HRS Focused Update of the 2008 Guidelines for Device-Based Therapy of Cardiac Rhythm Abnormalities: A Report of the American College of Cardiology Foundation/American Heart Association Task Force on Practice Guidelines," Circulation, **126**(14), pp. 1784–1800.
117. Owashi, K. P., Galli, E., Hubert, A., Donal, E., Hernandez, A. I., and Rolle, V. Le, 2020, "Patient-Specific Analysis of Myocardial Strains in Left Bundle Branch Block Based on Computational Models," Comput Cardiol, **2020-September**.

118. Owashi, K., Taconné, M., Courtial, N., Simon, A., Garreau, M., Hernandez, A., Donal, E., Le Rolle, V., and Galli, E., 2022, "Desynchronization Strain Patterns and Contractility in Left Bundle Branch Block through Computer Model Simulation," Journal of Cardiovascular Development and Disease 2022, Vol. 9, Page 53, **9**(2), p. 53.
119. Kerckhoffs, R. C. P., McCulloch, A. D., Omens, J. H., and Mulligan, L. J., 2009, "Effects of Biventricular Pacing and Scar Size in a Computational Model of the Failing Heart with Left Bundle Branch Block," Med Image Anal, **13**(2), pp. 362–369.
120. Finsberg, H., Balaban, G., Ross, S., Håland, T. F., Odland, H. H., Sundnes, J., and Wall, S., 2018, "Estimating Cardiac Contraction through High Resolution Data Assimilation of a Personalized Mechanical Model," J Comput Sci, **24**, pp. 85–90.
121. Kuijpers, N. H. L., Hermeling, E., Bovendeerd, P. H. M., Delhaas, T., and Prinzen, F. W., 2012, "Modeling Cardiac Electromechanics and Mechanoelectrical Coupling in Dyssynchronous and Failing Hearts: Insight from Adaptive Computer Models," J Cardiovasc Transl Res, **5**(2), pp. 159–169.
122. Okada, J. ichi, Washio, T., Nakagawa, M., Watanabe, M., Kadooka, Y., Kariya, T., Yamashita, H., Yamada, Y., Momomura, S. ichi, Nagai, R., Hisada, T., and Sugiura, S., 2017, "Multi-Scale, Tailor-Made Heart Simulation Can Predict the Effect of Cardiac Resynchronization Therapy," J Mol Cell Cardiol, **108**, pp. 17–23.
123. Isotani, A., Yoneda, K., Iwamura, T., Watanabe, M., Okada, J. ichi, Washio, T., Sugiura, S., Hisada, T., and Ando, K., 2020, "Patient-Specific Heart Simulation Can Identify Non-Responders to Cardiac Resynchronization Therapy," Heart Vessels, **35**(8), pp. 1135–1147.
124. Tuohy, C. V., Kaul, S., Song, H. K., Nazer, B., and Heitner, S. B., 2020, "Hypertrophic Cardiomyopathy: The Future of Treatment," Eur J Heart Fail, **22**(2), pp. 228–240.
125. Campbell, S. G., and McCulloch, A. D., 2011, "Multi-Scale Computational Models of Familial Hypertrophic Cardiomyopathy: Genotype to Phenotype," J R Soc Interface, **8**(64), pp. 1550–1561.
126. Wu, B., Wang, L., Liu, Q., and Luo, Q., 2012, "Myocardial Contractile and Metabolic Properties of Familial Hypertrophic Cardiomyopathy Caused by Cardiac Troponin I Gene Mutations: A Simulation Study," Exp Physiol, **97**(1), pp. 155–169.
127. Khalilimeybodi, A., Riaz, M., Campbell, S. G., Omens, J. H., McCulloch, A. D., Qyang, Y., and Saucerman, J. J., 2023, "Signaling Network Model of Cardiomyocyte Morphological Changes in Familial Cardiomyopathy," J Mol Cell Cardiol, **174**, pp. 1–14.
128. Syomin, F., Khabibullina, A., Osepyan, A., and Tsaturyan, A., 2020, "Hemodynamic Effects of Alpha-Tropomyosin Mutations Associated with Inherited Cardiomyopathies: Multiscale Simulation," Mathematics, **8**(7), p. 1169.
129. Filipovic, N., Sustersic, T., Milosevic, M., Milicevic, B., Simic, V., Prodanovic, M., Mijailovic, S., and Kojic, M., 2022, "SILICOFCM Platform, Multiscale Modeling of Left Ventricle from Echocardiographic Images and Drug Influence for Cardiomyopathy Disease," Comput Methods Programs Biomed, **227**, p. 107194.
130. Groenewegen, A., Rutten, F. H., Mosterd, A., and Hoes, A. W., 2020, "Epidemiology of Heart Failure," Eur J Heart Fail, **22**(8), pp. 1342–1356.
131. Ben Gal, T., Ben Avraham, B., Milicic, D., Crespo-Leiro, M. G., Coats, A. J. S., Rosano, G., Seferovic, P., Ruschitzka, F., Metra, M., Anker, S., Filippatos, G., Altenberger, J., Adamopoulos, S., Barac, Y. D., Chioncel, O., de Jonge, N., Elliston, J., Frigerio, M., Goncalvesova, E., Gotsman, I., Grupper, A., Hamdan, R., Hammer, Y., Hasin, T., Hill, L., Itzhakı Ben Zadok, O., Abuhazıra, M., Lavee, J., Mullens, W., Nalbantgıl, S., Pıepolı, M. F., Ponikowski, P., Potena, L., Ristic, A., Ruhparwar, A., Shaul, A., Tops, L. F., Tsui, S., Winnik, S., Jaarsma, T., and Gustafsson, F., 2021, "Guidance on the Management of Left Ventricular Assist Device (LVAD) Supported Patients for the Non-LVAD Specialist Healthcare Provider: Executive Summary," Eur J Heart Fail, **23**(10), pp. 1597–1609.
132. Presti, C. R., and Crenshaw, N. A., 2021, "Overview of Ventricular Assist Devices and the Total Artificial Heart," Dimensions of Critical Care Nursing, **40**(1), pp. 3–13.

133. Peirlinck, M., Sahli Costabal, F., Sack, K. L., Choy, J. S., Kassab, G. S., Guccione, J. M., De Beule, M., Segers, P., and Kuhl, E., 2019, "Using Machine Learning to Characterize Heart Failure across the Scales," Biomech Model Mechanobiol, **18**(6), pp. 1987–2001.
134. Han, J. J., Acker, M. A., and Atluri, P., 2018, "Left Ventricular Assist Devices.," Circulation, **138**(24), pp. 2841–2851.
135. Fraser, K. H., Taskin, M. E., Griffith, B. P., and Wu, Z. J., 2011, "The Use of Computational Fluid Dynamics in the Development of Ventricular Assist Devices," Med Eng Phys, **33**(3), pp. 263–280.
136. Bonnemain, J., I. Malossi, A. C., Lesinigo, M., Deparis, S., Quarteroni, A., and von Segesser, L. K., 2013, "Numerical Simulation of Left Ventricular Assist Device Implantations: Comparing the Ascending and the Descending Aorta Cannulations," Med Eng Phys, **35**(10), pp. 1465–1475.
137. Vandenberghe, S., Segers, P., Steendijk, P., Meyns, B., Dion, R. A. E., Antaki, J. F., and Verdonck, P., 2006, "Modeling Ventricular Function during Cardiac Assist: Does Time-Varying Elastance Work?," ASAIO Journal, **52**(1), pp. 4–8.
138. Cox, L. G. E., Loerakker, S., Rutten, M. C. M., De Mol, B. A. J. M., and Van De Vosse, F. N., 2009, "A Mathematical Model to Evaluate Control Strategies for Mechanical Circulatory Support," Artif Organs, **33**(8), pp. 593–603.
139. Lim, K. M., Constantino, J., Gurev, V., Zhu, R., Shim, E. B., and Trayanova, N. A., 2012, "Comparison of the Effects of Continuous and Pulsatile Left Ventricular-Assist Devices on Ventricular Unloading Using a Cardiac Electromechanics Model.," J Physiol Sci, **62**(1), pp. 11–9.
140. Sack, K. L., Baillargeon, B., Acevedo-Bolton, G., Genet, M., Rebelo, N., Kuhl, E., Klein, L., Weiselthaler, G. M., Burkhoff, D., Franz, T., and Guccione, J. M., 2016, "Partial LVAD Restores Ventricular Outputs and Normalizes LV but Not RV Stress Distributions in the Acutely Failing Heart in Silico," International Journal of Artificial Organs, **39**(8), pp. 421–430.
141. Sack, K. L., Dabiri, Y., Franz, T., Solomon, S. D., Burkhoff, D., and Guccione, J. M., 2018, "Investigating the Role of Interventricular Interdependence in Development of Right Heart Dysfunction During LVAD Support: A Patient-Specific Methods-Based Approach.," Front Physiol, **9**(MAY), p. 520.
142. Bakir, A. A., Al Abed, A., Stevens, M. C., Lovell, N. H., and Dokos, S., 2018, "A Multiphysics Biventricular Cardiac Model: Simulations with a Left-Ventricular Assist Device," Front Physiol, **9**(SEP), p. 391479.
143. Cardamone, L., Valentín, A., Eberth, J. F., and Humphrey, J. D., 2009, "Origin of Axial Prestretch and Residual Stress in Arteries," Biomech Model Mechanobiol, **8**(6), pp. 431–446.
144. Chuong, C. J., and Fung, Y. C., 1986, "On Residual Stresses in Arteries," J Biomech Eng, **108**(2), pp. 189–192.
145. Rajagopal, V., Chung, J.-H. H., Bullivant, D., Nielsen, P. M. F. F., and Nash, M. P., 2007, "Determining the Finite Elasticity Reference State from a Loaded Configuration," Int J Numer Methods Eng, **72**(12), pp. 1434–1451.
146. Sahli Costabal, F., Yang, Y., Perdikaris, P., Hurtado, D. E., and Kuhl, E., 2020, "Physics-Informed Neural Networks for Cardiac Activation Mapping," Front Phys, **8**, p. 510944.
147. Buoso, S., Joyce, T., and Kozerke, S., 2021, "Personalising Left-Ventricular Biophysical Models of the Heart Using Parametric Physics-Informed Neural Networks," Med Image Anal, **71**, p. 102066.
148. Dabiri, Y., Van der Velden, A., Sack, K. L., Choy, J. S., Guccione, J. M., and Kassab, G. S., 2020, "Application of Feed Forward and Recurrent Neural Networks in Simulation of Left Ventricular Mechanics," Scientific Reports 2020 10:1, **10**(1), pp. 1–11.

# Simulation of Heart Valve Cellular Micromechanics and Extracellular Matrix Modification

**Toni M. West, Gabriel Peery, Alex Khang, and Michael S. Sacks**

## Abbreviations

| | |
|---|---|
| AV | Aortic valve |
| AVIC | Aortic valve interstitial cell |
| CytoD | Cytochalasin D |
| ECM | Extracellular matrix |
| L-BFGS-B | Limited memory Broyden-Fletcher-Goldfarb-Shanno with bound constraints algorithm |
| MMP | Matrix metalloproteinase |
| MV | Mitral valve |
| PEG | Poly(ethylene) glycol |
| PV | Pulmonary valve |
| SF | Stress fiber |
| TV | Tricuspid valve |
| VEC | Valve endothelial cell |
| VIC | Valve interstitial cell |
| 3D-TFM | Three-dimensional traction force microscopy |
| αSMA | Alpha-smooth muscle actin |

T. M. West · G. Peery · A. Khang · M. S. Sacks (✉)
Department of Biomedical Engineering, Willerson Center for Cardiovascular Modelling and Simulation, Oden Institute for Computational Engineering and Sciences, University of Texas at Austin, Austin, TX, USA
e-mail: msacks@oden.utexas.edu

 
J. Liao, J. Y. Wong (eds.), *Integration and Bridging of Multiscale Bioengineering Designs and Tissue Biomechanics*, https://doi.org/10.1007/978-3-031-81743-4_13

# 1 Introduction

## *1.1 Cells and Extracellular Matrices as Important Links in the Multi-scale Modeling Chain*

Understanding the biomechanics of tissues in various pathological and physiological conditions requires the scaling of information from molecule to organism and even environment. Figure 1 depicts an example of the important aspect of scaling, as applied to a heart valve. In the applied setting, information is exchanged bidirectionally, where each tier influences the others. Therefore, it is imperative to comprehend the biomechanical scaling processes, encompassing both upward and downward trajectories to understand how tissues sustain physiologically normal biomechanical properties or undergo pathological remodeling [4, 37, 41, 54].

The content of this chapter centers on the links in the multi-scale modeling chain of the cells and their extracellular matrices (ECM). Since mammalian tissues are made up of cells and the ECM that the cells produce and maintain, understanding how the interrelatedness of these components feeds into the proper and pathological functionality of tissues is crucial to the understanding of tissue biomechanics in health and disease. Throughout the following pages, we will specifically focus on recent advances in both measuring and modeling the biomechanical properties of heart valve interstitial cells (VICs) and their surrounding ECM under varying

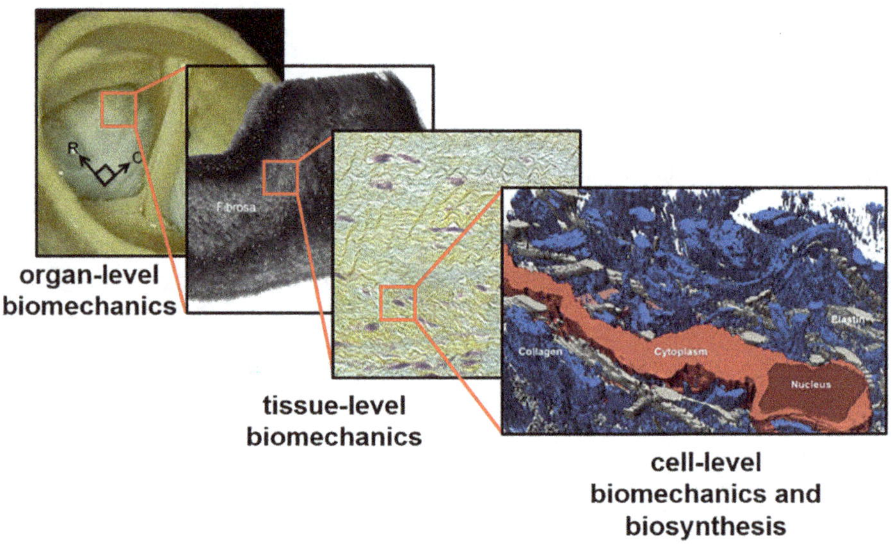


**Fig. 1** Multi-scale nature of biomechanical ques. from cell to organ in the heart valve. Phenomena that occur at the cellular level and at the organ level both affect the structure and function of the tissues that make up the aortic valve. (Images compiled from Merryman [38], Stella et al. [54], Ayoub et al. [4], Nicoara et al. [41])

conditions. VICs provide a fertile proving ground for developing and integrating biomechanical models into multi-scale models that are applicable to a variety of other cell and tissue types. This is because VIC deformation is a major factor in heart valve tissue physiology and pathological remodeling [3, 29, 57]. Furthermore, research indicates that the geometry, deformation, and biosynthetic behavior of VICs are greatly dependent on the biomechanical properties of the localized microenvironment [30, 42, 43, 46]. For instance, the deformation pattern applied to VICs affects their underlying cytoskeletal configurations and biosynthetic activity [8, 62]. Changes in VIC phenotype seen during disease are likely a response to return to a homeostatic biomechanical state through remodeling of the leaflet ECM [3, 46]. Given the complex spatial and nonlinear nature of cellular function, many of the determinant biomechanical factors of the system cannot be directly measured. Therefore, determining how the complex behaviors of cells and their surrounding ECM affect tissue dynamics, and vice versa, requires computational techniques based on integrated experiments [31]. Because of its pronounced mechanical attributes, heart valve tissue becomes a conducive medium for constructing computational models. These models, once tailored, can be extrapolated to different tissue types.

## *1.2 Heart Valves and Their Cells*

### 1.2.1 Heart Valves

Heart valves ensure the proper direction of blood flow through the circulatory system by preventing the backflow of blood into previous chambers of the heart. In the right heart, deoxygenated blood received from the body is prevented from backflow by the atrioventricular tricuspid valve (TV) and the semilunar pulmonary valve (PV) so that the blood will be pushed to the lungs upon contraction. In the left heart, oxygenated blood received from the lungs is prevented from backflow by the atrioventricular mitral valve (MV) and the semilunar aortic valve (AV) and is pumped to the body upon contraction. The job of heart valves is extremely demanding mechanically, with every beat of the heart causing the valves to experience tension, flexure, and shear stress [38]. For instance, research indicates peak strain rates of the aortic valve in the radial direction of 300%–400% per second and in the circumferential direction of 100%–130% per second [19, 20, 47, 48, 63]. These functional demands are repeated over three billion times in a typical lifetime. These complex mechanical demands highlight the constant demands on the physiological maintenance and pathological remodeling mechanisms that occur in valvular tissues.

Heart valve leaflets, when seen in cross-section, have distinct histological layers that function together as a single functionally graded material [7, 45]. Although there is some variation in the layers in different valve types, the aortic valve leaflet serves as a good example of the layering structure. The AV consists of three layers that contain VICs, with VECs forming a single-cell barrier surrounding the tri-layered tissue formed by the VICs. The tissue layer that faces the left ventricle is the ventricularis, which contains high concentrations of collagen and elastin fibers. The layer facing the aorta, the fibrosa, also contains large amounts of collagen. Sandwiched in between the ventricularis and fibrosa is the spongiosa. As its name suggests, the spongiosa has a lower small strain shear modulus than the other two layers because of its high water content due to large concentrations of proteoglycans and glycosaminoglycans [2, 49]. The detailed organization of the layers of ECM is controlled and maintained by the valve interstitial cells (VICs) that are embedded in the ECM.

### 1.2.2 Heart Valve Cells

Heart valves are primarily composed of two cell types: valve endothelial cells (VECs) and valve interstitial cells (VICs). VECs form a monolayer barrier between the valve tissue interior and the blood found within the chambers of the heart. The VECs therefore not only guard against unwanted chemical and immune cell invasion from the blood into the valve [15] but also absorb much of the impact from the shear stress that occurs on the surface of the valve due to blood flow [9, 10]. VICs are cells that are contained in the body of valve leaflets. They produce, mature, align, and maintain the extracellular matrix that makes up much of the volume of valves, with VICs existing relatively spread apart, similar to other connective tissues such as cartilage and ligaments [28].

VICs are drivers of heart valve remodeling during disease progression. A hallmark of early valve disease is a stenotic thickening and stiffening of the leaflets which occurs because of the increased production of collagen, elastin, glycosaminoglycan, and proteoglycan from the VICs [28]. At the same time that VICs increase the production and secretion of ECM components, they also increase the expression of their internal cytoskeletal components [36]. In particular, VICs increase the production of α-smooth muscle actin (αSMA) which gets incorporated into actin/myosin contractile stress fibers (SFs) [36]. This is important as the VIC SFs play a major role in VIC contractile behaviors, with local variations in SF expression and orientation affecting VIC mechanical function [21, 35, 61]. Moreover, ex vivo studies of diseased valves have shown that VIC activation levels, as assessed by expression levels of αSMA, are elevated in a variety of fibrotic and calcific valve diseases, including those that arise because of congenital defects as well as by exogenic or idiopathic factors [28, 36]. Thus, it is clear from previous studies that VIC SFs have large implications for the health and disease of heart valves.

### 1.3 2D and 3D Approaches

Producing trustworthy models that are accurate to biological phenomena is vitally important for studying cells in a way that will lead to efficacious interventions for patients. Although 2D cultures allow for a myriad of insightful assays to be performed on cells, they lack the complexity of the 3D mechanical environment that cells experience in vivo which leads to inaccurate readings of biomechanical responses. In environments such as those of heart valves, the large mechanical inputs that the cells encounter on a second-to-second basis magnify these inaccuracies. Therefore, understanding the full 3D nature of cells and their microenvironment is vital. In recent years, researchers have begun to perform experiments and develop models taking into account the 3D nature of tissues. Both looking at the cells in situ and developing 3D hydrogel in vitro culturing techniques have allowed for the measurement and development of models for cell mechanodynamics. Nonetheless, the development of models for understanding how cells remodel their matrices and modify their contractility remains challenging. Furthermore, the acumen of how biomechanical properties of cells and their ECM change must be refined. The upcoming section will cover recent measurement and modeling approaches that are being cultivated to address these current challenges to produce accurate understanding.

## 2 Studying Cells in 3D

### 2.1 Overview

Investigating the cells and ECM from native tissue samples has led to many fruitful understandings of physiology and pathology throughout the years. For instance, such studies have facilitated insight into the distribution of cells and different ECM components in heart valves in health and disease [28]. Some experiments have been performed in situ to determine the mechanical properties of VICs, elucidating through paired modeling that VIC contractile behavior greatly affects the mechanical properties of heart valve leaflets [6, 7, 46, 55]. That being said, mechanobiological experiments focusing on live cell readouts are notoriously challenging in native tissue samples because the opacity and density of the tissue make it difficult to image or assay the cells. The finest resolution with which the mechanobiological features have been accurately defined in the living dense connective tissue of heart valves is to a coarse cell membrane mesh (Fig. 2) [6, 7, 46, 55]. Furthermore, such experiments must be conducted expeditiously after extraction from the patient or animal to mitigate the risk of sample degradation and cell death. Because of these substantial hurdles, most contemporary mechanobiological research involving cells has focused on the implementation of isolated cells seeded within hydrogels.

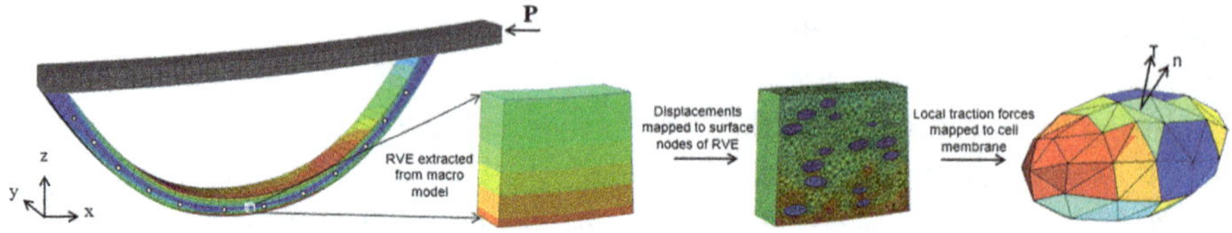


**Fig. 2** In situ experimentation and modeling elucidate that VIC contractile behavior greatly contributes to valve leaflet stiffness. Experiments of valve leaflet curvature when exposed to KCl or cytochalasin-D were developed into a representative volume element (RVE) model of the ECM. Cell locations from serial-stained histological analysis were then added to the model to determine cell surface traction forces that occur during contraction. (Images modified from Buchanan [6], Buchanan and Sacks [7])

Hydrogels have been shown to closely mimic the micromechanical environment seen by cells in situ [59]. Additionally, they are transparent, have similar refractive indices to water, and can have fluorescent microbead fiducial markers embedded in them, which enables high-resolution imaging of changes in both the cell and the gel during experimentation [17, 22]. Furthermore, hydrogels are tunable to varying stiffnesses based on the click chemistry concentrations, which allows researchers to study the effect of stenosis in a controlled manner [16, 26, 27]. Moreover, cells are easy to maintain for weeks within hydrogels utilizing standard cell culturing techniques. Together, these properties of hydrogels make them ideal for evaluating the biomechanical features of cells in 3D.

## 2.2 3D Traction Force Microscopy

3D traction force microscopy (3D-TFM) is a technique that is implemented to assess the geometry and contractility of cells under varying conditions [1, 5, 13, 26, 59]. We have developed an open-source software called FM-TRACK that tracks the deformations that occur in the cell and gel with high fidelity [32]. During experimentation, stained cells are embedded in a poly(ethylene) glycol (PEG)-based gel that contains fluorescent microspheres. A cell and its surrounding gel are imaged in the basal state, and then again after the drug [23]. The potent actin polymerization inhibitor, cytochalasin D (CytoD) can be introduced to analyze basal contractility, or agents such as endothelin-1 can be added to assess hypercontractility. A mask of the cell image produced through denoising, Gaussian filtering, binary closing, and hole filling is then employed to produce a finite element surface mesh with the Lewiner marching cubes algorithm. Smoothing and resampling of the cell surface mesh is then performed with the open-source software, Meshlab [12]. FM-TRACK is implemented to interpolate local field displacements onto the cell surface mesh nodes and produce a one-to-one node mapping of the basal and drug-

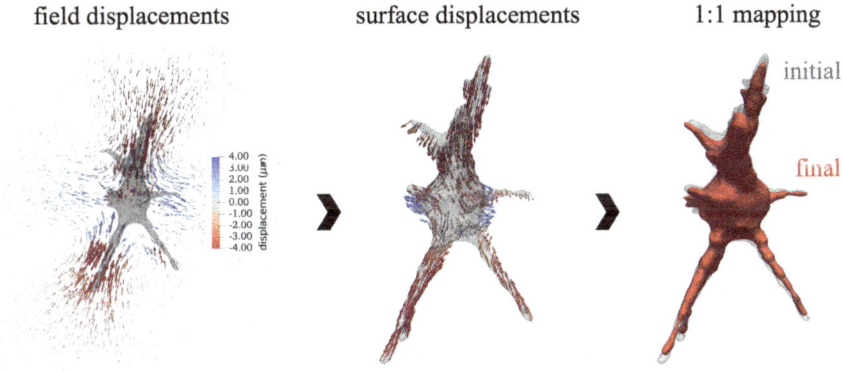


**Fig. 3** FM-TRACK software overview. FM-TRACK determines microsphere displacement within the gel with high fidelity and output cell-induced field displacements as well as 1:1 mapping of surface mesh nodes in the cellular geometry before and after drug treatment [25]

treated cell geometries (Fig. 3) [32]. This finite element model therefore allows for mathematical quantification of cell deformations that occur because of contractility.

The direction of a given displacement, either expansile or contractile, is then determined by finding the dot product between the displacement and the outward-facing node normal. Upon analysis of both porcine and human AVIC basal contractility, protrusions of the cells exhibit contractile behaviors, while the body of the cells portray expansile characteristics, with the cells being isovolumetric from state to state [25, 60]. Furthermore, hypercontractility in porcine AVICs indicates the same pattern. Additional exploration of the membrane kinematics of the protrusions of these porcine AVICs revealed that the length of a protrusion is correlated to its surface displacement magnitudes and that protrusions displace along their principal axes.

The above-mentioned methods do an excellent job of analyzing the deformations that occur in the cell and gel. However, prior calculations that mapped similar deformation measurements to possible traction forces in cells often assumed homogeneity in the hydrogel small strain shear modulus, typically measured by rheology. This assumption overlooks the well-documented evidence that cells must modify the hydrogel to maintain homeostasis [33, 34, 53, 56]. Micromechanical differences in the modulus of the gel can lead to substantial errors in the computed traction forces of the cell when homogeneity is assumed. However, experimentally measuring local differences in hydrogel mechanical characteristics is difficult at best. Approaches including magnetic bead rheology and atomic force microscopy can examine micromechanical differences in hydrogel behavior to an extent, but they lack the ability to map stiffness profiles in 3D to the resolution needed [11]. Therefore, mathematical modeling is needed to determine the micromechanical characteristics of the ECM.

## 2.3 Determination of How Cells Remodel Extracellular Matrices in 3D

By integrating experimental and computational methodologies, we have developed a process for calculating the spatial variation in stiffness that occurs because of cell-directed modification of the hydrogel [23, 25]. This integrated experimental-computational approach which allows for the calculation of varied hydrogel stiffness in full 3D is detailed in this section, which will include a discussion of both forward and inverse model development. The finite element model discussed in the previous section consists of discretized values of the displacement field. This FE model is used as input for the forward model, where the ECM is modeled as a Neo-Hookean nearly incompressible solid, as encoded in the form of the strain energy density of the system. Then, an inverse model that minimizes the objective function with respect to the gel stiffness field is developed. The degrees of freedom must be solved at each node of the finite element mesh, and a detailed adjoint expression must be determined in each iteration of the model. The high computational expense of such activities is mitigated by implementing the Dolfin-adjoint method extension within the FEniCS package for Python [39] to solve the nonlinear inverse problem by calculating the gradient of the objective function with respect to all degrees of freedom at each node. The application of this model on cell-specific data allows for information to be gleaned about how cells from different phenotypes or conditions form differing steady-state 3D micromechanical environments in culture. In addition, localized information on the stiffness of the ECM allows for much more accurate calculations of traction forces on the surface of the cell to be made, which is a crucial step to accurately determining stress fiber contractile force levels of the cell.

### 2.3.1 Extracellular Matrix Finite Element Forward Model

The domain for the model is first defined as a cell embedded in a cubic, homogenous, elastic, and isotropic hydrogel of 300 $\mu m^3$ with a cell as a hollow inclusion at its center. Measured bead displacements in the gel are interpolated into a displacement field and discretized into a finite element mesh made of evenly distributed tetrahedral elements (Fig. 4a). Surface displacements are prescribed as boundary conditions, where along the outer edges of the cube, it is assumed that no displacement occurs. The surface displacements of the cell at each cell surface mesh node, which have already been calculated as described in the previous section, act as the boundary conditions for the gel at the cell/gel interface, with the cell domain itself being depicted as a hollow inclusion. The 1:1 point correspondence of the cell surface nodes in basal and drug-treated states allows for simulation of contraction without the need for traction force values (Fig. 4b), which reduces the number of parameters that must be calculated (and therefore computational expense), which becomes especially important in the upcoming inverse model. For all points in the mesh, $\forall x_0 \in \Omega$, the deformation gradient, $\mathbf{F}$, is defined in Einstein notation as:

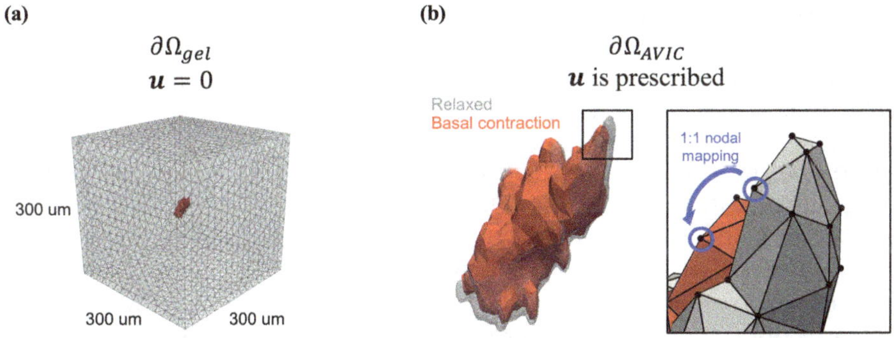


**Fig. 4** Hydrogel and cell as finite element meshes. (**a**) A finite element mesh of the ECM is composed of approximately 50,000 evenly distributed tetrahedral elements surrounding an imbedded cell surface mesh, forming a 300 $\mu m^3$ cube which has boundary conditions of zero displacement enforced at its outer surfaces. (**b**) The 1:1 nodal mapping between the cell surface meshes in the reference and deformed configurations are implemented to determine surface displacements at each surface node, which are set as the boundary conditions for the cell/gel interface and drive the forward model of gel stiffness [25]

$$\mathbf{F}_{ij} = \frac{\partial u_i}{\partial x_j^0}, i, j \in \{1, 2, 3\} \tag{1}$$

where $u_i$ is the displacement of a given point in the displacement field. The right Cauchy-Green deformation tensor, **C**, is given by

$$\mathbf{C} = \mathbf{F}^T\mathbf{F} \tag{2}$$

which quantifies how lengths change in the gel due to deformation. The strain energy density function of the hydrogel domain, $\Psi$, is modeled as a nearly incompressible Neo-Hookean solid:

$$\Psi = C_1 (I_1 - 3) - 2C_1 \ln J + D_1(\ln J)^2 \tag{3}$$

where $C_1$ is a material parameter that approaches half the shear modulus, $\mu$, in the small strain limit and can be experimentally measured by rheology. $I_1 = tr(\mathbf{C})$, as the first invariant of the right Cauchy-Green deformation tensor. $D_1$ is a material parameter that approaches one-half of Lame's first parameter, $\lambda$. $D_1/C_1$ is set to 10,000 to enforce the nearly incompressible condition so that $J = 1.00 \pm 0.02$ and therefore the volumetric term $D_1(\ln J)^2$ stays relatively small, and the equation maintains stability well under a variety of experimental conditions [22, 25]. Therefore, the hydrogel forward finite element model allows for a determination of strain energy within the system, which can be further analyzed with the implementation of the inverse model that is discussed henceforth.

### 2.3.2 Hydrogel Finite Element Inverse Model

Although previous mathematical models have assumed that the hydrogels of the cells are homogeneous, there is clear evidence that cells modify the gels as they grow. For instance, the formulation of hydrogels themselves must always include MMP-degradable crosslinkers so that the cells are able to grow into the gels; without cell-mediated MMP degradation of the gel, the cells will perish in their hydrogel cages [64]. Furthermore, fibroblasts and fibroblast-like cells such as VICs are the cell types that produce and remodel the ECM proteins such as collagen to form them into recognizable tissue types [28]. These modifications in the ECM need to be accounted for to accurately determine the traction forces that occur within the cell. Furthermore, determining the extent to which cells of different phenotypes modify their environments in 3D can be highly beneficial to understanding the biochemical differences that occur in the milieu of the ECM during health and disease.

To calculate the variation in stiffness of the ECM, a term that varies spatially over the nodes of the gel, $\alpha(\boldsymbol{x_0})$, is added to the strain energy density equation.

$$\Psi = \alpha\left(\boldsymbol{x_0}\right)\left[C_1\left(I_1 - 3\right) - 2C_1 \ \ln J\right] + D_1(\ln J)^2 \tag{4}$$

At first, values of all ones are input for $\alpha(\boldsymbol{x_0})$, and the existing finite element model of the gel is re-run to determine what the displacement values, $\boldsymbol{u_{sim}}$, in such a situation will be. These values are instituted to determine the initial simulated right Cauchy-Green deformation tensor, $\mathbf{C_{sim}}$. Concurrently, the right Cauchy-Green deformation tensor previously calculated from displacements extrapolated from the experimentally measured values is labeled as experimental, $\mathbf{C_{exp}}$. The goal of the inverse model is to effectively reduce the difference between the experimental and the simulated values:

$$\boldsymbol{\xi} = \mathbf{C}_{sim} - \mathbf{C}_{exp} \tag{5}$$

The gel is assumed to be unmodified by the cell near the outer gel boundary, and therefore defined $\alpha(\boldsymbol{x_0}) \equiv 1$ in the areas near the edges of the gel, with areas closer to the cell being assigned values relative to the outer gel boundary zone. Furthermore, to reduce computational expense, a more concentrated number of nodes are created at the cell surface and a reduced number of nodes at the gel boundary, as performed by gmsh [18]. Nevertheless, the computational rigor of this problem is large. To address this concern, the Dolfin-adjoint package is interfaced with scipy.optimize while using the Limited Memory Broyden-Fletcher-Goldfarb-Shanno with bound constraints (L-BFGS-B) algorithm [39, 40] to produce updated guesses for the new $\mathbf{C_{sim}}$, with the objective function, $\Phi_1$, being defined as:

$$\Phi_1 = \int_{\Omega} \boldsymbol{\xi} : \boldsymbol{\xi} \, d\Omega + \frac{\gamma}{2} \int_{\Omega} \nabla\alpha \bullet \nabla\alpha \, d\Omega \tag{6}$$

where : is the second rank tensor contraction, $d\Omega$ is an infinitesimal volume element, $\gamma$ is the regularization scaling parameter that is set to 0.001 $\mu$ m$^3$, and

$$\nabla\alpha \bullet \nabla\alpha = \left(\frac{\delta\alpha}{\delta x}\right)^2 + \left(\frac{\delta\alpha}{\delta y}\right)^2 + \left(\frac{\delta\alpha}{\delta z}\right)^2 \tag{7}$$

The regularization function is then iteratively evaluated with updated $\mathbf{C}_{sim}$ values until the function tolerance $1 \times 10^{-8}$ is reached or the number of iterations equals 5000. The resulting component values of $\alpha(\boldsymbol{x_0})$ therefore represent the degree to which the hydrogel stiffness is modulated by the cell at each point in the gel. Multiplying the components of $\alpha(\boldsymbol{x_0})$ by $2C_1$ subsequently provides the Pascal value of the gel at each node (Fig. 5).

### 2.3.3 AVICs Produce a Modulated Steady-State ECM When Cultured in 3D

The results from the hydrogel inverse model depict that there are significantly stiffened volumes localized near the tips of AVIC protrusions (Fig. 5a) and significantly degraded volumes that are more spatially uniform and dispersed around the AVIC (Fig. 5b). The ECM near AVIC protrusions displayed approximately 5 times greater stiffness than the unmodified hydrogel (~500 Pa vs 108 Pa). The distribution of the degradation is more uniform around the cell and reaches a value as low as 63 Pa. The model results correlate with experimental immunocytochemistry staining and imagery of collagen-I in 3D AVIC cultures, where the volume of locally deposited collagen (2624 μm$^3$) is similar to the volume of locally stiffened regions in the model (2718 μm$^3$; Fig. 6a). This approach therefore allows for a detailed quantification of the capacity of AVICs as well as other cell types to affect their local microenvironment. Furthermore, an accurate representation of the varied stiffness

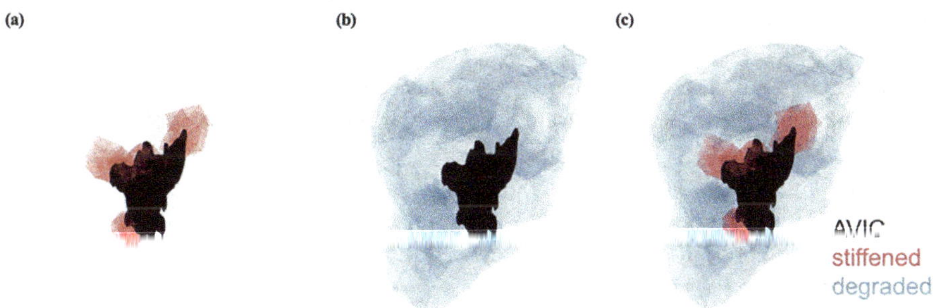


**Fig. 5** Inverse modeling results portray areas of stiffened and degraded ECM. (**a**) Significantly stiffened ECM ($\mu$ > 1 SD). (**b**) Significantly degraded ECM ($\mu$ < 1 SD). (**c**) Merge [25]

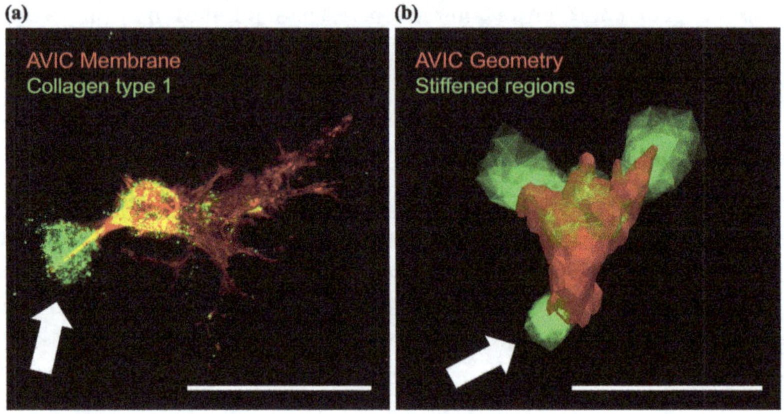


**Fig. 6** Experimental collagen-I volumes align with computational stiffening volumes. (**a**) Maximum intensity projection image of 3D immunocytochemistry for collagen-I. (**b**) Representative results from the inverse model that depicts areas of ECM are significantly stiffened. Arrows indicate the green 3D volumes that are quantified (2624 $\mu m^3$ and 2718 $\mu m^3$, respectively). Scale bars = 50 μm [25]

regions in the matrix surrounding the cell can be implemented for high-resolution and high-accuracy calculation of traction forces occurring because of contractility [60].

## 2.4 *Determination of 3D Cell Stress Fibers: Forces and Directions*

From the results of the above detailed inverse model of ECM stiffness, highly accurate and resolved traction forces of the cell can be computed to determine the structure and forces of stress fibers (SFs) within the cell, on a continuum modeling basis [24]. Again, as in the gel stiffness model, experimental cell surface displacements are employed, which reduces the number of parameters required for the simulation. Other needed parameters for the model are then applied from the literature [50–52], with the assumptions being made that the cell can be modeled on a continuum basis and that its components act as nearly incompressible solids. Moreover, it is assumed that SFs possess an intrinsic force per unit mass, leading to the local effective force being primarily influenced by the local orientation and mass fraction of SFs. [24].

### 2.4.1 Estimation of the 3D Aortic Valve Interstitial Cell Stress Fiber Force Levels

Since the focus of this model is on determining the properties of the stress fibers (SFs) in a cell that is embedded in a 3D matrix, the system is modeled as consisting of three compartments: the cytoplasm containing the SFs, the nucleus, and the hydrogel. The cell is assumed to act as a solid mixture that deforms together and contains the nucleus. Each component is modeled as a Neo-Hookean nearly incompressible solid, implementing Eq. 3 such that

$$\boldsymbol{\Psi} = \boldsymbol{\Psi}_{cyto+sf} + \boldsymbol{\Psi}_{nuc} + \boldsymbol{\Psi}_{gel} \tag{8}$$

A value of 750 Pa is defined for the shear modulus of the cytoplasm, and a value of 7500 Pa is defined for the shear modulus of the nucleus, as has previously been reported [50–52]. The inner boundary of the cell domain shares its nodes with the outer boundary of the nucleus, and the two are assumed to be perfectly bonded. Values for the nodes of $\boldsymbol{\Psi}_{gel}$ are collected from the outputs of the gel model described in the previous section, and again the inner boundary of the hydrogel shares its perfectly bonded nodes with the outer boundary of the cell. A total of about 50,000 tetrahedral elements are prescribed within the simulated space, which has been determined based on a mesh convergence analysis [25].

With finite deformations occurring concurrently with the adherence to the reference configuration, the second Piola-Kirchoff stress tensor, $\mathbf{S}$, is implemented as the traction force term. Furthermore, the nucleus and gel are assumed to be isotropic such that

$$\mathbf{S}^{nuc} = 2\frac{\delta\boldsymbol{\Psi}^{nuc}}{\delta\mathbf{C}^{nuc}} \tag{9}$$

$$\mathbf{S}^{gel} = 2\frac{\delta\boldsymbol{\Psi}^{gel}}{\delta\mathbf{C}^{gel}} \tag{10}$$

Meanwhile, the cell domain is modeled as a mixture of stress fibers and cytoplasm where there is a continuity of displacements of all elements, $\mathbf{u}$, between the two

$$\mathbf{u}^{cyto} = \mathbf{u}^{sf} \tag{11}$$

SF response is split into an active and passive component

$$\mathbf{S}^{sf} = \mathbf{S}^{a} + \mathbf{S}^{p} \tag{12}$$

Since stress fibers have directionality, both the active and passive components of the second Piola-Kirchoff stress include the stress fiber orientation unit vectors, $\mathbf{m}_0$, as part of their calculations [14, 44]. The passive component is calculated by

$$\mathbf{S}^{p} = 2\mu_{sf}\phi_{sf}H\left(I_4 - 1\right)\mathbf{m}_0 \otimes \mathbf{m}_0 \tag{13}$$

where $\phi_{sf}$ is the SF mass fraction. $\mu_{sf}$ is the SF stiffness which is set to 390 Pa, as previously reported [51]. $H$ is the Heaviside step function, which is enforced to ensure that only passive stretch from tension is calculated here. $I_4$ is the fourth invariant of the right Cauchy-Green deformation tensor which measures how much fiber is stretched or compressed relative to its initial configuration and equals the dot products of the unit vectors and the right Cauchy-Green deformation tensor, $\mathbf{m}_0 \bullet \mathbf{C} \bullet \mathbf{m}_0$. The active component of stress fibers is calculated as

$$\mathbf{S}^{a} = 2f_{sf}\phi_{sf}\mathbf{m}_0 \otimes \mathbf{m}_0 \tag{14}$$

where $f_{sf}$ is the SF contractile force per mass unit which is set to 1000 Pa based on previous work [50–52].

The FEniCS and Dolfin-adjoint packages are again employed here to run an inverse model that determines both the discrete adjoint and the tangent linear models of a cell's stress fibers, allowing for the determination of both concentration and directionality of the stress fibers [39]. Similar to the gel model described above, the goal of the inverse stress fiber model is to reduce the difference between the experimental and simulated right Cauchy green deformation tensors such that $\boldsymbol{\xi} = \mathbf{C}_{sim} - \mathbf{C}_{exp}$. The inverse iterations are deployed to determine the spatial variation in the parameters $\phi_{sf}$ and $\mathbf{m}_0$ until either a function tolerance of $1 \times 10^{-8}$ or iteration number of 5000 is reached. The objective function is defined as

$$\Phi_2 = \int_{\Omega_{gel}} \boldsymbol{\xi} : \boldsymbol{\xi}\, d\Omega_{gel}$$

$$+ \frac{\beta}{2}\int_{\Omega_{cyto+sf}} \nabla\phi_{sf} \bullet \nabla\phi_{sf}\, d\Omega_{cyto+sf}$$

$$+ \frac{\gamma}{2}\int_{\Omega_{cyto+sf}} \nabla\mathbf{m}_0 : \nabla\mathbf{m}_0\, d\Omega_{cyto+sf} \tag{15}$$

where $\beta$ and $\gamma$ are regularization scaling parameters for $\phi_{sf}$ and $\mathbf{m}_0$, respectively, which are both set to 0.001. In addition,

$$\nabla\phi_{sf} \bullet \nabla\phi_{sf} = \left(\frac{\delta\phi_{sf}}{\delta x}\right)^2 + \left(\frac{\delta\phi_{sf}}{\delta y}\right)^2 + \left(\frac{\delta\phi_{sf}}{\delta z}\right)^2 \tag{16}$$

and

$$\nabla\mathbf{m}_0 : \nabla\mathbf{m}_0 = \left(\frac{\delta m_0^x}{\delta x}\right)^2 + \left(\frac{\delta m_0^x}{\delta y}\right)^2 + \left(\frac{\delta m_0^x}{\delta z}\right)^2$$

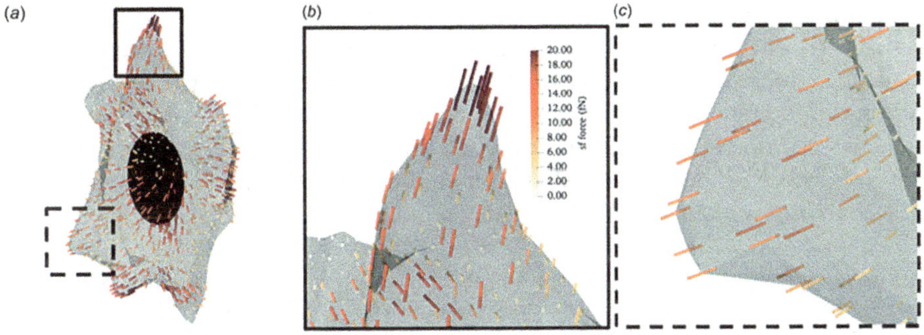


**Fig. 7** Stress fiber (SF) orientation and magnitude of force generation, as determined by the inverse stress fiber contraction model for (**a**) an entire cell, (**b**) a protrusion of the cell, and (**c**) a body region the cell [24]

$$+\left(\frac{\delta m_0^y}{\delta x}\right)^2+\left(\frac{\delta m_0^y}{\delta y}\right)^2+\left(\frac{\delta m_0^y}{\delta z}\right)^2$$

$$+\left(\frac{\delta m_0^z}{\delta x}\right)^2+\left(\frac{\delta m_0^z}{\delta y}\right)^2+\left(\frac{\delta m_0^z}{\delta z}\right)^2 \tag{17}$$

Model results allow for accurate predictions of SF orientations ($\mathbf{m}_0$), expression levels ($\phi_{sf}$), and stress ($\mathbf{S}^{sf}$), which are highly aligned and had higher values in the cell protrusions, as expected (Fig. 7). Validation studies depicted that in test problems with known values of the parameter fields, the model is able to recapitulate the values by an average difference of $0.004 \pm 0.01$ SD [24, 58].

## 3 Summary and Future Directions

In this chapter, methods for determining how contractile cells affect their local microenvironment through shape change, deformation, ECM deposition and remodeling, and traction forces in full 3D have been explored. These techniques that pair experimental testing with mathematical modeling can be used to identify intrinsic functional differences between cell types of varying phenotypes, as was exhibited throughout the chapter with AVICs from normal and diseased valves. This capability to detect functional disparities lays the groundwork for developing novel therapies for valvular heart disease and can be implemented in other disease models to do the same.

Future computational research focuses on robust mechanistic modeling of VICs within the 3D hydrogel systems. The tunable hydrogel system and the multi-scale, multi-resolution modeling framework collaborate to elucidate experimentally observed cell and system behavior, facilitating comparisons between synthetic and native VIC environments. Enhancements to the VIC-ECM model include more realistic cellular geometry, incorporation of additional subcellular features such as cell nuclei, a more complex strategy for capturing the cell-ECM attachment zone, and consideration of aleatoric uncertainty. These future enhancements not only can improve the knowledge about valvular disease but can be applied to other diseases as well to improve outcomes for patients.

**Acknowledgments** We thank Dr. Kristi Anseth's research group at the University of Colorado Boulder for help with hydrogel chemistry. We also thank Giovanni Ferrari's research group at Columbia University for providing human VICs for study. We acknowledge our funding sources: National Institutes of Health, United States, Ruth L. Kirschstein postdoctoral fellowships [grant no. F32 HL-007822 awarded to Alex Khang] and [grant no. F32 HL-167570 awarded to Toni West], Oden Institute of Computational Engineering and Sciences pre-doctoral fellowship in computational science, engineering, and mathematics [awarded to Gabriel Peery], and National Institutes of Health, United States, R01 grants [grant nos. R01 HL-157829, R01 HL-073021, and R01 HL-142504 awarded to Michael S. Sacks].

## References

1. Anseth, K., Bowman, C. & Brannon-Peppas, L. 1996. Mechanical properties of hydrogels and their experimental determination. *Biomaterials,* 17, 1647–1657.
2. Ayoub, S., Ferrari, G., Gorman, R. C., Gorman, J. H., Schoen, F. J. & Sacks, M. S. 2016. Heart Valve Biomechanics and Underlying Mechanobiology. *Compr Physiol,* 6, 1743–1780.
3. Ayoub, S., Lee, C.-H., Driesbaugh, K. H., Anselmo, W., Hughes, C. T., Ferrari, G., Gorman, R. C., Gorman, J. H. & Sacks, M. S. 2017. Regulation of Valve Interstitial Cell Homeostasis by Mechanical Deformation: Implications for Heart Valve Disease and Surgical Repair *Journal of the Royal Society Interface,* in-press.
4. Ayoub, S., Tsai, K. C., Khalighi, A. H. & Sacks, M. S. 2018. The Three-Dimensional Microenvironment of the Mitral Valve: Insights into the Effects of Physiological Loads. *Cell Mol Bioeng,* 11, 291–306.
5. Benton, J. A., Fairbanks, B. D. & Anseth, K. S. 2009. Characterization of valvular interstitial cell function in three dimensional matrix metalloproteinase degradable PEG hydrogels. *Biomaterials,* 30, 6593–603.
6. Buchanan, R. M. 2016. *An integrated computational-experimental approach for the in situ estimation of valve interstitial cell biomechanical state.* Ph.D., The University of Texas at Austin.
7. Buchanan, R. M. & Sacks, M. S. 2013. Interlayer micromechanics of the aortic heart valve leaflet. *Biomech Model Mechanobiol.*
8. Butcher, J. T. & Nerem, R. M. 2004. Porcine aortic valve interstitial cells in three-dimensional culture: comparison of phenotype with aortic smooth muscle cells. *J Heart Valve Dis,* 13, 478–85; discussion 485–6.
9. Butcher, J. T. & Nerem, R. M. 2006. Valvular Endothelial Cells Regulate the Phenotype of Interstitial Cells in Co-culture: Effects of Steady Shear Stress. *Tissue Engineering,* 12, 905–915.
10. Butcher, J. T. & Nerem, R. M. 2007. Valvular endothelial cells and the mechanoregulation of valvular pathology. *Philos Trans R Soc Lond B Biol Sci,* 362, 1445–57.
11. Canovic, E. P., Seidl, D. T., Polio, S. R., Oberai, A. A., Barbone, P. E., Stamenovic, D. & Smith, M. L. 2014. Biomechanical imaging of cell stiffness and prestress with subcellular resolution. *Biomech Model Mechanobiol,* 13, 665–78.

12. Cignoni, P., Callieri, M., Corsini, M., Dellepiane, M., Ganovelli, F. & Ranzuglia, G. 2008. MeshLab: an Open-Source Mesh Processing Tool. *In:* Scarano, V., De Chiara, R. & Erra, U. (eds.) *Eurographics Italian Chapter Conference.* The Eurographics Association.
13. Cushing, M. C., Liao, J. T. & Anseth, K. S. 2005. Activation of valvular interstitial cells is mediated by transforming growth factor-beta1 interactions with matrix molecules. *Matrix Biol,* 24, 428–37.
14. Feng, Y., Lee, C.-H., Sun, L., Ji, S. & Zhao, X. 2017. Characterizing white matter tissue in large strain via asymmetric indentation and inverse finite element modeling. *Journal of the mechanical behavior of biomedical materials,* 65, 490–501.
15. Ferrari, S. & Pesce, M. 2022. The complex interplay of inflammation, metabolism, epigenetics, and sex in calcific disease of the aortic valve. *Frontiers in Cardiovascular Medicine,* 8, 791646.
16. Fitzsimons, T. M. & Rosales, A. M. 2023. Leveraging mechanistic insight to design dynamic hydrogels. *Chem,* 9, 2051–2053.
17. Franck, C., Maskarinec, S. A., Tirrell, D. A. & Ravichandran, G. 2011. Three-dimensional traction force microscopy: a new tool for quantifying cell-matrix interactions. *PLoS One,* 6, e17833.
18. Geuzaine, C. & Remacle, J.-F. 2008. Gmsh: a three-dimensional finite element mesh generator with built-in pre-and post-processing facilities.
19. He, Z., Ritchie, J., Grashow, J. S., Sacks, M. S. & Yoganathan, A. P. 2005. In vitro dynamic strain behavior of the mitral valve posterior leaflet. *J Biomech Eng,* 127, 504–11.
20. He, Z., Sacks, M. S., Baijens, L., Wanant, S., Shah, P. & Yoganathan, A. P. 2003. Effects of papillary muscle position on in-vitro dynamic strain on the porcine mitral valve. *J Heart Valve Dis,* 12, 488–94.
21. Khang, A., Buchanan, R. M., Ayoub, S., Rego, B. V., Lee, C.-H., Ferrari, G., Anseth, K. S. & Sacks, M. S. 2018. Chapter 8—Mechanobiology of the heart valve interstitial cell: Simulation, experiment, and discovery. *In:* Verbruggen, S. W. (ed.) *Mechanobiology in Health and Disease.* Academic Press.
22. Khang, A., Gonzalez Rodriguez, A., Schroeder, M. E., Sansom, J., Lejeune, E., Anseth, K. S. & Sacks, M. S. 2019. Quantifying heart valve interstitial cell contractile state using highly tunable poly(ethylene glycol) hydrogels. *Acta Biomater,* 96, 354–367.
23. Khang, A., Nguyen, Q., Feng, X., Howsmon, D. P. & Sacks, M. S. 2023a. Three-dimensional analysis of hydrogel-imbedded aortic valve interstitial cell shape and its relation to contractile behavior. *Acta Biomater,* 163, 194–209.
24. Khang, A. & Sacks, M. S. 2023. An inverse modeling approach to estimate 3D aortic valve interstitial cell stress fiber force levels. *Journal Biomechanical Engineering,* Accepted.
25. Khang, A., Steinman, J., Tuscher, R., Feng, X. & Sacks, M. S. 2023b. Estimation of aortic valve interstitial cell-induced 3D remodeling of poly(ethylene glycol) hydrogel environments using an inverse finite element approach. *Acta Biomater,* 160, 123–133.
26. Kloxin, A. M., Kloxin, C. J., Bowman, C. N. & Anseth, K. S. 2010a. Mechanical properties of cellularly responsive hydrogels and their experimental determination. *Adv Mater,* 22, 3484–94.
27. Kloxin, A. M., Tibbitt, M. W. & Anseth, K. S. 2010b. Synthesis of photodegradable hydrogels as dynamically tunable cell culture platforms. *Nat Protoc,* 5, 1867–87.
28. Kodigepalli, K. M., Thatcher, K., West, T., Howsmon, D. P., Schoen, F. J., Sacks, M. S., Breuer, C. K. & Lincoln, J. 2020. Biology and Biomechanics of the Heart Valve Extracellular Matrix. *J Cardiovasc Dev Dis,* 7.
29. Lam, N. T., Muldoon, T. J., Quinn, K. P., Rajaram, N. & Balachandran, K. 2016. Valve interstitial cell contractile strength and metabolic state are dependent on its shape. *Integrative Biology,* 8, 1079–1089.
30. Lee, C. H., Carruthers, C. A., Ayoub, S., Gorman, R. C., Gorman, J. H., 3rd & Sacks, M. S. 2015. Quantification and simulation of layer-specific mitral valve interstitial cells deformation under physiological loading. *J Theor Biol,* 373, 26–39.
31. Lee, C. H., Zhang, W., Feaver, K., Gorman, R. C., Gorman, J. H., 3rd & Sacks, M. S. 2017. On the in vivo function of the mitral heart valve leaflet: insights into tissue-interstitial cell biomechanical coupling. *Biomech Model Mechanobiol.*

32. Lejeune, E., Khang, A., Sansom, J. & Sacks, M. S. 2020. FM-Track: A fiducial marker tracking software for studying cell mechanics in a three-dimensional environment. *Softwarex,* 11.
33. Li, S., van den Diepstraten, C., D'souza, S. J., Chan, B. M. & Pickering, J. G. 2003. Vascular smooth muscle cells orchestrate the assembly of type I collagen via alpha2beta1 integrin, RhoA, and fibronectin polymerization. *Am J Pathol,* 163, 1045–56.
34. Lindsey, M. L. 2004. MMP induction and inhibition in myocardial infarction. *Heart Fail Rev,* 9, 7–19.
35. Ma, H., Killaars, A. R., Delrio, F. W., Yang, C. & Anseth, K. S. 2017. Myofibroblastic activation of valvular interstitial cells is modulated by spatial variations in matrix elasticity and its organization. *Biomaterials,* 131, 131–144.
36. Mathieu, P., Bosse, Y., Huggins, G. S., Della Corte, A., Pibarot, P., Michelena, H. I., Limongelli, G., Boulanger, M. C., Evangelista, A., Bedard, E., Citro, R., Body, S. C., Nemer, M. & Schoen, F. J. 2015. The pathology and pathobiology of bicuspid aortic valve: State of the art and novel research perspectives. *J Pathol Clin Res,* 1, 195–206.
37. Merryman, W. D. 2007. *Mechanobiology of the aortic valve interstitial cell.* Ph.D. Doctoral Dissertation, University of Pittsburgh.
38. Merryman, W. D., Lukoff, H. D., Long, R. A., Engelmayr, G. C., Jr., Hopkins, R. A. & Sacks, M. S. 2007. Synergistic effects of cyclic tension and transforming growth factor-beta1 on the aortic valve myofibroblast. *Cardiovasc Pathol,* 16, 268–76.
39. Mitusch, S., Funke, S. & Dokken, J. 2019. dolfin-adjoint 2018.1: automated adjoints for FEniCS and Firedrake. *Journal of Open Source Software,* 4, 1292.
40. Moreira, M. & Fiesler, E. 1995. Neural networks with adaptive learning rate and momentum terms.
41. Nicoara, A., Popescu, W. M., Mathew, J., Swaminathan, M. & Ayoub, C. 2010. Assessment of left ventricular diastolic function. *Mathew JP, Swaminathan M, Ayoub CM. Clinical Manual and Review of Transesophageal Echocardiography. 2nd ed. New York, NY: McGraw-Hill,* 266–292.
42. Pierlot, C. M., Lee, J. M., Amini, R., Sacks, M. S. & Wells, S. M. 2014. Pregnancy-induced remodeling of collagen architecture and content in the mitral valve. *Ann Biomed Eng,* 42, 2058–71.
43. Pierlot, C. M., Moeller, A. D., Lee, J. M. & Wells, S. M. 2015. Pregnancy-induced remodeling of heart valves. *Am J Physiol Heart Circ Physiol,* 309, H1565–78.
44. Qiu, G. & Pence, T. 1997. Remarks on the behavior of simple directionally reinforced incompressible nonlinearly elastic solids. *Journal of Elasticity,* 49, 1–30.
45. Rego, B. V. & Sacks, M. S. 2017. A functionally graded material model for the transmural stress distribution of the aortic valve leaflet. *J Biomech,* 54, 88–95.
46. Rego, B. V., Wells, S. M., Lee, C. H. & Sacks, M. S. 2016. Mitral valve leaflet remodelling during pregnancy: insights into cell-mediated recovery of tissue homeostasis. *J R Soc Interface,* 13.
47. Sacks, M. S., Enomoto, Y., Graybill, J. R., Merryman, W. D., Zeeshan, A., Yoganathan, A. P., Levy, R. J., Gorman, R. C. & Gorman, J. H., 3rd 2006. In-vivo dynamic deformation of the mitral valve anterior leaflet. *Ann Thorac Surg,* 82, 1369–77.
48. Sacks, M. S., He, Z., Baijens, L., Wanant, S., Shah, P., Sugimoto, H. & Yoganathan, A. P. 2002. Surface strains in the anterior leaflet of the functioning mitral valve. *Ann Biomed Eng,* 30, 1281–90.
49. Sacks, M. S. & Yoganathan, A. P. 2007. Heart valve function: a biomechanical perspective. *Philos Trans R Soc Lond B Biol Sci,* 362, 1369–91.
50. Sakamoto, Y., Buchanan, R. M., Adams, J. S., Guilak, F. & Sacks, M. S. 2016a. On the contractile behavior of valve interstitial cell stress fibers. *Submitted to Biophysical Journal.*
51. Sakamoto, Y., Buchanan, R. M. & Sacks, M. S. 2016b. On intrinsic stress fiber contractile forces in semilunar heart valve interstitial cells using a continuum mixture model. *J Mech Behav Biomed Mater,* 54, 244–58.
52. Sakamoto, Y., Buchanan, R. M., Sanchez-Adams, J., Guilak, F. & Sacks, M. S. 2017. On the Functional Role of Valve Interstitial Cell Stress Fibers: A Continuum Modeling Approach. *J Biomech Eng,* 139.

53. Sander, E. A., Barocas, V. H. & Tranquillo, R. T. 2011. Initial fiber alignment pattern alters extracellular matrix synthesis in fibroblast-populated fibrin gel cruciforms and correlates with predicted tension. *Ann Biomed Eng,* 39, 714–29.
54. Stella, J. A., Liao, J. & Sacks, M. S. 2007. Time-dependent biaxial mechanical behavior of the aortic heart valve leaflet. *J Biomech,* 40, 3169–77.
55. Stella, J. A. & Sacks, M. S. 2007. On the biaxial mechanical properties of the layers of the aortic valve leaflet. *J Biomech Eng,* 129, 757-66.
56. Stock, U. A., Wiederschain, D., Kilroy, S. M., Shum-Tim, D., Khalil, P. N., Vacanti, J. P., Mayer, J. E., Jr. & Moses, M. A. 2001. Dynamics of extracellular matrix production and turnover in tissue engineered cardiovascular structures. *J Cell Biochem,* 81, 220-8.
57. Tandon, I., Razavi, A., Ravishankar, P., Walker, A., Sturdivant, N. M., Lam, N. T., Wolchok, J. C. & Balachandran, K. 2016. Valve interstitial cell shape modulates cell contractility independent of cell phenotype. *Journal of Biomechanics,* 49, 3289-3297.
58. Taylor, P. M., Batten, P., Brand, N. J., Thomas, P. S. & Yacoub, M. H. 2003. The cardiac valve interstitial cell. *Int J Biochem Cell Biol,* 35, 113-8.
59. Tibbitt, M. W. & Anseth, K. S. 2009. Hydrogels as extracellular matrix mimics for 3D cell culture. *Biotechnology and Bioengineering,* 103, 655-663.
60. Tuscher, R., Khang, A., West, T. M., Camillo, C., Ferrari, G. & Sacks, M. S. 2023. Functional differences in human aortic valve interstitial cells from patients with varying calcific aortic valve disease. *Frontiers in Physiology,* 14, 1168691.
61. Wang, H., Haeger, S. M., Kloxin, A. M., Leinwand, L. A. & Anseth, K. S. 2012. Redirecting Valvular Myofibroblasts into Dormant Fibroblasts through Light-mediated Reduction in Substrate Modulus. *Plos One,* 7.
62. West, T. M., Howsmon, D. P., Massidda, M. W., Vo, H. N., Janobas, A. A., Baker, A. B. & Sacks, M. S. 2023. The effects of strain history on aortic valve interstitial cell activation in a 3D hydrogel environment. *APL bioengineering,* 7.
63. Yap, C. H., Kim, H. S., Balachandran, K., Weiler, M., Haj-Ali, R. & Yoganathan, A. P. 2010. Dynamic deformation characteristics of porcine aortic valve leaflet under normal and hypertensive conditions. *Am J Physiol Heart Circ Physiol,* 298, H395–405.
64. Yu, J., Chen, F., Wang, X., Dong, N., Lu, C., Yang, G. & Chen, Z. 2016. Synthesis and characterization of MMP degradable and maleimide cross-linked PEG hydrogels for tissue engineering scaffolds. *Polymer Degradation and Stability,* 133, 312–320.

# Heterogeneity and Multi-Scale Modelling in Vascular Biomechanics

**Ankush Aggarwal, Alberto Coccarelli, and Sean McGinty**

## 1 Introduction

Blood vessels connect the heart to all parts of human body, in a cascading pattern. Starting with the largest artery closest to the heart, they bifurcate and reduce in radius and thickness as we move downstream (Fig. 1a). Their primary function is supplying nutrient rich blood to all organs and tissues. As a consequence, these vessels play a crucial role in many aspects of human health. For example, abnormally narrowed vessels increase the resistance to the flow and thereby decrease blood supply downstream. Vessel wall may also dilate locally and lead to the formation of an aneurysm that is at a high risk of dissection and rupture—both life-threatening scenarios. Vessel wall may also stiffen with age leading to

A. Aggarwal (✉)
Glasgow Computational Engineering Centre, University of Glasgow, Glasgow, United Kingdom

Division of Infrastructure and Environment, James Watt School of Engineering, University of Glasgow, Glasgow, United Kingdom
e-mail: ankush.aggarwal@glasgow.ac.uk

A. Coccarelli
Zienkiewicz Institute for Modelling, Data and AI, Faculty of Science and Engineering, Swansea University, Swansea, United Kingdom

Department of Mechanical Engineering, Faculty of Science and Engineering, Swansea University, Swansea, United Kingdom
e-mail: alberto.coccarelli@swansea.ac.uk

S. McGinty
Glasgow Computational Engineering Centre, University of Glasgow, Glasgow, United Kingdom

Division of Biomedical Engineering, James Watt School of Engineering, University of Glasgow, Glasgow, United Kingdom
e-mail: sean.mcginty@glasgow.ac.uk



J. Liao, J. Y. Wong (eds.), *Integration and Bridging of Multiscale Bioengineering Designs and Tissue Biomechanics*, https://doi.org/10.1007/978-3-031-81743-4_14

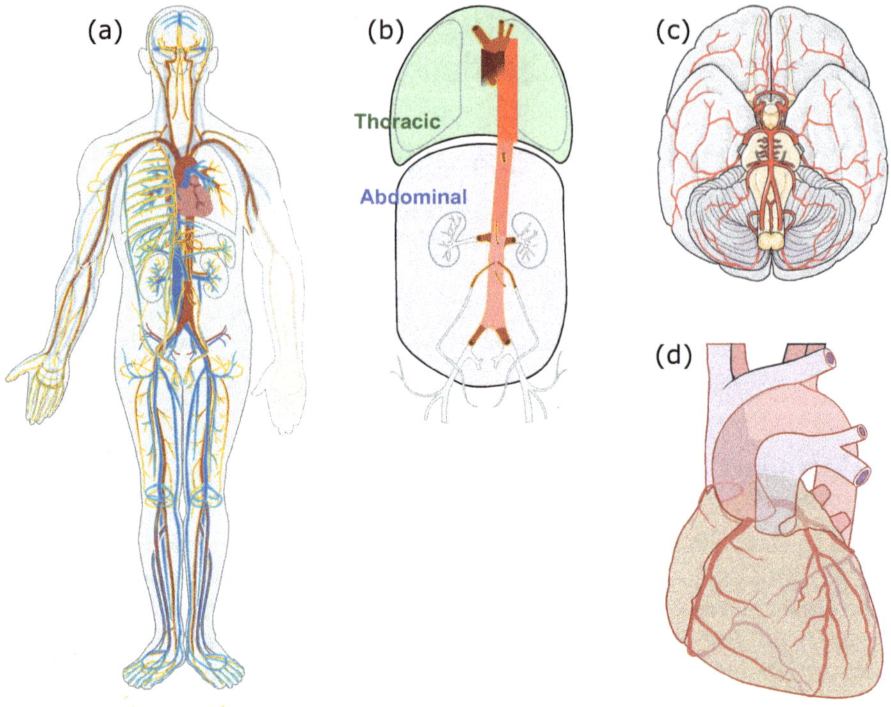


**Fig. 1** (**a**) The vascular network connects the heart to all parts of the human body in a cascading pattern with a large range of sizes. Three case studies are presented in this chapter: (**b**) biomechanical properties of the large aorta emanating from the heart; (**c**) flow regulation in small cerebral arteries; and (**d**) effect of drugs on coronary vessels

lower compliance, lower ability to store blood during diastole and higher pulse wave velocity, which can increase the diastolic blood pressure in the artery.

To support the high pulsatile blood pressure, ranging from 50 to 150 mm Hg each cardiac cycle, repeating nearly every 0.8 s, blood vessels contain a large amount of elastin and collagen fibres [1, 2]. These fibres endow the tissue with direction dependent, i.e. anisotropic, biomechanical properties. However, the amount of the elastin and collagen fibres in the vascular tissue can alter based on location. That is, not only the geometric parameters (radius and thickness) vary along the arterial tree, so do the tissue properties. Moreover, there is a natural turnover of these fibres carried out by smooth muscle cells embedded within the vascular tissue [3]. The deposition and degeneration of fibres follows a mechanobiological process, wherein the rates are affected by the stresses and strains being applied on the tissue and cells [4]. As a result, the tissue composition and fibrous components vary over long time scales, with age and during disease development.

The properties of vessels may also change in the intermediate time scales. For example, to ensure that sufficient blood is supplied during exercise, there are biochemical regulatory mechanisms in our body that alter the contractile properties of vascular smooth muscle cells and thereby control the vascular stiffness [5]. Another example is when, in order to treat diseases such as stenosis, drugs are used, these too have a biochemical effect on the vascular cells' proliferative properties and how they grow [6].

Evidently, there is a large amount of multi-factorial heterogeneity in vascular biomechanics—geometrical (radius/thickness), constituents (fibre components), and spatial scale (tissue layers and cellular). In addition, there are the biochemical pathways affecting the remodelling (such as generation/degeneration of fibres) and cell contractility and proliferation over several temporal scales (minutes/days/months). Thus, there is a need for multi-scale modelling to elucidate various aspects of vascular function and disease development. In this chapter, we report three case studies to demonstrate some recent approaches in this direction that cover the above-mentioned aspects. First, the quantification of heterogeneity in the biomechanical properties of aorta and the associated variation and uncertainty is examined using simplified models and Bayesian inference (Fig. 1b). Then, the pressure and flow regulation in cerebral arteries and arterioles are analyzed using a mathematical model for intracellular signalling and its coupling with vascular biomechanics (Fig. 1c). Lastly, advances in the mathematical model of the effect of drugs on coronary arteries and in-stent restenosis are discussed (Fig. 1d).

# 2 Variation and Heterogeneity in the Mechanics of Aorta

## *2.1 Background*

The aorta is the main blood vessel that starts at the outlet of left ventricle (Fig. 1b). The thoracic aorta consists of the starting section (aortic root), the proximal ascending aorta, the aortic arch, and the descending aorta. The descending aorta continues across the diaphragm and is known as the abdominal aorta. While these different sections of the aorta share some similarities, there are some notable variations that pose technical challenges in describing their mechanics. These challenges and tools that can be used to tackle them are presented next.

### 2.1.1 In Vivo Observations

The aortic vessels undergo repeated loading-unloading under the systolic-diastolic phases of the cardiac cycle. This leads to in vivo deformation that can be observed using clinical imaging modalities, such as ultrasound and magnetic resonance imaging (MRI). In a simplified analysis, vessel diameter is measured at the systole and diastole, $D_{sys}$ and $D_{dias}$, respectively. The corresponding pressures are commonly approximated with the cuff pressure measurements, $P_{sys}$ and $P_{dias}$. From

**Table 1** Characteristics of two commonly used in vivo measures for vessel mechanics

| Name | Distensibility ($\delta$) | $\beta$-stiffness index ($\beta$) |
|---|---|---|
| Definition[a] | $\dfrac{D_{\text{sys}} - D_{\text{dias}}}{D_{\text{dias}}} \dfrac{1}{P_{\text{sys}} - P_{\text{dias}}}$ | $\dfrac{D_{\text{dias}}}{D_{\text{sys}} - D_{\text{dias}}} \ln\left(\dfrac{P_{\text{sys}}}{P_{\text{dias}}}\right)$ |
| Measure | Compliance | Stiffness |
| Units | $(\text{Force/Area})^{-1}$ | Dimensionless |
| Depends on | Pulse pressure only | Systolic/diastolic pressures |
| Nominal order | $10^{-3}$ mm-Hg$^{-1}$ | 10 |
| With age | Decreases | Increases |
| Along the aortic length | Decreases | Increases |

[a] Sometimes area is used instead of the diameter, which, for small deformation, has a factor of two

these, two commonly derived "stiffness" (or its inverse, compliance) measures, and their comparisons, are listed in Table 1. For instance, in a large population-based in vivo analysis [7], in the ascending aorta, the median distensibility was found to be $\delta \approx 2.5 \times 10^{-3}$ mm-Hg$^{-1}$ and the median diastolic diameter was $D_{\text{dias}} \approx 30$ mm.

### 2.1.2 Mechanical Characteristics

It is important to recognise that the mechanical stiffness in vessels comes from two sources: material *and* geometry. The importance of the material (i.e. tissue) properties is obvious, since a stiffer/compliant tissue will make up a stiffer/compliant vessel. However, even if the tissue properties remain the same, changing the geometry, such as vessel's (diastolic) diameter and thickness, will lead to changes in the observed stiffness and compliance. Specifically, increasing the diameter or decreasing the thickness makes a vessel more compliant and vice versa.

To see these effects, tools of continuum mechanics can be employed by approximating the vessel as a perfect cylinder that deforms axisymmetrically under an internal pressure. If the reference configuration is denoted as $\Omega_{\text{R}}$, the aim is to find the deformed configuration denoted as $\Omega_{\text{D}}$ that is in equilibrium with the applied loads and boundary conditions. This problem can be framed in terms of the mapping between the two configurations. For an axisymmetric problem, this reduces to finding the mapping between undeformed radius $R$ and deformed radius $r$. Moreover, incompressibility implies that the deformed radius at any point can be expressed in terms of one (scalar) deformation measure, such as the internal radius [8]. That is, we can write the mapping as

$$\Omega_{\text{R}} \rightarrow \Omega_{\text{D}} : \quad r(R) = \sqrt{\left[R^2 - R_i^2\right] + r_i^2}, \tag{1}$$

where $R_i$ and $r_i$ are the reference and deformed inner (luminal) radii, respectively. Therefore, the deformation gradient is diagonal in cylindrical coordinates, i.e.

$$\mathbf{F} = \mathrm{diag}\,[\lambda_r, \lambda_\theta, \lambda_Z] = \mathrm{diag}\left[\frac{R}{r}, \frac{r}{R}, 1\right]. \tag{2}$$

Next, if the arterial tissue is considered to be hyperelastic, then the pressure acting on the artery can be related to the derivatives of strain energy density function $\Psi(\mathbf{F})$ as [9]:

$$P(r_i) - P_{\text{ext}} = -\int_{R_i}^{R_i+H} \frac{1}{\lambda_\theta \lambda_Z r}\left[\lambda_r \frac{\partial \Psi}{\partial \lambda_r} - \lambda_\theta \frac{\partial \Psi}{\partial \lambda_\theta}\right] \mathrm{d}R, \tag{3}$$

where $P(r_i)$ is the pressure acting on the internal radius, $P_{\text{ext}}$ is the external pressure acting on the outer surface of the vessel, and $H$ is the vessel thickness in the reference configuration.

In order to relate the above to the in vivo observation of distensibility presented in the previous section (Table 2), we note that if the applied pressure difference $P_{\text{sys}} - P_{\text{dias}}$ is infinitesimal, then one can arrive at the following approximation for the distensibility:

$$\frac{1}{\delta} \approx r_i \frac{\mathrm{d}P(r_i)}{\mathrm{d}r_i}. \tag{4}$$

Moreover, if this is calculated at the reference configuration and for a (locally) hyperelastic linear material $\Psi = \mu/2\left[\mathrm{tr}\left(\mathbf{F}^\top \mathbf{F}\right) - 3\right]$, the following expression can be derived from (3):

$$\left.\frac{\mathrm{d}P(r_i)}{\mathrm{d}r_i}\right|_{r_i=R_i} = 2\mu \frac{\bar{H}[2+\bar{H}]}{R_i[1+\bar{H}]^2}, \tag{5}$$

where $\bar{H} = H/R_i$ is the normalised thickness. The above expression clearly depends on the geometry ($R_i$ and $H$) as well as the shear modulus $\mu$. Thus, the derivative $\mathrm{d}P(r_i)/\mathrm{d}r_i$ increases with the increase of shear modulus $\mu$, but it also increases with increasing normalised thickness $\bar{H}$ and decreases with radius $R_i$. Combining Eqs. (4) and (5), the distensibility can be written as

$$\delta \approx \frac{[1+\bar{H}]^2}{2\mu \bar{H}[2+\bar{H}]}. \tag{6}$$

In other words, the observed variation in in vivo distensibility of vessels (e.g. along the arterial network) could be related to tissue properties and/or wall thickness/radius ratio. Critically, vessel thickness is highly challenging to measure in vivo; e.g. it is not visualised with commonly used MRI settings and it can be overestimated using ultrasound. Thus, there is a need to further improve our knowledge of the aortic in vivo thickness variation and how it relates to the variations in its radius/diameter and tissue stiffness. For example, could it be that thinner vessels compensate by increasing the fibrous components?

### 2.1.3 Ex Vivo Observations

In contrast to the in vivo observations, which are limited, there is a wealth of ex vivo studies and the observations reported from them. Arteries have a fibrous microstructure, which provides them nonlinear and anisotropic mechanical properties. Moreover, arterial tissue is multilayered, made up of intima, media, and adventitia layers (going from inside to outside) [10]. Importantly, the presence of residual stresses means that the "true" reference configuration is significantly different from in vivo. The residual stresses have been demonstrated through opening angle (O.A., $\omega$) and longitudinal stretch $\lambda_Z$. As a result, if we continue to assume axisymmetric deformation of a perfect cylinder, the mapping becomes [8]

$$\Omega_{\mathrm{R}} \rightarrow \Omega_{\mathrm{D}}: \quad r(R) = \sqrt{\frac{R^2 - R_i^2}{k_\omega \lambda_z} + r_i^2}, \tag{7}$$

where $k_\omega = 2\pi/(2\pi - \omega)$, and the deformation gradient becomes

$$\mathbf{F} = \operatorname{diag}\left[\lambda_r, \lambda_\theta, \lambda_Z\right] = \operatorname{diag}\left[\frac{R}{r k_\omega \lambda_Z}, \frac{r k_\omega}{R}, \lambda_Z\right]. \tag{8}$$

The pressure and deformation still follows the same relationship (3) but with modified stretches (8). It should be noted that $R_i$ and $H$ here represent the ex vivo radius and thickness, not in vivo, as the two can be significantly different.

In addition to the geometry, the mechanical response is also measured ex vivo, commonly using uniaxial, biaxial, and inflation-extension tests. Generally, the range of deformation in these ex vivo tests is much larger than in vivo, which provides a highly nonlinear response and a complex coupling of the deformation along different orientations. A large number of studies have investigated the arterial ex vivo properties, and several constitutive models have been proposed to describe the tissue behaviour. Amongst hyperelastic models, the following type is commonly used for describing the arterial tissues:

$$\Psi(\mathbf{F}) = \frac{\mu}{2}\left[I_1 - 3\right] + \sum_i \frac{k_1}{2k_2}\left[\exp\left(k_2 Q(I_1, I_{4i})\right) - 1\right], \tag{9}$$

where $I_1 = \operatorname{tr}\left(\mathbf{F}^\top \mathbf{F}\right)$ is the first strain invariant and $I_{4i} = \boldsymbol{M}_i \cdot \mathbf{F}^\top \mathbf{F} \boldsymbol{M}_i$ is the pseudo-invariant representing stretch squared along (fibre) direction $\boldsymbol{M}_i$. Different forms of the exponent function $Q(I_1, I_{4i})$ have been proposed:

$$Q(I_1, I_{4i}) = [I_{4i} - 1]^2 \qquad [10] \qquad (10a)$$

$$Q(I_1, I_{4i}) = \kappa\,[I_1 - 3]^2 + [1 - \kappa]\,[I_{4i} - 1]^2 \qquad [11] \qquad (10b)$$

$$Q(I_1, I_{4i}) = [\kappa\,[I_1 - 3] + [1 - 3\kappa]\,I_{4i} - 1]^2 \qquad [12]. \qquad (10c)$$

Here $\kappa$ is a dimensionless parameter representing the fibre dispersion.

In the ex vivo studies, a large variation in the properties, across artery location, age, and subjects, has been reported. For instance, the results from an extensive study detailing the properties of ascending and descending aorta in porcine samples [13] fitted to the model (9) with (10b) are summarised in Table 2. In this case, the tissue is assumed to be made up of two fibre families symmetrically arranged about the circumferential direction at angles $\pm\,\phi$, each with a direction vector $\boldsymbol{M}_i$. The variation in properties, in terms of both the standard deviation and also across layers and between ascending and descending aortas, is notable.

## 2.2 *Bridging Ex Vivo and In Vivo*

There are advantages and disadvantages to both in vivo and ex vivo observations. On one hand, in vivo observations can be made noninvasively and directly on the subject of interest (such as a patient for whom we are aiming to decide on the treatment). However, the range of deformation is small, and there are unknowns in terms of mechanical reference state. On the other hand, ex vivo observations provide more detailed picture of the nonlinear and anisotropic behaviour of the arteries. However, they are being observed in a nonfunctional environment and are often not from the subject of interest (such as the properties in Table 2 are from porcine aortas).

Due to the complementary nature of the two, it is not straightforward to use information from ex vivo studies in modelling in vivo situations. Bayes' theorem provides a framework to facilitate bridging the two and is described next.

### 2.2.1 Bayesian Framework

For two continuous random variables $u^{\mathcal{R}}$ and $v^{\mathcal{R}}$, let the joint prior probability density function be denoted by $\pi(u, v)$. Further, the prior marginal probability densities of $u^{\mathcal{R}}$ and $v^{\mathcal{R}}$ are denoted as $\pi(u)$ and $\pi(v)$, respectively. The posterior probability density of $u^{\mathcal{R}}$ given $v^{\mathcal{R}} = v$ (known as the conditional probability) is

**Table 2** Ex vivo properties of porcine aorta for intact and separated layers expressed as mean ± standard deviation (adopted from [13])

| | State | Ascending | Descending |
|---|---|---|---|
| $D_0$ [mm] | Intact | $11.25 \pm 1.3$ | $9.73 \pm 0.7$ |
| | Intima | $12 \pm 1.87$ | $9.93 \pm 0.93$ |
| | Media | $12.7 \pm 1.87$ | $10.79 \pm 1.06$ |
| | Adventitia | $13.74 \pm 2.46$ | $11.84 \pm 1.18$ |
| $H$ [mm] | Intact | $2.46 \pm 0.43$ | $1.57 \pm 0.2$ |
| | Intima | $0.29 \pm 0.04$ | $0.3 \pm 0.06$ |
| | Media | $1.46 \pm 0.27$ | $0.81 \pm 0.24$ |
| | Adventitia | $0.67 \pm 0.15$ | $0.55 \pm 0.03$ |
| O. A. [°] | Intact | $59.06 \pm 12.56$ | $74.43 \pm 14.80$ |
| | Intima | $228.96 \pm 80.06$ | $149.69 \pm 55.07$ |
| | Media | $64.47 \pm 22.62$ | $57.04 \pm 32.13$ |
| | Adventitia | $66.57 \pm 11.80$ | $111.99 \pm 42.65$ |
| $\lambda_Z$ [-] | Intact | $1.21 \pm 0.03$ | $1.43 \pm 0.07$ |
| | Intima | $1.19 \pm 0.06$ | $1.34 \pm 0.15$ |
| | Media | $1.26 \pm 0.11$ | $1.48 \pm 0.11$ |
| | Adventitia | $1.27 \pm 0.11$ | $1.44 \pm 0.08$ |
| $\mu$ [kPa] | Intact | $10 \pm 4$ | $10 \pm 5$ |
| | Intima | $9 \pm 8$ | $10 \pm 19$ |
| | Media | $13 \pm 2$ | $15 \pm 10$ |
| | Adventitia | $6 \pm 9$ | $5 \pm 12$ |
| $k_1$ [kPa] | Intact | $40 \pm 48$ | $67 \pm 56$ |
| | Intima | $52 \pm 89$ | $168 \pm 277$ |
| | Media | $34 \pm 19$ | $85 \pm 68$ |
| | Adventitia | $47 \pm 43$ | $128 \pm 146$ |
| $k_2$ [-] | Intact | $2.4 \pm 3$ | $10.1 \pm 8.6$ |
| | Intima | $1.8 \pm 1.4$ | $2.6 \pm 16.6$ |
| | Media | $1.2 \pm 0.3$ | $1.4 \pm 3.7$ |
| | Adventitia | $4.8 \pm 37.3$ | $4.3 \pm 20.3$ |
| $\kappa$ [-] | Intact | $0.112 \pm 0.204$ | $0.34 \pm 0.09$ |
| | Intima | $0.142 \pm 0.1745$ | $0.44 \pm 0.27$ |
| | Media | $0.565 \pm 0.25$ | $0.53 \pm 0.26$ |
| | Adventitia | $0.33 \pm 0.38$ | $0.46 \pm 0.17$ |
| $\phi$ [°] | Intact | $54.5 \pm 26.2$ | $71.6 \pm 10.3$ |
| | Intima | $51.1 \pm 20$ | $54 \pm 18.3$ |
| | Media | $74.4 \pm 12$ | $72.3 \pm 12.5$ |
| | Adventitia | $54.6 \pm 13.1$ | $66.1 \pm 19.8$ |

given by the Bayes' theorem:

$$\overbrace{\pi(u \mid v)}^{\text{Posterior}} = \frac{\overbrace{\pi(v \mid u)}^{\text{Likelihood}}\,\overbrace{\pi(u)}^{\text{Prior}}}{\underbrace{\pi(v)}_{\text{Normalization term}}}, \tag{11}$$

where $\pi(v \mid u)$ is the likelihood term. The denominator on the right-hand side is the normalisation term, i.e.

$$\pi(v) = \int \pi(u, v)\, du = \int \pi(v \mid u)\pi(u)\, du.$$

One approach of using Bayes theorem (11) for bridging ex vivo and in vivo observations is to: (1) first use ex vivo studies to construct a prior probability distribution of various quantities/parameters and (2) then use the in vivo properties as likelihood to obtain the posterior distribution. One could view the development of constitutive models as following the same steps: The models are developed/compared based on the ex vivo data before being used to simulate the in vivo situation, although without explicitly invoking the Bayes' theorem. We have recently proposed the use of Bayesian framework to compare constitutive models based on ex vivo data [14]. Taking a step further and employing the biomechanical models for in vivo deformation, such as Eq. (3), we can obtain the distribution of model parameters and other model outputs (such as stresses in the wall), based on in vivo observations (such as distensibility).

### 2.2.2 Methodology

Based on the ex vivo observations of ascending aortas (Table 2), we assume prior distributions as being independent and uniform in the range of reported mean ± twice the standard deviation values. These assumptions can be relaxed/refined if the full, detailed ex vivo data is available (for example, the parameters $k_1$ and $k_2$ are expected to be highly correlated [15]).

For any set of (assumed) parameters, the in vivo configuration can be simulated. We use the simplified axisymmetric model (3) to calculate the in vivo diameters at systole and diastole, $D_{\text{sys}}$ and $D_{\text{dias}}$, at pressures $P_{\text{sys}} = 120$ mm-Hg and $P_{\text{dias}} = 80$ mm-Hg, respectively. From the two diameters, the distensibility $\delta$ is calculated using definition in Table 1. Based on the in vivo observations in humans, we approximate the likelihood as

$$L(D_{\text{dias}}, \delta) = \mathcal{N}\left(D_{\text{dias}}^{\text{mean}} - D_{\text{dias}}, D_{\text{dias}}^{\text{sd}}\right)\mathcal{N}\left(\delta^{\text{mean}} - \delta, \delta^{\text{sd}}\right), \tag{12}$$

where

$$\mathcal{N}(m, \sigma) = \frac{1}{\sqrt{2\pi}\sigma} \exp\left(-\frac{m^2}{2\sigma^2}\right) \tag{13}$$

is based on the assumed Gaussian distribution of in vivo quantities. Specifically, $D_{\text{dias}}^{\text{mean}} = 30\,\text{mm}$, $D_{\text{dias}}^{\text{sd}} = 5\,\text{mm}$, $\delta^{\text{mean}} = 2.5 \times 10^{-3}\,\text{mm-Hg}^{-1}$, and $\delta^{\text{sd}} = 0.5 \times 10^{-3}\,\text{mm-Hg}^{-1}$. Similar to the ex vivo distributions, these likelihood functions can be refined if the full detailed data is available.

Two scenarios are modelled using the same approach: (1) artery is assumed to be made of a single, homogeneous layer, and (2) the artery is assumed to be made of three layers (intima, media, adventitia) each with different biomechanical properties. While the methodology remains the same, the second scenario entails three times as many model parameters, thus three times the associated uncertainty.

Using the above setup and the Bayes' theorem (11), a Markov-Chain Monte Carlo simulation is performed to sample the posterior distribution. In addition to the model parameters, the stress tensor across arterial wall is also recorded. From the stress tensor, the three normal stress components $\sigma_{rr}$, $\sigma_{\theta\theta}$, and $\sigma_{zz}$, as well as the first principal stress $\sigma_1$ and von-Misès stress $\sigma_{\text{vM}}$ are also extracted. From the samples, histograms, correlations, and variations of the parameters as well as the corresponding stress measures are constructed. Only correlations with absolute values above 0.05 are considered not insignificant. Variations are depicted in terms of the standard deviation.

### 2.2.3 Results

Histograms and cross-correlation of various arterial parameters are shown in Fig. 2. Remarkably, there is a large range of parameters that can match the expected in vivo observations. This signifies the uncertainty related to our biomechanical parameters. Moreover, the correlations are generally weak. Notably, the opening angle $\omega$ is not correlated to any other parameters. Thus, the coarse in vivo measurements are not able to establish any relations between these parameters.

The variations of stresses in the vessel wall at systole are shown in Fig. 3. There is a relatively small variation in the radial stress $\sigma_{rr}$; this is expected since the radial stress primarily depends on the applied pressure, which is assumed constant here. The variation in circumferential stress $\sigma_{\theta\theta}$ is considerable, while that in the longitudinal stress is massive. The variations increase even further when we consider the three-layered structure of the artery. The correlations of different stress measures with the model parameters as well as in vivo measurements are also listed in Fig. 3. Interestingly, the longitudinal stress is strongly correlated to fibre angle $\phi$, while the circumferential stress is strongly correlated to the diastolic diameter. Moreover, the correlations of various stress measures with distensibility $\delta$ are relatively small.

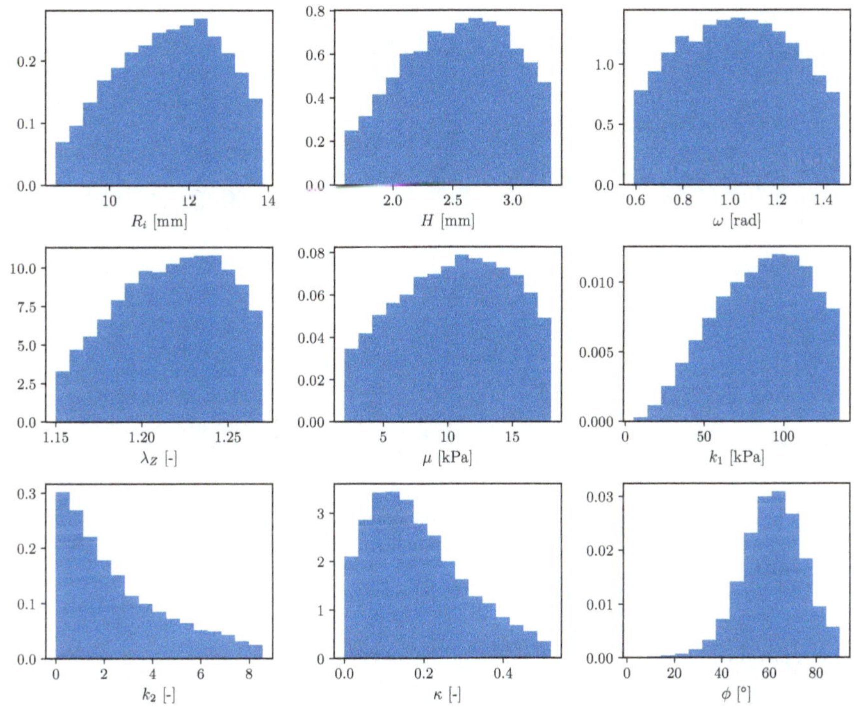

| | $R_i$ [mm] | $H$ [mm] | $\omega$ [rad] | $\lambda_Z$ [-] | $\mu$ [kPa] | $k_1$ [kPa] | $k_2$ [-] | $\kappa$ [-] | $\phi$ [°] |
|---|---|---|---|---|---|---|---|---|---|
| $R_i$ [mm] | | 0.07 | | 0.11 | | 0.22 | 0.12 | | -0.17 |
| $H$ [mm] | 0.07 | | | | | -0.11 | | 0.07 | |
| $\omega$ [rad] | | | | | | | | | |
| $\lambda_Z$ [-] | 0.11 | | | | | -0.09 | | 0.11 | 0.05 |
| $\mu$ [kPa] | | | | | | -0.08 | | | |
| $k_1$ [kPa] | 0.22 | -0.11 | | -0.09 | -0.08 | | 0.15 | -0.06 | 0.16 |
| $k_2$ [-] | 0.12 | | | | | 0.15 | | -0.31 | 0.23 |
| $\kappa$ [-] | | 0.07 | | 0.11 | | -0.06 | -0.31 | | -0.09 |
| $\phi$ [°] | -0.17 | | | 0.05 | | 0.16 | 0.23 | -0.09 | |

**Fig. 2** Artery parameters from Bayesian inference: (top) normalised histogram of each parameter and (bottom) their correlations (only the correlations above 0.05 or below $-0.05$ are shown)

## 2.3 *Discussion and Future Outlook*

Due to the variations in observed arterial properties, both in vivo and ex vivo, it is nontrivial to link them in a deterministic way. Bayes' theorem provides a systematic framework to bridge the ex vivo and in vivo observations, while also accounting for the uncertainties in both. Here, the framework was applied to the ascending aorta. The ex vivo properties were used as a prior knowledge. The in vivo deformation was approximated using a simplified model to obtain the diastolic and systolic states of the artery. The parameter values that matched the observed in vivo diameter and distensibility were sampled using MCMC algorithm.

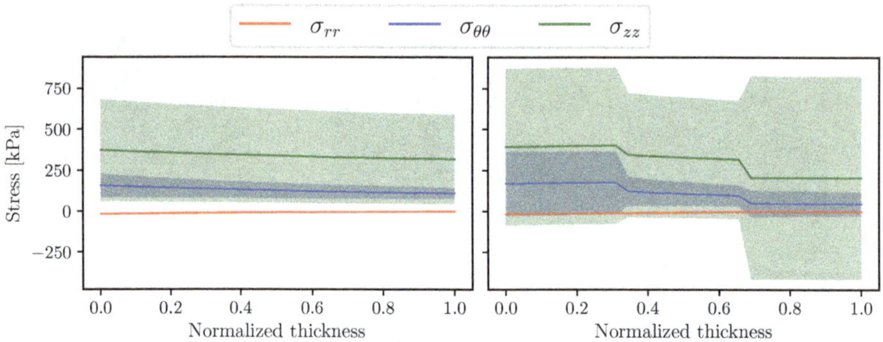


| | $R_i$ [mm] | $H$ [mm] | $\omega$ [rad] | $\lambda_Z$ [-] | $\mu$ [kPa] | $k_1$ [kPa] | $k_2$ [-] | $\kappa$ [-] | $\phi$ [°] | $D_{\text{dias}}$ | $\delta$ |
|---|---|---|---|---|---|---|---|---|---|---|---|
| $\sigma_{rr}$ | | | | | | | | -0.13 | | | 0.21 |
| $\sigma_{\theta\theta}$ | -0.07 | -0.47 | | -0.09 | | -0.30 | -0.33 | -0.13 | 0.30 | 0.77 | |
| $\sigma_{zz}$ | | -0.16 | | 0.22 | -0.05 | 0.16 | 0.50 | -0.33 | 0.58 | 0.13 | 0.09 |
| $\sigma_1$ | | -0.17 | | 0.21 | -0.05 | 0.12 | 0.48 | -0.32 | 0.56 | 0.16 | 0.09 |
| $\sigma_{\text{vM}}$ | | -0.17 | | 0.21 | -0.05 | 0.12 | 0.47 | -0.31 | 0.54 | 0.15 | 0.09 |

**Fig. 3** Stress distribution at systole in single-layered and three-layered artery (top) and correlation between maximum stress and various arterial parameters (below, only the correlations above 0.05 or below $-0.05$ are shown)

The results illustrate that large uncertainty exists in arterial mechanics. Due to the lack of detailed ex vivo data, we had to assume that there was no correlation between the biomechanical parameters a priori, i.e. the prior distributions were assumed to be independent. Moreover, the observed in vivo diameter and distensibility do not necessitate any significant correlation between the parameters either. Thus, there is a need for more detailed dataset, from which improved, nonindependent prior distributions and correlations can be inferred.

The stresses in the arterial tissue are considered as one of the main quantities of interest, since these are expected to play an important role in the arterial remodelling and degradation of tissue. However, with the dataset used, we see massive uncertainty in the longitudinal stress, which gets even bigger when we consider multilayered model. Importantly, these variations do not show any significant reduction if the in vivo observations are not known more accurately (i.e. with a smaller standard deviation used in the likelihood). Thus, to gain higher confidence in the stresses, we need to either improve our knowledge of ex vivo observations or add more in vivo observations. For example, could the in vivo longitudinal stretch or thickness be measured accurately?

Lastly, the choice of model should be carefully considered. Here, even with the simplest model of a uniform cylinder, we found large uncertainties, which further increased by simply adding multilayered structure. The recent advances in in vivo image-based analysis can help more realism to our models. While it is tempting to increase the model complexity to include detailed three-dimensional shape characteristics using a finite element model, it might add even further uncertainty. Thus, the model complexity must be balanced with more data, such as detailed in vivo strain estimates [16]. Similarly, the question of constitutive model development and selection remains an active area of research [14]. Data-based models are gaining particular attention, which provide better fidelity to the observed ex vivo datasets [17]. While these advanced approaches provide an advantage, the associated uncertainty needs to be accounted for, by applying them to large datasets.

## 3 Small Cerebral Arteries and Their Pressure and Flow Regulation

Small arteries and arterioles play a fundamental role in the regulation of blood flow and pressure in many organs such as heart, brain, stomach, and kidney. Due to their small diameter (50–300 $\mu$m), these resistance vessels are responsible for the major blood pressure drop across the circulatory system. In line with the Hagen-Poiseuille equation, the pressure drop across a vessel is inversely proportional to the fourth power of its radius. Thus, a tiny variation in diameter of one of these vessels could lead, if not controlled, to substantial flow and pressure changes in both upstream and downstream vascular territories. Diameter in small blood vessels is regulated through a combination of local and systemic control mechanisms that operate muscular apparatus made of smooth muscle cells (SMCs). This enables blood vessels to develop tone across their wall thickness and ultimately to adjust their inner diameter upon different mechanical stimuli. From an hierarchical point of view, pressure-induced/myogenic tone is considered to be the lowest key control mechanism of luminal diameter as it directly depends on the local pressure level. This mechanism enables the vessel to limit diameter variations and stabilise flow in face of significant pressure changes by developing tone within its muscle layer. This means that for an upstream pressure increase (decrease), the vessel will reduce (expand) its diameter to keep the volumetric flow relatively constant. The presence of basal vascular tone/contractility is also essential for tissue metabolic control, which operates by relaxing the vasculature and favouring blood flow to the regions where there is an increased metabolic activity [18].

Since tone regulation dysfunction is associated with the development of different pathological conditions (such hypertension, stroke, heart failure, and others), there is a pressing need to accelerate our understanding on the causative links between the vascular components constituting the contractile machinery. Despite some recent attempts [8, 19, 20], the modelling of these key regulatory mechanisms is still at its infant stage and requires further development and validation. In the following we discuss our recently proposed modelling strategy [5] that aims to recapitulate the key processes involved in pressure-induced tone development in small cerebral vessels. Although this framework was developed by considering only experimental data from rat middle cerebral arteries, we expect that this can also describe, with a reasonable level of accuracy, regulatory mechanisms governing other cerebral vessels (such as pial arteries) in rats as well as in humans. Vascular tone development and resulting contractility are captured through two sub-models: a signalling network that describes the cellular pathways that are activated upon mechanical loading and a multi-scale model that integrates the contractile units mechanics into tissue tone and deformation. The key equations and assumptions of the proposed methodology are reported below, but we recommend to see [5] for more details on the modelling assumptions and solution procedure.

## 3.1 Intracellular Signalling

When smooth muscle cells are subjected to normal loading, different cascades of intracellular processes are activated. Among these, the pathways involving ion $Ca^{2+}$, Rho-kinase/ROCK and Protein Kinase C (PKC) appear to be the major contributors to tone development. Some authors [21] suggest that these pressure-induced pathways are not sequentially activated, but they can independently and simultaneously contribute to tone generation across a wide and overlapping pressure load ranges. Figure 4 summarises the key cellular pathways that contribute to vascular tone development. The pressure-induced whole cell signalling is represented through a system of ordinary differential equations, which was devised by following the logic-based approach by Irons and Humphrey [22]. The nodes of the network are the independent variables that can represent either the concentration/content of an ion/protein or the phosphorylation level of a protein. The processes characterising the interactions between nodes are described through a set of logistic functions ($\chi_0$, $\chi_1$, $\chi_2$, $\chi_3$, $\chi_4$, $\chi_5$, $\chi_6$, $\chi_7$, $\chi_8$). The adopted logic-based formalism leads to the following system of ordinary differential equations:

$$\frac{d\xi_0}{dt} = \frac{1}{\tau_0}\left[\chi_0(\bar{P}) - \xi_0\right], \tag{14a}$$

$$\frac{d\xi_1}{dt} = \frac{1}{\tau_1}\left[\chi_1(\bar{P}) - \xi_1\right], \tag{14b}$$

$$\frac{d\xi_2}{dt} = \frac{1}{\tau_2}\left[\chi_2(\bar{P}) - \xi_2\right], \tag{14c}$$

$$\frac{d\xi_3}{dt} = \frac{1}{\tau_3}\left[\chi_3(\xi_1) - \xi_3\right], \tag{14d}$$

$$\frac{d\xi_4}{dt} = \frac{1}{\tau_4}\left[\chi_4(\xi_1) - \xi_4\right], \tag{14e}$$

$$\frac{d\xi_5}{dt} = \frac{1}{\tau_5}\{\chi_5(\xi_0)\,[1 - \xi_5] - [1 - \chi_6(\xi_3)]\,\xi_5\}, \tag{14f}$$

$$\frac{d\xi_6}{dt} = \frac{1}{\tau_6}\{[1 - \chi_7(\xi_4)] + [1 - \chi_8(\xi_2)] - [1 - \chi_7(\xi_4)]\,[1 - \chi_8(\xi_2)] - \xi_6\}. \tag{14g}$$

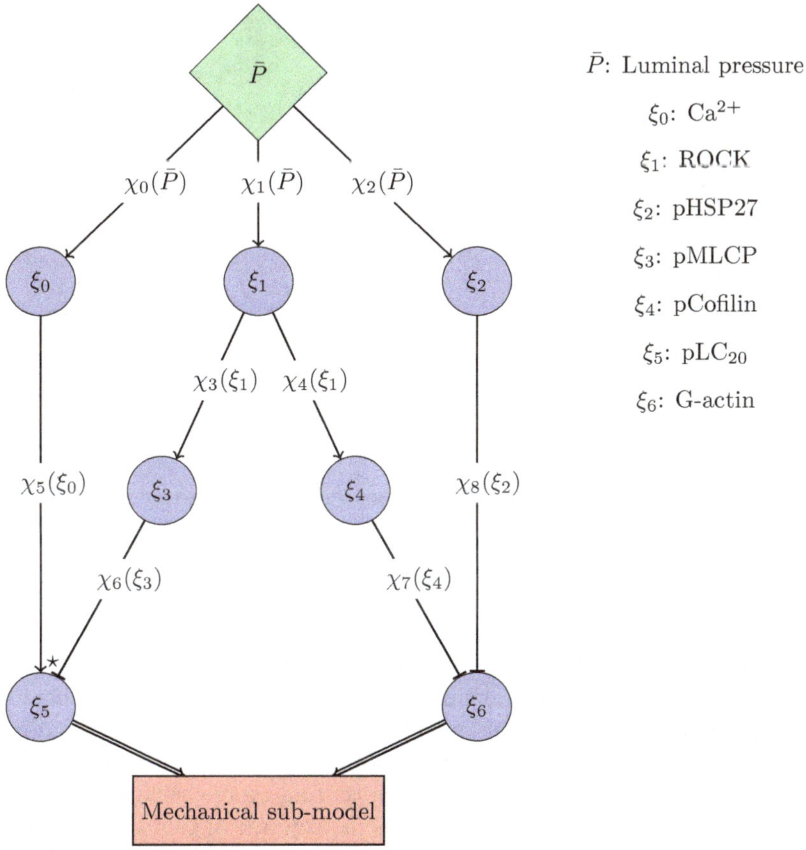


**Fig. 4** Signalling network proposed in [5] for describing pressure-induced cellular pathways contributing to tone development. The considered intracellular variables for this network are the normalised values (with respect to their maximum) of $Ca^{2+}$ concentration ($\xi_0$), ROCK activity level ($\xi_1$), HSP27 phosphorylation level ($\xi_2$), MLCP phosphorylation level ($\xi_3$), Cofilin phosphorylation level ($\xi_4$), $LC_{20}$ phosphorylation level ($\xi_5$), and G-actin content ($\xi_6$). The symbols $\leftarrow$ and $\vdash$ represent activation and inhibition processes, respectively. The symbol $\star$ indicates that the dynamics of $\xi_5$ is described through an ad hoc equation that does not follow the logic-based rules of the remaining part of the network, whilst $\Leftarrow$ implies that the associated variable serves as an input for the mechanical sub-model

In [5] we identified the coefficients of the nine logistic functions ($\chi_0$,...,$\chi_8$) by fitting the model results against measured intracellular quantities for a wide range of pressures. Here we report how the variables describing the signalling network respond to pressure variations over time (Fig. 5). We consider the intracellular signalling under control conditions and with (partial) inhibition of the $Ca^{2+}$ pathway ($\xi_0$ set constant equal to 0.1 irrespective of the pressure level). The latter condition is generally achieved in laboratory by removing the $Ca^{2+}$ ions from the extracellular space.

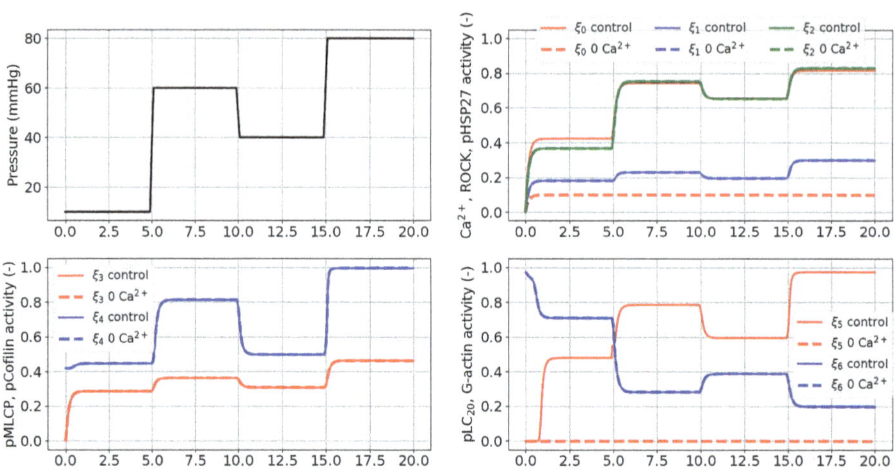


**Fig. 5** Dynamic response of intracellular variables to luminal pressure changes for the control and 0 extracellular $Ca^{2+}$ cases. The time step is set 0.1 s whilst all the other model parameters are taken as in [5]

With the exception of the level G-actin content ($\xi_6$), all variables increased and decreased in line with pressure variation. The variable representing the level of Cofilin phosphorylation ($\xi_4$) exhibits the most significant variations across the considered pressure range, but this might be different if we considered a different pressure level. Upon extracellular $Ca^{2+}$ removal, the level of phosphorylated $LC_{20}$ is nullified, despite the ROCK pathway is intact (indeed the behaviour of $\xi_4$ is the same as in the control case). This highlights the key role of $LC_{20}$ phosphorylation level ($\xi_5$) on SMC tone development. Indeed, the described intracellular dynamics converts mechanical stimuli into changes of actin-myosin activity (producing power stroke) and cytoskeleton stiffness, represented by G-actin content $\xi_6$. In the proposed framework, these intracellular quantities serve as input for the multi-scale mechanical model for tone development.

### 3.2 *Tissue Tone Development*

The intracellular signalling activates the SMC contractile machinery, which in turns drives tissue tone generation and eventual deformation. This is achieved by coupling the chemical model described above with the mechanical model proposed by Murtada et al. [23]. The latter turns sub-cellular mechanical interactions into an emergent response at tissue level through an active strain energy function.

### 3.2.1 Sub-Cellular Mechanics

The contractile machinery accounts for actin-myosin interaction leading to cross-bridges (XBs) formation within contractile units (CUs) and cytoskeleton contribution. Passive surrounding elements constituting the actin cortex are connected with contractile units, arranged in series to form tissue contractile fibres (CFs). In the proposed methodology it is assumed that only phosphorylated cross-bridges contribute to active tone generation. The total stiffness ($k_{tCU}$) associated with a number of CUs ($N_{CU}$) is directly related to the level of $LC_{20}$ phosphorylation via

$$k_{tCU} = \frac{\bar{L}_{fo} L_m \xi_5 n_{XBmax} k_{XB}}{2\delta_m N_{CU}}, \tag{15}$$

where $L_m$ is the average length of myosin filaments, $n_{XBmax}$ is the maximum phosphorylation rate, $\delta_m$ is the average distance between myosin monomers heads, $k_{XB}$ represents the elastic stiffness of the cross-bridge, and $\bar{L}_{fo}$ is a normalised function representing the overlap between thin and thick filaments. The stiffness of the actin cortex ($k_{AC}$) is evaluated as a function of the F-actin content level

$$k_{AC} = k_{ACmax} \frac{\xi_7^{n_{AC}}}{\xi_7^{n_{AC}} + K_{AC}^{n_{AC}}}, \tag{16}$$

where $k_{ACmax}$ is the maximum stiffness under loading conditions, whilst $n_{AC}$ and $K_{AC}$ are the coefficients of the associated activation function. The circumferential stretch ($\lambda_\theta$) can be written as

$$\lambda_\theta = 1 + 2N_{CU}(\bar{u}_{CB} + \bar{u}_{fs}) + 2\bar{u}_{FA}, \tag{17}$$

where $\bar{u}_{CB}$, $\bar{u}_{fs}$, and $\bar{u}_{FA}$ are the normalised (with respect to the SMC length) average XB elongation, the relative filament sliding, and the average surrounding passive elements elongation, respectively. The resulting reaction force due to the resistance from $N_{CU}$ in series with an F-actin element at each extremity is given as total displacement by the resulting stiffness of the system:

$$F_a = (\lambda_\theta - 1 - 2N_{CU}\bar{u}_{fs}) \frac{k_{tCU} k_{AC}}{2k_{tCU} + k_{AC}}, \tag{18}$$

whilst the average driving force generated from the cycling XBs results

$$F_c = \bar{L}_{fo} \frac{L_m}{\delta_m} \xi_5 n_{XBmax} k_{XB} u_{PS}, \tag{19}$$

where $u_{\mathrm{PS}}$ is the average displacement associated to power stroke. The kinetics of the filament sliding induced by the XB cycling is described via

$$\frac{d\bar{u}_{\mathrm{fs}}}{dt} = \beta_{\mathrm{c}}(F_{\mathrm{a}} - F_{\mathrm{c}}), \tag{20}$$

where $\beta_{\mathrm{c}}$ is the time constant associated to the filament sliding process.

### 3.2.2 Tissue Mechanical Model Component

The generated tone from CFs can be quantified through the average first Piola-Kirchhoff stress $P_{\mathrm{a}}$ of the vascular tissue, $P_{\mathrm{a}} = N_{\mathrm{CF}} F_{\mathrm{a}}$, where $N_{\mathrm{CF}}$ is the surface density of the contractile fibres. The considered vascular tissue, consisting of different SMC layers, is modelled as an axisymmetric homogeneous hyperelastic thick-walled tube as described in Eq. (3) with a mapping given in Eq. (7). The strain energy density function $\Psi$ is now decomposed into active and passive components $\Psi = \Psi_{\mathrm{a}} + \Psi_{\mathrm{p}}$, with the former evaluated as

$$\Psi_{\mathrm{a}} = \int P_{\mathrm{a}}\, d\lambda_{\theta} = \frac{N_{\mathrm{CF}}}{2} \frac{k_{\mathrm{tCU}} k_{\mathrm{AC}}}{2k_{\mathrm{tCU}} + k_{\mathrm{AC}}} \left[\lambda_{\theta} - 1 - 2N_{\mathrm{CU}}\bar{u}_{\mathrm{fs}}\right]^2 . \tag{21}$$

The Holzapfel-Gasser-Ogden model [24] (Eq. (9) with (10a)) is used for representing the passive behaviour $\Psi_{\mathrm{p}}$.

Figure 6 shows how the pressure-induced intracellular signalling can affect the vessel geometry as well as the CU dynamics for both control and 0 extracellular $Ca^{2+}$ cases. As expected, the control (active) and 0 extracellular $Ca^{2+}$ (passive) cases present a different diameter for the pressure (60 mmHg). A decrease in pressure (to 40 mmHg) caused a significant diameter change in the passive case, whereas for the active case the variation is much more limited. When the vessel is inflated at high pressure (80 mmHg), the difference in behaviour is even more evident: The control vessel further reduces its diameter, whilst the counterpart expands to a size that is bigger than the initial one. On the other hand, filaments sliding for the 0 extracellular $Ca^{2+}$ case are almost nullified, whilst the control case is associated with negative values, indicating that filaments shorten their relative distance. Together, these results give an informative picture on how intraluminal pressure is converted into SMC signalling that can alter the vascular tissue tone. We expect that this regulatory effect will be even more evident when integrated into blood flow simulations across vessel networks. Therefore, future modelling studies assessing the impact of the myogenic control on the haemodynamics across vascular networks are warranted.

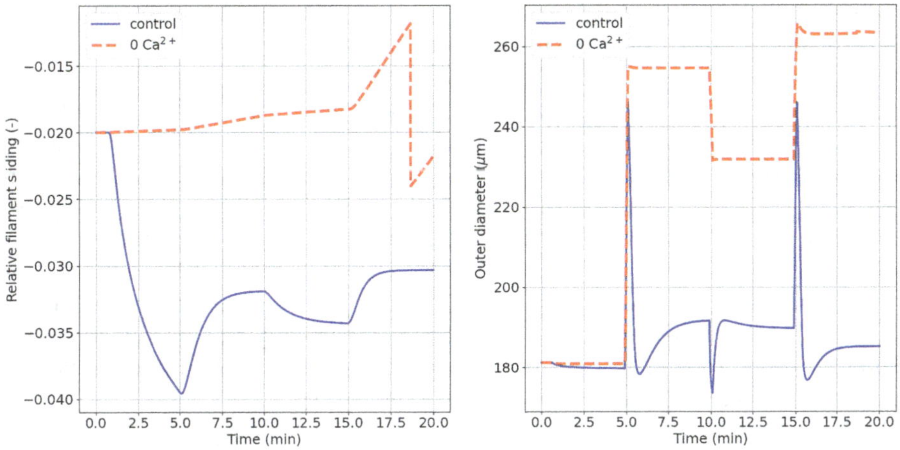


**Fig. 6** Variation of relative filament sliding and vessel diameter in time for different pressure levels for the control and 0 extracellular $Ca^{2+}$ cases. Model parameters were taken as in [5], whilst the load-free radius was $R_o = 0.1$ mm

# 4 Coronary Vessels and the Effect of Drugs

## *4.1 Background*

The coronary vasculature fulfils the primary role of supplying oxygenated blood to the heart muscle. If the supply is disrupted, ischaemia may occur, a condition where the heart muscle can become deprived of oxygen. Angina (chest pain) is the most common side effect of ischaemia. However, in the most serious cases, myocardial infarction (MI, also known as a heart attack) can occur. The main cause is atherosclerosis—a build-up of plaque that narrows the lumen of coronary arteries and restricts blood flow. Atherosclerosis is a complex progressive disease that develops over a number of years, with key risk factors including unhealthy diet, smoking, alcohol intake, and hereditary factors. Given the high prevalence of the disease, and potential severity of the outcome for patients, there has been an explosion of research over the past few decades to better understand both the disease progression and treatment. This section focusses on the latter with particular emphasis on computational approaches to improve treatment of obstructive coronary artery disease.

Percutaneous coronary intervention (PCI) is the most common clinical procedure undertaken to relieve the symptoms of ischaemic heart disease, by improving blood flow in coronary arteries. Nowadays, PCI typically involves the minimally invasive implantation of a scaffolding device (a stent) to widen the artery lumen and restore blood flow. These devices have evolved over the years from permanent bare metal stents (BMS) to drug-eluting stents (DES), with the incorporation of antiproliferative drug becoming a revolutionary step that reduced the incidence of the main complication (renarrowing known as restenosis) from around 30% to 10%. While the success of DES is unquestionable, stubborn repeat revascularisation rates have driven the development of alternative therapies, with a primary aim of "leaving nothing behind". These include bioresorbable scaffolds (BVS), similar to DES, but whose structure is entirely resorbed over a period of time and, most recently, drug-coated balloons (DCB) where a temporary inflation of a balloon at high pressure widens the artery and delivers drug simultaneously. A comprehensive description of the evolution of endovascular devices in the context of coronary artery disease treatment is provided in [25].

Clearly, there are many aspects of PCI that are amenable to computational modelling, with the aim of better understanding the underlying physical, chemical, and biological processes in pursuit of optimal device design and/or optimisation of procedural aspects. These primarily relate to studies of:

- Design of stents focussing on improving mechanical properties for deliverability and biocompatibility
- Altered arterial fluid dynamics in the presence of stent implants, to better understand the role of shear stress and other related fluid dynamic quantities in disease progression and onset of complications
- Device deployment, tissue damage, growth and remodelling, to enable prediction of the time course and severity of restenosis
- Drug release from endovascular devices and subsequent transport and retention in the arterial wall, with a view to optimising the drug type, dose, and release rate.

These aspects have been studied both in isolation and in tandem, but a fully comprehensive modelling framework is still elusive. This section focusses specifically on the role of drug delivery.

## 4.2 The Role of Antiproliferative Drugs

Deployment of stent and balloon devices injures the arterial wall and initiates an inflammatory response (Fig. 7). Most notably, damage (often denudation) of the endothelial layer of cells lining the vessel exposes the arterial wall to circulating growth factors which promote differentiation of the key cellular component—smooth muscle cells (SMCs)—from a contractile to a proliferative synthetic phenotype. Through proliferation, deposition of extracellular matrix (ECM), and migration towards the lumen, tissue growth ensues until such times as the endothelium has recovered. Excessive growth leads to restenosis, with the potential requirement for reintervention.

The most common compounds used on DES and DCB are paclitaxel (PTX) and sirolimus (SIR) or analogues such as zotarolimus, everolimus, and biolimus. Notably, the modes of action of PTX and SIR are different. Specifically, they exert their effect with greater affinity at different checkpoints through the cell cycle

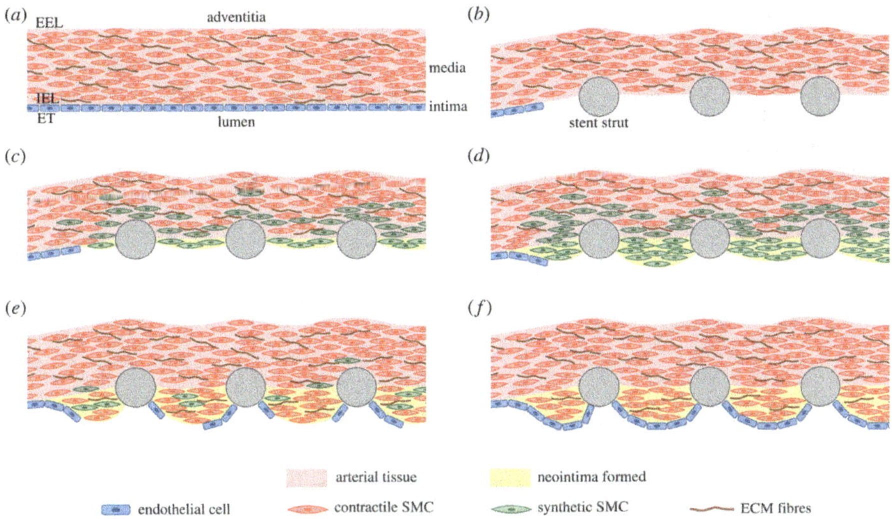


**Fig. 7** Schematic of arterial damage and healing process following stent deployment. (**a**) Healthy artery with intact endothelium and media layer populated by contractile, quiescent smooth muscle cells (SMCs). (**b**) Endothelial injury and denudation following stent deployment. (**c**) Vascular injury exposes the arterial wall to circulating growth factors which promote differentiation of cells to their active, synthetic phenotype. (**d**) SMC proliferation, ECM deposition, and cellular migration lead to tissue growth. (**e**) As endothelium recovers, SMCs begin differentiating back to their dormant phenotype. (**f**) Vessel healed with fully regenerated endothelial layer. EEL: external elastic lamina. IEL: internal elastic lamina. ET: endothelium. Image reproduced with permission from [26]

(Fig. 8). While both drugs are known to enter cells via passive diffusion, PTX binds to the $\beta$ subunit of tubulin, stabilising microtubules, whereas SIR binds to a protein named FKBP12. In each case, inhibition of SMC proliferation is achieved in a dose-dependent manner, meaning that the rate of drug delivery and the exposure concentration at the cell surface play an important role in the effectiveness of the therapy. These aspects are amenable to mathematical and computational modelling.

## 4.3 *The Need for Heterogeneous and Multi-Scale Modelling*

In many ways, the modelling methodologies and level of detail required to understand and optimise drug delivery to the arterial wall really depend on the question(s) to be addressed. But before one begins to consider this, it is essential to have a basic understanding of the structure of the arterial wall (Fig. 9) and of the different spatial and temporal scales at play. At the spatial scale of the tissue, the healthy arterial wall displays heterogeneity in the sense that it is a multi-layer structure with three distinct layers: the intima containing the endothelial layer of

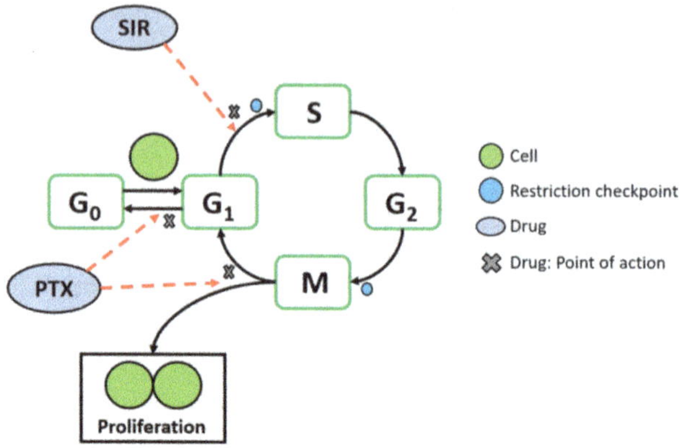


**Fig. 8** Illustration of mode of action of sirolimus (SIR) and paclitaxel (PTX). Obtained from McQueen et al. [27] under Creative Commons licence https://creativecommons.org/licenses/by/4.0/

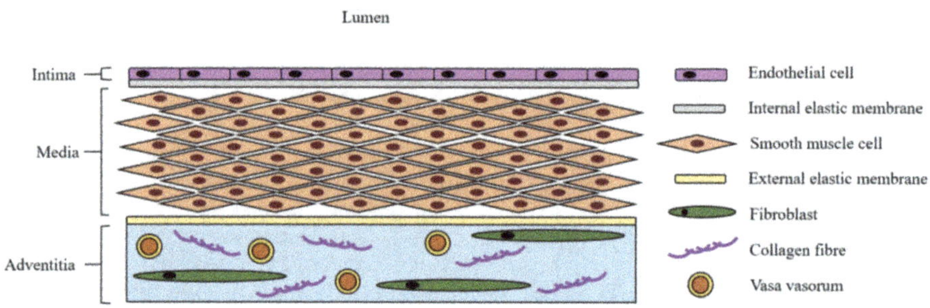


**Fig. 9** Illustration of the structure of the healthy coronary arterial wall. Acquired from McQueen [28] with permission

cells, the media primarily composing SMCs, collagen and elastin and the adventitia composed of fibroblast cells, collagen fibres and a vascular network named the *vasa vasorum*. The arterial wall is porous with voids filled with plasma driven by a pressure gradient from the inner lumen to the outer wall. The picture in disease is quite different and is addressed in Sect. 4.7. At this tissue scale, bulk transfer phenomena such as advection and diffusion play an important role in the distribution of dissolved drug. However, the ultimate goal of the therapy is to delivery drug to interrupt cellular proliferation, which happens on the scale of an individual cell. Yet, the main complication of the therapy—restenosis—is quantified on the tissue scale as a bulk reduction in the lumen of the vessel. There are also vastly different temporal scales to consider—DES typically deliver their payload over a period of weeks to months, DCBs are in contact with the arterial wall for around 1 minute, drug binding processes and cellular processes such as proliferation and migration happen over sub-seconds to hours, while restenosis typically occurs over a period of months to years. From a modelling perspective, these processes have been roughly grouped into one of three categories, focussing on either (i) drug delivery, (ii) drug transport, and retention in tissue or (iii) the role of drug in modulating tissue growth.

### 4.4 Mathematical and Computational Modelling of Drug Delivery from DES and DCB

There is a large literature on mathematical modelling of drug delivery from DES, dating back over 20 years [29]. The primary aim of these models is to understand how varying design aspects influence the time course and quantity of drug delivered (known as the drug release profile) with a view to optimising the drug release profile. Typically, these models assume sink conditions, reflecting common in vitro testing of these devices where drug release is studied in vials or containers with a large volume of fluid, often stirred to homogenise the drug concentration. This, coupled with the assumption of uniform initial drug concentration, often enables a one-dimensional model to be justified, with analytical solutions obtainable in some cases (e.g. [30]).

The vast majority of models consider durable polymer coatings on the devices containing drug at some uniform initial concentration. Such models are based on linear diffusion:

$$\frac{\partial c}{\partial t} = D_p \nabla^2 c, \quad \text{in } \Omega_p, \quad t > 0, \tag{22}$$

where $c$ is the volume averaged concentration of drug in the polymer domain $\Omega_p$, $D_p$ is the effective drug diffusion coefficient in the polymer, assumed to be isotropic and constant, and $t$ is time. Typical boundary conditions are zero flux at the inner surface in contact with the stent struts and sink conditions ($c = 0$) at the outer surface. Bi-phasic diffusion models have also been considered, where two distinct routes for drug out of the coating (a fast and a slow route represented by different diffusion coefficients and initial conditions) are assumed as a result of the underlying microstructure [31, 32]. Following solution of these models, drug release profiles can be calculated as follows:

$$M_{rel}(t) = M_0 - \int_{\Omega_p} c \, d\Omega_p, \tag{23}$$

where $M_0$ is the initial drug mass loaded on the device. These profiles are typically normalised by $M_0$ and expressed as a fraction or percentage. Such models have been shown to accurately capture drug release from commercial DES and prototype DES that have undergone pre-clinical trial (Fig. 10). Extensions have introduced dissolution mechanisms to account for slow release in the case of poorly soluble drugs. Concerns regarding the biocompatibility of the polymer used on original DES drove the development of biodegradable polymer coatings and polymer-free DES, each requiring more sophisticated models. In the case of biodegradable polymers, models of drug release tend to be based on coupled nonlinear reaction-diffusion equations describing hydrolysis and evolution of bi-products and transport properties (see, e.g. [33]). Polymer-free stents typically require drug to be loaded in solid form, so that dissolution can be used to slow the release. Models describing

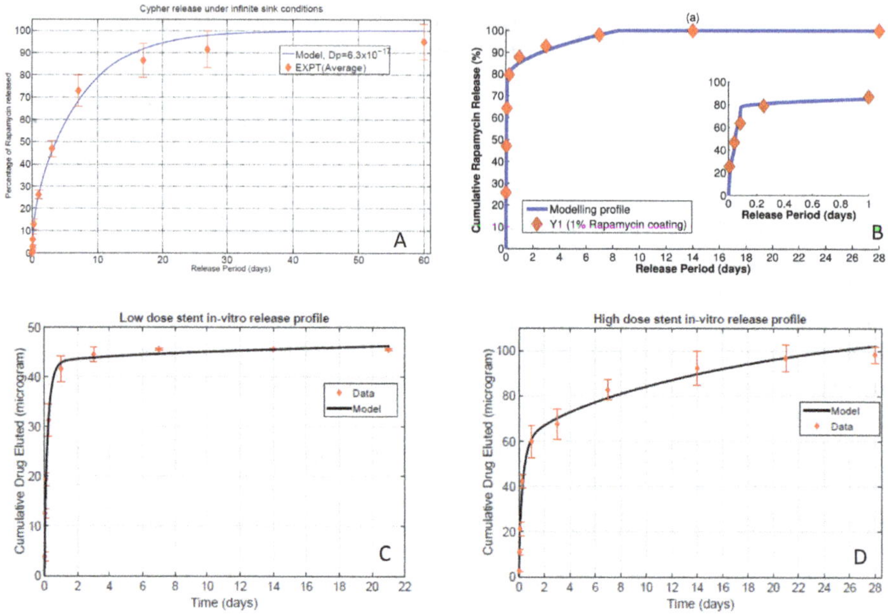


**Fig. 10** Comparison between diffusion-based drug release models and in vitro experimental data of drug release from stents. (**a**) Simple diffusion model captures rapamycin release data from the commercial Cypher DES (based on data reported in [30]). (**b**) Dissolution-diffusion moving-boundary model describes rapamycin release from a polymer-free stent (obtained from Vo et al. [34] under Creative Commons licence https://creativecommons.org/licenses/by/4.0/). (**c**)–(**d**) Bimodal diffusion model captures drug release data from prototype low dose (**c**) and high dose (**d**) sirolimus-coated stents (obtained from McKittrick et al. [31] under Creative Commons licence https://creativecommons.org/licenses/by/4.0/). Note that rapamycin is an alternative name for sirolimus

these systems, therefore, have tended to be based on moving-boundary problems, where a moving front is introduced to separate dissolved from undissolved drugs [34, 35]. By comparison, the literature on modelling drug release from DCBs is significantly smaller. However, the increasing clinical and commercial interest in these devices is driving a resurgence in research and development. To date, drug release from DCBs has typically been modelled as a constant source for the application period, or via a flux condition derived from experiential testing [36].

## *4.5 Mathematical and Computational Modelling of Drug Transport in Arterial Tissue*

While having an understanding of the rate of drug delivery from a device is clearly important, a key limitation of the aforementioned models is their assumption of sink conditions and a lack of coupling with the arterial environment. Thus, they do not address the key questions of how drug distributes in tissue and what spatial concentrations are observed over what time course. It is evident that dissolved drug will traverse the arterial wall as a result of various transport phenomena, most notably diffusion and advection, and that the complex arterial structure dictates that the nature of these will vary on a layer-specific basis. However, early attempts to model drug delivery to arterial tissue made a number of simplifying assumptions. For example, the wall was modelled as a continuum, assumed to be a single-layer medium with a focus on transport in the radial direction, enabling lower-dimensional models with constant and isotropic diffusion coefficients governing the speed of drug transport [30]. Subsequent efforts uncovered the potentially important role of advection due to plasma flow, concluding that the balance of advection and diffusion dictates the spatiotemporal transport of drug within tissue [37]. Experimental evidence indicating preferential diffusion in circumferential and axial directions compared with the radial direction turned attention towards anisotropic diffusion, necessitating higher dimensional models, while data indicating varying diffusion coefficients within each layer paved the way for multi-layer models [38]. Thus layer-specific advection-diffusion equations of the following form are now typically employed [39, 40]:

$$\frac{\partial c_i}{\partial t} + \frac{\gamma_i}{\phi_i}\mathbf{u}_i \cdot \nabla c_i = \nabla \cdot (\mathbf{D}_i \nabla c_i), \quad \text{in } \Omega_i, \quad i = \{\text{intima, media, adventitia}\}, \tag{24}$$

with $\mathbf{D}_i$ anisotropic in general and $\gamma_i$, $\phi_i$ denoting the hindrance coefficient and porosity, respectively. The fluid velocity field, $\mathbf{u}_i$, is typically computed through consideration of Darcy's law and the continuity equation:

$$\mathbf{u}_i = -\frac{K_i}{\mu}\nabla p_i, \quad \nabla \cdot \mathbf{u}_i = 0, \quad \text{in } \Omega_i, \quad i = \{\text{intima, media, adventitia}\}, \tag{25}$$

where $K_i$ is the Darcian permeability, $\mu$ is the dynamic viscosity of plasma, and $p_i$ is the pressure field. Equations (24)–(25) must be supplemented with initial conditions ($c_i = 0$) and interface conditions between and at the extremities of the layers. These typically take the form of Kedem-Katchalsky equations [41] with varying simplifications, depending on the complexity of the model.

### 4.5.1 Mechanical Influences on Drug Transport in Arterial Tissue

It is important to note that the deployment of DES and DCB can have a long lasting or temporary influence on arterial drug transport properties. For example, device deployment mechanically deforms the porous arterial tissue, leading to changes in tortuosity and therefore changes in effective diffusion coefficients. An initial effective diffusion coefficient $D^0$ should be modified to $D$ after compression, via the expression [40]:

$$D = D^0 [1 + \varepsilon], \tag{26}$$

where $\varepsilon$ is the strain induced through stent deployment and may be calculated through a stent deployment simulation, with appropriate mechanical properties assigned to the stent and the arterial tissue. Nowadays, healthy arterial tissue is typically modelled as an anisotropic hyperelastic and incompressible material, reinforced by two families of collagen fibres. For example, Escuer et al. [40] described the response of the arterial wall using the strain energy density function proposed by Gasser [42], and fitted parameters to experimental data in [43] on a layer-specific basis (Eq. (9) with (10c)).

### 4.5.2 The Crucial Role of Drug Binding

While advection-diffusion models are capable of describing the spatiotemporal distribution of drug, they are unable to account for prolonged retention of drug in tissue. Experimental studies have uncovered that the physicochemical properties of drugs have a strong influence on arterial drug transport and retention [44]. Specifically, lipophilic drugs like PTX and SIR are strongly retained within tissue, while hydrophilic drugs like heparin are rapidly cleared through. These findings indicated that existing advection-diffusion models were inadequate and that more complex models that incorporated drug retention were required to better understand and capture drug retention within tissue. This led to the development of advection-diffusion-reaction models of varying complexity. Initially, reactions were described through the inclusion of linear "sink" terms in equations of the form (24), but evolved to incorporate linear first-order reactions [30], often approximated through equilibrium partition coefficients, which can be estimated experimentally via studies of drug uptake into bulk tissue. Through greater appreciation of the mode of action of the drugs, models have evolved to consider drug binding reactions, introducing reversible linear kinetics and, most recently, the nonlinear saturable nature of drug binding. Nowadays, the state-of-the-art models incorporate two distinct phases of bound drug (specific and non-specific), which appear as "sink terms" in the governing equations (24) [31, 32]:

$$\frac{\partial b^j}{\partial t} = k^j_{on} c_{\text{media}} \left[ B^j - b^j \right] - k^j_{off}, \quad \text{in } \Omega_{\text{media}}, \quad j = \{\text{specific, non-specific}\}, \tag{27}$$

where $b^j$ refers to the concentration of drug bound within each phase, $B^j$ the local density of binding sites, and $k_{on}$, $k_{off}$ denoting the binding-on and binding-off rates, respectively. These equations are typically applied to only the media layer, although binding of both types likely occurs throughout the arterial wall. The specific nature of drug binding refers to drug molecules associating with target receptors within cells, while the non-specific component accounts for drug binding to general ECM sites. The former has been strongly linked to efficacy of the therapy [31, 32].

### 4.5.3 Key Drug Performance Indices

The literature indicates that saturation of SMC specific receptors ($RS$) with drug for the duration of healing period is desired for maximum efficacy [32]. $RS$ is therefore a key index that may be calculated from models but is impractical to measure in vivo:

$$RS = \int_{\Omega_{\mathrm{media}}} \frac{b^{\mathrm{specific}}}{B^{\mathrm{specific}}} \, d\Omega_{\mathrm{media}}. \tag{28}$$

However, simply ramping up the drug dose to maximise $RS$ is unlikely to yield satisfactory results. If the drug concentration is too high, toxicity or other undesired effects may occur. Moreover, it is evident that the rate of drug delivery is important to ensure $RS$ is maintained for the duration of the healing process, thought to be of the order of 1 month. An important quantity related to safety of the therapy is drug content in tissue, $DC$, essentially the $\mu$g drug per g tissue:

$$\begin{aligned} DC = {} & \sum_i \frac{MW}{\rho V_i} \int_{\Omega_i} c_i \, d\Omega_i \quad i = \{\text{intima, adventitia}\} \\ & + \frac{MW}{\rho V_{\mathrm{media}}} \int_{\Omega_{\mathrm{media}}} \left[ c_{\mathrm{media}} + b_{\mathrm{media}}^{\mathrm{specific}} + b_{\mathrm{media}}^{\mathrm{non\text{-}specific}} \right] d\Omega_{\mathrm{media}}, \end{aligned} \tag{29}$$

where $MW$ is the molecular weight of the drug, $\rho$ is the density of wet arterial tissue, and $V$ represents the volume of the tissue layer. This quantity is readily measurable in vivo in animal models and provides a basis for comparison of alternative therapies and formulations. Despite the importance of these measures, a key challenge in the field is the sparsity of experimentally derived estimates of the binding parameters, which are restricted to a few studies involving animal tissue, with parameter values reported in [32]. Figure 11 compares predictions from a 1D model, parametrised on in vitro stent drug release data, and available literature tissue parameter data, with data obtained from an in vivo animal study for prototype low-dose and high-dose sirolimus coated stents. Despite the simplicity of the model, the predictions broadly capture the experimental data and, moreover, highlight that despite similar levels of drug content in tissue, differential $RS$ levels dictate device efficacy.

To summarise, the state-of-the-art models of drug transport in arterial tissue incorporate advection-diffusion-reaction mechanisms and have the following features: (i) 2D axisymmetric or 3D geometries, (ii) multi-layer tissue with layer-specific properties including anisotropic diffusion, (iii) advection in the porous tissue described via Darcy's law, either through a prescribed physiological pressure

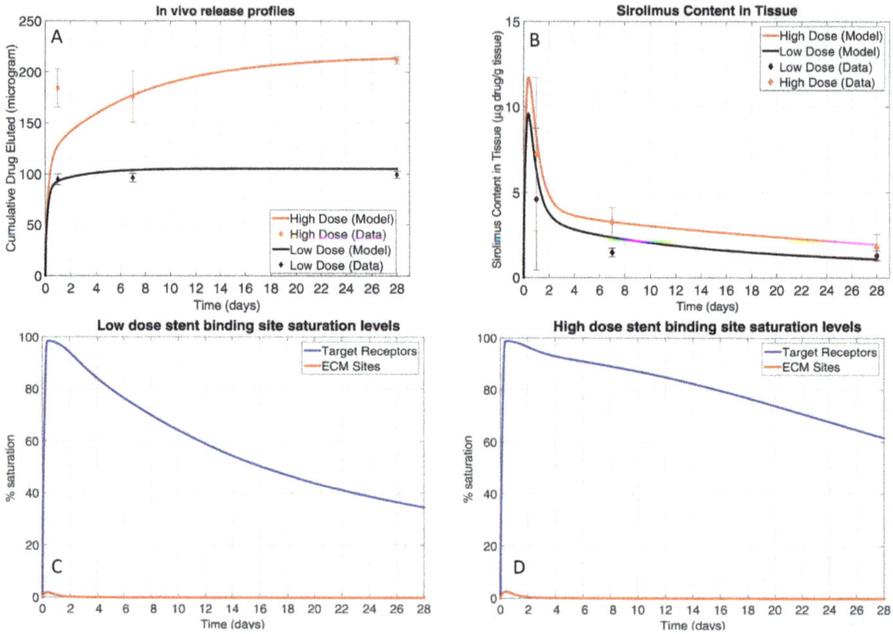


**Fig. 11** Comparison between model simulations and experiential data from an in vivo porcine study, comparing low-dose and high-dose sirolimus stents [31]. (**a**) In vivo release profile. (**b**) Drug content in tissue. (**c**)–(**d**): Binding site saturation levels for low-dose and high-dose stents, respectively. Obtained from McKittrick et al. [31] under Creative Commons licence https://creativecommons.org/licenses/by/4.0/

gradient or coupled to equations describing fluid flow in the lumen, and (iv) two distinct phases of drug binding.

## 4.6 *Modelling the Effect of Drug on Growth and Remodelling*

While models of drug transport and retention in tissue enable calculation of drug performance metrics, which may be correlated with device efficacy, a key limitation is that they almost exclusively consider a *static* artery. In other words, they do not account for cellular proliferation, ECM deposition, and, ultimately, tissue growth—excessive levels of which drive the main complication with the therapy. Thus, in order to be able to optimise drug delivery from DES and DCB, it is imperative to model the effect of drug on tissue growth. While several approaches to arterial remodelling following stent and balloon deployment have been developed, encompassing primarily discrete models, until recently these have either entirely neglected the effect of drug or do not account for the state-of-the-art understanding of how the drugs act on cells.

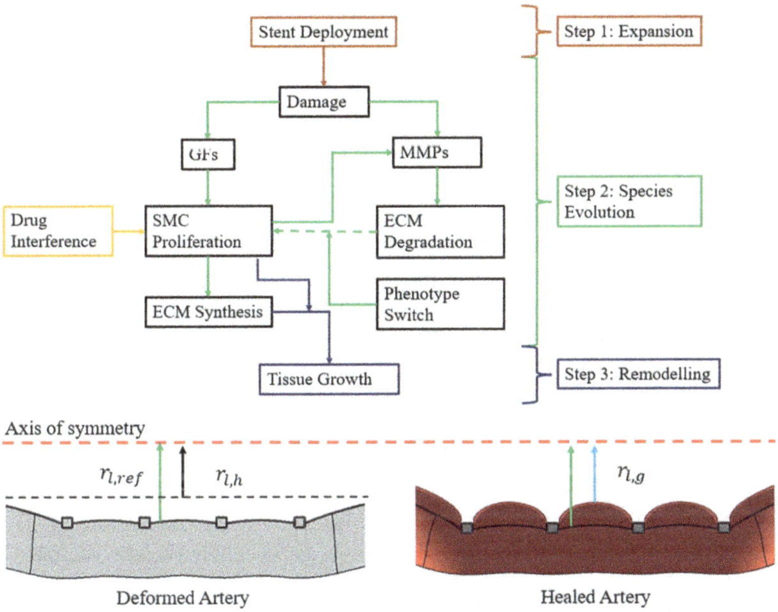


**Fig. 12** Schematic depicting the model developed by McQueen et al. [6]. Step 1 involved deploying stent struts into an idealised artery ("Deformed Artery"), from which the Von Mises stress is calculated and linked to a damage variable. Step 2 involved a system of nonlinear coupled partial differential equations (PDEs) to describe species evolution subject to drug delivery. Step 3 calculated the level of arterial narrowing ("Healed Artery") through consideration of the volume increase due to SMC and EC proliferation and ECM production. Obtained from McQueen et al. [6] under Creative Commons licence https://creativecommons.org/licenses/by/4.0/

This void has been addressed in two recent publications. McQueen et al. [6] developed a comprehensive modelling framework that incorporates the four stages of the process within a 2D-axisymmetric setting: (i) deployment of a DES, (ii) drug delivery, (iii) drug transport and retention, and (iv) arterial remodelling (Fig. 12).

The work considered an idealised deployment scenario, consisting of prescribing a displacement to stent struts to mimic different levels of overexpansion within a pressurised idealised straight artery. The resulting von Mises stress map ($\sigma_{vm}(r, z)$) is linked to a variable termed "damage", which decays over time following first-order kinetics as the artery heals. Stress is assumed to trigger a series of biological events, described though a species evolution model comprising multiple coupled nonlinear partial differential equations. The key species considered in the process are growth factors and metalloproteinases, which each undergo production and decay, SMC which may differentiate between contractile and synthetic types, with the latter able to proliferate and deposit ECM, and EC. Drug transport within the stent struts is assumed to be governed by diffusion, and drug kinetics in the tissue are described via reaction-diffusion equations incorporating the state-of-the-art two-phase nonlinear saturable binding model. A key advantage of this framework is that the local concentration of free and bound drug may be used to directly influence local SMC proliferation (and therefore ECM production), key contributors to tissue growth. Specifically, SMC proliferation, $P$, is calculated via

$$P = g\ c_{ssmc}\ c_{gf}\left[1 - \frac{c_{ecm}}{K_{ecm}}\right]\left[1 - \frac{b^{\text{specific}}}{B^{\text{specific}}}\right], \tag{30}$$

where $c_{ssmc}$ is the concentration of synthetic SMCs of intrinsic growth rate $g$; $c_{gf}$ is the concentration of growth factors; and $c_{ecm}$ is the concentration of ECM which synthesises up to some limit $K_{ecm}$. Tissue growth is calculated as a stationary problem where the species evolution model informs the volumetric growth of tissue at the time of interest. Specifically:

$$\frac{\partial \rho_i^0}{\partial t} = \Pi_i - \nabla \cdot \mathbf{M}_i, \tag{31}$$

where $i$ refers to the species contributing to growth, with concentration $\rho_i$ in the reference configuration; $\Pi_i$ describes the balance of source and sink terms (e.g. proliferation, degradation, apoptosis); and $\mathbf{M}_i$ are the mass fluxes of the species. Assuming locally isotropic volumetric changes a growth deformation tensor may be defined as $\mathbf{F}_i = \rho_i^0/\rho_i^{init}\mathbf{I}$, where $\rho_i^{init}$ is the initial density of species $i$ in the reference configuration. Under the small strain approximation, isotropic growth is then described through increases in volume of cellular components and density of non-cellular components:

$$\nabla \cdot \nu = \frac{\partial \Delta c_{ec}}{\partial t} V_{ec} + \frac{\Delta c_{smc}}{\partial t} V_{smc} + \frac{1}{\rho_{ecm}} \frac{\Delta c_{ecm}}{\partial t}, \tag{32}$$

where $\Delta c_i$ defines the change in species concentration. The level of restenosis may then be calculated as

$$\%\text{Area Restenosis} = 100\left[1 - \frac{r_{l,g}^2}{r_{l,ref}^2}\right], \tag{33}$$

where $r_{l,g}$ is the minimum radius of the lumen within the treated region and $r_{l,ref}$ is the reference radius proximal to the stented region (Fig. 12).

McQueen et al. [6] explored variations in stent design features such as inter-strut spacing and strut thickness, as well as the overexpansion ratio, and achieved results consistent with experiential and clinical studies. The most important finding was an intricate interplay between initial drug mass loaded on the stent, the drug release rate and restenosis. In particular, it was found that simply increasing the drug dose or extending the period of drug release was not sufficient to minimise restenosis and that a "sweet spot" appeared to exist in terms of drug loading mass and release rate, suggesting that these aspects are amenable to optimisation [6].

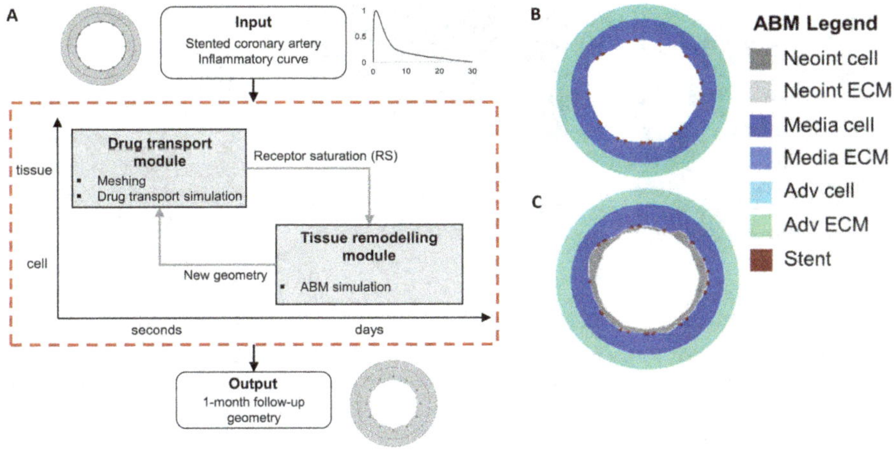


**Fig. 13** (**a**) Multi-scale arterial remodelling framework proposed by Corti et al. [45]. (**b**) Agent-based model of a patient-specific coronary artery cross-section. (**c**) Agent-based model showing neointimal tissue growth after 30 days. Obtained from McQueen et al. [45] under Creative Commons licence https://creativecommons.org/licenses/by/4.0/

In an alternative approach, Corti et al. [45] developed a hybrid continuum-agent-based modelling approach to consider the effect of drug release on restenosis formation. The drug release and transport model employed were similar to that those in McQueen et al. [6]. However, an agent-based model of cellular dynamics was employed to simulate remodelling. Although adopting a 2D arterial cross-section geometrical framework, a proof-of-concept application to a patient-specific geometry was considered. Figure 13 displays the modelling framework and shows the interaction between distinct drug transport and remodelling modules. Importantly, the stimulus was considered to be an inflammation curve, derived from clinical data, rather than a stress-induced damage field following deployment. The RS curve was used to mediate cellular activities (division, apoptosis, ECM production) through probability equations. The framework proved capable of capturing different responses in terms of levels of restenosis depending on different initial drug loading masses and drug release rates.

## 4.7 *Future Outlook—Incorporation of Plaque Composition*

Up until now, all of the approaches described to model the effect of drugs on coronary vessels have assumed idealised healthy arteries. Even if patient-specific arterial geometries have been derived from imaging, the wall's drug transport and retention properties have not accounted for disease composition. Unfortunately, at the point of intervention, a given patient's artery will not look like the idealised representation in Fig. 9. In reality, heterogeneous plaque is likely to be present, com-

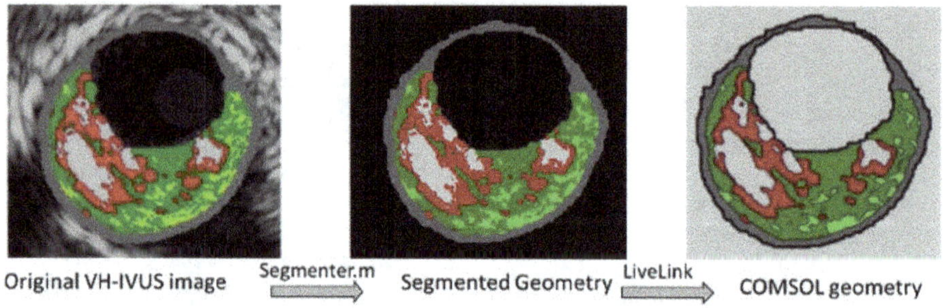


**Fig. 14** Proposed workflow—from image to simulation—for an example VH-IVUS image. Left: The image is segmented following an automatic routine. Middle: A computational geometry is automatically created in COMSOL Multi-physics. Right: A drug transport and retention simulation is performed. The different colours refer to different components of plaque. Red—necrotic core. Green—Fibrous. Yellow—fibrofatty. White—calcium

prising a variety of different components with varying material properties, across varying scales. Figure 14 shows a cross-section of a patient artery derived from Virtual Histology—Intravascular Ultrasound (VH-IVUS) imaging. In the future, imaging such as VH-IVUS or alternatives such as Optical Coherence Tomography (OCT) should be utilised to create the most realistic patient geometries from which diseased material properties (both mechanical and solute transport) can be assigned. A suggested workflow is provided in Fig. 14 based upon 2D cross-sectional VH-IVUS images. Clearly, this approach could be extended to produce 3D diseased geometries and adapted to suit different types of images from different intravascular imaging systems. A key challenge however is the assignment of accurate material properties. In the case of drug transport and retention, these parameters are primarily derived from experimental testing in healthy animal tissues, or from diseased animal tissue where the plaque complexity is not comparable to that in humans. If this challenge can be overcome, then it paves the way for lesion-specific drug delivery optimisation and potentially better outcomes for patients.

## 5 Conclusion

Through the three case studies presented in this chapter, we highlight how highly complex blood vessels are. Their native, healthy biomechanical properties are heterogeneous and uncertain, and large detailed ex vivo and in vivo datasets are required to improve our understanding. The biochemical regulation mechanisms are important for maintaining sufficient blood flow to various body parts. However, we are only beginning to understand how these couple with the mechanics via multi-scale models. More experimental data are needed to prove/disprove different hypotheses on how these regulation mechanisms work. Lastly, to understand the effect of drugs during treatment requires a multi-physics description, incorporating diffusion and advection of the drug through the tissue and reaction of the drug with the vascular cells. Lastly, the inclusion of disease further complicates our investigations by introducing heterogeneities associated with the unhealthy vascular components (such as calcium and lipidic plaques). For all the challenges above, a judicious combination of further experimentation and mathematical and statistical modelling is required to disentangle the various scales and phenomenon that dictate our vascular system.

## References


1. Harkness, M.L.R., Harkness, R.D., McDonald, D.A., Franklin, K.J.: The collagen and elastin content of the arterial wall in the dog. Proceedings of the Royal Society of London. Series B—Biological Sciences **146**(925), 541–551 (1957) https://royalsocietypublishing.org/doi/pdf/10.1098/rspb.1957.0029. https://doi.org/10.1098/rspb.1957.0029
2. Fung, Y.C.: Mechanical Properties of Blood Vessels, pp. 261–301. Springer, New York, NY (1981). https://doi.org/10.1007/978-1-4757-1752-5_8
3. Strauss, B.H., Chisholm, R.J., Keeley, F.W., Gotlieb, A.I., Logan, R.A., Armstrong, P.W.: Extracellular matrix remodeling after balloon angioplasty injury in a rabbit model of restenosis. Circulation Research **18**(4), 650–658 (1994). https://www.ahajournals.org/doi/abs/10.1161/01.RES.75.4.650
4. Tilahun, H.G., Mullagura, H.N., Humphrey, J.D., Baek, S.: A biochemomechanical model of collagen turnover in arterial adaptations to hemodynamic loading. Biomechanics and Modeling in Mechanobiology **22**(6), 2063–2082 (2023). https://doi.org/10.1007/s10237-023-01750-1
5. Coccarelli, A., Polydoros, I., Drysdale, A., Harraz, O.F., Kadapa, C.: A new computational model for quantifying blood flow dynamics across myogenically-active cerebral arterial networks. arXiv:2411.09046
6. McQueen, A., Escuer, J., Schmidt, A.F., Aggarwal, A., Kennedy, S., McCormick, C., Oldroyd, K., McGinty, S.: An intricate interplay between stent drug dose and release rate dictates arterial restenosis. Journal of Controlled Release **349**, 992–1008 (2022). https://doi.org/10.1016/j.jconrel.2022.07.037
7. Bai, W., Suzuki, H., Huang, J., Francis, C., Wang, S., Tarroni, G., Guitton, F., Aung, N., Fung, K., Petersen, S.E., et al.: A population-based phenome-wide association study of cardiac and aortic structure and function. Nature medicine **26**(10), 1654–1662 (2020). https://doi.org/10.1038/s41591-020-1009-y
8. Coccarelli, A., Edwards, D.H., Aggarwal, A., Nithiarasu, P., Parthimos, D.: A multiscale active structural model of the arterial wall accounting for smooth muscle dynamics. J R Soc Interface **15**(139) (2018). https://doi.org/10.1098/rsif.2017.0732
9. Coccarelli, A., Carson, J.M., Aggarwal, A., Pant, S.: A framework for incorporating 3d hyperelastic vascular wall models in 1d blood flow simulations. Biomech Model Mechanobiol **20**(4), 1231–1240 (2021). https://doi.org/10.1007/s10237-021-01437-3
10. Holzapfel, G.A., Gasser, T.C., Ogden, R.W.: A new constitutive framework for arterial wall mechanics and a comparative study of material models. Journal of Elasticity and the Physical Science of Solids **61**(1), 1–48 (2000). https://doi.org/10.1023/A:1010835316564
11. Holzapfel, G.A., Sommer, G., Gasser, C.T., Regitnig, P.: Determination of layer-specific mechanical properties of human coronary arteries with nonatherosclerotic intimal thickening and related constitutive modeling. American Journal of Physiology-Heart and Circulatory Physiology **289**(5), 2048–2058 (2005). https://doi.org/10.1152/ajpheart.00934.2004

12. Gasser, T.C., Ogden, R.W., Holzapfel, G.A.: Hyperelastic modelling of arterial layers with distributed collagen fibre orientations. Journal of the Royal Society Interface **3**(6), 15–35 (2006). https://doi.org/10.1098/rsif.2005.0073
13. Peña, J.A., Martínez, M.A., Peña, E.: Layer-specific residual deformations and uniaxial and biaxial mechanical properties of thoracic porcine aorta. Journal of the Mechanical Behavior of Biomedical Materials **50**, 55–69 (2015). https://doi.org/10.1016/j.jmbbm.2015.05.024
14. Aggarwal, A., Hudson, L.T., Laurence, D.W., Lee, C.-H., Pant, S.: A Bayesian constitutive model selection framework for biaxial mechanical testing of planar soft tissues: Application to porcine aortic valves. Journal of the Mechanical Behavior of Biomedical Materials **138**, 105657 (2023). https://doi.org/10.1016/j.jmbbm.2023.105657
15. Aggarwal, A.: An improved parameter estimation and comparison for soft tissue constitutive models containing an exponential function. Biomechanics and Modeling in Mechanobiology **16**(4), 1309–1327 (2017). https://doi.org/10.1007/s10237-017-0889-3
16. Aggarwal, A., Mortensen, P., Hao, J., Kaczmarczyk, Ł., Cheung, A.T., Al Ghofaily, L., Gorman, R.C., Desai, N.D., Bavaria, J.E., Pouch, A.M.: Strain estimation in aortic roots from 4d echocardiographic images using medial modeling and deformable registration. Medical Image Analysis **87**, 102804 (2023). https://doi.org/10.1016/j.media.2023.102804
17. Aggarwal, A., Jensen, B.S., Pant, S., Lee, C.-H.: Strain energy density as a Gaussian process and its utilization in stochastic finite element analysis: Application to planar soft tissues. Computer Methods in Applied Mechanics and Engineering **404**, 115812 (2023). https://doi.org/10.1016/j.cma.2022.115812
18. Longden, T.A., Dabertrand, F., Koide, M., Gonzales, A.L., Tykocki, N.R., Brayden, J.E., Hill-Eubanks, D., Nelson, M.T.: Capillary k(+)-sensing initiates retrograde hyperpolarization to increase local cerebral blood flow. Nat Neurosci **20**(5), 717–726 (2017). https://doi.org/10.1038/nn.4533
19. Daher, A., Payne, S.: A dynamic multiscale model of cerebral blood flow and autoregulation in the microvasculature. Applied Mathematical Modelling **123**, 213–240 (2023). https://doi.org/10.1016/j.apm.2023.06.035
20. Uhlmann, K., Balzani, D.: Chemo-mechanical modeling of smooth muscle cell activation for the simulation of arterial walls under changing blood pressure. Biomech Model Mechanobiol **22**(3), 1049–1065 (2023). https://doi.org/10.1007/s10237-023-01700-x
21. Lidington, D., Schubert, R., Bolz, S.S.: Capitalizing on diversity: an integrative approach towards the multiplicity of cellular mechanisms underlying myogenic responsiveness. Cardiovasc Res **97**(3), 404–12 (2013). https://doi.org/10.1093/cvr/cvs345
22. Irons, L., Humphrey, J.D.: Cell signaling model for arterial mechanobiology. PLoS Comput Biol **16**(8), 1008161 (2020). https://doi.org/10.1371/journal.pcbi.1008161
23. Murtada, S.I., Lewin, S., Arner, A., Humphrey, J.D.: Adaptation of active tone in the mouse descending thoracic aorta under acute changes in loading. Biomech Model Mechanobiol **15**(3), 579–92 (2016). https://doi.org/10.1007/s10237-015-0711-z
24. Holzapfel, G.A., Gasser, T.C., Ogden, R.W.: A new constitutive framework for arterial wall mechanics and a comparative study of material models. J Elast (2000)
25. Hassan, S., Ali, M.N., Ghafoor, B.: Evolutionary perspective of drug eluting stents: from thick polymer to polymer free approach. Journal of Cardiothoracic Surgery **17**(1), 1–20 (2022)
26. Escuer, J., Martínez, M.A., McGinty, S., Peña, E.: Mathematical modelling of the restenosis process after stent implantation. Journal of the Royal Society Interface **16**(157), 20190313 (2019). https://doi.org/10.1098/rsif.2019.0313
27. McQueen, A., Escuer, J., Aggarwal, A., Kennedy, S., Mccormick, C., Oldroyd, K., McGinty, S.: Do we really understand how drug eluted from stents modulates arterial healing? International Journal of Pharmaceutics **601**, 120575 (2021). https://doi.org/10.1016/j.ijpharm.2021.120575
28. McQueen, A.: Understanding how anti–proliferative drug modulates arterial healing following stent deployment. PhD thesis, University of Glasgow (2022)
29. McGinty, S.: A decade of modelling drug release from arterial stents. Mathematical Biosciences **257**, 80–90 (2014). https://doi.org/10.1016/j.mbs.2014.06.016. Multiscale models and methods in biomedicine

30. McGinty, S., McKee, S., McCormick, C., Wheel, M.: Release mechanism and parameter estimation in drug-eluting stent systems: analytical solutions of drug release and tissue transport. Mathematical Medicine and Biology: A Journal of the IMA **32**(2), 163–186 (2014). https://doi.org/10.1093/imammb/dqt025
31. McKittrick, C.M., McKee, S., Kennedy, S., Oldroyd, K., Wheel, M., Pontrelli, G., Dixon, S., McGinty, S., McCormick, C.: Combining mathematical modelling with in vitro experiments to predict in vivo drug-eluting stent performance. Journal of Controlled Release **303**, 151–161 (2019). https://doi.org/10.1016/j.jconrel.2019.03.012
32. Tzafriri, A.R., Groothuis, A., Price, G.S., Edelman, E.R.: Stent elution rate determines drug deposition and receptor-mediated effects. Journal of Controlled Release **161**(3), 918–926 (2012). https://doi.org/10.1016/j.jconrel.2012.05.039
33. Prabhu, S., Hossainy, S.: Modeling of degradation and drug release from a biodegradable stent coating. Journal of Biomedical Materials Research Part A **80A**(3), 732–741 (2007). https://doi.org/10.1002/jbm.a.31053
34. Vo, T.T.N., Morgan, S., McCormick, C., McGinty, S., McKee, S., Meere, M.: Modelling drug release from polymer-free coronary stents with microporous surfaces. International Journal of Pharmaceutics **544**(2), 392–401 (2018). https://doi.org/10.1016/j.ijpharm.2017.12.007. Advances in Drug Delivery Related Biosensors and Medical Devices
35. McGinty, S., Vo, T.T.N., Meere, M., McKee, S., McCormick, C.: Some design considerations for polymer-free drug-eluting stents: A mathematical approach. Acta Biomaterialia **18**, 213–225 (2015). https://doi.org/10.1016/j.actbio.2015.02.006
36. Escuer, J., Schmidt, A.F., Peña, E., Martínez, M.A., McGinty, S.: Mathematical modelling of endovascular drug delivery: Balloons versus stents. International Journal of Pharmaceutics **620**, 121742 (2022). https://doi.org/10.1016/j.ijpharm.2022.121742
37. Hwang, C.-W., Wu, D., Edelman, E.R.: Physiological transport forces govern drug distribution for stent-based delivery. Circulation **104**(5), 600–605 (2001). https://doi.org/10.1161/hc3101.092214
38. Hwang, C.-W., Edelman, E.R.: Arterial ultrastructure influences transport of locally delivered drugs. Circulation Research **90**(7), 826–832 (2002). https://doi.org/10.1161/01.RES.0000016672.26000.9E
39. Bozsak, F., Chomaz, J.-M., Barakat, A.: Modeling the transport of drugs eluted from stents: Physical phenomena driving drug distribution in the arterial wall. Biomechanics and Modeling in Mechanobiology **13** (2014). https://doi.org/10.1007/s10237-013-0546-4
40. Escuer, J., Cebollero, M., Peña, E., McGinty, S., Martínez, M.A.: How does stent expansion alter drug transport properties of the arterial wall? Journal of the Mechanical Behavior of Biomedical Materials **104**, 103610 (2020). https://doi.org/10.1016/j.jmbbm.2019.103610
41. Kedem, O., Katchalsky, A.: Thermodynamic analysis of the permeability of biological membranes to non-electrolytes. Biochimica et Biophysica Acta **27**, 229–246 (1958). https://doi.org/10.1016/0006-3002(58)90330-5
42. Gasser, T.C., Ogden, R.W., Holzapfel, G.A.: Hyperelastic modelling of arterial layers with distributed collagen fibre orientations. Journal of the Royal Society Interface **3**(6), 15–35 (2006). https://doi.org/10.1098/rsif.2005.0073
43. Holzapfel, G.A., Stadler, M., Gasser, T.C.: Changes in the Mechanical Environment of Stenotic Arteries During Interaction with Stents: Computational Assessment of Parametric Stent Designs . Journal of Biomechanical Engineering **127**(1), 166–180 (2005). https://doi.org/10.1115/1.1835362
44. Chao-Wei Hwang, D.W., Edelman, E.R.: Impact of transport and drug properties on the local pharmacology of drug-eluting stents. International Journal of Cardiovascular Interventions **5**(1), 7–12 (2003). https://doi.org/10.1080/14628840304614. PMID: 12623559
45. Corti, A., McQueen, A., Migliavacca, F., Chiastra, C., McGinty, S.: Investigating the effect of drug release on in-stent restenosis: A hybrid continuum–agent-based modelling approach. Computer Methods and Programs in Biomedicine **241**, 107739 (2023). https://doi.org/10.1016/j.cmpb.2023.107739

# Multiscale Biomechanics of the Right Ventricle: The Relation of Tissue Structure, Mechanics, and Physiology

**Kristen LeBar, Lik Chuan Lee, and Zhijie Wang**

## 1 Introduction

The right ventricle (RV) is historically an understudied organ compared to the left compartment of the heart. However, RV failure is prevalent in the United States and worldwide and is a key contributor to the mortality of multiple cardiovascular diseases. During RV failure development, the tissue undergoes significant structural remodeling and mechanical changes, which lead to the alteration of overall organ function. Significant advancements have been achieved recently in the understanding of cell and tissue mechanical behavior, as well as RV function, using both experimental and computational approaches. This chapter aims to discuss the current methodologies and findings to expand our understanding of RV biomechanics and pathophysiology at multiple scales.

K. LeBar
Department of Mechanical Engineering, Colorado State University, Fort Collins, CO, USA

L. C. Lee
Department of Mechanical Engineering, Michigan State University, East Lansing, MI, USA

Z. Wang (✉)
Department of Mechanical Engineering, Colorado State University, Fort Collins, CO, USA

School of Biomedical Engineering, Colorado State University, Fort Collins, CO, USA
e-mail: Zhijie.Wang@colostate.edu

 
J. Liao, J. Y. Wong (eds.), *Integration and Bridging of Multiscale Bioengineering Designs and Tissue Biomechanics*, https://doi.org/10.1007/978-3-031-81743-4_15

## 2 Classification of Material Mechanical Behavior and Mechanical Type of Right Ventricles

### *2.1 Nonlinear Elasticity*

Elasticity describes the material's ability to deform and store energy under loadings. In other words, the applied forces are converted to mechanical energy and stored, and this energy can be fully released after the removal of external forces. Elasticity can be derived from the slope of the stress-strain curve. This is referred to as the elastic modulus or Young's modulus (Fig. 1a). There are two forms of elastic behavior: linear and nonlinear. A linear elastic material displays the same degree of elasticity regardless of the loading (stress) or deformation (strain) (Fig. 1a). In contrast, a nonlinear elastic material displays altered elasticity that depends on deformation (strain) (Fig. 1b). Like all biological tissues, the RV exhibits a nonlinear elastic behavior [1–4].

### *2.2 Viscoelasticity*

Viscoelasticity refers to the material's ability to store and dissipate energy under loadings. In other words, the applied forces cause both elastic and viscous resistances, and there is a certain loss of mechanical energy after the removal of external forces. The elastic resistance originates from the atomic bond energy of molecules and contributes to energy storage during deformation. On the other hand, viscous resistance originates from friction arising from the sliding of molecular bonds, and it contributes to the energy dissipation during deformation. The viscoelastic behavior can be evident or measured in three ways: creep (Fig. 2a), stress relaxation (Fig. 2b), and the hysteresis loop during cyclic loadings (Fig. 2c).

Three forms of viscoelastic behavior exist: linear viscoelastic, quasilinear viscoelastic, and fully nonlinear viscoelastic. A linear viscoelastic material exhibits a linear elastic response, as well as a linear time-dependent (viscous) response. In other words, both elasticity and viscosity are independent of strain or strain

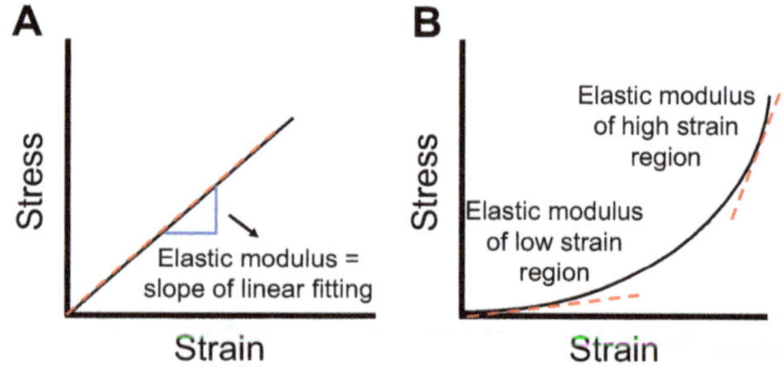


**Fig. 1** Stress-strain relations of a linear (**a**) and a nonlinear (**b**) elastic material

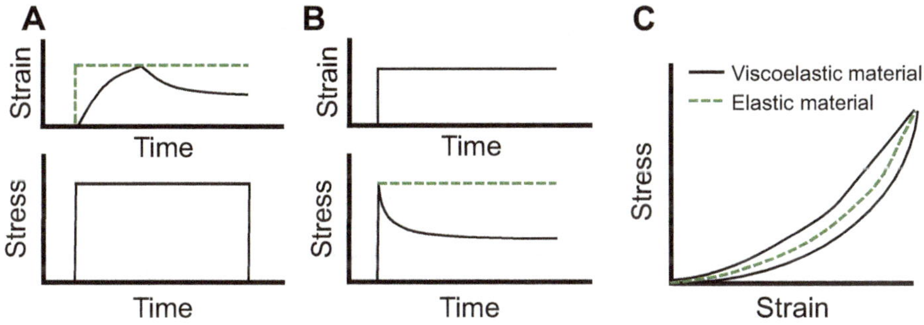


**Fig. 2** Creep (**a**), stress relaxation (**b**), and cyclic loading (**c**) behavior of a viscoelastic (black solid line) material. The response of an elastic (green dash line) material is shown to illustrate the difference from the viscoelastic response

**Fig. 3** Stress-strain (**a**) and stress relaxation (**b**) measurements of a linear viscoelastic material

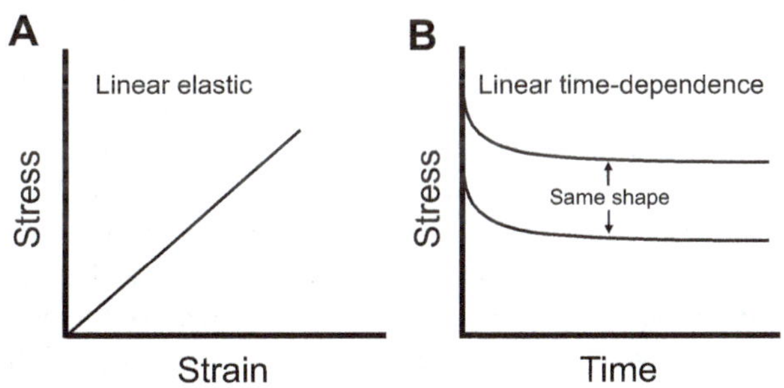


**Fig. 4** Stress-strain (**a**) and stress relaxation (**b**) measurements of a quasilinear viscoelastic material

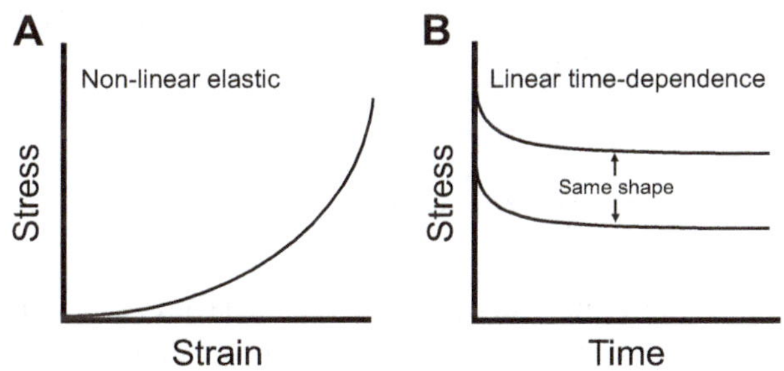


rate (Fig. 3). A quasilinear material exhibits a nonlinear elastic response but a linear viscous response (Fig. 4). In the cases of both linear and quasilinear viscoelasticity, the elasticity and viscosity of the material are separable. In a fully nonlinear viscoelastic material, both the elastic and the viscous behaviors are nonlinear (strain- and strain rate-dependent) (Fig. 5). In this case, the viscosity and elasticity are non-separable (inter-dependent).

Fung developed the quasilinear viscoelastic theory in order to characterize the nonlinear stress-strain relationship of the tissue [5], and the formulation has been adopted in numerous studies of cardiac tissue mechanics [6–8]. Under a certain range of strains and strain rates, the RV tissue can be treated as a quasilinear

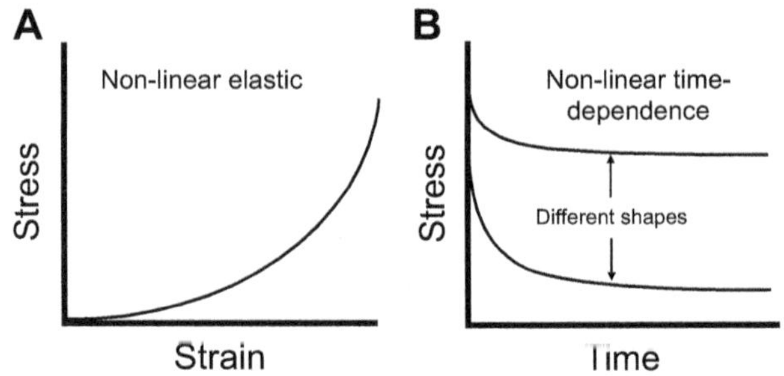


**Fig. 5** Stress-strain (**a**) and stress relaxation (**b**) measurements of a fully nonlinear viscoelastic material

viscoelastic material to characterize the mechanical behavior at a lower computational cost. Based on our recent studies of ovine hearts, however, the RV tissue exhibited nonlinear viscoelastic behavior in both longitudinal (apex-to-base axis) and circumferential directions [4].

In myocardial tissues, elasticity and viscoelasticity are enhanced with elevated stretch rates [9–12]. The strain rate and strain dependency of RV tissue mechanics indicate the importance of measuring the tissue mechanical behavior under physiological conditions to capture the *in vivo* behavior. This means that in small animal hearts, a higher strain rate should be adopted than that of large animal hearts to mimic the higher heart rate. In most mechanical testing devices and/or protocols, the physiological strains can be readily achieved. However, the physiological strain rates, as well as the data acquisition capability of the system, are not available to enable a physiological viscoelastic measurement in rodent hearts. This limit has impeded the complete understanding of the dynamic (viscoelastic) mechanical behavior of the RV [12].

## 2.3 *Isotropy and Anisotropy*

In an isotropic material, its mechanical properties are independent of direction. That is, the mechanical behavior is the same in all directions/orientations. On the other hand, an anisotropic material's mechanical properties depend on the direction/orientation. One needs to note that such classification is different than the definition of homogenous and heterogeneous materials, which refers to the uniformity of the material properties throughout the volume. A homogenous material has the same mechanical behavior everywhere, whereas a heterogeneous material has different mechanical behaviors at distinct locations of the specimen.

The RV myocardium is anisotropic as a result of the amount and structural orientation of the tissue constituents, such as myo- and collagen fibers [13]. In general, the RV is treated as an orthotropic material, which is a subset of anisotropic materials. This means that its material properties are different in three perpendicular directions [3, 14, 15]. However, the RV is sometimes assumed and treated as a transversely anisotropic material for simplicity.

## 3 Mechanical Assessments of RV at Cell and Tissue Levels

Currently, RV mechanics have been investigated at various scales, ranging from isolated cardiomyocytes, trabeculae, papillary muscles, and even the entire RV free wall. These methodologies have their own advantages and disadvantages. For instance, the loading of individual cells or papillary muscles can only be performed at one dimension, whereas the loading of RV free wall tissue can be performed at one (by uniaxial or indentation test), two (by biaxial tests), or three (by triaxial tests) dimensions. Thus, the latter measurements offer more comprehensive mechanical characterization of the myocardial wall. Alternatively, cell-level mechanics can be assessed in both active and passive states of the cardiac muscle, whereas tissue-level mechanics are mainly assessed in the passive state of the tissue. The active mechanical behavior (from the biomechanical engineering perspective) of myocardium is sometimes referred to as the contractile behavior or contractility of the muscle (from the physiology perspective). In this section, we will summarize current *ex vivo* mechanical testing methodologies at cellular and tissue levels.

### *3.1 Ex Vivo, Multi-axial Mechanical Tests*

Cardiomyocytes (CM) are a major cell population of the myocardium and contribute significantly to the passive and active mechanical behaviors of the tissue. The CM consists of a cytoskeletal network, made up of the sarcomeric cytoskeleton such as actin, myosin, and non-sarcomeric cytoskeleton such as titin, microtubules, and intermediate filaments [16]. These cytoskeletons are predominantly aligned axially in the cell, with some transverse filaments interspersed [17]. Despite the 3D structure of the cell, only uniaxial (axial) cell mechanics are measured and reported. The RV free wall, however, is a complex tissue comprising myofibers (CM), collagen, and other cells and extracellular matrix proteins, which contribute to the anisotropic mechanical behavior of the tissue. To date, uniaxial, biaxial, and triaxial mechanical tests have all been reported for ventricular free wall tissues. Particularly, biaxial and triaxial mechanical tests provide more comprehensive measurements of the myocardium. Additionally, they better mimic the *in vivo* mechanical loading of the RV [2, 13, 18–22]. In such tests, a different degree of deformation or loading [15, 23–25] or the same degree of deformation or loading [9, 26–34] can be applied in each direction.

Because of the unique physiological motions (mainly planar deformation) and the thin wall of the RV, planar (biaxial) tests are often used to assess the anisotropic behavior of the tissue. Conventionally, the orientation of the tissue is defined by main fiber and cross-fiber/sheet orientations of the heart [2, 3, 13, 15, 35]. But recent studies have oriented the tissue based on anatomic orientations. That is, the testing axes are defined by the apex-to-outflow tract direction and the perpendicular direction [10, 28, 30, 34].

Biaxial and uniaxial mechanical tests apply normal tensile or compressive forces to the specimen, and only planar deformation is induced in these tests. However, a few studies have suggested transmural changes in RV free wall microstructure and mechanical properties [1, 3]. To account for this, triaxial mechanical tests are performed to measure the shear and/or orthotropic properties of the tissue [14, 15, 36, 37]. These tests are more common for the left ventricle (LV). Some studies have used the combination of biaxial and shear tests to characterize the 3D mechanical behavior of the ventricle [38, 39].

## *3.2 Sample Preparation*

Multiple preparation steps must be taken to maintain the physiological behavior of cell or tissue samples. This includes harvest time and storage. Typically, fresh samples should be used (testing within several hours of harvest), and the sample is kept on ice prior to the test. To preserve cell viability, media such as Tyrode's solution is used in isolated CM studies [40–42], and phosphate buffer or cardioplegic solutions are used in tissue studies [32, 40]. Moreover, to capture the passive behavior of the tissue, a relaxant solution is often used to break the actin-myosin cross-bridge in cardiac muscle cells, such as 2,3-butanedione monoxime [23, 30].

Before data acquisition, the samples should undergo a series of preconditioning cycles to achieve a stable mechanical response. This protocol was proposed and developed by Fung [5]. When the sample is excised, it moves from a loaded to an unloaded state, which may cause microstructural changes in the tissue and, consequently, affects the mechanical behavior. After it is mounted and reloaded for testing, the preconditioning cycles enable the tissue to re-adapt to reset the microstructural properties to those under *in vivo* loading conditions. A typical response to the preconditioning cycles is presented in Fig. 6. It often takes a few (~10–15) cycles of loading for the specimen to reach an equilibrium state.

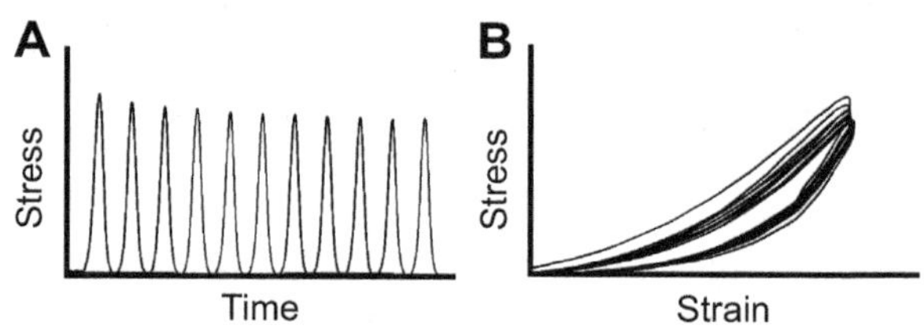


**Fig. 6** The stress-time (**a**) and stress-strain (**b**) curves obtained from preconditioning arise from viscoelastic behavior of the material

## 3.3 *Passive Mechanical Measurements*

### 3.3.1 Cyclic Tensile Test

The cyclic tensile mechanical test is a common method to measure the passive, elastic behavior of CM [43–45] and free wall [26–30, 32, 33] of the RV. In isolated cell tests, the degree of length extension is typically set by the resting (slack) and extended (tensioned) sarcomere length, for instance, from 1.4 μm to 2.0 μm, respectively [46]. In studies of RV free wall mechanics, the tissue is deformed to an *in vivo* strain level that is derived from medical images at end-systole and end-diastole. Thus, a range of 10–20% of strain is used depending on the animal species [3, 13, 28–30, 47].

Cyclic mechanical tests can also be used to measure the viscoelastic behavior of the cell or tissue. To derive the material elastic property, the elastic modulus or the stored energy is quantified from the hysteresis loop (Fig. 2c) [12, 15, 30, 48, 49]. The viscous property can be derived from various parameters, such as the loop width, loop area, or dissipated energy [12, 16].

### 3.3.2 Nanoindentation Test

Atomic force microscopy (AFM) indentation testing has been used for measuring cell [42, 50–58] and tissue [59–62] mechanics. In such tests, a cantilever is actuated to push a nano-indenter to a certain depth, which deforms the cell. The material stiffness is derived as Young's modulus, using mathematical models such as the Hertz model [42, 63–66]. The Young's modulus obtained using such tests is considered a tensile mechanical property, although the deformation is compressive. Another important assumption in AFM measurement is that the material is isotropic and homogeneous [67], which is not the case in cells or tissues. Moreover, the nanoindentation tests only provide the *local* (microscopic) mechanical behavior of the specimen, while the normal compressive or tensile mechanical test allows the experimentalist to obtain the *global* (macroscopic) mechanical behavior of the specimen. Thus, mechanical behaviors obtained from indentation tests are highly dependent on the spatial selection of the sample and may vary significantly from place to place in heterogenous materials. However, some groups have performed indentations at multiple locations in the RV tissue to obtain the mechanical behavior across the entirety of the material in an effort to overcome this limitation [60, 61].

AFM indentation can also be used to quantify the viscoelastic behavior of myocardium tissues. In these tests, the speed of compression is slow (0.6 μm/s) [55, 59]. However, for viscoelastic materials such as ventricle walls, the mechanical behavior is dependent on the deformation rate, and it is appropriate to adopt the physiological stretch rates of the species to obtain physiologically relevant mechanical properties. To account for this, Caporizzo et al. applied a wide range of indentation speeds (0.1 μm/s–150 μm/s) and measured the elastic modulus of the cardiomyocyte at each speed. They found that the elastic modulus increased with increasing indentation speed, indicating a viscoelastic behavior of the cell (strain rate-dependent) [42]. Moreover, they attributed the low-speed indentation measures to a static (elastic) mechanical property, and the high-speed indentation measures to the dynamic (viscoelastic) mechanical behavior of the cell.

### 3.3.3 Stress Relaxation

Finally, stress relaxation is another mechanical testing method from which the viscoelastic mechanical properties of the material can be measured. In these tests, the material is quickly deformed to a specified length, and then held constant to observe the viscous decay (Fig. 2b) [4, 10, 15, 16, 56]. As in the cyclic mechanical tests, a physiological strain level is applied to the material to better mimic the *in vivo* mechanical behavior. However, the loading scheme applied in these tests is certainly not physiological as the material is held stretched for prolonged periods of time. From the stress relaxation curve, relaxation modulus is derived as the stress at a given time divided by the applied strain [9, 10, 56]. Additionally, normalized stress is derived as the stress at a given time divided by the maximal stress and has been used previously as a metric of viscosity [10, 30]. As from the hysteresis loops, Liu et al. has derived stored and dissipated energies from the stress relaxation curve as metrics of elasticity and viscosity, respectively [9, 10, 34]. In this method, elasticity is defined as the product of the input strain and the area beneath the raw stress relaxation curve, and the dissipated energy as the product of the input strain and the area between the raw stress relaxation curve and the theoretical purely elastic curve. We do not include studies/details of creep tests in quantifying RV viscoelasticity. This is partly because the viscoelastic behavior of heart tissues is often negligible when quantified by such approaches.

## *3.4 Active Mechanical Measurements*

In general, there are two techniques used to measure active force in isolated CM or cardiac muscles: the measurement of force under electrical stimulation [43, 68], and the measurement of force under altered calcium concentration [69]. Applying either stimulus, isometric contraction can be achieved at a fixed sarcomere length. Then, the length-dependent activation (Frank-Starling mechanism) of the cell can then be measured at varying sarcomere lengths. A more detailed review of cell active mechanics and organ function is provided here [70].

The measurement of active force in a larger scale of myocardium such as the entire free wall, though, is less established. When measuring the active behavior of myocardial tissue, though, many groups obtain these measurements in isolated papillary muscles, not integrated ventricular free walls [71–75]. However, Neelakantan et al. recently presented a method measuring the active mechanics of mouse left ventricular tissue [76]. LV free wall tissue was passively stretched up to 20% of strain and then submerged in an activation solution, in which the trabeculae or papillary muscles fully contract and the active force could be measured. Overall, the measurement of RV free wall active mechanics using *ex vivo* techniques is underdeveloped.

# 4 *In Silico* Models of Myocardial Mechanics

*In silico* models of cardiac mechanics at cellular or tissue levels are valuable tools for understanding the mechanical behavior and the structure-function relations of the material. Such models have constitutive relations based on physical elements or mathematical theory, and they can be used to model elastic or viscoelastic mechanical responses and passive or active mechanical behavior. In this section, we will provide a general picture of the modeling of myocardial (cellular and tissue) in passive and active states, with a focus on RV-relevant research and recent findings in RV viscoelastic modeling. For additional reviews of *in silico* modeling in general biomechanics or ventricular biomechanics, please refer to the following references [77, 78].

## *4.1 Hyperelastic Models*

From cyclic mechanical tests, constitutive models have been used to fit the stress-strain curve to describe the elastic behavior of the RV. The traditional Fung-type model has been used extensively for the fitting of hyperelastic curves of the RV and other cardiovascular tissues [23, 32, 79, 80]. Studies have reported strong fitting capabilities of the Fung-type model, as well as the ability of the model to capture the anisotropic behavior of the RV [23, 79, 81]. In this formulation, tissue elasticity is derived using the Fung exponential strain energy density function.

While the Fung-type model captures the anisotropic behavior of the tissue, it does not have the ability to distinguish the contributions of myo- and collagen fibers to the total elastic behavior. Thus, Avazmohammadi et al. incorporated the collagen and myofiber structure, including content and orientation, in a hyperelastic model of the RV [1, 3]. The model provided accurate fitting of the hyperelastic curve and elucidated distinct contributions of myo- and collagen fibers to the total elastic behavior of the RV. Additionally, this model attributed the low-strain elastic modulus to the contribution of myofibers, and the high-strain elastic modulus to the contribution of collagen fibers, as described previously [2, 13]. Kakaletsis et al. implemented a similar approach in a modified Holzapfel model. The Holzapfel model incorporates tissue microstructural information (fiber orientation and dispersion). However, according to Kakaletsis et al., higher fidelity microstructural information in the Holzapfel model greatly improved its predictive capability [36].

While the fiber orientation is a principal factor to incorporate in a constitutive model to predict the elastic behavior of the anisotropic RV, Kakaletsis et al. also included the area fraction of myo- and collagen fibers, as well as the ground matrix to delineate the microstructural contributions to RV elasticity, as done recently by Li et al. [31].

## 4.2 *Viscoelastic Models*

While tremendous efforts have been made and remain to be devoted in the hyperelastic modeling of the 3D, anisotropic composite material of myocardium, it is imperative to develop viscoelastic models at multiple scales and improve the understanding of the organ that is constantly under dynamic motions. The establishment of various viscoelastic models, including the phenomenological and constitutive models to describe the quasilinear viscoelastic (QLV) or fully nonlinear viscoelastic (NLV) behaviors, has been achieved in broad fields of materials science or biological tissues. Nevertheless, the application in myocardial cells or tissues, particularly in the context of the RV, has remained limited. Therefore, we will start our discussion with more simplistic approaches of viscoelastic modeling using the Maxwell and Kelvin-Voigt models and move to the more complex and robust QLV and NLV models.

### 4.2.1 Phenomenological Models

A simplistic approach to describe the viscoelastic phenomenon of a material is to represent the elastic and viscous properties using spring and dashpot elements. Depending on the number of components and the relation between elastic and viscous components, the Maxwell and Kelvin-Voigt models have been used to model myocardial viscoelastic behavior. The Maxwell model simulates the material as a spring and dashpot in series, and the formulation for stress in the material is described as

$$\sigma_{\text{total}} = \eta \dot{\varepsilon} - \frac{\eta}{E} \dot{\sigma}, \tag{1}$$

where $\sigma_{\text{total}}$ is the total stress in the material, $\eta$ is the viscosity of the material, $\dot{\varepsilon}$ is the strain rate, $E$ is the relaxation modulus (stiffness), and $\dot{\sigma}$ is rate of stress applied. In this model, the stress applied to the entire system is applied to both the spring and dashpot elements, but the total strain is the sum of the strain in the dashpot and in the spring elements. Unlike models of hyperelasticity, the elastic element is a *linear* spring, which is too simple in capturing the nonlinear elastic behavior of the RV.

The Kelvin-Voigt model simulates the material as a spring and dashpot in parallel, and the formulation for the total stress in the material is described as

$$\sigma_{\text{total}} = E\varepsilon + \eta\dot{\varepsilon}, \tag{2}$$

where $\varepsilon$ is the applied strain. In this model, the strain applied to the entire system is equivalent to the strain in the spring and dashpot elements, while the total stress is the sum of the stress of the spring and dashpot elements.

Many groups have combined the Kelvin-Voigt and Maxwell models to improve accuracy of simulating the physical system. These models are referred to as standard linear solid (SLS) models. Such models have been used to describe stress relaxation or cyclic loading mechanical data in papillary muscles [75, 82] and ventricular tissue [14, 19, 35, 60, 83]. At the cell level, Caporizzo et al. modeled isolated CM stress relaxation and cyclic loading response as a combined Maxwell-Kelvin-Voigt model, with a spring and dashpot in parallel with a spring-dashpot series element [42]. Each element was related to a different portion of the loading regime: In stress relaxation, the individual dashpot element represented the ramp phase, the spring-dashpot series element represented the early relaxation phase, and the individual spring element represented the end-stage relaxation phase (equilibrium). Alternatively, in the cyclic loading regime, the individual spring element represents the low-strain-rate stiffness where viscous stress is negligible, the spring-dashpot series element represents the transition-strain-rate stiffness, and the dashpot element represents the high-strain-rate stiffness when viscous stress dominates the mechanical behavior. The rheological elements are added to better describe the physiological and rate-dependent response of the cell or tissue [77]. At the tissue level, Rubiano et al. has applied an SLS model to AFM indentation test data and predicted the viscous and elastic behavior of rodent myocardium [60]. Early uses of the Maxwell and Kelvin-Voigt models used linear elements, but the formulations have been improved by using nonlinear spring and dashpot elements to capture the nonlinear viscoelastic behavior of the myocardium [75, 84–86].

### 4.2.2 Quasilinear and Nonlinear Viscoelastic Constitutive Models

Unlike phenomenological models, constitutive models employ known physical and mathematical relations to formulate the governing equations. Depending on the type of viscoelasticity, various models have been used to predict the mechanical behavior of the myocardium. For a QLV material, the main governing equation is given as

$$\boldsymbol{S}(\boldsymbol{E}, t) = \boldsymbol{S}_0 + \int_0^t \boldsymbol{G}(t-\tau)\,\frac{\partial \boldsymbol{S}^e}{\partial \boldsymbol{E}} : \frac{\partial \boldsymbol{E}(\tau)}{\partial \tau} d\tau, \tag{3}$$

where **E** is the Green strain tensor, t is time, $\mathbf{S_0}$ is the initial stress obtained from the force in the zero-stress configuration, **G**(t) is the reduced relaxation function, $\mathbf{S^e}$ is the instantaneous elastic stress, and τ is a time variable of integration. Tissue viscosity is given by the reduced relaxation function, **G**(t), which is defined using a Prony series:

$$G(t) = G_\infty + \sum_{n=1}^{3} G_n e^{-t/\tau_n} \tag{4}$$

In the QLV model, the sum of the relaxation constants, $G_n$, is sometimes constrained to be 1 [4, 9, 34]. However, whether this constraint is necessary and whether varying numbers of $G_n$ values can be used are unclear. The above Prony series was adopted because it provided optimal fitting to ovine and rat RV stress relaxation experimental data. A similar approach has been applied to human myocardium tissues by Nordsletten et al. [35], but it is unclear whether the model would work well for the RV and the exact Prony series is not specified.

The traditional Fung-type QLV fitting scheme (Eq. 3), though, has offered strong fitting capabilities of stress relaxation raw data of both ovine [4] and rodent [9, 34] RVFW tissue. Moreover, the QLV model elucidates further information regarding the viscous relaxation of the tissue that can be obtained via experimental approaches, such as the time-dependent relaxation strength at various time points. Liu et al. observed a stronger relaxation strength at late (30 seconds) rather than earlier (0.3 and 3) relaxation time points of the healthy ovine RV [4]. Alternatively, our own group observed equivalent relaxation strengths between the early (0.3 seconds) and late (30 seconds) relaxation time points of the healthy rodent RV [9].

As we introduced previously, the viscosity of a QLV material is independent of strain or strain history. Thus, the relaxation function (viscosity) and elastic stress (elasticity) are separated. However, this is not the case for nonlinear viscoelastic materials. The QLV formulation was unable to fit RV stress relaxation raw data at multiple strain levels, suggesting a nonlinear viscoelastic response. Thus, to model a NLV material, the main governing equation becomes:

$$\boldsymbol{S}\left(\boldsymbol{E}(t), t\right) = \boldsymbol{S}_0 + \int_0^t \boldsymbol{M}\left(\boldsymbol{E}\left(\tau\right), t-\tau\right) : \frac{\partial \boldsymbol{E}\left(\tau\right)}{\partial \tau} d\tau, \tag{5}$$

where **M** is the strain-dependent relaxation modulus. Due to the high computational cost, the first NLV model of the RV was developed with the assumptions of a step increase in strain (Heaviside function) and a power-law behavior in the relaxation modulus.

Liu et al. found that this NLV model had a higher capacity in predicting the stress relaxation at multiple strain levels than the QLV model; thus, the NLV model formulation is better suited for describing RV viscoelasticity [4, 7]. This was the first study to treat the RV as a fully nonlinear viscoelastic material, and the study only investigated healthy ovine tissue. Future work should move toward the applications in other species or states and the use of a fully NLV model to capture the viscoelastic behavior of the RV.

### 4.2.3 Fractional Viscoelastic Models

Many of the viscoelastic models presented thus far consider the viscous and elastic components to be additive, which implies that the elasticity (spring) and viscosity (dashpot) are separate elements. But these models have their own disadvantages [87]. Recently, a new category of viscoelastic models (fractional viscoelastic models) has been developed to overcome these limitations. With the introduction of fractional calculus as a mathematical tool, it is possible to replace the convolution integrals used in a QLV modeling (see Sect. 4.2.2) with a Caputo derivative, with the assumption of isotropic, power-law behavior materials. The application of fractional derivatives to represent viscoelastic behavior was introduced by Suki et al. in a mechanical study of lung tissue [88]. The power-law assumption is mostly valid as it has been shown that in a wide range of materials, including biological materials, the stress relaxation response follows a power-law signature, and this has been attributed to the complex microstructure of the material (please refer to the review of Bonfanti et al. [87]). As a result, the governing constitutive equation and the number of model parameters can be simplified or reduced [87]:

$$\sigma(t) = c_\beta \frac{d^\beta \varepsilon(t)}{dt^\beta} \tag{6}$$

where $c_\beta$ is a constant determined by the elastic and/or viscous properties and often interpreted as the "firmness" of a material, and the fractional derivative is also called the Caputo derivative.

In the fractional viscoelastic model, a new spring-pot element (defined by the Caputo derivative) is introduced, in which the viscous and elastic components are not necessarily separate, additive components. This element can also be combined with spring or dashpot elements, and such models are referred to as generalized fractional viscoelastic models (see Table 1 in [87]). Depending on the power exponent $\beta$ ($0 \leq \beta \leq <1$), the fractional element (spring-pot) behaves as a spring ($\beta = 0$) or a dashpot ($\beta = 1$) and the stress in this element is derived using a Caputo derivative [77]. Recently, Nordsletten and colleagues have compared and evaluated over 4000 viscoelastic models across different test conditions (uniaxial, biaxial, pure shear, simple shear, torsion tests with sinusoidal or sawtooth wave functions, as well as relaxation and creep tests) [35, 77]. Although all models are limited to isotropic materials, they found that different viscoelastic models should be chosen for different testing conditions, and the fractional Zener model is the most robust model for a wide range of tests and frequencies. To our knowledge, the use of fractional viscoelastic modeling on myocardial tissue is limited to the LV only [35]. It would be interesting to assess this model in describing the viscoelastic behavior of the RV.

## 4.3 Active Cardiac Mechanics Models

At the subcellular and cellular levels, models incorporate myofilament mechanics, geometry, and interactions to capture the active mechanical behavior of isolated CM [70], such as the Huxley model [89]:

$$F(t) = k \sum_{-\infty}^{\infty} n(x, t)\, x dx \tag{7}$$

where $F$ is the contractile force, $k$ is the stiffness of the actin-myosin cross-bridge, and the distribution function $n(x, t)$ describes the rate of connections between actin filaments and myosin heads [90]. Alternatively, Rice et al. developed a phenomenological model to predict CM activation and force generation using myofilament behaviors [91]. Using an experimental approach, they determined 8 key modulators of CM active mechanics, including steady-state force-sarcomere length relations, steady-state force-calcium relations, steady-state sarcomere length-calcium relations, as well as the effect of sarcomere length on the intracellular calcium transients. The model accurately fit the predictions of the well-established Hill model. Additionally, studies have incorporated calcium activation [92], as well as sarcomere dynamics, including calcium binding and length changes [93]. For example, Lumens et al. derived myofiber active mechanical stress using sarcomere length-dependent activation and the total active stress of the tissue [94]:

$$\sigma_{f,\mathrm{act}} = \sigma_{\mathrm{act}} C \left(L_{sc} - L_{sc0}\right) \frac{L_s - L_{sc}}{L_{se,iso}}, \tag{8}$$

where $\sigma_{\mathrm{act}}$ is the total active stress in the tissue, $C$ is the mechanical activation, $L_{sc}$ is the contractile element length, $L_{sc0}$ is the contractile element length with zero active stress, $L_s$ is the sarcomere length, and $L_{se,\,iso}$ is the length of an isometrically stressed series elastic element. The time derivative of $C$ is heuristically obtained, with separate terms describing the rise and decay of mechanical activation $C$:

$$\frac{dC}{dt} = \frac{1}{\tau_R} C_L \left(L_{sc}\right) F_{\mathrm{rise}}(t) + \frac{1}{\tau_D} \frac{C_{\mathrm{rest}} - C}{1 + e^{(T(L_{sc}) - t)/\tau_D}} \tag{9}$$

where $\tau_R$ and $\tau_D$ are scaling rise and decay time parameters, respectively. Functions $C_L$, $F_{\mathrm{rise}}$, and $T$ refer to increase of activation with sarcomere length, rise of mechanical activation, and decrease of activation duration with decrease of sarcomere length, respectively. Finally, symbols $t$ and $C_L$ denote the time and diastolic resting level of activation, respectively.

All these above models have been validated by the cell-level experiments, and thus can be used to delineate the independent role of each factor (i.e., sarcomere length control) and to predict the active force at certain hypothetical conditions. However, except for the Hill model that incorporates the interactions of the muscle to extracellular component, the other models are microscopic models of muscle contraction and do not take extracellular factors into consideration. For instance, the interactions of the CM and extracellular matrix proteins, the orientation and network of fibrillar and non-fibrillar components in the ventricular wall, as well as the electrical conductivity of the tissue, all play a critical role to the overall contractive behavior of the tissue or entire organ. More importantly, the cell-level modeling is only 1D and does not have the capacity to capture the multi-dimensional feature of active constriction. As a result, the macroscopic active force at the tissue level or organ level is not well understood, and the total force or stress is a simple summation of passive and active mechanical force, mostly in a scaler format rather than a tensor format.

## 5 RV Mechanics in Physiology and Pathophysiology

Heart failure (HF) is the leading cause of death in both the United States and worldwide [95]. Ventricular dysfunction is the most common type of HF, including right-sided HF secondary to pulmonary hypertension (PH), heart failure with preserved ejection fraction (HFpEF), and congenital heart disease [96–98]. During HF failure development, significant cell and extracellular matrix (ECM) remodeling takes place, which contributes to overall changes in tissue mechanics and function. The initial remodeling occurs to compensate for altered hemodynamic loads (adaptive remodeling). However, in chronic disease, the tissue will reach a maladaptive remodeling state, and the structure and function of the cells and organ will be impaired. The mechanism of transitioning from an adaptive to a maladaptive state is not completely understood.

Historically, the RV was considered the "forgotten" chamber, and the knowledge obtained from the LV is often considered true for the RV. However, the LV and RV are distinct in their physiology, pathology, and mechanical behavior. The systemic circulation operates in a high pressure, high resistance, and low compliance system, while the pulmonary circulation operates in a low pressure, low resistance, and high compliance system [23]. Thus, the LV and RV experience drastically different hemodynamic loads. Furthermore, in the case of pressure overload, the RV experiences a significantly greater increase in afterload compared to the LV: fivefold vs 1.5-fold, respectively [99–101]. At the cell level, the Ca2+ sensitivity and handling of CM are different between the LV and RV [102], and more discussions of CM function between these chambers can be found in [103, 104].

In addition, it has been shown that, unlike the LV, the passive mechanics of the RV are a key determinant of organ function [2, 33], as RV elasticity is closely correlated to the degree of RV failure and is associated with the prediction of systolic function [105–107]. Thus, it is imperative to investigate the RV biomechanics in disease development. The focus of this section is on the structural remodeling and alterations of tissue mechanics in RV failure progression. We will discuss biomechanical mechanisms of the RV at multiple scales (from molecules and cells to tissue and organ), which includes the alteration of RV mechanical properties and the implications in RV malfunction.

## 5.1 *Intracellular Components of the RV*

The CM serves as the contractile machine of the myocardium. Microtubules, actin, and myosin are the three cytoskeletal filaments of CM, and the microtubules are the largest and stiffest of these three filaments [17]. Titin is a stiff, large sarcomeric protein, and is often viewed as the dominant protein determining cardiomyocyte stiffness. Thus, for the purposes of this review, we will focus on microtubules and titin.

### 5.1.1 Microtubules

#### Structure and Assembly

Microtubules comprise two subunits: α- and β-tubulin, and the assembly and disassembly of microtubules are extremely dynamic, with a turnover of α- and β-tubulin that is dependent on the degree and rate of stretch [108]. Without loading, the microtubules grow at a rate of 0.2–0.4 μm/s, with a much higher shrinking rate [109]. Individual microtubules with crosslinkers interspersed between them form what is called the microtubule network. Under stretch, the crosslinkers engage, which causes the microtubules to become more aligned and induces a nonlinear stiffening behavior of the microtubule network in the cell [17]. Additionally, there are two processes that will alter its structure and mechanics: polymerization and tyrosination. These processes are heavily determined by temperature, rate of stretch, and can be altered via pharmaceutical treatment [16, 17, 42, 110–112].

#### Changes in RV Failure Progression

In the pressure-overloaded RV, microtubule density is increased in the hypertrophied CM, with elevated expressions in both α- and β-tubulin [17, 40, 113, 114]. Furthermore, during heart failure progression, microtubule network stability is weakened [115, 116]. It has been observed, too, that the microtubule network becomes disorganized in pressure-overloaded RV [113], and there is an upregulation of genetic expression associated with abnormal microtubule assembly [117, 118]. In addition, the accumulation of microtubules in PH patients is not the same between males and females. Xiong et al. reported significant elevations of α- and β-tubulin with PH in male rat RVs, while no significant change in α- and β-tubulin expression or microtubule polymerization was observed in female rat RVs [119].

Role in RV Mechanics

Cooper and his colleagues have reported significant impact of MT removal on CM in the contractile function of CM or papillary muscles from healthy and pressure-overloaded LVs and RVs [120–122]. But the mechanics of the cells or muscles have not been quantified. This is the first evidence that suggests a linkage between cell mechanics and function, but the translation at the tissue level is unknown. Caporizzo et al. observed that depolymerization of the microtubule network via colchicine treatment reduced cardiomyocyte viscoelasticity and enhanced the extent of shortening [16, 42], which indicates the impact of microtubule on cell mechanics and contractile function. But this behavior was only observed in human CM from healthy and failing LVs. Recently, we applied a similar approach at the tissue level by exposing healthy and pressure-overloaded RV samples to colchicine treatment. In the healthy RV, we observed significant reductions in tissue viscosity and elasticity (from stress relaxation) in both longitudinal (outflow tract) and circumferential directions after depolymerization of the microtubules [9]. Moreover, the impact was different at different strains and directions of the tissue. First, these changes were more substantial at lower level of strain that corresponds to the early diastolic phase, where myofibers (CM) dominate RV mechanical behavior [1, 2]. Second, the reductions were stronger in the longitudinal than circumferential direction, indicating an anisotropic impact of microtubules. In the pressure-overloaded RV, we observed similar but stronger effects of colchicine than in healthy RVs [34]. The degree of reduction in elasticity and viscosity was not equivalent, though, as the ratio of viscosity to elasticity was significantly decreased. These findings indicate distinct molecular mechanisms of RV viscoelasticity.

### 5.1.2 Titin

Structure and Assembly

Titin is a stiff, large sarcomeric protein, with its Z-disk connected to the M-band region of the sarcomere [123, 124]. Titin comprises multiple segments along its structure, each acting as a distinct spring element [125]. There are multiple isoforms of titin, with two main titin isoforms expressed in cardiac muscle: N2A and N2BA [126]. N2A is the stiff isoform, and N2BA is the compliant isoform of titin [127, 128]. Like microtubules, titin's structure and mechanical property are heavily influenced by mechanical load. When the cell is stretched to a sarcomere length greater than 2.1 μm, the Z-disk and the myofilaments become more aligned and increase the stiffness of titin and, therefore, the entire cell [110, 129, 130].

Changes in RV Failure Progression

Titin undergoes significant remodeling with RV failure progression. Rain et al. reported reduced titin phosphorylation in rats with mild RV dysfunction, but the reduction was not significant in rats with severe RV dysfunction [127]. However, they did not find marked change in the total expression of titin. Moreover, multiple studies have observed a shift in titin isoform expression, with alterations in the N2BA/N2Aratio [127, 128, 131–136]. These changes are dependent on the severity of disease, as well as pathology. For example, in PH, the N2BA/N2A ratio is increased in the RV, suggesting a more compliant muscle in the chamber wall [127, 128], whereas in dilated cardiomyopathy and tachycardia, the ratio is reduced, suggesting a shift to a stiffer muscle [131–133]. Interestingly, though, in end-stage dilated cardiomyopathy, an increase in the N2BA/N2A ratio has been reported [134–136]. While the increase in N2BA/N2A ratio is viewed as a compensatory mechanism, significant myofiber stiffening is still observed in severe RV dysfunction [127], which may be due to other molecular mechanisms. Thus, the alteration in titin is unlikely to account for the increase in RVFW stiffness.

Role in RV Mechanics

In isolated muscle fiber bundles from the LV, Loescher et al. cleaved titin via HaloTag-Tobacco Etch Virus protease and observed marked elastic *and* viscous force reductions at multiple strain levels (5–25%) [137]. The contribution of titin was more marked at a higher strain level, with an approximate 73% reduction in elastic force at 10% of strain. After the breakdown of titin via KCl/KI, Granzier et al. observed a 70% reduction in passive tension, and this response was stronger with a higher sarcomere slack length [138]. Interestingly, there was no difference in the response of the trabeculae from the LV or RV. Moreover, Rain et al. investigated the role of titin phosphorylation on myocardial stiffness in the pressure-overloaded RV [127]. In the muscle with phosphorylated titin, myocardial stiffness was significantly reduced in both moderate and severe RV dysfunction samples. Similarly, in skinned rat RV muscles, Fukuda et al. observed a decrease in magnitude of peak force (elasticity) and a weakened stress relaxation response (viscosity) after titin phosphorylation [139]. Although it is unclear how phosphorylation alters the cellular structure and then mechanics, these studies suggest a significant role of titin in RV myocardial mechanics. Unfortunately, the contribution of titin to the tissue-level mechanical behavior of the RV has not yet been elucidated.

## 5.2 *Extracellular Components of the RV*

Collagen and fibronectin are two of the most predominant ECM proteins in the ventricular free wall [140]. The content of these two proteins is heavily influenced by mechanical load [141–143]. This regulation occurs via the mechanosensing cardiac fibroblast, which is responsible for the increased synthesis of collagen and fibronectin under elevated loading environment [140, 144, 145]. Due to the lack of data in the role of fibronectin in RV mechanics, we will mainly discuss the findings related to collagen in this section.

### 5.2.1 Structure and Assembly

Collagen is the most prevalent structural ECM protein in the RV. It is formed from the base unit called tropocollagen [146]. Tropocollagens twist together, making a triple helix structure, which forms the procollagen fibril. Procollagen fibrils then crosslink to form mature collagen fibers, bundles, or sheets [148]. To date, over 25 types of collagen have been reported in biological tissues, and type I and type III collagen are the most predominant type in the heart to display a unique structure and mechanical behavior [147].

### 5.2.2 Changes in RV Failure Progression

In pressure-overloaded RVs, elevated collagen content (fibrosis) is reported in small and large animal models, as well as in early- and late-stage RV failure [102, 148–150]. Both Bogaard et al. and Borgdorff et al. have reported a marked fibrotic response of RVs in experimental models of PH [151, 152]. It is not well understood, though, whether certain collagen subtypes are more upregulated in diseased states than others. Golob et al. reported that, in a mouse model of PH (Sugen + hypoxia), the ratio of collagen type I to collagen type III was increased, and this alteration was primarily contributed by type I collagen [150]. Rain et al. observed the same trend in a rat model of PH (pulmonary arterial banding) [127].

In addition to collagen deposition, alterations in collagen morphology and alignment have also been reported in diseased RVs. Rain et al. observed increased collagen crosslinking in the RVs four weeks after PH development, and this increase was correlated with elevated diastolic pressures [127]. But the crosslinking formation may not form in early remodeling as Baicu et al. did not observe a change in insoluble collagen expression or diastolic pressures two weeks post-pulmonary arterial banding [153]. However, collagen content was elevated, suggesting new collagen fibers were deposited but not yet crosslinked at this time point. These results further suggest that the increased diastolic pressure is associated with collagen crosslinking, not content. It is speculated that the increased crosslinking stiffens the RV myocardium and prevents overstretch of the tissue, which leads to a fatal pathological condition [154].

Finally, alterations in myofiber and collagen alignment in the RV have been reported in PH development. Fibers have been found to become more aligned (organized), and their preferred direction is altered toward the long axis (apex-to-outflow tract direction) in several rodent studies, including our own work [1, 3, 13].

Interestingly, though, there is no significant change in fiber orientation in an ovine model of RV failure secondary to PH [155], with a preferred alignment of 60° in the healthy ovine RV. Park et al., though, observed significant fiber orientation change in pressure-overloaded rat RVs at both the endocardial and epicardial sides of the free wall [156]. Moreover, the transmural change in the fiber orientation of the pressure-overloaded RV was significantly smaller than that of the healthy RV, suggesting a small transmural fiber orientation distribution with PH progression. It has also been shown that RV collagen fiber preferred orientation moves toward the circumferential direction during growth and development [157, 158], although these measurements were performed only in porcine RV samples. Whether the collagen fiber orientation of human RV changes with growth and disease progression is unknown.

### 5.2.3 Role in RV Mechanics

In the pressure-overloaded rat RV, Hill et al. reported no change in individual collagen fiber stiffness with PH progression, which was quantified using a constitutive model [13]. Hence, their findings suggest that it is the remodeling of collagen structure and morphology, not the intrinsic mechanical properties of the protein that contributes to the changes in mechanical behavior in PH development. In general, fibrillar (type I) collagen is more dominant in high-strain regions, as it recruits during deformation. In an ovine model of PH, Liu et al. reported that type III collagen, not myofiber hypertrophy, is strongly correlated with the longitudinal low-strain elastic modulus of the RV, whereas the total collagen content is correlated with the high-strain elastic modulus [23]. We speculate the discrepant contributions between low- and high-strain regions are most likely influenced by the different recruitment and mechanical behaviors of type I and III collagen. Fibrillar type I collagen is stiffer compared to the non-fibrillar, mesh-like type III collagen [159]. Typically, the collagen fibers are thought to be recruited at larger strains and thus contribute to the nonlinear elastic behavior of the RV. The findings in the ovine PH RVs suggest a new role of type III collagen in the low-strain RV mechanical behavior, which needs further investigations.

# 6 Translation from RV Mechanics to Organ Physiology

## *6.1 Implications of RV Mechanics to Organ Function from Ex Vivo Studies*

As previously discussed, there are marked biological remodeling and mechanical alterations in the RV with disease progression. While many *ex vivo* studies have been performed to thoroughly characterize the mechanical behavior of the RV tissue, only a few directly examined the physiological relevance of the measured mechanical properties. For instance, Trip et al. and Rain et al. found that the RV diastolic stiffness measured by the end-diastolic pressure-volume relation (EDPVR) was related with clinical progression of both healthy and pulmonary arterial hypertension human patients, as an elevated stiffness was associated with worse prognosis [107]. Additionally, Jang et al. reported that the rat longitudinal RV elastic modulus was correlated with the EDPVR, indicating a correlation of *ex vivo* and *in vivo* measurements [2]. Later, Liu et al. found in an ovine model of RV failure that the longitudinal elastic modulus at high strain was positively and significantly correlated with ejection time, a metric of systolic function [33].

Furthermore, as the viscoelastic behavior of ventricular wall raises more attention in recent studies [4, 9, 12, 15, 34, 60, 160], the changes in RV viscoelastic properties with disease progression and their implications in RV dysfunction have been noted. In a rat model of PH, elevated RV tissue elasticity and viscosity, as well as enhanced anisotropy in both parameters, were reported with disease progression [10, 34]. In both studies, it was observed that the relative changes of viscosity and elasticity were different, evidenced by an impaired damping capacity (ratio of dissipated energy to total energy) [10] as well as a reduced ratio of dissipated energy over stored energy (i.e., V/E ratio) [34] with PH. These findings suggest distinct viscous and elastic responses of the RV in disease progression. Furthermore, it was observed that RV's circumferential damping capacity was significantly correlated with acceleration time and the ratio of acceleration time/ejection time, both of which are metrics of systolic function [10]. These findings suggest the roles of both RV elasticity and viscosity in organ performance. It is unclear, though, whether a homeostasis of these two parameters should be maintained in compensated RV hypertrophy.

The regulation of mechanical behavior via the microtubule network may offer some insights into the role of tissue viscoelasticity in organ function. Hancock et al. depolymerized the microtubules via colchicine in LV myocardial slices and observed improved diastolic performance by assessment of *ex vivo* work loops [161]. Moreover, in a mouse model of hypertrophic cardiomyopathy, Pietsch et al. observed improved LV systolic performance after microtubule destabilization using *in vivo* pressure-volume measurements [162]. In both studies, an increased loop width (stroke volume) was observed after the removal of microtubule network. Considering the Frank-Starling mechanism, it can be inferred that the reduced viscoelastic resistance by the destabilization of the microtubule network leads to this enhanced stroke volume and improved performance. Additionally, the effect of chronic microtubule destabilization on RV function in a rat model of PH has been investigated by Prins et al. RV performance was significantly improved after microtubule depolymerization, although this result is attributed to t-tubule architecture enhancement and reduced adverse pulmonary vascular remodeling [113]. Overall, a complete assessment of the dependence of RV function on tissue mechanics remains a knowledge gap. Moreover, these prior studies did not investigate the separate roles of tissue viscosity and elasticity but treated the "viscoelastic resistance" as a whole to impact the ventricular function. Since Liu et al. have observed distinct changes in RV viscosity and elasticity with PH progression [10], it is possible that these mechanical properties may contribute differently to RV failure development.

## 6.2 Implications of RV Mechanics to Organ Function from In Vivo Studies

The advancement in medical imaging has enabled *in vivo* assessments of RV from various imaging modalities. The utilization of these methods offers different mechanical and physiological metrics, as we will briefly discuss below. Moreover, the imaging data can be used to develop shape models of the RV, which are useful inputs for multiscale computational models as discussed in Sect. 6.3.

### 6.2.1 Echocardiography

Echocardiography is a common, noninvasive approach to assess ventricular performance, often used in clinical settings. This technique uses imaging data to derive certain metrics of contractility, such as fractional shortening (FS), acceleration time (AT), ejection time (ET), and tricuspid annular plane systolic excursion (TAPSE). Echocardiography also offers information about ventricular and atrial geometry, which provides insight into the health of the organ. Using 3D echocardiography, strain mapping can be performed to track tissue stretch and contraction during the cardiac cycle [163]. This method is referred to as speckle-tracking echocardiography (STE). However, STE has mainly been performed in the LV and is limited in the RV due to its irregular geometry and thin wall. Recently, though, 2D STE of the RV has been performed and the strain parameters were found to be prognostic of the mortality of HF with reduced ejection fraction (HFrEF) [164].

### 6.2.2 Cardiac MRI

Cardiac magnetic resonance (CMR) imaging is another noninvasive method for measuring ventricular and vascular geometry, as well as flow parameters and myocardial relaxation [165]. RV strain quantification has been performed from CMR images. Using a hyperelastic warping method, which is a finite element method, RV strains were derived in a single ventricular model [166] as well as a bi-ventricular model [167]. Strain-encoded (SENC) MRI is another effective tool in strain mapping of the RV [168, 169], and it has been validated against conventional tagged images [170]. The RV strains derived from CMR have been shown to be indicative of systolic function [168–170].

### 6.2.3 CT

Computed tomography (CT) is an X-ray imaging technique in which X-ray beams are directed at the patient to compile 3D images of the area of interest. Cardiac CT is a powerful tool in cardiovascular disease diagnosis. Additionally, strain analysis can be performed from CT images. As described by Xiong et al., strain tensors are derived from subsequent 3D images throughout the cardiac cycle [171]. Strain mapping from CT scans, though, is not standard. The time resolution of CT is low, which results in an inaccurate strain-time wave form, though the spatial resolution is high. Thus, strain measurements from echocardiography or CMR are better options.

## 6.3 *Implications of RV Mechanics to Organ Function from Multiscale Modeling*

While current medical imaging modalities offer great tools in assessing RV geometry and function, the evaluation of RV mechanics with *in vivo* methods is limited compared to the *ex vivo* measurements. Furthermore, the experimental approaches are often time- and cost-expensive, and it is difficult to rule out individual factors' influence on the overall organ performance. Recent development in multiscale computational modeling has the potential to overcome these obstacles and provide novel insights on the translation from RV mechanics to organ physiology.

Multiscale computational models of the cardiovascular system, including both ventricular and vascular components, can incorporate multiple physical and physiological processes at different scales, such as (vascular) hemodynamic loads, ventricular geometry, tissue mechanical behavior, and cardiac muscle cell contractility, into the organ-level performance and function (Fig. 7). The researcher can adjust the model by choosing the necessary components and processes and varying from 0D to 3D modeling for each component. Such computational frameworks offer a powerful tool capable of delineating specific modulators (e.g., RV mechanical behavior) and their involvement in multiscale processes as well as predicting the ventricular performance in specific processes and pathologies [172–175]. Below, we will provide a summary of the model components (Fig. 7) and their applications in RV research. More general reviews on the multiscale modeling of the RV and/or pulmonary circulation system can be found here [176–178].

### 6.3.1 Shape Model as Inputs for Multiscale Modeling

Accurate modeling of ventricular geometry is critical in the development of multiscale computational models, as chamber geometry and myocardial strain information are direct inputs to the passive mechanical and vascular hemodynamic components of the framework. The ventricular chamber is often simulated using a so-called shape model that includes anatomy and geometry. A well-known shape model, CircAdapt, developed by Arts et al. [179], is a 4-chamber model coupled to the vasculature that can simulate beat-to-beat dynamics. The CircAdapt model can also simulate geometric and structural remodeling of the ventricles in response to mechanical load. The independence of the two ventricles, though, was not

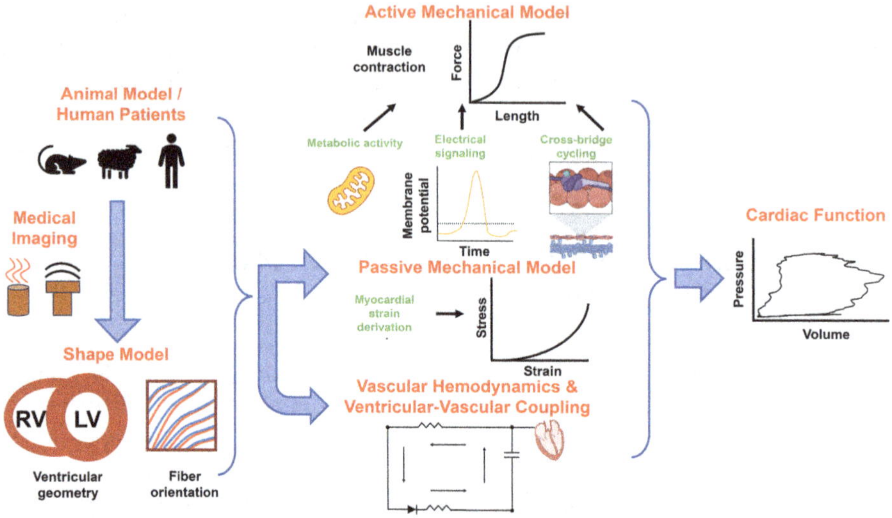


**Fig. 7** Functional diagram of multiscale computational modeling framework inputs in organ function prediction. (The figure was partially generated using BioRender)

considered as the ventricular shapes are simulated as one large outer chamber encapsulating a small inner chamber. The septum was also not considered. The TriSeg model was, thus, developed to counter these limitations. Developed by Lumens et al., the TriSeg model simulates the LV, RV, and septum as three thick-walled segments [94] and has been used previously in studies of the RV [180, 181]. Both the TriSeg and CircAdapt models are considered bi-ventricular models. The TriSeg and CircAdapt models assume a homogeneous material, which is a simplistic approach to modeling the myocardium that only includes the overall chamber geometry. Pourmodheji et al., though, developed a single chamber shape model of the RV with an assumed geometry of a quarter prolate spheroid [173].

While these shape models use an idealized geometry of the ventricles, medical imaging modalities can be used to construct patient-specific 3D finite element (FE) models of the heart. For example, Xi et al. developed a patient-specific bi-ventricular model of a healthy and pressure-overloaded RV to assess the ventricular interactions during disease progression [182]. The FE mesh was developed from CINE MRI scans in healthy and PAH patients, and fit contours of the LV and RV endocardium, and epicardium at various short- and long-axis views to obtain 3D endocardial and epicardial surfaces of the LV and RV. Similarly, Gomez et al. obtained the geometry for a bi-ventricular FE model from CINE MRI scans of healthy adult rabbits [183]. The study was interested in the effect of myofiber alignment during PH progression. Thus, DT-MRI scans were performed to incorporate transmural myofiber architecture and obtain different FE meshes at the healthy and pressure-overloaded state. Similarly, Avazmohammadi et al. and Finsburg et al. have developed patient-specific bi-ventricular FE models in rat and human [184, 185]. Because the FE modeling framework incorporates the morphological and mechanical changes of the RV, it is widely considered the most versatile modeling method for predicting myocardial mechanics [178, 186, 187].

While the development of FE models of the heart is more computationally demanding than using an idealized geometry, the former offers the ability to incorporate microstructural information into the framework, which is critical in understanding RV remodeling. Recently, though, the MultiPatch module was developed by Walmsley et al. to allow modeling of the heterogeneity in the RV tissue, which improves its capabilities of capturing structural properties of the RV free wall [188].

### 6.3.2 Active Mechanics Modeling (Contraction Model)

Many models of myocardial active mechanics incorporated in multiscale computational frameworks are at cellular levels and only one-dimensional. We have previously reviewed the constitutive models of cellular contractility in Sect. 4.3, and such models have been applied to the multiscale modeling framework. Common formulations of active stress include those presented by Kim et al. and Lumens et al., where active stress is a function of sarcomere mechanics (length under zero tension and under stress, shortening velocity) and mechanical activation [94, 180, 189]. Alternatively, instead of using a description of sarcomere kinetics, Tözeren et al. proposed a modified Hill model to describe the muscle active tension as a function of intracellular calcium concentration and length-dependent calcium sensitivity [190, 191]. Tossas-Betancourt applied this model in a bi-ventricular model of pulmonary arterial hypertension to describe the active stress in a single myocardial muscle fiber of the RV [192].

The modeling of tissue-level active mechanics is not a common approach. Thus, unlike the passive mechanics modeling of the RV, the active mechanics of the tissue is often reduced to 1D that aligns with the muscle fiber's long axis. When the total stress (i.e., the summation of the passive and active stress) is derived, the passive stress in the fiber direction is used [3, 15, 173]. The lack of a 2D or 3D description of the active mechanical behavior of the tissue remains a major knowledge gap in the field.

### 6.3.3 Passive Mechanics Modeling

As we discussed in Sects. 4.1 and 4.2, a variety of constitutive formulations are established to describe the passive stress within the RV. Some of them have been adapted to the multiscale modeling of the RV. For example, Colebank et al. modeled the passive stress (1D) generated as a sum of stresses in the ECM ($G_{ECM}$) and titin ($G_{titin}$), using a structurally informed approach [181], where the stress in the ECM and titin is a function of sarcomere length, $L_s$:

$$G_{\text{pas}} = G_{\text{ECM}}(L_s) + G_{\text{titin}}(L_s) \tag{10}$$

Similarly, in a bi-ventricular model of HFpEF, Kim et al. treated the passive stress (1D) as a function of the time-dependent sarcomere length based on myocardial strain ($L_i$) and the sarcomere length at zero active stress ($L_{c,0}$) [180]:

$$\sigma_{\text{pas},i} = \left(\nu_L \left(L_i - L_{c,0}\right)\right)^{\gamma} \tag{11}$$

where $\nu_L$ is a unit conversion factor. The model accurately predicted the RV pressure-volume relation and revealed the role of RV compensation in the progression of LV failure.

The constitutive relations used in the above studies derive the passive stress at the sarcomeric level and is only one-dimensional. However, other groups determine the passive mechanics at tissue level based on material parameters of the entire tissue. For example, in a bi-ventricular mode of pediatric pulmonary arterial hypertension, Tossas-Betancourt et al. derived the RV passive stress (3D) from a modified Ogden-Holzapfel strain energy density function [182, 192]:

$$W = \frac{1}{2}c\left(e^{Q} - 1\right) \tag{12}$$

where $Q$ is a quadratic function of the strain components defined in the material directions:

$$Q = b_{ff}E_{ff}^2 + b_{xx}\left(E_{ss}^2 + E_{nn}^2 + E_{sn}^2 + E_{ns}^2\right) + b_{fx}\left(E_{fn}^2 + E_{nf}^2 + E_{fs}^2 + E_{sf}^2\right) \tag{13}$$

where $E_{ij}$ is the Green-Lagrange strain tensor with components $(i,j) \in (f, s, n)$ that denote the myocardial fiber ($f$), sheet ($s$), and sheet normal ($n$) directions, and $b_{ff}$, $b_{xx}$, and $b_{fx}$ are material constants. In this model, the ventricular and vascular hemodynamics (i.e., pressures and flows in the RV and pulmonary vessels) were predicted, and several model-derived metrics were strongly correlated with clinically assessed disease severity. Gomez et al. applied the same formulation for passive stress in a patient-specific bi-ventricular model of PH and found that the longitudinal alignment of myofibers was related to an enhanced RV ejection fraction [183]. This response was more pronounced in the pressure-overloaded RV. Thus, the re-orientation of the myofibers to the long axis of the RV may serve as a compensatory mechanism. Such findings provide critical knowledge of RV pathophysiology that cannot be obtained *in vivo*.

In addition, Pourmodheji et al. incorporated a growth and remodeling (G&R) component of RV in pulmonary arterial hypertension patients with the same strain energy density function shown in Eq. 12 [173]. Then, the RV passive fiber stress (1D) was given as:

$$\sigma_{RV,p}(s,t) = 2C(s)b_f E_f(s,t)\lambda_f^2(s,t)\,e^{Q(s,t)} \tag{14}$$

where $Q$, $C$, and $b_f$ are material parameters associated with the strain energy density function. Because this is a model of G&R, the passive stress is given as a function of time, where $t$ refers to the short timescale (defined by the duration of a single heartbeat), and $s$ refers to the long timescale (defined by the duration of time during disease progression). This is one of the first computational models to consider the G&R of tissues in the context of coupling between the pulmonary circulation and the RV in a closed-loop circulation. It was predicted that progressive RV stiffening led to a large increase in RV end-diastolic pressure but a small increase in end-diastolic volume. Moreover, RV stiffening resulted in an enhanced pulmonary arterial pressure and an upregulation of pulmonary arterial mass.

In another bi-ventricular model of PH, Odeigah et al. applied the Ogden strain energy density function to derive the passive mechanical behavior of the RV [193]. A shape model was developed using a FE framework to match experimentally measured RV wall thickness and inflated both ventricles to match the cavity volumes. The fiber angle architecture was prescribed to vary transmurally from +60° at the endocardium to −60° at the epicardium. The model predicted an increase in myocardial wall stress in the fiber, cross-fiber, and apex-to-outflow directions, which was consistent with previous experimental findings. Moreover, the authors attributed these increases to the enhanced wall thickening of the RV during PH progression.

Pourmodheji et al., Tossas-Betancourt et al., and Odeigah et al. all applied a transversely isotropic form of the hyperelastic strain energy density function to derive the passive stress of the RV. However, an orthotropic model may better represent RV mechanics, as the RV tissue has shown transmural changes in mechanical behavior. Moreover, recent works have suggested a significant viscoelastic behavior of the RV, and thus, a viscoelastic model would more comprehensively capture the dynamic mechanical behavior than the hyperelastic model. Future modeling work may consider an improvement in these aspects.

### 6.3.4 Vascular Hemodynamics and Ventricular-Vascular Coupling

In a multiscale model, the vasculature component is often incorporated using a lumped parameter approach. The hemodynamic loads are represented by the resistance and compliance of the vessels (arteries and veins), and both proximal and distal vessels can be included. Next, the hemodynamics are determined by simple physical laws (e.g., resistance = mean arterial pressure / mean flow, compliance = volume change / pressure change) and then used as the input pressure and volume of the RV chamber. A common approach using simple flow calculations to derive the arterial, venous, and valvular pressure and volumes has been documented in previous studies [173, 181, 192–194]. The lumped parameter approach is also called 0D modeling. Using more sophisticated computational fluid dynamic (CFD) modeling, 1D to 3D hemodynamics from the vasculature can be obtained and used for the multiscale model, as done previously [195, 196]. A thorough review of pulmonary vasculature modeling can be found in Vélez-Rendón et al. [197, 198]. In this chapter, we will focus on the coupling of vasculature with the RV using the lumped parameter approach.

Typically, clinical (human) or preclinical (large animal) data are used to determine baseline ventricular and vasculature pressure [193], or volume [173]. The lumped parameter model is then connected to a model of a single or two ventricular chambers. This method is typically used in shape models of idealized ventricular geometry. In FE models of the RV, the pulmonary circulation can be simulated using a lumped parameter model, or the vasculature can be included as the boundary conditions of the FE simulation [199]. Alternatively, Xi et al. have used direct hemodynamic measures of the RV and LV instead of using 0D modeling [182]. Pressure and volume waveforms were obtained via pressure-volume loop acquisition, and the waveforms were then synchronized with CINE MR images of the LV and RV, using the end-diastolic stage as the reference point. Thus, the true *in vivo* hemodynamic condition was used in this study.

The ability to combine developed models of the ventricle and the vascular system allows us to better understand the ventricular-vascular coupling (VVC) in RV pathophysiology. Experimentally, the VVC is quantified via a single parameter, defined as the ratio of end-systolic elastance ($E_{es}$) to effective arterial elastance ($E_a$) ($E_{es}/E_a$). However, the coupling mechanism may involve more aspects than what a single parameter can measure. In a computational model, the VVC can be comprehensively assessed by a metric of model parameters and easily obtained in different pathologies [172, 192] and at multiple time points of disease progression [173]. Moreover, multiscale models provide further insight into the relationship between vascular and ventricular remodeling [94, 173].

### 6.3.5 Future Work to Link RV Mechanics to Organ Function

The incorporation of RV mechanics in multiscale modeling frameworks is imperative in accurately predicting ventricular function. Current computational frameworks primarily incorporate the *elastic* behavior of the RV tissue and often only in 1D. However, as we have discussed previously, the RV exhibits marked viscoelastic and anisotropic mechanical behavior, and the neglect of these key features will impede a full understanding of the translational role of RV mechanics to organ function. Furthermore, the distinct alterations of tissue viscosity and elasticity during PH progression and their contributions to RV dysfunction have yet to be elucidated. Finally, a transmural mechanical difference has been reported in the ovine septum [32], and septal wall remodeling and geometrical changes with PH progression have been observed widely. A further investigation incorporating the shape and mechanics of the septum tissue will be valuable to shed new lights on the bi-ventricular interactions and RV failure.

## 7 Conclusions

RV mechanical behavior, including both elastic and viscous properties, is a critical factor in the progression of RV failure. Our review highlights current experimental and *in silico* approaches to characterize passive and active mechanics of the RV myocardium. In addition, we have shown a significant contribution of both intra- and extracellular components to the cell- and tissue-level viscoelasticity, the distinct alterations of viscosity and elasticity in disease progression, and implications of the RV mechanics to organ function. While the delineation of these multi-temporal, multiscale remodeling processes and hemodynamics/mechanics are not always feasible via experimental approaches, the implementation of multiscale modeling opens new ways to incorporate multiple or individual mechanical factors that will inform RV function. Further achievement in the multiscale computational framework will offer a promising tool to capture the biomechanical mechanism of the RV in different pathologies.

## References


1. R. Avazmohammadi, M. R. Hill, M. A. Simon, W. Zhang, and M. S. Sacks, 'A novel constitutive model for passive right ventricular myocardium: evidence for myofiber–collagen fiber mechanical coupling', *Biomech Model Mechanobiol*, vol. 16, no. 2, pp. 561–581, Apr. 2017, doi: https://doi.org/10.1007/S10237-016-0837-7.
2. S. Jang *et al.*, 'Biomechanical and hemodynamic measures of right ventricular diastolic function: Translating tissue biomechanics to clinical relevance', *J Am Heart Assoc*, vol. 6, no. 9, Sep. 2017, doi: https://doi.org/10.1161/JAHA.117.006084.
3. R. Avazmohammadi, M. Hill, M. Simon, and M. Sacks, 'Transmural remodeling of right ventricular myocardium in response to pulmonary arterial hypertension', *APL Bioeng*, Dec. 2017.
4. W. Liu *et al.*, 'Strain-dependent stress relaxation behavior of healthy right ventricular free wall', *Acta Biomater*, vol. 152, pp. 290–299, Oct. 2022, doi: https://doi.org/10.1016/J.ACTBIO.2022.08.043.
5. Y. C. Fung, 'Biomechanics: Mechanical Properties of Living Tissues', *J Appl Mech Eng*, vol. 49, no. 2, pp. 464–465, Jun. 1982.
6. K. M. Labus and C. M. Puttlitz, 'Viscoelasticity of brain corpus callosum in biaxial tension', *J Mech Phys Solids*, vol. 96, pp. 591–604, Nov. 2016, doi: https://doi.org/10.1016/j.jmps.2016.08.010.
7. K. L. Troyer and C. M. Puttlitz, 'Nonlinear viscoelasticity plays an essential role in the functional behavior of spinal ligaments', *J Biomech*, vol. 45, no. 4, pp. 684–691, Feb. 2012, doi: https://doi.org/10.1016/j.jbiomech.2011.12.009.
8. K. L. Troyer and C. M. Puttlitz, 'Human cervical spine ligaments exhibit fully nonlinear viscoelastic behavior', *Acta Biomater*, vol. 7, no. 2, pp. 700–709, 2011, doi: https://doi.org/10.1016/j.actbio.2010.09.003.
9. K. LeBar, W. Liu, A. J. Chicco, and Z. Wang, 'Role of Microtubule Network in the Passive Anisotropic Viscoelasticity of Healthy Right Ventricle ', *J Biomech Eng*, 2023.
10. W. Liu *et al.*, 'Alterations of biaxial viscoelastic properties of the right ventricle in pulmonary hypertension development in rest and acute stress conditions', *Front Bioeng Biotechnol*, vol. 11, 2023, doi: https://doi.org/10.3389/FBIOE.2023.1182703.

11. W. Liu *et al.*, 'Effects of Pulmonary Hypertension on Biaxial Viscoelastic Properties of the Right Ventricle at Rest and Exercise Conditions', in *Experimental Biology*, Philadelphia: The Federation of American Societies for Experimental Biology, Apr. 2022.
12. K. Roth, W. Liu, K. LeBar, M. Ahern, and Z. Wang, 'Establishment of a Biaxial Testing System for Characterization of Right Ventricle Viscoelasticity Under Physiological Loadings', *Cardiovasc Eng Technol*, pp. 1–13, Mar. 2024, doi: https://doi.org/10.1007/S13239-024-00722-5/METRICS.
13. M. R. Hill, M. A. Simon, D. Valdez-Jasso, W. Zhang, H. C. Champion, and M. S. Sacks, 'Structural and Mechanical Adaptations of Right Ventricle Free Wall Myocardium to Pressure Overload', *Ann Biomed Eng*, vol. 42, no. 12, pp. 2451–2465, Nov. 2014, doi: https://doi.org/10.1007/s10439-014-1096-3.
14. O. Gültekin, G. Sommer, and G. A. Holzapfel, 'An orthotropic viscoelastic model for the passive myocardium: continuum basis and numerical treatment', *Comput Methods Biomech Biomed Engin*, vol. 19, no. 15, pp. 1647–1664, Nov. 2016, doi: https://doi.org/10.1080/10255842.2016.1176155.
15. G. Sommer *et al.*, 'Biomechanical properties and microstructure of human ventricular myocardium', *Acta Biomater*, vol. 24, pp. 172–192, Sep. 2015, doi: https://doi.org/10.1016/j.actbio.2015.06.031.
16. M. A. Caporizzo, C. Y. Chen, K. Bedi, K. B. Margulies, and B. L. Prosser, 'Microtubules Increase Diastolic Stiffness in Failing Human Cardiomyocytes and Myocardium', *Circulation*, vol. 141, no. 11, pp. 902–915, Mar. 2020, doi: https://doi.org/10.1161/CIRCULATIONAHA.119.043930.
17. M. A. Caporizzo and B. L. Prosser, 'The microtubule cytoskeleton in cardiac mechanics and heart failure', *Nat Rev Cardiol*, vol. 19, no. 6, pp. 364–378, Jun. 2022, doi: https://doi.org/10.1038/S41569-022-00692-Y.
18. M. S. Sacks and C. J. Chuong, 'Biaxial mechanical properties of passive right ventricular free wall myocardium', *J Biomech Eng*, vol. 115, no. 2, pp. 202–205, May 1993.
19. L. L. Demer and F. C. Yin, 'Passive biaxial mechanical properties of isolated canine myocardium.', *J Physiol*, Jun. 1983.
20. M. S. Sacks, 'Biaxial Mechanical Evaluation of Planar Biological Materials', 2000.
21. H. Ghaemi, K. Behdinan, and A. D. Spence, 'In vitro technique in estimation of passive mechanical properties of bovine heart. Part I. Experimental techniques and data', *Med Eng Phys*, vol. 31, no. 1, pp. 76–82, Jan. 2009, doi: https://doi.org/10.1016/j.medengphy.2008.04.008.
22. M. Nguyen-Truong and Z. Wang, 'Biomechanical Properties and Mechanobiology of Cardiac ECM', in *Cardiac Extracellular Matrix. Advances in Experimental Medicine and Biology*, vol. 1098, E. Shmuck, P. Hematti, and A. Raval, Eds., Springer, Cham, 2018, pp. 1–19.
23. W. Liu *et al.*, 'Multiscale Contrasts Between the Right and Left Ventricle Biomechanics in Healthy Adult Sheep and Translational Implications', *Front Bioeng Biotechnol*, vol. 10, Apr. 2022, doi: https://doi.org/10.3389/FBIOE.2022.857638.
24. M. S. Sacks and C. J. Chuong, 'Orthotropic Mechanical Properties of Chemically Treated Bovine Pericardium', 1998.
25. M. S. Sacks and W. Sun, 'Multiaxial mechanical behavior of biological materials', *Annual Review of Biomedical Engineering*, vol. 5. pp. 251–284, 2003. doi: https://doi.org/10.1146/annurev.bioeng.5.011303.120714.
26. M. S. Sacks and C. J. Chuong, 'Biaxial Mechanical Properties of Passive Right Ventricular Free Wall Myocardium', *J Biomech Eng*, vol. 115, no. 2, pp. 202–205, May 1993.
27. D. Valdez-Jasso, Marc. A. Simon, H. C. Champion, and M. S. Sacks, 'A murine experimental model for the mechanical behaviour of viable right-ventricular myocardium', *J Physiol*, Aug. 2012.
28. D. Vélez-Rendón, E. R. Pursell, J. Shieh, and D. Valdez-Jasso, 'Relative Contributions of Matrix and Myocytes to Biaxial Mechanics of the Right Ventricle in Pulmonary Arterial Hypertension', *J Biomech Eng*, vol. 141, no. 9, Sep. 2019, doi: https://doi.org/10.1115/1.4044225.

29. C. Witzenburg, R. Raghupathy, S. M. Kren, D. A. Taylor, and V. H. Barocas, 'Mechanical changes in the rat right ventricle with decellularization', *J Biomech*, vol. 45, no. 5, pp. 842–849, Mar. 2012, doi: https://doi.org/10.1016/J.JBIOMECH.2011.11.025.
30. W. Liu, M. Nguyen-Truong, M. Ahern, K. M. Labus, C. M. Puttlitz, and Z. Wang, 'Different Passive Viscoelastic Properties between the Left and Right Ventricles in Healthy Adult Ovine', *J Biomech Eng*, vol. 143, no. 12, Dec. 2021, doi: https://doi.org/10.1115/1.4052004.
31. D. S. Li, E. A. Mendiola, R. Avazmohammadi, F. B. Sachse, and M. S. Sacks, 'A multi-scale computational model for the passive mechanical behavior of right ventricular myocardium', *J Mech Behav Biomed Mater*, vol. 142, Jun. 2023, doi: https://doi.org/10.1016/j.jmbbm.2023.105788.
32. M. Nguyen-Truong *et al.*, 'The Interventricular Septum Is Biomechanically Distinct from the Ventricular Free Walls', *Bioengineering 2021, Vol. 8, Page 216*, vol. 8, no. 12, p. 216, Dec. 2021, doi: https://doi.org/10.3390/BIOENGINEERING8120216.
33. W. Liu *et al.*, 'Correlations between the right ventricular passive elasticity and organ function in adult ovine', *J Integr Cardiol*, vol. 6, pp. 1–6, 2020, doi: https://doi.org/10.15761/JIC.1000294.
34. K. LeBar, W. Liu, J. Pang, A. Chicco, and Z. Wang, 'Role of the Microtubule Network in the Passive Anisotropic Viscoelasticity of Right Ventricle with Pulmonary Hypertension Progression', *Acta Biomater*, 2024.
35. D. Nordsletten *et al.*, 'A viscoelastic model for human myocardium', *Acta Biomater*, vol. 135, pp. 441–457, Nov. 2021, doi: https://doi.org/10.1016/j.actbio.2021.08.036.
36. S. Kakaletsis *et al.*, 'Right ventricular myocardial mechanics: Multi-modal deformation, microstructure, modeling, and comparison to the left ventricle', *Acta Biomater*, vol. 123, pp. 154–166, Mar. 2021, doi: https://doi.org/10.1016/j.actbio.2020.12.006.
37. D. S. Li *et al.*, 'Insights into the passive mechanical behavior of left ventricular myocardium using a robust constitutive model based on full 3D kinematics', *J Mech Behav Biomed Mater*, vol. 103, Mar. 2020, doi: https://doi.org/10.1016/j.jmbbm.2019.103508.
38. S. Dokos, B. H. Smaill, A. A. Young, and I. J. LeGrice, 'Shear properties of passive ventricular myocardium', *Heart and Circulatory Physiology*, vol. 283, no. 6, pp. 2650–2659, Dec. 2002.
39. R. Avazmohammadi *et al.*, 'An integrated inverse model-experimental approach to determine soft tissue three-dimensional constitutive parameters: application to post-infarcted myocardium', *Biomech Model Mechanobiol*, vol. 17, no. 1, pp. 31–53, Feb. 2018, doi: https://doi.org/10.1007/S10237-017-0943-1.
40. C. Y. Chen *et al.*, 'Suppression of detyrosinated microtubules improves cardiomyocyte function in human heart failure', *Nat Med*, vol. 24, no. 8, pp. 1225–1233, Aug. 2018, doi: https://doi.org/10.1038/s41591-018-0046-2.
41. G. M. Roth, D. M. Bader, D. M. Bader, and E. R. Pfaltzgraff, 'Isolation and physiological analysis of mouse cardiomyocytes', *Journal of Visualized Experiments*, no. 91, Jul. 2014, doi: https://doi.org/10.3791/51109.
42. M. A. Caporizzo, C. Y. Chen, A. K. Salomon, K. B. Margulies, and B. L. Prosser, 'Microtubules Provide a Viscoelastic Resistance to Myocyte Motion', *Biophys J*, vol. 115, no. 9, pp. 1796–1807, Nov. 2018, doi: https://doi.org/10.1016/j.bpj.2018.09.019.
43. G. Iribe, M. Helmes, and P. Kohl, 'Force-length relations in isolated intact cardiomyocytes subjected to dynamic changes in mechanical load', *Am J Physiol Heart Circ Physiol*, vol. 292, no. 3, Mar. 2007, doi: https://doi.org/10.1152/ajpheart.00909.2006.
44. W. K. K. Weiwad, W. A. Linke, and M. H. P. Wussling, 'Sarcomere length-tension relationship of rat cardiac myocytes at lengths greater than optimum', *J Mol Cell Cardiol*, vol. 32, no. 2, pp. 247–259, 2000, doi: https://doi.org/10.1006/jmcc.1999.1069.
45. X. Zhijie Wang, J. R. Patel, D. A. Schreier, T. A. Hacker, R. L. Moss, and N. C. Chesler, 'Organ-level right ventricular dysfunction with preserved Frank-Starling mechanism in a mouse model of pulmonary arterial hypertension', *J Appl Physiol*, vol. 124, pp. 1244–1253, 2018, doi: https://doi.org/10.1152/japplphysiol.00725.2017.-Pulmonary.
46. C. Crocini and M. Gotthardt, 'Cardiac sarcomere mechanics in health and disease', doi: 10.1007/s12551-021-00840-7/Published.

47. D. Rappaport, D. Adam, P. Lysyansky, and S. Riesner, 'Assessment of myocardial regional strain and strain rate by tissue tracking in B-mode echocardiograms', *Ultrasound Med Biol*, vol. 32, no. 8, pp. 1181–1192, Aug. 2006, doi: https://doi.org/10.1016/j.ultrasmedbio.2006.05.005.
48. Z. Wang, R. S. Lakes, M. Golob, J. C. Eickhoff, and N. C. Chesler, 'Changes in Large Pulmonary Arterial Viscoelasticity in Chronic Pulmonary Hypertension', *PLoS One*, Nov. 2013.
49. Z. Wang and N. C. Chesler, 'Role of collagen content and cross-linking in large pulmonary arterial stiffening after chronic hypoxia', *Biomech Model Mechanobiol*, vol. 11, no. 1–2, pp. 279–289, Jan. 2012, doi: https://doi.org/10.1007/s10237-011-0309-z.
50. J. C. Benech *et al.*, 'Diabetes increases stiffness of live cardiomyocytes measured by atomic force microscopy nanoindentation', *J Physiol Cell Physiol*, vol. 307, pp. 910–919, 2014, doi: https://doi.org/10.1152/ajpcell.00192.2013.-Stiffness.
51. L. Lou, A. Sesena Rubfiaro, J. He, and A. Agarwal, 'Understanding spatiotemporal mechanical behavior, viscoelasticity, and functions of stem cell-derived cardiomyocytes', *Nanoscale*, vol. 15, no. 24, pp. 10360–10370, Jun. 2023, doi: https://doi.org/10.1039/d3nr01553j.
52. B. W. Lee *et al.*, 'Adult human cardiomyocyte mechanics in osteogenesis imperfecta', *Am J Physiol Heart Circ Physiol*, vol. 325, no. 4, pp. H814–H821, Oct. 2023, doi: https://doi.org/10.1152/ajpheart.00391.2023.
53. D. Borin, I. Pecorari, B. Pena, and O. Sbaizero, 'Novel insights into cardiomyocytes provided by atomic force microscopy', *Seminars in Cell and Developmental Biology*, vol. 73. Elsevier Ltd, pp. 4–12, Jan. 01, 2018. doi: https://doi.org/10.1016/j.semcdb.2017.07.003.
54. J. Zhu, T. Sabharwal, A. Kalyanasundaram, L. Guo, and G. Wang, 'Topographic mapping and compression elasticity analysis of skinned cardiac muscle fibers in vitro with atomic force microscopy and nanoindentation', *J Biomech*, vol. 42, no. 13, pp. 2143–2150, Sep. 2009, doi: https://doi.org/10.1016/j.jbiomech.2009.05.031.
55. S. C. Lieber, N. Aubry, J. Pain, G. Diaz, S. J. Kim, and S. F. Vatner, 'Aging increases stiffness of cardiac myocytes measured by atomic force microscopy nanoindentation', *Am J Physiol Heart Circ Physiol*, vol. 287, no. 2 56-2, pp. 645–651, Aug. 2004, doi: https://doi.org/10.1152/AJPHEART.00564.2003/ASSET/IMAGES/LARGE/ZH40080431910007.JPEG.
56. S. Deitch, B. Z. Gao, and D. Dean, 'Effect of Matrix on Cardiomyocyte Viscoelastic Properties in 2D Culture'.
57. R. Peyronnet *et al.*, 'Simultaneous assessment of radial and axial myocyte mechanics by combining atomic force microscopy and carbon fibre techniques', *Philosophical Transactions of the Royal Society B: Biological Sciences*, vol. 377, no. 1864, Nov. 2022, doi: https://doi.org/10.1098/rstb.2021.0326.
58. C. Zhang, W. Wang, W. He, N. Xi, Y. Wang, and L. Liu, 'Dynamic Model for Characterizing Contractile Behaviors and Mechanical Properties of a Cardiomyocyte', *Biophys J*, vol. 114, no. 1, pp. 188–200, Jan. 2018, doi: https://doi.org/10.1016/j.bpj.2017.11.002.
59. J. C. Benech and G. Romanelli, 'Atomic force microscopy indentation for nanomechanical characterization of live pathological cardiovascular/heart tissue and cells', *Micron*, vol. 158, Jul. 2022, doi: https://doi.org/10.1016/j.micron.2022.103287.
60. A. Rubiano, Y. Qi, D. Guzzo, K. Rowe, C. Pepine, and C. Simmons, 'Stem cell therapy restores viscoelastic properties of myocardium in rat model of hypertension', *J Mech Behav Biomed Mater*, vol. 59, pp. 71–77, Jun. 2016, doi: https://doi.org/10.1016/j.jmbbm.2015.11.041.
61. I. Andreu *et al.*, 'Heterogeneous micromechanical properties of the extracellular matrix in healthy and infarcted hearts', *Acta Biomater*, vol. 10, no. 7, pp. 3235–3242, 2014, doi: https://doi.org/10.1016/j.actbio.2014.03.034.
62. K. Genovese, A. Montes, A. Martínez, and S. L. Evans, 'Full-surface deformation measurement of anisotropic tissues under indentation', *Med Eng Phys*, vol. 37, no. 5, pp. 484–493, May 2015, doi: https://doi.org/10.1016/j.medengphy.2015.03.005.

63. T. Lanzicher *et al.*, 'The cardiomyopathy lamin A/C D192G mutation disrupts whole-cell biomechanics in cardiomyocytes as measured by atomic force microscopy loading-unloading curve analysis', *Sci Rep*, vol. 5, Sep. 2015, doi: https://doi.org/10.1038/srep13388.
64. E. Dague *et al.*, 'Atomic force and electron microscopic-based study of sarcolemmal surface of living cardiomyocytes unveils unexpected mitochondrial shift in heart failure', *J Mol Cell Cardiol*, vol. 74, pp. 162–172, 2014, doi: https://doi.org/10.1016/j.yjmcc.2014.05.006.
65. J. R. Tangney *et al.*, 'Novel role for vinculin in ventricular myocyte mechanics and dysfunction', *Biophys J*, vol. 104, no. 7, pp. 1623–1633, Apr. 2013, doi: https://doi.org/10.1016/j.bpj.2013.02.021.
66. X. Shi *et al.*, 'Elasticity of cardiac cells on the polymer substrates with different stiffness: An atomic force microscopy study', *Physical Chemistry Chemical Physics*, vol. 13, no. 16, pp. 7540–7545, Apr. 2011, doi: https://doi.org/10.1039/c1cp20154a.
67. L. Qian and H. Zhao, 'Nanoindentation of soft biological materials', *Micromachines*, vol. 9, no. 12. MDPI AG, Dec. 11, 2018. doi: https://doi.org/10.3390/mi9120654.
68. G. Iribe, T. Kaneko, Y. Yamaguchi, and K. Naruse, 'Load dependency in force-length relations in isolated single cardiomyocytes', *Prog Biophys Mol Biol*, vol. 115, no. 2–3, pp. 103–114, Aug. 2014, doi: https://doi.org/10.1016/j.pbiomolbio.2014.06.005.
69. A. Borbély *et al.*, 'Cardiomyocyte stiffness in diastolic heart failure', *Circulation*, vol. 111, no. 6, pp. 774–781, Feb. 2005, doi: https://doi.org/10.1161/01.CIR.0000155257.33485.6D.
70. M. Golob, R. L. Moss, and N. C. Chesler, 'Cardiac Tissue Structure, Properties, and Performance: A Materials Science Perspective', *Annals of Biomedical Engineering*, vol. 42, no. 10. Kluwer Academic Publishers, pp. 2003–2013, Oct. 01, 2014. doi: https://doi.org/10.1007/s10439-014-1071-z.
71. T. Shibata *et al.*, 'Metabolic and functional consequences of barium-induced contracture in rabbit myocardium'.
72. Y. Saeki, K. Sagawa, and H. Suga, 'Transient tension responses of heart muscle in Ba2+ contracture to step length changes'.
73. E. Marban, H. Kusuoka, D. T. Yue, M. L. Weisfeldt, and W. G. Wier, 'Maximal Ca2+-activated force elicited by tetanization of ferret papillary muscle and whole heart: mechanism and characteristics of steady contractile activation in intact myocardium.', *Circ Res*, vol. 59, no. 3, pp. 262–269, 1986.
74. T. Shibata, W. C. Hunter, and K. Sagawa, 'Dynamic Stiffness of Barium-Contractured Cardiac Muscles With Different Speeds of Contraction'. [Online]. Available: http://ahajournals.org
75. J. D. Stroud, C. F. Baicu, M. A. Barnes, F. G. Spinale, and M. R. Zile, 'Viscoelastic properties of pressure overload hypertrophied myocardium: effect of serine protease treatment', https://doi.org/10.1152/ajpheart.00711.2001, vol. 282, no. 6 51-6, 2002, doi: https://doi.org/10.1152/AJPHEART.00711.2001.
76. S. Neelakantan *et al.*, 'Multiscale characterization of left ventricle active behavior in the mouse', *Acta Biomater*, vol. 162, pp. 240–253, May 2023, doi: https://doi.org/10.1016/j.actbio.2023.03.022.
77. W. Zhang, A. Capilnasiu, and D. Nordsletten, 'Comparative Analysis of Nonlinear Viscoelastic Models Across Common Biomechanical Experiments', *J Elast*, vol. 145, no. 1–2, pp. 117–152, Aug. 2021, doi: https://doi.org/10.1007/s10659-021-09827-7.
78. R. Avazmohammadi, J. S. Soares, D. S. Li, S. S. Raut, R. C. Gorman, and M. S. Sacks, 'A Contemporary Look at Biomechanical Models of Myocardium', *Annu Rev Biomed Eng*, vol. 21, pp. 417–442, 2019, doi: https://doi.org/10.1146/ANNUREV-BIOENG-062117-121129.
79. S. Javani, M. Gordon, and A. N. Azadani, 'Biomechanical Properties and Microstructure of Heart Chambers: A Paired Comparison Study in an Ovine Model', *Ann Biomed Eng*, vol. 44, no. 11, pp. 3266–3283, Nov. 2016, doi: https://doi.org/10.1007/s10439-016-1658-7.
80. T. Matsumoto *et al.*, 'Biaxial Tensile Properties of Thoracic Aortic Aneurysm Tissues', *Journal of Biomechanical Science and Engineering*, vol. 4, no. 4, pp. 518–529, 2009, doi: https://doi.org/10.1299/JBSE.4.518.
81. D. Pfau *et al.*, 'Angiotensin Receptor Neprilysin Inhibitor Attenuates Myocardial Remodeling and Improves Infarct Perfusion in Experimental Heart Failure', *Sci Rep*, vol. 9, no. 1, Dec. 2019, doi: https://doi.org/10.1038/S41598-019-42113-0.

82. L. Loeffler and K. Sagawa, 'A one dimensional viscoelastic model of cat heart muscle studied by small length perturbations during isometric contraction', *Circ Res*, vol. 36, no. 4, pp. 498–512, 1975, doi: https://doi.org/10.1161/01.RES.36.4.498.
83. A. Propp, A. Gizzi, F. Levrero-Florencio, and R. Ruiz-Baier, 'An orthotropic electro-viscoelastic model for the heart with stress-assisted diffusion', *Biomech Model Mechanobiol*, vol. 19, no. 2, pp. 633–659, Apr. 2020, doi: https://doi.org/10.1007/s10237-019-01237-y.
84. O. M. Hess, J. Schneider, R. Koch, C. Bamert, J. Grimm, and H. P. Krayenbuehl, 'Diastolic function and myocardial structure in patients with myocardial hypertrophy. Special reference to normalized viscoelastic data', *Circulation*, vol. 63, no. 2, pp. 360–371, 1981, doi: https://doi.org/10.1161/01.CIR.63.2.360
85. C. O. Olsen, D. D. Glower, K. L. Lee, P. A. McHale, and J. S. Rankin, 'Diastolic anisotropic properties of the left ventricle in the conscious dog', *Circ Res*, vol. 69, no. 3, pp. 765–778, 1991, doi: https://doi.org/10.1161/01.RES.69.3.765.
86. H. Pouleur, J. S. Karliner, M. M. Lewinter, and J. W. Covell, 'Diastolic Viscous Properties of the Intact Canine Left Ventricle', 1979. [Online]. Available: http://ahajournals.org
87. A. Bonfanti, J. L. Kaplan, G. Charras, and A. Kabla, 'Fractional viscoelastic models for power-law materials', *Soft Matter*, vol. 16, no. 26. Royal Society of Chemistry, pp. 6002–6020, Jul. 14, 2020. doi: 10.1039/d0sm00354a.
88. B. Suki, K. R. Lutchen, and A.-L. Barabhsi, 'Lung tissue viscoelasticity: a mathematical framework and its molecular basis'.
89. J. M. Guccione and A. D. McCulloch, 'Mechanics of Active Contraction in Cardiac Muscle: Part I—Constitutive Relations for Fiber Stress That Describe Deactivation', *J Biomech Eng*, vol. 115, no. 1, pp. 72–81, Feb. 1993, doi: https://doi.org/10.1115/1.2895473.
90. B. Milićević, M. Ivanović, B. Stojanović, M. Milošević, M. Kojić, and N. Filipović, 'Huxley muscle model surrogates for high-speed multi-scale simulations of cardiac contraction', *Comput Biol Med*, vol. 149, Oct. 2022, doi: https://doi.org/10.1016/j.compbiomed.2022.105963.
91. J. J. Rice, F. Wang, D. M. Bers, and P. P. De Tombe, 'Approximate model of cooperative activation and crossbridge cycling in cardiac muscle using ordinary differential equations', *Biophys J*, vol. 95, no. 5, pp. 2368–2390, Sep. 2008, doi: https://doi.org/10.1529/biophysj.107.119487.
92. N. A. Trayanova and J. J. Rice, 'Cardiac electromechanical models: From cell to organ', *Front Physiol*, vol. 2 AUG, 2011, doi: https://doi.org/10.3389/fphys.2011.00043.
93. T. Washio, J. I. Okada, S. Sugiura, and T. Hisada, 'Approximation for cooperative interactions of a spatially-detailed cardiac sarcomere model', *Cell Mol Bioeng*, vol. 5, no. 1, pp. 113–126, Mar. 2012, doi: https://doi.org/10.1007/s12195-011-0219-2.
94. J. Lumens, T. Delhaas, B. Kirn, and T. Arts, 'Three-wall segment (TriSeg) model describing mechanics and hemodynamics of ventricular interaction', *Ann Biomed Eng*, vol. 37, no. 11, pp. 2234–2255, Nov. 2009, doi: https://doi.org/10.1007/s10439-009-9774-2.
95. S. S. Virani and Connie. W. Tsao, 'Heart Disease and Stroke Statistics-2020 Update: A Report from the American Heart Association', *Circulation*, vol. 141, no. 9, Jan. 2020.
96. C. Yancy, '2013 ACCF/AHA Guideline for the Management of Heart Failure: Executive Summary: A Report of the American College of Cardiology Foundation/American Heart Association Task Force on Practice Guidelines', *J Am Coll Cardiol*, vol. 128, no. 16, 2013.
97. A. Dhingra *et al.*, 'Epidemiology of Heart Failure with Preserved Ejection Fraction', *Curr Heart Fail Rep*, vol. 11, pp. 354–365, Sep. 2014.
98. O. E. Theophilus, D. O. Hodge, R. M. Herges, S. J. Jacobsen, V. L. Roger, and M. M. Redfield, 'Trends in Prevalence and Outcome of Heart Failure with Preserved Ejection Fraction', *N Engl J Med*, vol. 355, pp. 251–259, Jul. 2006.
99. F. L. Abel, 'Effects of alterations in peripheral resistance on left ventricular function', *Proceedings of the Society of Experimental Biology and Medicine*, vol. 120, no. 1, pp. 52–56, Oct. 1965.
100. S. Andersen, J. E. Nielsen-Kudsk, A. Vonk Noordegraaf, and F. S. de Man, 'Right Ventricular Fibrosis', *Circulation*, vol. 139, no. 2, pp. 269–285, Jan. 2019.

101. M. Pahuja and D. Burkhoff, 'Right Ventricular Afterload Sensitivity Has Been on My Mind', *Circ Heart Fail*, vol. 12, no. 9, Sep. 2019.
102. W. Liu and Z. Wang, 'Current understanding of the biomechanics of ventricular tissues in heart failure', *Bioengineering*, vol. 7, no. 1, Mar. 2020, doi: https://doi.org/10.3390/BIOENGINEERING7010002.
103. J. Bernal-Ramirez *et al.*, 'Exploring Functional Differences between the Right and Left Ventricles to Better Understand Right Ventricular Dysfunction', *Oxid Med Cell Longev*, vol. 2021, 2021, doi: https://doi.org/10.1155/2021/9993060.
104. C. E. Molina, J. Heijman, and D. Dobrev, 'Differences in Left Versus Right Ventricular Electrophysiological Properties in Cardiac Dysfunction and Arrhythmogenesis', *Arrhthm Electrophysiol Rev.*, vol. 5, no. 1, pp. 14–19, May 2016.
105. S. Rain *et al.*, 'Right ventricular diastolic impairment in patients with pulmonary arterial hypertension', *Circulation*, vol. 128, no. 18, pp. 2016–2025, Oct. 2013, doi: https://doi.org/10.1161/CIRCULATIONAHA.113.001873.
106. M. Murayama *et al.*, 'Simple and noninvasive method to estimate right ventricular operating stiffness based on echocardiographic pulmonary regurgitant velocity and tricuspid annular plane movement measurements during atrial contraction', *International Journal of Cardiovascular Imaging*, vol. 35, no. 10, pp. 1871–1880, Oct. 2019, doi: https://doi.org/10.1007/s10554-019-01637-2.
107. P. Trip *et al.*, 'Clinical relevance of right ventricular diastolic stiffness in pulmonary hypertension', *European Respiratory Journal*, vol. 45, no. 6, pp. 1603–1612, Jun. 2015, doi: https://doi.org/10.1183/09031936.00156714.
108. G. A. Buxton, S. L. Siedlak, G. Perry, and M. A. Smith, 'Mathematical modeling of microtubule dynamics: Insights into physiology and disease', *Prog Neurobiol*, vol. 92, no. 4, pp. 478–483, Dec. 2010, doi: https://doi.org/10.1016/J.PNEUROBIO.2010.08.003.
109. A. J. Zwetsloot, G. Tut, and A. Straube, 'Measuring microtubule dynamics', *Essays Biochem.*, vol. 62, no. 6, pp. 725–735, Oct. 2018.
110. M. A. Caporizzo and B. L. Prosser, 'Need for Speed: The Importance of Physiological Strain Rates in Determining Myocardial Stiffness', *Front Physiol*, vol. 12, Jul. 2021, doi: https://doi.org/10.3389/FPHYS.2021.696694.
111. H. Tsutsui, K. Ishihara, and G. Cooper IV, 'Cytoskeletal role in the contractile dysfunction of hypertrophied myocardium', *Science*, vol. 260, no. 5108, pp. 682–687, 1993, doi: https://doi.org/10.1126/SCIENCE.8097594.
112. H. Sato *et al.*, 'Microtubule Stabilization in Pressure Overload Cardiac Hypertrophy', *J Cell Biol*, vol. 139, no. 4, pp. 963–973, 1997, Accessed: Sep. 04, 2023. [Online]. Available: http://www.jcb.org
113. K. W. Prins, L. Tian, D. Wu, T. Thenappan, J. M. Metzger, and S. L. Archer, 'Colchicine Depolymerizes Microtubules, Increases Junctophilin-2, and Improves Right Ventricular Function in Experimental Pulmonary Arterial Hypertension', *J Am Heart Assoc*, vol. 6, no. 6, Jun. 2017, doi: https://doi.org/10.1161/JAHA.117.006195.
114. M. Schuldt *et al.*, 'Proteomic and Functional Studies Reveal Detyrosinated Tubulin as Treatment Target in Sarcomere Mutation-Induced Hypertrophic Cardiomyopathy', *Circ Heart Fail*, vol. 14, no. 1, p. E007022, Jan. 2021, doi: https://doi.org/10.1161/CIRCHEARTFAILURE.120.007022.
115. C. Zhang *et al.*, 'Microtubule-mediated defects in junctophilin-2 trafficking contribute to myocyte transverse-tubule remodeling and Ca2+ handling dysfunction in heart failure', *Circulation*, vol. 129, no. 17, pp. 1742–1750, Apr. 2014, doi: https://doi.org/10.1161/CIRCULATIONAHA.113.008452.
116. B. S. Scopacasa, V. P. A. Teixeira, and K. G. Franchini, 'Colchicine attenuates left ventricular hypertrophy but preserves cardiac function of aortic-constricted rats', *J Appl Physiol*, vol. 94, no. 4, pp. 1627–1633, Apr. 2003, doi: https://doi.org/10.1152/japplphysiol.00744.2002.
117. J. Luo *et al.*, 'Integrative analyses of gene expression profile reveal potential crucial roles of mitotic cell cycle and microtubule cytoskeleton in pulmonary artery hypertension', *BMC Med Genomics*, vol. 13, no. 1, p. 86, Jun. 2020, doi: https://doi.org/10.1186/s12920-020-00740-x.

118. Q. Li, L. B. Meng, and D. P. Liu, 'Screening and Identification of Therapeutic Targets for Pulmonary Arterial Hypertension Through Microarray Technology', *Front Genet*, vol. 11, Jul. 2020, doi: https://doi.org/10.3389/fgene.2020.00782.
119. P. Y. Xiong, S. L. Archer, and K. W. Prins, 'Sex Hormones Directly Regulate Microtubules: New Insights Into Sex-Differences in Right Ventricular Function in Pulmonary Arterial Hypertension', *Circulation*, vol. 140, no. 1, Nov. 2019.
120. G. Cooper, 'Cytoskeletal networks and the regulation of cardiac contractility: microtubules, hypertrophy, and cardiac dysfunction', *Heart and Circulatory Physiology*, vol. 291, no. 3, pp. 1003–1014, Sep. 2006.
121. H. Tsutsui *et al.*, 'Role of microtubules in contractile dysfunction of hypertrophied cardiocytes', *Circulation*, vol. 90, no. 1, pp. 533–555, 1994, doi: https://doi.org/10.1161/01.CIR.90.1.533.
122. M. R. Zile *et al.*, 'Role of microtubules in the contractile dysfunction of hypertrophied myocardium', *J Am Coll Cardiol*, vol. 33, no. 1, pp. 250–260, Jan. 1999, doi: https://doi.org/10.1016/S0735-1097(98)00550-6.
123. M. M. LeWinter, Y. Wu, S. Labeit, and H. Granzier, 'Cardiac titin: Structure, functions and role in disease', *Clinica Chimica Acta*, vol. 375, nó. 1–2. pp. 1–9, Jan. 2007. doi: https://doi.org/10.1016/j.cca.2006.06.035.
124. M. M. Lewinter and H. Granzier, 'Cardiac titin: A multifunctional giant', *Circulation*, vol. 121, no. 19. pp. 2137–2145, May 18, 2010. doi: https://doi.org/10.1161/CIRCULATIONAHA.109.860171.
125. M. Helmes *et al.*, 'Mechanically Driven Contour-Length Adjustment in Rat Cardiac Titin's Unique N2B Sequence Titin Is an Adjustable Spring', 1999. [Online]. Available: http://www.circresaha.org
126. W. A. Linke, 'Titin Gene and Protein Functions in Passive and Active Muscle', 2017, doi: https://doi.org/10.1146/annurev-physiol-021317-121234.
127. S. Rain *et al.*, 'Right ventricular myocardial stiffness in experimental pulmonary arterial hypertension', *Circ Heart Fail*, vol. 9, no. 7, Jul. 2016, doi: https://doi.org/10.1161/CIRCHEARTFAILURE.115.002636.
128. K. Stam *et al.*, 'Cardiac remodelling in a swine model of chronic thromboembolic pulmonary hypertension: comparison of right vs. left ventricle', *Journal of Physiology*, vol. 597, no. 17, pp. 4465–4480, Jun. 2019.
129. M. R. Zile *et al.*, 'Myocardial stiffness in patients with heart failure and a preserved ejection fraction contributions of collagen and titin', *Circulation*, vol. 131, no. 14, pp. 1247–1259, 2015, doi: https://doi.org/10.1161/CIRCULATIONAHA.114.013215.
130. W. A. Linke and J. M. Fernandez, 'Cardiac titin: molecular basis of elasticity and cellular contribution to elastic and viscous stiffness components in myocardium', *Mechanics of Elastic Biomolecules*, vol. 23, pp. 483–497, 2003, doi: https://doi.org/10.1007/978-94-010-0147-2_9.
131. S. P. Bell, L. Nyland, M. D. Tischler, M. Mcnabb, H. Granzier, and M. M. Lewinter, 'Alterations in the Determinants of Diastolic Suction During Pacing Tachycardia Integrative Physiology', 2000. [Online]. Available: http://www.circresaha.org
132. Y. Wu *et al.*, 'Changes in titin isoform expression in pacing-induced cardiac failure give rise to increased passive muscle stiffness', *Circulation*, vol. 106, no. 11, pp. 1384–1389, Sep. 2002, doi: https://doi.org/10.1161/01.CIR.0000029804.61510.02.
133. W. A. Jaber *et al.*, 'Titin isoforms, extracellular matrix, and global chamber remodeling in experimental dilated cardiomyopathy: functional implications and mechanistic insight.', *Circ Heart Fail*, vol. 1, no. 3, pp. 192–199, 2008, doi: https://doi.org/10.1161/CIRCHEARTFAILURE.108.768465.
134. S. F. Nagueh *et al.*, 'Altered titin expression, myocardial stiffness, and left ventricular function in patients with dilated cardiomyopathy', *Circulation*, vol. 110, no. 2, pp. 155–162, Jul. 2004, doi: https://doi.org/10.1161/01.CIR.0000135591.37759.AF.
135. C. Neagoe *et al.*, 'Titin isoform switch in ischemic human heart disease', *Circulation*, vol. 106, no. 11, pp. 1333–1341, Sep. 2002, doi: https://doi.org/10.1161/01.CIR.0000029803.93022.93.

136. I. Makarenko *et al.*, 'Passive stiffness changes caused by upregulation of compliant titin isoforms in human dilated cardiomyopathy hearts', *Circ Res*, vol. 95, no. 7, pp. 708–716, Oct. 2004, doi: https://doi.org/10.1161/01.RES.0000143901.37063.2f.
137. C. M. Loescher *et al.*, 'Titin governs myocardial passive stiffness with major support from microtubules and actin and the extracellular matrix', *Nature Cardiovascular Research*, vol. 2, no. 11, pp. 991–1002, Nov. 2023, doi: https://doi.org/10.1038/s44161-023-00348-1.
138. H. L. Granzier and T. C. Irving, 'Passive Tension in Cardiac Muscle: Contribution of Collagen, Titin, Microtubules, and Intermediate Filaments', 1995.
139. N. Fukuda, Y. Wu, P. Nair, and H. L. Granzier, 'Phosphorylation of titin modulates passive stiffness of cardiac muscle in a titin isoform-dependent manner', *Journal of General Physiology*, vol. 125, no. 3, pp. 257–271, Mar. 2005, doi: https://doi.org/10.1085/jgp.200409177.
140. D. Mackenna, S. R. Summerour, and F. J. Villarreal, 'Role of mechanical factors in modulating cardiac fibroblast function and extracellular matrix synthesis', 2000. [Online]. Available: www.elsevier.com/locate/cardioreswww.elsevier.nl/locate/cardiores
141. F. J. Villarreal and W. H. Dillmann, 'Cardiac hypertrophy-induced changes in mRNA levels for TGF-P, fibronectin, and collagen', 1992.
142. F. J. Villarreal, N. N. Kim, and F. J. Villareal, 'Regulation of Myocardial Extracellular Matrix Components by Mechanical and Chemical Growth Factors'.
143. W. Carver, L. Terracio, and T. K. Borg, 'Expression and accumulation of interstitial collagen in the neonatal rat heart', *Anat Rec*, vol. 236, no. 3, pp. 511–520, 1993, doi: https://doi.org/10.1002/ar.1092360311.
144. K. LeBar and Z. Wang, 'Extracellular Matrix in Cardiac Tissue Mechanics and Physiology: Role of Collagen Accumulation', *Extracellular Matrix - Developments and Therapeutics [Working Title]*, Mar. 2021, doi: https://doi.org/10.5772/INTECHOPEN.96585.
145. D. A. MacKenna, S. M. Vaplon, and A. D. McCulloch, 'Microstructural model of perimysial collagen fibers for resting myocardial mechanics during ventricular filling', 1997.
146. J. Parvizi and G. K. Kim, 'Collagen', in *High Yield Orthopaedics*, 2010, pp. 107–109.
147. J. Rodriguez-Feo, J. Sluijter, D. Kleijn, and G. Pasterkamp, 'Modulation of Collagen Turnover in Cardiovascular Disease', *Curr Pharm Des*, vol. 11, no. 19, pp. 2501–2514, Jul. 2005, doi: https://doi.org/10.2174/1381612054367544.
148. Z. Wang, D. A. Schreier, T. A. Hacker, and N. C. Chesler, 'Progressive right ventricular functional and structural changes in a mouse model of pulmonary arterial hypertension', *Physiol Rep*, vol. 1, no. 7, p. e00184, Dec. 2013, doi: https://doi.org/10.1002/PHY2.184.
149. L. Todorovich-Hunter, D. J. Johnson, P. Ranger, and M. Rabinovitch, 'Altered Elastin and Collagen Synthesis Associated with Progressive Pulmonary Hypertension Induced by Monocrataline. A Biochemical and Ultrastructural Study', *Lab Invest.*, vol. 58, no. 2, pp. 184–195, 1988.
150. M. J. Golob, Z. Wang, A. J. Prostrollo, T. A. Hacker, and N. C. Chesler, 'Limiting collagen turnover via collagenase-resistance attenuates right ventricular dysfunction and fibrosis in pulmonary arterial hypertension', *Physiol Rep*, vol. 4, no. 11, 2016, doi: https://doi.org/10.14814/phy2.12815.
151. H. J. Bogaard *et al.*, 'Chronic pulmonary artery pressure elevation is insufficient to explain right heart failure', *Circulation*, vol. 120, no. 20, pp. 1951–1960, Nov. 2009, doi: https://doi.org/10.1161/CIRCULATIONAHA.109.883843.
152. M. A. J. Borgdorff *et al.*, 'Distinct loading conditions reveal various patterns of right ventricular adaptation', *Am J Physiol Heart Circ Physiol*, vol. 305, pp. 354–364, 2013, doi: https://doi.org/10.1152/ajpheart.00180.2013.-Right.
153. C. F. Baicu *et al.*, 'Time course of right ventricular pressure-overload induced myocardial fibrosis: relationship to changes in fibroblast postsynthetic procollagen processing', *Am J Physiol Heart Circ Physiol*, vol. 303, pp. 1128–1134, 2012, doi: https://doi.org/10.1152/ajpheart.00482.2012.-Myocardial.
154. A. J. Woodiwiss *et al.*, 'Reduction in Myocardial Collagen Cross-Linking Parallels Left Ventricular Dilatation in Rat Models of Systolic Chamber Dysfunction', 2001. [Online]. Available: http://www.circulationaha.org

155. M. Nguyen-Truong *et al.*, 'Establishment of adult right ventricle failure in ovine using a graded, animal-specific pulmonary artery constriction model', *Animal Model Exp Med*, vol. 3, no. 2, pp. 182–192, Jun. 2020, doi: https://doi.org/10.1002/AME2.12124.
156. D. W. Park, A. Sebastiani, C. H. Yap, M. A. Simon, and K. Kim, 'Quantification of Coupled Stiffness and Fiber Orientation Remodeling in Hypertensive Rat Right-Ventricular Myocardium Using 3D Ultrasound Speckle Tracking with Biaxial Testing', *PLoS One*, vol. 11, no. 10, Oct. 2016, doi: https://doi.org/10.1371/JOURNAL.PONE.0165320.
157. F. Ahmad, S. Soe, J. Albon, R. Errington, and P. Theobald, 'Quantifying the microstructural and biomechanical changes in the porcine ventricles during growth and remodelling', *Acta Biomater*, vol. 171, pp. 166–192, Nov. 2023, doi: https://doi.org/10.1016/j.actbio.2023.09.044.
158. M. Ren, C. W. Ong, M. L. Buist, and C. H. Yap, 'Biventricular biaxial mechanical testing and constitutive modelling of fetal porcine myocardium passive stiffness', *J Mech Behav Biomed Mater*, vol. 134, Oct. 2022, doi: https://doi.org/10.1016/j.jmbbm.2022.105383.
159. F. H. Silver, I. Horvath, and D. J. Foran, 'Mechanical implications of the domain structure of fiber-forming collagens: Comparison of the molecular and fibrillar flexibilities of the α1-chains found in types I-III collagen', *J Theor Biol*, vol. 216, no. 2, pp. 243–254, 2002, doi: https://doi.org/10.1006/jtbi.2002.2542.
160. F. Ahmad *et al.*, 'Biomechanical properties and microstructure of neonatal porcine ventricles', *Journal of the Mechanical Behavior of Biomedical Materials*, vol. 88, pp. 18–28, Dec. 2018.
161. E. N. Hancock, B. M. Palmer, and M. A. Caporizzo, 'Microtubule destabilization with colchicine increases the work output of myocardial slices', *Journal of Molecular and Cellular Cardiology Plus*, vol. 7, p. 100066, Mar. 2024, doi: https://doi.org/10.1016/j.jmccpl.2024.100066.
162. C. Sanyal, N. Pietsch, S. Ramirez Rios, L. Peris, L. Carrier, and M. J. Moutin, 'The detyrosination/re-tyrosination cycle of tubulin and its role and dysfunction in neurons and cardiomyocytes', *Seminars in Cell and Developmental Biology*, vol. 137. Elsevier Ltd, pp. 46–62, Mar. 15, 2023. doi: 10.1016/j.semcdb.2021.12.006.
163. W. M. Torres *et al.*, 'Regional and temporal changes in left ventricular strain and stiffness in a porcine model of myocardial infarction', *Am J Physiol Heart Circ Physiol*, vol. 315, no. 4, pp. H958–H967, Oct. 2018, doi: https://doi.org/10.1152/ajpheart.00279.2018.
164. I. J. Lundorff *et al.*, 'Imaging-Tissue Doppler, Speckle Tracking and Strain Imaging Right ventricular speckle tracking in patients with heart failure-a comparison of right ventricular measures'. [Online]. Available: https://academic.oup.com/eurheartj/article/41/Supplement_2/ehaa946.0145/6003664
165. N. Kawel-Boehm *et al.*, 'Reference ranges ("normal values") for cardiovascular magnetic resonance (CMR) in adults and children: 2020 update', *Journal of Cardiovascular Magnetic Resonance*, vol. 22, no. 1. BioMed Central Ltd, Dec. 01, 2020. doi: https://doi.org/10.1186/s12968-020-00683-3.
166. A. I. Veress, J. A. Weiss, G. J. Klein', and G. T. Gullberg', 'Quantification of 3D Left Ventricular Deformation using Hyperelastic Warping: Comparisons between MRI and PET Imaging'.
167. H. Zou *et al.*, 'Three-dimensional biventricular strains in pulmonary arterial hypertension patients using hyperelastic warping', *Comput Methods Programs Biomed*, vol. 189, Jun. 2020, doi: https://doi.org/10.1016/j.cmpb.2020.105345.
168. A. Youssef, E.-S. H. Ibrahim, G. Korosoglou, R. Abraham, R. G. Weiss, and N. F. Osman, 'Strain-encoding cardiovascular magnetic resonance for assessment of right-ventricular regional function', 2008, doi: https://doi.org/10.1186/1532-429X-10-33.
169. G. Korosoglou *et al.*, 'Real-time fast strain-encoded magnetic resonance imaging to evaluate regional myocardial function at 3.0 Tesla: Comparison to conventional tagging', *Journal of Magnetic Resonance Imaging*, vol. 27, no. 5, pp. 1012–1018, May 2008, doi: https://doi.org/10.1002/jmri.21315.

170. Z. A. Fayad *et al.*, 'Right ventricular regional function using MR tagging: Normals versus chronic pulmonary hypertension', *Magn Reson Med*, vol. 39, no. 1, pp. 116–123, 1998, doi: https://doi.org/10.1002/mrm.1910390118.
171. G. Xiong *et al.*, 'Comprehensive Modeling and Visualization of Cardiac Anatomy and Physiology from CT Imaging and Computer Simulations', *IEEE Trans Vis Comput Graph*, vol. 23, no. 2, pp. 1014–1028, Feb. 2017, doi: https://doi.org/10.1109/TVCG.2016.2520946.
172. S. M. Shavik *et al.*, 'Computational Modeling Studies of the Roles of Left Ventricular Geometry, Afterload, and Muscle Contractility on Myocardial Strains in Heart Failure with Preserved Ejection Fraction', *J Cardiovasc Transl Res*, vol. 14, pp. 1131–1145, Apr. 2021.
173. R. Pourmodheji *et al.*, 'Computational modelling of multi-temporal ventricular–vascular interactions during the progression of pulmonary arterial hypertension', *Journal of Royal Society Interface*, vol. 19, Nov. 2022.
174. M. Lei, M. Franz, J. A. Fraser, H. Zhang, and I. Adeniran, 'Abnormal calcium homeostasis in heart failure with preserved ejection fraction is related to both reduced contractile function and incomplete relaxation: an electromechanically detailed biophysical modeling study', *Frontiers in Physiology* |www.frontiersin.org, vol. 6, p. 78, 2015, doi: https://doi.org/10.3389/fphys.2015.00078.
175. M. Pagani *et al.*, 'Relationship of Transmural Variations in Myofiber Contractility to Left Ventricular Ejection Fraction: Implications for Modeling Heart Failure Phenotype With Preserved Ejection Fraction', *Frontiers in Physiology* |www.frontiersin.org, vol. 1, p. 1003, 2018, doi: https://doi.org/10.3389/fphys.2018.01003.
176. O. O. Odeigah, D. Valdez-Jasso, S. T. Wall, and J. Sundnes, 'Computational models of ventricular mechanics and adaptation in response to right-ventricular pressure overload', *Frontiers in Physiology*, vol. 13. Frontiers Media S.A., Aug. 24, 2022. doi: https://doi.org/10.3389/fphys.2022.948936.
177. D. Sharifi Kia, K. Kim, and M. A. Simon, 'Current Understanding of the Right Ventricle Structure and Function in Pulmonary Arterial Hypertension', *Front Physiol*, vol. 12, May 2021, doi: https://doi.org/10.3389/FPHYS.2021.641310.
178. K. L. Sack, N. H. Davies, J. M. Guccione, and T. Franz, 'Personalised computational cardiology: Patient-specific modelling in cardiac mechanics and biomaterial injection therapies for myocardial infarction', *Heart Fail Rev*, vol. 21, no. 6, pp. 815–826, Nov. 2016, doi: https://doi.org/10.1007/s10741-016-9528-9.
179. T. Arts, T. Delhaas, P. Bovendeerd, X. Verbeek, and F. W. Prinzen, 'Adaptation to mechanical load determines shape and properties of heart and circulation: the CircAdapt model', *Am J Physiol Heart Circ Physiol*, vol. 288, pp. 1943–1954, 2005, doi: https://doi.org/10.1152/ajpheart.00444.
180. S. M. Kim, E. B. Randall, F. Jezek, D. A. Beard, and N. C. Chesler, 'Computational modeling of ventricular-ventricular interactions suggest a role in clinical conditions involving heart failure', *Front Physiol*, vol. 14, 2023, doi: https://doi.org/10.3389/fphys.2023.1231688.
181. M. J. Colebank and N. C. Chesler, 'An in-silico analysis of experimental designs to study ventricular function: A focus on the right ventricle', *PLoS Comput Biol*, vol. 18, no. 9, Sep. 2022, doi: https://doi.org/10.1371/journal.pcbi.1010017.
182. C. Xi *et al.*, 'Patient-Specific Computational Analysis of Ventricular Mechanics in Pulmonary Arterial Hypertension', *J Biomech Eng*, vol. 138, no. 11, Nov. 2016, doi: https://doi.org/10.1115/1.4034559.
183. A. D. Gomez, H. Zou, M. E. Bowen, X. Liu, E. W. Hsu, and S. H. McKellar, 'Right ventricular fiber structure as a compensatory mechanism in pressure overload: A computational study', *J Biomech Eng*, vol. 139, no. 8, Aug. 2017, doi: https://doi.org/10.1115/1.4036485.
184. X. Henrik Finsberg *et al.*, 'Computational quantification of patient-specific changes in ventricular dynamics associated with pulmonary hypertension', *Am J Physiol Heart Circ Physiol*, vol. 317, pp. 1363–1375, 2019, doi: https://doi.org/10.1152/ajpheart.00094.
185. R. Avazmohammadi *et al.*, 'A Computational Cardiac Model for the Adaptation to Pulmonary Arterial Hypertension in the Rat', *Ann Biomed Eng*, vol. 47, no. 1, pp. 138–153, Jan. 2019, doi: https://doi.org/10.1007/s10439-018-02130-y.

186. R. Shimkunas *et al.*, 'Left ventricular myocardial contractility is depressed in the borderzone after posterolateral myocardial infarction', *Annals of Thoracic Surgery*, vol. 95, no. 5, pp. 1619–1625, May 2013, doi: https://doi.org/10.1016/j.athoracsur.2013.02.005.
187. J. F. Wenk *et al.*, 'A novel method for quantifying the in-vivo mechanical effect of material injected into a myocardial infarction', *Annals of Thoracic Surgery*, vol. 92, no. 3, pp. 935–941, Sep. 2011, doi: https://doi.org/10.1016/j.athoracsur.2011.04.089.
188. J. Walmsley *et al.*, 'Fast Simulation of Mechanical Heterogeneity in the Electrically Asynchronous Heart Using the MultiPatch Module', *PLoS Comput Biol*, vol. 11, no. 7, Jul. 2015, doi: https://doi.org/10.1371/journal.pcbi.1004284.
189. S. M. Shavik, L. Zhong, X. Zhao, and L. C. Lee, 'In-silico assessment of the effects of right ventricular assist device on pulmonary arterial hypertension using an image based biventricular modeling framework', *Mech Res Commun*, vol. 97, pp. 101–111, Apr. 2019, doi: https://doi.org/10.1016/j.mechrescom.2019.04.008.
190. A. Tozeren, 'CONTINUUM RHEOLOGY OF MUSCLE CONTRACTION AND ITS APPLICATION TO CARDIAC CONTRACTILITY'.
191. J. M. Guccione, L. K. Waltlman, and A. Mem, 'Mechanics of Active Contraction in Cardiac Muscle: Part II-Cylindrical Models of the Systolic Left Ventricle', 1991. [Online]. Available: http://asmedigitalcollection.asme.org/biomechanical/article-pdf/115/1/82/5766440/82_1.pdf
192. C. Tossas-Betancourt *et al.*, 'Data-driven computational models of ventricular-arterial hemodynamics in pediatric pulmonary arterial hypertension', *Front Physiol*, vol. 13, Sep. 2022, doi: https://doi.org/10.3389/fphys.2022.958734.
193. O. O. Odeigah, E. D. Kwan, K. M. Garcia, H. Finsberg, D. Valdez-Jasso, and J. Sundnes, 'A computational study of right ventricular mechanics in a rat model of pulmonary arterial hypertension', *Front Physiol*, vol. 15, 2024, doi: https://doi.org/10.3389/fphys.2024.1360389.
194. S. G. Tewari, S. M. Bugenhagen, K. C. Vinnakota, J. J. Rice, P. M. L. Janssen, and D. A. Beard, 'Influence of metabolic dysfunction on cardiac mechanics in decompensated hypertrophy and heart failure', *J Mol Cell Cardiol*, vol. 94, pp. 162–175, May 2016, doi: https://doi.org/10.1016/j.yjmcc.2016.04.003.
195. V. Vedula, J. Lee, H. Xu, C. C. J. Kuo, T. K. Hsiai, and A. L. Marsden, 'A method to quantify mechanobiologic forces during zebrafish cardiac development using 4-D light sheet imaging and computational modeling', *PLoS Comput Biol*, vol. 13, no. 10, Oct. 2017, doi: https://doi.org/10.1371/journal.pcbi.1005828.
196. J. Lee *et al.*, 'Moving Domain Computational Fluid Dynamics to Interface with an Embryonic Model of Cardiac Morphogenesis', *PLoS One*, vol. 8, no. 8, Aug. 2013, doi: https://doi.org/10.1371/journal.pone.0072924.
197. D. Vélez-Rendón and D. Valdez-Jasso, 'Multiscale modeling of ventricular–vascular dysfunction in pulmonary arterial hypertension', *Current Opinion in Biomedical Engineering*, vol. 11. Elsevier B.V., pp. 68–75, Sep. 01, 2019. doi: https://doi.org/10.1016/j.cobme.2019.09.004.
198. J. W. Gerringer, J. C. Wagner, D. Vélez-Rendón, and D. Valdez-Jasso, 'Lumped-parameter models of the pulmonary vasculature during the progression of pulmonary arterial hypertension', *Physiol Rep*, vol. 6, no. 3, Feb. 2018, doi: https://doi.org/10.14814/phy2.13586.
199. C. Stevens, E. Remme, I. LeGrice, and P. Hunter, 'Ventricular mechanics in diastole: Material parameter sensitivity', *J Biomech*, vol. 36, no. 5, pp. 737–748, May 2003, doi: https://doi.org/10.1016/S0021-9290(02)00452-9.

# Part V
# Multiscale Characterization Methodologies

# Multi-scale Light-Sheet Imaging for Volumetric Assessment of Cardiovascular Architecture and Function

**Milad Almasian, Xinyuan Zhang, Alireza Saberigarakani, and Yichen Ding**

## 1 Introduction

The volumetric imaging of developmental and regenerative processes of entire organs at cellular resolution presents a significant optical challenge. Traditional optical microscopes encounter limitations such as restricted tissue penetration, limited working distance, and slow imaging speed, leading to major barriers in the time-lapse imaging of live specimens and three-dimensional (3-D) imaging of thick and opaque models. Mechanical sectioning required by conventional optical microscopes for the imaging of 3-D structure also leads to the distortion of the inherent tissue integrity, and under sampling during the subsequent volumetric reconstruction. While techniques such as μPET [1, 2], μCT [3, 4], μMRI [5, 6], and bioluminescence [7, 8] imaging can generate 3-D images from live samples, their spatial resolution and imaging speed often prove inadequate for accurately capturing organ morphogenesis in small animal models [9–13]. To address these challenges, the emergence of light-sheet microscopy (LSM) [14, 15] has marked a significant breakthrough in multi-scale imaging. This innovative technique enables the observation of the cardiovascular system and circulation ranging from live zebrafish embryos [16] to adult mice [17], with exceptional spatiotemporal resolution. In the

M. Almasian · X. Zhang · A. Saberigarakani
Department of Bioengineering, The University of Texas at Dallas, Richardson, TX, USA

Y. Ding (✉)
Department of Bioengineering, The University of Texas at Dallas, Richardson, TX, USA

Center for Imaging and Surgical Innovation, The University of Texas at Dallas, Richardson, TX, USA

Hamon Center for Regenerative Science and Medicine, UT Southwestern Medical Center, Dallas, TX, USA
e-mail: yichen.ding@utdallas.edu

 
J. Liao, J. Y. Wong (eds.), *Integration and Bridging of Multiscale Bioengineering Designs and Tissue Biomechanics*, https://doi.org/10.1007/978-3-031-81743-4_16

meanwhile, its rapid data acquisition becomes imperative to mitigate the effects of photobleaching and phototoxicity in specimens [18–20]. Thus, the transformative LSM allows us to delve deeper into cardiac morphogenesis and remodeling with unprecedented clarity and accuracy.

While the concept of LSM was initially proposed by Siedentopf and Zsigmondy in 1902 [21], the practical implementation of LSM remained unfeasible until the advent of laser and high-speed data acquisition facilitated by fast-rate charge-coupled devices (CCD) /scientific complementary metal-oxide-semiconductor (sCMOS) cameras. The modern LSM systems, orthogonal-plane fluorescence optical sectioning (OPFOS), and thin laser light-sheet microscope (TLSM) were developed for the imaging of the internal architecture of the cochlea in 1993 [22] and microbial oceanography in 2002 [23], respectively. The design of decoupled illumination and detection led to the first widely recognized method termed selective plane illumination microscopy (SPIM) (Fig. 1a) [24], enabling the in vivo imaging of Medaka fish with the light-sheet generated by the cylindrical lens. The orthogonally positioned illumination and detection lenses allow for faster scanning rates and reduced photo-damage in contrast to conventional methods. To minimize artifacts such as stripes induced by photon scattering and absorption, variations including multidirectional selective plane illumination microscopy (mSPIM) were proposed to scan the light-sheet across the entire specimen [25]. Following that, the integration of SPIM with other methods including tissue clearing [26] and optogenetic control [27] has further advanced the study of neuronal circuits in intact mouse brains and cardiac function in zebrafish, respectively.

To enhance image contrast, an alternative method known as a digital scanned laser microscope (DSLM) generates a virtual sheet by rapidly scanning the laser beam along the axis perpendicular to the illumination and detection axes [28]. This implementation provides uniform illumination across scanning axis and enables control of the illumination intensity in highly scattering specimens, while SPIM illumination pattern follows the Gaussian distribution. Due to the usage of Gaussian beams in both SPIM and DSLM, a fundamental trade-off between a larger field of view (FOV) and a higher axial resolution remains to be addressed. Thus, a remote focusing-based method [29] termed axially swept light-sheet microscopy (ASLM) was proposed [30–32]. Unlike SPIM and DSLM, ASLM involves sweeping the well-focused beam waist axially along the laser illumination and synchronizing it with rolling pixels on the sCMOS to create a cross-sectional image. This technique decouples the axial resolution and FOV, resulting in high axial resolution and uniform optical sectioning across the large specimen [33–36]. However, the improvements in ASLM come at the cost of reduced temporal resolution and increased photo-damage in contrast to SPIM.

Due to photon scattering and absorption, the image quality of LSM is compromised as the thickness of light-sheet increases. Thus, capturing images from complementary views, achieved by either rotating the specimen or incorporating multiple illumination/detection modules, was proposed to significantly enhance the physical coverage of light-sheet [37]. The first multi-view light-sheet image was computationally fused 36 views of mulberry pollen grain [37], leading to an

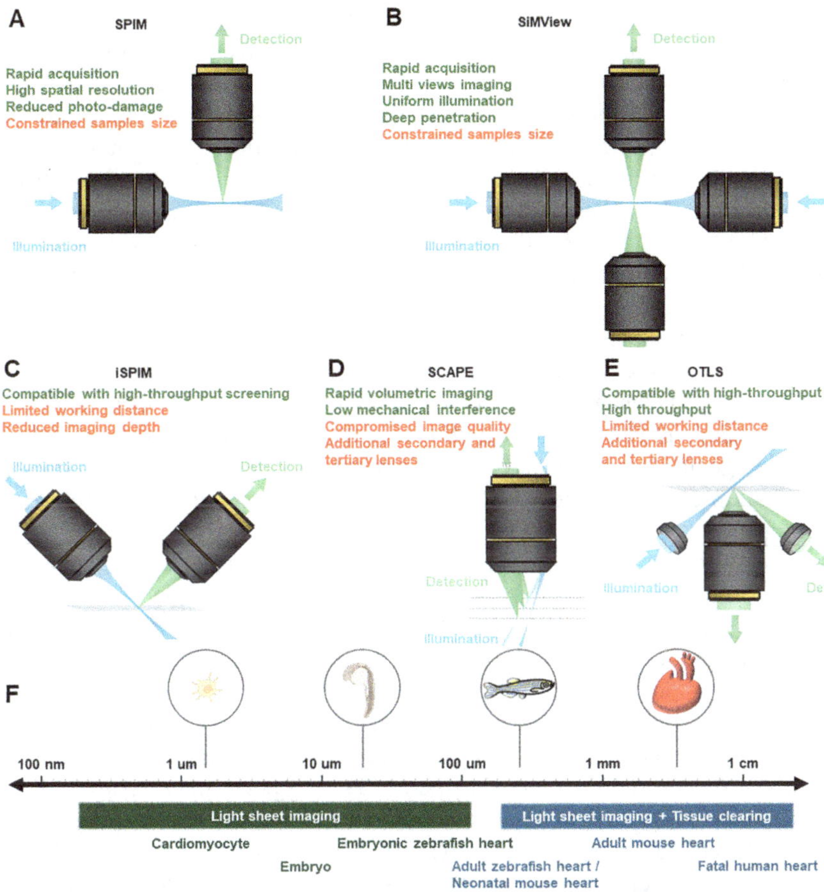


**Fig. 1** Representative designs of LSM for illumination and detection. (**a**) The fundamental concept of SPIM involves independent illumination and detection pathways. The waist of laser beam precisely aligns with the focal plane of the detection objective lens for optical sectioning. (**b**) Two illumination and detection modules enable concurrent fluorescence excitation and collection in the SiMView. (**c**) The iSPIM system is compatible with conventional sample holders while maintaining the orthogonal design of illumination and detection. (**d**) Single-objective lens is utilized for generating inclined illumination and detection, while remote focusing is employed to accurately project the angled light-sheet onto a camera. (**e**) The hybrid OTLS, containing three objectives, is designated for both orthogonal dual-objective and non-orthogonal dual-objective imaging. (**f**) The LSM strategy enables multi-scale imaging, ranging from micrometer to centimeter levels, covering a spectrum from single-cell studies to intact rodent hearts

improved spatial resolution in 3-D microscopy. It is important to note that the lateral resolution of LSM is primarily determined by the numerical aperture (NA) of the detection objective, whereas the axial resolution is additionally affected by the thickness of the light sheet. Through the implementation of multi-view approaches, isotropic resolution has been achieved in fruit flies [38] and intact neonatal mouse hearts [39]. However, sequential data acquisition and image registration from multiple views are not only time-consuming in image post-processing but also hinder the investigation of dynamic processes, posing additional challenges to be addressed. Simultaneous multi-view (SiMView) [40] (Fig. 1b), isotropic multi-view (IsoView) [41], and multi-view selective plane illumination microscopy (MuVi-SPIM) [42] are representative optimized multi-view systems that have been proposed to capture dynamics such as nuclei tracking and cell shape changes during embryonic development. Briefly, these microscopes incorporate at least two sets of illumination/detection modules, concurrently shining a laser onto the specimen and capturing images from opposing views. This strategy ensures high-quality images while minimizing the time consumption during data acquisition.

In addition to optics and computation, another trade-off in sample preparation and working distance remains to be addressed. Biological specimens are typically prepared in Petri dishes, multiwell plates, or on slides and coverslips for high-throughput screening and imaging purposes. However, this limits the wide application of conventional LSM systems that require unique chambers for sample mounting and refractive index matching. Inverted selective illumination microscopy (iSPIM) [43] (Fig. 1c), followed by dual inverted selective plane illumination microscopy (diSPIM) [44], opens a new chapter by successfully accommodating conventional sample holders while maintaining the orthogonal design of illumination and detection in LSM imaging. However, this design often comes at the cost of an effective working distance between the specimen and objectives, limiting its use to small animal models such as *Caenorhabditis elegans*. In this context, another alternative solution is to use the single-objective lens for both illumination and detection, including oblique plane microscopy (OPM) [45], single-objective selective plane illumination microscopy (SoPi) [46], swept confocally aligned planar excitation (SCAPE) [47, 48] (Fig. 1d), and DaXi [49]. In these configurations, selective illumination is introduced at an inclined angle, and fluorescence is subsequently collected from a non-orthogonal angle using the same lens. This unique design allows for implementation on conventional microscopes with standard sample preparation while retaining the minimal photobleaching and phototoxicity advantages of LSM, as well as swift volumetric imaging and effective functional imaging. However, as the light-sheet traverses the focal volume diagonally, the introduction of additional optical elements, such as secondary and tertiary lenses, becomes necessary to ensure distortion-free perspectives. This, in turn, increases the complexity of the system design. Another limitation of single-objective LSM is its restriction in customizing axial and lateral resolutions, resulting in compromised quality across the imaged plane. Thus, the most recent open-top LSM (OTLS) has been developed to integrate high tolerance to refractive index mismatch and maintain the entire working distance of a high NA objective lens in its new design [50] (Fig. 1e).

Collectively, LSM has emerged as an indispensable tool for multi-scale imaging and analysis (Fig. 1f), allowing us to explore structural changes and functional dynamics from the sub-cellular to organ levels. LSM is particularly useful for imaging both small organisms due to their transparency and thick models with the aid of tissue clearing. Specifically, we summarize the applications of the LSM concept in cardiac and vascular systems in this review, focusing on the following aspects. (a) Visualizing the micro-structure of intact hearts from zebrafish larvae to adult mice, and even fetal human hearts, (b) mapping the vascular branching network for the assessment of hemodynamics in combination with computational fluid dynamics, (c) tracking contractile function in its entirety within multiple cardiac cycles, and (d) investigating developmental and regenerative processes of cardiovascular systems. Overall, the advance of multi-scale LSM imaging has proven valuable in deepening our understanding of cardiovascular systems, particularly cardiac morphogenesis, and disease pathogenesis, across applications ranging from zebrafish and rodents to clinical investigations.

## 2 Applications of Light-Sheet Microscopy in Cardiovascular Studies

The applications of LSM are diverse, ranging from time-lapse imaging of cardiac development to dynamic imaging of cardiac contractile function to volumetric imaging of myocardial micro-structure. Generally, LSM is capable of both ex vivo imaging of tissue-cleared specimens, such as rodent models, and in vivo imaging of live models, such as zebrafish larvae.

### *2.1 Ex Vivo Imaging*

The integration of numerous tissue clearing methods [51–53] with LSM has emerged as a powerful framework for the imaging of opaque specimens, such as mouse hearts, offering unprecedented opportunities for 3-D visualization of intact cardiac architecture [39, 54, 55]. Briefly, tissue clearing techniques render organs or thick tissue optically transparent by removing lipids and other light-scattering elements, enabling deeper photon penetration [56]. The combination of LSM and tissue clearing approaches enhances the imaging depth and allows for the assessment of intricate structures including cardiac trabeculation [39] and nervous system [31, 33] in their entirety. The rationale underlying this integration lies in the complementary strengths of each technique; LSM allows for rapid imaging with high spatial resolution and minimal photo damage, while tissue clearing enhances optical transparency in thick tissue. The synergistic application of these methods holds great promise for advancing our understanding of cardiac morphogenesis and regenerative processes [57–59], paving the way for new discoveries in various scientific disciplines.

### 2.1.1 Tissue Clearing

Tissue clearing encompasses a variety of methods that enhance tissue transparency while preserving their original structure [60]. Generally, lipids and proteins in tissue have a refractive index ranging from 1.45 to 1.47, while the cytosol, which is the aqueous interior of cells, has a refractive index of 1.33, similar to that of water [61–63]. The mismatch in refractive indices within the tissue leads to photon scattering, resulting in the opaque appearance of the tissue [64]. Thus, the basic concept in tissue clearing approaches is to establish a uniform refractive index, achieved through methods such as the removal of lipids, to minimize diffusion and render the tissue transparent. Recent progress demonstrates that three representative approaches are prevalent in tissue clearing, namely hydrophobic, hydrophilic, and hydrogel techniques [65] (Fig. 2).

Hydrophobic approaches, such as BABB [26], iDISCO [66], and PEGASOS [67], involve dehydrating tissue, removing lipids with a water-immiscible solvent,

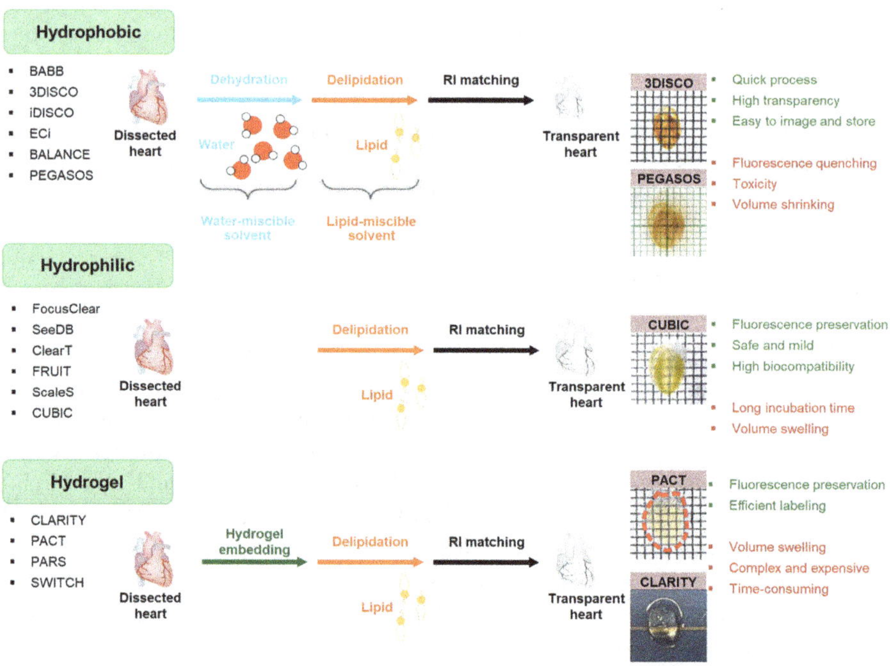


**Fig. 2** Schematics of three primary categories of optical clearing techniques. Hydrophobic approaches involve thorough tissue dehydration, followed by lipid removal and refractive index matching using organic solvents. Hydrophilic methods involve the use of water-soluble solutions, typically offering greater biosafety and compatibility compared to hydrophobic methods. Hydrogel-based methods utilize either monomer and initiator molecules to create a synthetic gel or polyepoxide to establish a reinforced scaffold, improving tissue robustness during the clearing process [65]. (Reproduced with permission from Ref 59, 67, 78, 79)

and matching refractive indices. These methods are fast, robust, and compatible with immunostaining. However, tissue volume tends to shrink, and the organic solvent removes water molecules from the chromophore of fluorescent proteins, leading to fluorescence quenching during the dehydration step [66, 68, 69]. In contrast, hydrophilic methods, including ClearT [70] and CUBIC [71], primarily focus on detergent-based lipid removal and refractive index matching, providing significantly lower hazards compared to other clearing protocols. While this group of methods is generally compatible with immunostaining and preserves endogenous fluorescence, concerns arise regarding the longer incubation time and tissue swelling required for complete transparency [72, 73]. Hydrogel-embedding techniques, encompassing CLARITY [72], PACT [74], and SWITCH [75], establish robust hydrogel scaffolds to protect the tissue structure from damage during the clearing process. While these approaches involve hydrogel-embedding, delipidation, and refractive index matching, they effectively preserve endogenous fluorescence, albeit at the cost of time-consuming procedures [65, 76, 77].

### 2.1.2 Integration of LSM and Tissue Clearing in Cardiac Applications

In contrast to other organs, the heart possesses a notable autofluorescence signal, particularly from the myocardium [80]. Thus, autofluorescence provides inherent contrast to reflect the myocardial structure [39]. Integrating the multi-view LSM with Adipo-clear [81], we have extensively explored the 3-D myocardial trabeculation and compaction across the entire heart with an isotropic resolution of 5 μm (Fig. 3a) [39]. A notable LSM application showcasing changes in myocardial structure has been characterized by calculating helix angles and global helix angle transmurality, indicating a significant reduction in global helix angle transmurality in the ischemic group compared to the control group [82]. In addition, other studies have delved into the interactions between lymphatic and immune cells in regulating cardiac recovery following myocardial infarction [83], as well as exploring the effects of CD47 inhibition on enhancing the phagocytosis of cardiac cell debris by neutrophils after acute myocardial reperfusion injury [84].

In addition, this framework has also demonstrated superiority over traditional histological methods in investigating carotid artery ligation injury and remodeling models (Fig. 3b) [85]. Volumetric images of immunolabeled carotid arteries reveal the presence of puncta exclusively in the ligated arteries, concentrated around the ligation site and adjacent to the neovascular plexus. Besides cardiac research, the combination of LSM and CUBIC clearing method facilitates high-resolution imaging of vascular network and enables 3-D visualization of its remodeling in response to coronary disease [86]. Another similar study created a 3-D model depicting the vascular structure of collateral arteries from neonatal and adult mouse hearts to the fetal human heart (Fig. 3c) [87]. This study also implemented computational fluid dynamics into the 3-D model, enabling the delineation of the spatial architecture of collateral arteries and the prediction of blood flow and hemodynamic forces through these arteries in both neonatal and adult mouse hearts

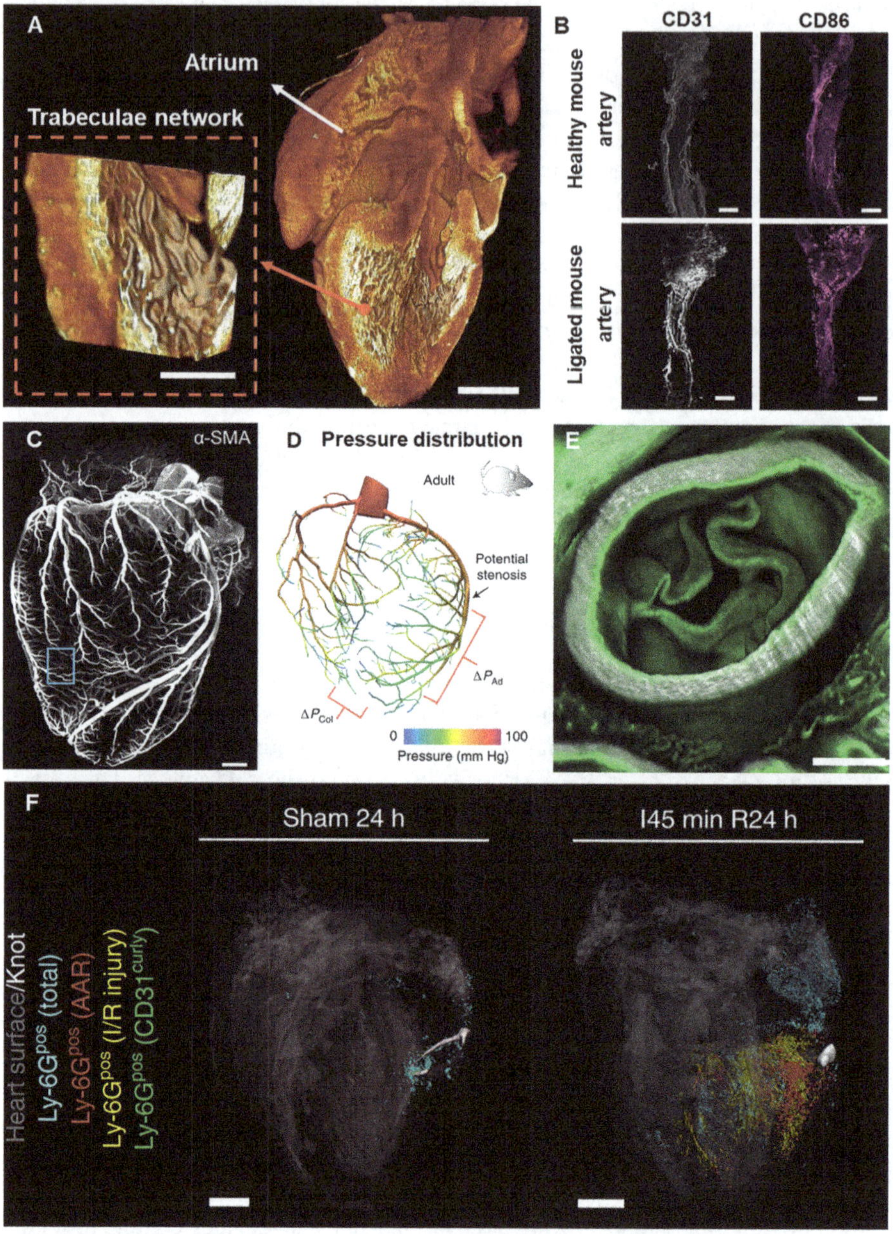


**Fig. 3** Integration of LSM and tissue clearing in cardiovascular study. (**a**) Multi-view LSM enhanced by image stitching enables investigation of the intricate trabecular network in the whole heart with a near isotropic resolution of 5 μm [39]. (**b**) Immunolabeled carotid arteries and macrophage in both healthy and ligated mouse carotids [85]. (**c**) Imaging of entire coronary arteries in fetal human heart immunolabeled with α-SMA [87]. (**d**) Integration of LSM with computational fluid dynamics depicts the pressure distribution pattern and hemodynamics within coronary models in adult mouse hearts [87]. (**e**) 3-D reconstruction of the aortic valve based on autofluorescence and immunolabeling [88]. (**f**) Analyzing immune cell infiltration in an ischemia/reperfusion injury model across the entire mouse heart [88]. Scale bars: (**a**) 500 μm; for red dashed boxes 100 μm, (**b**) 500 μm, (**c**) 1000 μm, (**e**) 300 μm, and (**f**) 1000 μm. (Reproduced with permission from Ref 39, 85, 87, 88)

(Fig. 3d) [87]. These findings revealed that neonatal collaterals form more rapidly and are more numerous, larger in diameter, and more effective at restoring blood flow compared to adult collaterals. This difference is attributed to the expansion of coronary arteries during postnatal growth, leading to alterations in pressure distributions in adults.

This 3-D integrative strategy also enables the investigation of acute and late ischemia/reperfusion injury models (Figs. 3e, f) [88]. By combining autofluorescence and immunostaining, the integrated results depict cardiac structure and vascular integrity, unveiling acute and late damage zones, inflammatory responses, and affirming that the integration of LSM and tissue clearing can accurately assess the extent of damage and recovery processes. Furthermore, this framework also holds great potential to characterize the impact of small molecule compounds during therapy. For example, a preclinical study of captopril on late-stage post-myocardial infarction remodeling suggests that administering captopril after a delay effectively mitigated left ventricle dilation and hypertrophy, resulting in an enhanced ejection fraction, all without influencing the infarct size [89].

## 2.2 *In Vivo Imaging*

Due to reduced photobleaching and phototoxicity, LSM has transformed the realm of live imaging of cardiac development and contractile function across the entire ventricle and atrium, particularly in embryonic and adult zebrafish. The utility of zebrafish as models for cardiovascular research is underscored by their rapid development and translucent embryos, making them ideal for exploring vertebrate cardiac development and remodeling [90–93]. Additionally, their electrocardiogram patterns exhibit remarkable similarity to those of human hearts, enabling real-time investigation of structural and functional changes in the heart (Fig. 4a) [16, 94–96]. The superb temporal resolution of LSM allows for high-speed imaging of contracting zebrafish hearts at the cellular resolution, offering significant insights into cardiac contractile function and electrical conduction. Following the first reconstruction of 4-D (3-D spatial +1-D temporal) cardiac cycles in zebrafish in 2005 [97], the development of retrospective synchronization algorithms-based LSM enables us to study the intricacies of cardiac contractions within the beating zebrafish heart (Fig. 4b) [15, 95]. However, redundant images of each slice from multiple cardiac cycles have to be collected, leading to a deluge of dataset of the 4-D contracting heart and increased photo damage. A further phase-lock algorithm for prospective optical gating has been developed for longitudinal imaging of cardiac development in zebrafish (Fig. 4c) [94].

Besides mechanical contraction and developmental process of the heart, LSM has shown immense capability in functional imaging such as mapping electrical conduction across the entire atrium and ventricle [96]. When using transgenic zebrafish that express GCaMP fluorescence, the electrical excitation of individual cardiomyocytes spreading the entire heart has been captured for in vivo 3-D optical

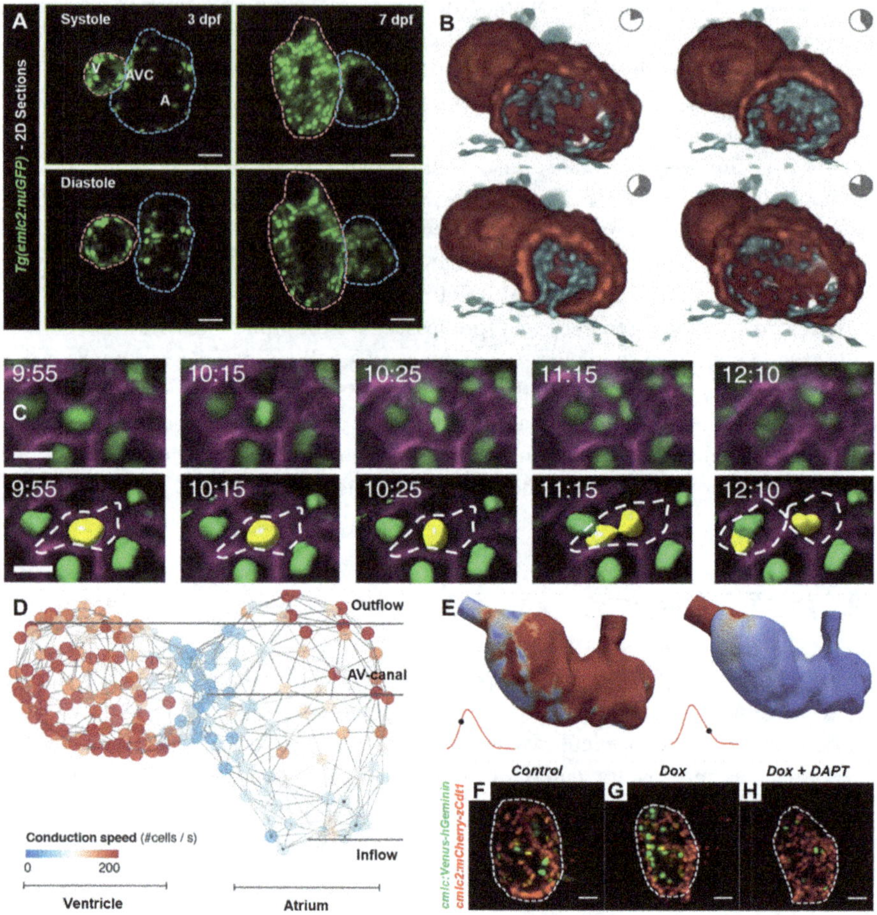


**Fig. 4** In vivo LSM imaging of cardiac development and contractile function. (**a**) LSM empowered by retrospective synchronization algorithm allows for the imaging of nuclei of cardiomyocytes in transgenic *Tg(cmlc2:nucGFP)* zebrafish from systole to diastole between 3 and 7 days post-fertilization (dpf) [16]. (**b**) Myocardium (red) and vasculature (cyan) in an open atrium of transgenic *Tg(myl7:DsRed, kdrl:GFP)* zebrafish are captured from diastole to systole [95]. (**c**) Migration and division of cardiomyocytes during myocardial trabeculation are recorded in the time-lapse LSM imaging. Mitotic stages such as interphase, prophase, metaphase, and anaphase (top row), and typical karyokinesis along with noticeable changes in nuclear volume (yellow) are depicted (bottom row) [94]. (**d**) Single-cell level optical mapping delineates calcium transients across the entire heart in zebrafish embryos at 52 hpf [96]. (**e**) Integration of LSM with computational fluid dynamics reveals wall shear stress patterns in the endocardium of zebrafish embryos during cardiac development [98]. (**f–h**) Study of chemotherapy-induced cardiac injury and remodeling by using LSM for tracking proliferating and non-proliferating cardiomyocytes in *FUCCI* zebrafish larvae at 5 dpf [99]. (Scale bars: (**a**) 30 μm, (**c**) 10 μm, and (**f–h**) 25 μm. Reproduced with permission from Ref 16, 94–96, 98, 99)

mapping at the single cell level (Fig. 4d) [96]. Additionally, mechanobiological forces can be extracted from the integration of LSM and computational fluid dynamics. Representative examples in zebrafish reveal that hemodynamic shear stress regulates Notch signaling in the endocardium, controlling trabeculation and cardiac function during cardiac development [15, 98]. Further study of drug-induced cardiac defects has also been proposed for the in vivo exploration of susceptible myocardial segments and remodeling processes in response to the treatment of small molecule compounds in zebrafish larvae (Fig. 4e–h) [99].

Despite the recent progress, LSM for live cardiac imaging still meets challenges that necessitate ongoing research and optimization. While its superb spatial resolution is sufficient for distinguishing individual cardiomyocytes within zebrafish, challenges persist due to anisotropic resolutions that can affect the quality of volumetric images. While strategies like multi-view fusion and deconvolution have proven effective in studying cardiac development [37, 100], their feasibility in capturing instantaneous dynamics such as contractility, hemodynamics, and neural activity remains to be determined. Thus, the emerging technique of light-field imaging and its integration into LSM for rapid volumetric imaging holds great potential for the study of cardiac dynamics, especially when leveraging innovative computational approaches such as deep learning [101–103]. In addition, utilizing advanced computing strategies, such as parallel computation involving multi-core CPUs and GPUs, has improved 4-D reconstruction efficiency of the retrospective synchronization algorithm by around ten-fold [16, 104].

To facilitate the in vivo study of mammalian development, advanced LSM has been developed to investigate mouse embryo development [105]. This groundbreaking work allows for time-lapse imaging of cardiac development until embryonic day 9.5 (E9.5), but it is still limited in capturing the entire term of cardiac development. In summary, despite numerous ongoing research efforts, LSM has set a new benchmark for high-resolution and real-time cardiac imaging, underlining its significance in deepening our understanding of cardiac abnormalities and contractile dysfunction.

## 2.3 Data Processing, Visualization, and Management

LSM has provided unprecedented details in both spatial and temporal domains for cardiovascular research but at the expense of handling a substantial volume of data for processing, visualization, and management. Generally, a single 2048 * 2048 pixel image with 16-bit depth occupies approximately 8 MB of data for transfer and storage. Therefore, considering a standard frame rate of 100 frames per second for most current sCMOS cameras, LSM easily generates nearly 800 MB of data per second. Consequently, an IsoView microscope is able to generate data at rates of up to 10 TB per hour [41]. In this context, an efficient framework for LSM image post-processing is critical to address these challenging issues in processing, visualization, and management.

Open-source software such as ImageJ [106], Fiji [107], and 3D Slicer [108], along with commercial software like Imaris and Amira [109], is widely used in microscopy image rendering, segmentation, tracking, and quantitative analysis, providing comprehensive solutions for tasks such as image import/export, editing, processing, and analysis. While these are suitable for LSM images, the availability of numerous plugins in ImageJ/Fiji tailored to LSM images enhances their potential for post-processing. For example, image registration is required for multi-view LSM images to align multiple views, correct sample drift, and stitch image tiles (Fig. 5a). Custom-made plugins for Fiji have offered solutions for point-based linear and affine registration [110–112], content-based registration [40, 113], and fusion [37, 110], resulting in multi-view deconvolution with isotropic resolution (Fig. 5b) [38, 41].

Among these, image segmentation and classification stand out as the cornerstone of customized methods for emerging imaging results. It delineates regions of interest to extract binary or multi-level data for quantitative analysis. Besides established methods such as Watershed and Voronoi [114], various methods have been developed for nuclei segmentation and cell tracking in fruit fly, zebrafish, and mouse models [114–119]. These methods typically utilize techniques such as gradient diffusion, optical flow, and local adaptive thresholding. Due to the significant enhancement of computational power, machine learning-based approaches including 3DeeCellTracker (Fig. 5c) [121], u-track3D [122], and LLSM5DTools [123] are also rapidly emerging to provide accurate segmentation and tracking results in a time-efficient manner. Specifically for cardiovascular studies, an open-source software termed SimVascular (Fig. 5d) [124, 125] provides a comprehensive cardiovascular modeling pipeline, from image segmentation to computational fluid dynamics simulation and analysis. Other machine learning-based image segmentation methods tailored to cardiac trabeculation have also been reported [118, 119, 126, 127].

Due to the unparalleled structural details and functional dynamics uncovered by LSM, advanced image visualization and reconstruction of 3-D and 4-D data for analysis and interpretation are also unmet needs. While there are commercial software options like Imaris and Amira (Fig. 5e) available for rendering and visualizing data, the immense volume of LSM data makes data analysis challenging. Big Data Viewer [128], a plugin in Fiji, has attempted to address the issue of large datasets by providing solutions for image compression, transferring, and processing [113]. To address the sample shift issue during the time-lapse imaging, a method called ClearVolume [129] has been developed. In addition to tracking sample drift, this method allows for image sharpness estimation, enhancing real-time and multi-view visualization, multicolor rendering, as well as instant rewinding and replay of time-lapse recordings. To interpret the intricate 3-D structure captured in LSM and enhance the interactive analysis, virtual reality (Fig. 5f) and augmented reality-based image analysis have also been proposed for recapitulating developmental cardiac mechanics and analyzing cardiac contraction in a user-directed microenvironment [130]. In this context, a generic framework aimed at addressing LSM system construction and control, image processing, analysis, visualization, and

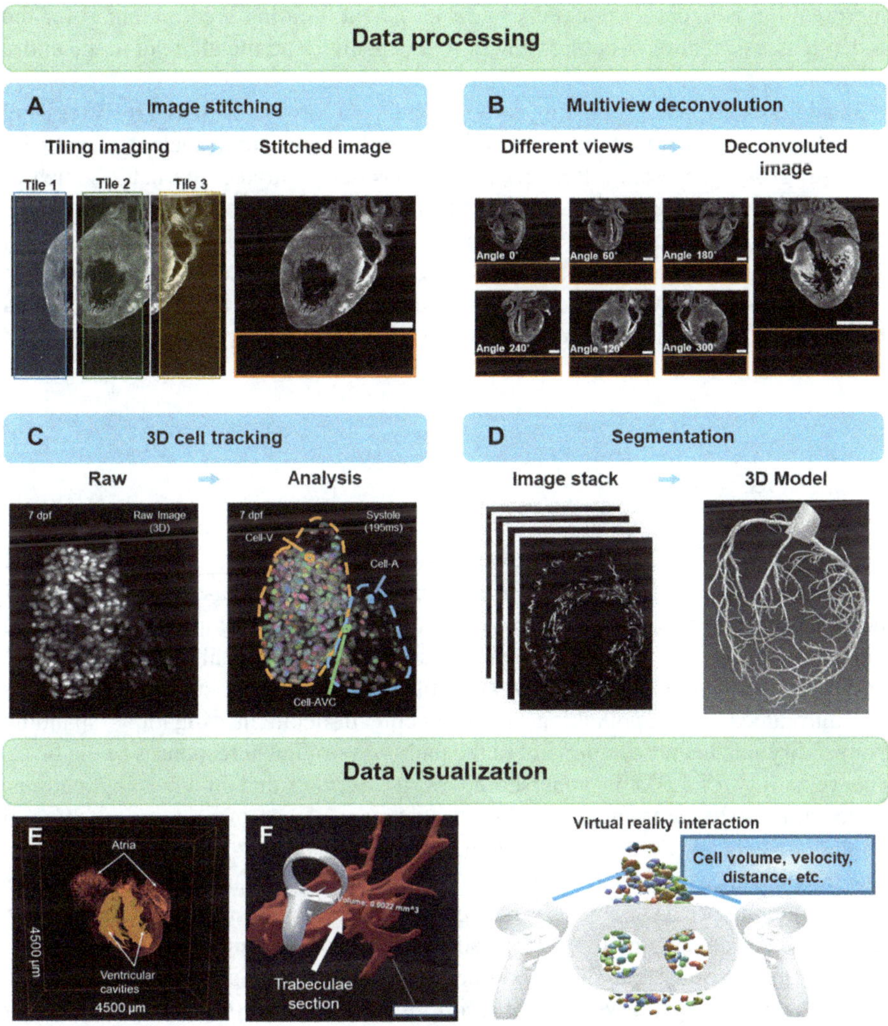


**Fig. 5** A typical framework for computational analysis of LSM images includes both data processing and visualization. (**a**) Image stitching of LSM images aligns and combines multiple individual images to create a seamless, larger image that covers a wider field of view [39]. (**b**) Multiview deconvolution merges various views of an object to enhance spatial resolution and imaging depth [39]. (**c**) 3DeeCellTracker is a machine learning-based approach that provides cellular segmentation and tracking in a time-efficient manner, facilitating the study of dynamic cellular processes [16]. (**d**) SimVascular provides a comprehensive pipeline from image segmentation to computational fluid dynamics simulation and analysis, offering specialized tools for detailed examination of cardiovascular structures and dynamics [87]. (**e**) Conventional volume rendering and (**f**) emerging extended reality-based methods allow for the data visualization and interpretation of 3-D and 4-D cardiac LSM image stacks. The latter advances an immersive and interactive environment for exploration and analysis. (Scale bars: (**a**) 500 μm, (**b**) 500 μm, and (**f**) 5 μm. Reproduced with permission from Ref 16, 39, 87)

interpretation is undergoing extensive development, holding great potential for the in-depth assessment of structure, function, mechanics, and metabolism in the entire heart.

In addition, concerns regarding conventional data formats such as TIFF are rising due to their design limitations, especially when handling large-scale datasets with single-threaded processing. Recent progress has demonstrated alternatives such as Zarr [123] and OME-Zarr [36, 131, 132] formats, is designed for multi-dimensional data, and parallels read-write operations over networks. These formats also boast comprehensive metadata structures. Overall, the integration of LSM with efficient computational frameworks enables enhanced image quality and streamlined data interpretation, providing unprecedented insights into cardiac morphogenesis and disease pathogenesis.

## 3 Conclusion

In this review, we have outlined recent advancements in the emerging field of multi-scale LSM imaging and its implementation in cardiovascular studies. This method provides profound insights into cardiovascular development, injury, and regenerative processes, covering both structural and functional aspects. We have elucidated the development of LSM systems along with image analysis techniques and introduced their applications in advancing in-depth investigations spanning from embryonic heart development to the pathophysiological responses of the heart. The versatility of LSM is underscored by its ex vivo and in vivo applications, demonstrating its outstanding ability to overcome trade-offs in spatiotemporal resolution, field of view, depth of field, and photon budget. This framework significantly enhances imaging capabilities for studying cardiovascular morphogenesis and remodeling processes, with broad applications in both fundamental research on cardiovascular development and translational studies addressing cardiac diseases.

Compared to other imaging techniques for cardiovascular study, LSM stands out due to its unique capability for rapid recording of the contracting heart and vascular network while maintaining high spatial resolution. The numerous tissue clearing methods and extended field of view provided by remote focusing and image stitching further allow for the investigation of the entire heart or cardiovascular system from sub-cellular to organ level. In combination with computational fluid dynamics, the LSM strategy is able to provide mechanical insights into hemodynamics across the entire circulation system. The advancement of image processing, visualization, and management further improves the assessment of multi-scale volumetric LSM datasets with high spatiotemporal resolution.

The limitations of LSM for studying the cardiovascular system are also evident. Firstly, the in vivo imaging depth is restricted by photon scattering and absorption. Most LSM applications in live models are currently limited to *C.elegans*, fruit fly, and zebrafish. Specifically for cardiac research, zebrafish is the only vertebrate model that is used for in vivo investigation. Despite the emergence of tissue clearing methods, new techniques that can improve transparency in live models remain unmet needs. Secondly, most LSM systems are customized for specific groups and applications, thus raising the barrier for wider adoption by groups with less engineering expertise. While commercial solutions are available on the market, the complexity of these systems and their operation typically necessitate experienced staff for handling in most imaging centers. Thirdly, the throughput of most LSM systems is limited, and therefore, the full extent of cardiovascular discoveries enabled by this strategy has yet to be fully realized. With rapidly emerging machine learning-based analysis tools, there is a growing expectation for high-throughput LSM systems to further advance research in this field. In addition, false-positive signals resulting from pigment or residual hemoglobin in cardiovascular systems remain challenging. Unmixing fluorophores based on hyperspectral microscopy [133, 134] holds great promise for improving cardiovascular applications.

Collectively, the unique strengths of LSM, in conjunction with advanced biochemical and computational methodologies, have opened up new landscapes for exploring the entire cardiac system and vascular networks across multiple scales, offering unprecedented insights into their structure and function with remarkable spatiotemporal resolution. By enabling comprehensive investigations spanning from sub-cellular interactions to organ-level dynamics, this transformative synergy has not only deepened our understanding of normal physiological processes but has also shed light on the complex pathophysiological mechanisms underlying cardiovascular diseases.

**Acknowledgments** We appreciate all the constructive comments provided by D-incubator members at UT Dallas. This work was supported by NIH R00HL148493, R01HL162635, NSF 2326628, and UT Dallas STARS program.

## References

1. Miyaoka, R. S., Lewellen, T. K., Yu, H. & McDaniel, D. L. Design of a depth of interaction (DOI) PET detector module1. *IEEE Trans. Nucl. Sci.***45**, 1069–1073 (1998).
2. Tai, Y. C. *et al.* MicroPET II: design, development and initial performance of an improved microPET scanner for small-animal imaging. *Phys. Med. Biol.***48**, 1519 (2003).
3. Paxton, R. & Ambrose, J. The EMI scanner. A brief review of the first 650 patients. *Brit. J. Radiol.***47**, 530–565 (2014).
4. Prokop, M. General principles of MDCT. *Eur. J. Radiol.***45**, S4–S10 (2003).
5. Kelly, K. A. *et al.* Detection of Vascular Adhesion Molecule-1 Expression Using a Novel Multimodal Nanoparticle. *Circ. Res.***96**, 327–336 (2005).
6. Van Der Graaf, M. In vivo magnetic resonance spectroscopy: Basic methodology and clinical applications. *Eur. Biophys. J.***39**, 527–540 (2010).
7. Ntziachristos, V. Fluorescence Molecular Imaging. *Annu. Rev. Bio. Eng.***8**, 1–33 (2006).
8. Li, C. *et al.* Proapoptotic protein Bim attenuates estrogen-enhanced survival in lymphangioleiomyomatosis. *JCI Insight***1**, 86629 (2016).
9. Patel, M. R. *et al.* ACCF Proposed Method for Evaluating the Appropriateness of Cardiovascular Imaging. *J. Am. Coll. Cardiol.***46**, 1606–1613 (2005).

10. Slavin, G. S. & Bluemke, D. A. Spatial and Temporal Resolution in Cardiovascular MR Imaging: Review and Recommendations. *Radiology* **234**, 330–338 (2005).
11. Vuletic, I. *et al.* Real-time volumetric reconstruction of biological dynamics with light-field microscopy and deep learning. *Sci. Rep.* **3**, 4114–4120 (2018).
12. Sanz, J. & Fayad, Z. A. Imaging of atherosclerotic cardiovascular disease. *Nature* **451**, 953–957 (2008).
13. Plana, J. C. *et al.* Expert consensus for multimodality imaging evaluation of adult patients during and after cancer therapy: a report from the American Society of Echocardiography and the European Association of Cardiovascular Imaging. *J. Am. Soc. Echocardiogr.* **15**, 1063–1093 (2014).
14. Fei, P. *et al.* Cardiac light-sheet fluorescent microscopy for multi-scale and rapid imaging of architecture and function. *Sci. Rep.* **6**, 1–12 (2016).
15. Lee, J. *et al.* 4-Dimensional light-sheet microscopy to elucidate shear stress modulation of cardiac trabeculation. *J. Clin. Invest.* **126**, 1679–1690 (2016).
16. Zhang, X. *et al.* 4D Light-sheet imaging and interactive analysis of cardiac contractility in zebrafish larvae. *APL Bioeng.* **7**, 26112 (2023).
17. Ding, Y. *et al.* Light-sheet fluorescence imaging to localize cardiac lineage and protein distribution. *Sci. Rep.* **7**, 42209 (2017).
18. Huisken, J. & Stainier, D. Y. R. Selective plane illumination microscopy techniques in developmental biology. *Development* **136**, 1963–1975 (2009).
19. Power, R. M. & Huisken, J. A guide to light-sheet fluorescence microscopy for multiscale imaging. *Nat. Methods* **14**, 360–373 (2017).
20. Jemielita, M., Taormina, M. J., Delaurier, A., Kimmel, C. B. & Parthasarathy, R. Comparing phototoxicity during the development of a zebrafish craniofacial bone using confocal and light sheet fluorescence microscopy techniques. *J. Biophotonics* **6**, 920–928 (2013).
21. Siedentopf, H. & Zsigmondy, R. Uber Sichtbarmachung und Größenbestimmung ultramikoskopischer Teilchen, mit besonderer Anwendung auf Goldrubingläser. *Ann. Phys.* **315**, 1–39 (1902).
22. Voie, A. H., Burns, D. H. & Spelman, F. A. Orthogonal-plane fluorescence optical sectioning: Three-dimensional imaging of macroscopic biological specimens. *J. Microsc.* **170**, 229–236 (1993).
23. Fuchs, E., Long, R. A., Jaffe, J. S. & Azam, F. Thin laser light sheet microscope for microbial oceanography. *Opt. Express.* **10**, 145–154 (2002).
24. Huisken, J., Swoger, J., Del Bene, F., Wittbrodt, J. & Stelzer, E. H. K. K. Optical sectioning deep inside live embryos by selective plane illumination microscopy. *Science* **305**, 1007–1009 (2004).
25. Huisken, J. & Stainier, D. Y. R. Even fluorescence excitation by multidirectional selective plane illumination microscopy (mSPIM). *Opt. Lett.* **32**, 2608–2610 (2007).
26. Dodt, H.-U. *et al.* Ultramicroscopy: three-dimensional visualization of neuronal networks in the whole mouse brain. *Nat. Methods* **4**, 331–336 (2007).
27. Arrenberg, A. B., Stainier, D. Y. R., Baier, H. & Huisken, J. Optogenetic control of cardiac function. *Science* **330**, 971–974 (2010).
28. Keller, P. J., Schmidt, A. D., Wittbrodt, J. & Stelzer, E. H. K. Reconstruction of zebrafish early embryonic development by scanned light sheet microscopy. *Science* **322**, 1065–1069 (2008).
29. Botcherby, E. J., Juškaitis, R., Booth, M. J. & Wilson, T. An optical technique for remote focusing in microscopy. *Opt. Commun.* **281**, 880–887 (2008).
30. Dean, K. M., Roudot, P., Welf, E. S., Danuser, G. & Fiolka, R. Deconvolution-free subcellular imaging with axially swept light sheet microscopy. *Biophys J.* **108**, 2807–2815 (2015).
31. Chakraborty, T. *et al.* Light-sheet microscopy of cleared tissues with isotropic, subcellular resolution. *Nat. Methods* **16**, 1109–1113 (2019).
32. Dean, K. M. *et al.* Isotropic imaging across spatial scales with axially swept light-sheet microscopy. *Nat. Protoc.* **17**, 2025–2053 (2022).
33. Voigt, F. F. *et al.* The mesoSPIM initiative: open-source light-sheet microscopes for imaging cleared tissue. *Nat. Methods* **16**, 1105–1108 (2019).

34. Vladimirov, N. *et al.* Benchtop mesoSPIM: a next-generation open-source light-sheet microscope for cleared samples. *Nat. Commun.***15**, 1–16 (2024).
35. Liu, Y., Rollins, A. M. & Jenkins, M. W. CompassLSM: axially swept light-sheet microscopy made simple. *Biomed. Opt. Express***12**, 6571–6589 (2021).
36. Glaser, A. *et al.* Expansion-assisted selective plane illumination microscopy for nanoscale imaging of centimeter-scale tissues. *Elife***12**, RP91979 (2023).
37. Stelzer, E. H. K., Huisken, J., Swoger, J., Greger, K. & Verveer, P. Multi-view image fusion improves resolution in three-dimensional microscopy. *Opt. Express***15**, 8029–8042 (2007).
38. Preibisch, S. *et al.* Efficient Bayesian-based multiview deconvolution. *Nat. Methods***11**, 645–648 (2014).
39. Sodimu, O. *et al.* Light sheet imaging and interactive analysis of the cardiac structure in neonatal mice. *J. Biophotonics***16**, e202200278 (2023).
40. Tomer, R., Khairy, K., Amat, F. & Keller, P. J. Quantitative high-speed imaging of entire developing embryos with simultaneous multiview light-sheet microscopy. *Nat. Methods***9**, 755–763 (2012).
41. Chhetri, R. K. *et al.* Whole-animal functional and developmental imaging with isotropic spatial resolution. *Nat. Methods***12**, 1171–1178 (2015).
42. Krzic, U., Gunther, S., Saunders, T. E., Streichan, S. J. & Hufnagel, L. Multiview light-sheet microscope for rapid in toto imaging. *Nat. Methods***9**, 730–733 (2012).
43. Wu, Y. *et al.* Inverted selective plane illumination microscopy (iSPIM) enables coupled cell identity lineaging and neurodevelopmental imaging in Caenorhabditis elegans. *Proc. Natl. Acad. Sci. U. S. A.***108**, 17708–17713 (2011).
44. Wu, Y. *et al.* Spatially isotropic four-dimensional imaging with dual-view plane illumination microscopy. *Nat. Biotechnol.***31**, 1032–1038 (2013).
45. Dunsby, C. Optically sectioned imaging by oblique plane microscopy. *Opt. Express***16**, 20306–20316 (2008).
46. Kumar, M., Kishore, S., Nasenbeny, J., McLean, D. L. & Kozorovitskiy, Y. Integrated one- and two-photon scanned oblique plane illumination (SOPi) microscopy for rapid volumetric imaging. *Opt. Express***26**, 13027 (2018).
47. Bouchard, M. B. *et al.* Swept confocally-aligned planar excitation (SCAPE) microscopy for high-speed volumetric imaging of behaving organisms. *Nat. Photonics***9**, 113–119 (2015).
48. Voleti, V. *et al.* Real-time volumetric microscopy of in vivo dynamics and large-scale samples with SCAPE 2.0. *Nat. Methods***16**, 1054–1062 (2019).
49. Yang, B. *et al.* DaXi—high-resolution, large imaging volume and multi-view single-objective light-sheet microscopy. *Nat. Methods***19**, 461–469 (2022).
50. Ku, T. *et al.* Multiplexed and scalable super-resolution imaging of three-dimensional protein localization in size-adjustable tissues. *Nat. Biotechnol.***34**, 973–981 (2016).
51. Richardson, D. S. & Lichtman, J. W. Clarifying Tissue Clearing. *Cell***162**, 246–257 (2015).
52. Richardson, D. S. & Lichtman, J. W. SnapShot: Tissue Clearing. *Cell***171**, 496-496.e1 (2017).
53. Richardson, D. S. *et al.* Tissue clearing. *Nat. Rev. Methods Prim.***1**, 1–24 (2021).
54. Almasian, M., Saberigarakani, A., Zhang, X., Lee, B. & Ding, Y. Light-Sheet Imaging to Reveal Cardiac Structure in Rodent Hearts. *J. Vis. Exp.* (2024) https://doi.org/10.3791/66707.
55. Tomer, R., Ye, L., Hsueh, B. & Deisseroth, K. Advanced CLARITY for rapid and high-resolution imaging of intact tissues. *Nat. Protoc.***9**, 1682–1697 (2014).
56. Molbay, M., Kolabas, Z. I., Todorov, M. I., Ohn, T. & Ertürk, A. A guidebook for DISCO tissue clearing. *Mol. Syst. Biol.***17**, 9807 (2021).
57. Ding, Y. *et al.* Multiscale light-sheet for rapid imaging of cardiopulmonary system. *JCI Insight***3**, e121396 (2018).
58. Ding, Y. *et al.* Light-Sheet Imaging to Elucidate Cardiovascular Injury and Repair. *Curr. Cardiol. Rep.***20**, 35 (2018).
59. Sereti, K. I. *et al.* Analysis of cardiomyocyte clonal expansion during mouse heart development and injury. *Nat. Commun.***9**, 754 (2018).
60. Ueda, H. R. *et al.* Whole-Brain Profiling of Cells and Circuits in Mammals by Tissue Clearing and Light-Sheet Microscopy. *Neuron***106**, 369–387 (2020).

61. Susaki, E. A. & Ueda, H. R. Whole-body and Whole-Organ Clearing and Imaging Techniques with Single-Cell Resolution: Toward Organism-Level Systems Biology in Mammals. *Cell Chem. Biol.***23**, 137–157 (2016).
62. Feuchtinger, A., Walch, A. & Dobosz, M. Deep tissue imaging: a review from a preclinical cancer research perspective. *Histochem. Cell Biol.***146**, 781–806 (2016).
63. Seo, J., Choe, M. & Kim, S. Y. Clearing and Labeling Techniques for Large-Scale Biological Tissues. *Mol. Cells***39**, 439–446 (2016).
64. Costa, E. C., Silva, D. N., Moreira, A. F. & Correia, I. J. Optical clearing methods: An overview of the techniques used for the imaging of 3D spheroids. *Biotechnol. Bioeng.***116**, 2742–2763 (2019).
65. Ueda, H. R. *et al.* Tissue clearing and its applications in neuroscience. *Nat. Rev. Neurosci.***21**, 61–79 (2020).
66. Renier, N. *et al.* iDISCO: a simple, rapid method to immunolabel large tissue samples for volume imaging. *Cell***159**, 896–910 (2014).
67. Jing, D. *et al.* Tissue clearing of both hard and soft tissue organs with the PEGASOS method. *Cell Res.***28**, 803–818 (2018).
68. Azaripour, A. *et al.* A survey of clearing techniques for 3D imaging of tissues with special reference to connective tissue. *Prog. Histochem. Cytochem.***51**, 9–23 (2016).
69. Kirchner, K. N. *et al.* A hydrophobic tissue clearing method for rat brain tissue. *J. Vis. Exp.***166**, e61821 (2020).
70. Kuwajima, T. *et al.* ClearT: a detergent- and solvent-free clearing method for neuronal and non-neuronal tissue. *Development***140**, 1364–1368 (2013).
71. Susaki, E. A. *et al.* Whole-brain imaging with single-cell resolution using chemical cocktails and computational analysis. *Cell***157**, 726–739 (2014).
72. Chung, K. *et al.* Structural and molecular interrogation of intact biological systems. *Nature***497**, 332–337 (2013).
73. Tainaka, K. *et al.* Chemical Landscape for Tissue Clearing Based on Hydrophilic Reagents. *Cell Rep.***24**, 2196-2210.e9 (2018).
74. Yang, B. *et al.* Single-cell phenotyping within transparent intact tissue through whole-body clearing. *Cell***158**, 945–958 (2014).
75. Murray, E. *et al.* Simple, Scalable Proteomic Imaging for High-Dimensional Profiling of Intact Systems. *Cell***163**, 1500–1514 (2015).
76. Sideris, E. *et al.* Particle Hydrogels Based on Hyaluronic Acid Building Blocks. *ACS. Biomater. Sci. Eng.***2**, 2034–2041 (2016).
77. Chung, K. & Deisseroth, K. CLARITY for mapping the nervous system. *Nat. Methods***10**, 508–513 (2013).
78. Xu, J., Ma, Y., Yu, T. & Zhu, D. Quantitative assessment of optical clearing methods in various intact mouse organs. *J. Biophotonics***12**, e201800134 (2019).
79. Sung, K. *et al.* Simplified three-dimensional tissue clearing and incorporation of colorimetric phenotyping. *Sci Rep.***6**, 30736 (2016).
80. Kolesová, H., Olejníčková, V., Kvasilová, A., Gregorovičová, M. & Sedmera, D. Tissue clearing and imaging methods for cardiovascular development. *iScience***24**, 102387 (2021).
81. Chi, J., Crane, A., Wu, Z. & Cohen, P. Adipo-clear: a tissue clearing method for three-dimensional imaging of adipose tissue. *J. Vis. Exp.***137**, e58271 (2018).
82. Lee, S. E. *et al.* Three-dimensional Cardiomyocytes Structure Revealed By Diffusion Tensor Imaging and Its Validation Using a Tissue-Clearing Technique. *Sci. Rep.***8**, 1–11 (2018).
83. Houssari, M. *et al.* Lymphatic and Immune Cell Cross-Talk Regulates Cardiac Recovery After Experimental Myocardial Infarction. *Arterioscler. Thromb. Vasc. Biol.***40**, 1722–1737 (2020).
84. Haj-Yehia, E. *et al.* CD47 blockade enhances phagocytosis of cardiac cell debris by neutrophils. *Int. J. Cardiol. Hear. Vasc.***48**, 101269 (2023).
85. Buglak, N. E. *et al.* Light sheet fluorescence microscopy as a new method for unbiased three-dimensional analysis of vascular injury. *Cardiovasc. Res.***117**, 520–532 (2021).
86. Nehrhoff, I., Gómez-Gaviro, M. V., Ripoll, J., Samaniego, R. & Desco, M. Looking inside the heart: a see-through view of the vascular tree. *Biomed. Opt. Express.***8**, 3110–3118 (2017).

87. Anbazhakan, S. *et al.* Blood flow modeling reveals improved collateral artery performance during the regenerative period in mammalian hearts. *Nat. Cardiovasc. Res.***1**, 775–790 (2022).
88. Merz, S. F. *et al.* Contemporaneous 3D characterization of acute and chronic myocardial I/R injury and response. *Nat. Commun.***10**, 1–14 (2019).
89. Roostalu, U. *et al.* Effect of captopril on post-infarction remodelling visualized by light sheet microscopy and echocardiography. *Sci. Rep.***11**, 1–13 (2021).
90. González-Rosa, J. M. Zebrafish Models of Cardiac Disease: From Fortuitous Mutants to Precision Medicine. *Circ. Res.***130**, 1803–1826 (2022).
91. Moossavi, M., Lu, X., Herrmann, J. & Xu, X. Molecular mechanisms of anthracycline induced cardiotoxicity: Zebrafish come into play. *Front. Cardiovasc. Med.***10**, 1080299 (2023).
92. Dvornikov, A. V., de Tombe, P. P. & Xu, X. Phenotyping Cardiomyopathy in Adult Zebrafish. *Prog. Biophys. Mol. Biol.***138**, 116 (2018).
93. Liu, J. & Stainier, D. Y. R. Zebrafish in the Study of Early Cardiac Development. *Circ. Res.***110**, 870 (2012).
94. Taylor, J. M. *et al.* Adaptive prospective optical gating enables day-long 3D time-lapse imaging of the beating embryonic zebrafish heart. *Nat. Commun.***10**, 1–15 (2019).
95. Mickoleit, M. *et al.* High-resolution reconstruction of the beating zebrafish heart. *Nat. Methods***11**, 919–922 (2014).
96. Weber, M. *et al.* Cell-accurate optical mapping across the entire developing heart. *Elife***6**, e28307 (2017).
97. Liebling, M., Forouhar, A. S., Gharib, M., Fraser, S. E. & Dickinson, M. E. Four-dimensional cardiac imaging in living embryos via postacquisition synchronization of nongated slice sequences. *J. Biomed. Opt.***10**, 54001 (2005).
98. Vedula, V. *et al.* A method to quantify mechanobiologic forces during zebrafish cardiac development using 4-D light sheet imaging and computational modeling. *PLoS Comput. Biol.***14**, e1006482 (2018).
99. Chen, J. *et al.* Displacement analysis of myocardial mechanical deformation (DIAMOND) reveals segmental susceptibility to doxorubicin-induced injury and regeneration. *JCI insight***4**, e125362 (2019).
100. Schmid, B. & Huisken, J. Real-time multi-view deconvolution. *Bioinformatics***31**, 3398–3400 (2015).
101. Wang, Z. *et al.* Real-time volumetric reconstruction of biological dynamics with light-field microscopy and deep learning. *Nat. Methods***18**, 551–556 (2021).
102. Wang, Z. *et al.* A hybrid of light-field and light-sheet imaging to study myocardial function and intracardiac blood flow during zebrafish development. *PLoS Comput. Biol.***17**, e1009175 (2021).
103. Wagner, N. *et al.* Deep learning-enhanced light-field imaging with continuous validation. *Nat. Methods***18**, 557–563 (2021).
104. Zhang, X. *et al.* 4D Light-sheet Imaging of Zebrafish Cardiac Contraction. *J. Vis. Exp.***203**, e66263 (2024).
105. Wan, Y. *et al.* Single-cell reconstruction of emerging population activity in an entire developing circuit. *Cell***179**, 355 (2019).
106. Schindelin, J., Rueden, C. T., Hiner, M. C. & Eliceiri, K. W. The ImageJ ecosystem: An open platform for biomedical image analysis. *Mol. Reprod. Dev.***82**, 518–529 (2015).
107. Schindelin, J. *et al.* Fiji: an open-source platform for biological-image analysis. *Nat. Methods***9**, 676–682 (2012).
108. Fedorov, A. *et al.* 3D Slicer as an image computing platform for the Quantitative Imaging Network. *J. Magn. Reson. Imaging***30**, 1323–1341 (2012).
109. Stalling, D., Westerhoff, M., Hege, H.-C. C. & others. Amira: A highly interactive system for visual data analysis. *Vis. Handb.***38**, 749–767 (2005).
110. Preibisch, S., Saalfeld, S., Schindelin, J. & Tomancak, P. Software for bead-based registration of selective plane illumination microscopy data. *Nat. Methods***7**, 418–419 (2010).

111. Hörl, D. *et al.* BigStitcher: reconstructing high-resolution image datasets of cleared and expanded samples. *Nat. Methods***16**, 870–874 (2019).
112. Preibisch, S., Saalfeld, S., Rohlfing, T. & Tomancak, P. Bead-based mosaicing of single plane illumination microscopy images using geometric local descriptor matching. *Proc. SPIE***7259**, 926–935 (2009).
113. Amat, F. *et al.* Efficient processing and analysis of large-scale light-sheet microscopy data. *Nat. Protoc.***10**, 1679–1696 (2015).
114. Jones, T. R., Carpenter, A. & Golland, P. Voronoi-based segmentation of cells on image manifolds. *CVBIA***3765**, 535–543 (2005).
115. Li, G. *et al.* 3D cell nuclei segmentation based on gradient flow tracking. *BMC Cell Biol.***8**, 40 (2007).
116. Krämer, P. *et al.* Comparison of segmentation algorithms for the zebrafish heart in fluorescent microscopy images. *Adv. Vis. Comput.***5876**, 1041–1050 (2009).
117. Amat, F., Myers, E. W. & Keller, P. J. Fast and robust optical flow for time-lapse microscopy using super-voxels. *Bioinformatics***29**, 373–380 (2013).
118. Sommer, C., Straehle, C., Kothe, U. & Hamprecht, F. A. Ilastik: Interactive learning and segmentation toolkit. *Proc. Int. Symp. Biomed. Imaging***1109**, 230–233 (2011).
119. Strnad, P. *et al.* Inverted light-sheet microscope for imaging mouse pre-implantation development. *Nat. Methods***13**, 139–142 (2015).
120. Packard, R. R. S. *et al.* Automated segmentation of light-sheet fluorescent imaging to characterize experimental doxorubicin-induced cardiac injury and repair. *Sci. Rep.***7**, 1–11 (2017).
121. Wen, C. *et al.* 3deecelltracker, a deep learning-based pipeline for segmenting and tracking cells in 3d time lapse images. *Elife***10**, e59187 (2021).
122. Roudot, P. *et al.* u-track3D: Measuring, navigating, and validating dense particle trajectories in three dimensions. *Cell Rep. Methods***3**, 100655 (2023).
123. Ruan, X. *et al.* Image processing tools for petabyte-scale light sheet microscopy data. *bioRxiv***2023.12.31**, 573734 (2024).
124. Updegrove, A. *et al.* SimVascular: An Open Source Pipeline for Cardiovascular Simulation. *Ann. Biomed. Eng.***45**, 525 (2017).
125. Lan, H. *et al.* A Re-Engineered Software Interface and Workflow for the Open-Source SimVascular Cardiovascular Modeling Package. *J. Biomech. Eng.***140**, (2018).
126. Ding, Y. *et al.* Saak Transform-Based Machine Learning for Light-Sheet Imaging of Cardiac Trabeculation. *IEEE Trans. Biomed. Eng.***68**, 225–235 (2021).
127. Jiao, Y., Yuan, J., Sodimu, O. M., Qiang, Y. & Ding, Y. Deep Neural Network-Aided Histopathological Analysis of Myocardial Injury. *Front. Cardiovasc. Med.***8**, 724183 (2021).
128. Pietzsch, T., Saalfeld, S., Preibisch, S. & Tomancak, P. BigDataViewer: visualization and processing for large image data sets. *Nat. Methods***12**, 481–483 (2015).
129. Royer, L. A. *et al.* ClearVolume: open-source live 3D visualization for light-sheet microscopy. *Nat. Methods***12**, 480–481 (2015).
130. Abiri, A. *et al.* Simulating Developmental Cardiac Morphology in Virtual Reality Using a Deformable Image Registration Approach. *Ann. Biomed. Eng.***46**, 2177–2188 (2018).
131. Moore, J. *et al.* OME-NGFF: a next-generation file format for expanding bioimaging data-access strategies. *Nat. Methods***18**, 1496–1498 (2021).
132. Moore, J. *et al.* OME-Zarr: a cloud-optimized bioimaging file format with international community support. *Histochem. Cell Biol.***160**, 223–251 (2023).
133. Jahr, W., Schmid, B., Schmied, C., Fahrbach, F. O. & Huisken, J. Hyperspectral light sheet microscopy. *Nat. Commun.***6**, 1–7 (2015).
134. Lu, G. & Fei, B. Medical hyperspectral imaging: a review. *J. Biomed. Opt.***19**, 10901 (2014).

# High-Frequency Ultrasound Techniques for Measurement of Cell and Tissue Mechanics

**Seungyeop Lee, Heidi Kim, Moon Hwan Lee, Yongsoon Eun, and Jae Youn Hwang**

High-frequency ultrasound has emerged as a versatile tool in both medicine and biology. Operating at frequencies exceeding ~20 MHz, this technology excels in delivering superior resolution, albeit at the expense of imaging depth. Consequently, it finds utility in scrutinizing superficial structures such as the skin, eyes, and organs, offering detailed insights into tissue anatomy [1, 2].

Incorporating high-frequency ultrasound facilitates precise visualization of structures, enriching diagnostic capabilities. Moreover, its integration with ultrasound elastography methods, such as acoustic radiation force impulse and shear wave elastography, enables quantitative assessment of tissue elasticity, distinguishing normal from pathological conditions [3–5].

Moreover, the advancement of highly focused high-frequency ultrasound microbeams has expanded the scope of cellular biomechanics studies. These microbeams, generated through precise focusing, enable the measurement of cellular mechanical properties via acoustic trapping and radiation force impulse techniques. However, to augment the accuracy of cell mechanics measurements, a synergistic approach combining optical imaging techniques with high-frequency ultrasound microbeam methodologies has been adopted. This approach enhances the precision and comprehensiveness of cellular biomechanics analyses, fostering advancements in both medical diagnostics and fundamental biological research [6–8].

S. Lee · M. H. Lee · Y. Eun · J. Y. Hwang (✉)
Department of Electrical Engineering and Computer Science, DGIST (Daegu Gyeongbuk Institute of Science and Technology), Daegu, South Korea
e-mail: jyhwang@dgist.ac.kr

H. Kim
Department of Psychology, University of Texas at Austin, Austin, TX, USA

 
J. Liao, J. Y. Wong (eds.), *Integration and Bridging of Multiscale Bioengineering Designs and Tissue Biomechanics*, https://doi.org/10.1007/978-3-031-81743-4_17

# 1 High-Frequency Ultrasound Techniques for Quantification of Cell Mechanics

The mechanical properties of cells, pivotal in various cellular functions such as proliferation, migration, and gene expression, can serve as key biomarkers for disease detection and cell phenotype identification, necessitating the development of biophysical tools for quantifying cell mechanics. Numerous tools, including atomic force microscopy (AFM), optical tweezers, magnetic tweezers, and acoustic tweezers, have been devised for probing cell mechanics. Among these, acoustic tweezers offer several advantages over optical tweezers, providing higher trapping forces with minimal sample damage. This capability allows their application in manipulating relatively large biosamples across a vast length scale [9, 10].

Acoustic tweezers include standing wave-based and single-beam configurations. In particular, single-beam acoustic tweezers have shown significant advancements over the past decades due to their simple experimental setup compared to standing wave counterparts. The basic working principle, based on ray tracing mechanisms, is illustrated in Fig. 1. Here, a highly focused Gaussian ultrasound beam undergoes the refraction of ultrasound rays upon encountering a cell. Rays near the central beam axis exhibit greater intensity than those at the periphery. Refraction conserves momentum but alters the direction of ultrasound rays, generating a gradient force that acts on the cell. This force, generated from refracted ray momentum, manifests as both forward and lateral components. $F_1$ and $F_2$ depict the repulsive forces exerted on the cell by $R_1$' and $R_2$', respectively, with the net force ($F_{net}$) being the vector sum of these individual forces. This net force enables transverse trapping of cells, displacing them toward the ultrasound focus along the lateral direction as depicted in Fig. 1 [11].

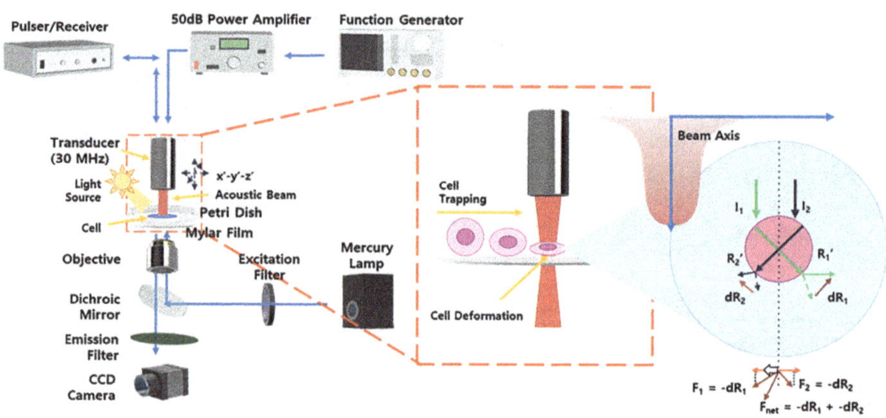


**Fig. 1** Single-beam acoustic tweezers: (1) Experimental setup, (2) single-beam acoustic trapping mechanism. (Modified from Hwang et al. [14])

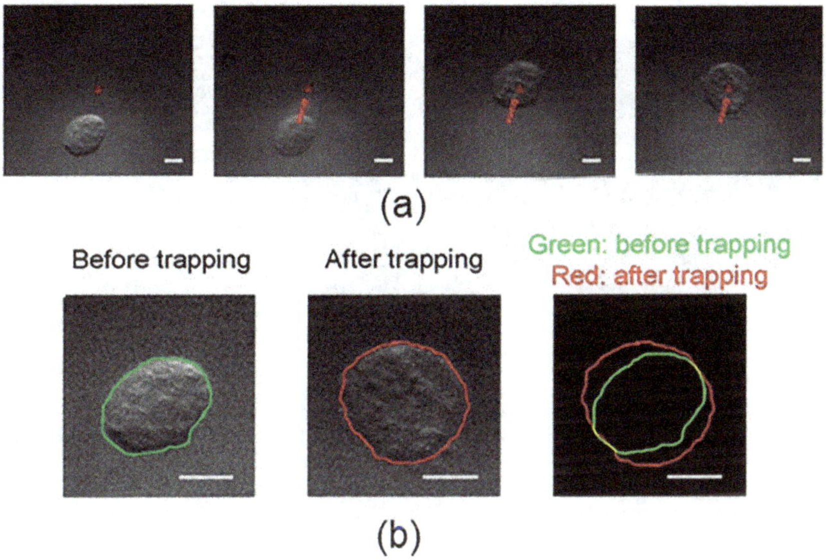


**Fig. 2** Deformation of the acoustically trapped cell: (**a**) Cell movements due to trapping force. (**b**) Changes in the size of the cell before and after acoustic trapping at 1.64 MPa. (Reprinted from Hwang et al. [6], CC BY 4.0 licensed figure. License information: http://creativecommons.org/licenses/by/4.0/)

The single-beam acoustic tweezers have been applied to manipulate and characterize the properties of cells. They allow us to characterize the membrane tension, mechanical properties, and chemotherapeutic response of various cells [6, 8, 12]. The single-beam acoustic tweezers have the potential to quantify the mechanics of cells by measuring the deformability of acoustically trapped cells. The deformation of acoustically trapped cells is likely caused by the acoustic radiation force, while a highly focused ultrasound beam is applied to trap the cells. Figure 2 shows the deformation of a cell before and after trapping. The size of the cell after trapping becomes larger than that before trapping. The red color indicates the boundary of the cell after trapping, whereas the green color indicates the boundary of the cell before trapping.

Moreover, through precise quantification of cellular deformability, distinct cell phenotypes can be discerned. Notably, highly invasive breast cancer cells such as MDA-MB-231 demonstrate markedly greater deformability compared to their weakly invasive counterparts, MCF-7 and BT-474. The calculated normalized diameter of the MDA-MB-231 cell is much larger than those of the MCF-7 and BT-474 cells at 1.22 MPa. These findings underscore that the single-beam acoustic tweezers have the potential as a valuable biophysical tool for probing the intricate mechanics of cellular structures (Fig. 3).

In addition to the acoustic tweezer technique, which is widely used to quantify the mechanical properties of cells, other non-contact high-frequency ultrasound

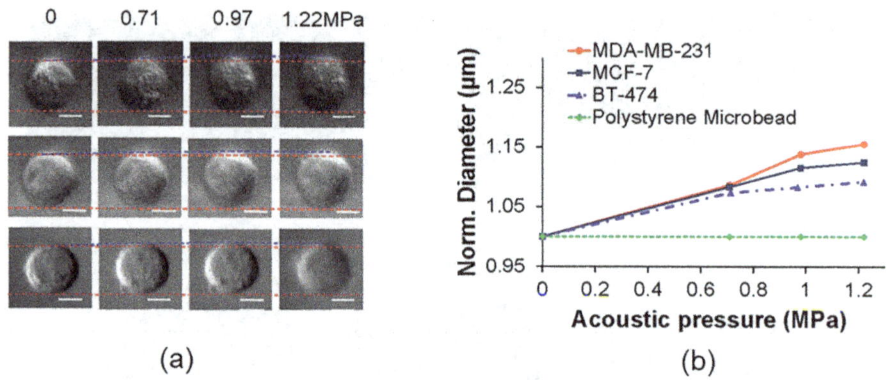


**Fig. 3** Comparisons of deformation of breast cancer cells with different invasiveness by acoustic trapping: (a) Deformation of breast cancer cells with different invasiveness (MDA-MB-231(upper), MCF-7 (middle), and BT-474 (lower) by acoustic trapping at different acoustic pressures (0, 0.71, 0.97, and 1.22 MPa) and (**b**) changes in the normalized diameters of the different trapped cells at the indicated acoustic pressures. (Reprinted from Hwang et al. [6], CC BY 4.0 licensed figure. License information: http://creativecommons.org/licenses/by/4.0/)

techniques traditionally used for the measurement of tissue mechanics are now being adapted for cell-level analysis. One particularly promising technique is acoustic radiation force impulse (ARFI) imaging. ARFI imaging is a non-contact and non-invasive method that measures the viscoelastic properties of tissues. It can be realized by emitting short-duration acoustic radiation forces that induce localized displacements within the tissue, which can then be measured to determine mechanical properties [13].

Based on the principles of ARFI imaging, a novel tool, non-contact acoustic radiation force impulse microscopy via photoacoustic detection (PA-ARFI), has been developed. This innovative technique is designed to measure the mechanical properties of individual cells with high precision. The PA-ARFI microscopy technique utilizes functionalized carbon nanotubes (FCNTs), which are attached to the membrane of target cells, serving as contrast agents. These nanotubes produce strong photoacoustic signals, which are crucial for the temporal measurement of cell membrane displacement induced by ARFI [14].

When ARFI is applied to a target cell, it generates localized displacements in the cell membrane. The FCNTs enhance the detection of these displacements due to their robust photoacoustic response. By tracking the photoacoustic signals over time, the mechanical properties of the cell membrane can be accurately measured. The development of PA-ARFI microscopy represents a significant advancement in the field of cellular mechanics (Fig. 4). This technique opens up new possibilities for studying the mechanical properties of individual cells, which could have profound implications for understanding various physiological and pathological processes. Its potential applications include cancer, stem cell, and drug development research.

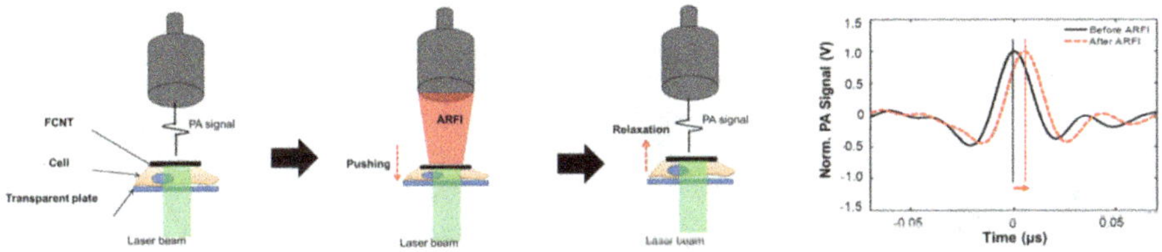


(a) (b)

**Fig. 4** Working principle of PA-ARFI microscopy: (**a**) Sequential procedures of PA-ARFI. After positioning a target cell labeled with FCNTs at the center of the image field of view (left), acoustic radiation force impulses are applied to the cells (middle), followed by tracking of cell membrane displacements (right). (**b**) PA signals from the cell membrane before and after ARFI application to the cell. (Reprinted with permission from Hwang et al. [14] © Optical Society of America)

The PA-ARFI microscopy technique was utilized to measure the mechanical properties of human breast cancer cells with varying metastatic potentials. The cell lines studied included MDA-MB-231 (highly metastatic), SKBR3 (weakly metastatic), and MCF-7 (weakly metastatic). To generate photoacoustic signals from functionalized carbon nanotubes (FCNTs) attached to the cell membranes, a low-intensity pulsed light with a short duration of 25 ms and a total fluence of less than 9 mJ/cm$^2$ was applied to the target cells. Concurrently, while measuring the photoacoustic signals from the FCNTs, ARFI was applied to induce cell membrane displacements.

The displacements of the breast cancer cells with different metastatic potentials were recorded and compared. Figure 5 illustrates the displacements of the various breast cancer cells in response to ARFI. The weakly metastatic breast cancer cells, SKBR3 and MCF-7, exhibited lower membrane displacements compared to the highly metastatic MDA-MB-231 cells. The MDA-MB-231 cells demonstrated a wider distribution of displacements, with more frequent occurrences of displacements greater than 3 μm upon ARFI application.

These results indicate that the SKBR3 and MCF-7 cells are mechanically stiffer than the MDA-MB-231 cells. This suggests that highly invasive breast cancer cells are likely to be mechanically softer, which aligns with the hypothesis that softer cells may have a higher metastatic potential [15]. The findings demonstrate that the non-contact PA-ARFI microscopic system is a valuable tool for measuring and differentiating the mechanical properties of individual cells. This technique has significant potential for cancer research and diagnostic applications.

In conclusion, PA-ARFI microscopy offers a non-invasive, precise method for assessing the mechanical characteristics of single cells, with significant applications in cancer research and beyond.

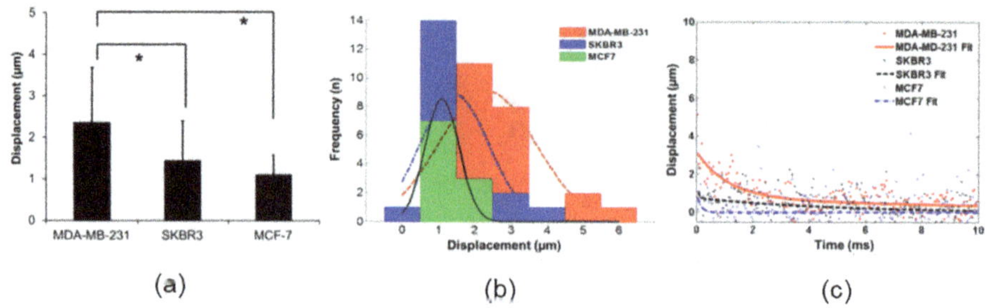


**Fig. 5** Measurement of breast cancer cells with different invasiveness using PA-ARFI: (**a**) Mean cell membrane displacement of MDA-MB-231, SKBR3, and MCF-7 cells (*: $p$-value <0.01). (**b**) Histogram of membrane displacements of each cell type (red: MDA-MB-231, blue: SKBR3, green: MCF-7). The dotted curves represent Gaussian distributions. (**c**) Temporal changes in the displacement of FNTs, which are attached to the membrane of target cells, due to ARFI. (Reprinted with permission from Hwang et al. [14] © Optical Society of America)

## 2 Directional Shear Wave Ultrasound Imaging for Measurement of Cardiac Tissues

Cardiovascular diseases are the leading cause of death worldwide [16]. Despite advancements in diagnostic techniques in recent years, classic clinical imaging methods are generally only used to identify the degree of cardiac damage after cardiac dysfunction has already occurred [17]. This suggests a potential gap in the early detection and prevention of cardiac issues, which could benefit from advances in diagnostic methods and the development of diagnostic instruments.

Evaluating tissue mechanics is a key diagnostic indicator for the diagnosis of diseases because tissue mechanics are closely related to the pathological state of the tissue [18]. For example, cancerous tissue is stiffer than normal tissue. Ultrasound shear wave imaging (USWI) has been studied for tissue diagnosis because USWI can measure tissue stiffness [19–21]. In USWI, the shear waves occur from the user or natural shear wave source such as a heartbeat, then the propagated shear wave through the human tissue is observed by an ultrasound transducer [22]. The observed shear wave speeds are used for diagnosis because the shear wave speed and the stiffness (Young's modulus) of human tissues are related by the following equation:

$$\text{Young's modulus } (E) = 3\rho V^2 \tag{1}$$

In this equation, $E$ , $\rho$ , and $V$ represent Young's modulus, the density of the tissue, and the shear wave speed, respectively. This equation is obtained from a simplified version of Hooke's law for isotropic and incompressible media. Since the density of the tissue is almost constant in tissue, the shear wave speed is transformed to Young's modulus by Eq. 1. Thus, observing the shear wave speed can be used to diagnose tissues.

However, in cardiac tissue, the arrangement of cardiac tissue fibers causes transverse isotropy, leading to directional mechanics in which shear wave speed varies depending on the shear propagation direction [23]. In transversely isotropic tissues, shear wave speed is described by the following equation which is from a simplified version of Hooke's law for transverse isotropic and incompressible:

$$V(\theta) = \sqrt{\frac{\mu_{\|}\cos^2(\theta) + \mu_{\perp}\sin^2(\theta)}{\rho}} \tag{2}$$

In this equation, $\theta$ is the angle between the direction in which the shear wave propagates and fiber orientation, $V(\theta)$ is the shear wave speed, $\mu_{\|}$ and $\mu_{\perp}$ are elastic modulus, and $\rho$ is the density of the tissue. To estimate elastic modulus, the shear wave speed should be measured at several different points and $\theta$ needs to be known. In other words, most shear wave ultrasound imaging process based on Eq. 1 is insufficient to work well as a diagnostic indicator in cardiac tissue. From Eq. 2, to utilize shear wave ultrasound imaging as a diagnostic indicator for cardiac tissue, it is necessary to measure shear wave speed in at least two different directions. In other words, a novel shear wave ultrasound imaging process to evaluate directional mechanics is necessary.

This section introduces a novel shear wave ultrasound imaging system, which assesses directional mechanics, as a diagnostic indicator for cardiac tissue. In addition, a novel metric to quantify directional mechanics considering viscosity or geometry which can affect the propagation of the shear wave is described.

A new type of shear wave ultrasound imaging system, called a directional shear wave ultrasound imaging system, has been developed to measure directional mechanics [24]. The directional shear wave imaging system comprises a mechanical shaker to generate strong displacement, a high-frequency ultrasound transducer for high-resolution imaging, and a rotation stage for imaging at different angles. The entire directional shear wave ultrasound imaging system is depicted in Fig. 6.

The directional shear wave ultrasound imaging system measures a 2-dimensional (2D) shear wave image within an angle range from 0° to 360° with an increment of 3° to know the shear wave speed depending on the angle ($\theta$ ). The 2D shear wave images are acquired by the following processes. The shear wave is generated by stimulation from acoustic radiation force or mechanical shaker, which propagates into the tissue and makes displacement. The ultrasound transducer receives RF signal for a few milliseconds at a point after the shear wave generation to measure the displacement of tissue that changes over time. After data acquisition, the received RF signals at each tracking point are converted to displacement data by the cross-correlation technique [25]. A directional filtering technique is applied to the displacement data to reduce reverberation at tissue boundaries before estimating shear wave speed [26]. Finally, the shear wave speed is estimated by applying time-of-flight techniques to the displacement data at several different points. The shear

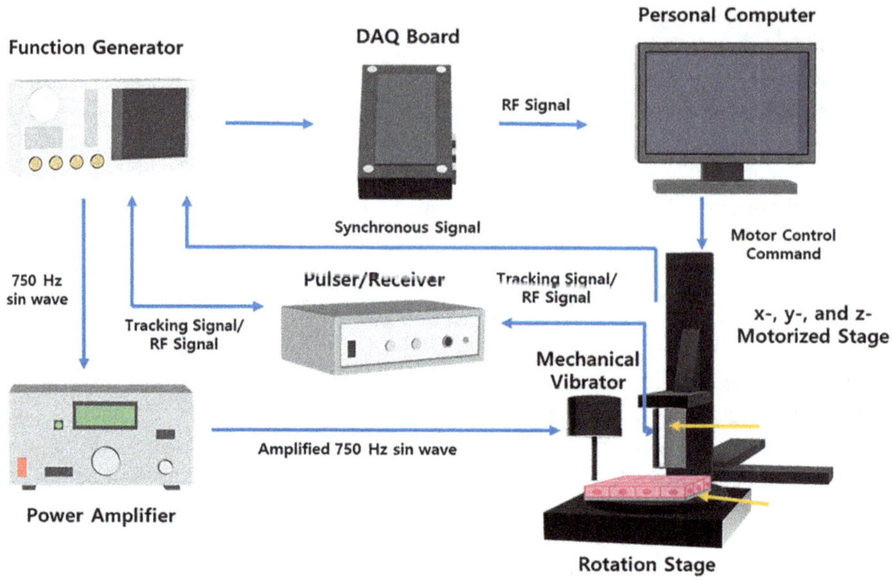


**Fig. 6** Directional shear wave ultrasound imaging system. (Modified from Lee et al. [24])

wave speeds estimated at entire points are combined into a 2D shear wave speed map of the region of interest. The obtained 2D shear wave ultrasound images within an angle range from 0° to 360° with an increment of 3° are combined into the directional shear wave image.

The directional shear wave ultrasound image on the cardiac tissue ex vivo is illustrated in Fig. 7, which shows that the shear wave speed on the cardiac tissue varies depending on the direction in which the shear wave propagates. This result demonstrates that shear wave speed, varying on the direction a wave propagates, can be measured by the directional shear wave ultrasound imaging system.

The directional shear wave speed can be quantified using fractional anisotropy (FA), which is a method to quantify tissue mechanics in transversely isotropic tissue, for the diagnosis of cardiac tissue [27]. However, this system does not consider frequency dispersion due to the viscosity or geometry of the tissue. Frequency dispersion underestimates the shear wave speed when the tissue has viscosity, or when the thickness of the tissue is lower than the wavelength [28, 29]. This phenomenon may cause a misunderstanding of FA. Therefore, the value obtained by applying FA to the directional shear wave ultrasound imaging is insufficient to diagnose cardiac tissue.

To overcome the above problems, a metric, named maximum cosine similarity (MaxCosim), can be used to quantify the directional mechanics of cardiac tissues by evaluating the periodicity of the directional shear wave speed [30].

The periodicity of the shear wave speed can be compared by contrasting it to an ideal cosine wave that has a period of 180° because the shear wave speed is

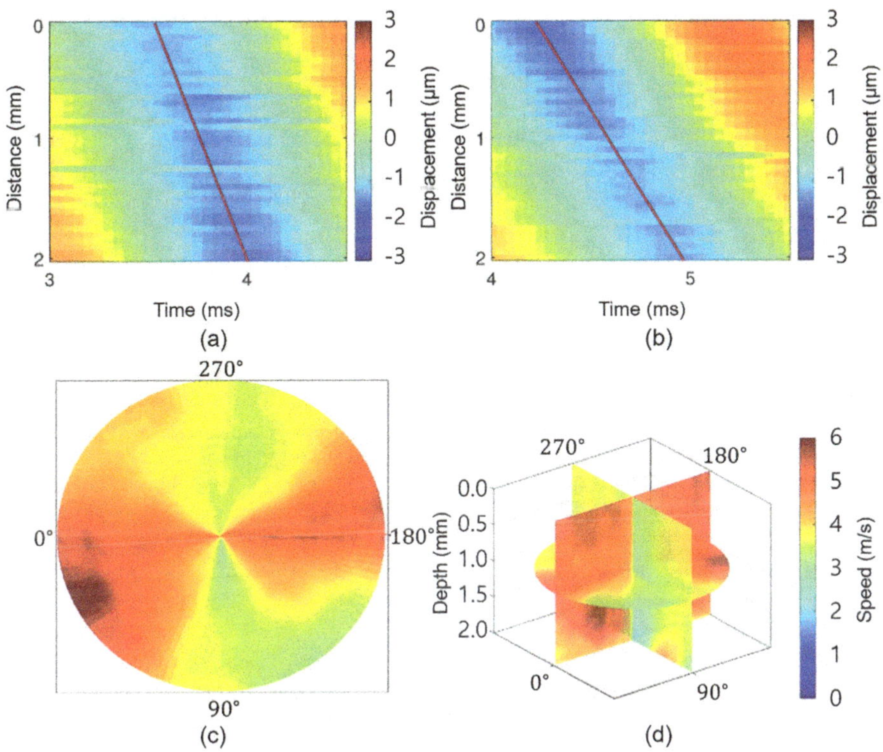


**Fig. 7** Heart image: directional dependence of the heart at the middle myocardium. The slope of the red line indicates the wave speed. (**a**) displacement map: circumferential direction, (**b**) displacement map: longitudinal direction, (**c**) heart image: 0 mm depth, and (**d**) heart image: 360° (A depth of 0 mm does not mean the surface, but the point at which the ultrasound transducer started measuring inside of the tissue). (Reprinted from Lee et al. [24], CC BY 4.0 licensed figure. License information: http://creativecommons.org/licenses/by/4.0/)

periodic with a period of 180°. To evaluate the periodicity of shear wave speed, cosine similarity is adopted as a metric. Cosine similarity measures the similarity between two signals as follows:

$$Cosim(A) = \frac{\int_0^{2\pi} A(\theta)\cos(\theta)\,d\theta}{\sqrt{\int_0^{2\pi} A(\theta)\,d\theta}\,\sqrt{\int_0^{2\pi} \cos^2(\theta)\,d\theta}} \quad (3)$$

In this equation, $A$ is the shear wave speed profile depending on the measurement angle. This equation represents the similarity between $A$ and an ideal cosine waveform. Since cosine similarity does not consider the phase shift, cosine similarity is insufficient when the phase of wave speed profile $A$ does not coincide with that of the ideal cosine wave. Therefore, they adopted a phase sweep to find the maximum value of the Cosim, which would then correspond to the phase of signal $A$ that best

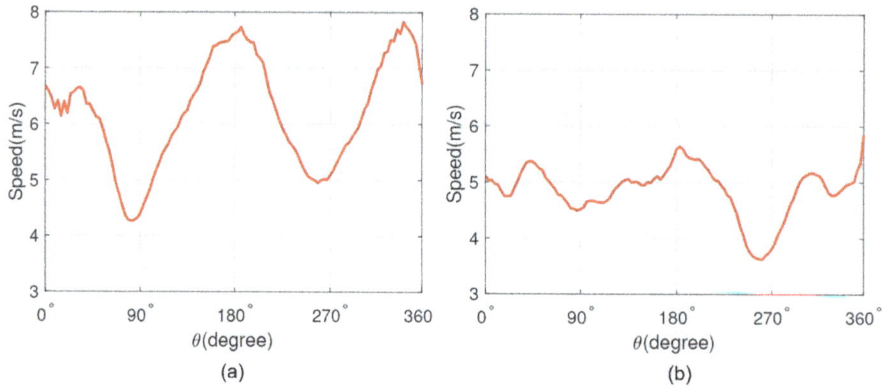


**Fig. 8** Wave speed profile for different wave traveling directions parameterized by the angle (the wave speed profile for each group is obtained from a representative individual). (**a**) Control group. (**b**) DOX treatment group. (Reprinted with permission from Lee et al. [30] © IEEE)

matches the phase of the ideal cosine waveform. MaxCosin is calculated by the following equation:

$$MaxCosim(A) = \max_{0\leq\phi\leq 2\pi} \frac{\int_0^{2\pi} A(\theta)\cos(\theta+\phi)\,d\theta}{\sqrt{\int_0^{2\pi} A(\theta)\,d\theta}\sqrt{\int_0^{2\pi}\cos^2(\theta+\phi)\,d\theta}} d\phi \tag{4}$$

From the equation, MaxCosim represents the degree of similarity between the pattern of the wave speed profile and a sinusoidal waveform. The MaxCosim value close to 1 implies that the tissue is transversely anisotropic material.

The feasibility of MaxCosim as a quantification metric was validated by animal experiments on rats. The example of shear wave speed profile in control group and doxorubicin (DOX) treatment group is shown in Fig. 8. The shear wave speed profile is disrupted in DOX treatment groups. The tissues were classified as normal or cardiomyopathy based on the histopathological evaluation performed by two pathologists. The distribution of the MaxCosim values based on histological changes is plotted in Fig. 9. A decrease in the MaxCosim value is observed in the cardiomyopathy group. To evaluate MaxCosim as a diagnostic indicator, a *t*-test is performed between the values of MaxCosim in cardiomyopathy and normal tissues. The *p*-value is calculated as 0.000002, which implies that evaluating transverse anisotropy can help in the diagnosis of a heart. From these results, the MaxCosim value below 0.7 indicates the possibility of damage in heart tissues.

Therefore, this section implies that the direction-dependent shear wave speed measured by the directional shear wave ultrasound imaging and quantified by MaxCosim can be a useful diagnostic indicator for cardiac diseases.

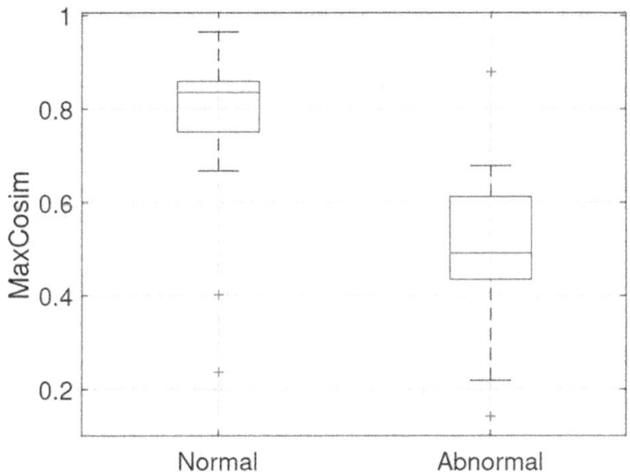


**Fig. 9** Box plot of MaxCosim for normal hearts and hearts with cardiomyopathy. (Reprinted with permission from Lee et al. [30] © IEEE)

## 3 Acoustic Radiation Force Impulse Method for Tumor Detection

Cancer is the second-leading cause of death in the United States overall and a major public health problem worldwide [31]. Despite significant technological advancements related to the diagnosis and treatment of cancer, there have been limited improvement in the overall survival rate due to the heterogeneous patterns of cancer [32]. Therefore, ultrasound imaging and optical imaging are being developed to diagnose tumors because of some advantages, such as non-invasiveness, time-efficiency, and inexpensiveness [33, 34].

Multispectral imaging is highly useful as a non-invasive imaging modality for the characterization and delineation of tumors and therefore has been explored for various clinical and biological applications [35, 36]. However, the use of spectral imaging is limited to shallow penetration depths with imaged tissues because light cannot penetrate deep inside tissues. High-frequency ultrasound B-mode imaging and ARFI imaging may complement the limited capability of spectral imaging for the detection of tumors deep inside excised tissue due to the permeability of ultrasound waves in tissues. The high-frequency ultrasound B-mode image offers structural information about the tissues of interest in more detail because the high-frequency ultrasound provides high axial and lateral resolution [37]. However, high-frequency B-mode imaging often fails to detect a malignant tumor because it may appear isoechoic with surrounding non-cancerous tissue [38]. Thus, an additional reliable method is needed to detect tumors more accurately.

Recently, high-frequency acoustic radiation force impulse (ARFI) imaging has shown to be a viable alternative in overcoming the shortcomings of B-mode imaging by overlaying a map of the stiffness of tissues with an ultrasound B-mode image [39]. ARFI imaging is an advanced sonographic technique to diagnose tissue by qualitatively evaluating tissue mechanics. ARFI imaging can obtain qualitative tissue stiffness by measuring displacement response to a force because the displacements are typically inversely proportional to local tissue stiffness. For instance, displacements in response to a compressional force in stiff tissues are lower than those in softer tissues. This relationship between displacement and stiffness helps diagnose tissue by observing its displacement response to force. Based on these characteristics, ARFI imaging has been combined with multispectral imaging for tumor detection, which is called a multimodal biomicroscopic system [40].

ARFI imaging utilizes short-duration acoustic radiation forces (ARF) to induce localized displacements in tissue. Such displacements in the order of ten microns are then tracked using ultrasound. These relative displacements in response to ARF provide relative information about tissue mechanics, which can be used to detect tumors.

In ARFI imaging, the short-duration ARF is exerted in the direction of acoustic wave propagation by a transfer of momentum from the acoustic wave to a target medium. The momentum transfer causes a mechanical force to the target. The exerted force can be expressed with the following equation:

$$F = \frac{W_{\text{absorbded}}}{c} = \frac{2\alpha I}{c} \tag{5}$$

where $F$ is acoustic radiation force, $W_{\text{absorbed}}$ is the power absorbed by the medium at a given spatial location, $c$ is the sound speed in the medium, $\alpha$ is the absorption coefficient of the medium, and $I$ is the temporal average intensity at a given point in space [41]. The applied force on tissues causes tissue displacement at a region of the acoustic beam, and the displacement is monitored both spatially and temporally by an ultrasound transducer. Represented properties, such as the maximum displacement in response to ARF, the time it took for the tissue to reach its maximum displacement, or the recovery time of tissue can be used to differentiate tumors [42]. In the multimodal biomicroscopic system, maximum displacement in response to ARF is used for ARFI imaging.

The block diagram of the multimodal biomicroscopic system that combines high-frequency ultrasound B-mode imaging, ARFI imaging, and multispectral imaging is shown in Fig. 10. The system consisted of two parts for high-frequency B-mode/ARFI and multispectral imaging. For high-frequency B-mode and ARFI imaging, a 45 MHz ultrasound transducer was fabricated and attached to x-, y-, and z-axis motorized stages, which allows both two- and three-dimensional ultrasound imaging. The transducer is driven by a pulser-receiver to perform high-frequency B-mode and ARFI imaging simultaneously. After the transducer is activated by a pulse from the pulser-receiver, echo signals from the target are detected by the transducer and then amplified in the pulser-receiver. Subsequently, they are recorded in order to obtain B-mode and ARFI images. For ARFI imaging, 12 trigger signals from the first channel of a function generator are input to the pulser-receiver with a pulse repetition frequency of 20 kHz to acquire reference and tracking signals for the measurement of displacements of the target due to the acoustic radiation force

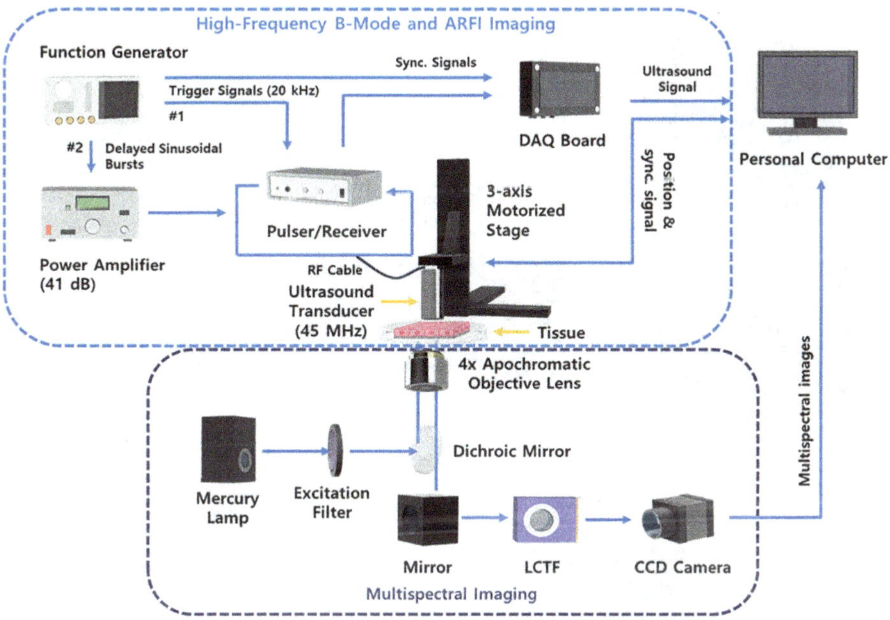


**Fig. 10** Block diagram of a multimodal biomicroscopic system. (Modified from Kim et al. [40])

applied to the target. In addition, the acquisition frequency per each scan line for ARFI imaging was 100 Hz. 45 MHz sinusoidal bursts (number of cycles: 10,000) from the second channel of the function generator amplified by a power amplifier are input to the transducer to generate acoustic radiation force. The sinusoidal bursts are input to the transducer after acquisition of a reference signal. For the synchronization of all the equipment utilized for ARFI imaging, synchronization signals from an analog-to-digital converter are input to both the pulser-receiver and the function generator. Since this section focuses on ultrasound imaging techniques, the details of multispectral imaging are not treated (Please refer to Ref. [40]).

To validate the feasibility of ARFI imaging for tumor detection, extensive multimodal imaging of colon tissues including tumors excised from six cancer patients was performed. Figure 11a and b show multispectral images within the wavelength range of 420 nm to 700 nm with a step size and a bandwidth of 10 nm as well as the normalized spectral signatures of colorectal tumors and normal colon tissues. The normalized spectral signatures of the normal colon tissues and colorectal tumors exhibited similar peaks around 550 nm. However, the spectral signatures for the colorectal tumors were broader than those for the normal colon tissues. Moreover, the intensity of the former was approximately 700 nm lower than that of normal tissues. Colorectal tumor regions on the surfaces of the colon tissues were clearly distinguished from normal tissues by spectral classification with these spectral signatures normalized as a reference, as shown in the spectral classified image (Fig. 11c). Figure 11d and e indicate high-resolution B-mode

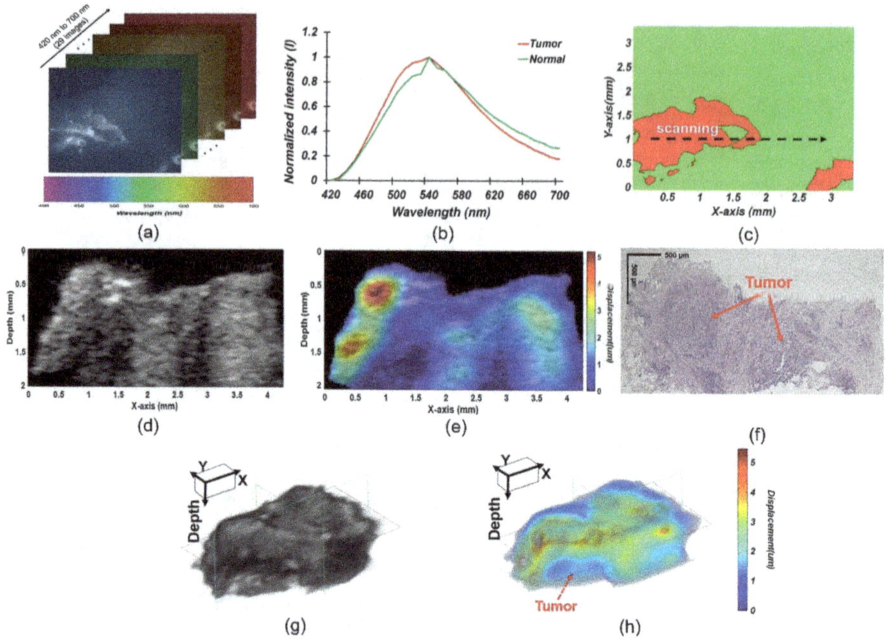

**Fig. 11** Multimodal imaging of colorectal tumors ex vivo using the proposed system. (**a**) Multispectral images obtained at the indicated wavelengths. (**b**) Normalized spectral signatures of colorectal tumors and normal colon tissues. (**c**) Multispectral classified image. The black dotted arrow indicates the scanning line. (**d**) A high-frequency B-mode image. (**e**) A high-frequency ARFI image overlaid with the corresponding B-mode image. (**f**) H&E-stained image corresponding to the B-mode and ARFI image. The red arrows indicate tumor regions. (**g**) Three-dimensional B-mode and (**f**) ARFI image (2 mm × 2.8 mm × 4 mm in x, y, and depth). The pseudo-colored map indicates tissue displacements. (Reprinted from Kim et al. [12], CC BY 4.0 licensed figure. License information: http://creativecommons.org/licenses/by/4.0/)

and ARFI images of the excised colon tissues including tumors. In the B-mode image, tumor regions exhibit slightly darker intensity levels than the normal colon regions, but they are not clearly distinguished. In contrast, in the ARFI overlaid on a B-mode image, the colorectal tumors, shown in blue, could be identified from the normal colon tissues. As expected, tumor regions typically exhibited greater stiffness compared to normal regions. Figure 11f shows the H&E-stained tissue image in the longitudinal direction corresponding to the B-mode and ARFI images. The regions pointed by the red arrows in the H&E image that were found to be stiffer in the ARFI image were indeed the tumor regions, whereas the regions at the other part of the H&E image were the normal regions. Furthermore, Fig. 11g and h show the tumor's three-dimensional B-mode and ARFI imaging, representing the potential of three-dimensional B-mode and ARFI imaging for tumor detection.

Finally, statistical analysis of reference spectral signatures for multispectral imaging and mean displacement of the tumors and normal tissues induced by acoustic radiation force were carried out. Figure 12a presents the normalized mean

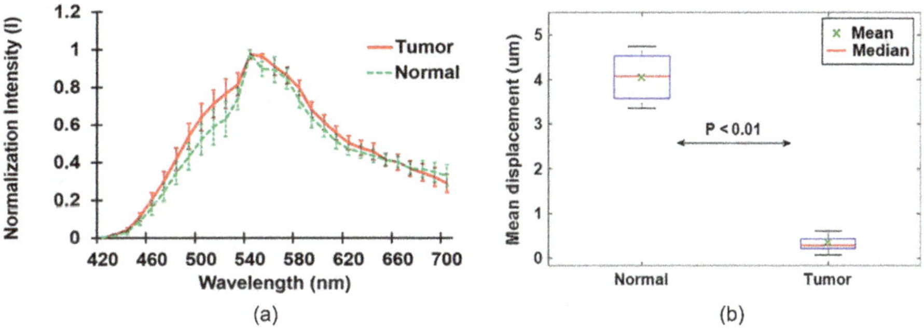


**Fig. 12** Statistical analysis of reference spectral signatures for multispectral imaging and mean displacements of tumors and normal regions induced by acoustic radiation force. (**a**) Normalized mean spectrum with the standard deviation (sample number = 6). (**b**) Box and whisker plot of the displacements of ARFI images (sample number = 6). (Reprinted from Kim et al. [12], CC BY 4.0 licensed figure. License information: http://creativecommons.org/licenses/by/4.0/)

spectrum with the standard deviation of six samples of tumors and normal tissues. It was observed that the spectral signature of the tumors was distinguished from that of the normal colon tissues. The spectral signature of the tumors clearly exhibited higher intensity than that of the normal colon tissues at 520 nm. On the other hand, Fig. 12b illustrates the mean displacements of the ARFI images from the six samples. In this case, it was found that the mean displacements between the normal tissues and tumors were significantly different ($p$-value<0.01). These results thus demonstrated that the proposed system has the potential to discriminate between tumors and normal regions in excised human colorectal tissue samples.

Furthermore, the total acquisition time for ARFI, B-mode, and 29 multispectral images was 3 min 15 s. Note that the total time for examinations of excised tumors using our proposed system is still much shorter than that of examinations of human colorectal excised tumors using histopathological methods (preparation time: typically, over ~24 h with the fixation of tissues). Therefore, these results suggested that this system has the potential for rapid tumor characterizations compared to the histopathological methods.

**Acknowledgments** This book chapter is supported by the Technology Innovation Program funded by the Ministry of Trade, Industry and Energy (Korea) under Grant No. 20014214.

## References

1. Lockwood, G., et al., *Beyond 30 MHz [applications of high-frequency ultrasound imaging].* IEEE Engineering in Medicine and Biology Magazine, 1996. **15**(6): p. 60–71.
2. Shung, K.K., *High frequency ultrasonic imaging.* Journal of medical ultrasound, 2009. **17**(1): p. 25–30.

3. Zhang, J., et al., *High-frequency ultrasound elastography to assess the nonlinear elastic properties of the cornea and ciliary body.* IEEE transactions on ultrasonics, ferroelectrics, and frequency control, 2022. **69**(9): p. 2621–2629.
4. Lay, F.-Y., et al., *Ex vivo evaluation of mouse brain elasticity using high-frequency ultrasound elastography.* IEEE Transactions on Biomedical Engineering, 2019. **66**(12): p. 3426–3435.
5. Botar-Jid, C.M., et al., *Assessment of cutaneous melanoma by use of very-high-frequency ultrasound and real-time elastography.* American Journal of Roentgenology, 2016. **206**(4): p. 699–704.
6. Hwang, J.Y., et al., *Cell deformation by single-beam acoustic trapping: a promising tool for measurements of cell mechanics.* Scientific reports, 2016. **6**(1): p. 27238.
7. Hwang, J.Y., et al., *Non-contact high-frequency ultrasound microbeam stimulation for studying mechanotransduction in human umbilical vein endothelial cells.* Ultrasound in medicine & biology, 2014. **40**(9): p. 2172–2182.
8. Youn, S., et al., *Acoustic trapping technique for studying calcium response of a suspended breast cancer cell: Determination of its invasion potentials.* IEEE Transactions on Ultrasonics, Ferroelectrics, and Frequency Control, 2019. **66**(4): p. 737–746.
9. Li, Z., et al., *The forbidden band and size selectivity of acoustic radiation force trapping.* Iscience, 2021. **24**(1).
10. Lam, K.H., et al., *Multifunctional single beam acoustic tweezer for non-invasive cell/organism manipulation and tissue imaging.* Scientific reports, 2016. **6**(1): p. 37554.
11. Lee, J., et al., *Transverse acoustic trapping using a Gaussian focused ultrasound.* Ultrasound in medicine & biology, 2010. **36**(2): p. 350–355.
12. Kim, M.G., et al., *Label-free analysis of the characteristics of a single cell trapped by acoustic tweezers.* Scientific reports, 2017. **7**(1): p. 14092.
13. Nightingale, K., *Acoustic Radiation Force Impulse (ARFI) Imaging: a Review.* Curr Med Imaging Rev, 2011. **7**(4): p. 328–339.
14. Hwang, J.Y., et al., *Non-contact acoustic radiation force impulse microscopy via photoacoustic detection for probing breast cancer cell mechanics.* Biomedical optics express, 2015. **6**(1): p. 11–22.
15. Swaminathan, V., et al., *Mechanical stiffness grades metastatic potential in patient tumor cells and in cancer cell lines.* Cancer research, 2011. **71**(15): p. 5075–5080.
16. WHO, *The Top 10 Causes of Death. World Health Organization.* 2020, WHO Geneva, Switzerland.
17. Konofagou, E., *Cardiovascular elastography*, in *Tissue Elasticity Imaging*. 2020, Elsevier. p. 67–107.
18. Chen, W.L.K. and C.A. Simmons, *Lessons from (patho) physiological tissue stiffness and their implications for drug screening, drug delivery and regenerative medicine.* Advanced drug delivery reviews, 2011. **63**(4-5): p. 269–276.
19. Sandrin, L., et al., *Transient elastography: a new noninvasive method for assessment of hepatic fibrosis.* Ultrasound in medicine & biology, 2003. **29**(12): p. 1705–1713.
20. Youk, J.H., H.M. Gweon, and E.J. Son, *Shear-wave elastography in breast ultrasonography: the state of the art.* Ultrasonography, 2017. **36**(4): p. 300.
21. Taljanovic, M.S., et al., *Shear-wave elastography: basic physics and musculoskeletal applications.* Radiographics, 2017. **37**(3): p. 855–870.
22. Santos, P., et al., *Natural shear wave imaging in the human heart: normal values, feasibility, and reproducibility.* IEEE transactions on ultrasonics, ferroelectrics, and frequency control, 2018. **66**(3): p. 442–452.
23. Lee, W.-N., et al., *Mapping myocardial fiber orientation using echocardiography-based shear wave imaging.* IEEE transactions on medical imaging, 2011. **31**(3): p. 554–562.
24. Lee, S., et al., *Ex vivo evaluation of mechanical anisotropic tissues with high-frequency ultrasound shear wave Elastography.* Sensors, 2022. **22**(3): p. 978.
25. Rouze, N.C., et al., *Parameters affecting the resolution and accuracy of 2-D quantitative shear wave images.* IEEE transactions on ultrasonics, ferroelectrics, and frequency control, 2012. **59**(8): p. 1729–1740.

26. Deffieux, T., et al., *On the effects of reflected waves in transient shear wave elastography.* IEEE transactions on ultrasonics, ferroelectrics, and frequency control, 2011. **58**(10): p. 2032–2035.
27. Tseng, W.Y.I., et al., *Imaging myocardial fiber disarray and intramural strain hypokinesis in hypertrophic cardiomyopathy with MRI.* Journal of Magnetic Resonance Imaging: An Official Journal of the International Society for Magnetic Resonance in Medicine, 2006. **23**(1): p. 1–8.
28. Chen, S., M. Fatemi, and J.F. Greenleaf, *Quantifying elasticity and viscosity from measurement of shear wave speed dispersion.* The Journal of the Acoustical Society of America, 2004. **115**(6): p. 2781–2785.
29. Li, G.-Y., et al., *Guided wave elastography of layered soft tissues.* Acta biomaterialia, 2019. **84**: p. 293–304.
30. Lee, S., et al., *New metric to evaluate cardiac anisotropic mechanics by directional high frequency ultrasound-based transverse wave elastography.* IEEE Transactions on Ultrasonics, Ferroelectrics, and Frequency Control, 2023.
31. Siegel, R.L., A.N. Giaquinto, and A. Jemal, *Cancer statistics, 2024.* CA: a cancer journal for clinicians, 2024. **74**(1): p. 12–49.
32. Partridge, M., et al., *Detection of minimal residual cancer to investigate why oral tumors recur despite seemingly adequate treatment.* Clinical cancer research, 2000. **6**(7): p. 2718–2725.
33. Lee, W.-N., et al., *Theoretical quality assessment of myocardial elastography with in vivo validation.* ieee transactions on ultrasonics, ferroelectrics, and frequency control, 2007. **54**(11): p. 2233–2245.
34. Balas, C., *Review of biomedical optical imaging—a powerful, non-invasive, non-ionizing technology for improving in vivo diagnosis.* Measurement science and technology, 2009. **20**(10): p. 104020.
35. Behrooz, A., et al., *Multispectral open-air fluorescence-guided imaging and detection of tumors using a hands-free translational platform with liquid crystal tunable filters (LCTF).* Cancer Research, 2016. **76**(14_Supplement): p. 4246–4246.
36. Yang, V.X., et al., *A multispectral fluorescence imaging system: Design and initial clinical tests in intra-operative Photofrin-photodynamic therapy of brain tumors.* Lasers in surgery and medicine, 2003. **32**(3): p. 224–232.
37. CC, H., *High frequency, high resolution B-scan ultrasound in the assessment of skin tumours.* Br J Dermatol, 1993. **128**: p. 525–532.
38. Saarenmaa, I., et al., *The effect of age and density of the breast on the sensitivity of breast cancer diagnostic by mammography and ultasonography.* Breast cancer research and treatment, 2001. **67**: p. 117–123.
39. Park, J., et al., *Acoustic radiation force impulse (ARFI) imaging of zebrafish embryo by high-frequency coded excitation sequence.* Annals of biomedical engineering, 2012. **40**: p. 907–915.
40. Kim, J., et al., *A multimodal biomicroscopic system based on high-frequency acoustic radiation force impulse and multispectral imaging techniques for tumor characterization ex vivo.* Scientific Reports, 2017. **7**(1): p. 17518.
41. Nightingale, K., et al., *Acoustic radiation force impulse imaging: in vivo demonstration of clinical feasibility.* Ultrasound in medicine & biology, 2002. **28**(2): p. 227–235.
42. Nightingale, K., *Acoustic radiation force impulse (ARFI) imaging: a review.* Current medical imaging, 2011. **7**(4): p. 328–339.

# Non-obstructive Methods to Detect Sarcomere Remodeling of Live In Vitro Cardiomyocytes

**Nicholas Rogozinski, Sarah Velez, Yi Hong, and Huaxiao Yang**

## 1 Introduction

Heart failure (HF) is a leading cause of death globally within the twenty-first century, accounting for about 64 million individuals annually [1]. Therefore, a way to further characterize the pathogenesis of myocardial diseases that lead to effective treatment strategies is one of the highest priorities. Various cardiomyopathic conditions that lead to the development of HF such as hypertrophic cardiomyopathy (HCM), dilated cardiomyopathy (DCM), and arrhythmic cardiomyopathy (ACM) are the focus of many therapeutics and disease models aiming to reduce mortality rates and treat heart failure. However, not all aspects of these diseases are sufficiently understood, and further light needs to be shed on the contractile protein apparatus, the sarcomere. The sarcomere is the smallest functional unit of cardiomyocytes (CMs) for contraction, and sarcomere dysfunction may cause various cardiac-related complications [2]. For example, an overwhelming majority of HCM cases are associated with the genetic inheriting of sarcomere mutation and result in structural and functional alteration of the sarcomere [3]. The ability to accurately understand the sarcomere remodeling of cardiac microfibers in CMs will significantly help us delineate the molecular pathogenesis of various cardiomyopathies [4].

The sarcomere consists of two major components: the thick filaments and the thin filaments. These components are arranged into a hexagonal pattern, with the thin actin-based filaments attaching in parallel at the Z-line, and the thick myosin-

N. Rogozinski · S. Velez · H. Yang (✉)
Department of Biomedical Engineering, University of North Texas, Denton, TX, USA
e-mail: huaxiao.yang@unt.edu

Y. Hong
Department of Bioengineering, University of Texas at Arlington, Arlington, TX, USA

 
J. Liao, J. Y. Wong (eds.), *Integration and Bridging of Multiscale Bioengineering Designs and Tissue Biomechanics*, https://doi.org/10.1007/978-3-031-81743-4_18

containing filaments spanning the middle of the sarcomere unit [5, 6]. Sliding filament theory, the basis for which our current understanding of muscle contraction comes, dates to 1954 when it was first proposed. The theory encompasses both chemical and physical interactions between the two filament types [7, 8]. In short, each myosin-actin contact point contains two myosin heads, which serve as the active site for hydrolyzation and the formation of cross bridges. Allowing for the physical dragging or sliding of these filaments to occur. The proper functioning of the sarcomere is crucial for the CM to respond to its ever-changing environment [9]. Human hemodynamic conditions can be rapidly altered in response to a variety of stimuli, resulting in changes to ventricular pressure and physical demands of the CM [10, 11]. In response, the length of the sarcomere is often remodeled to directly impact force output, a concept known as the Frank-Sterling mechanism [12]. However, sarcomere pathologic remodeling can occur in a variety of instances, such as the addition and subtraction to existing myofibrils resulting in length changes [13], as well as organizational disarray which has been correlated with severe prognostic outcomes [14]. Regardless of occurring during natural development, or the genesis of pathology, sarcomere remodeling highlights the importance of the myofibril structure-function relationship [15, 16].

The difficulty of tracking sarcomere remodeling has been realized due to its small dimension at 1 ~ 2 μm and the fast dynamics of sarcomere shortening [17, 18]. Typically, sarcomere observations have primarily centered around histological and immunostaining techniques. These typically involve labeling specific sarcomere components with fluorescently active compounds that are conjugated to antibodies or other delivery molecules. The samples are usually derived from cryo-sectioned tissue slices or whole cells in culture and then are subjected to a chemical agent that effectively fixes the sarcomere structure at the end point of experiments. The preserved cells can then be labeled with a high degree of sensitivity and functionality. However, chemical fixation is only able to produce an image of a single selected timepoint, making it very difficult to piece together what is occurring during complex and dynamic biological processes of sarcomere remodeling. With the advancement of increasingly complex analysis techniques, the ability to measure the dynamic changes in sarcomere structure and function by various non-obstructive live-cell imaging modalities, such as fluorescence resonance energy transfer (FRET) [19, 20] and phase contrast [21], to track sarcomere remodeling as it occurs in real time in the process of cardiogenesis and pathogenesis of cardiomyopathies. This can allow for a more detailed explanation of how and why sarcomere changes occur from genetic or environmental stressors. In this chapter, we introduced the current state of the art in live-cell sarcomere imaging techniques, focusing on the advantages and challenges of imaging live CMs in non-obstructive ways for studying sarcomere remodeling in various in vitro culture models. We further highlighted the applications of understanding of sarcomerogenesis and sarcomere remodeling, evaluating cardiac function and mechanotransduction, disease modeling of cardiomyopathies, evaluating pharmaceutical efficacy and cardiotoxicity, and cardiac regenerative medicine. Last, we discussed the current limitations and future perspectives of studying sarcomere remodeling with more advancements in sources, maturation, and sub-types of human pluripotent stem cell (hPSC)-derived CMs, and more potential non-obstructive live imaging technologies of sarcomeres.

## 2 Non-obstructive Methods to Detect Sarcomere in Live CMs

### *2.1 Genetically Encoded Fluorescence-Based Imaging*

Live-cell transfections are a simple and instant method to introduce fluorescently reporting molecules via the insertion of a plasmid containing the tagging gene, to a component of interest within the cell. Fluorescent proteins like green fluorescent protein (GFP) or mCherry are used broadly to label sarcomeres for fluorescence visible by fluorescence microscopic imaging. Transfections will enable the expression of these proteins as a conjugate to the protein of interest. For example, a study applied GFP-alpha-actinin labeled embryonic heart cells to track Z-line structure and its early development [22, 23], and found the alignment of new sarcomeres during the formation of new myofibrils. The simplicity of this technique allows for high degrees of image clarity and use without specialized laboratory equipment, making the use of transfection-based GFP expression very accessible. Conversely, the simplicity of the system leaves room for variance to occur. This inconsistency can stem from numerous steps in the process such as the efficiency of the transfection, duration of expression, fluorescence photobleaching after long-time continuous imaging, and altered configuration and function of labeled sarcomere unit. Regardless, with Z-lines and other sarcomere components visible via fluorescence labeling, researchers can extract useful information on sarcomerogenesis, and sarcomere shortening and remodeling [24].

Similarly, encoded reporter gene systems (ERGS) are powerful analytic tools allowing for detecting the expressions of fluorescent proteins [25, 26]. However, a major difference between live-cell transfections and ERGS is that cells harnessing ERGS have been permanently engineered to innately express fluorescent proteins. This method also seeks to avoid concerns over the use of transfection agents including cytotoxicity, transfection efficiency, dilution of fluorescence upon cell dividing, stability, and duration of fluorescence expression that are associated with transfected cells [27]. Due to the limited capability of the proliferation of CMs, ERGS are often generated via genetic editing technologies like CRISPR/Cas9, TALEN, and zinc-finger nucleases (ZNFs) in dividing cell lines such as hPSCs. These cells can then be passaged and further differentiated into hPSC-derived CMs, which produce permanent expression of fluorescence protein-tagged on the targeted sarcomere proteins. The fluorescence protein-tagged sarcomere protein will be visualized by a fluorescence microscope for further studying sarcomere structure and remodeling. However, ERGS has notable limitations, for example, the disruption of a healthy allele [28], loci-dependent efficiency [29, 30], and off-target modifications of genomic editing [31, 32] are an ever-present possibility. Regardless, both methods offer a simple and accessible approach to tracking live-cell sarcomere remodeling [33].

## 2.2 Fluorescence Resonance Energy Transfer (FRET)

FRET is an imaging technique that has seen widespread usage in biomedical research, including the characterization of sarcomere remodeling [34]. FRET works if there is a transfer of energy between two chromophore probes, known as donor-acceptor pairs. To function properly, these pairs must have an overlap in their absorption spectra, resulting in the emission of fluorescence in a distance-dependent manner with a remarkably high degree of accuracy, therefore making the determination of dynamic biological processes accomplishable. FRET probes come in many categories including fluorescence proteins, organic fluorescent dyes, lanthanides [35], and quantum dots [36]. This unique advantage has allowed researchers to measure sarcomere activity both spatially and temporally [37, 38].

The accuracy of FRET is very high that subtle nuances in sarcomere dynamics can be detected, such as the presence of a "twitch" that occurs on the cardiac troponin C (cTnC) interface [39]. This was determined by replacing intrinsic TNC with the FRET biosensor and observing the transient of sarcomere-activating ligands in real time. In another example, researchers have taken advantage of FRET-based biosensors to develop imaging techniques that simultaneously measure conformational changes in the sarcomere protein actomyosin inhibitory troponin I (cTnI), as well as intracellular calcium transients [40] at various sarcomere lengths. Similarly, FRET-based bioprobes have been demonstrated to measure the changes in sarcomere lengths directly using α-actinin integrated fluorescent set of bioprobes, providing highly accurate Z-disk spacing measurements in neonatal rat CMs during active contraction cycles [41]. Additionally, calcium transients can be observed, further underscoring the potential of FRET to simultaneously record sarcomere structural changes with other dynamic processes. This means researchers can make unique comparisons between sarcomere remodeling and other key cellular mechanisms (e.g., calcium transient) related to cardiac development and pathology.

While FRET offers the unique features of spatial and temporal characterization of sarcomere in live CMs, this technique has several limitations. The first is the reliance on transfection efficiency to produce viable probes, this is especially relevant with the use of difficult-to-transfect primary CMs. Secondly, highly specialized equipment is required to excite the fluorophore pair and make the necessary genetic modifications of cell lines. It is also worth pointing out that FRET is susceptible to a phenomenon known as "spectral crossover," where spectral overlap has become too great and results in a very high background signal coupled with a weak emission signal [42]. Despite these limitations, FRET-based imaging offers tremendous potential for the precise and accurate tracking of sarcomere remodeling in real time.

## 2.3 Second Harmonic Generation (SHG) Microscopy

SHG is unlike the previously mentioned methods, which all rely on some form of excitable compounds, probes, or genetic modifications. The underlying principle of SHG centers around the doubling in the frequency of emitted photons from laser light, when passing through a non centrosymmetric object, making sarcomere-related proteins an ideal candidate for imaging without specific labeling [43, 44]. One thing to note is that SHG is a non-absorptive process and therefore is not subject to photobleaching. However, it is also important to consider that SHG microscopy generally requires the use of a high-powered laser, which can potentially cause thermal and photo damage to cells with prolonged exposure [45]. Traditionally, a shortcoming of SHG has been a low signal-to-noise ratio due to intrinsically weak signals. However, it has been found that the additional incorporation of photon counting detection has aided in overcoming this disadvantage [46]. SHG boasts an extreme level of precision, allowing the localization of single myofibrils for observing sarcomere-level changes occurring in real time. Particular regions of the sarcomere can be imaged by SHG, including the "A-band" [47, 48], the most central region of the contractile apparatus. SHG provides accurate metrics of sarcomere length measurement with reduced variability in quantification [49]. Moreover, the in situ and simultaneous SHG and a two-photon confocal microscope were achieved by us previously to observe myofibrillogenesis and sarcomere addition in primary neonatal rat CMs under mechanical stretching [50]. This research offers insightful revelations into the molecular dynamics at play during cardiac hypertrophy. Due to the understood strengths and special attention to this strategy, SHG holds promise in determining the underlying causes of CM stress, a root basis of contraction-related sarcomere cardiomyopathies and cardiac eletrophysiology [51, 52].

## 2.4 Phase-Contrast Imaging

Phase-contrast imaging is one of the most minimally invasive techniques that exploits the differences in refractive index between the cellular and subcellular components and the surrounding background [53–55]. Recently, the specific application of this technique to image myofilaments has gained more attractions as the demand for non-intrusive modes of cardiac sarcomere quantification has increased. A typical phase-contrast microscopy apparatus uses an emission bulb that passes light through the sample. The light will either pass through some areas undisturbed or is diffracted to varying degrees. These deviations can be interpreted to generate an image with a high degree of clarity. To take advantage of the high degree of contrast generated from these images, SarcLen, an imaging module was developed by IonOptix, for quantifying sarcomere changes to very high degrees of resolution, and without the need for additional labeling [56–58]. The module uses a fast Fourier transform (FFT) to analyze repeating patterns in contrast and determine sarcomere lengths in live CMs. For example, one study analyzed the force-tension profiles and sarcomere lengths of primary rat adult CMs post-exposure to protein kinase A, a method to incite β-adrenergic stimulation and in chronic instances cause structural and functional alterations to heart tissue [59]. Phase-contrast imaging, and more specifically SarcLen, was able to determine that sarcomere lengths dictate force development rates, and that β-adrenergic stimulation was able to shift the length-tension profile of CMs.

While phase-contrast-based image analysis programs have become a viable and efficient way to increase the throughput of data and provide meaningful results, phase-contrast imaging also has its intrinsic limitations. Notably, phase-contrast imaging is highly specialized for thin samples and cellular monolayers. The thick 3D tissues, such as engineered heart tissue and organoids, may not be easy to let light transmit through for efficient microscopic imaging as the occurring phase shifts may not be uniform or blocked [60, 61]. Additionally, low-image resolution and artifacts within the image have been reported among users [62, 63].

## *2.5 Additional Imaging Techniques for Live-Cell Sarcomere Imaging*

To date, there is a clear and pressing need to develop new techniques in the field of live-cell sarcomere imaging. This presents the goal to overcome the challenges associated with fluorescence probes and keep pace with the rapid advancements in image analysis. The development of such techniques could potentially offer more accurate, reliable, and safer methods for characterizing sarcomere remodeling, thereby pushing the boundaries of what is currently achievable in the field. This is an exciting prospect that could open new avenues for research and discovery.

Borrowing from the field of analytical chemistry comes Raman imaging, which traditionally has been utilized to determine chemical species [64, 65]. This imaging method is based on the concept of detectable light scattering, the Raman scattering, and the resulting shift of photon frequencies. Raman imaging of live cells in its traditional capacity has not been used to directly quantify sarcomere changes. However, it can be used on imaging sarcomere indirectly. For instance, Raman imaging was used to investigate the spectral properties of both rod- and rounded-shaped CMs. The rod-shaped CMs have more organized and aligned sarcomeres in the adult CMs in comparison with the neonatal or immature CMs, which are either under dedifferentiation from adult CMs or immature neonatal CMs [66, 67]. Interestingly, spectral differences were discernible between the two morphologies, as well as between the sarcomere-containing region of the rod-shaped CMs and the surrounding cytosolic area [68]. It is also worth noting that this method employed the use of a low-powered laser, avoiding the potential of photobleaching, this being a major limitation of traditional probe imaging methods. Additionally, Raman spectroscopy has the unique advantage of chemical specificity, giving the technique tremendous potential to evaluate contractile mechanisms like the shortening of sarcomeres via the phosphorylation of myosin [69, 70]. However, like phase-contrast imaging, Raman imaging is also susceptible to the presence of low resolution and potential image artifacts. This, coupled with the generally low signal produced by Raman scattering, potentially makes the automated processing of images more difficult to achieve and may be why the technique has not seen widespread adoption [71, 72]. Despite these limitations, there is tremendous interest in the potential of Raman spectroscopy.

### 2.6 *Comparison of Each Live-Cell Sarcomere Imaging Technique*

Collectively ranging from genetically encoded probes to detecting intrinsic physical properties of protein in sarcomere to derive sarcomere images, live-cell imaging has demonstrated minimally intrusive quantification of the dynamic process of sarcomere remodeling with unique advantages and technical disadvantages as summarized in Table 1, which provides suggestive pros and cons to guide us how to apply various live-cell imaging modalities in sarcomere remodeling-relevant studies and applications.

## 3 Applications of Live-Cell Imaging Sarcomere Remodeling in CMs

We explored significant applications in better understanding new sarcomere formation, evaluating cardiac contractile function, characterizing cardiac mechanotransduction, modeling cardiopathies, and seeking possibilities of cardiac regenerative medicine as summarized in Fig. 1.

### 3.1 *Better Understanding of Sarcomerogenesis and Sarcomere Remodeling*

In recent years, numerous in vitro engineered cardiac culture platforms have been established for recapitulating in vivo cardiac pathophysiological conditions, such as micropatterned CMs for sarcomere alignment and mechanical coupling of CMs, and mechanical stretching CMs for modeling hypertrophied sarcomerogenesis [77, 79]. These engineering techniques provide better and more accurate observation of the sarcomere development and remodeling in a more in vivo-*like* state. Coupled with the unique advantages of live-cell imaging, engineered tissue platforms have paved

**Table 1** Summary table of popular quantitative methods used to track sarcomere changes in live CMs. The pros and cons of the techniques are addressed to discuss the state of the available technologies

| | Live-cell transfections [24] | Encoded reporter gene systems (ERGS) [27, 33, 73–76] | Fluorescence resonance energy transfer (FRET) [34, 35, 36, 40–42] | S,econd harmonic generation (SHG) microscopy [47, 48, 50, 77] | Phase-contrast imaging [59, 78] | Raman imaging [68] |
|---|---|---|---|---|---|---|
| High accessibility | Yes | No | No | No | Yes | No |
| High temporal/spatial resolution | Yes | Yes | Yes | Yes | No | No |
| Long-term imaging | No | Yes | No | Yes | Yes | Yes |
| Photobleaching susceptible | Yes | Yes | Yes | No | No | No |
| Potentially cytotoxic | Yes | Yes | Yes | No | No | No |
| Image noise of concern | No | No | Yes | No | Yes | Yes |
| Need for laser excitation | No | Yes | Yes | Yes | No | Yes |
| Allele disruption | No | Yes | No | No | No | No |

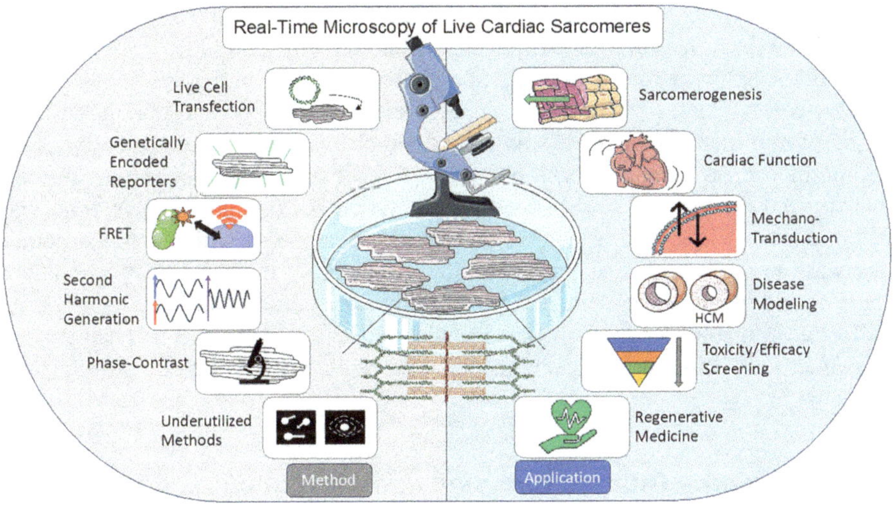


**Fig. 1** Summary of current live-cell sarcomere imaging methods and corresponding applications. Figure 1 was recreated and modified based on the illustrations from Servier Medical Art (https://smart.servier.com/)

the way for more detailed and improved exploration of how sarcomeres are formed in developmental stages and stress response.

The focus on studying the process of sarcomere development and maturation in hPSC-CMs has been at the forefront of many research efforts [80–82]. Live-cell imaging looks to make the most of these techniques by dynamically tracking how sarcomere structure and function can improve in response to the tailored conditions for enhancing cardiac maturation. The hPSC-CM was labeled with RFP on the actin filament by adenovirus rAV CAG-LifeActTagRFP transfection and then was subjected to certain aspect ratios from 1:1 to 7:1 in rectangular shape by microcontact printing to study the sarcomere alignment and cardiac maturation. The method has seen widespread usage for its accessibility and has proven to be an easy and effective way to promote CM maturation and align myofibrils for easy quantification. The RFP-labeled actin in 7:1 hPSC-CMs demonstrated better sarcomere shortening and forces with enhanced maturity in comparison with other aspect ratios by in situ live-cell sarcomere imaging [76].

Another notable example is the use of SHG to investigate sarcomere formations in response to CMs subjected to static stretching for simulating mechanical load as seen in the hypertrophied myocardium. Here, neonatal rat CMs were placed on a custom apparatus that consisted of an adjustable frame and an elastomeric cell substrate. Without the use of probes and enacting minimal damage to live cells, SHG imaging allowed for precise tracking of sarcomere formation and addition due to mechanical stretching [77]. Similarly, microfluidic devices aim to emulate the flow of nutrients, exert mechanical fluid forces, and provide precise control of the culture microenvironment [83–85]. Pressure overload hypertrophic cardiomyopathy, and its downstream effects which result in increased myofibril density [86], is a prime example of how in vitro modeling can be potentially combined with live-cell imaging to capture unique characteristics of the native myocardium.

Live-cell tracking of sarcomeres has been employed to both explain the restructuring of myofilaments in developing CMs and characterize 3D engineered tissues within microfluidic devices. Both points were successfully addressed by one study that utilized GFP-expressing hPSC-CMs to visualize α-actinin in real time and determine the length and orientation of cardiac sarcomeres within a 3D structure that was improved with vascularization [87]. The ability to characterize sarcomere changes and quantify the effectiveness of engineered culture platforms in real time will offer valuable insight into the development and remodeling mechanisms of sarcomeres in CMs that were previously unobtainable by imaging of fixed samples.

## 3.2 *Evaluating Cardiac Function and Mechanotransduction*

Acting as the primary motor of CMs, sarcomeres are responsible for both active and passive forces. The sliding of the thick filaments is what generates active force and contributes to work done at the tissue level [88]. Tracking the forces generated by sarcomeres and how these changes over time can be key to determining critical steps in the remodeling process of sarcomere and following cardiac function. While cardiac function is historically measured on the tissue or even organ level by techniques such as echocardiogram [89], measuring contractile properties on the sarcomere level can give even more clues as to how orientation and density play a larger role in overall cardiac output. Interestingly, one study was able to observe CMs actively expressing α-actinin–GFP within live mouse hearts using a custom imaging setup [90]. This study shows the immense potential of imaging to determine cardiac function, as a positive correlation between sarcomere shortening ratio and diastolic dysfunction was able to be demonstrated. However, the limitations of using animal models persist within this study. Therefore, the use of in vitro imaging is critical to developing methods that are accessible, ethical, and cost-efficient to obtain this information. hPSC-CMs and primary animal CMs have remained at the forefront of these studies, demonstrating feats such as live recording of contractile velocity and relaxation impairment of sarcomeres [91, 92]. Determination of individual sarcomere length has also been an investigated metric, as length is known to be directly correlated to cardiac force output. Guinea Pig CMs isolated at different regions in the heart were used to track sarcomere lengths across several contraction cycles in real time, and it was found that sarcomere length variability was present in end-diastole and end-systole contraction states in comparison between CMs from subepi- or subendocardial regions via phase-contrast live-cell imaging [93]. Sarcomere length nonuniformity, and more importantly a reliable method to evaluate it, could prove useful in the development of new cardiac function assessment assays and further refining sarcomere length-force relationships.

Passive forces in CM are static tension on the sarcomeres, which is generated by certain sarcomere proteins, mainly titin [94, 95]. The protein titan has a profound effect on the sarcomere as it spans almost the entirety of its length, from the M-line to the Z-disc [96]. Titin also has been known to play a pivotal role in the mechanotransduction pathway [97], acting as a signal conductor from the sarcomere to the cytoskeleton, out to the ECM, and to adjacent CMs [98, 99]. Most notably, titin plays the same role in the reverse direction, interpreting mechanical signals or stressors from the environment and relaying them to the sarcomere for remodeling to occur. A variety of titin-mediated pathways have been investigated such as the MLP-TCAP-Tinin complex, which is associated with stretch sensing and heart failure [100]. Given the sarcomeres' ability to sense mechanical stimuli and remodel via feedback, the use of live-cell imaging to track changes in sarcomeres has serious potential to further investigate pathology [101]. One example of a study looking to capitalize on the knowledge surrounding mechanotransduction used GFP-tagged α-actinin biosensors to track sarcomere changes when cells were cultured on PDMS substrates of various stiffness with nanopillar arrays [102]. They noted that sarcomeres had an increased rate of shortening as well as a greater capacity to do so when cultured on the softer PDMS. Another study looked to characterize Titin and its greater role in pathogenesis. Using live-cell-fluorescent microscopy in combination with other conventional techniques, it was found that truncated titin molecules result in impaired sarcomerogenesis [103]. Additionally, it was observed that sarcomerogenesis initiates at sites of cell-extracellular matrix adhesion within hPSC-CMs. Findings such as this provide critical information on how cellular substrate and the mechanotransduction process dictate how remodeling can occur in maturation, and when disrupted, in pathology. Future studies may look to modulate surfaces to determine which mechanisms perturb sarcomere structures and their remodeling in specific ways.

## *3.3 Disease Modeling of Cardiomyopathies*

While many of the genetic mutations and clinical manifestations of cardiomyopathies are well accounted for, their individual effects on CM sarcomeres are less known. The combination of iPSC technology and current live imaging techniques has resulted in a new era of in vitro disease modeling of human cardiac diseases, making it possible to better understand the molecular basis of pathogenesis on the sarcomere level. This possibility is crucial for improving several areas such as for developing targeted therapies [104, 105], improving clinical diagnostic tools, generating more precise disease models, and an overall better understanding of the mechanism of sarcomere remodeling responsible. Recent efforts have seen significant progress in creating accurate and informative disease models. For example, hPSC-CMs with various sarcomere mutations (e.g., myosin heavy chain 7 (MYH7), myosin binding protein C3 (MYBPC3), and cardiac troponin T (TNNT2) were applied to better understand the diastolic dysfunction in HCM patients.

These genetically modified models were able to recapitulate impaired diastolic function and shortened diastolic sarcomere length, qualities observed in patient-derived iPSC-CMs with the same mutations [106]. Thus, via live-cell imaging of overexpressing GFP-actin, the investigators were able to determine novel cellular mechanisms of diastolic dysfunction and clues into its origin.

Given that many cardiomyopathies result in extracellular matrix remodeling [107, 108], the ability to track sarcomere changes in response to stimuli can prove useful in deciphering the mechanisms of cardiomyopathies. For example, tracking sarcomere remodeling in response to stiffness changes in substrates is critical to developing accurate models for cardiomyopathies such as HCM [109]. One specific study used actin-labeled iPSC-CMs on higher stiffness hydrogel substrates to induce myofibril defects and decrease the overall mechanical output of the cells [110]. Such progress in the understanding of sarcomere dynamics and how remodeling can occur from increased substrate stiffness, as seen in various cardiomyopathies, furnish an essential framework for delving into the workings of the heart and its ailments on the microscale.

## 3.4 Evaluating Pharmaceutical Efficacy and Cardiotoxicity

Live-cell imaging of sarcomere structure holds immense promise in developing the next generation of pharmaceutical characterization assays. An immediate beneficiary from the advantages of real-time tracking of sarcomere function and structure are screens that evaluate the potency of novel therapeutics, looking to more quickly and accurately determine the cardiotoxic potential of various compounds. In a recent study, an engineered cell line expressing the MYH6-mScarlet reporter demonstrated successful tracking of sarcomeres when dosed with known cardiotoxic compounds, many of which alter contraction profile or initiate sarcomere damage [75]. Over 4 days, they recorded various responses in real time. This included apoptosis induced by doxorubicin exposure [111], and sarcomere deterioration and disarray from compounds like sunitinib and amiodaron. Various other studies have seen the usage of genetically encoded reporter lines to track sarcomere dynamics in real time, even incorporating high-throughput automated image analysis to discriminate between sarcomere toxic effects [73, 74]. The potential for drug screening increases many folds with further automation. In one study utilizing the SarcLen software with phase-contrast imaging, researchers observed sarcomere shortening in the presence of various pharmacological stimuli, such as β-adrenergic stimulation, PDE3 inhibition, and increased calcium sensitivity [78]. SarcLen has been employed to evaluate sarcomere contraction [112] and the effects of various drugs resulting from environmental toxins [113], among other applications [114, 115]. Notably, the capacity for real-time sarcomere monitoring in living cells has yielded insights into the mechanisms behind diverse toxic effects. This paves the way for further exploration into the toxicological characteristics of novel medical treatments and presents a distinct chance to leverage such knowledge in the creation of advanced pharmaceuticals.

### 3.5 Cardiac Regenerative Medicine

Cardiac regenerative medicine encompasses a wide range of diagnostic and therapeutic approaches aimed at restoring or rejuvenating heart health. One powerful tool within this field is real time in vitro imaging, which allows us to observe biological function within its natural structural context [116, 117]. As we strive to translate stem cell therapies into pre-clinical applications, appropriate imaging techniques play a crucial role in elucidating the function of these therapies. The current regenerative medicine landscape is defined as the primary strategy for reinitiating CM proliferation [118, 119]. Given the human heart's limited regenerative abilities [120], the importance of stimulating the proliferation of CMs cannot be understated. To quantify this process in vitro, the real-time tracking of sarcomeres can prove beneficial, as it is known that CMs reentering the cell cycle will undergo sarcomere disassembly [121, 122]. One recent study was able to achieve this feat by observing Halo tagged α-actinin in real time, here they revealed that during early prophase rapid disassembly occurs, and at a hasty rate [123]. Moreover, there have been several instances of dedifferentiation and reentry into the cell cycle among the in vitro isolated CMs. These instances have been triggered primarily by miRNA [124, 125] or via drug treatment of a cocktail containing the small molecules Phenylephrine, Baricitinib, Harmine, Vo-topic Trihydrate, and AZD3965 [126]. Another example used live imaging and fluorescently tagged TNNT2, a thin filament protein, to perform a screen and determine the efficacy of various pro-proliferative compounds and their cell cycle-inducing activities [127]. Combined with a fluorescent cell cycle indicator, the investigators were able to suggest that clonidine has regenerative potential for mammalian hearts. This technology lays the groundwork for further characterization of sarcomere dynamics during the activation of CM proliferation and may give us further modulating capabilities to promote more effective cardiac repair in patients.

A second route of approach is to induce the body's innate ability to repair tissue through exogenous stimulation from stem cells. Considerable evidence has emerged to back the concept that stem cells primarily deliver their healing and regenerative benefits by secreting active biological compounds that function practically to influence neighboring cells [128, 129]. The coculturing of cardiac progenitor cells with CMs has also been demonstrated to disorganize sarcomere structure in vitro [130]. Here, the use of live-cell sarcomere tracking techniques would allow for a more accurate and precise understanding of how sarcomeres are remodeled by stem cells in preparation for proliferation. Even though the research area is in its infancy, there have been clear instances where live-sarcomere tracing has been instrumental in visualizing essential functions for the success of regenerative medicine. Utilizing imaging technology, specifically real-time tracking of sarcomere dynamics will allow scientists to delve into the efficacy and underlying processes of these regenerative approaches and provide hope for creating therapies to aid those suffering from serious cardiomyopathies.

## 4 Challenges and Perspectives

The advancements in non-obstruction or live-cell imaging technologies have significantly improved the characterization and detection of sarcomeres in live CMs. The emergence of these minimally invasive techniques has led to the development of highly specific quantification methods of sarcomere dynamics in a temporospatial manner. These advancements have been well developed, but they are accompanied by limitations and corresponding challenges regarding sarcomere remodeling. A main constraint is the limited number of labeling modalities available to researchers, mainly characterized by all having limited imaging depth, and either a lack of high resolution or a cytotoxic element. Additionally, the source of CM is a limiting factor as well. Both primary CMs and those derived from hPSCs have been the only reliable way to obtain cells for experimentation and are compatible with many non-invasive imaging types. This restricts the accessibility and flexibility of many experiments, as the choice of cell source can greatly influence the outcome and impact of the study due to tradeoffs in cell availability (human primary CMs) and maturity (hPSC-CM).

Furthermore, only a finite number of reporter cell hPSC lines have been successfully generated for applications of the reviewed methods. These cell lines typically have taken the form of engineered hPSC lines by genome editing and primary CMs. However, the viability of primary CMs, harvested from both healthy and diseased hosts, is much more limited than those derived from hPSC lines [131]. Additionally, the lack of human homology about specific protein structures must also be considered. While the recent advancements in technology have greatly improved our ability to study sarcomere remodeling, there are still significant limitations that need to be addressed. For example, the finite number of reporter cell lines must dramatically be expanded upon as sarcomere is a dense construct, consisting of over 200 various proteins [17]. Therefore, while determining remodeling-induced changes of a select few sarcomere proteins can yield valuable applications, it only offers a piece of the puzzle. Future research should then focus on developing new labeling methods and focus on utilizing automated sarcomere structure analysis pipelines [74], allowing for more flexibility in experimental design and the generation of more reporter cell lines, increasing throughput, and ultimately leading to a better understanding of sarcomere remodeling.

hPSC-CMs hold the potential for advancing cardiac sarcomere research through *in vitro* modeling and imaging, providing more efficient methods for characterizing sarcomere remodeling compared to primary CMs. However, hPSC-CMs have their limitation in lack of sarcomere maturity showing irregular sarcomere organization and overall lower contractile force. The immature hPSC-CMs have sarcomeres localized to the periphery of the cell and are generally less orderly [132, 133], only requiring minimal $Ca^{2+}$ release to initiate contraction [134, 135]. The immaturity of hPSC-CMs in these aspects means that they may not accurately represent the function of mature heart cells, potentially dampening the results of studies. To address this, novel bioengineering methods have been created to improve the maturity of cardiac sarcomeres in hPSC-CMs. For example, various micro-/nano-patterning techniques have been implemented to replicate this alignment of hPSC-CMs, which further induce the alignment of sarcomere and formation of myofibrils [136–138]. Moreover, electrospinning of extracellular matrix (ECM)-like fibers uses an electric field to draw charged threads of polymer solutions to create aligned fibrous structures, which further align the sarcomere in hPSC-CMs culturing on those ECM-like fibers [139, 140]. Other methods to induce greater myofibril alignment and overall maturation include the 3D bioprinting of engineered heart tissues [141, 142]. This unique tissue engineering method aims to take advantage of its ability to spatially distribute cells with high precision and encapsulate them in ECM-like materials to result in enhanced maturation signaling and sarcomere organization. Additionally, seemingly less complex methods such as pulsed electrical stimulation have been proven to enhance contractile profile [143, 144], the expression of key sarcomere proteins such as α-actinin, cardiac troponin I (cTnI), and calcium handling machinery such as SERCA2 [145].

Another limitation of hPSC-CMs to consider is the sub-type of CMs, referring to the chamber or region of origin within the heart (i.e., atrial, ventricular, and pacemaker). While CMs are found throughout most of the heart, they can differ in functional optimizations. CMs of ventricular origin have a higher density of myofilaments and efficient calcium handling mechanisms, unlike CMs of atrial origin which have generally less organized sarcomeres and are generally smaller in size [5, 146]. Unsurprisingly, cells of the His-Purkinje system are more specialized for electrical signal conduction, having contractile minimal machinery that reflects this specialization [147]. Although all CMs possess the ability to cycle calcium and contract, the stark differences that arise among cell types need proper distinguishing for mechanistic interrogation of region-specific cardiomyopathies to become an effective practice [148].

Given the discussed drawbacks of current live-cell imaging techniques and cell culture practices, the development of new and more powerful tools for improving sarcomere remodeling detection is necessary to make meaningful insights and develop next-generation therapeutics for various sarcomere-related cardiomyopathies. One such potential tool is the use of multiphoton confocal microscopy (MCM) [149]. MCM is less phototoxic and works on the principle of near-infrared laser excitation of two/three fluorophore photons with a living biological sample, allowing for the collection of 2D and 3D images with high spatial resolution [150]. Most notably, MCM is capable of surpassing tissue depth barriers experienced by conventional single-photon live imaging techniques [151, 152]. Sarcomere proteins in the deep cardiac tissue could be imaged due to the strengths of MCM. Additionally, utilizing MCM would open the door for the use of 3D-engineered cardiac tissues such as bio wires [153] and organoids [154] to help understand sarcomere remodeling on a molecular scale.

Regardless of current limitations, live-cell sarcomere tracking offers an exciting opportunity to accelerate the clinical translation of therapies for sarcomere-related cardiomyopathies. In addition to treatment, the methodology for diagnosis of various cardiomyopathies may change as well. This means a patient-derived CMs could now be screened on new and more specific sarcomere criteria as an understanding of cardiomyopathy etiologies. The use of machine learning in particular stands to bolster the capabilities of researchers, making extensive data sets more navigable and helping to visualize previously unnoticeable correlations. It is equally important to consider the impacts of improving sarcomere remodeling tracking tools outside the sphere of obtaining better results. The ethical advantages of utilizing live tracking of sarcomeres go hand in hand with improving metrics of modeling systems. With a decreased need to rely on animal models, their usage can be significantly reduced, and hPSC-derived CMs take their place albeit with enhanced maturity alongside this fact.

## 5 Conclusions

The advent of live-cell imaging for cardiomyocytes, and more specifically for their sarcomeres, marks a significant breakthrough in the field of cardiac research, providing critical insights into the dynamic activities that dictate cardiac function. This innovative approach permits the direct monitoring of living cells, offering a glimpse into the complex operations of CMs during their developmental stages and in various pathologies. Its application has been crucial for deciphering the underlying mechanisms of cardiac disorders and shaping new therapeutic avenues. The unique advantages of live-cell imaging empower researchers to investigate CMs in their physiological context, bolstering progress in fields such as regenerative medicine and pharmacological innovation. Although promising data have been generated and novel correlations have been established, it is important to emphasize the need to develop further reporting methodologies to better complement existing techniques and allow for the field to increase in scale, taking full advantage of the throughput capabilities of machine learning.

**Funding Supports** This work was partly supported by the NIH R15HD108720 (HXY), Harry Moss Heart Foundation (HXY), start-up from the University of North Texas (UNT) Biomedical Engineering (HXY), Research Seed Grants (2021 and 2023) from UNT Research and Innovation Office (HXY), and American Heart Association Established Investigator Award (# 959644, YH).

**Conflicts of Interest** None.

## References


1. Savarese, G. *et al.* Global burden of heart failure: a comprehensive and updated review of epidemiology. *Cardiovasc Res***118**, 3272–3287 (2023).
2. Hamdani, N. *et al.* Sarcomeric dysfunction in heart failure. *Cardiovasc Res***77**, 649–658 (2007).
3. Topriceanu, C.-C., Pereira, A. C., Moon, J. C., Captur, G. & Ho, C. Y. Meta-Analysis of Penetrance and Systematic Review on Transition to Disease in Genetic Hypertrophic Cardiomyopathy. *Circulation***149**, 107–123 (2024).
4. Kimura, A. Molecular genetics and pathogenesis of cardiomyopathy. *J Hum Genet***61**, 41–50 (2016).
5. Burbaum, L. *et al.* Molecular-scale visualization of sarcomere contraction within native cardiomyocytes. *Nat Commun***12**, 4086 (2021).
6. Au, Y. The muscle ultrastructure: a structural perspective of the sarcomere. *Cellular and Molecular Life Sciences***61**, 3016–3033 (2004).
7. Cooke, R. The Sliding Filament Model. *J Gen Physiol***123**, 643–656 (2004).
8. Squire, J. M. Muscle contraction: Sliding filament history, sarcomere dynamics and the two Huxleys. *Glob Cardiol Sci Pract***2016**, (2016).
9. Hanft, L. M., Korte, F. S. & McDonald, K. S. Cardiac function and modulation of sarcomeric function by length. *Cardiovasc Res***77**, 627–636 (2007).
10. Nadruz, W. Myocardial remodeling in hypertension. *J Hum Hypertens***29**, 1–6 (2015).
11. Weiner, R. B. & Baggish, A. L. Exercise-Induced Cardiac Remodeling. *Prog Cardiovasc Dis***54**, 380–386 (2012).
12. Holubarsch, C. *et al.* Existence of the Frank-Starling Mechanism in the Failing Human Heart. *Circulation***94**, 683–689 (1996).
13. Yang, H. *et al.* Dynamic Myofibrillar Remodeling in Live Cardiomyocytes under Static Stretch. *Sci Rep***6**, 20674 (2016).
14. Chen, B. *et al.* Disruption of a GATA4/Ankrd1 Signaling Axis in Cardiomyocytes Leads to Sarcomere Disarray: Implications for Anthracycline Cardiomyopathy. *PLoS One***7**, e35743 (2012).
15. Sehnert, A. J. *et al.* Cardiac troponin T is essential in sarcomere assembly and cardiac contractility. *Nat Genet***31**, 106–110 (2002).
16. Russell, B., Motlagh, D. & Ashley, W. W. Form follows function: how muscle shape is regulated by work. *J Appl Physiol***88**, 1127–1132 (2000).
17. Crocini, C. & Gotthardt, M. Cardiac sarcomere mechanics in health and disease. *Biophys Rev***13**, 637–652 (2021).
18. Janssen, P. M. L. Kinetics of cardiac muscle contraction and relaxation are linked and determined by properties of the cardiac sarcomere. *American Journal of Physiology-Heart and Circulatory Physiology***299**, H1092–H1099 (2010).
19. Padilla-Parra, S. & Tramier, M. FRET microscopy in the living cell: Different approaches, strengths and weaknesses. *BioEssays***34**, 369–376 (2012).
20. Jares-Erijman, E. A. & Jovin, T. M. FRET imaging. *Nat Biotechnol***21**, 1387–1395 (2003).
21. Wang, Z. *et al.* Advantages of intermediate X-ray energies in Zernike phase contrast X-ray microscopy. *Biotechnol Adv***31**, 387–392 (2013).
22. Dabiri, G. A., Turnacioglu, K. K., Sanger, J. M. & Sanger, J. W. Myofibrillogenesis visualized in living embryonic cardiomyocytes. *Proceedings of the National Academy of Sciences***94**, 9493–9498 (1997).
23. Ribeiro, M. C. *et al.* A cardiomyocyte show of force: A fluorescent alpha-actinin reporter line sheds light on human cardiomyocyte contractility versus substrate stiffness. *J Mol Cell Cardiol***141**, 54–64 (2020).
24. Dabiri, G. A., Turnacioglu, K. K., Sanger, J. M. & Sanger, J. W. Myofibrillogenesis visualized in living embryonic cardiomyocytes. *Proceedings of the National Academy of Sciences***94**, 9493–9498 (1997).

25. Chanthra, N. *et al.* A Novel Fluorescent Reporter System Identifies Laminin-511/521 as Potent Regulators of Cardiomyocyte Maturation. *Sci Rep***10**, 4249 (2020).
26. Sharma, A., Toepfer, C. N., Schmid, M., Garfinkel, A. C. & Seidman, C. E. Differentiation and Contractile Analysis of GFP-Sarcomere Reporter hiPSC-Cardiomyocytes. *Curr Protoc Hum Genet***96**, (2018).
27. Ansari, A. M. *et al.* Cellular GFP Toxicity and Immunogenicity: Potential Confounders in in Vivo Cell Tracking Experiments. *Stem Cell Rev Rep***12**, 553–559 (2016).
28. Mueller, K. E., Wolf, K. & Fields, K. A. Gene Deletion by Fluorescence-Reported Allelic Exchange Mutagenesis in Chlamydia trachomatis. *mBio***7**, (2016).
29. Xu, Z., Song, Y. & Wang, F. Rational design of genetically encoded reporter genes for optical imaging of apoptosis. *Apoptosis***25**, 459–473 (2020).
30. Mikkelsen, N. S. & Bak, R. O. Enrichment strategies to enhance genome editing. *J Biomed Sci***30**, 51 (2023).
31. Zhang, X.-H., Tee, L. Y., Wang, X.-G., Huang, Q.-S. & Yang, S.-H. Off-target Effects in CRISPR/Cas9-mediated Genome Engineering. *Mol Ther Nucleic Acids***4**, e264 (2015).
32. Cho, S. W. *et al.* Analysis of off-target effects of CRISPR/Cas-derived RNA-guided endonucleases and nickases. *Genome Res***24**, 132–141 (2014).
33. Acharya, A. *et al.* Live-Cell Imaging of the Contractile Velocity and Transient Intracellular Ca2+ Fluctuations in Human Stem Cell-Derived Cardiomyocytes. *Cells***11**, 1280 (2022).
34. Schobesberger, S. *et al.* T-tubule remodelling disturbs localized β2-adrenergic signalling in rat ventricular myocytes during the progression of heart failure. *Cardiovasc Res***113**, 770–782 (2017).
35. Pilch-Wrobel, A., Kotulska, A. M., Lahtinen, S., Soukka, T. & Bednarkiewicz, A. Engineering the Compositional Architecture of Core-Shell Upconverting Lanthanide-Doped Nanoparticles for Optimal Luminescent Donor in Resonance Energy Transfer: The Effects of Energy Migration and Storage. *Small***18**, (2022).
36. Grazon, C. *et al.* The quantum dot *vs.* organic dye conundrum for ratiometric FRET-based biosensors: which one would you chose? *Chem Sci***13**, 6715–6731 (2022).
37. Miyawaki, A. Visualization of the Spatial and Temporal Dynamics of Intracellular Signaling. *Dev Cell***4**, 295–305 (2003).
38. Sprenger, J. U., Bork, N. I., Herting, J., Fischer, T. H. & Nikolaev, V. O. Interactions of Calcium Fluctuations during Cardiomyocyte Contraction with Real-Time cAMP Dynamics Detected by FRET. *PLoS One***11**, e0167974 (2016).
39. Vetter, A. D., Martin, A. A., Thompson, B. R., Thomas, D. D. & Metzger, J. M. Sarcomere integrated biosensor detects myofilament-activating ligands in real time during twitch contractions in live cardiac muscle. *J Mol Cell Cardiol***147**, 49–61 (2020).
40. Li, K.-L., Rieck, D., Solaro, R. J. & Dong, W. In Situ Time-Resolved FRET Reveals Effects of Sarcomere Length on Cardiac Thin-Filament Activation. *Biophys J***107**, 682–693 (2014).
41. Hanft, L. M. & McDonald, K. S. Length dependence of force generation exhibit similarities between rat cardiac myocytes and skeletal muscle fibres. *J Physiol***588**, 2891–2903 (2010).
42. CHEN, Y., MAULDIN, J. P., DAY, R. N. & PERIASAMY, A. Characterization of spectral FRET imaging microscopy for monitoring nuclear protein interactions. *J Microsc***228**, 139–152 (2007).
43. Aghigh, A. *et al.* Second harmonic generation microscopy: a powerful tool for bio-imaging. *Biophys Rev***15**, 43–70 (2023).
44. Zoumi, A., Yeh, A. & Tromberg, B. J. Imaging cells and extracellular matrix *in vivo* by using second-harmonic generation and two-photon excited fluorescence. *Proceedings of the National Academy of Sciences***99**, 11014–11019 (2002).
45. Li, B., Li, J., Zeb, J., Yuan, Q. & Gan, W. The effect of laser repetition rate in second harmonic generation imaging and intensity detection. *Chem Phys***572**, 111955 (2023).
46. Jiang, J., Eisenthal, K. B. & Yuste, R. Second Harmonic Generation in Neurons: Electro-Optic Mechanism of Membrane Potential Sensitivity. *Biophys J***93**, L26–L28 (2007).
47. Awasthi, S. *et al.* Multimodal SHG-2PF Imaging of Microdomain Ca $^{2+}$ -Contraction Coupling in Live Cardiac Myocytes. *Circ Res***118**, (2016).

48. Lookin, O. *et al.* Cardiomyocyte sarcomere length variability: Membrane fluorescence versus second harmonic generation myosin imaging. *Journal of General Physiology* **155**, (2023).
49. Lookin, O. *et al.* Cardiomyocyte sarcomere length variability: Membrane fluorescence versus second harmonic generation myosin imaging. *Journal of General Physiology* **155**, (2023).
50. Liu, H. *et al.* Myosin filament assembly onto myofibrils in live neonatal cardiomyocytes observed by TPEF-SHG microscopy. *Cardiovasc Res* **97**, 262–270 (2013).
51. Ruwhof, C. Mechanical stress-induced cardiac hypertrophy: mechanisms and signal transduction pathways. *Cardiovasc Res* **47**, 23–37 (2000).
52. Seo, K. *et al.* Relevance of cardiomyocyte mechano-electric coupling to stretch-induced arrhythmias: Optical voltage/calcium measurement in mechanically stimulated cells, tissues and organs. *Prog Biophys Mol Biol* **115**, 129–139 (2014).
53. Masters, B. R. Phase Microscopy to Enhance Contrast. **227**, 213–228 (2020).
54. Xu, W., Jericho, M. H., Meinertzhagen, I. A. & Kreuzer, H. J. Digital in-line holography for biological applications. *Proceedings of the National Academy of Sciences* **98**, 11301–11305 (2001).
55. Bhaduri, B. *et al.* Cardiomyocyte Imaging Using Real-Time Spatial Light Interference Microscopy (SLIM). *PLoS One* **8**, e56930 (2013).
56. Nishimura, S. *et al.* Single cell mechanics of rat cardiomyocytes under isometric, unloaded, and physiologically loaded conditions. *American Journal of Physiology-Heart and Circulatory Physiology* **287**, H196–H202 (2004).
57. Charo, D. N. *et al.* Endogenous regulation of cardiovascular function by apelin-APJ. *American Journal of Physiology-Heart and Circulatory Physiology* **297**, H1904–H1913 (2009).
58. Nishimura, S. *et al.* Expression of Green Fluorescent Protein Impairs the Force-Generating Ability of Isolated Rat Ventricular Cardiomyocytes. *Mol Cell Biochem* **286**, 59–65 (2006).
59. Barbagallo, F. *et al.* Genetically Encoded Biosensors Reveal PKA Hyperphosphorylation on the Myofilaments in Rabbit Heart Failure. *Circ Res* **119**, 931–943 (2016).
60. Rogozinski, N., Yanez, A., Bhoi, R., Lee, M.-Y. & Yang, H. Current methods for fabricating 3D cardiac engineered constructs. *iScience* **25**, (2022).
61. Shi, H. *et al.* Organoid intelligence: Integration of organoid technology and artificial intelligence in the new era of in vitro models. *Med Nov Technol Devices* **21**, 100276 (2024).
62. Nguyen, T. H. *et al.* Halo-free Phase Contrast Microscopy. *Sci Rep* **7**, 44034 (2017).
63. Yin, Z., Kanade, T. & Chen, M. Understanding the phase contrast optics to restore artifact-free microscopy images for segmentation. *Med Image Anal* **16**, 1047–1062 (2012).
64. Cialla-May, D. *et al.* Raman Spectroscopy and Imaging in Bioanalytics. *Anal Chem* **94**, 86–119 (2022).
65. Jones, R. R., Hooper, D. C., Zhang, L., Wolverson, D. & Valev, V. K. Raman Techniques: Fundamentals and Frontiers. *Nanoscale Res Lett* **14**, 231 (2019).
66. Yang, H., Borg, T. K., Wang, Z., Ma, Z. & Gao, B. Z. Role of the Basement Membrane in Regulation of Cardiac Electrical Properties. *Ann Biomed Eng* **42**, 1148–1157 (2014).
67. Yang, H., Borg, T. K., Liu, H. & Gao, B. Z. Interactive relationship between basement-membrane development and sarcomerogenesis in single cardiomyocytes. *Exp Cell Res* **330**, 222–232 (2015).
68. Brazhe, N. A. *et al.* Mapping of Redox State of Mitochondrial Cytochromes in Live Cardiomyocytes Using Raman Microspectroscopy. *PLoS One* **7**, e41990 (2012).
69. Kampourakis, T., Sun, Y.-B. & Irving, M. Myosin light chain phosphorylation enhances contraction of heart muscle via structural changes in both thick and thin filaments. *Proceedings of the National Academy of Sciences* **113**, (2016).
70. van Manen, H.-J., Lenferink, A. & Otto, C. Noninvasive Imaging of Protein Metabolic Labeling in Single Human Cells Using Stable Isotopes and Raman Microscopy. *Anal Chem* **80**, 9576–9582 (2008).
71. Byrne, H. J., Knief, P., Keating, M. E. & Bonnier, F. Spectral pre and post processing for infrared and Raman spectroscopy of biological tissues and cells. *Chem Soc Rev* **45**, 1865–1878 (2016).

72. Butler, H. J. *et al.* Using Raman spectroscopy to characterize biological materials. *Nat Protoc* **11**, 664–687 (2016).
73. Schwach, V. *et al.* A safety screening platform for individualized cardiotoxicity assessment. *iScience* **27**, 109139 (2024).
74. Cao, L. *et al.* Automated Sarcomere Structure Analysis for Studying Cardiotoxicity in Human Pluripotent Stem Cell-Derived Cardiomyocytes. *Microscopy and Microanalysis* **29**, 254–264 (2023).
75. Maria Cherian, R. *et al.* Fluorescent hiPSC-derived MYH6-mScarlet cardiomyocytes for real-time tracking, imaging, and cardiotoxicity assays. *Cell Biol Toxicol* **39**, 145–163 (2023).
76. Ribeiro, A. J. S. *et al.* Multi-Imaging Method to Assay the Contractile Mechanical Output of Micropatterned Human iPSC-Derived Cardiac Myocytes. *Circ Res* **120**, 1572–1583 (2017).
77. Yang, H. *et al.* Dynamic Myofibrillar Remodeling in Live Cardiomyocytes under Static Stretch. *Sci Rep* **6**, 20674 (2016).
78. Gao, B. *et al.* Assessment of sarcomere shortening and calcium transient in primary human and dog ventricular myocytes. *J Pharmacol Toxicol Methods* **123**, 107278 (2023).
79. Strimaityte, D. *et al.* Contractility and Calcium Transient Maturation in the Human iPSC-Derived Cardiac Microfibers. *ACS Appl Mater Interfaces* **14**, 35376–35388 (2022).
80. Tu, C., Chao, B. S. & Wu, J. C. Strategies for Improving the Maturity of Human Induced Pluripotent Stem Cell-Derived Cardiomyocytes. *Circ Res* **123**, 512–514 (2018).
81. Bedada, F. B., Wheelwright, M. & Metzger, J. M. Maturation status of sarcomere structure and function in human iPSC-derived cardiac myocytes. *Biochimica et Biophysica Acta (BBA) - Molecular Cell Research* **1863**, 1829–1838 (2016).
82. Feyen, D. A. M. *et al.* Metabolic Maturation Media Improve Physiological Function of Human iPSC-Derived Cardiomyocytes. *Cell Rep* **32**, 107925 (2020).
83. Marsano, A. *et al.* Beating heart on a chip: a novel microfluidic platform to generate functional 3D cardiac microtissues. *Lab Chip* **16**, 599–610 (2016).
84. Aung, A. *et al.* 3D cardiac μtissues within a microfluidic device with real-time contractile stress readout. *Lab Chip* **16**, 153–162 (2016).
85. Zhang, F. *et al.* Design and fabrication of an integrated heart-on-a-chip platform for construction of cardiac tissue from human iPSC-derived cardiomyocytes and in situ evaluation of physiological function. *Biosens Bioelectron* **179**, 113080 (2021).
86. Tagawa, H. *et al.* Cytoskeletal Mechanics in Pressure-Overload Cardiac Hypertrophy. *Circ Res* **80**, 281–289 (1997).
87. Arslan, U. *et al.* Vascularized hiPSC-derived 3D cardiac microtissue on chip. *Stem Cell Reports* **18**, 1394–1404 (2023).
88. Powers, J. D., Malingen, S. A., Regnier, M. & Daniel, T. L. The Sliding Filament Theory Since Andrew Huxley: Multiscale and Multidisciplinary Muscle Research. *Annu Rev Biophys* **50**, 373–400 (2021).
89. Gopal, A. S. *et al.* Assessment of Cardiac Function by Three-dimensional Echocardiography Compared With Conventional Noninvasive Methods. *Circulation* **92**, 842–853 (1995).
90. Kobirumaki-Shimozawa, F. *et al.* Nano-imaging of the beating mouse heart in vivo: Importance of sarcomere dynamics, as opposed to sarcomere length per se, in the regulation of cardiac function. *Journal of General Physiology* **147**, 53–62 (2016).
91. Acharya, A. *et al.* Live-Cell Imaging of the Contractile Velocity and Transient Intracellular Ca2+ Fluctuations in Human Stem Cell-Derived Cardiomyocytes. *Cells* **11**, 1280 (2022).
92. Sala, L. *et al.* MUSCLEMOTION. *Circ Res* **122**, (2018).
93. Lookin, O. *et al.* Contractile State Dependent Sarcomere Length Variability in Isolated Guinea-Pig Cardiomyocytes. *Front Physiol* **13**, (2022).
94. Granzier, H., Kellermayer, M., Helmes, M. & Trombitás, K. Titin elasticity and mechanism of passive force development in rat cardiac myocytes probed by thin-filament extraction. *Biophys J* **73**, 2043–2053 (1997).
95. Lyon, R. C., Zanella, F., Omens, J. H. & Sheikh, F. Mechanotransduction in Cardiac Hypertrophy and Failure. *Circ Res* **116**, 1462–1476 (2015).

96. Wang, K., McClure, J. & Tu, A. Titin: major myofibrillar components of striated muscle. *Proceedings of the National Academy of Sciences***76**, 3698–3702 (1979).
97. Wu, Y. *et al.* Changes in Titin Isoform Expression in Pacing-Induced Cardiac Failure Give Rise to Increased Passive Muscle Stiffness. *Circulation***106**, 1384–1389 (2002).
98. Knöll, R. *et al.* The Cardiac Mechanical Stretch Sensor Machinery Involves a Z Disc Complex that Is Defective in a Subset of Human Dilated Cardiomyopathy. *Cell***111**, 943–955 (2002).
99. Herzog, W. The multiple roles of titin in muscle contraction and force production. *Biophys Rev***10**, 1187–1199 (2018).
100. Knöll, R. *et al.* The Cardiac Mechanical Stretch Sensor Machinery Involves a Z Disc Complex that Is Defective in a Subset of Human Dilated Cardiomyopathy. *Cell***111**, 943–955 (2002).
101. Miller, L. & Penta, R. Investigating the effects of microstructural changes induced by myocardial infarction on the elastic parameters of the heart. *Biomech Model Mechanobiol***22**, 1019–1033 (2023).
102. Pandey, P. *et al.* Cardiomyocytes Sense Matrix Rigidity through a Combination of Muscle and Non-muscle Myosin Contractions. *Dev Cell***44**, 326-336.e3 (2018).
103. Chopra, A. *et al.* Force Generation via β-Cardiac Myosin, Titin, and α-Actinin Drives Cardiac Sarcomere Assembly from Cell-Matrix Adhesions. *Dev Cell***44**, 87-96.e5 (2018).
104. Mohr, E., Thum, T. & Bär, C. Accelerating cardiovascular research: recent advances in translational 2D and 3D heart models. Eur J Heart Fail**24**, 1778–1791 (2022).
105. Ribas, J. *et al.* Cardiovascular Organ-on-a-Chip Platforms for Drug Discovery and Development. *Appl In Vitro Toxicol***2**, 82–96 (2016).
106. Wu, H. *et al.* Modelling diastolic dysfunction in induced pluripotent stem cell-derived cardiomyocytes from hypertrophic cardiomyopathy patients. *Eur Heart J***40**, 3685–3695 (2019).
107. Fan, D., Takawale, A., Lee, J. & Kassiri, Z. Cardiac fibroblasts, fibrosis and extracellular matrix remodeling in heart disease. *Fibrogenesis Tissue Repair***5**, 15 (2012).
108. Perestrelo, A. R. *et al.* Multiscale Analysis of Extracellular Matrix Remodeling in the Failing Heart. *Circ Res***128**, 24–38 (2021).
109. Ariyasinghe, N. R., Lyra-Leite, D. M. & McCain, M. L. Engineering cardiac microphysiological systems to model pathological extracellular matrix remodeling. *American Journal of Physiology-Heart and Circulatory Physiology***315**, H771–H789 (2018).
110. Ribeiro, A. J. S. *et al.* Contractility of single cardiomyocytes differentiated from pluripotent stem cells depends on physiological shape and substrate stiffness. *Proceedings of the National Academy of Sciences***112**, 12705–12710 (2015).
111. Zhao, L. & Zhang, B. Doxorubicin induces cardiotoxicity through upregulation of death receptors mediated apoptosis in cardiomyocytes. *Sci Rep***7**, 44735 (2017).
112. Hulbert, J., Elger, C. E., Meyer, R. & Surges, R. Antiepileptic Drugs Impair Shortening of Isolated Cardiomyocytes. *Front Neurol***8**, (2017).
113. Luo, J. *et al.* Mechanisms of acrolein-induced myocardial dysfunction: implications for environmental and endogenous aldehyde exposure. *American Journal of Physiology-Heart and Circulatory Physiology***293**, H3673–H3684 (2007).
114. Peterson, P., Kalda, M. & Vendelin, M. Real-time determination of sarcomere length of a single cardiomyocyte during contraction. *American Journal of Physiology-Cell Physiology***304**, C519–C531 (2013).
115. Korte, F. S. & McDonald, K. S. Sarcomere length dependence of rat skinned cardiac myocyte mechanical properties: dependence on myosin heavy chain. *J Physiol***581**, 725–739 (2007).
116. Wang, Y., Shyy, J. Y. J. & Chien, S. Fluorescence Proteins, Live-Cell Imaging, and Mechanobiology: Seeing Is Believing. *Annu Rev Biomed Eng***10**, 1–38 (2008).
117. Scarfe, L. *et al.* Preclinical imaging methods for assessing the safety and efficacy of regenerative medicine therapies. *NPJ Regen Med***2**, 28 (2017).
118. Tao, Z. *et al.* Angiopoietin-1 enhanced myocyte mitosis, engraftment, and the reparability of hiPSC-CMs for treatment of myocardial infarction. *Cardiovasc Res***117**, 1578–1591 (2021).
119. Yu, D., Wang, X. & Ye, L. Cardiac Tissue Engineering for the Treatment of Myocardial Infarction. *J Cardiovasc Dev Dis***8**, 153 (2021).

120. Foglia, M. J. & Poss, K. D. Building and re-building the heart by cardiomyocyte proliferation. *Development* **143**, 729–740 (2016).
121. Yuan, Q. *et al.* Sarcomere Disassembly and Transfection Efficiency in Proliferating Human iPSC-Derived Cardiomyocytes. *J Cardiovasc Dev Dis* **9**, 43 (2022).
122. Jopling, C. *et al.* Zebrafish heart regeneration occurs by cardiomyocyte dedifferentiation and proliferation. *Nature* **464**, 606–609 (2010).
123. Fan, X. *et al.* Dynamic Alterations to α-Actinin Accompanying Sarcomere Disassembly and Reassembly during Cardiomyocyte Mitosis. *PLoS One* **10**, e0129176 (2015).
124. Ali, H., Braga, L. & Giacca, M. Cardiac regeneration and remodelling of the cardiomyocyte cytoarchitecture. *FEBS J* **287**, 417–438 (2020).
125. Huang, W. *et al.* Loss of microRNA-128 promotes cardiomyocyte proliferation and heart regeneration. *Nat Commun* **9**, 700 (2018).
126. Powers, N. & Huang, G. N. Mending a broken heart with novel cardiogenic small molecules. *Cell Regeneration* **11**, 20 (2022).
127. Murganti, F. *et al.* FUCCI-Based Live Imaging Platform Reveals Cell Cycle Dynamics and Identifies Pro-proliferative Compounds in Human iPSC-Derived Cardiomyocytes. *Front Cardiovasc Med* **9**, (2022).
128. Leri, A., Kajstura, J. & Anversa, P. Cardiac Stem Cells and Mechanisms of Myocardial Regeneration. *Physiol Rev* **85**, 1373–1416 (2005).
129. Duran, J. M. *et al.* Bone-Derived Stem Cells Repair the Heart After Myocardial Infarction Through Transdifferentiation and Paracrine Signaling Mechanisms. *Circ Res* **113**, 539–552 (2013).
130. Floy, M. E. *et al.* Direct coculture of human pluripotent stem cell-derived cardiac progenitor cells with epicardial cells induces cardiomyocyte proliferation and reduces sarcomere organization. *J Mol Cell Cardiol* **162**, 144–157 (2022).
131. Onódi, Z. *et al.* Systematic transcriptomic and phenotypic characterization of human and murine cardiac myocyte cell lines and primary cardiomyocytes reveals serious limitations and low resemblances to adult cardiac phenotype. *J Mol Cell Cardiol* **165**, 19–30 (2022).
132. Guo, Y. & Pu, W. T. Cardiomyocyte Maturation. *Circ Res* **126**, 1086–1106 (2020).
133. Skwarek-Maruszewska, A., Hotulainen, P., Mattila, P. K. & Lappalainen, P. Contractility-dependent actin dynamics in cardiomyocyte sarcomeres. *J Cell Sci* **122**, 2119–2126 (2009).
134. Lee, Y.-K. *et al.* Calcium Homeostasis in Human Induced Pluripotent Stem Cell-Derived Cardiomyocytes. *Stem Cell Rev Rep* **7**, 976–986 (2011).
135. Satin, J. *et al.* Calcium Handling in Human Embryonic Stem Cell-Derived Cardiomyocytes. *Stem Cells* **26**, 1961–1972 (2008).
136. Salick, M. R. *et al.* Micropattern width dependent sarcomere development in human ESC-derived cardiomyocytes. *Biomaterials* **35**, 4454–4464 (2014).
137. Sun, S. *et al.* Progressive Myofibril Reorganization of Human Cardiomyocytes on a Dynamic Nanotopographic Substrate. *ACS Appl Mater Interfaces* **12**, 21450–21462 (2020).
138. Kim, D.-H. *et al.* Nanopatterned cardiac cell patches promote stem cell niche formation and myocardial regeneration. *Integrative Biology* **4**, 1019 (2012).
139. Han, J., Wu, Q., Xia, Y., Wagner, M. B. & Xu, C. Cell alignment induced by anisotropic electrospun fibrous scaffolds alone has limited effect on cardiomyocyte maturation. *Stem Cell Res* **16**, 740–750 (2016).
140. Schoen, B. *et al.* Electrospun Extracellular Matrix: Paving the Way to Tailor-Made Natural Scaffolds for Cardiac Tissue Regeneration. *Adv Funct Mater* **27**, (2017).
141. Liu, J. *et al.* Rapid 3D bioprinting of in vitro cardiac tissue models using human embryonic stem cell-derived cardiomyocytes. *Bioprinting* **13**, e00040 (2019).
142. Maiullari, F. *et al.* A multi-cellular 3D bioprinting approach for vascularized heart tissue engineering based on HUVECs and iPSC-derived cardiomyocytes. *Sci Rep* **8**, 13532 (2018).

143. Ruan, J.-L. *et al.* Mechanical Stress Conditioning and Electrical Stimulation Promote Contractility and Force Maturation of Induced Pluripotent Stem Cell-Derived Human Cardiac Tissue. *Circulation***134**, 1557–1567 (2016).
144. Chan, Y.-C. *et al.* Electrical Stimulation Promotes Maturation of Cardiomyocytes Derived from Human Embryonic Stem Cells. *J Cardiovasc Transl Res***6**, 989–999 (2013).
145. Llucià-Valldeperas, A. *et al.* Physiological conditioning by electric field stimulation promotes cardiomyogenic gene expression in human cardiomyocyte progenitor cells. *Stem Cell Res Ther***5**, 93 (2014).
146. Brandenburg, S., Arakel, E. C., Schwappach, B. & Lehnart, S. E. The molecular and functional identities of atrial cardiomyocytes in health and disease. *Biochimica et Biophysica Acta (BBA) - Molecular Cell Research***1863**, 1882–1893 (2016).
147. Kane, C. & Terracciano, C. M. N. Concise Review: Criteria for Chamber-Specific Categorization of Human Cardiac Myocytes Derived from Pluripotent Stem Cells. *Stem Cells***35**, 1881–1897 (2017).
148. Zhao, Y. *et al.* Towards chamber specific heart-on-a-chip for drug testing applications. *Adv Drug Deliv Rev***165–166**, 60–76 (2020).
149. König, K. Multiphoton microscopy in life sciences. *J Microsc***200**, 83–104 (2000).
150. Dunn, K. W. & Young, P. A. Principles of Multiphoton Microscopy. *Nephron Exp Nephrol***103**, e33–e40 (2006).
151. Levene, M. J., Dombeck, D. A., Kasischke, K. A., Molloy, R. P. & Webb, W. W. In Vivo Multiphoton Microscopy of Deep Brain Tissue. *J Neurophysiol***91**, 1908–1912 (2004).
152. Andresen, V. *et al.* Infrared multiphoton microscopy: subcellular-resolved deep tissue imaging. *Curr Opin Biotechnol***20**, 54–62 (2009).
153. Nunes, S. S. *et al.* Biowire: a platform for maturation of human pluripotent stem cell–derived cardiomyocytes. *Nat Methods***10**, 781–787 (2013).
154. Lewis-Israeli, Y. R. *et al.* Self-assembling human heart organoids for the modeling of cardiac development and congenital heart disease. *Nat Commun***12**, 5142 (2021).

GPSR Compliance
The European Union's (EU) General Product Safety Regulation (GPSR) is a set of rules that requires consumer products to be safe and our obligations to ensure this.

If you have any concerns about our products, you can contact us on

ProductSafety@springernature.com

In case Publisher is established outside the EU, the EU authorized representative is:

Springer Nature Customer Service Center GmbH
Europaplatz 3
69115 Heidelberg, Germany

www.ingramcontent.com/pod-product-compliance
Lightning Source LLC
Chambersburg PA
CBHW050717010525
26019CB00026B/33
* 9 7 8 3 0 3 1 8 1 7 4 2 7 *